# Electrodynamics

Third Edition

# Electrodynamics

**Third Edition**

Harald J W Müller-Kirsten

University of Kaiserslautern, Germany

World Scientific

NEW JERSEY • LONDON • SINGAPORE • BEIJING • SHANGHAI • HONG KONG • TAIPEI • CHENNAI • TOKYO

*Published by*

World Scientific Publishing Co. Pte. Ltd.
5 Toh Tuck Link, Singapore 596224
*USA office:* 27 Warren Street, Suite 401-402, Hackensack, NJ 07601
*UK office:* 57 Shelton Street, Covent Garden, London WC2H 9HE

**British Library Cataloguing-in-Publication Data**
A catalogue record for this book is available from the British Library.

**ELECTRODYNAMICS**
**Third Edition**



ISBN 978-981-98-2559-2 (hardcover)
ISBN 978-981-98-2599-8 (paperback)
ISBN 978-981-98-2560-8 (ebook for institutions)
ISBN 978-981-98-2561-5 (ebook for individuals)

For any available supplementary material, please visit
https://www.worldscientific.com/worldscibooks/10.1142/14654#t=suppl

Desk Editors: Eshak Nabi Akbar Ali/Rhaimie Wahap

Typeset by Stallion Press
Email: enquiries@stallionpress.com

# Preface to Third Edition

The main incentive that led to a revision of the Second Edition was the desire to make the practical physics behind the formulas in the topics of currents and transmission cables more transparent. In the course of this endeavor the entire text was thoroughly revised and numerous comments and new exercises have been added, and occasionally minor changes have been made throughout the text.

H.J.W. Müller–Kirsten

# Preface to Second Edition

The text has been completely and thoroughly revised. Throughout errors and omissions which have been detected have been corrected, and at numerous points additional explanatory comments have been inserted. To simplify cross referencing most equations have been given numbers, and in many cases the units are stated alongside in order to assist a concurrent familiarization with these. Numerous references to literature serve not only the purpose of citing specialized literature, but also to record sources consulted in the preparation and revision of the text. A main change compared with the first edition is the addition of exercises. Further additions in the text are the topic of 4-terminal networks included in Chapter 8 and an introduction to diffraction in Chapter 11.

H.J.W. Müller–Kirsten

# Preface to First Edition

Electrodynamics is one of the pillars of physics with arms reaching through classical and atomic physics far into high energy particle physics and present day string theory. With its unification of electric and magnetic phenomena and subsequently with classical mechanics in Special Relativity it pointed towards a unification of all physical theories in a future theory of "everything".

Classical electrodynamics is the electrodynamics of macroscopic phenomena and is summarized essentially in the four Maxwell equations which, however, permeate in one way or another into microscopic phenomena. Although beset with tantalizing divergence problems, the quantized form of Maxwell's electrodynamics, that is quantum electrodynamics, has so far defied all mathematical purists by yielding highly precise values of such vital physical quantities as the Lamb shift and magnetic moments. In its nonabelian extension this quantized electrodynamics led to quantum chromodynamics and beyond, thereby permitting its unification with strong and weak interactions. It is expected that the ultimate theory, possibly a form of higher dimensional string theory, will one day permit its unification also with quantum gravity. This theoretical endeavor, starting with Maxwell's unification of electric and magnetic phenomena and now reaching as far as cosmology, is one of the most amazing intellectual achievements of mankind.

A countless number of treatises has been written on electrodynamics, including applications in all directions. These cover the entire range from elementary introductions to authoritative reference tomes. The present text is but one more, and like any other is imprinted with the author's likes and preferences. In repeatedly teaching electrodynamics the author set himself the task of making the presentation as easily comprehensible as possible, and to direct the view beyond the traditional domain of pure macroscopic Maxwell theory.

Rightly or not, and irrespective of what others think, the author claims that the companion text of a first course in a basic subject must be readable, must explain the issues at hand in detail and, not the least, must present any nontrivial calculational steps. The author remembers from his time as a student the relief derived from a book or script which explains what others passed over (knowingly or not), if only as a reassurance that one's own understanding is correct. A student's struggle with a book may flatter the vanity of the author, but will more likely result in the book being put aside. The present text therefore attempts to avoid as far as possible the "*it–can–easily–be–shown–jumps*" familiar from everyday literature. With more than a hundred mostly worked and typical examples interspersed, the text might also serve as a "*Teach Yourself Course*" in electrodynamics. To what extent this succeeds only the reader can decide.

The further attempt pursued in the text (frequently in examples) is to open some views to other domains which are usually excluded in traditional presentations of electrodynamics. This applies in particular to intimately related quantum effects (mostly of Schrödinger equation or harmonic oscillator type), electric–magnetic duality, reference to higher dimensions and to gravity, all of which the student of today encounters elsewhere much earlier than in the past, and hence might expect some mention in the text.[1]

I thank Dr. K.K. Yim of World Scientific for his untiring help and a countless number of suggestions for improving the manuscript.

H.J.W. Müller–Kirsten

[1]Only textbook literature consulted repeatedly at various times in the course of preparation of this text is cited in the bibliography. Many other sources have, of course, also been used at various points but their specific relevance has been lost track of as the years passed by. Hence no attempt is made to even refer to all important and standard works on the subject. An extensive list of this literature can be found in the book of Jackson [73]. Literature related to very specific topics or sources is cited in footnotes at appropriate points in the text.

# About the Author

**Harald J.W. Müller-Kirsten**, born in Germany, attended High School and First Year University in Sydney, obtained 1st class Honours degree and Ph.D. at the University of W. Aust., was post-doc with Habilitation at L.M. University of Munich, NATO-Fellow at Lawrence Radiation Laboratory (Berkeley), Max-Kade-Foundation-Fellow at SLAC (Stanford), assistant professor at American University of Beirut, finally professor at Technische Universität Kaiserslautern/University of Kaiserslautern (TUKL).

# Contents

# Chapter 1

# Introduction

## 1.1 What Is Electrodynamics?

*Electrodynamics* as a theory deals with electric charges which move in space and with electromagnetic fields that are produced by the charges and again interact with charges. Electrodynamics is described by Maxwell's equations whose most important consequences are: (a) the electromagnetic nature of light (Maxwell), (b) the emission of electromagnetic waves by an oscillating dipole (H. Hertz), and (c) the unification of electric and magnetic forces. Unlike in Maxwell's days one does not refer to an ether anymore — it was Einstein who concluded that this does not exist — and instead one uses the concept of fields in space. Einstein's Special Theory of Relativity unifies electrodynamics with classical mechanics.

## 1.2 Presentation of Macroscopic Electrodynamics

In general a first course in electrodynamics, or electricity and magnetism as it is also called, is preceded by a course in classical mechanics which even today is not always combined with the Special Theory of Relativity (as would be desirable). In addition the concept of a field in space and the distinction between directly observable field quantities and at best indirectly observable field potentials is at that stage still too vague to permit an immediate relativistic field theoretic approach to appear plausible. Moreover, in general the term electrodynamics is usually restricted to macroscopically observable phenomena, so that a quantized treatment with operator-valued fields is beyond its scope, the latter being dealt with in quantum electrodynamics. Thus the classical fields are c-numbers. As a consequence of this restriction, and also in order to establish classical electrodynamics as a theory in its own right, the fundamental equations of this theory, Maxwell's equations, are usually not derived from Hamilton's principle as a prior course in methods of classical mechanics might suggest. In addition the law of force between moving charges is considerably more complicated than that between pointlike masses so that an analogous procedure is not immediately advisable. Nonetheless Hamilton's variational principle is of such basic significance that it permits, of course, the derivation of Maxwell's equations as the Euler–Lagrange equations (and their consequences) of an appropriate variational principle. Thus here we do not adopt this procedure immediately (i.e. till Chapter 18). What are the other methods which suggest themselves? Perusing the literature, one observes two main procedures. The approach with emphasis on logic starts from an axiomatic presentation of Maxwell's equations, whereas the other more historical and phenomenological approach abstracts these from observations. A textbook which adopts the former procedure is that of Sommerfeld [139] who follows Hertz in this respect, but whereas Hertz starts from the differential form of Maxwell's equations, Sommerfeld chooses the vector integral form. Other texts which follow this procedure are, for instance, the books of Ferrari [53] and Lim [88]. Books which choose the second procedure are those of Jackson [73] and Greiner [61].

In the axiomatic formulation of classical or macroscopic electrodynamics the *two principal axioms* are Faraday's law of induction and Ampère's flux theorem. In integral form and in units of the internationally agreed system of units (ISU), these are

$$\frac{d}{dt}\int_F \mathbf{B}\cdot d\mathbf{F} = -\oint_{C(F)} \mathbf{E}\cdot d\mathbf{l} \tag{1.1}$$

and

$$\frac{d}{dt}\int_F \mathbf{D}\cdot d\mathbf{F} + \int_F \mathbf{j}\cdot d\mathbf{F} = \oint_{C(F)} \mathbf{H}\cdot d\mathbf{l}, \tag{1.2}$$

where for a given fixed surface of integration $F$,

$$\frac{d}{dt}\int_F \mathbf{D}\cdot d\mathbf{F} = \int_F \frac{\partial \mathbf{D}}{\partial t}\cdot d\mathbf{F},$$

since (by assumption) $F$ does not change with $t$.[1]

---

[1]Recall, for instance the analogy, that as shown in Whittaker and Watson [162], p. 67,

$$\frac{d}{d\alpha}\int_a^b f(x,\alpha)dx = f(b,\alpha)\frac{db}{d\alpha} - f(a,\alpha)\frac{da}{d\alpha} + \int_a^b \frac{\partial f(x,\alpha)}{\partial \alpha}dx,$$

which shows the $\alpha$-dependence of boundary values.

In the first or *Faraday's law*, **B** is the magnetic induction and **E** the electric field strength. The law says that every change of the magnetic flux[2] with time through an open, double-sided area $F$ (a plane is single-sided) with boundary given by the closed curve $C(F)$ — this flux being the number of lines of force in Faraday's considerations — generates an equal but oppositely directed circuit potential, the *electromotive force* $\oint \mathbf{E}\cdot d\mathbf{l}$, along the boundary $C(F)$. An open surface can, for instance, be visualized as a container without a lid, and a closed surface as one closed with the lid. In the former case the rim of the opening corresponds to the curve $C(F)$. The word *double-sided* implies that the surface possesses direction normals directed towards inside or outside regions. A closed surface has no boundary, but we can imagine on it a closed curve $C(F)$ which divides the surface into two regions.

In the second or *Ampère's law*, **D** is the so-called dielectric displacement ($\mathbf{D} = \epsilon_0\mathbf{E} + \mathbf{P}, \epsilon_0$ the dielectric constant of vacuum, **P** the polarization vector or dipole moment per unit volume of the medium), **j** is the density of the electron current and **H** the magnetic field strength. The law says that analogous to Faraday's law the time change of an electric flux through an area $F$ is equal to a magnetic circuit potential in the boundary curve $C(F)$ in the same direction. The expression $\partial\mathbf{D}/\partial t$ obviously has the dimension of a current density. The expression

$$\frac{d}{dt}\int_F \mathbf{D}\cdot d\mathbf{F}$$

is called *Maxwell's displacement current.*

The two principal axioms are supplemented by *two subsidiary axioms* which are consequences of the principal axioms with implementation of some empirical findings. We arrive at these by considering a boundary curve $C(F)$ enclosing the area $F$, and by permitting $C(F)$ to shrink to zero, so that the line integrals vanish, i.e.

$$\lim_{C(F)\to 0}\frac{d}{dt}\int_F \mathbf{B}\cdot d\mathbf{F} = 0, \tag{1.3}$$

$$\lim_{C(F)\to 0}\left\{\frac{d}{dt}\int_F \mathbf{D}\cdot d\mathbf{F} + \int_F \mathbf{j}\cdot d\mathbf{F}\right\} = 0. \tag{1.4}$$

The area $F$ thus loses its boundary and becomes closed. In the case of magnetic dipoles (irrespectively of whether one considers magnets or current loops) every line of force leaving the area $F$ is accompanied by a

---

[2]Flux is in general the scalar product of a vector field with an area.

corresponding line of force towards it. The integral extended over the entire closed surface thus yields the value zero, i.e.

$$\int_{F_{\text{closed}}} \mathbf{B} \cdot d\mathbf{F} = 0. \tag{1.5}$$

This is a general result in view of the empirical fact that no isolated magnetic poles exist and hence no regions with only ingoing or only outgoing lines of force. With the definition of the divergence

$$\boldsymbol{\nabla} \cdot \mathbf{A} \equiv \text{div } \mathbf{A}(\mathbf{r}) = \lim_{V \to 0} \frac{1}{V} \int_{F(V)} \mathbf{A} \cdot d\mathbf{F} \tag{1.6}$$

we obtain the differential form of Eq. (1.5), i.e.

$$\text{div } \mathbf{B}(\mathbf{r}) \equiv \boldsymbol{\nabla} \cdot \mathbf{B}(\mathbf{r}) = 0. \tag{1.7}$$

One says: $\mathbf{B}(\mathbf{r})$ is *divergenceless.*

Equation (1.4) permits an analogous consideration but with different consequences. Equation (1.2), or rather (1.4), applies to cases in which the electron current

$$\int_{F(V)} \mathbf{j} \cdot d\mathbf{F}$$

leaves the surface $F$ enclosing the volume $V(F)$. This implies that the volume with total charge $q$ loses per second charge $q$ at the rate

$$\int_{F(V)} \mathbf{j} \cdot d\mathbf{F} = -\frac{dq}{dt}. \tag{1.8}$$

The minus sign indicates that $V(F)$ is losing charge. Put differently, as

$$\frac{dq}{dt} + \int_{F(V)} \mathbf{j} \cdot d\mathbf{F} = 0, \tag{1.9}$$

the equation says that the amount of charge in $V(F)$ remains constant (this is called charge conservation). Here one has to be careful, because if we write

$$\frac{dq}{dt} = + \int_F \mathbf{j} \cdot d\mathbf{F},$$

then $q$ represents the amount of charge which passes through the area $F$ per second *with no reference to* $V$. We can now write Eq. (1.4) as

$$\lim_{C(F) \to 0} \left\{ \frac{d}{dt} \int_F \mathbf{D} \cdot d\mathbf{F} - \frac{dq}{dt} \right\} = 0, \tag{1.10}$$

which implies

$$\int_{F_{\text{closed}}} \mathbf{D} \cdot d\mathbf{F} = q. \tag{1.11}$$

We could have added a constant of integration on the right side. But for charge $q = 0$ the field $\mathbf{D} = 0$, and thus this constant must be zero. The differential form of Eq. (1.11) again follows with Eq. (1.6); i.e.

$$\boldsymbol{\nabla} \cdot \mathbf{D} = \rho, \quad \rho = \frac{q}{V}. \tag{1.12}$$

Here $\rho$ is the charge density. Equations (1.11) and (1.12) are known as the Gauss law of electrodynamics. We can now write Eq. (1.8):

$$\frac{d}{dt} \int_V \rho dV = -\int_{F(V)} \mathbf{j} \cdot d\mathbf{F}. \tag{1.13}$$

Going to an infinitesimal volume and using the Gauss divergence theorem, this implies

$$\int_V \boldsymbol{\nabla} \cdot \mathbf{j} dV = \int_{F(V)} \mathbf{j} \cdot d\mathbf{F} \tag{1.14}$$

and hence the *equation of continuity*

$$\frac{\partial \rho(\mathbf{r}, t)}{\partial t} + \boldsymbol{\nabla} \cdot \mathbf{j} = 0. \tag{1.15}$$

A method of deriving electrodynamics essentially from this equation and the Lorentz force (the force exerted on a charge in an electromagnetic field) — as an alternative approach to electrodynamics — can be found in a paper of Bopp.[3]

Returning to comments at the beginning we mention that attempts have repeatedly been made to derive Maxwell's equations from the Coulomb law and the Special Theory of Relativity. Since it is known that the General Theory of Relativity contains the generalization of Newton's law of gravitation, i.e. a law which has the same form as the Coulomb law for charges, such an attempt[4] cannot succeed without additional assumptions.[5] A discussion of this topic can be found in the book of Jackson [73]. An entirely different approach which sees the only logical foundation of

[3]Bopp [20].

[4]See e.g. Cook [33], pp. 448–452.

[5]Some differences are: The masses in Newton's law are always positive, whereas the charges in Coulomb's law can be positive or negative. Also there is no screening of the gravitational force, no analogue of the Faraday cage.

Maxwell's theory in a relativistic treatment of radiation theory is the book of Page and Adams [109]. In the following we adopt the historical and phenomenological approach starting with electrostatics, since this seems to be more suitable for an understanding of the physics of electrodynamics, particularly in a first course on the subject.

## 1.3 On the Choice of Units

In textbooks on electrodynamics nowadays two systems of units are in use: The more common internationally used MKSA-system of units with meter (m), kilogramme (kg), second (s) and ampere (A), and the still frequently used system of Gaussian units which is based on the c.g.s. units, i.e. centimeter (cm), gramme (g) and second (s) (the current is then given in statamperes, 1 A $= 3\times 10^9$ statamperes). Either system has its advantages — the former has been agreed upon internationally and is therefore used particularly in applications, and the latter, the Gaussian system, is somewhat theory–oriented and therefore frequently used in considerations of singly charged particles. Here we employ the first system but will refer occasionally also to the other system. We add the following comments.

In discussions on units the word "dimension" is frequently referred to. The dimension of a physical quantity is nothing absolute. It is possible to choose units (like the so-called *natural units*) in which Planck's action quantum constant $\hbar$ and the velocity of light $c$ in vacuum have the value 1 and are dimensionless. Since velocity$=$ distance/time, one can then express lengths in units of time, i.e. seconds, or more commonly time intervals in meters, and mass in reciprocal meters. This arbitrariness was already pointed out by Planck. Some authors even amuse themselves today about dimensional considerations.[6] The example shows that also the number of fundamental units is a matter of choice (in "natural units" every quantity can be expressed in a power of length). In earlier days the units used in the literature on electrodynamics were the so-called electrostatic units (e.s.u.) and electromagnetic units (e.m.u.) which today are indirectly contained in the Gaussian system. In essence, in the Gaussian system of units electric quantities are expressed in e.s.u. and magnetic quantities in e.m.u. while the change to Gaussian units requires a factor $c$ or $1/c$ ($c$ the velocity of

[6]See e.g. O'Rahilly, Vol. I [122], p. 65, where the author says: "*Maxwell invented a second system... But this system is never employed, it merely occurs in those pages of textbooks which profess to deal with something called 'dimensions'.*" See also p. 68.

light in vacuum), which can be determined experimentally (cf. Appendix B). For instance

$$\mathbf{B}_{\text{e.s.u.}} = \frac{1}{c}\mathbf{B}_{\text{e.m.u.}}. \tag{1.16}$$

In particular one has in these units for the dielectric constant of the vacuum $\epsilon_0$ and for the magnetic permeability of the vacuum $\mu_0$:

$$\epsilon_0(\text{e.s.u.}) = 1, \quad \epsilon_0(\text{e.m.u.}) = \frac{1}{c^2}, \quad \mu_0(\text{e.s.u.}) = \frac{1}{c^2}, \quad \mu_0(\text{e.m.u.}) = 1. \tag{1.17}$$

We do not enter into a deeper discussion of these units here (sometimes it is not clear whether a quantity is electric or magnetic).

The choice of a system of units in electrodynamics depends on the choice of the magnitude as well as the dimension of two arbitrary constants, as an exhaustive investigation of Maxwell's equations shows.[7] The appearance of one constant $k$ can immediately be seen by looking at the Coulomb law in electrostatics, which determines the force $\mathbf{F}_{12}$ between two point charges $q_1, q_2$, separated by a distance $\mathbf{r}$, i.e.

$$\mathbf{F}_{12} = kq_1q_2\frac{\mathbf{r}}{r^3}. \tag{1.18}$$

For length, mass and time, we choose here as agreed upon internationally the ISU (international system of units, internationally abbreviated SI, see Appendix B) and thus the units meter (m), kilogramme (kg), second (s) (correspondingly in the Gaussian system centimeter (cm), gramme (g), second (s)). Depending on the choice of dimension and magnitude of $k$ we obtain different units for the charge $q$. The electric field strength $\mathbf{E}(\mathbf{r})$ at the distance $\mathbf{r}$ away from the charge $q$ is

$$\mathbf{E}(\mathbf{r}) = k\frac{q}{r^3}\mathbf{r}. \tag{1.19}$$

Thus the field $\mathbf{E}(\mathbf{r})$ can be defined as force per unit charge.

In the consideration of magnetic phenomena we have to deal with currents. Currents $I$ are defined with respect to charges $q$, $I = dq/dt$. The connection between electric and magnetic phenomena or between $\mathbf{E}$ and $\mathbf{B}$ introduces another constant for the choice of units of $\mathbf{B}$. This constant can

[7]See Jackson [73], p. 811.

be introduced in a number of ways. With the help of the theorem of Stokes (see later), Eq. (1.1) can be written as

$$\nabla \times \mathbf{E} + k^* \frac{\partial \mathbf{B}}{\partial t} = 0, \tag{1.20}$$

where $k^*$ is chosen to be 1 and dimensionless in the MKSA-system (the constants $k$ and $k^*$ can be chosen arbitrarily in magnitude and dimension).

In accordance with the ISU, i.e. the MKSA-system of units, the unit of current is taken as the *ampere* (the spelling in English being this, ampere, which is not part of SI). One ampere is defined as that amount of current which runs through two parallel, straight, infinitely long, thin conductors which are one meter apart when the force between these is $2 \times 10^{-7}$ newtons/m (per meter in length) (the expression for this attractive force is derived later, cf. Eq. (5.34)). Since current $\times$ time is charge, we obtain the definition of the unit of charge:

$$1 \text{ coulomb (C)} = 1 \text{ ampere} \cdot \text{second (A·s)}.$$

It remains to determine the dimensions of $\mathbf{D}$ and $\mathbf{H}$. For a large number of different materials the following "*matter*" or "*material equations*" are valid:

$$\mathbf{D} = \epsilon \mathbf{E}, \quad \mathbf{B} = \mu \mathbf{H}. \tag{1.21}$$

Here $\epsilon$ is the *dielectric constant*, also called *permittivity* of the material and $\mu$ its *magnetic permeability*. Both have magnitudes and dimensions depending on the chosen system of units. The vacuum values $\epsilon_0, \mu_0$ obey, as we shall see, the important relation

$$\epsilon_0 \mu_0 = \text{const.} \tag{1.22}$$

In the Gaussian system of units $\epsilon_0, \mu_0$ are taken as 1 and dimensionless. The constant is then 1. In the MKSA-system the constant is found to be $1/c^2$, where $c$ is the velocity of light in vacuum, i.e.

$$c = 2.998 \times 10^8 \text{ m/s}. \tag{1.23}$$

In this case ($\epsilon/\epsilon_0$ being the *relative permittivity*)

$$\epsilon_0 = \frac{10^7}{4\pi c^2} \text{ farads/meter} \tag{1.24}$$

with the dimension of current$^2$ time$^4$ mass$^{-1}$ length$^{-3}$, or $\epsilon_0 = 8.854 \times 10^{-12}$ A·s/(V·m) or $C^2$/(joule·meter), where 1 volt (V) = 1 A$^{-1}$ m$^2$ kg s$^{-2}$ and

1 joule = 1 newton · meter (N·m). In addition

$$\mu_0 = 4\pi \times 10^{-7} \text{ newton} \cdot \text{ampere}^{-2} \; (\text{N}\cdot\text{A}^{-2}) \text{ or henries/meter,}$$

or $\mu_0 = 1.257 \times 10^{-6}$ with dimension of mass length current$^{-2}$ time$^{-2}$.

In the *Gaussian system* the constants $k, k^*$ are chosen as follows:

$$k = 1 \quad (\text{dimensionless}), \qquad k^* = \frac{1}{c} \quad (\text{dimension time/length}).$$

In the MKSA-*system* the vacuum constants $k, k^*$ are chosen as

$$k = \frac{1}{4\pi\epsilon_0} = 10^{-7}c^2 \tag{1.25}$$

with dimension kg m$^3$ s$^{-4}$ A$^{-2}$ and

$$k^* = 1 \tag{1.26}$$

(dimensionless). We see that in comparison with the Gaussian system the MKSA-system requires the use of $\epsilon_0$, $\mu_0$. The units of the electric field intensity **E** are V/m (or N/C) and those of **D**, also called *electric flux density*, C/m$^2$ (or FV/m$^2$). The units of the *magnetic flux density* **B** and the magnetic field intensity **H** are respectively Wb/m$^2$ and A/m.

# Chapter 2

# Electrostatics: Basic Aspects

## 2.1 Introductory Remarks

In this chapter the *Coulomb law* is introduced, and various static consequences are investigated, such as electrical screening and diverse applications. The latter require also the introduction of essential mathematics, in particular that of the *delta distribution*, originally called *delta function*.

## 2.2 The Coulomb Law

Electrostatics is the theory of static charges, which means that moving charges, i.e. currents, are not considered. The fundamental phenomenological law which is the basis of electrostatics is *Coulomb's law*, which determines the force acting between two point charges $q_1, q_2$ and which in vacuum is given by[1]

$$\mathbf{F}_{12} = k\frac{q_1 q_2}{r^3}\mathbf{r} \quad \text{newtons (N)}. \tag{2.1}$$

Here $\mathbf{r} = \mathbf{r}_1 - \mathbf{r}_2$ is the vectorial separation of the charges $q_i$ at positions $\mathbf{r}_i$. The parameter $k$ is the constant fixed by the choice of units. In the ISU, as explained in Chapter 1 and in Appendix A with symbols of units in upright type,

$$k = \frac{1}{4\pi\epsilon_0} \quad \text{meters/farad (m/F)} \tag{2.2}$$

[1]Observe that we assume this here as given phenomenologically as extracted from observation.

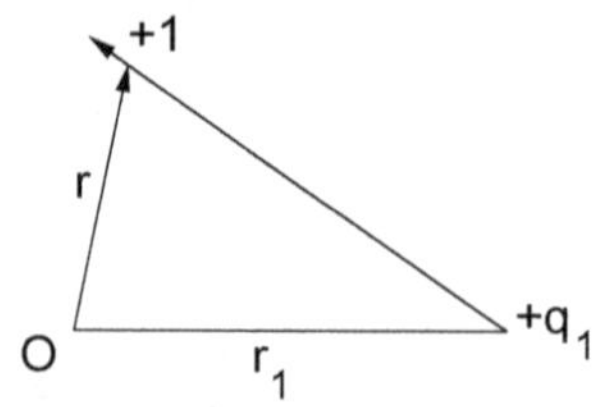


**Fig. 2.1.** Point charges $q_1$ and $+1$.

in vacuum. The force is given in newtons (N), the charge in coulombs (C) and the separation in meters (m). The electric field strength $\mathbf{E}(\mathbf{r})$ at a point $\mathbf{r}$ as in Fig. 2.1 is defined as the Coulomb force acting on a unit charge (+1C) at this point, i.e.

$$\mathbf{E}(\mathbf{r}) = k\frac{q_1(\mathbf{r}-\mathbf{r}_1)}{|\mathbf{r}-\mathbf{r}_1|^3} \quad \text{newtons/coulomb.} \tag{2.3}$$

In the ISU or MKSA-system of units the electric field strength is given in units of N/C or V/m. In the *Gaussian system* the charge is given in *statcoulombs*, the distance in cm, the force in dynes, and the electric field strength in statvolt/cm (1 C $=$ 3 $\times$ $10^9$ statcoulombs, 1 statcoulomb $=$ 10 e.s.u.; 1 newton $=$ $10^5$ dynes).

It should be kept in mind that a charge $q$ can be positive or negative. Conventionally electric lines of force are drawn as leaving positive charges and directed towards negative charges. In this sense positive charges are regarded as sources of the electric fields and negative charges as sinks.

One distinguishes between discrete charge distributions (charges at distinct points) and continuous charge distributions (charges spread over spatial domains). Correspondingly one defines a charge distribution or charge density $\rho(\mathbf{r})$ as charge per unit volume. Considering an infinitesimal volume element, we can use the above expression for the field strength at a point $\mathbf{r}$ due to a point charge $q_1$ at a point $\mathbf{r}_1$ to obtain the field strength at a point $\mathbf{r}$ resulting from superposition, i.e. summation or integration of the contributions of point or pointlike charges in a domain $V$ of space, i.e.

$$\mathbf{E}(\mathbf{r}) = k\int_V \rho(\mathbf{r}')\frac{(\mathbf{r}-\mathbf{r}')}{|\mathbf{r}-\mathbf{r}'|^3}d\mathbf{r}' \quad \text{newtons/coulomb,} \quad dV' \equiv d\mathbf{r}'. \tag{2.4}$$

Here $V$ is a volume which does not enclose the observation point at $\mathbf{r}$. The lines of force of $\mathbf{E}$, as shown in Fig. 2.2, are continuous for one and the same medium (having the same $\epsilon$). Since $E \propto k$, this does not apply in the case of different media. But if we define $D := E/k$, called *dielectric induction* or

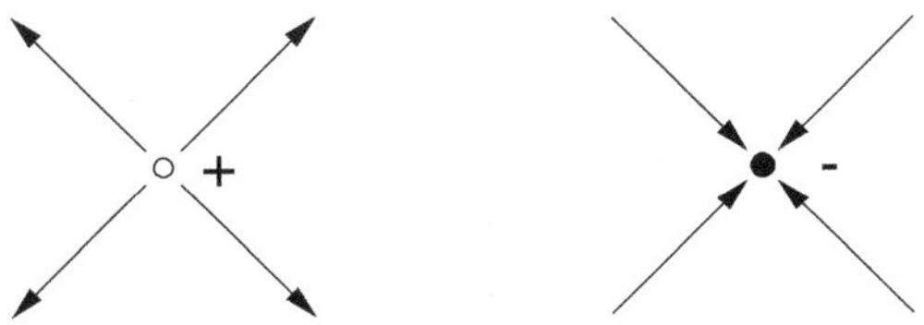

**Fig. 2.2.** Charges and lines of force.

*displacement*, the lines of $D$ are continuous at interfaces between different media. The law of force

$$F_{12} \propto \frac{q_1 q_2}{r^2}$$

for point charges $q_i$ was discovered experimentally by Coulomb, who published his observations in 1785.

Later it became known that the law had been observed earlier by Cavendish who, however, never published his findings. Coulomb used for his observations a so-called torsion balance consisting of a vertical rod or fibre with another fibre attached horizontally at its lower end. At one end of this horizontal fibre a small chargeable sphere was mounted and at the other end a piece of paper to stabilize the system in a rotation. If another charge of the same magnitude is brought close to the mounted charge, at a distance $r$ say, the latter experiences a repulsion evidenced by a rotation of the horizontal fibre through an angle $\theta$. The corresponding torque is proportional to $\theta$. By selecting various values of $r$ $(nr, n = 1, 2, 3, \ldots)$ Coulomb arrived at the law now named after him for the case of repulsive charges. For charges which attract each other Coulomb modified his experiment. His experiments did not establish the proportionality to the charges themselves. This was more or less assumed in analogy to masses in Newton's law of gravitation. It was Gauss who later recognized that Coulomb's law permits the definition of charge. In recent times the Coulomb law has, of course, been subjected to much more critical investigations. Thus the power of $-2$ of $r$ has been verified up to deviations of less than $10^{-9}$.

Cavendish, on the other hand, assumed a law of force of the form

$$F_{12} \propto \frac{1}{r^n}, \quad E = k\frac{q}{r^n}, \tag{2.5}$$

and then designed his experiment to determine $n$. His experimental setup consisted of two concentric conducting spherical shells connected by a conducting wire from one to the other. He charged the exterior shell and then removed the wire together with the exterior shell. The interior shell then provided no evidence of a field in its neighborhood. From this Cavendish concluded that $n = 2$, because in this case the interior of a charged spherical shell has $E = 0$ (thus in the case $n = 2$ the charge covered the exterior shell, as we explain later; see also Example 2.2).

## 2.3 The Electrostatic Potential

We now demonstrate that the electric field $\mathbf{E}$ can be derived from a potential, the *electrostatic potential.* Since (with $r = |\mathbf{r}| = \sqrt{x^2 + y^2 + z^2}$)

$$\nabla_{\mathbf{r}} \frac{1}{|\mathbf{r} - \mathbf{r}'|} = -\frac{(\mathbf{r} - \mathbf{r}')}{|\mathbf{r} - \mathbf{r}'|^3}, \quad \nabla = \left(\frac{\partial}{\partial x}, \frac{\partial}{\partial y}, \frac{\partial}{\partial z}\right), \tag{2.6}$$

we obtain from Eq. (2.4)

$$\mathbf{E}(\mathbf{r}) = -k \int \rho(\mathbf{r}') \nabla_{\mathbf{r}} \frac{1}{|\mathbf{r} - \mathbf{r}'|} d\mathbf{r}' = k \int d\mathbf{r}' \frac{\rho(\mathbf{r}')(\mathbf{r} - \mathbf{r}')}{|\mathbf{r} - \mathbf{r}'|^3} \equiv -\nabla\phi, \tag{2.7}$$

i.e.

$$\mathbf{E} = -\nabla\phi \quad \text{newtons/coulomb}, \tag{2.8}$$

where $\phi$ is the electrostatic potential

$$\phi(\mathbf{r}) = k \int \frac{\rho(\mathbf{r}')}{|\mathbf{r} - \mathbf{r}'|} d\mathbf{r}' \quad \text{volts}. \tag{2.9}$$

Since *curl grad* $= 0$, it follows that

$$\nabla \times \mathbf{E} = 0, \tag{2.10}$$

i.e. the electric field is *curl-free* (or whirl- or vortex-free; *curl* is also often written *rot* for "rotation") .

Next we consider the *flux* $\mathbf{E} \cdot d\mathbf{F}$ of the vector field $\mathbf{E}$ through the surface area element $d\mathbf{F}(\mathbf{r})$ at $\mathbf{r}$ of a surface $F$ surrounding the charges. Then, cf. Fig. 2.3,

$$\mathbf{E} \cdot d\mathbf{F}(\mathbf{r}) = k \int_V \rho(\mathbf{r}') \frac{(\mathbf{r} - \mathbf{r}') \cdot d\mathbf{F}(\mathbf{r})}{|\mathbf{r} - \mathbf{r}'|^3} d\mathbf{r}'. \tag{2.11}$$

Let $V$ be the volume $V(F)$ enclosed by $F$; the case of $V \neq V(F)$ has to be dealt with separately. It follows that

$$\int_F \mathbf{E} \cdot d\mathbf{F} = k \int_{V(F)} \rho(\mathbf{r}') d\mathbf{r}' \int_F \frac{(\mathbf{r} - \mathbf{r}') \cdot d\mathbf{F}(\mathbf{r})}{|\mathbf{r} - \mathbf{r}'|^3}. \tag{2.12}$$

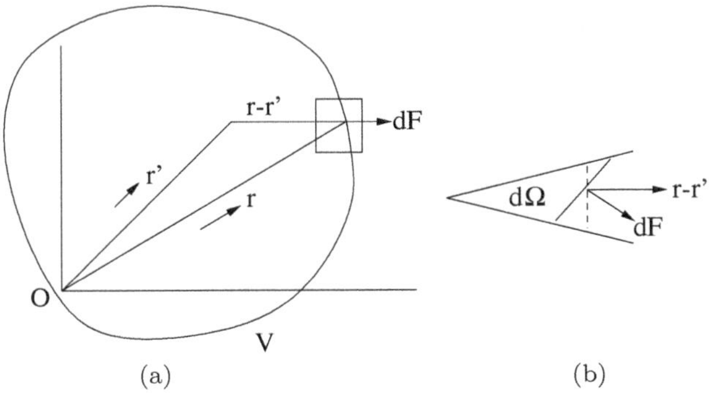


**Fig. 2.3.** (a) The surface and (b) the solid angle.

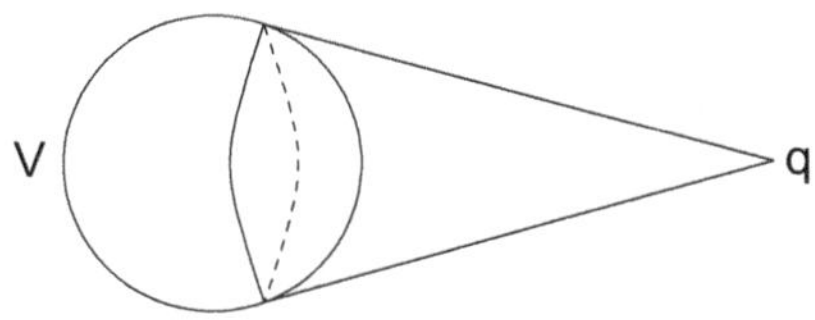


**Fig. 2.4.** Charge $q$ outside of $V$.

But

$$|\mathbf{r}-\mathbf{r}'|^2 d\Omega = \frac{(\mathbf{r}-\mathbf{r}')\cdot d\mathbf{F}(\mathbf{r})}{|\mathbf{r}-\mathbf{r}'|}. \tag{2.13}$$

Hence

$$\int_F \mathbf{E}\cdot d\mathbf{F} = k\int_{V(F)} \rho(\mathbf{r}')d\mathbf{r}' \int_F d\Omega = 4\pi k \int_V \rho(\mathbf{r}')d\mathbf{r}', \tag{2.14}$$

provided the point $\mathbf{r}'$ (charge element $\rho(\mathbf{r}')$) is within $V$. This relation is the integral form of the Gauss law $\boldsymbol{\nabla}\cdot\mathbf{E} = 4\pi k\rho$. If $\mathbf{r}'$ is outside of $V$, that is if we wish to know the flux through a closed surface $F$ in the case when the field is due to charges outside of $F$, we have

$$\int_F d\Omega = \Omega_1 - \Omega_2 = 0, \tag{2.15}$$

as we can see from Fig. 2.4, since the consideration of the solid angles of a closed volume such as that of a sphere subtended at a charge $q$ implies:

$$\Omega_1 + (-\Omega_1) = 0.$$

In this consideration the volume $V$ is a purely mathematical volume, so to speak. If the material of the volume $V(F)$ is such that the lines of force of

$\mathbf{E}$ penetrate into $V$, this result implies that every ingoing line must also leave the volume again. If the closed surface $F$, which does not enclose a charge, is made of conducting material (see later), then the field $\mathbf{E}$ in the interior is zero. With the help of the *Gauss divergence theorem* (here assumed known) we obtain from Eq. (2.14) (the delta function is explained below):

$$\int_F \mathbf{E} \cdot d\mathbf{F} = \int_V \boldsymbol{\nabla} \cdot \mathbf{E} d\mathbf{r} = 4\pi k \int_V \rho(\mathbf{r}) d\mathbf{r} = 4\pi k \int_V \Sigma_i q_i \delta(\mathbf{r} - \mathbf{r}_i) d\mathbf{r}, \quad (2.16)$$

depending on whether the charge distribution is continuous or discrete. It follows that

$$\int_V (\boldsymbol{\nabla} \cdot \mathbf{E} - 4\pi k \rho(\mathbf{r})) d\mathbf{r} = 0. \quad (2.17)$$

Since this is valid for any volume $V$, i.e. $\int_V = \int_{V'} = \cdots = 0$, it follows that

$$\boldsymbol{\nabla} \cdot \mathbf{E} = 4\pi k \rho(\mathbf{r}). \quad (2.18)$$

This relation is known as the *Gauss law.* Thus, *charges in space are sources* of the electric field.

## 2.4 The Equations of Electrostatics

We have now obtained the *three important equations of electrostatics*:

$$\begin{aligned} &\mathbf{E} = -\boldsymbol{\nabla}\phi, \\ \boldsymbol{\nabla} \cdot \mathbf{E} = 4\pi k \rho, \quad \text{or} \quad &\int \mathbf{E} \cdot d\mathbf{F} = 4\pi k \Sigma_i q_i, \\ &\boldsymbol{\nabla} \times \mathbf{E} = 0. \end{aligned} \quad (2.19a)$$

Later we shall assume that these equations are also valid when $\rho, \mathbf{E}, \phi$ are time-dependent.[2] From the first two equations we obtain the *Poisson equation*,

$$\triangle\phi = -4\pi k \rho(\mathbf{r}). \quad (2.19b)$$

The equation for $\rho = 0$, i.e. $\triangle\phi = 0$, is known as the *Laplace equation.*

[2]As Konopinski [79] remarks (p. 13), the most important reason for putting the Maxwell equation $\boldsymbol{\nabla} \cdot \mathbf{E} = 4\pi k \rho$ at the basis of the theory is that it is "more safely generalizable to nonstatic situations" than other formulations (see there for further discussion).

We recapitulate first the following important three-dimensional formulas which are needed in various applications, also in this text[3]:

(a) In *Cartesian coordinates* $(x, y, z)$:

$$\nabla \cdot \mathbf{E} = \frac{\partial E_x}{\partial x} + \frac{\partial E_y}{\partial y} + \frac{\partial E_z}{\partial z}, \tag{2.20a}$$

$$\triangle\phi = \nabla \cdot \nabla\phi = \frac{\partial^2 \phi}{\partial x^2} + \frac{\partial^2 \phi}{\partial y^2} + \frac{\partial^2 \phi}{\partial z^2}, \tag{2.20b}$$

(b) In *cylindrical coordinates* $(r, \theta, z)$:

$$\nabla \cdot \mathbf{E} = \frac{1}{r}\frac{\partial}{\partial r}(rE_r) + \frac{1}{r}\frac{\partial E_\theta}{\partial \theta} + \frac{\partial E_z}{\partial z}, \tag{2.21a}$$

$$\triangle\phi = \frac{1}{r}\frac{\partial}{\partial r}\left(r\frac{\partial \phi}{\partial r}\right) + \frac{1}{r^2}\frac{\partial^2 \phi}{\partial \theta^2} + \frac{\partial^2 \phi}{\partial z^2}, \tag{2.21b}$$

(c) In *spherical polar coordinates* (spherical coordinates) $(r, \theta, \varphi)$[4]:

$$\nabla \cdot \mathbf{E} = \frac{1}{r^2}\frac{\partial}{\partial r}(r^2 E_r) + \frac{1}{r\sin\theta}\frac{\partial}{\partial \theta}(\sin\theta E_\theta) + \frac{1}{r\sin\theta}\frac{\partial E_\varphi}{\partial \varphi}, \tag{2.22a}$$

$$\triangle\phi = \frac{1}{r^2}\frac{\partial}{\partial r}\left(r^2\frac{\partial \phi}{\partial r}\right) + \frac{1}{r^2\sin\theta}\frac{\partial}{\partial \theta}\left(\sin\theta\frac{\partial \phi}{\partial \theta}\right) + \frac{1}{r^2\sin\theta}\frac{\partial^2 \phi}{\partial \varphi^2}. \tag{2.22b}$$

Thus e.g. the potential $\phi$ outside a uniformly charged sphere (i.e. where $\rho(\mathbf{r}) = 0$ in Eq. (2.19b)) is given by (since independent of $\theta$ and $\varphi$)

$$\frac{\partial}{\partial r}\left(r^2\frac{\partial \phi}{\partial r}\right) = 0, \quad r^2\frac{\partial \phi}{\partial r} = \text{const.}, \quad \phi = -\frac{\text{const.}}{r}. \tag{2.22c}$$

**Example 2.1 (Discrete and continuous charge distributions).** Let the following potential be given:

$$\phi(r) = kq\frac{e^{-\alpha r}}{r}\left(1 + \frac{\alpha r}{2}\right) \quad \text{volts.} \tag{2.23}$$

Show that there are both discrete and continuous charge distributions with a vanishing total charge.

[3] Magnus and Oberhettinger [90], pp. 144–146.

[4] For $\varphi$ we also use in the text $\psi$ or $\phi$.

**Solution:** In the case of the given spherically symmetric potential $\phi(r)$ or electric field $E_r$, the Gauss law (2.18) is

$$\frac{1}{r^2}\frac{\partial}{\partial r}(r^2 E_r) = 4\pi k\rho(r), \tag{2.24}$$

so that

$$r^2 E_r = \int_0^r 4\pi k\rho(r)r^2 dr \equiv kQ(r), \tag{2.25}$$

where $Q(r)$ is the charge enclosed in a sphere of radius $r$. With $E_r = -d\phi(r)/dr$, it follows that

$$Q(r) = -\frac{r^2}{k}\frac{d\phi(r)}{dr}. \tag{2.26}$$

For the given potential $\phi(r)$ we have

$$\begin{aligned} Q(r) &= -4\pi\epsilon_0 r^2 \frac{d\phi(r)}{dr} = qe^{-\alpha r}\left[1 + \alpha r + \frac{1}{2}\alpha^2 r^2\right] \\ &= -qe^{-\alpha r}\left[\frac{\alpha r}{2} - (1+\alpha r)\left(1 + \frac{\alpha r}{2}\right)\right]. \end{aligned} \tag{2.27}$$

For $r \to 0$ we obtain the discrete charge

$$Q(r) \to q \quad \text{coulombs}, \tag{2.28}$$

whereas for $r \to \infty$

$$Q(r) \to 0.$$

The continuous charge distribution is given by $\rho(r)$, for which we obtain from Eq. (2.25)

$$4\pi\rho(r)r^2 = \frac{dQ(r)}{dr}, \tag{2.29a}$$

i.e.

$$\begin{aligned} \rho(r) &= \frac{1}{4\pi r^2}\frac{dQ(r)}{dr} = \frac{qe^{-\alpha r}}{4\pi r^2}\left[-\alpha\left(1 + \alpha r + \frac{1}{2}\alpha^2 r^2\right) + \alpha + \alpha^2 r\right] \\ &= -\frac{1}{8\pi}q\alpha^3 e^{-\alpha r}. \end{aligned} \tag{2.29b}$$

This can be verified by computing the sum of the distribution, i.e.

$$\int_0^\infty \rho(r)4\pi r^2 dr = -q. \tag{2.30}$$

The vanishing of the sum of results (2.28) and (2.30) implies that the total charge in space is zero. Applied to the neutral hydrogen atom, we can interpret the discrete charge as that of the proton and the continuous charge distribution as that of the electron.

**Example 2.2 (Theoretical explanation of Cavendish's experiment).** Assuming that the electric field $\mathbf{E}$ of a point charge $q$ is given by the radially symmetric expression

$$E = \frac{1}{4\pi\epsilon_0}\frac{q}{r^n} \quad \text{newtons/coulomb}, \tag{2.31}$$

show that the field $\mathbf{E}$ at a point $r$ without and correspondingly within a spherical shell of radius $r_0$ and with surface charge density $\sigma$ are given by the following expressions:

$$\begin{aligned} r > r_0: \quad E &= \frac{\sigma r_0}{2\epsilon_0 r^2(3-n)(1-n)}[(r+r_0)^{2-n}\{r(2-n)-r_0\} \\ &\quad - (r-r_0)^{2-n}\{r(2-n)+r_0\}], \end{aligned} \tag{2.32a}$$

$$\begin{aligned} r < r_0: \quad E &= \frac{\sigma r_0}{2\epsilon_0 r^2(3-n)(1-n)}[(r+r_0)^{2-n}\{r(2-n)-r_0\} \\ &\quad + (r-r_0)^{2-n}\{r(2-n)+r_0\}]. \end{aligned} \tag{2.32b}$$

What can be deduced from these results for the cases $n < 2, n = 2, n > 2$?

**Solution:** Let $\sigma = q/4\pi r_0^2$ be the charge per unit area of the spherical shell with radius $r_0$. The quantity to be calculated is the electric field at the point $P$ in Fig. 2.5. The charge on an infinitesimal element of area $dF$ is $\sigma dF = \sigma r_0^2 \sin\theta d\theta d\psi$. From the formula for the assumed law of force

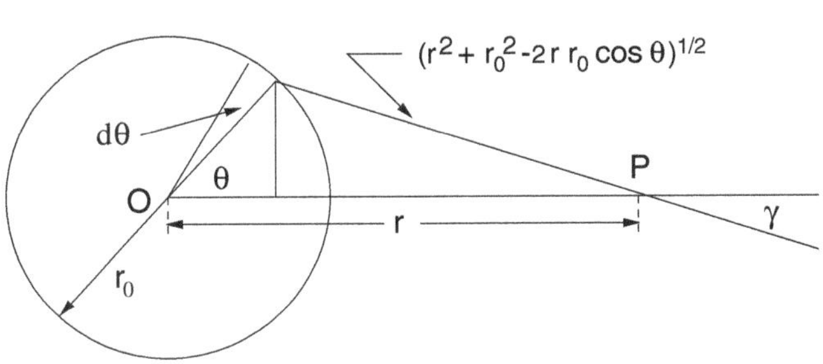


**Fig. 2.5.** Field point $P$ outside the spherical shell.

Eq. (2.31): Field at $P$ equals

$$E_P = \text{charge}/4\pi\epsilon_0(\text{distance})^n, \tag{2.33}$$

we deduce that here

$$dE_P = \frac{\sigma r_0^2 \sin\theta d\theta d\psi}{4\pi\epsilon_0(r^2 + r_0^2 - 2rr_0\cos\theta)^{n/2}}. \tag{2.34}$$

Integrating over the entire sphere we obtain the total field strength at $P$. From the symmetry of the geometry we infer that contributions perpendicular to OP cancel each other so that only the component along OP remains. We obtain this component by multiplying $dE_P$ by $\cos\gamma$ where (cf. Fig. 2.5):

$$\cos\gamma = \frac{r - r_0\cos\theta}{\sqrt{r^2 + r_0^2 - 2rr_0\cos\theta}}.$$

We thus obtain for the field at $P$:

$$E_P = \int_0^{2\pi}\int_0^{\pi} \frac{\sigma r_0^2(r - r_0\cos\theta)\sin\theta d\theta d\psi}{4\pi\epsilon_0(r^2 + r_0^2 - 2rr_0\cos\theta)^{(n+1)/2}}. \tag{2.35}$$

With $z = \cos\theta$ the integral is of the form

$$\int_{-1}^{1} \frac{(a - bz)dz}{(a^2 + b^2 - 2abz)^{(n+1)/2}}.$$

But[5]

$$\int \frac{dz}{(A + Bz)^{(n+1)/2}} = -\frac{2}{B(n-1)}\frac{1}{(A + Bz)^{(n-1)/2}}, \tag{2.36a}$$

and using this integral we can evaluate:

$$\begin{aligned}\int \frac{zdz}{(A + Bz)^{(n+1)/2}} &= \frac{1}{B}\int \frac{(A + Bz - A)dz}{(A + Bz)^{(n+1)/2}} \\ &= -\frac{A}{B}\int \frac{dz}{(A + Bz)^{(n+1)/2}} + \frac{1}{B}\int \frac{dz}{(A + Bz)^{(n-1)/2}}.\end{aligned} \tag{2.36b}$$

If $E_P$ is evaluated with the help of these integrals, one obtains the expressions for the field given at the beginning. We then see that for $n = 2$, the field $E_P$ vanishes if $r < r_0$. This is the theoretical explanation of Cavendish's

[5]Dwight [42], formula 90, p. 17.

experiment. We also observe that if we take in Eqs. (2.32a) and (2.32b) the limits of first $n = 2$ and then $r = r_0$ we obtain the field on the surface of the sphere as

$$E = \frac{\sigma}{\epsilon_0} \quad \text{newtons/coulomb}, \tag{2.37}$$

an expression which will reappear several times later, as in Eq. (2.104).

## 2.5 Dirac's Delta Distribution

A solution of the Poisson equation $\triangle\phi = -4\pi k\rho$ is the Poisson integral (already encountered in Eq. (2.9)):

$$\phi(\mathbf{r}) = k\int \frac{\rho(\mathbf{r}')d\mathbf{r}'}{|\mathbf{r}-\mathbf{r}'|}. \tag{2.38}$$

Before we show that this expression satisfies the Poisson equation, we introduce the so-called *δ-distribution* (an introduction is given in Appendix A). This quantity was introduced by Dirac and was originally called *delta function*. Actually it is an improper function which is today called *distribution*, and is defined as a functional by the property

$$f(a) = \int_{-\infty}^{\infty} f(x)\delta(x-a)dx. \tag{2.39}$$

The quantity $\delta(x)$ can be looked at as the limit of a function which vanishes everywhere except at $x = 0$, where it is singular. Thus e.g.

$$\begin{aligned}\delta(x) &= \lim_{\epsilon\to 0}\frac{1}{2\pi}\int_{-\infty}^{\infty} dk e^{-\epsilon|k|}e^{ikx} \\ &= \lim_{\epsilon\to 0}\frac{1}{\pi}\frac{\epsilon}{\epsilon^2+x^2} \\ &= \begin{cases} 0 & : \quad x \neq 0, \\ \infty & : \quad x = 0, \end{cases}\end{aligned} \tag{2.40}$$

or

$$\begin{aligned}\delta(x) &= \lim_{a\to 0}\frac{1}{2\pi}\int_{-\infty}^{\infty} dk e^{-ak^2}e^{ikx} \\ &= \lim_{a\to 0}\frac{1}{2\sqrt{\pi a}}e^{-x^2/4a}.\end{aligned} \tag{2.41}$$

The delta function has the integral representation

$$\delta(x) = \frac{1}{2\pi}\int_{-\infty}^{\infty} dk e^{ikx}. \tag{2.42}$$

Important properties of $\delta(x)$ are:

$$\begin{aligned}
\delta(x) &= \delta(-x), \\
\delta'(x) &= -\delta'(-x), \\
x\delta(x) &= 0, \\
x\delta'(x) &= -\delta(x), \\
\delta(ax) &= \frac{1}{a}\delta(x), \quad (a > 0).
\end{aligned} \tag{2.43}$$

These relations can be verified by multiplication by a continuous differentiable function and subsequent integration (as in Appendix A). Furthermore the following relations can be shown to hold:

$$\begin{aligned}
\delta(x^2 - a^2) &= \frac{1}{2a}[\delta(x-a) + \delta(x+a)], \quad (a > 0), \\
f(x)\delta(x) &= f(0)\delta(x), \\
\theta'(x) &= \delta(x),
\end{aligned} \tag{2.44}$$

where the $\theta$-function or *Heaviside step function* is defined by

$$\theta(x) = \begin{cases} 1 & : \quad x > 0, \\ 0 & : \quad x < 0. \end{cases} \tag{2.45}$$

Equations (2.44) are explicitly verified in Appendix A. We observe that

$$\delta(\mathbf{r}) \equiv \delta(x)\delta(y)\delta(z). \tag{2.46}$$

A solution of the inhomogeneous equation

$$\triangle G(\mathbf{r}) = \delta(\mathbf{r}) \tag{2.47}$$

is called *Green's function* (we shall consider such functions in detail later). Once $G(\mathbf{r})$ is known, the inhomogeneous solution $\phi(\mathbf{r})$ of $\triangle\phi = -4k\pi\rho$ follows from the relation

$$\phi(\mathbf{r}) = -k\int G(\mathbf{r}-\mathbf{r}')4\pi\rho(\mathbf{r}')d\mathbf{r}' \tag{2.48}$$

apart from a solution of the homogeneous equation $\triangle\phi = 0$ which could be added subject to the same boundary conditions as for $G$. The relation (2.48) is readily verified, because if we apply the Laplace operator $\triangle$ to this, we obtain

$$\triangle\phi(\mathbf{r}) = -k\int \delta(\mathbf{r}-\mathbf{r}')4\pi\rho(\mathbf{r}')d\mathbf{r}' = -4k\pi\rho(\mathbf{r}). \tag{2.49}$$

For $\phi(\mathbf{r})$ of Eq. (2.38) to satisfy the integral relation (2.48) we must have

$$G(\mathbf{r}-\mathbf{r}') = -\frac{1}{4\pi|\mathbf{r}-\mathbf{r}'|}, \tag{2.50}$$

i.e.

$$\triangle_{\mathbf{r}}\frac{1}{|\mathbf{r}|} = -4\pi\delta(\mathbf{r}). \tag{2.51}$$

This expression makes sense only for $\mathbf{r} \neq 0$. Now,

$$r = \sqrt{x^2+y^2+z^2}, \quad \triangle = \frac{\partial^2}{\partial x^2} + \frac{\partial^2}{\partial y^2} + \frac{\partial^2}{\partial z^2} \tag{2.52}$$

and

$$\frac{\partial}{\partial x}\left(\frac{1}{r}\right) = -\frac{x}{r^3}, \quad \frac{\partial^2}{\partial x^2}\left(\frac{1}{r}\right) = -\frac{1}{r^3} + \frac{3x^2}{r^5}, \tag{2.53}$$

so that

$$\triangle\frac{1}{r} = -\frac{3}{r^3} + \frac{3r^2}{r^5} = 0. \tag{2.54}$$

We can see by integration that the factor $-4\pi$ in Eq. (2.51) is correct:

$$\begin{aligned}\int_V \triangle_{\mathbf{r}}\frac{1}{r}d\mathbf{r} &= \int_V \boldsymbol{\nabla}\cdot\left(\boldsymbol{\nabla}\frac{1}{r}\right)d\mathbf{r} = \int_F \left(\boldsymbol{\nabla}\frac{1}{r}\right)\cdot d\mathbf{F} \\ &= -\int_F \frac{\mathbf{r}}{r^3}\cdot d\mathbf{F} = -\int\frac{dF}{r^2} = -\int d\Omega = -4\pi. \end{aligned} \tag{2.55}$$

This verifies Eq. (2.51). Equation (2.50) shows that the Green's function is the potential of a negative unit charge at the source point multiplied by $\epsilon_0$

since with

$$\rho(\mathbf{r}') = -\epsilon_0 \delta(\mathbf{r}') \tag{2.56}$$

we have

$$\phi(\mathbf{r}) = -k \int G(\mathbf{r} - \mathbf{r}') 4\pi \rho(\mathbf{r}') d\mathbf{r}' = G(\mathbf{r}). \tag{2.57}$$

We see also that the Green's function $G$ is the response of a system (here the system is represented by the Laplacian, but this could also be some other second order differential operator, such as that of an equation of motion in classical mechanics or of a Schrödinger equation in quantum mechanics) to a delta function-like source. One can also refer to output and input. In the expression for the potential $\phi(\mathbf{r})$ as an integral over the Green's function $G$ the values of $-4k\pi\rho(\mathbf{r}')$ at points over which one integrates play the role of weight factors. The overall expression thus represents the sum or superposition of the responses of the system thus weighted to delta function-like sources or inputs.

**Example 2.3 (The Planck mass).** In Chapter 1 natural units were referred to. Show that in a world with $n$ spatial dimensions Newton's gravitational constant $G$ raised to the power $1/(1-n)$ has in natural units the dimension of a mass, which defines the so-called *Planck mass* $M_P$, i.e.

$$M_P := G^{1/(1-n)}, \tag{2.58}$$

for $\hbar = 1, c = 1$ and is dimensionless.

**Solution:** In $n$ spatial dimensions the Laplacian is the operator

$$\triangle = \sum_{i=1}^{n} \frac{\partial^2}{\partial x_i^2}, \tag{2.59}$$

and Newton's gravitational potential $\psi$ is just like the Coulomb potential given by the solution of $\triangle\psi = 0$ with $r = \sqrt{\sum_{i=1}^{n} x_i^2} \neq 0$. For $\psi = 1/r^m$ we have

$$\frac{\partial \psi}{\partial x_i} = -\frac{m x_i}{r^{m+2}}, \quad \sum_{i=1}^{n} \frac{\partial^2 \psi}{\partial x_i^2} = -\frac{mn}{r^{m+2}} + \frac{m(m+2)}{r^{m+2}}. \tag{2.60}$$

For $\triangle\psi$ to be zero, we must have $m = n-2$. The gravitational constant $G$ is now defined by the potential $V = -G \times \text{mass}/r^{n-2}$ per unit mass. In natural units this potential per unit mass (here $V/\text{mass}$) is dimensionless (recall Einstein's formula energy $=$ mass times $c^2$). Then $G$ has the dimension of

$(\text{mass})^{-(n-1)}$, or one can say the mass varies as $G^{-1/(n-1)}$. In the case of $n = 3$ the mass $M_P$ varies as $1/\sqrt{G}$ as in Table B.2 of Appendix B.

**Example 2.4 (The delta function in various coordinates).** What is $\delta(\mathbf{r})$ in three dimensional cylindrical and spherical polar coordinates?

**Solution:** Since $\int d\mathbf{r}'\delta(\mathbf{r}-\mathbf{r}') = 1$ and in cylindrical coordinates $\rho, \psi, z$ and spherical polar coordinates $r, \theta, \varphi$ the volume element $d\mathbf{r} = \rho d\rho dz d\psi$ and $d\mathbf{r} = r^2 dr \sin\theta d\theta d\varphi$ respectively, we have

$$\delta(\mathbf{r}-\mathbf{r}')|_{\text{cyl}} = \frac{1}{\rho'}\delta(\rho-\rho')\delta(\psi-\psi')\delta(z-z') \tag{2.61}$$

and

$$\begin{aligned}\delta(\mathbf{r}-\mathbf{r}')|_{\text{spher}} &= \frac{1}{r'^2}\delta(r-r')\delta(\cos\theta-\cos\theta')\delta(\varphi-\varphi') \\ &= \frac{1}{r'^2\sin\theta'}\delta(r-r')\delta(\theta-\theta')\delta(\varphi-\varphi'). \end{aligned}\tag{2.62}$$

**Example 2.5 (Representations of delta functions).** By appropriate evaluation of the integral

$$\delta(\mathbf{x}) = \lim_{\epsilon\to 0}\frac{1}{(2\pi)^n}\int d^n k e^{-\epsilon|\mathbf{k}|}e^{i\mathbf{k}\cdot\mathbf{x}} \tag{2.63}$$

for $n = 1$ and 2 determine representations of the one- and two-dimensional delta functions.

**Solution:** The case $n = 1$ is simple:

$$\frac{1}{2\pi}\int_{-\infty}^{\infty} dk e^{-\epsilon|k|}e^{ikx} = \frac{1}{\pi}\int_0^{\infty} dk e^{-\epsilon k}\cos kx = \frac{1}{\pi}\frac{\epsilon}{\epsilon^2+x^2}, \quad (\epsilon > 0). \tag{2.64}$$

We thus obtain the following representation for the delta function in one dimension:

$$\delta(x) = \lim_{\epsilon\to 0}\frac{1}{2\pi}\int_{-\infty}^{\infty} dk e^{-\epsilon|k|}e^{ikx} = \lim_{\epsilon\to 0}\frac{1}{\pi}\frac{\epsilon}{\epsilon^2+x^2}. \tag{2.65}$$

Higher dimensional cases are not quite so easy. In two dimensions we have

$$\begin{aligned}\frac{1}{(2\pi)^2}\int dk_x dk_y e^{-\epsilon\sqrt{k_x^2+k_y^2}}e^{i(k_x x+k_y y)} &= \frac{1}{(2\pi)^2}\int_0^{\infty} kdk\int_{-\pi}^{\pi} d\theta e^{-\epsilon k}e^{ik|\mathbf{x}|\cos\theta} \\ &= \frac{1}{(2\pi)^2}\int_{-\pi}^{\pi} d\theta\left(-\frac{d}{d\epsilon}\right)\frac{1}{\epsilon - i|\mathbf{x}|\cos\theta}. \end{aligned}\tag{2.66}$$

The integral

$$\int_{-\pi}^{\pi} \frac{d\theta}{\epsilon - ix\cos\theta}$$

can be evaluated with the help of Tables of Integrals and with caution in the two cases $\epsilon^2 \lessgtr \mathbf{x}^2$. In both cases one obtains $2\pi/\sqrt{\epsilon^2 + \mathbf{x}^2}$. Hence the representation that follows is

$$\delta(\mathbf{x}) = \lim_{\epsilon\to 0} \frac{1}{(2\pi)^2} \int d^2k e^{-\epsilon|\mathbf{k}|} e^{i\mathbf{k}\cdot\mathbf{x}} = \lim_{\epsilon\to 0} \frac{\epsilon}{2\pi(\epsilon^2 + \mathbf{x}^2)^{3/2}}. \tag{2.67}$$

Representations of the type discussed here are frequently useful in applications.

**Example 2.6 (Further representations of the delta function).** Show that[6]

$$K(\mathbf{x}) = \lim_{x_0\to 0} \tilde{K}(\mathbf{x}), \quad \tilde{K}(\mathbf{x}) = c\frac{x_0^n}{[x_0^2 + \sum_{j=1}^n x_j^2]^n}, \tag{2.68}$$

is a delta function with coefficient 1 if the constant $c$ is chosen suitably.

**Solution:** In order to establish the result, one has to show that (a) $K(\mathbf{x}) = 0$ for $x_j \neq 0$, and (b) $\int d\mathbf{x} K(\mathbf{x}) = 1$ and independent of $x_0$. We limit ourselves here exercise-wise to a few remarks. First of all we convince ourselves that a rescaling of $x_i \to x_i' = x_i/x_0$ verifies that the integral over $K$ is independent of $x_0$:

$$\int \tilde{K}(\mathbf{x}) d\mathbf{x} = c\int \frac{x_0^n d\mathbf{x}}{[x_0^2 + \mathbf{x}^2]^n} = c\int \frac{d\mathbf{x}'}{[1 + \mathbf{x}'^2]^n}. \tag{2.69}$$

By going to polar coordinates it is not too difficult to verify that for example in the case $n = 2 : c = 1/\pi$, in the case $n = 3 : c = 4/\pi^2$, and so on. Also the following representation can be verified:

$$\delta(\mathbf{x}) = \lim_{x_0\to 0} \frac{(2h_+ - 1)! x_0^{-2h_-}}{\pi(2h_+ - 2)!} \left[\frac{x_0}{x_0^2 + \mathbf{x}^2}\right]^{2h_+} \equiv \lim_{x_0\to 0} K(\mathbf{x}),$$

$$\text{with} \quad h_+ + h_- = 1. \tag{2.70}$$

[6]See also Witten [164].

For example we have for $n = 2$ (in this case $h_+ = 1, h_- = 0$):

$$\delta(\mathbf{x}) = \lim_{x_0 \to 0} \frac{x_0^2}{\pi(x_0^2 + \mathbf{x}^2)^2}. \tag{2.71}$$

We shall encounter in particular this case $n = 2$ in Chapter 21 in the regularization of the field of a magnetic monopole.

**Example 2.7 (The Coulomb potential in higher dimensions).** Determine the Coulomb potential $\phi$ in $N > 3$ space dimensions.

**Solution:** One way to solve this problem is the following. In three dimensions from Eq. (2.18):

$$\nabla \cdot \mathbf{E} = \frac{\rho}{\epsilon_0}, \quad \nabla = \left(\mathbf{e}_r \frac{\partial}{\partial r}, \mathbf{e}_\varphi \frac{1}{r \sin\theta} \frac{\partial}{\partial \varphi}, \mathbf{e}_\theta \frac{1}{r} \frac{\partial}{\partial \theta}\right). \tag{2.72}$$

For charge $e$ at $\mathbf{r}' = \mathbf{r}$ we have (cf. Example 2.4)

$$\rho(\mathbf{r} - \mathbf{r}') = e\delta(\mathbf{r} - \mathbf{r}') = \frac{e}{r'^2}\delta(r - r')\delta(\cos\theta - \cos\theta')\delta(\varphi - \varphi'). \tag{2.73}$$

Thus (since the field is radial),

$$E_r = \int^r \frac{\rho(r')}{\epsilon_0} dr' \tag{2.74}$$

and with further integrating out of the angles

$$\begin{aligned} &\int_0^{2\pi} d\varphi' \int_0^{-1} d\cos\theta' E_r \\ &= \int dr' \frac{e}{\epsilon_0 r'^2} \delta(r - r')\delta(\varphi - \varphi')\delta(\cos\theta - \cos\theta') d\varphi' d\cos\theta', \end{aligned} \tag{2.75}$$

i.e.

$$-4\pi E_r = \frac{e}{\epsilon_0 r^2}, \quad E_r = -\frac{e}{4\pi\epsilon_0 r^2}, \quad \phi = -\int E_r dr = -\frac{e}{4\pi\epsilon_0 r}. \tag{2.76}$$

For the case of $N$ space dimensions we require the $N$-dimensional generalization, in particular

$$\begin{aligned} d\mathbf{r} &= r^{N-1}(\sin\theta_1)^{N-2}(\sin\theta_2)^{N-3}\dots(\sin\theta_{N-2})^2 dr d\theta_1 \dots d\theta_{N-1} \\ &\equiv r^{N-1} dr d\Omega_{N-1}. \end{aligned} \tag{2.77}$$

The solid angle $\Omega_{N-1}$ can be deduced from the equality of the following integrals:

$$\int d^N x e^{-\frac{1}{2}\sum_{i=1}^N x_i^2} = \int_0^\infty dr r^{N-1} d\Omega_{N-1} e^{-\frac{1}{2}r^2}, \tag{2.78}$$

and so from[7]

$$(2\pi)^{N/2} = \frac{2^{N/2}}{N}\left(\frac{N}{2}\right)!\,\Omega_{N-1}. \tag{2.79}$$

Hence

$$\Omega_{N-1} = \frac{N\pi^{N/2}}{(N/2)!}. \tag{2.80}$$

For $N = 3$ with $(-1/2)! = \sqrt{\pi}$ (see gamma functions and factorials in Tables of Special Functions), we obtain $\Omega_2 = 4\pi$, as expected. Hence

$$E_r = -\frac{e}{\Omega_{N-1}\epsilon_0 r^{N-1}} \quad \text{and} \quad \phi(r) = -\frac{e}{(N-2)\Omega_{N-1}\epsilon_0 r^{N-2}}. \tag{2.81}$$

**Example 2.8 (The Coulomb potential in space dimensions 1 and 2).** What is the space dependence of the Coulomb potential in a world with one space dimension? What is it in the case of 2 space dimensions?

**Solution:** We have $\nabla \cdot \mathbf{E} = 4\pi k\rho$ with $\mathbf{E} = -\nabla\phi$; hence in the case of one space dimension (keeping the same symbols on the right originally defined for three dimensions)

$$\frac{d^2\phi}{dx^2} = -4\pi k\rho = -\frac{\rho}{\epsilon_0}. \tag{2.82}$$

With the Green's function $G(x)$, the solution is

$$\phi(x) = \int dx' G(x - x')\left(-\frac{\rho(x')}{\epsilon_0}\right), \quad \frac{d^2 G(x)}{dx^2} = \delta(x). \tag{2.83}$$

The Green's function may be obtained by "trial and error" or, better, by contour integration which will be explained later. The result is

$$G(x) = \frac{1}{2\pi}\int \frac{e^{ikx}}{-k^2} dk = \frac{1}{2}|x|, \tag{2.84}$$

[7]Here we use $\int_0^\infty e^{-\alpha x^2} dx = \sqrt{\pi/\alpha}/2$, $\int_0^\infty x^n e^{-\alpha x^2} dx = \{[(n-1)/2]!/2\}\alpha^{-(n+1)/2}$.

as one can verify by differentiation. Thus it follows that in particular for $\rho(x') = e\delta(x' - x_0)$:

$$\phi(x) = -\frac{1}{\epsilon_0}\int dx'\rho(x')|x - x'|, \quad \text{or} \quad \phi(x) = -\frac{e}{\epsilon_0}|x - x_0|. \tag{2.85}$$

Such a linear potential is known as a *confinement potential*, since particles (like quarks) bound together by such a potential cannot be separated with finite energy. (See also Example 18.5).

In the case of 2 space dimensions one can go to polar coordinates $r, \varphi$, so that

$$\left(\frac{\partial^2}{\partial r^2} + \frac{1}{r}\frac{\partial}{\partial r} + \frac{1}{r^2}\frac{\partial^2}{\partial \varphi^2}\right)\phi = -\frac{\rho(r)}{\epsilon_0}. \tag{2.86}$$

In a similar way one finds that $\phi = \text{const.} \ln r$.

## 2.6 Potential Energy of Charges in Electric Fields

Let a charge $q$ be moved from $A$ to $B$ *against* the field $\mathbf{E}$. The work done in this process is

$$\begin{aligned} W &= -\int_A^B q\mathbf{E}(\mathbf{r}) \cdot d\mathbf{r} = q\int_A^B \boldsymbol{\nabla}\phi \cdot d\mathbf{r} \\ &= q\int_A^B d\phi = q(\phi(B) - \phi(A)). \end{aligned} \tag{2.87}$$

The minus sign indicates that the force exerted on the charge $q$ is directed opposite to the field $\mathbf{E}$. The final result shows that the work done is independent of the path from A to B, that is, $q\mathbf{E}$ is a *conservative* field, i.e. $\boldsymbol{\nabla} \times \mathbf{E} = 0$. We also observe that the potential difference $\phi(B) - \phi(A)$ is the work done or energy change per unit test charge. Thus the energy of a charge $q$ in a potential $\phi$ is $q\phi$ joules (see Example 2.16).

## 2.7 The Electric Field at Charged Surfaces

We now consider the behavior of the electric field at charged surfaces or films.[8] Following Faraday it is very instructive to draw lines of force starting

[8]We are not yet considering macroscopic electrostatics and therefore have not yet introduced the dielectric displacement $\mathbf{D}$. Nonetheless a comment seems appropriate at this point. Since the expression for $\mathbf{E}$ contains the factor $1/\epsilon_0$ or $1/\epsilon$, the displacement $\mathbf{D}$ as the product $\epsilon\mathbf{E}$ is independent of the dielectric constant, and this implies the continuity of $\mathbf{D}$ at the interface between media. One should note the difference between this and the case to be considered here, i.e. that of a charged surface in the same medium.

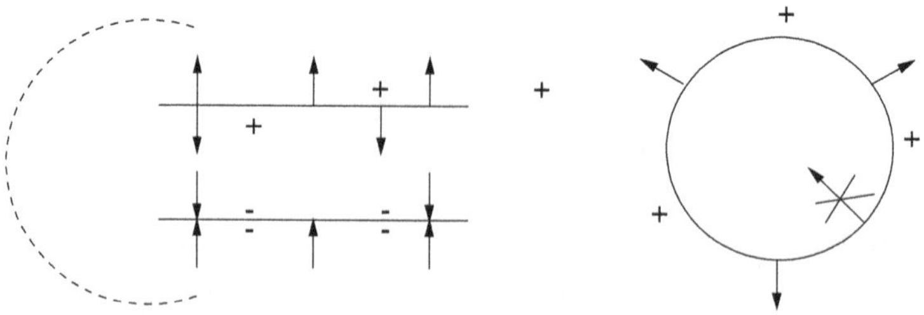


**Fig. 2.6.** Charged surfaces and lines of force.

from positive charges and aiming at negative charges. Already in drawing these lines of force, as in Fig. 2.6, one observes that the electric field suffers a discontinuity at the charged surface, because, as the term "surface" implies (with the same medium on either side), its charge is the same on both sides, so that the vector $\mathbf{E}$ on one side is opposite to that on the other side. Since we wish to consider charged surfaces, we first define as *charge per unit area* the expression $\sigma(\mathbf{r})$ in the following relation, in which $q$ is the total charge of a surface area $\triangle F$, i.e.

$$q = \int_{\triangle F} \sigma(\mathbf{r}) dF \quad \text{coulombs.} \tag{2.88}$$

If $\mathbf{E}_1$ and $\mathbf{E}_2$ are as in Fig. 2.6 the vectors $\mathbf{E}$ at a point on the surface but on either sides and directed away as for a positive charge, then we write the difference of their components along the surface normals $E_{2n} - E_{1n}$ and the difference along a surface tangent $E_{2t} - E_{1t}$. We let $\mathbf{n}$ be a unit vector along the surface normal on the side with index "2" of the charged surface. In the first place we consider a surface of thickness $d$ and volume $\triangle V = d\triangle F$. We apply the Gauss law to this volume and then allow $d$ to approach zero, i.e. we consider

$$\begin{aligned} \int_{F(\triangle V), d\to 0} \mathbf{E} \cdot d\mathbf{F} = 4k\pi q &= 4k\pi \int_{\triangle V, d\to 0} \sigma(\mathbf{r}) dF \\ &= 4k\pi\sigma\triangle F. \end{aligned} \tag{2.89}$$

For thickness $d \to 0$ we have

$$\int_{F(\triangle V), d\to 0} \mathbf{E} \cdot d\mathbf{F} = (\mathbf{E}_2 + \mathbf{E}_1) \cdot \triangle \mathbf{F} = (\mathbf{E}_2 - \mathbf{E}_1) \cdot \mathbf{n} \triangle F,$$

where $\mathbf{n}$ is a unit vector directed along the normal on side "2" of the surface. It follows that

$$(\mathbf{E}_2 - \mathbf{E}_1) \cdot \mathbf{n} = 4k\pi\sigma(\mathbf{r}). \tag{2.90}$$

Thus in passing through the charged surface of vanishing thickness the normal component of the electric field $\mathbf{E}$ jumps from one value to another depending on $\sigma(\mathbf{r})$.

We consider two particular cases. As the first we consider the case of a *large charged plate*, whose boundary effects (distortions of lines of force) may be neglected (hence the specification 'large'). In this case we have as depicted in Fig. 2.7,

$$\mathbf{E}_2 \cdot \mathbf{n} = E, \quad \mathbf{E}_1 \cdot \mathbf{n} = -E, \tag{2.91}$$

so that

$$2E = 4\pi k\sigma, \quad E = \frac{\sigma}{2\epsilon_0} \quad \text{newtons/coulomb.} \tag{2.92}$$

Thus the field has the same value $2\pi k\sigma$ on both sides of the plate, but is in each case directed away from the plate; the difference is $4\pi k\sigma$.

These results, in particular formula (2.90), are of considerable importance and will be used in many examples. In the second particular case which we consider, we assume that the plate is made of an *ideal conducting material*,[9] i.e. a material in which the electrons can move about freely without doing any work. We let the thickness of the plate be $d \neq 0$. In this case the electrostatic potential $\phi =$ const., so that inside the plate the electric field vanishes, i.e. $E_1 = 0$, and the field *above* the plate is $E = 4\pi k\sigma$.

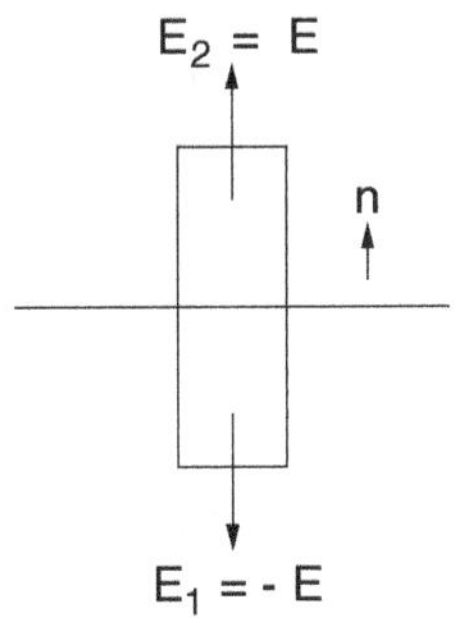


**Fig. 2.7.** Large charged plate.

[9]Conductors will be treated in detail later.

The discontinuity formula is valid at every surface. If $\mathbf{E}$ has the same value on both sides of the surface, then $\sigma = 0$ on this surface. Conversely if $\mathbf{E}_1 \cdot \mathbf{n}, \mathbf{E}_2 \cdot \mathbf{n}$ are different on both sides, the surface carries a charge. This is generally valid for the refraction of light at an interface.

In the preceding we considered the fields $\mathbf{E}_1, \mathbf{E}_2$ as originating solely from the one and only charged surface considered. We can drop this restriction now and allow charges and surfaces anywhere in space. These have the effect of distorting the lines of force of $\mathbf{E}_1$ and $\mathbf{E}_2$, and to change their magnitudes. This does not change anything in the previous considerations since these involve only the difference

$$\mathbf{E}_1 \cdot \mathbf{n}_1 + \mathbf{E}_2 \cdot \mathbf{n}_2 = E_{1n} - E_{2n}. \tag{2.93}$$

Finally we consider the *tangential components*. Since $\mathbf{E}$ is a conservative field, it follows that $\oint \mathbf{E} \cdot d\mathbf{r} = 0$ (along a closed path). If we choose a path along just above the surface to just below as in Fig. 2.8, then

$$(\mathbf{E}_{2t} - \mathbf{E}_{1t}) \cdot \triangle \mathbf{r} = 0, \tag{2.94}$$

i.e. the tangential components along an arbitrary direction remain unchanged:

$$\mathbf{E}_{2t} = \mathbf{E}_{1t}. \tag{2.95}$$

In the preceding we have been concerned with the electric field $\mathbf{E}$ which is also a measure of its field strength. Later we will in addition be concerned with the magnetic field strength $\mathbf{H}$. Electromagnetic radiation which is our main subject here has these two fields as components, and these both contribute to the energy of the overall electromagnetic field and to its propagation in the form of waves. In order to make this propagation serve practical purposes (e.g. in transmission lines) one naturally wants to be able to manipulate the field components and their energy in the lines and thus e.g. in electric circuits with a battery of potential $V$ and a current $I$. For this purpose one introduces two constants known as capacitance $C$ and inductance $L$. The capacitance $C$ (represented by a condenser) couples voltage $V$ to the energy stored in the electric field (see Example 2.25), and the inductance $L$ (represented by an inductor) couples the current $I$ to the

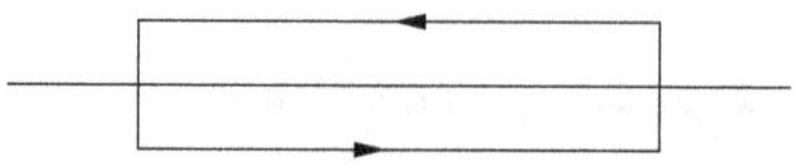

**Fig. 2.8.** The closed path above and below the surface.

energy stored in the magnetic field. The study of examples with calculation of the constant $C$ is therefore the object of our next section; the inductance $L$ will be treated correspondingly later in connection with magnetic fields.

We close with a comment. The basis of our considerations so far was the experimentally determined Coulomb law. If one recalls that the Coulomb law has the same form as Newton's gravitational law, one may enquire about further analogies between electrostatics and gravitation theory. Indeed a relation analogous to that for the electric field at a charged surface applies also for the gravitational field at a surface of mass. For applications we refer to the literature.[10]

## 2.8 Examples

The following examples are needed later. We therefore treat them in some detail. Most of these examples consider condensers which are combinations of conductors[11] separated by an insulating material (i.e. a dielectric) and store electric energy (i.e. in the electric field). However, the second conductor is not always apparent as such, e.g. the walls of a room in which an insulated conductor is placed (see e.g. Example 3.13). The parallel plate condenser we consider first is a prototype; we shall see an application later in transducers which convert acoustic energy into electromagnetic energy. Such a condenser consisting of two elements or components with a dielectric medium in-between is incorporated into an electric circuit by connecting one element (i.e. one plate or cylinder) to one part of the circuit and the other element to another part, or simply to the other open end if the circuit has been cut somewhere, the entire circuit, of course, also involving e.g. a battery, resistances (representing also the resistance of the conductors) and inductances and other condensers.

**Example 2.9 (The parallel plate condenser or capacitor).** Two parallel plates made of conductor material and separated by a distance $d$ are given, one with charge $+q$, the other with charge $-q$, as depicted in Fig. 2.9. Determine the capacity of the condenser or parallel plate capacitor.

**Solution:** Let $\sigma$ be the charge per unit area, i.e. $\sigma = q/F$ C/m$^2$. We use the coordinate system $O(x, y)$ with origin at the center of the condenser and the $y$-axis perpendicular to the plates. The lines of force (in the direction of $\mathbf{E}$)

[10]Blau, Guendelman and Guth [11]; see in particular Eqs. (3.3), (3.4), (4.11) and (4.27).

[11]This term, not yet defined here, will be specified in Section 2.9 and further at various other points in the text.

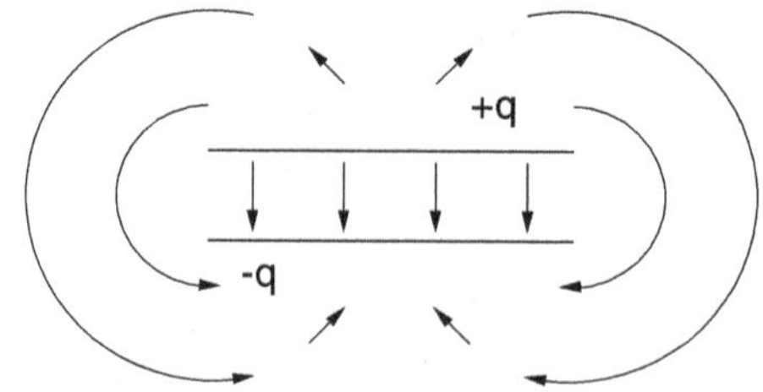


**Fig. 2.9.** The parallel plate condenser.

proceed from $+q$ to $-q$. We apply the Gauss law (2.19a) to the cylindrical volume with cross sectional area $\triangle F$, i.e.

$$\oint \mathbf{E} \cdot d\mathbf{F} = 4\pi k q. \tag{2.96}$$

Since the total charge contained in the cylindrical volume is zero, i.e. $\sigma\triangle F + (-\sigma)\triangle F = 0$, and since the tangential components of the field (along to the condenser plates) vanish, we obtain

$$E(y) - E(-y) = 0. \tag{2.97}$$

This relation expresses the symmetry of the field on both sides above and below the condenser plates. In the case of a *finite condenser* it is difficult to compute the field in the external domains. The reason is the *deformation of the lines of force at the open ends* as indicated in Fig. 2.9. However, we can approximate this finite condenser by a segment of a large spherical condenser consisting of two concentric shells. As we shall see in Example 2.10 the field outside is zero. In some books it is simply assumed that the field outside is zero. We see that this is an approximation. We now apply Gauss' law to a single plate and make use of the approximations just explained (i.e. that the field outside is zero); then in Eq. (2.90) we have to put $\mathbf{E}_1 = 0$. Then it follows with $\mathbf{E}_2 \equiv \mathbf{E}$ that

$$(\mathbf{E} - \mathbf{0}) \cdot \mathbf{n}\triangle F = 4\pi k\sigma\triangle F, \tag{2.98}$$

i.e. the field in the condenser is given by

$$E = 4\pi k\sigma \text{ N/C}, \quad \mathrm{k} = 1/4\pi\epsilon_0, \quad \mathrm{E} = \frac{\sigma}{\epsilon_0} \text{ N/C}. \tag{2.99}$$

The difference $V$ of the potentials between the two plates is defined as

$$V = \phi_2 - \phi_1 = \int_0^d \mathbf{E} \cdot d\mathbf{r} = Ed = 4\pi k\sigma d \quad \text{volts}. \tag{2.100}$$

The *capacity* (American usage: capacitance) $C$ of a condenser is defined as (with $q = \sigma F$)

$$C = \frac{q}{V} = \frac{q}{4\pi k\sigma d} = \frac{F}{4\pi dk} \quad \text{farads.} \tag{2.101}$$

Capacity is a measure of the ability of a system to store electric energy (that of the field $\mathbf{E}$). We saw previously that in MKSA units the constant $k$ has a complicated dimension. The unit of capacity in this system is the farad (F), i.e. 1 farad = 1 coulomb per volt. In the Gaussian system (see Section 1.3) the capacity C is given in cm with 1 farad $\Leftrightarrow 9 \times 10^{11}$ cm. Since a farad is a huge quantity, it is customary to employ micro-farads with 1 $\mu$F $= 10^{-6}$ F. Note that the two plates of the condenser carry opposite charges. Thus there is a Coulomb attraction between the latter and hence, at least theoretically an oppositely directed force is needed to maintain the system in equilibrium (see Example 2.23). A similar phenomenon is provided by the magnetic force acting between electromagnetic circuits; cf. Eq. (8.46).

**Example 2.10 (The spherical condenser).** The condenser consists of two conducting, concentric, spherical surfaces with internal radius $r_1$ and external radius $r_2$ and charge $+q$ on the inner shell and $-q$ on the outer shell, as indicated in Fig. 2.10. Determine the capacity $C$ of the condenser.

**Solution:** In view of the spherical symmetry of the condenser, the field components $\mathbf{E}_\theta, \mathbf{E}_\varphi = 0$. We consider a spherical surface of radius $r_1 - \epsilon$, $\epsilon > 0$. Then, according to the Gauss law (2.19a),

$$\int_F \mathbf{E} \cdot d\mathbf{F} = 4\pi k \int_{V(F)} \rho(\mathbf{r}')d\mathbf{r}' = 0, \tag{2.102}$$

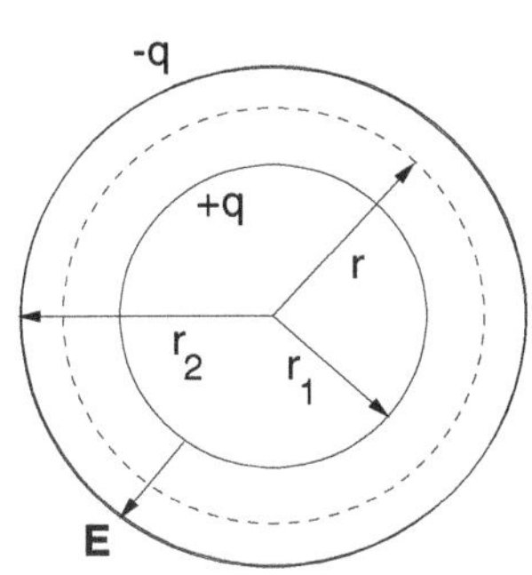


**Fig. 2.10.** The spherical condenser.

since the volume $V(r_1 - \epsilon)$ does not enclose a charge. Hence

$$\int_F \mathbf{E} \cdot d\mathbf{F} = 4\pi r_1^2 E_{r_1} = 0, \quad E_{r_1} = 0. \tag{2.103}$$

This means that the field in the interior of a charged closed surface is zero, i.e. this is field-free.[12] Thus the field of the condenser originates from the inner shell. We now consider a sphere of radius $r > r_1$:

$$\int \mathbf{E} \cdot d\mathbf{F} = 4\pi r^2 E_r = 4\pi kq,$$

so that (with $\sigma = q/4\pi r^2$, V upright the symbol for volts)

$$E_r \equiv E = \frac{4\pi kq}{4\pi r^2} = k\frac{q}{r^2} = \frac{\sigma}{\epsilon_0}\ \mathrm{N/C}, \quad \phi = \frac{kq}{r}\ V, \tag{2.104}$$

as for a point charge $+q$ at the origin. Outside of $r_2$, i.e. for $r > r_2$, the field again vanishes, because there the field contributions of the two equal and opposite charge distributions cancel each other (one could say, the field there corresponds to the sum of the fields of point charges $+q$ and $-q$ at the origin). With this we obtain for the potential $V$ of the condenser

$$V = \phi_2 - \phi_1 = qk\int_{r_1}^{r_2} \frac{dr}{r^2} = kq\left(\frac{1}{r_1} - \frac{1}{r_2}\right) \quad \text{volts}, \tag{2.105}$$

and for the capacity ($V$ sloping type (italics) meaning potential)[13]

$$C = \frac{q}{V} = \frac{r_1 r_2}{k(r_2 - r_1)} \quad \text{farads}. \tag{2.106}$$

With reference to a *single* spherical shell we can define its capacity as

$$\mathrm{C} = \text{charge/potential},$$

or by taking $r_2 \to \infty$, so that in the case of the inner sphere

$$C = \frac{q}{\phi_1} = \frac{qr_1}{kq}, \quad \text{i.e.} \quad C = 4\pi\epsilon_0 r_1 \quad \text{farads}. \tag{2.107}$$

The spherical condenser like the spherical conductor and the spherical charge distribution are easy to deal with since the sphere is a surface with no boundary. A charged disc is already much more difficult to handle. See Example 2.24.

---

[12]This was the observation of Cavendish in his experiment as discussed in Example 2.2.
[13]The space between the surface of the earth and the ionosphere (see Chapter 13) constitutes roughly a condenser of this type for radio waves.

**Example 2.11 (The van de Graaff generator).** A van de Graaff generator[14] has a spherically shaped dome made of conducting material which receives charges at the rate of a current of 1 μA. The dome has a radius of 1 m. At a critical electric field strength of $3 \times 10^6$ V/m the surrounding air produces sparks. How many seconds after the beginning of the charge transfer to the dome does this effect appear? ($4\pi\epsilon_0 = 10^7/c^2$ farads/meter).

**Solution:** In the van de Graaff generator charge is sprayed on a transmission belt made of insulator material. The belt transports the charge to the conducting dome of radius $R$ as depicted in Fig. 2.11. For a single spherical shell the capacity $C$ is defined by the ratio of charge to potential, as we saw previously, so that with potential $\phi = kq/R$ the capacity is given by

$$C = \frac{q}{\phi} = \frac{qR}{kq} = 4\pi\epsilon_0 R \quad \text{farads} \tag{2.108}$$

for a charge $q$. If charge is continuously added, the potential $\phi$ changes accordingly, i.e.

$$C\frac{d\phi}{dt} = \frac{dq}{dt} = I \quad \text{amperes}, \tag{2.109}$$

where $I$ is the current. We thus have the relation

$$\frac{d\phi}{dt} = \frac{I}{C}, \quad \phi = \frac{I}{C}t \tag{2.110}$$

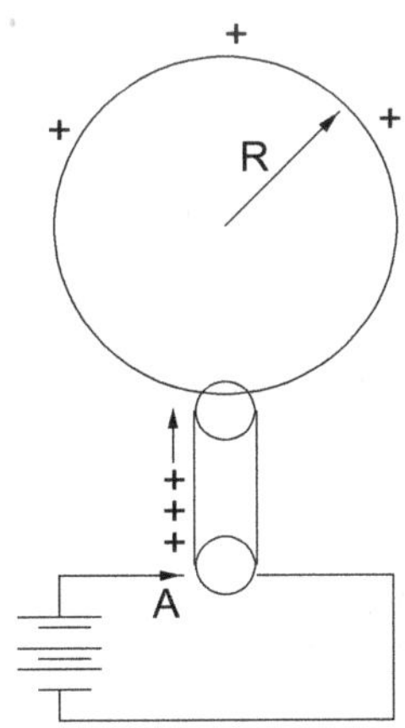


**Fig. 2.11.** The van de Graaff generator.

[14]For a more detailed but elementary description see e.g. Nelkon [104], pp. 16–17.

with $\phi = 0$ at time $t = 0$. The potential or voltage increases until the corresponding field $\mathbf{E}$ around the sphere is strong enough to knock electrons out of the surrounding air — the corresponding atomic transitions then become visible in the form of sparks. In air this stage is attained when $E_{\text{crit}} = 3\times 10^6$ V/m. The field $\mathbf{E}$ in the neighborhood of the dome is given by

$$E = k\frac{q}{r^2}, \quad r > R. \tag{2.111}$$

For $r \sim R$:

$$E = k\frac{q}{R^2} = \frac{\phi}{R}, \tag{2.112}$$

or $\phi = ER$, so that $\phi_{\text{crit}} = E_{\text{crit}}R$. For $\phi_{\text{crit}} = 3 \times 10^6$ V, we have $R = 1$ m and

$$C = 4\pi\epsilon_0 R = \frac{10^7}{c^2} \times 1 = \frac{10^7}{(3 \times 10^8)^2} \quad \text{farads.} \tag{2.113}$$

This implies that

$$C = \frac{1}{9} \text{ nF (nanofarads).} \tag{2.114}$$

Thus the appropriate current $I$ is: $I = \phi_{\text{crit}}C/t$, i.e.

$$I = \frac{3 \times 10^6 V}{9 \times 10^9 \text{F}}\frac{1}{t} \text{ A} = \frac{0.3 \times 10^{-3}}{t} \text{ A.}$$

For $I = 1 \times 10^{-6}$ A:

$$1 \times 10^{-6} \text{ A} = \frac{0.3 \times 10^{-3}}{t} \text{ A}$$

we obtain $t = 0.3 \times 10^3$ seconds. Thus if the dome is charged at the rate of a current of 1 micro-ampere, it takes about 300 seconds for the potential to reach its maximal value.

**Example 2.12 (The cylindrical condenser).** This condenser consists of two coaxial cylinders of height $h$ and radii $r_1, r_2, r_2 > r_1$. The inner cylinder carries charge $q$, the outer cylinder charge $-q$. What is the capacity $C$ of the condenser?

**Solution:** Clearly one uses cylindrical coordinates $\theta, \rho, z$ in order to exploit the cylindrical symmetry. We consider, as in Fig. 2.12, a cylindrical surface

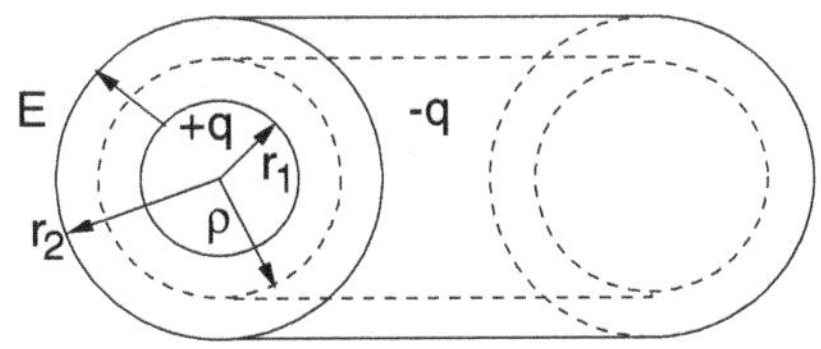


**Fig. 2.12.** The cylindrical condenser.

of radius $\rho, r_2 > \rho > r_1$. In view of the cylindrical symmetry $E_\theta = E_z = 0, E_\rho = E$. The Gauss law (2.19a) implies

$$\oint \mathbf{E} \cdot d\mathbf{F} = E_\rho \cdot 2\pi\rho h = 4\pi k q, \tag{2.115}$$

so that

$$E_\rho = \frac{2qk}{\rho h} = \frac{q}{2\pi\epsilon_0 h\rho} \quad \text{newtons/coulomb.} \tag{2.116}$$

We thus obtain for the potential $V$:

$$V = \int_{r_1}^{r_2} E_\rho d\rho = \frac{2qk}{h}\int_{r_1}^{r_2} \frac{d\rho}{\rho} = \frac{2qk}{h}\ln\frac{r_2}{r_1} \quad \text{volts,} \tag{2.117}$$

and for the capacity

$$C = \frac{q}{V} = \frac{h/k}{2\ln(r_2/r_1)} \quad \text{farads.} \tag{2.118}$$

The finite cylindrical condenser can be approximated by a segment of a toroidal condenser, like the parallel plate condenser can be approximated by a segment of the spherical condenser.

**Example 2.13 (The dipole).** A dipole is generally defined as a system of two spatially separated charges $+q, -q$. Obviously an infinitesimally imagined parallel plate condenser is of this form. We now want to obtain the potential of such an arrangement of charges, and we shall define in this context the important concept of a dipole moment.

**Solution:** We consider first two infinitesimal parallel plates with surface areas $dF', dF''$, and charge densities $+\sigma(\mathbf{r}'), -\sigma(\mathbf{r}')$, which are elements of surfaces $S', S''$ with charges $+\sigma, -\sigma$ as depicted in Fig. 2.13. The potential

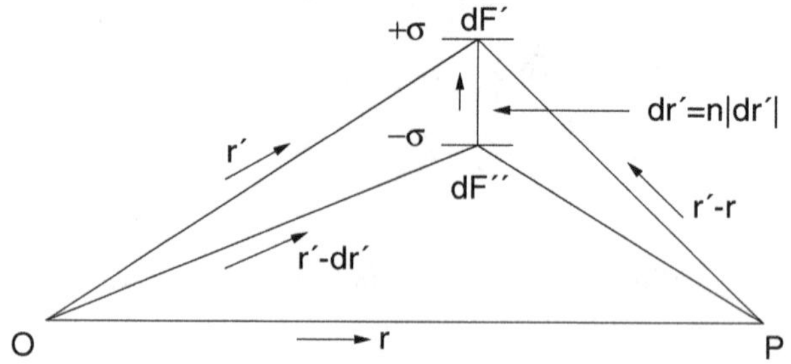


**Fig. 2.13.** The dipole as an infinitesimal parallel plate condenser.

at the point $P$ is given by

$$\phi(\mathbf{r}) = k\int\frac{\sigma(\mathbf{r}')dF'}{|\mathbf{r}-\mathbf{r}'|} - k\int\frac{\sigma(\mathbf{r}')dF'}{|\mathbf{r}-\mathbf{r}'+d\mathbf{r}'|} \quad \text{volts.} \tag{2.119}$$

In the following one should not confuse the line element $d\mathbf{r}$ with the volume element $d\mathbf{r}$ (the distinction should be clear from the context). We have

$$\begin{aligned}\frac{1}{|\mathbf{r}-\mathbf{r}'+d\mathbf{r}'|} &= \frac{1}{\{(\mathbf{r}-\mathbf{r}')^2+2d\mathbf{r}'\cdot(\mathbf{r}-\mathbf{r}')+(d\mathbf{r}')^2\}^{1/2}}\\ &\simeq \frac{1}{|\mathbf{r}-\mathbf{r}'|}\left(1-\frac{(\mathbf{r}-\mathbf{r}')\cdot d\mathbf{r}'}{(\mathbf{r}-\mathbf{r}')^2}+\cdots\right).\end{aligned} \tag{2.120}$$

It follows that

$$\phi(\mathbf{r}) \simeq k\int\frac{\sigma(\mathbf{r}')(\mathbf{r}-\mathbf{r}')\cdot d\mathbf{r}'}{(\mathbf{r}-\mathbf{r}')^3}dF'. \tag{2.121}$$

We put

$$D(\mathbf{r}') \equiv \lim_{d\mathbf{r}'\to 0}\sigma(\mathbf{r}')|d\mathbf{r}'|. \tag{2.122}$$

The quantity $D$ is called *dipole density* (i.e. we consider a case in which the potential between the surface elements $dF', dF''$ remains constant, when $d\mathbf{r}'\to 0$ and $\sigma(\mathbf{r}')\to\infty$). An *infinitesimal dipole surface* (area $\triangle F$) in the limit $\to 0$ is sometimes also described as a *point dipole.* Then (in the limit $d\mathbf{r}'\to 0$)

$$\phi(\mathbf{r}) \simeq k\int\frac{D(\mathbf{r}')\mathbf{n}\cdot(\mathbf{r}-\mathbf{r}')}{(\mathbf{r}-\mathbf{r}')^3}dF'. \tag{2.123}$$

Let $d\Omega$ be the solid angle subtended by the area $dF'$ at P. Let $\mathbf{n}$ be a unit vector perpendicular to $dF'$ and directed to the outside (i.e. $\mathbf{n}$ parallel to $d\mathbf{r}'$).

It then follows from Fig. 2.13 that

$$\frac{\mathbf{n}\cdot(\mathbf{r}'-\mathbf{r})dF'}{(\mathbf{r}-\mathbf{r}')^3}=d\Omega \tag{2.124}$$

and hence

$$\phi(\mathbf{r})=-k\int_F D(\mathbf{r}')d\Omega \tag{2.125}$$

(the minus sign appears because $d\Omega$ involves the vector $(\mathbf{r}'-\mathbf{r})$ and not $(\mathbf{r}-\mathbf{r}')$ like $\phi$). Replacing $F$ by $\triangle F$ we obtain *the potential of a point dipole*:

$$\begin{aligned}\phi(\mathbf{r}') &= -k\int_{\triangle F} D(\mathbf{r}')d\Omega = -k\int_{\triangle F}\frac{\sigma(\mathbf{r}')|d\mathbf{r}'|dF'\mathbf{n}\cdot(\mathbf{r}'-\mathbf{r})}{(\mathbf{r}-\mathbf{r}')^3}\\ &= -k\frac{\mathbf{n}\cdot(\mathbf{r}'-\mathbf{r})}{(\mathbf{r}-\mathbf{r}')^3}|d\mathbf{r}'|\int_{\triangle F}\sigma(\mathbf{r}')dF' = -k\mathbf{p}\cdot\frac{(\mathbf{r}'-\mathbf{r})}{(\mathbf{r}-\mathbf{r}')^3},\end{aligned} \tag{2.126}$$

where

$$\mathbf{p}=\mathbf{n}|d\mathbf{r}'|Q,\quad Q=\int_{\triangle F}\sigma(\mathbf{r}')dF'. \tag{2.127}$$

The expression $\mathbf{p}$ is called *dipole moment* of the point dipole; this is a vector quantity with *direction from the negative charge to the positive charge.* The quantity $Q$ is the charge.

## 2.9 Conductors and Electrical Screening

A *conductor* is a macroscopic object in which charges are contained which can move about freely (i.e. without doing work). Inside the conductor the electric field $\mathbf{E}$ therefore vanishes, because only then a charge can move about freely in the conductor. It follows that if the conductor is given a charge, this charge must be located in a thin layer at its surface. This must be so because if we consider an arbitrary closed Gaussian surface wholly within the conductor material, this cannot contain charge since everywhere on the surface $\mathbf{E}=0$. Thus inside the conductor the charge is zero at every point (the charges of the conductor material itself averaging out to zero). It follows that the surface of the conductor is an *equipotential surface.*

**Example 2.14 (The charged sphere).** Let a spherically shaped conductor be given with charge $q$. Explain with the help of potential considerations that the charge accumulates on the outer surface of the conductor.

**Solution:** We consider a sphere with external radius $b$ and an internal hollow space of radius $r_0 = a < b$. We subdivide the intermediate region into concentric spherical shells of radii $r_1, r_2, \ldots$. Each shell has initially charge zero. Since the sphere is a conductor, each of its parts has the same constant potential. If we give e.g. the sphere with radius $r_2$ the charge $q$, the potential changes for all shells with radius $r_i, i > 3$. Thus the gradient of a potential arises. Obviously the energy of the system becomes minimal when the charge $q$ is located on the outermost shell (recall that the modulus of the Coulomb potential decreases with increase of the distance from the charge).

The electric field inside arbitrarily shaped conducting bodies whose surfaces are charged but which do not contain any enclosed charges is always zero. This is so, because if we apply to the interior the integral form of the law of Gauss, we obtain

$$\int_{F'} \mathbf{E} \cdot d\mathbf{F} = 0. \tag{2.128}$$

The right hand side is zero, since the surface $F'$ does not enclose charges. If there were inside the given positively charged, closed surface a field $E \neq 0$ then the field would be directed from the surface towards the interior. All contributions to the integral would then be positive. The integral would be a sum of positive contributions. All these would therefore have to vanish, implying that the field inside is zero. This result is apparently also valid for homogeneous charge distributions on the surface. Such closed bodies which do not contain charges are called *Faraday cages* because Faraday crept into such a container in an attempt to measure the field therein.[15] Containers of this type are used when field-free regions are needed. That the field inside closed charged bodies has to be zero can also be seen by the impossibility to sketch lines of force in the interior which travel from positive charges to negative charges.[16]

A conductor is said to be *earthed* if its potential is the same as at infinity, i.e. zero. This effect is achieved in practice by connection with the ground, i.e. the "earth".

If a body, e.g. a solid sphere, is made of a nonconducting material, that is a dielectric medium, which is polarized in an applied electric field, then one part of the body can be positively charged and another negatively. These cases will be treated later, but see Example 2.36.

[15] See e.g. Zahn [167], pp. 77–78.

[16] For a discussion of conductors of a general, i.e. nonspherical shape, see Anderson [3], pp. 40–42.

Finally we consider some *systems of conductors and their electric screening*.[17] Above we obtained the Poisson equation for the electric potential $\phi$. Assuming also the *superposition principle*, we can treat the case of $n$ conductors with potentials $\phi_i$ and charges $q_i, i = 1, 2, \ldots n$, in vacuum or air by writing

$$\phi_i = \sum_{j=1}^{n} p_{ij} q_j \tag{2.129}$$

or inversely

$$q_i = \sum_{j=1}^{n} C_{ij}\phi_j, \quad \sum_{j=1}^{n} C_{ij} p_{jk} = \delta_{ik}. \tag{2.130}$$

The coefficients $C_{ij}$ are described as *capacities*, and the coefficients $p_{ij}$ as *induction coefficients*. Obviously these coefficients depend on the geometry of the conductors and their arrangement in space. In the special case of two conductors with charges $q$ and $-q$ (as in preceding examples), we obtain

$$\phi_1 = (p_{11} - p_{12})q, \quad \phi_2 = (p_{21} - p_{22})q, \tag{2.131}$$

or

$$\phi_1 - \phi_2 = (p_{11} - p_{12} - p_{21} + p_{22})q. \tag{2.132}$$

Hence

$$V = \phi_1 - \phi_2 = \frac{q}{C} \quad \text{volts}, \tag{2.133}$$

where

$$C = \frac{1}{(p_{11} - p_{12} - p_{21} + p_{22})} \quad \text{farads}. \tag{2.134}$$

Expressed in terms of the coefficients $C_{ij}$, one obtains

$$C = \frac{C_{11} + C_{12} + C_{21} + C_{22}}{C_{11}C_{22} - C_{12}C_{21}} \quad \text{farads}. \tag{2.135}$$

It can be shown that[18]

$$C_{ij} = C_{ji}, \quad p_{ij} = p_{ji}. \tag{2.136}$$

[17]For considerations similar to those below see e.g. Cheston [30], pp. 99–103.
[18]See below the consequence of Eq. (2.145) or e.g. Chirgwin, Plumpton and Kilmister [31], pp. 82–84.

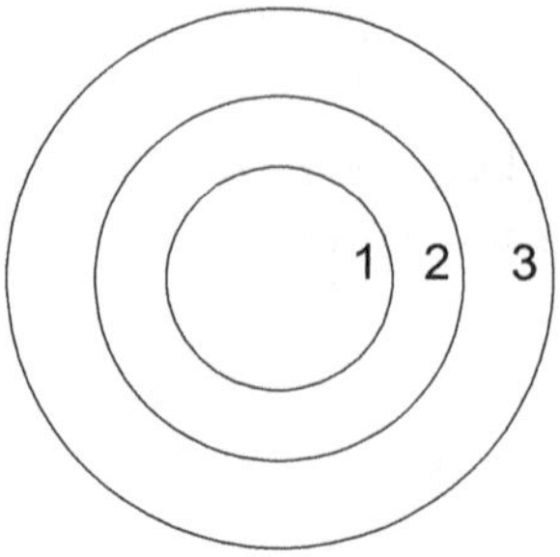


**Fig. 2.14.** Conducting shells screening the inner ones.

*Electrical screening* results if one conductor is completely surrounded by another conductor. Instead of considering the most general case, we consider here the example of 3 concentric spherical shells which provide such enclosures. We will find that, as illustrated in Fig. 2.14, conductor 1 is screened off by conductor 3 and other external conductors in the sense that the coefficient $C_{13} = 0$.

We consider the three concentric spherical conductor shells with charges $q_i$ and radii $r_i, i = 1, 2, 3$ and $r_1 < r_2 < r_3$. We know that in the case of a single sphere with charge $q$ and radius $r_0$:

$$\begin{aligned} \phi &= k\frac{q}{r_0} \quad \text{for } r < r_0 \ \text{(there } E = 0), \\ \phi &= k\frac{q}{r} \quad \text{for } r > r_0 \ \text{(there } E \neq 0). \end{aligned} \tag{2.137}$$

It follows that

$$\begin{aligned} \phi_1 &= k\left[\frac{q_1}{r_1} + \frac{q_2}{r_2} + \frac{q_3}{r_3}\right], \\ \phi_2 &= k\left[\frac{q_1}{r_2} + \frac{q_2}{r_2} + \frac{q_3}{r_3}\right], \\ \phi_3 &= k\left[\frac{q_1}{r_3} + \frac{q_2}{r_3} + \frac{q_3}{r_3}\right]. \end{aligned} \tag{2.138}$$

From these relations we deduce that if $\phi_i = \sum_j p_{ij} q_j$, as in Eq. (2.129), we have

$$\frac{k}{r_2} = p_{12} = p_{21} = p_{22}, \quad p_{13} = p_{31} = p_{23} = p_{33} = p_{32} = \frac{k}{r_3}. \tag{2.139}$$

The matrix $(C_{ij})$ is the inverse of the matrix $(p_{ij})$ (and one can show that the inverse exists, i.e. $\det(p_{ij}) \neq 0$). According to the rule for establishing the elements of an inverse matrix we have for example in the case of the $(3 \times 3)$-matrix

$$A = \begin{pmatrix} a_1 & b_1 & c_1 \\ a_2 & b_2 & c_2 \\ a_3 & b_3 & c_3 \end{pmatrix} \tag{2.140a}$$

the inverse[19]

$$A^{-1} = \frac{1}{\det A} \begin{pmatrix} |b_2c_3| & -|b_1c_3| & |b_1c_2| \\ -|a_2c_3| & |a_1c_3| & -|a_1c_2| \\ |a_2b_3| & |a_1b_3| & |a_1b_2| \end{pmatrix}, \tag{2.140b}$$

where

$$|b_2c_3| \equiv b_2c_3 - b_3c_2,$$

and so on. Applied to the matrix $(C_{ij})$, this implies (corresponding to $|b_1c_2|$) $C_{13} \propto |p_{12}p_{23}|$, i.e.

$$C_{13} \propto p_{12}p_{23} - p_{22}p_{13} = 0 \tag{2.141}$$

according to the above relations. This means the charge $q_1$ (cf. Eq. (2.130)) remains unaffected by the field $\phi_3$, or external conductors. This is what one calls electrical screening. In practice the conductor which is used for screening purposes is frequently "*earthed*", i.e. in the previous example one would have put $\phi_2 = 0$.

It is convenient at this point to introduce one form of a theorem known as the *reciprocity theorem*, also called *Green's reciprocity theorem*, here applied to charges and conductors (in Chapter 8 to currents and circuits). Assume we have charge densities $\rho$ and $\rho'$ which give rise to potentials $\phi$ and $\phi'$ respectively. Then we can prove that

$$\int \rho\phi' dV = \int \rho'\phi dV \tag{2.142}$$

[19]See e.g. Aitken [1], p. 52, or any other book on determinants and matrices.

when integrated over all space. In order to verify this claim we take the left hand side and use Eqs. (2.18) and (2.8). Then

$$\begin{aligned}\int \rho\phi' dV &= \frac{1}{4\pi k}\int \boldsymbol{\nabla}\cdot\mathbf{E}\phi' dV\\ &= \frac{1}{4\pi k}\int [\boldsymbol{\nabla}\cdot(\phi'\mathbf{E}) - \mathbf{E}\cdot(\boldsymbol{\nabla}\phi')]dV\\ &= \frac{1}{4\pi k}\int dV\mathbf{E}\cdot\mathbf{E}' \qquad (2.143)\end{aligned}$$

(the integral over the divergence becoming with Gauss' divergence theorem a surface integral at infinity where the fields are zero). Since the result is symmetric in primed and unprimed quantities the claim (2.142) is verified. In the special case of charges $q_i$ and conductors with potentials $\phi_i$ Eq. (2.142) becomes

$$\sum_{i=1}^{n} q_i\phi_i' = \sum_{i=1}^{n} q_i'\phi_i. \qquad (2.144)$$

Inserting here for $q_i$ the relation (2.130) we obtain

$$\sum_{i,j=1}^{n} C_{ij}\phi_j\phi_i' = \sum_{i,j=1}^{n} C_{ij}\phi_j'\phi_i \equiv \sum_{i,j=1}^{n} C_{ji}\phi_i'\phi_j. \qquad (2.145)$$

It follows that for all $i, j{:}C_{ij} = C_{ji}$. As an application of this reciprocity theorem consider a charge $q$ placed at a distance $d$ away from the center of a conducting sphere. We ask: What is the potential of the sphere? In this case of two objects Eq. (2.144) is

$$q_1\phi_1' + q_2\phi_2' = q_1'\phi_1 + q_2'\phi_2. \qquad (2.146)$$

We have two situations — one involving unprimed quantities, and the other involving primed quantities — in both cases the index 1 referring to position 1 (the sphere) and the index 2 to position 2 (charge), $\phi_1$ being the potential to be found. Thus in the first situation we have $q_1$ = charge of the sphere = $Q = 0, q_2 = q$. In the second situation we take the sphere as charged, $q_1' = Q' \neq 0$ and $q_2' = 0$, i.e. no charge at $d$. Inserting these values into Eq. (2.146) we obtain

$$0 + q\phi_2' = Q'\phi_1 + 0.$$

Here $\phi_2'$ is the potential at position 2, i.e. $d$, when the sphere has charge $Q'$ and there is no charge outside, i.e. at position $d$. Clearly $\phi_2' = kQ'/d$.

Hence

$$q\frac{kQ'}{d} = Q'\phi_1, \quad \phi_1 = q\frac{k}{d}. \tag{2.147}$$

**Example 2.15 (The two hemispherical shells).** A conductor has the shape of a spherical shell with radius $a$. The shell is divided diametrically into two hemispherical parts. Calculate the force which is needed to keep the hemispheres together if the sphere carried originally the total charge $q$ ($q = 4\pi a^2\sigma$).

**Solution:** The hemispheres are separated. Thus each has on both sides a field $\mathbf{E} \neq 0$. In this case (as we saw above, Eq. (2.92)) the electric field at the surface is given by

$$E = \frac{\sigma}{2\epsilon_0} = \frac{q}{\epsilon_0 8\pi a^2} \quad \text{newtons/coulomb}. \tag{2.148}$$

We now calculate the force acting in the direction of $z$ (see Fig. 2.15). To this end we compute the force acting on the surface charge $\sigma$ in the general direction $\theta$, take the component along $z$ of this (multiplication by $\cos\theta$), and add all force components in the direction of $z$ (integration over $\theta$ for one hemisphere). We thus obtain for the force in the direction of $z$

$$\begin{aligned} F_z &= \int_0^{\pi/2} 2\pi a \sin\theta \, a d\theta \sigma \frac{q}{\epsilon_0 8\pi a^2} \cos\theta \\ &= 2\pi a^2 \frac{q}{4\pi a^2} \frac{q}{\epsilon_0 8\pi a^2} \int_0^{\pi/2} \sin\theta\cos\theta d\theta \\ &= \frac{q^2}{16\pi a^2 \epsilon_0} \int_0^{\pi/2} \sin\theta d(\sin\theta) \\ &\equiv \frac{q^2}{32\pi a^2\epsilon_0} \equiv k\frac{q^2}{8a^2} \quad \text{newtons}. \end{aligned} \tag{2.149}$$

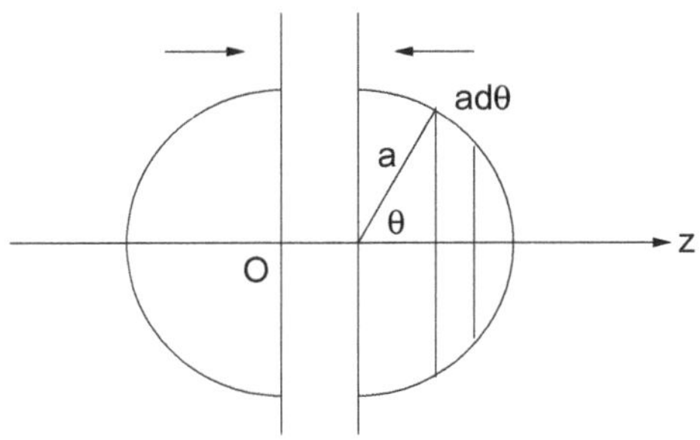


**Fig. 2.15.** The sphere divided into hemispheres.

If one considers a spherical condenser consisting of two concentric spherical shells with radii $a$ and $b$ ($a > b$) and one divides this diametrically into two parts, and if $q$ and $-q$ are the charges on the spheres originally, then the following force is required to keep the two parts together[20]

$$k\frac{q^2}{8}\left(\frac{1}{b^2}-\frac{1}{a^2}\right) \quad \text{newtons} \tag{2.150}$$

(difference of two forces so that the electric field remains between the two spheres).

**Example 2.16 (Two connected spherical shells).** Two spherically shaped conductors with radii $a$ and $b$ and respectively charges $q_1$ and $q_2$ are separated from one another by a distance $r \gg a, b$. The conductors are connected by a thin conducting wire as indicated in Fig. 2.16. Calculate by what amount the energy of the system is thereby lowered.

**Solution:** Connecting the spheres by a wire alters the equipotential surfaces of the system. The charges rearrange themselves such that the energy is minimized. We first calculate the energy of the conductors when not connected. For this we calculate the potential of each of the conductors. These are in general determined by both charge distributions. The potential of the first conductor follows from a contribution $q_1/4\pi\epsilon_0 a$, originating from the charge of this conductor, and an additional contribution due to the charge $q_2$. Since $r \gg a, b$, we can approximate the latter by $q_2/4\pi\epsilon_0 r$. Hence the potential of the first conductor is approximately

$$\phi_1 = \frac{1}{4\pi\epsilon_0}\left[\frac{q_1}{a}+\frac{q_2}{r}\right] \quad \text{volts.} \tag{2.151}$$

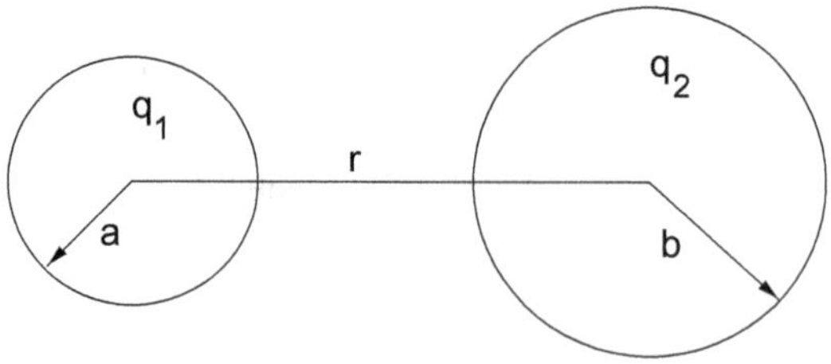


**Fig. 2.16.** Two spherical shells connected by a wire.

[20]See problem 1, p. 54 of Chambers [27].

Analogously we obtain for the potential of the other conductor

$$\phi_2 = \frac{1}{4\pi\epsilon_0}\left[\frac{q_2}{b} + \frac{q_1}{r}\right] \quad \text{volts.} \tag{2.152}$$

If the potential of a conductor is $\phi$ volts and its charge is $q$ coulombs, its energy is $q\phi$ joules. The energy of the system of separated conductors is therefore

$$\begin{aligned} W &= q_1\phi_1 + q_2\phi_2 \\ &= \frac{q_1}{4\pi\epsilon_0}\left[\frac{q_1}{a} + \frac{q_2}{r}\right] + \frac{q_2}{4\pi\epsilon_0}\left[\frac{q_2}{b} + \frac{q_1}{r}\right] \\ &= \frac{1}{4\pi\epsilon_0}\left[\frac{q_1^2}{a} + \frac{q_2^2}{b} + \frac{2q_1q_2}{r}\right] \quad \text{joules.} \end{aligned} \tag{2.153}$$

When the conductors are connected by a conducting wire they form an equipotential surface. The charges redistribute themselves to minimize the potential. Let $q_1', q_2'$ be the charges of the conductors in the new state of equilibrium. Since no charge is lost, we must have

$$(a) \qquad q_1' + q_2' = q_1 + q_2. \tag{2.154}$$

Since both spheres now have the same potential we have

$$(b) \quad \phi_1' = \frac{1}{4\pi\epsilon_0}\left[\frac{q_1'}{a} + \frac{q_2'}{r}\right] = \frac{1}{4\pi\epsilon_0}\left[\frac{q_2'}{b} + \frac{q_1'}{r}\right] = \phi_2'. \tag{2.155}$$

This implies that

$$\frac{q_1'}{a} + \frac{q_2'}{r} = \frac{q_2'}{b} + \frac{q_1'}{r}, \tag{2.156}$$

i.e.

$$q_1' = q_2'R, \quad q_2' = \frac{q_1'}{R}, \quad \text{where} \quad R = \left(\frac{1/b - 1/r}{1/a - 1/r}\right). \tag{2.157}$$

With (a) we have

$$q_1' = (q_1 + q_2 - q_1')R$$

and so

$$q_1' = \frac{(q_1 + q_2)R}{1 + R}. \tag{2.158}$$

Similarly

$$q_2' = \frac{(q_1 + q_2)(1/R)}{1 + 1/R}. \tag{2.159}$$

With these expressions we obtain the new energy of the system as

$$W' = q_1'\phi_1{}' + q_2'\phi_2{}', \tag{2.160}$$

i.e.

$$W' = \frac{1}{4\pi\epsilon_0}\left[\frac{q_1'^2}{a} + \frac{q_2'^2}{b} + \frac{2q_1'q_2'}{r}\right]. \tag{2.161}$$

The amount by which the *energy is lowered* is therefore $W - W'$; for $a = b$, this is

$$(W - W')\Big|_{a=b} = \frac{(q_1 - q_2)^2}{8\pi\epsilon_0}\left[\frac{1}{a} - \frac{1}{r}\right] \quad \text{joules.} \tag{2.162}$$

This is the result.[21]

**Example 2.17 (The charges of connected spherical shells).** Two conducting spherical shells have radii $R_1$ and $R_2$. Their centers are a distance $D, D \gg R_1 + R_2$ apart. The sphere with radius $R_1$ is given a charge $Q_1$ and the sphere with radius $R_2$ the charge $Q_2$. A thin conducting wire is now added to connect the two spheres. Calculate the resulting surface charge densities on both spheres, as well as the electric field strengths. Without further calculations discuss the effect that appears when the system with very small radius $R_2$ is subjected to a high voltage (i.e. why high voltage systems avoid sharp points like that of the small sphere).

**Solution:** It is easiest to obtain the potential of the spheres from the Gauss law (2.19a),

$$\int \mathbf{E}\cdot d\mathbf{F} = 4\pi k\int \rho(\mathbf{r}')d\mathbf{r}' = 4\pi k Q, \tag{2.163}$$

i.e. for $r > R$:

$$4\pi r^2 E_r = 4\pi k Q, \quad E_r = k\frac{Q}{r^2}. \tag{2.164}$$

In general the charge of the distant sphere contributes to the potential of the other sphere (ignoring the former is not always a good approximation). We can assume for simplicity here that at the other sphere the potential

[21] See also Chambers [27], Problem 8, p. 55.

of the distant sphere does not vary much with the radius of the sphere considered, so that

$$\phi_1 = k\left[\frac{Q_1}{R_1} + \frac{Q_2}{D}\right] \quad \text{and} \quad \phi_2 = k\left[\frac{Q_2}{R_2} + \frac{Q_1}{D}\right]. \tag{2.165}$$

The wire connecting the spheres ensures that both spheres together form an equipotential surface with potential

$$\phi_0 = k\left[\frac{Q_1{}'}{R_1} + \frac{Q_2{}'}{D}\right] = k\left[\frac{Q_2{}'}{R_2} + \frac{Q_1{}'}{D}\right], \tag{2.166}$$

where $Q_1{}'$ and $Q_2{}'$ are the new charges, in other words the charges rearrange themselves on the spheres in such a way that both spheres have the same potential $\phi_0$. Since in this process charge is neither created nor annihilated, we must have

$$Q_1 + Q_2 = Q_1{}' + Q_2{}'. \tag{2.167}$$

From this relation and Eq. (2.166) we can calculate $Q_1{}'$ and $Q_2{}'$:

$$Q_1{}' = \frac{R_1}{R_2}Q_2{}'\left[1 + O\left(\frac{1}{D}\right)\right] \tag{2.168}$$

for $D$ large. By neglecting contributions of the order of $1/D$ we obtain

$$Q_1 + Q_2 \simeq \frac{R_1}{R_2}Q_2' + Q_2', \tag{2.169}$$

i.e.

$$Q_2' \simeq \frac{R_2}{R_1 + R_2}(Q_1 + Q_2), \quad Q_1' \simeq \frac{R_1}{R_1 + R_2}(Q_1 + Q_2). \tag{2.170}$$

The potential $\phi_0$ is therefore

$$\phi_0 = k\frac{Q_1 + Q_2}{R_1 + R_2}\left[1 + O\left(\frac{1}{D}\right)\right]. \tag{2.171}$$

Now, surface charge density = charge/area, so that

$$\sigma_1 \simeq \frac{Q_1'}{4\pi R_1^2} = \frac{Q_1 + Q_2}{4\pi R_1(R_1 + R_2)}, \tag{2.172}$$

and

$$\sigma_2 \simeq \frac{Q_2'}{4\pi R_2^2} = \frac{Q_1 + Q_2}{4\pi R_2(R_1 + R_2)}. \tag{2.173}$$

According to Eq. (2.164) the corresponding field strengths are given by

$$E_1 \simeq k\frac{Q_1'}{R_1^2}, \quad E_2 \simeq k\frac{Q_2'}{R_2^2}, \tag{2.174}$$

i.e.

$$E_1 \simeq 4\pi k\sigma_1, \quad E_2 \simeq 4\pi k\sigma_2 \quad \text{newtons/coulomb}. \tag{2.175}$$

When $R_2$ is very small, $E_2$ becomes very large. If a high voltage is applied, for $E_2 > E_{2,\text{critical}}$, this electric field is so large that it can eject electrons from atoms of the surrounding air. The resulting reordering of atomic states leads to the emission of light which can be observed in the form of sparks.

## 2.10 Energy of Charge Distributions

The *potential energy of a point charge* $q_i$ at the point $\mathbf{r}_i$ in the electric field $\mathbf{E} = -\boldsymbol{\nabla}\phi$ of a point charge $q_k$ at point $\mathbf{r}_k$ is the *Coulomb potential energy*

$$k\frac{q_i q_k}{|\mathbf{r}_i - \mathbf{r}_k|} \quad \text{joules}, \quad i \neq k. \tag{2.176}$$

The potential energy of $N$ charges $q_1, q_2, \ldots, q_N$ is correspondingly

$$W = \frac{1}{2}k\sum_i\sum_{k\neq i}\frac{q_i q_k}{|\mathbf{r}_i - \mathbf{r}_k|} \quad \text{joules}, \tag{2.177}$$

since $\sum_i\sum_{k\neq i}$ sums over $i$ and $k$, i.e. *twice.*

For *continuous charge distributions* $\rho(\mathbf{r})$ we have analogously

$$W = \frac{1}{2}k\int dV\int dV'\frac{\rho(\mathbf{r})\rho(\mathbf{r}')}{|\mathbf{r} - \mathbf{r}'|} = \frac{1}{2}\int dV\rho(\mathbf{r})\phi(\mathbf{r}) \quad \text{joules}, \tag{2.178}$$

$$\phi(\mathbf{r}) = k\int\frac{dV'\rho(\mathbf{r}')}{|\mathbf{r} - \mathbf{r}'|} \quad \text{volts}. \tag{2.179}$$

The expression for $W$ does not explicitly exclude the point where $\mathbf{r} - \mathbf{r}' = 0$, so that it is possible that the expression involves divergent contributions. This is a problem connected with the Coulomb potential and requires a correction which we cannot enter into here (it is concerned with the problem dealt with under the term "*self-energy*"), which, however, is to be understood as a screening effect and replaces effectively the Coulomb potential $\sim 1/r$

by a type of Yukawa potential $\sim e^{-\mu r}/r$. In the case of surface charges we have correspondingly

$$W = \frac{1}{2}k \int_F dF(\mathbf{r}) \int_F dF(\mathbf{r}') \frac{\sigma(\mathbf{r})\sigma(\mathbf{r}')}{|\mathbf{r} - \mathbf{r}'|}. \tag{2.180}$$

For example, $F$ can consist of two condenser plates with areas $F_1$ and $F_2$ as illustrated in Fig. 2.17. We can also express $W$ in terms of $\mathbf{E}$: We saw that $\rho(\mathbf{r})$ is the source function of the potential $\phi$, i.e. (cf. Eq. (2.19b))

$$\nabla^2\phi \equiv \Delta\phi = -4\pi k\rho(\mathbf{r}). \tag{2.181}$$

Hence with Eq. (2.178) the energy becomes

$$W = -\frac{1}{8\pi k}\int \phi\nabla^2\phi dV = -\frac{1}{8\pi k}\int \phi\Delta\phi dV. \tag{2.182}$$

However,

$$\nabla \cdot (\phi\nabla\phi) = \phi\Delta\phi + (\nabla\phi)^2, \tag{2.183}$$

so that

$$W = -\frac{1}{8\pi k}\int dV[\nabla \cdot \{\phi(\nabla\phi)\} - (\nabla\phi)^2]. \tag{2.184}$$

But with Gauss' divergence theorem

$$\int dV\nabla \cdot (\phi\nabla\phi) = \int d\mathbf{F} \cdot (\phi\nabla\phi). \tag{2.185}$$

Now, $\phi \sim 1/r, \nabla\phi \sim 1/r^2$. If we integrate over an infinitely large surface (i.e. with $r \to \infty$), the integral vanishes, since $dF \propto r^2 d\Omega$ which implies

$$\int d\mathbf{F} \cdot (\phi\nabla\phi) \sim \frac{1}{r}\int d\Omega \to 0 \tag{2.186}$$

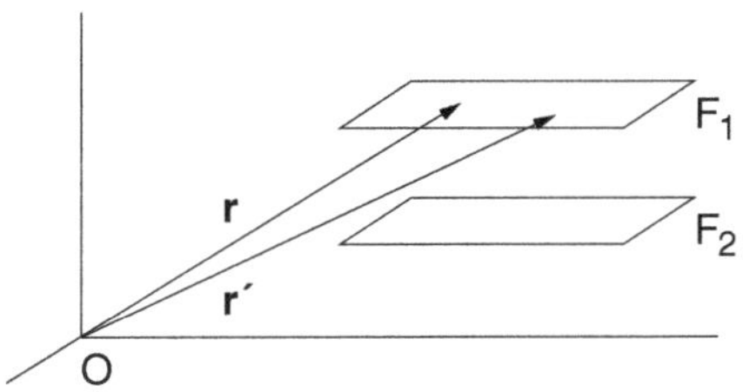


**Fig. 2.17.** Parts of a single surface.

for $r \to \infty$. It follows that (since $\mathbf{E} = -\boldsymbol{\nabla}\phi$)

$$W = \frac{1}{8\pi k}\int_{V_\infty} dV \mathbf{E}^2 \quad \text{joules.} \tag{2.187}$$

Thus the expression $\mathbf{E}^2/8\pi k$ represents the *energy density of the electric field.* Thus the energy in the capacitor is stored in its electric field. One should observe: For large values of $r$ the integrand of $W$ behaves as $dr/r^2$, but from this we cannot conclude that it vanishes, since the integral is the sum of many nonvanishing positive contributions. On the contrary, inserting the Coulomb potential for a single charge at $r = 0$ one obtains $W = \infty$.

**Example 2.18**[22] **(The gravitational potential).** Consider a static particle with charge $Q$. With the help of Einstein's formula *energy* $=$ *mass* $\cdot\, c^2$ calculate the mass density which is equivalent to the energy density of the electric field of the charge, and then insert this together with the contribution of a "bare" (i.e. fieldless) mass $m_0$ of the particle into the equation for Newton's gravitational potential $\phi$, i.e.

$$\triangle\phi = 4\pi G\rho_{\text{total}}. \tag{2.188}$$

Taking into account an additional contribution to $\rho$, which makes $\phi$ its own source (similar to the case in electrostatics), solve the equation with the help of the ansatz ($c$ being the velocity of light)

$$\psi(\phi) = e^{\phi/2c^2} \tag{2.189}$$

and the substitution $u = 1/r$. Then calculate $m_0$.

**Solution:** At a distance $r$ from the charge $Q$ the electric field is given by Eq. (2.3) ($k = 1/4\pi\epsilon_0$):

$$\mathbf{E} = \frac{Q}{4\pi\epsilon_0}\frac{\mathbf{r}}{r^3} \quad \text{newtons/coulomb.} \tag{2.190}$$

The corresponding electromagnetic energy density is (here only the electric part as given by Eq. (2.187)):

$$\rho_{\text{elm}} = \frac{1}{8\pi k}|\mathbf{E}|^2 = \frac{1}{8\pi}\frac{1}{4\pi\epsilon_0}\frac{Q^2}{r^4}. \tag{2.191}$$

[22] See Visser [155].

According to Einstein's energy-mass relation (see Eq. (17.166)) the electric energy density corresponds to a mass density

$$\rho_m = \frac{1}{8\pi k c^2}|\mathbf{E}|^2 \quad \text{kg/m}^3.$$

Let $m_0$ be the so-called bare mass of the particle (like the electron) with charge (say) $Q$ (i.e. not taking into account the radiation field of the particle). Then the total mass density is ($\int d\mathbf{r}\delta(\mathbf{r}) = 1$)

$$\rho_m = m_0\delta(\mathbf{r}) + \frac{1}{8\pi k c^2}|\mathbf{E}|^2 \quad \text{kg/m}^3. \tag{2.192}$$

This mass density contributes to the gravitational potential in space (like any other mass). Newton's gravitational potential $\phi$ is the solution of Eq. (2.188) (where $G$ is the gravitational constant, $G = 6.7 \times 10^{-11}\text{m}^3\text{kg}^{-1}\text{s}^{-2}$). The density $\rho_{\text{total}}$ is the total mass density contributing to the gravitational potential which now includes the electromagnetic energy density (in this way gravity couples to any kind of energy density). One contribution is apparently the contribution $\rho$ of Eq. (2.192). Another contribution has its origin in the gravitational field itself. Like the electromagnetic energy density $\rho_{\text{elm}}$ the (Newtonian) gravitational field also possesses an energy density which is given by the analogous expression (with minus sign in view of the attractive nature of the gravitational field)

$$\rho_{\text{Grav}} = -\frac{1}{8\pi G c^2}|\boldsymbol{\nabla}\phi|^2.$$

Adding this contribution also to $\rho_m$ we obtain a nonlinear self-interaction of the gravitational field which could be looked at as a neo-Newtonian version which incorporates model-like the nonlinear contributions in Einstein's theory. We obtain therefore

$$\Delta\phi = 4\pi G\left[m_0\delta(\mathbf{r}) + \frac{1}{8\pi c^2}\frac{1}{4\pi\epsilon_0}\frac{Q^2}{r^4}\right] - \frac{1}{2c^2}|\boldsymbol{\nabla}\phi|^2. \tag{2.193}$$

In solving the equation we set (observe the dimension of $\phi$ is that of $c^2$)

$$\psi(\phi) = e^{\phi/2c^2}, \tag{2.194}$$

so that

$$\boldsymbol{\nabla}\psi(\phi) = \frac{1}{2c^2}(\boldsymbol{\nabla}\phi)\psi(\phi) \tag{2.195}$$

and

$$\triangle\psi = \nabla\cdot\nabla\psi(\phi) = \frac{1}{2c^2}\nabla\cdot\{(\nabla\phi)\psi(\phi)\} = \frac{1}{2c^2}\{\nabla\psi\cdot\nabla\phi + \psi(\phi)\triangle\phi\}$$
$$= \frac{1}{2c^2}\left\{\frac{1}{2c^2}(\nabla\phi)^2 + \triangle\phi\right\}\psi(\phi). \tag{2.196}$$

Replacing here $\triangle\phi$ by the expression of Eq. (2.188), we have

$$\triangle\psi = \left[\frac{2\pi G m_0}{c^2}\delta(\mathbf{r}) + \frac{G}{4c^2 4\pi\epsilon_0 c^2}\frac{Q^2}{r^4}\right]\psi. \tag{2.197}$$

The right side is spherically symmetric. This suggests the separation of $\triangle$ in spherical polar coordinates $r, \theta, \varphi$. Then setting $u = 1/r$, one obtains

$$\triangle\psi = u^4\frac{\partial^2}{\partial u^2}\psi, \tag{2.198}$$

with

$$\frac{\partial^2}{\partial u^2}\psi = \frac{GQ^2}{4c^2 4\pi\epsilon_0 c^2}\psi, \quad \frac{\partial^2\psi}{\partial u^2} - \frac{\tilde{Q}^2}{4}\psi = 0 \tag{2.199}$$

with $\tilde{Q} = \kappa Q, \kappa = \sqrt{G}/\sqrt{4\pi\epsilon_0}c^2$. It should be noted that because $r^4\delta(\mathbf{r}) = 0$, the other source term has disappeared (but will be regained by consideration of $r \to 0$, which is here excluded). The resulting equation can now be solved provided we know the boundary condition that $\psi(\phi)$ ought to satisfy. Since $\phi$ is to vanish for $r \to \infty$, we require (cf. Eq. (2.194))

$$r \to \infty: \quad \psi(\phi) \to 1,$$

and hence

$$\psi \propto \cosh\left(\frac{\tilde{Q}u}{2}\right), \quad \psi = \frac{\cosh(A - \tilde{Q}/2r)}{\cosh A}, \quad A = \text{const.} \tag{2.200}$$

However, for $r \to \infty$,

$$\phi \overset{(2.194)}{=} 2c^2\ln\psi = 2c^2\ln\left\{\frac{\cosh(A - \tilde{Q}/2r)}{\cosh A}\right\}$$
$$= 2c^2\ln\left(1 - \frac{1}{2r}\tilde{Q}\tanh(A) + \cdots\right)$$
$$= -\frac{1}{r}\tilde{Q}c^2\tanh(A) + \cdots. \tag{2.201}$$

Since this expression represents the gravitational potential (for $r \to \infty$) the quantity $\tilde{Q}c^2 \tanh(A)$ has to be interpreted as gravitational mass. With the above expression for $\psi$ we have in addition, as given

$$\phi = 2c^2 \ln\left(\frac{\cosh(A - \tilde{Q}/2r)}{\cosh A}\right), \tag{2.202}$$

and with this ($d$coshx/dx = sinhx)

$$\nabla\phi = \tilde{Q}c^2 \tanh\left(A - \frac{\tilde{Q}}{2r}\right)\frac{\mathbf{r}}{r^3}, \tag{2.203}$$

and, using $d$tanhx/dx = 1/cosh$^2$x, (see below for the first term)

$$\triangle\phi = -4\pi\tilde{Q}c^2\delta(\mathbf{r}) + \frac{\tilde{Q}^2c^2}{2r^4\cosh^2(A - \tilde{Q}/2r)}. \tag{2.204}$$

The *first contribution* on the right of Eq. (2.204) is obtained as follows. For $r \neq 0$ we have $(\partial(1/r^3)/\partial x = -3x/r^5)$

$$\nabla\cdot\left(\frac{\mathbf{r}}{r^3}\right) = \frac{1}{r^3}\nabla\cdot\mathbf{r} + \left(\nabla\frac{1}{r^3}\right)\cdot\mathbf{r} = \frac{3r^2 - 3r^2}{r^5} = 0. \tag{2.205}$$

For $r \to 0$ we have

$$\lim_{r\to 0}\tanh\left(A - \frac{\tilde{\tilde{Q}}}{2r}\right) = \tanh(-\infty) = -1. \tag{2.206}$$

Then, returning to Eq. (2.203), for $r \to 0$: $\nabla\phi \simeq -\tilde{Q}c^2\mathbf{r}/r^3$ and

$$\begin{aligned}\int\nabla\cdot\nabla\phi d\mathbf{r} &= -\tilde{Q}c^2\int_V\nabla\cdot\left(\frac{\mathbf{r}}{r^3}\right)d\mathbf{r}\\ &= -\tilde{Q}c^2\int_{F(V)}\frac{\mathbf{r}}{r^3}\cdot d\mathbf{F} = -\tilde{Q}c^2\int\frac{\mathbf{r}}{r^3}(d\Omega r^2)\frac{\mathbf{r}}{r}\\ &= -4\pi\tilde{Q}c^2 \equiv -4\pi\tilde{Q}c^2\int\delta(\mathbf{r})d\mathbf{r}.\end{aligned} \tag{2.207}$$

Hence $\triangle\phi$ acquires the contribution $-4\pi\tilde{Q}c^2\delta(\mathbf{r})$. Comparison of the Eqs. (2.188) and (2.197) then yields

$$m_0 = -\tilde{Q}c^2/G = -Q/\sqrt{4\pi\epsilon_0 G} \quad \text{kg}. \tag{2.208}$$

The *second contribution* on the right side of Eq. (2.204) follows from the sum of the other two contributions on the right side of Eq. (2.193) with $1/\cosh^2 x = 1 - \tanh^2 x$.

**Example 2.19 (Energy of the spherical condenser).** Show that for the spherical condenser (radii $a$ and $b, a < b$, with charges $q, -q$) the electrostatic energy is $W = q^2/2C$. Calculate the electrostatic pressure on the outer spherical shell and compare the result with the product of charge/area of the outer shell with the electric field strength at the latter.

**Solution:** The charge on the inner sphere (radius $a$, charge $+q$) acts like concentrated at the center $O$. Therefore the field $E_r$ at a distance $r$ away from $O$ is

$$E_r = k\frac{q}{r^2} \quad \text{newtons/coulomb}, \quad \text{a} < \text{r} < \text{b}.$$

Hence the total electrostatic energy is (cf. Eq. (2.187))

$$\begin{aligned} W &= \frac{1}{8\pi k}\int d\mathbf{r}E_r^2 = \frac{1}{8\pi k}\int 4\pi r^2 dr E_r^2 = \frac{kq^2}{2}\int_a^b \frac{dr}{r^2} \\ &= \frac{kq^2}{2}\left(\frac{1}{a} - \frac{1}{b}\right) \quad \text{joules.} \end{aligned} \tag{2.209}$$

Since according to Example 2.10, Eq. (2.106), the capacity $C$ is

$$C = \frac{1}{k(1/a - 1/b)} \quad \text{farads,} \tag{2.210}$$

we obtain $W = q^2/2C$. The force acting on the external spherical shell is

$$-\frac{\partial W}{\partial b} = -\frac{kq^2}{2b^2} \quad \text{newtons,} \tag{2.211}$$

so that the pressure on this shell, force/area, is

$$-\frac{kq^2}{2b^2 4\pi b^2} = -\frac{kq^2}{8\pi b^4} \quad \text{newtons/m}^2 \tag{2.212}$$

(toward the center). On the other hand the charge density $\sigma$ of the outer shell is $-q/4\pi b^2$. Thus

$$\sigma E_{r=b} = -\frac{q}{4\pi b^2}\frac{kq}{b^2} = -\frac{kq^2}{4\pi b^4} \quad \text{newtons/m}^2. \tag{2.213}$$

This is not the pressure, which acts on the outer shell, since the surface charge on the rest of the outer shell exerts a force in the opposite direction and thus compensates partially the force due to the charge on the inner sphere (cf. Eq. (2.92)).

**Example 2.20 (A charged particle inside a charge density[23]).** A spherically shaped volume is uniformly charged with charge density, charge per unit volume, $q_v$. A particle with mass $m$ and charge $-q$ is placed into this spherical volume. Show that the mass executes harmonic oscillations and calculate the frequency of these.

**Solution:** We have here

$$\int \mathbf{E}\cdot d\mathbf{F} = 4\pi k\sum_i q_i, \quad \text{i.e.} \quad E_r 4\pi r^2 = 4\pi k\frac{4}{3}\pi r^3 q_v, \tag{2.214}$$

and hence

$$E_r = \frac{4}{3}\pi r q_v k, \tag{2.215}$$

so that the equation of motion of the particle is

$$m\ddot{r} = -qE_r = -\frac{4}{3}\pi r q q_v k \quad \text{or} \quad m\ddot{r} + m\omega^2 r = 0 \tag{2.216}$$

with

$$\omega^2 = \frac{4\pi k q q_v}{3m}, \quad T = \frac{2\pi}{\omega} = 2\pi\sqrt{\frac{3m}{4\pi q q_v k}} \quad \text{seconds.} \tag{2.217}$$

**Example 2.21 (A multivalued potential).** Determine the equipotential surfaces of the potential

$$\phi = \tan^{-1}\frac{y}{x}. \tag{2.218}$$

What is the value of a contour integral $\int \mathbf{\nabla}\phi\cdot d\mathbf{l}$ along a closed path A, which does not encircle the $z$-axis, and what is the value of a contour integral along a closed path B, which does encircle the $z$-axis? (Remark: The function $\phi$ will be discussed further, in particular in Section 8.3).

**Solution:** Since $\phi = \tan^{-1}(y/x) = \theta \pm n\pi$, n a positive integer or zero, the equipotential surfaces are the radial planes with $\theta =$ const. These intersect along the $z$-axis. These planes at different values of $\phi$ therefore violate the physically expected single valuedness of the potential, and the $z$-axis is the seat of singularities.

---

[23]A similar problem based on Newton's potential can naturally be found in books on Classical Mechanics, see e.g. Müller–Kirsten [101].

The gradient is

$$\nabla\phi = -\frac{y}{x^2+y^2}\mathbf{e}_x + \frac{x}{x^2+y^2}\mathbf{e}_y, \quad \text{so that} \quad \nabla\phi\cdot d\mathbf{l} = \frac{xdy - ydx}{x^2+y^2}. \tag{2.219}$$

For a circle around the origin in the plane of $x$ and $y$, and $a =$ const., one has $x = a\cos\theta, y = a\sin\theta, xdy - ydx = a^2 d\theta$ and thus $\int \nabla\phi\cdot d\mathbf{l} = \int d\theta$. This expression is independent of $a$ and is therefore valid for an arbitrary contour. Since the direction of $\nabla\phi$ is that of $\mathbf{e}_\theta$, contour components along $\mathbf{e}_z$ and $\mathbf{e}_r$ do not contribute. In the case of contour A we clearly have $\int d\theta = [\theta - \theta] = 0$. Also curl grad $\phi = 0$. In the case of contour B, however, the value of the integral depends on the endpoint. The potential is multivalued, i.e. each time O is encircled its value changes by $\pm 2\pi$. Such a contour cannot be shrunk to zero without cutting the singularity at O.

## 2.11 Exercises

**Example 2.22 (Transformation to other coordinates).** The coordinates $x, y, z$ are related to another set of coordinates $x', y', z'$ through the Jacobian $J(x, y, z, x', y', z')$. Show that

$$\delta(x)\delta(y)\delta(z) = \frac{1}{|J|}\delta(x')\delta(y')\delta(z'). \tag{2.220}$$

**Example 2.23 (Mechanical force versus electric force).** Use the principle of virtual work familiar from classical mechanics to obtain the mechanical force $F$ between the oppositely charged plates of a parallel plate condenser which are a distance $d$ apart and in static equilibrium.

(Answer: $F = \epsilon_0 SE^2/2$ newtons, where $S$ is the area of a plate).

**Example 2.24 (Potential of a uniformly charged disc).** Determine the electrostatic potential $\phi$ of a circular disc of radius $a$ and charge $\sigma$ per unit area at a height $z$ on the perpendicular through the center of the disc. What is the electric field strength $E$ there?

(Answer: $\phi = 2\pi k\sigma z(1 - \cos\alpha)$,
$E = 2\pi k\sigma(1 - \cos\alpha), \cos\alpha = z/\sqrt{a^2 + z^2}$).

The problem is not so easy if the point of interest does not lie on the central perpendicular. As Konopinski [79] (p. 33) remarks, any circular shape looks like an elliptical one when viewed from a point off axis. Hence the general problem requires evaluation of elliptic integrals (see Example 3.26). A different problem is to obtain the potential at the rim of a uniformly charged *conducting* disc (constant potential). In this case it is convenient to consider an oblate spheroid and to take the limit of the axial length shrunk to zero. This is how this problem was treated by Bradley [22] (pp. 57–59) by solving the Laplace equation in oblate spheroidal coordinates. The potential on the surface and hence at the rim was thus obtained as (see Example 3.20)

$$\phi(a) = \frac{\pi a \sigma}{8\epsilon_0} \quad \text{volts.} \tag{2.221}$$

The problem is also referred to by Jackson [73] (p. 125) in the context of rather complicated dual integral equations.

**Example 2.25 (Energy stored in a capacitor).** Considering the rate at which work $W$ is done in the process of charging a capacitor of capacity $C$ ($W = VdQ/dt, V$ the voltage) show that the energy of the system when raised to a voltage $V_0$ is

$$U = \frac{1}{2} C V_0^2 \quad \text{joules.} \tag{2.222}$$

This is for instance the energy transferred from the battery and to be stored in the condenser.[24]

**Example 2.26 (Electrostatic pressure).** Consider a charged condenser composed of two concentric spheres with radii $a$ and $b, b > a$, and its capacity $C$ and energy $U$. By considering a virtual increase in the radius $b$ of the outer sphere calculate from the resultant virtual change in energy the electrostatic pressure acting on the outer sphere. Can this pressure alternatively be determined by multiplying the surface charge density on the outer sphere by the electric field strength $E$ created by the charge on the inner sphere? If not, why not?

**Example 2.27 (Determination of electric energy).** Two equal and opposite charges are separated by a distance $2s$. By vectorial addition of the fields due to the charges, e.g. by the parallelogram of forces or otherwise, determine the electrostatic energy density at a distance $r$ from the midpoint of the line joining the charges when $r$ makes an angle $\theta$ with this line.

**Example 2.28 (Coulomb potential in two space dimensions).** Derive the Coulomb potential in two space dimensions (cf. Example 2.8).

**Example 2.29 (Conducting plates in parallel).** A capacitor consists of four parallel plates arranged in pairs in the middle of ladder-like rungs

[24] For discussion see Ferrari [53], pp. 41–44.

or cross-bars with charges $+q, -q, +q, -q$. What is the capacity of the capacitor? Such capacitors (for which the capacities add) are said to be "*in parallel*". What is a corresponding arrangement which is said to be "*in series*" and what is the reciprocal of its capacity? See Example 8.13. For further discussion see Clemmow [32], pp. 83–84 and Bradley [22], p. 45.

**Example 2.30 (Electric and gravitational forces acting on a particle).** A particle of mass $m$ and charge $q$ is subjected to oppositely directed forces of gravity and an electric field. Is there a position of equilibrium?

**Example 2.31 (Mutual potential energy of 3 charges).** The charges $q, -q, -q$ are placed at the vertices of an isosceles triangle with side lengths $a, a, 2a$, the charge $q$ being placed opposite to the side of length $2a$. What is the mutual potential energy of the three charges?

(Answer: $-3kq^2/2a$ joules).

**Example 2.32 (Plate condenser as limit of spherical condenser).** Show that in the case of a spherical condenser with radii $r_1, r_2$ and $r_2 - r_1 = d$ finite, the capacity becomes that of the parallel plate condenser in the limit $r_1, r_2 \to \infty$.

**Example 2.33 (One-dimensional delta function properties).** Prove the following relations:

$$\frac{1}{2}\frac{d^2}{dx^2}|x| = \delta(x), \tag{2.223}$$

$$\delta(x) = \lim_{\lambda\to 0}\frac{1}{\sqrt{\pi\lambda}}\exp(-x^2/\lambda), \tag{2.224}$$

$$\delta(x) = \lim_{a\to 0}\frac{a}{\pi(x^2+a^2)}, \tag{2.225}$$

$$\delta(x) = \lim_{a\to\infty}\frac{a}{\sqrt{\pi}}e^{-a^2x^2}. \tag{2.226}$$

**Example 2.34 (The electrostatic loudspeaker).** An electrostatic loudspeaker uses the attractive electrostatic force of the electrodes. Show that in the case of a parallel plate capacitor with voltage $V$ and separation $d$ of the plates, the pressure $p$ on the plates is given by

$$p = \frac{1}{2}\epsilon\frac{V^2}{d^2}\quad \mathrm{N/m^2}, \tag{2.227}$$

where $\epsilon$ is the dielectric constant.

**Example 2.35 (Capacity of a van de Graaff generator).** The dome of a van de Graaff generator has a radius of 0.15 m. What is its capacity?

(Answer: 16.67 pf).

**Example 2.36 (Charges inside and outside a capacitor).** The (internal) surfaces of a parallel plate capacitor facing each other hold equal and opposite charges as a result of Coulomb's law. What are these charges if one plate or element is given charge $Q_1$ and the other charge $Q_2$? What are the charges on the outer surfaces?

(Answers: $\pm(Q_1 - Q_2)/2; (Q_1 + Q_2)/2$).

## Chapter 3

# Applications of Electrostatics

### 3.1 Introductory Remarks

In the following we investigate further extensions of electrostatics as well as more complicated applications, thereby also emphasizing the methodical differences. *Green's theorems* and *Fourier transforms* are introduced, and finally the *multipole expansion* of the potential of a charge distribution is derived.

### 3.2 Method I: The Gauss Law

The problems we considered thus far were mostly treated with the help of the integral form of the Gauss law, and so with Eq. (2.19a), i.e.

$$\int \mathbf{E}\cdot d\mathbf{F} = 4\pi k \sum_i q_i. \tag{3.1}$$

We now investigate other methods.

### 3.3 Method II: Poisson and Laplace Equations

In Chapter 2 we encountered the Poisson equation as the equation (cf. Eq. (2.19a))

$$\boldsymbol{\nabla}^2\phi = -4\pi k\rho, \quad k = \frac{1}{4\pi\epsilon_0}. \tag{3.2}$$

In solving this equation we require boundary conditions. These are:

1. $\phi =$ const. *on conducting surfaces* (in an ideal conductor the electrons do not perform work) with the discontinuity for the derivatives as for the

electric field strength, i.e. (cf. (2.90))

$$-\left(\frac{\partial \phi}{\partial n}\right)_2 + \left(\frac{\partial \phi}{\partial n}\right)_1 = 4\pi k\sigma, \tag{3.3}$$

2. *otherwise* $\phi$ *continuous* (this is no contradiction with the difference of the potential on both sides of a dipole layer; the potential of a dipole $\mathbf{p}$ is, as we shall see in Section 3.9, $\mathbf{p}\cdot\mathbf{r}/r^3$, and this is continuous in $\mathbf{r}$ (no step function $\theta(x)$). If $\phi$ were discontinuous anywhere, i.e. if $\phi$ contained a step-function $\theta(x - x_0)$, then at this point $\mathbf{E} = -\boldsymbol{\nabla}\phi$ would be proportional to $\delta(x - x_0)$, i.e. would be infinite, which would be meaningless and unphysical.

**Example 3.1 (The cylindrical condenser).** Calculate the potential difference $V$ of a cylindrical condenser.

**Solution:** We consider again the cylindrical condenser without boundary effects, and charges $q$ and $-q$ on the surfaces of radii $r_1$ and $r_2 > r_1$ and height $h$. In the region between the cylindrical surfaces we have no charge, so that there $\nabla^2\phi = 0$. In view of the cylindrical symmetry $\partial\phi/\partial\theta = 0 = \partial\phi/\partial z$, so that separation of the Laplace operator yields in cylindrical coordinates

$$\frac{1}{r}\frac{\partial}{\partial r}\left(r\frac{\partial \phi}{\partial r}\right) = 0, \tag{3.4}$$

i.e. $\partial\phi/\partial r = \alpha/r, \alpha = \text{const}$. Hence $E = -\partial\phi/\partial r = -\alpha/r$. At $r = r_1$ we have (see Eq. (2.99)) $E = 4\pi k\sigma = 4\pi kq/F = 4\pi qk/2\pi h r_1$, so that $\alpha = -2qk/h$, and for $E$ and the potential difference $V$ we obtain as before

$$E = \frac{2qk}{h}\frac{1}{r} \quad \text{newtons/coulomb}, \quad V = \int_{r_1}^{r_2} E dr = \frac{2qk}{h}\ln\frac{r_2}{r_1} \quad \text{volts}. \tag{3.5}$$

The result is positive as a consequence of its physical interpretation as the work done *in the direction of* $\mathbf{E}$. Note the difference in the calculation compared with that of Example 2.12.

As a more advanced application we consider a proportional counter.[1]

**Example 3.2 (The proportional counter).** Calculate the potential of a proportional counter. An important part of a proportional counter is its coaxial circular cylindrically shaped anode of radius $a$ made of wire

[1]Sipila, Vanha-Honko and Bergqvist [135].

and a corresponding circular cylindrical cathode with radius $b > a$. The intermediate space is filled with an ionizable mixture of gases. In practice one frequently has the situation that the volume charge density $\rho$ differs significantly from zero only along a length $L$(with $|z| \leq L/2$) of the cylinders. Show that the potential $\phi(r, z)$ in this intermediate space obtained with the following boundary conditions

$$\phi(a, z) = V, \quad \phi(b, z) = 0, \quad \rho = \begin{cases} \text{const.} & \text{for} \quad |z| \leq L/2, \\ 0 & \text{for} \quad |z| > L/2, \end{cases} \tag{3.6}$$

is given by — where $K_n = -J_0(\mu_n a)/Y_0(\mu_n a)$ —

$$\begin{aligned} \phi(r, z) = & \frac{V}{\ln(a/b)} \ln(r/b) + \rho \sum_n \frac{2B_n}{\mu_n} [J_0(\mu_n r) + K_n Y_0(\mu_n r)] \\ & \times \begin{cases} 1 - e^{-\mu_n L/2} \cosh(\mu_n z) & \text{for} \quad |z| \leq L/2, \\ \sinh(\mu_n L/2) e^{-\mu_n |z|} & \text{for} \quad |z| > L/2, \end{cases} \end{aligned} \tag{3.7}$$

where $\mu_n$ satisfies the following relation involving the special functions $J$ and $Y$ known as Bessel and Neumann functions:

$$J_0(\mu_n a) Y_0(\mu_n b) - J_0(\mu_n b) Y_0(\mu_n a) = 0.$$

Which condition determines the coefficients $B_n$?

**Solution:** For the detailed solution we refer this time to Sipila *et al.* [135].

## 3.4 Method III: Direct Integration

In this case one integrates either *vectorially*, as in the formula

$$\mathbf{E} = k \int \frac{\mathbf{r} dq}{r^3} \quad \text{newtons/coulomb,} \quad \mathrm{dq} = \rho \mathrm{dV} \quad \text{or} \quad \sigma \, \mathrm{ds}, \tag{3.8}$$

or *scalarly* as in the formula

$$\phi = k \int \frac{dq}{r} \quad \text{volts,} \quad \mathrm{dq} = \rho \mathrm{dV} \quad \text{or} \quad \sigma \, \mathrm{ds}. \tag{3.9}$$

In the latter case the integral is evaluated by integrating over a charged surface $S$, and $\mathbf{r}$ is the vector from the element of area $ds$ to the point $P$ at which the potential is to be determined. We illustrate the method again by application to some important examples.

**Example 3.3 (The finite, charged, thin rod).** Calculate the field surrounding a charged, thin rod of finite length.

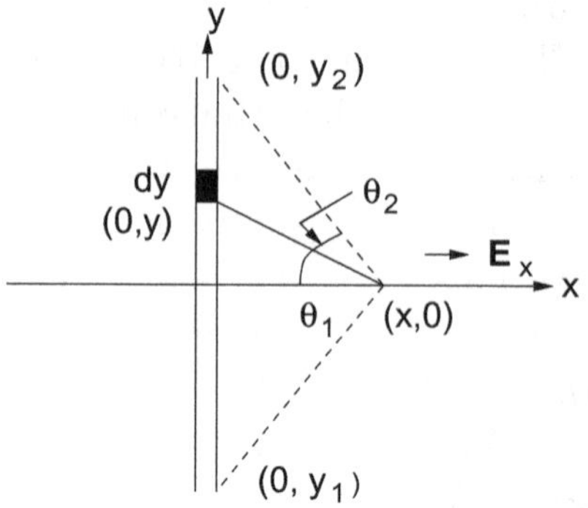


**Fig. 3.1.** The charged rod.

**Solution:** The word “thin” implies that the rod is to be considered as a line. Let $q$ be its charge per unit length. We choose the coordinates as in Fig. 3.1.

The potential energy at the point $(x, 0)$ in Fig. 3.1 is

$$\begin{aligned}\phi(x) &= kq\int_{y_1}^{y_2}\frac{dy}{\sqrt{x^2+y^2}} = qk\left[\sinh^{-1}\frac{y}{x}\right]_{y_1}^{y_2}\\ &= qk\left\{\sinh^{-1}\frac{y_2}{x} - \sinh^{-1}\frac{y_1}{x}\right\} \quad \text{volts.}\end{aligned} \tag{3.10}$$

Hence the electric field at the point $(x, 0)$ is

$$\begin{aligned}E_x &= -\frac{\partial\phi}{\partial x} = \frac{qk}{x}\left\{\frac{y_2}{\sqrt{y_2^2+x^2}} - \frac{y_1}{\sqrt{y_1^2+x^2}}\right\}\\ &= \frac{qk}{x}\{\sin\theta_2 - (-\sin\theta_1)\} \quad \text{newtons/coulomb.}\end{aligned} \tag{3.11}$$

In the limit $y_2, -y_1 \to \infty$: $\theta_1, \theta_2 \to \pi/2$, i.e. $E_x \to 2qk/x$ as for a point charge at the origin. One can show that the equipotential surfaces are ellipses with foci at the ends of the rod. Above we calculated as a matter of simplicity only the $x$-component of the field. We can deduce from Eq. (3.10) what the potential is. To obtain this, we use the expansions[2]

$$\begin{aligned}\sinh^{-1}x &= \ln(2x) + \frac{1}{4x^2} + O\left(\frac{1}{x^4}\right) \quad \text{for } x > 1,\\ \sinh^{-1}x &= -\ln|2x| - \frac{1}{4x^2} + O\left(\frac{1}{x^4}\right) \quad \text{for } x < -1.\end{aligned} \tag{3.12}$$

[2]See e.g. Dwight [42], formula 706, p. 156.

Inserting the dominant contributions into Eq. (3.10) we obtain

$$\phi(x) \simeq qk\left\{\ln\left(\frac{2y_2}{x}\right) + \ln\left|\frac{2y_1}{x}\right|\right\} = -2qk\ln x + qk\ln|4y_2y_1|,$$
$$\left(\frac{y_2}{x} > 1, \frac{y_1}{x} < 1\right). \tag{3.13}$$

Replacing $x$ by $r$ leads to the expression

$$\phi(r) = -\frac{q}{2\pi\epsilon_0}\ln r + \text{const.}, \quad r > 0, \tag{3.14}$$

in which the constant can be infinite. In the computation of the field strength, however, the contribution of the constant drops out. With the help of the last relation and the product formula[3]

$$\sinh x = x\prod_{n=1}^{\infty}\left(1 + \frac{x^2}{n^2\pi^2}\right) \tag{3.15}$$

one can calculate the potential of a system of parallel wires, a distance $b$ apart, and each carrying the same charge. One finds

$$\phi(r) = -\frac{q}{2\pi\epsilon_0}\ln\sinh\left(\frac{\pi r}{b}\right) + \text{const.} \quad \text{volts}, \tag{3.16}$$

which for $b \to \infty$ yields Eq. (3.14). Systems of this or a similar kind are used in particle detectors.[4]

## 3.5 Method IV: Kelvin Method of Image Charges

In an electrostatic problem the potential determines in a unique way quantities like $\mathbf{E}$ and $\sigma$. It is therefore possible to imagine certain potential distributions replaced by fictitious charges called "*image charges*" and to calculate the field of these. An example is shown in Fig. 3.2. The effect of the earthed conductor (i.e. with potential $\phi = 0$, this alone is our definition of "*earthed*") can also be achieved by a fictitious charge.

The earthed conductor implies a *boundary condition*: $\phi = 0 = \phi(r = \infty)$. The field $\mathbf{E}$ then results as the effect of both charges, $+q$ and $-q$, in the region *above* the conductor. A similar example is illustrated in Fig. 3.3; again the field of case (a) can be calculated from the situation of case (b).[5]

---

[3] Gradshteyn and Ryzhik [60], formula 1.431(2), p. 37.
[4] See e.g. Killian [76].
[5] See also Ferrari [53], pp. 179–180.

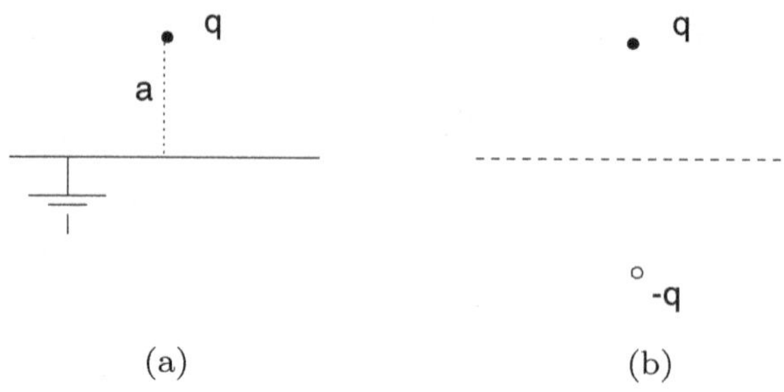


**Fig. 3.2.** (a) The earthed conductor and (b) its fictitious replacement.

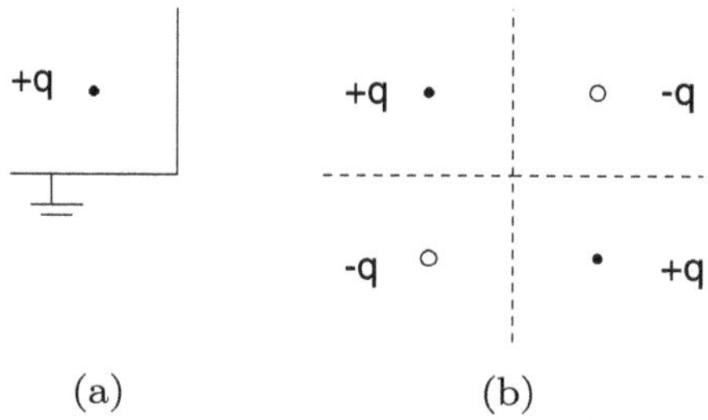


**Fig. 3.3.** (a) An earthed conductor and (b) its fictitious replacement.

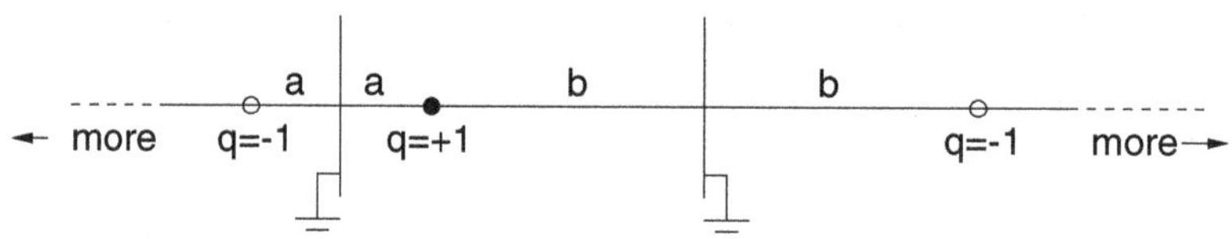


**Fig. 3.4.** Charge $q$ between earthed conductor plates.

In the case of two earthed conducting plates and a single charge between them we have to introduce a large number of image charges as shown in Fig. 3.4.

The potential at an arbitrary point between the plates enclosing the charge (in Fig. 3.4 with charge +1 coulomb), is then the same as that of the given charge together with the image charges. An analogous situation is obtained for the earthed conductor plates arranged as in Fig. 3.5.

Although the method of image charges is used in applications, it is not always treated in relevant texts. Some texts which include it are those of Stiefel [145], Ferrari [53] (particularly pp. 176–185), Zahn [167] and in particular Schelkunoff [129]. The method of image charges finds application

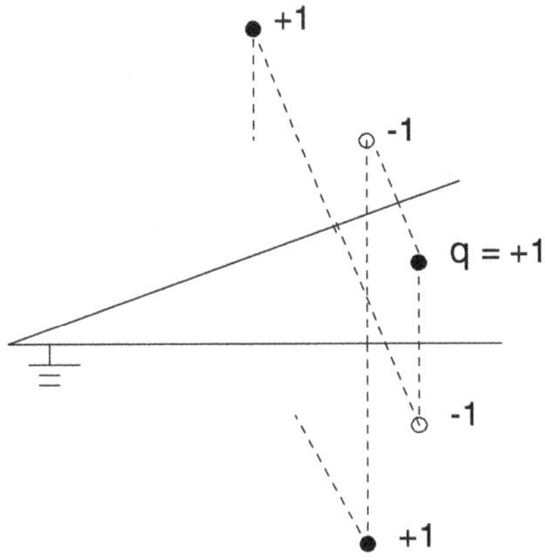


**Fig. 3.5.** Conducting plates with angular orientation.

for instance in the calculation of the electric field in particle detectors.[6] The method can be extended to include electric and magnetic current elements.[7]

**Example 3.4 (The oscillating charge).** A pointlike mass $m$ with charge $q$ at the end of a string of length $a$ oscillates in the gravitational field of the earth above an earthed horizontal metal plate. The mass oscillates like a pendulum in a plane. Determine the equation of motion of the pendulum expressed in terms of the angular deflection $\theta$. What is the period in the case of small oscillations (of course, a moving charge generates a magnetic field but this is of no significance here)?

**Solution:** Using the method of image charges as sketched in Fig. 3.6, we place a charge $-q$ in the position of a mirror image of the charge $+q$, so that in the plane of the earthed conductor the potential $\phi$ is zero. In this way we can ignore the plate (or its boundary condition) in the evaluation of the electric field. It is easiest to use the principle of conservation of energy $E$. The total energy of the charge $+q$ is ($k = 1/4\pi\epsilon_0$)

$$E = \frac{1}{2}m(a\dot{\theta})^2 + mga(1 - \cos\theta) - \frac{kq^2}{(2d + 2a - 2a\cos\theta)} \quad \text{joules}, \tag{3.17}$$

where $d$ is the distance shown in Fig. 3.6. With

$$\frac{d}{d\theta}\left(\frac{1}{2}\dot{\theta}^2\right) = \ddot{\theta}, \quad \frac{dE}{d\theta} = 0,$$

[6]See e.g. Weihs and Zech [160].
[7]See Schelkunoff [129], p. 171.

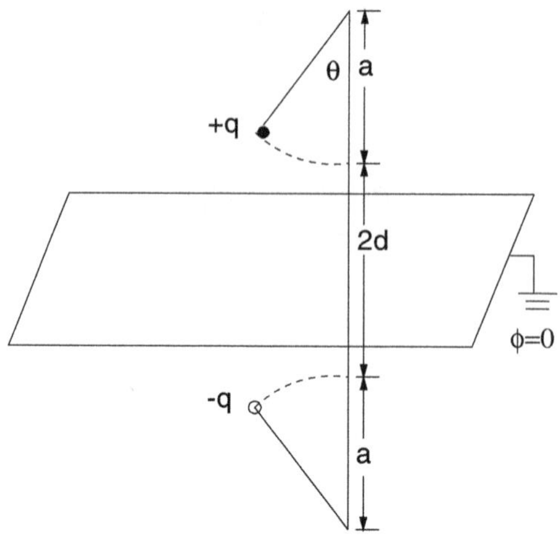


**Fig. 3.6.** The oscillating charge.

it follows that

$$ma^2\ddot{\theta} + mga\sin\theta + \frac{kq^2 a\sin\theta}{(d+a-a\cos\theta)^2} = 0. \qquad (3.18)$$

For $\theta$ small we have

$$\ddot{\theta} + \Omega^2\theta \simeq 0, \quad \Omega^2 = \left(\frac{mg + kq^2/d^2}{ma}\right). \qquad (3.19)$$

Thus the desired period is $T = 2\pi/\Omega$ seconds.

## 3.6 Theoretical Aspects of Image Charges

### 3.6.1 *The induced charge*

Before we investigate general theoretical aspects of the method, we consider an example: A charge $q$ placed outside an earthed spherical shell of radius $a$. Different from the cases discussed above, in which we considered either point charges *or* charge distributions, we now have the example of a mixed case. Here "*earthed*" or "*grounded*" means potential $\phi$ as on the ground, and this means as at $r = \infty$. Everywhere on the spherical surface the potential is to be $\phi = 0$. Apart from this, $\phi$ is a continuous function (so that $\mathbf{E} = -\boldsymbol{\nabla}\phi$ is nowhere infinite).

*The first question* is therefore: Where do I have to put which image charge so that at any point on the sphere one has $\phi = 0$? Inside the conductor, i.e. on the sphere, the charges (electrons) can move about freely (performing no

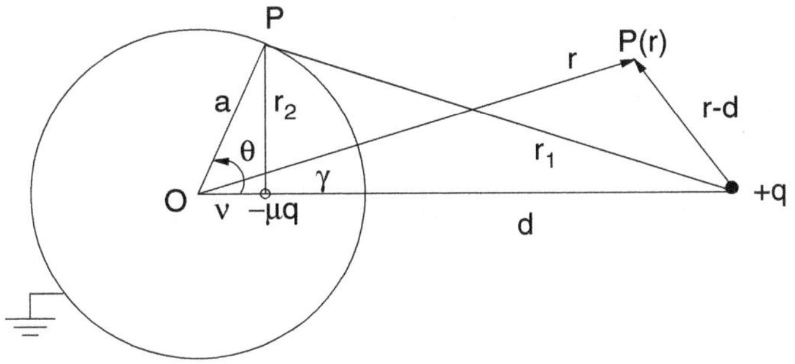


**Fig. 3.7.** The earthed spherical shell.

work, since $\phi = 0$, also $\mathbf{E} = 0$). As indicated in Fig. 3.7 we put a fictitious charge $-\mu q$ into the interior of the sphere. In order that $\phi = 0$ at a point $P$ on the shell, we must have

$$0 = \frac{q}{r_1} - \frac{\mu q}{r_2}, \tag{3.20}$$

i.e. $r_2 = \mu r_1$. From the geometry of Fig. 3.7 we obtain

$$r_1^2 = d^2 + a^2 - 2ad\cos\theta = d^2\left(1 + \frac{a^2}{d^2} - 2\frac{a}{d}\cos\theta\right), \tag{3.21}$$

and

$$r_2^2 = a^2 + \nu^2 - 2a\nu\cos\theta = a^2\left(1 + \frac{\nu^2}{a^2} - 2\frac{\nu}{a}\cos\theta\right). \tag{3.22}$$

It follows that $r_2$ is proportional to $r_1$ if

$$\nu = \frac{a^2}{d}. \tag{3.23}$$

Then $r_2 = ar_1/d$, i.e. $\mu = a/d$. In other words, a fictitious charge $-aq/d$ has to be placed at a point, called the "*image point*", a distance $a^2/d$ away from $O$, so that $\phi = 0$ at all points on the sphere.

*The second question*: What is the potential $\phi(\mathbf{r})$ at an arbitrary point $P(\mathbf{r})$ outside the earthed sphere, where $r > a$? Outside the sphere in vacuum

$$\mathbf{E} \equiv \mathbf{E}_2 = -\boldsymbol{\nabla}\phi \neq 0.$$

At an arbitrary point $\mathbf{r}$ outside the sphere in vacuum

$$\begin{aligned} 4\pi\epsilon_0\phi(\mathbf{r}) &= \left(\frac{q}{|\mathbf{r}-\mathbf{d}|} - \frac{qa}{d|\mathbf{r}-\boldsymbol{\nu}|}\right) \\ &\stackrel{r\geq a}{=} \frac{q}{(r^2+d^2-2rd\cos\gamma)^{1/2}} - \frac{aq}{d(r^2+\nu^2-2r\nu\cos\gamma)^{1/2}}, \end{aligned} \tag{3.24}$$

and this vanishes for $r = a$. Hence inside the sphere $\mathbf{E} = 0$, i.e. $\phi$ constant. This constant is zero since on the sphere $\phi = 0$. (If the entire sphere is filled with conductor material, the potential is everywhere constant or zero as on the boundary surface). The relation (2.90) we derived earlier for the passage of the electric field through a charged surface (now with $\mathbf{E}_1 = 0$) is

$$(\mathbf{E}_2 - \mathbf{E}_1)\cdot\mathbf{n} = 4\pi k\sigma, \quad \text{or} \quad -\left(\frac{\partial\phi}{\partial n}\right)_2 + \left(\frac{\partial\phi}{\partial n}\right)_1 = 4\pi k\sigma \tag{3.25}$$

where $\partial\phi/\partial n = (\partial\phi/\partial\mathbf{r})\cdot\mathbf{n}$. This allows us to compute the surface charge density $\sigma$. The corresponding charge on the sphere is called *induced charge.* This charge follows from the fact that the charge outside repels charges of the same polarity in the conductor and attracts those of opposite polarity.

*The third question*: What is the induced charge? The induced charge is a real charge which we can calculate. This charge is given by

$$Q = \int_S \sigma(\gamma)a^2 d\Omega \quad \text{coulombs}, \quad d\Omega = 2\pi d(\cos\gamma) \tag{3.26}$$

(a circular annulus about the horizontal axis in Fig. 3.7 at angular height $\gamma$ and of infinitesimal width $ad\gamma$ has area $(ad\gamma)2\pi(a\sin\gamma)$; hence the element of area under the integral).

We compute first $\sigma$. Since $\mathbf{E} = -\boldsymbol{\nabla}\phi = \mathbf{E}_2$, we obtain from Eq. (3.25)

$$\mathbf{E}_2\cdot\mathbf{n} = -\nabla\phi\cdot\mathbf{n} = 4\pi k\sigma,$$

i.e.

$$-\frac{1}{4\pi}\frac{\partial\phi}{\partial n}\bigg|_{r=a} = -\frac{1}{4\pi}\frac{\partial\phi}{\partial r}\bigg|_{r=a} = k\sigma. \tag{3.27}$$

Since $S \equiv$ sphere, it follows that

$$4\pi\epsilon_0\frac{\partial\phi}{\partial r} \stackrel{(3.24)}{=} -\frac{2q(r-d\cos\gamma)}{2(r^2+d^2-2rd\cos\gamma)^{3/2}} + \frac{2aq(r-\nu\cos\gamma)}{2d(r^2+\nu^2-2r\nu\cos\gamma)^{3/2}}. \tag{3.28}$$

For $\nu = a^2/d$ and $r = a$ we obtain

$$\underbrace{4\pi\epsilon_0}_{1/k} \frac{\partial\phi}{\partial r}\bigg|_{r=a} = -\frac{q(a - d\cos\gamma)}{(a^2 + d^2 - 2ad\cos\gamma)^{3/2}} + \frac{aq\{a - (a^2/d)\cos\gamma\}}{d(a^2 + a^4/d^2 - 2(a^3/d)\cos\gamma)^{3/2}}$$
$$= \frac{q(d^2/a)(1 - a^2/d^2)}{(a^2 + d^2 - 2ad\cos\gamma)^{3/2}}. \tag{3.29}$$

We observe that this expression giving the derivative of the potential on the sphere varies with the angle $\gamma$, the potential itself being zero there. We note that $(\phi)_a = 0$, $(\partial\phi/\partial r)_a \neq 0$. Hence

$$k\sigma(\gamma) = -\frac{1}{4\pi}\frac{\partial\phi}{\partial r}\bigg|_{r=a} = -k\frac{q}{4\pi}\frac{d^2}{a}\frac{(1 - a^2/d^2)}{(a^2 + d^2 - 2ad\cos\gamma)^{3/2}}$$
$$= -k\frac{q}{4\pi ad}\frac{(1 - a^2/d^2)}{(1 + a^2/d^2 - (2a/d)\cos\gamma)^{3/2}}. \tag{3.30}$$

We see that the surface charge density $\sigma$ varies with the angle $\gamma$; with $q > 0$ it is negative and largest at $\gamma = 0$ and smallest at $\gamma = \pi$, implying an accumulation of negative charge at the point closest to the point charge $+q$. The total charge on the sphere is therefore

$$Q = \int \sigma(\gamma)a^2 \cdot 2\pi d(\cos\gamma)$$
$$= -\frac{qa}{2d}\left(1 - \frac{a^2}{d^2}\right)\int_{-1}^{1}\frac{d(\cos\gamma)}{(1 + a^2/d^2 - (2a/d)\cos\gamma)^{3/2}}$$
$$= -\frac{qa}{2d}\left(1 - \frac{a^2}{d^2}\right)\left[\frac{(-2)(-d/2a)}{(1 + a^2/d^2 - (2a/d)\cos\gamma)^{1/2}}\right]_{-1}^{1}$$
$$= -\frac{qa}{2d}\left(1 - \frac{a^2}{d^2}\right)\frac{d}{a}\left[\frac{1}{1 - a/d} - \frac{1}{1 + a/d}\right]$$
$$= -q\frac{a}{d} \quad \text{coulombs.} \tag{3.31}$$

This is precisely the fictitious or image charge. Thus the fictitious charge ensures that $\phi = 0$ on the surface of the spherical conductor, but the conductor itself, that is the sphere, has total charge $-qa/d$. If we want to achieve that the (earthed or unearthed, insulated) sphere carries the total charge zero, we have to supply the sphere with extra charge $q' = +qa/d$ in which case the potential of the sphere becomes $kq'/a$. Summarizing, we can say: The charge induced on the sphere has the same effect as the charge $-qa/d$ at $\nu$.

**Example 3.5 (The spherical shell hanging in the gravitational field).** A conductor in the form of a homogeneous spherical shell of radius $R$, is attached to one end of an elastic string, with string constant $k_0$ (not to be confused with $k = 1/4\pi\epsilon_0$). With the other end, at which a charge $+q$ is placed, the sphere is hanging in the gravitational field of the earth. Establish the equation from which the extension of the string can be calculated (of course, a moving charge generates a magnetic field but this can here be ignored).

**Solution:** We let $Q$ be the charge of the sphere (if not zero) and $x_0$ the unextended length of the string. The conductor with the constant potential can effectively be replaced by an image charge $-q'$, whereby $q'$ and the position $d$, see Fig. 3.8, of the image charge assume the values we calculated previously. We then obtain for the equation of motion of the spherical shell of mass $m$ (observe that the rigid body "spherical shell" can be treated like a pointlike particle)

$$m\ddot{x} = mg - k_0(x - x_0) - \frac{kqq'}{(x_0 + d + (x - x_0))^2} + \frac{kq(Q + q')}{|x + R|^2}. \tag{3.32}$$

As we explained previously: The unearthed sphere carries the total charge $Q$.

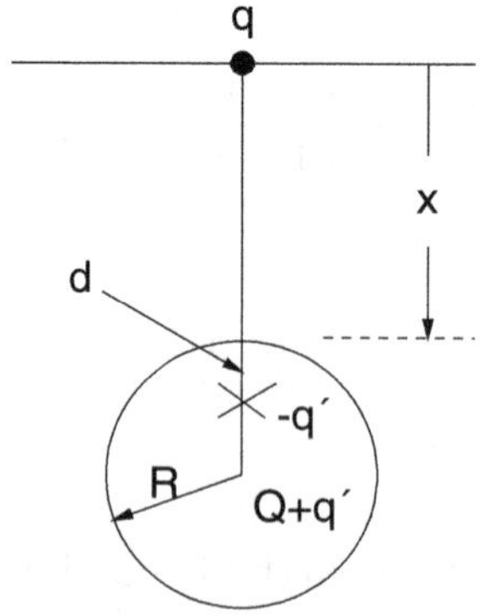


**Fig. 3.8.** The hanging spherical shell.

### 3.6.2 Green's theorems

The previous example illustrates the point in the following formulation of a problem. We had the relation (where in vacuum $k = 1/4\pi\epsilon_0$)

$$\phi(\mathbf{r}) = k \int_V \frac{\rho(\mathbf{r}')}{|\mathbf{r} - \mathbf{r}'|} d\mathbf{r}', \tag{3.33}$$

i.e. if $\rho(\mathbf{r}')$ is given, then this relation yields the potential $\phi(\mathbf{r})$. In electrostatics one frequently encounters a different type of problem: $\phi(\mathbf{r})$ is known on certain surfaces, i.e. boundary conditions supplementing the Poisson equation are given, but not $\rho(\mathbf{r})$, and it is required to determine for instance $\mathbf{E}$. The solution of such boundary problems is simplified with the help of *Green's Theorems*[8]:

Theorem (1):

$$\int_V [\varphi \Delta \psi + (\nabla \varphi) \cdot (\nabla \psi)] \, dV = \int_F \varphi (\nabla \psi) \cdot d\mathbf{F}, \tag{3.34}$$

Theorem (2):

$$\int_V [\varphi \Delta \psi - \psi \Delta \varphi] \, dV = \int_F [\varphi \nabla \psi - \psi \nabla \varphi] \cdot d\mathbf{F}. \tag{3.35}$$

We *verify* these theorems: Set

$$\mathbf{A} = \varphi \nabla \psi. \tag{3.36}$$

Then

$$\nabla \cdot \mathbf{A} = \nabla \cdot (\varphi \nabla \psi) = \varphi \Delta \psi + (\nabla \varphi) \cdot (\nabla \psi). \tag{3.37}$$

But with Gauss' divergence theorem

$$\int_V \nabla \cdot \mathbf{A} dV = \int_F \mathbf{A} \cdot d\mathbf{F} = \int_F \mathbf{A} \cdot \mathbf{n} dF. \tag{3.38}$$

Moreover

$$\mathbf{A} \cdot \mathbf{n} = \varphi (\nabla \psi) \cdot \mathbf{n} \equiv \varphi \frac{\partial \psi}{\partial n}, \tag{3.39}$$

where $\mathbf{n}$ is a unit vector pointing vertically out of the surface $F$. We then have

$$\int \nabla \cdot \mathbf{A} dV = \int_V [\varphi \Delta \psi + (\nabla \varphi) \cdot (\nabla \psi)] \, dV = \int_F \varphi (\nabla \psi) \cdot d\mathbf{F}, \tag{3.40}$$

---

[8] See e.g. Anderson [3], pp. 20–26.

which is the first theorem. If we interchange in this $\varphi$ and $\psi$, we obtain

$$\int_V [\psi\triangle\varphi + (\boldsymbol{\nabla}\psi)\cdot(\boldsymbol{\nabla}\varphi)]\, dV = \int_F \psi(\boldsymbol{\nabla}\varphi)\cdot d\mathbf{F}. \tag{3.41}$$

Subtracting this equation from that of the first theorem, we obtain the second theorem, i.e.

$$\int_V [\varphi\triangle\psi - \psi\triangle\varphi]\, dV = \int_F [\varphi\boldsymbol{\nabla}\psi - \psi\boldsymbol{\nabla}\varphi]\cdot d\mathbf{F}. \tag{3.42}$$

An application which will be needed later is the case of

$$\psi = \frac{1}{|\mathbf{r}-\mathbf{r}'|}, \quad \varphi = \phi, \tag{3.43}$$

so that, with Eq. (2.51),

$$\triangle\psi = \triangle_{\mathbf{r}}\frac{1}{|\mathbf{r}-\mathbf{r}'|} = -4\pi\delta(\mathbf{r}-\mathbf{r}'), \quad \triangle\varphi = \triangle\phi(\mathbf{r}) = -4\pi k\rho(\mathbf{r}). \tag{3.44}$$

The second of Green's theorems then implies from Eqs. (3.42) and (3.43)

$$\begin{aligned} -4\pi\int_V \phi(\mathbf{r}')\delta(\mathbf{r}-\mathbf{r}')dV' &+ 4k\pi\int_V \frac{\rho(\mathbf{r}')}{|\mathbf{r}-\mathbf{r}'|}dV' \\ &= \int_F\left[\phi\frac{\partial}{\partial n'}\left(\frac{1}{|\mathbf{r}-\mathbf{r}'|}\right) - \frac{1}{|\mathbf{r}-\mathbf{r}'|}\frac{\partial\phi}{\partial n'}\right]dF'. \end{aligned} \tag{3.45}$$

In particular, *for* $\mathbf{r}$ *inside* $V$:

$$\phi(\mathbf{r}) = k\int_V \frac{\rho(\mathbf{r}')}{|\mathbf{r}-\mathbf{r}'|}dV' + \frac{1}{4\pi}\int_F\left[\frac{1}{|\mathbf{r}-\mathbf{r}'|}\frac{\partial\phi}{\partial n'} - \phi\frac{\partial}{\partial n'}\frac{1}{|\mathbf{r}-\mathbf{r}'|}\right]dF'. \tag{3.46}$$

Allowing here the area of integration to become infinite, we have

$$\begin{aligned} &dF' \sim r'^2 d\Omega, \quad \phi \propto \frac{1}{r}, \quad \frac{\partial\phi}{\partial n} \propto \frac{1}{r^2}, \quad \frac{1}{|\mathbf{r}-\mathbf{r}'|}\frac{\partial\phi(\mathbf{r}')}{\partial n'} \sim \frac{1}{r'^3}, \\ &\frac{dF'}{r'^3} \sim \frac{d\Omega}{r'} \overset{r',F'\to\infty}{\longrightarrow} 0. \end{aligned} \tag{3.47}$$

We thus obtain the result familiar from Chapter 2, i.e.

$$\phi(\mathbf{r}) = k\int_{V_\infty} \frac{\rho(\mathbf{r}')}{|\mathbf{r}-\mathbf{r}'|}dV'. \tag{3.48}$$

On the other hand, *if the volume of integration V does not enclose charges*, we have $\rho = 0$ and

$$\phi(\mathbf{r})\bigg|_{(\mathbf{r}\epsilon V)} = \frac{1}{4\pi}\int_F \left[\frac{1}{|\mathbf{r}-\mathbf{r}'|}\frac{\partial\phi(\mathbf{r}')}{\partial n'} - \phi\frac{\partial}{\partial n'}\frac{1}{|\mathbf{r}-\mathbf{r}'|}\right] dF'. \tag{3.49}$$

Consequently the potential $\phi$ at $\mathbf{r}$ is determined by values and derivatives of $\phi$ on the boundary $F$ of $V$.

One describes as *Dirichlet boundary conditions* those that specify the potential on the boundary (not necessarily as zero!). One describes as *Neumann boundary conditions* those which specify the normal derivative on the boundary, i.e. $\partial\phi/\partial n$. By a specification of Dirichlet *or* Neumann boundary conditions *the potential is uniquely determined.* This can be seen as follows. Setting $u = \phi_1 - \phi_2$, with $\phi_1, \phi_2$ assumed to be two different solutions (but single-valued on the boundary, i.e. we do not assume multi-valued boundary conditions), then

$$\triangle u = 0, \quad \text{since} \quad \triangle\phi_{1,2} = -4\pi k\rho. \tag{3.50}$$

Setting in the first of Green's theorems $\varphi = \psi = u$, we have

$$\int_V (u\triangle u + \nabla u\cdot\nabla u)dV = \int_F u\nabla u\cdot d\mathbf{F}. \tag{3.51}$$

In the case of Dirichlet boundary conditions $u$ vanishes on the boundary of $F$ (given uniquely), in the case of Neumann boundary conditions $\nabla u$. Thus, in either of the two cases $\int_F = 0$ and hence with Eq. (3.50),

$$\int_V (\nabla u)^2 dV = 0, \quad \text{i.e.} \quad \nabla u = 0,$$

implying $u$ inside $V =$ const. In the case of Dirichlet boundary conditions $u$ is zero on the boundary; hence the constant is zero, i.e. $u = 0$. In the case of Neumann boundary conditions the constant can be $\neq 0$, and so $\phi_1$ and $\phi_2$ differ at most by an insignificant constant.

### 3.6.3 *Green's function and image potential*

We saw earlier with Eq. (2.51) that

$$\nabla^2_{\mathbf{r}}\frac{1}{|\mathbf{r}-\mathbf{r}'|} = -4\pi\delta(\mathbf{r}-\mathbf{r}'). \tag{3.52}$$

(The expression $1/|\mathbf{r}-\mathbf{r}'|$ is at $\mathbf{r}$ effectively the potential of a point charge at $\mathbf{r}'$). The function $1/|\mathbf{r}-\mathbf{r}'|$, however, is *only one class* of functions $G$, for which Eq. (3.52) is valid, i.e. we can have

$$\triangle_{\mathbf{r}} G_{\mathbf{r}} \equiv \boldsymbol{\nabla}^2_{\mathbf{r}} G(\mathbf{r}, \mathbf{r}') = -4\pi\delta(\mathbf{r} - \mathbf{r}') \tag{3.53}$$

(for the moment we call $G$ Green's function *in spite of* the factor $-4\pi$; here this is only a matter of names!) with (cf. Eq. (3.67) below)

$$G(\mathbf{r}, \mathbf{r}') = \frac{1}{|\mathbf{r} - \mathbf{r}'|} + F(\mathbf{r}, \mathbf{r}'), \quad \text{where} \quad \triangle_{\mathbf{r}} F(\mathbf{r}, \mathbf{r}') = 0. \tag{3.54}$$

The function $F$ is called *image potential.* Solutions of the equation $\triangle F = 0$ (called Laplace equation) are known as *harmonic functions.* Previously we obtained the expression (3.46) as a general integral representation of the electrostatic potential. In order to obtain an expression which takes also the image potential into account, we proceed as follows. In the second of Green's theorems, Eq. (3.35), we replace $\varphi$ by $\phi$ and $\psi$ by $G(\mathbf{r}, \mathbf{r}')$:

$$\begin{aligned} &\int_V (\phi \triangle G(\mathbf{r}, \mathbf{r}') - G(\mathbf{r}, \mathbf{r}') \triangle \phi) \, dV(\mathbf{r}') \\ &\quad = \int_F \left( \phi \frac{\partial G(\mathbf{r}, \mathbf{r}')}{\partial n'} - G(\mathbf{r}, \mathbf{r}') \frac{\partial \phi}{\partial n'} \right) dF(\mathbf{r}'), \end{aligned} \tag{3.55}$$

and hence we obtain

$$\begin{aligned} \phi(\mathbf{r}) &= k \int_V \rho(\mathbf{r}') G(\mathbf{r}, \mathbf{r}') dV' \\ &\quad + \frac{1}{4\pi} \int_{F(V)} \left[ G(\mathbf{r}, \mathbf{r}') \frac{\partial \phi(\mathbf{r}')}{\partial n'} - \phi(\mathbf{r}') \frac{\partial G(\mathbf{r}, \mathbf{r}')}{\partial n'} \right] dF' \end{aligned} \tag{3.56}$$

(since $\triangle G(\mathbf{r}, \mathbf{r}') = -4\pi\delta(\mathbf{r} - \mathbf{r}')$, $\mathbf{r} \epsilon V$, $\triangle \phi = -4k\pi\rho$). This expression still involves both $\phi(F)$ and $(\partial\phi/\partial n)_F$ (i.e. the potential on the boundary and its normal derivative at the boundary). With a suitable choice of $G(\mathbf{r}, \mathbf{r}')$, i.e. $F(\mathbf{r}, \mathbf{r}')$, one or the other surface integral can be eliminated, so that an expression for $\phi(\mathbf{r})$ results with *either Dirichlet or Neumann boundary conditions.* In the case of a *Dirichlet boundary condition* we choose

$$G_D(\mathbf{r}, \mathbf{r}') \equiv G(\mathbf{r}, \mathbf{r}') \Big|_{\mathbf{r}' \text{ on } F} = 0. \tag{3.57}$$

Then

$$\phi(\mathbf{r}) = k\int_V \rho(\mathbf{r}')G_D(\mathbf{r},\mathbf{r}')dV' - \frac{1}{4\pi}\int_F \phi(\mathbf{r}')\frac{\partial G(\mathbf{r},\mathbf{r}')}{\partial n'}dF'$$
$$= k\int_V \rho(\mathbf{r}')G_D(\mathbf{r},\mathbf{r}')dV', \quad \text{if} \quad \phi(F) = 0. \tag{3.58}$$

In Eq. (3.56) we make the following substitution for $G$ (this is explained in more detail in Section 3.6.4):

$$G \to G_D(\mathbf{r},\mathbf{r}') = \frac{1}{|\mathbf{r}-\mathbf{r}'|} + F(\mathbf{r},\mathbf{r}'). \tag{3.59}$$

Substituting this into Eq. (3.56), we obtain

$$\begin{aligned}\phi(\mathbf{r})\Big|_{(\mathbf{r}\epsilon V)} &= k\int_V \rho(\mathbf{r}')\left\{\frac{1}{|\mathbf{r}-\mathbf{r}'|} + F(\mathbf{r},\mathbf{r}')\right\}dV' \\ &\quad + \frac{1}{4\pi}\int_F \left[\left\{\frac{1}{|\mathbf{r}-\mathbf{r}'|} + F(\mathbf{r},\mathbf{r}')\right\}\frac{\partial\phi(\mathbf{r}')}{\partial n'}\right. \\ &\quad \left. - \phi(\mathbf{r}')\frac{\partial}{\partial n'}\left\{\frac{1}{|\mathbf{r}-\mathbf{r}'|} + F(\mathbf{r},\mathbf{r}')\right\}\right]dF',\end{aligned} \tag{3.60}$$

and using Eq. (3.46) this equation becomes

$$0 = k\int_V \rho(\mathbf{r}')F(\mathbf{r},\mathbf{r}')dV' + \frac{1}{4\pi}\int_F \left[F(\mathbf{r},\mathbf{r}')\frac{\partial\phi(\mathbf{r}')}{\partial n'} - \phi(\mathbf{r}')\frac{\partial F(\mathbf{r},\mathbf{r}')}{\partial n'}\right]dF'. \tag{3.61}$$

This equation represents a relation between the image potential $F(\mathbf{r},\mathbf{r}')$ and the boundary conditions. If $\phi(\mathbf{r}) = 0$ on the boundary of $V$ (Dirichlet condition), then the following relation

$$0 = k\int_V \rho(\mathbf{r}')F(\mathbf{r},\mathbf{r}')dV' + \frac{1}{4\pi}\int_F F(\mathbf{r},\mathbf{r}')\frac{\partial\phi(\mathbf{r}')}{\partial n'}dF' \tag{3.62}$$

is the equation from which, in principle, the image potential $F(\mathbf{r},\mathbf{r}')$ is to be determined. It equates a volume integral with volume charge density $\rho$ to a surface integral with surface charge density $\sigma$ (cf. Eq. (3.27)). The validity of the relation (3.62) can be verified in the case of the spherically shaped conductor treated above, although such a check is nontrivial. In the case of Neumann boundary conditions, which we shall not pursue further here, one has to choose (see e.g. Jackson [73], p.44)

$$\partial G_N(\mathbf{r},\mathbf{r}')/\partial n' = -4\pi/F \quad \text{for} \quad \mathbf{r}' \,\text{on}\, F. \tag{3.63}$$

### 3.6.4 *The image potential in an example*

We ask ourselves now: What do the preceding considerations imply in the concrete case of the earlier problem of a charge $q$ outside an earthed sphere? We refer to Fig. 3.9. The Dirichlet boundary condition is

$$\phi(\mathbf{r}=\mathbf{a}) = \phi(\mathbf{r}=\infty) = 0.$$

The charge density (in $V$) is

$$\rho(\mathbf{r}') = q\delta(\mathbf{r}' - \mathbf{y}). \tag{3.64}$$

The function $G_D(\mathbf{r},\mathbf{r}')$ as solution of $\triangle G = -4\pi\delta(\mathbf{r}-\mathbf{r}')$ with boundary condition $G = 0$ on the boundary, where $r' = a$, is given by the following expression, as we verify below:

$$G_D(\mathbf{r},\mathbf{r}') = \frac{1}{|\mathbf{r}-\mathbf{r}'|} + \frac{q'}{q|\mathbf{r}-(a^2/r'^2)\mathbf{r}'|}, \quad \text{i.e.} \quad F(\mathbf{r},\mathbf{r}') = \frac{q'}{q|\mathbf{r}-(a^2/r'^2)\mathbf{r}'|}, \tag{3.65}$$

where $q' = -qa/r'$. In Eq. (3.65), the first term is the inhomogeneous solution of the equation for $G$. We see immediately, that $G_D$ vanishes for $r' = a$, since

$$G_D(\mathbf{r},\mathbf{r}')\bigg|_{r'=a} = \left[\frac{1}{|\mathbf{r}-\mathbf{r}'|} + \frac{q'}{q|\mathbf{r}-(a^2/r'^2)\mathbf{r}'|}\right]_{r'=a} = 0. \tag{3.66}$$

We can verify that the expression $G_D$ satisfies its differential equation. Assume $\mathbf{r}' \neq \mathbf{r}''$. Then

$$\begin{aligned}\triangle_{\mathbf{r}}\left(\frac{a}{|\mathbf{r}-\mathbf{r}'|} + \frac{b}{|\mathbf{r}-\mathbf{r}''|}\right) &= \frac{\partial^2}{\partial(\mathbf{r}-\mathbf{r}')^2}\left(\frac{a}{|\mathbf{r}-\mathbf{r}'|} + \frac{b}{|\mathbf{r}-\mathbf{r}'+\mathbf{r}'-\mathbf{r}''|}\right)\\ &\overset{(2.51)}{=} -4\pi a\delta(\mathbf{r}-\mathbf{r}').\end{aligned} \tag{3.67}$$

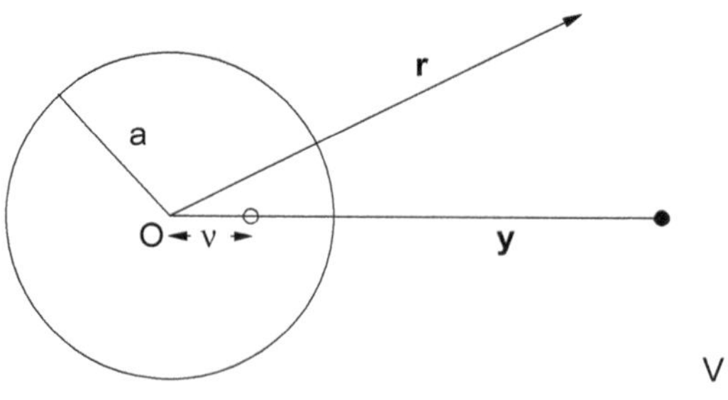


**Fig. 3.9.** The sphere as boundary surface.

The following expression (with $\rho(\mathbf{r}') = q\delta(\mathbf{r}' - \mathbf{y})$) arises from $F$, i.e. this is the contribution to the potential at $\mathbf{r}$ arising from the image charge at $\boldsymbol{\nu}$ in our example above:

$$\begin{aligned} k\int_V \rho(\mathbf{r}')F(\mathbf{r},\mathbf{r}')dV' &= k\int_V q\delta(\mathbf{r}'-\mathbf{y})F(\mathbf{r},\mathbf{r}')dV' \\ &= kqF(\mathbf{r},\mathbf{y}) \overset{(3.65)}{=} k\frac{q'(r'=y)}{|\mathbf{r}-(a^2/y^2)\mathbf{y}|} \\ &\overset{(3.23)}{=} k\frac{q'(r'=y)}{|\mathbf{r}-\boldsymbol{\nu}|}, \end{aligned} \tag{3.68}$$

where $q'(r'=y) = -aq/y$. The expression (see Eq. (3.62)):

$$-\frac{1}{4\pi}\int_{F(V)} F(\mathbf{r},\mathbf{r}')\frac{\partial\phi(\mathbf{r}')}{\partial n'}dF' \tag{3.69}$$

represents the same, but calculated as the potential of the equivalent induced charge distribution $\sigma$ on the sphere.[9] In order to verify this, we would have to show that (with $F(\mathbf{r},\mathbf{r}')|_{r'=a} = -1/|\mathbf{r}-\mathbf{r}'|_{r'=a}$) the second term in Eq. (3.62),

$$\begin{aligned} &-\frac{1}{4\pi}\int_{r'=a}\frac{1}{|\mathbf{r}-\mathbf{r}'|}\frac{\partial\phi(\mathbf{r}')}{\partial r'}dF' \\ &\overset{(3.29)}{=} -\frac{1}{4\pi}\int_0^\pi \frac{2\pi a\sin\gamma' a d\gamma'}{\sqrt{r^2+a^2-2ra\cos(\gamma'-\gamma)}}\cdot\frac{kq(y^2/a)(1-a^2/y^2)}{(\sqrt{a^2+y^2-2ay\cos\gamma'})^3} \end{aligned} \tag{3.70}$$

is equal to minus the expression in Eq. (3.68),

$$\frac{-kqa}{y|\mathbf{r}-(a^2/y)(\mathbf{y}/y)|} = \frac{-kqa}{y\sqrt{r^2+a^4/y^2-2(a^2r/y)\cos\gamma}}.$$

For a point charge $q$ at $\mathbf{r}' = \mathbf{y}$ (implying $\rho(\mathbf{r}') = q\delta(\mathbf{r}'-\mathbf{y})$), we obtain from $\phi(\mathbf{r})$ (with $F(\mathbf{r},\mathbf{r}')$ from Eq. (3.65) and see Eq. (3.58))

$$\phi(\mathbf{r}) = qk\left[\frac{1}{|\mathbf{r}-\mathbf{y}|} + \frac{q'}{q|\mathbf{r}-(a^2/y^2)\mathbf{y}|}\right], \tag{3.71}$$

i.e. as obtained from the charge and the image charge.

We do not investigate further examples here. We only remind that in the case of a conducting sphere in a homogeneous electric field we replace the

[9] See Jackson [73], p. 45.

homogeneous electric field by two opposite but equal charges, initially with the same separations from the sphere, but finally removed to infinity.

**Example 3.6 (The image charge potential).** Point charges $q_i$ at $(r_i, \theta_i, \varphi_i)$ produce at point $(r, \theta, \varphi)$ the potential $\Phi(r, \theta, \varphi)$. Show that the image charges $q_i' = (a/r_i)q_i$ of a conducting sphere of radius $a$ at the points $(a^2/r_i, \theta_i, \varphi_i)$ generate at $(r, \theta, \varphi)$ the potential

$$\Phi(r, \theta, \varphi) = \frac{a}{r}\Phi\left(\frac{a^2}{r}, \theta, \varphi\right). \tag{3.72}$$

**Solution:** The solution follows immediately from the preceding considerations (compare with the image potential contribution in Section 3.6.1; the ratio $a^2/r$ corresponds to $\nu$ there).

**Example 3.7 (A point charge below a spherical shell).** A charge $q$ with mass $m$ is placed at a point distance $z$ below the center of a fixed, conducting and earthed spherical shell of radius $a$. Verify that the charge induced on the sphere can be replaced by an image charge $-qa/z$ at a distance $a^2/z$ (cf. Eq. (3.23)) below the center of the sphere (this part of the Example has been dealt with in Section 3.6.1). Taking into account also the gravitational field of the Earth calculate the potential energy of the charge. Finally assume that the charge $q$ is allowed to fall (from rest) and obtain its velocity at a depth $z_2$ below the center of the sphere (at $z = z_1 = 0$).

**Solution:** The attractive force between the charge $q$ and the image charge $-qa/z$ is

$$q\frac{qa}{z}\frac{k}{(z - a^2/z)^2} = \frac{q^2az}{(z^2 - a^2)^2}k \quad \text{newtons.} \tag{3.73}$$

The electrostatic potential energy is

$$\begin{aligned}\int \text{force} \cdot dz &= q^2ak\int_{\infty}^{z}\frac{zdz}{(z^2 - a^2)^2}\\ &= \frac{1}{2}q^2ak\int_{\infty}^{z^2}\frac{d(z^2)}{(z^2 - a^2)^2} = -\frac{k}{2}\frac{q^2a}{(z^2 - a^2)} \quad \text{volts.}\end{aligned} \tag{3.74}$$

The gravitational potential energy is $= -mgz$ ("minus", since measured downward). Hence energy conservation implies (energy at $z$ = energy at $z_2$)

$$\left[\frac{1}{2}mv^2 - mgz - \frac{k}{2}\frac{q^2a}{(z^2 - a^2)}\right]_z^{z_2} = 0,$$

and therefore

$$v = \sqrt{2g(z_2 - z)\left\{1 - \frac{kq^2a(z_2 + z)}{2mg(z_2^2 - a^2)(z^2 - a^2)}\right\}} \quad \text{meters/second.} \tag{3.75}$$

**Example 3.8 (A point charge in front of a conducting plate).** A point charge $+q$ is placed in front of an infinitely large conducting plane at shortest separation $d$. The plane is earthed. The perpendicular from the charge to the plane meets the latter at a point $O$ which is the center of a circle in the plane of the conductor with radius $R_0$. Calculate the ratio $Q_0/q$, where $Q_0$ is the charge induced on the area of the circle with radius $R_0$.

**Solution:** The plane is a conductor which means that the electrons can move about freely without doing work. The potential $\phi$ on the entire conductor is therefore constant, i.e. zero. We therefore put $\phi = 0$ on the plane (earthed). We can achieve the same condition on the potential by placing a charge $-q$ at a distance $d$ on the other side of the plane, as depicted in Fig. 3.10. We now consider an arbitrary point $P(\mathbf{R})$ on the same side as charge $+q$. We use $O$ as origin of coordinates and the line connecting the charges as $x$-axis. Let $\mathbf{R} = (X, Y, Z)$. The potential at $P$ is then (with $k = 1/4\pi\epsilon_0$)

$$\begin{aligned}\phi(\mathbf{R}) &= k\left[\frac{q}{R_+} - \frac{q}{R_-}\right] \\ &= qk\left[\frac{1}{\sqrt{(X-d)^2 + Y^2 + Z^2}} - \frac{1}{\sqrt{(X+d)^2 + Y^2 + Z^2}}\right].\end{aligned} \tag{3.76}$$

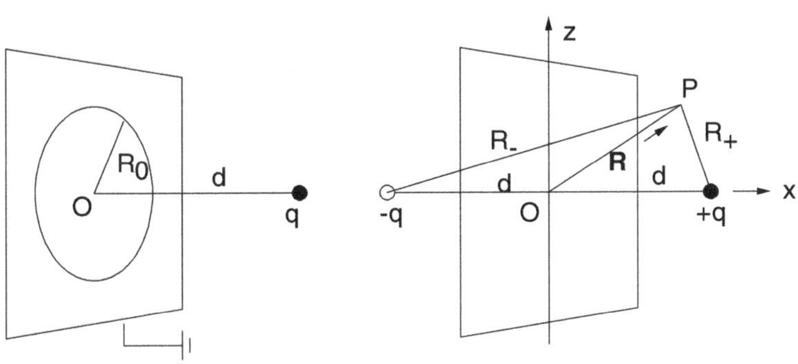


**Fig. 3.10.** Charge $q$ in front of a conducting plane.

We have

$$\frac{\partial\phi}{\partial X} = qk\left[\frac{-(X-d)}{[(X-d)^2+Y^2+Z^2]^{3/2}} - \frac{-(X+d)}{[(X+d)^2+Y^2+Z^2]^{3/2}}\right]. \tag{3.77}$$

For $X \to 0$, it follows that

$$\left(\frac{\partial\phi}{\partial X}\right)_{X\to 0} = \frac{2qdk}{(d^2+Y^2+Z^2)^{3/2}}. \tag{3.78}$$

We apply the Gauss law to the volume $V$ enclosed by the plane to the left, where $\mathbf{E} \to \mathbf{E}_1$, and obtain:

$$\int \mathbf{E}\cdot d\mathbf{F} = 4\pi k \sum(\text{charges}) = 0, \tag{3.79}$$

since no real charge is enclosed to the left. This means $E_{1x} = 0$, i.e. $E_x = -\partial\phi/\partial x = 0$ to the left of the infinitely large conducting plane, as well as inside. We obtain the charge density per unit area, $\sigma$, from the relation (2.90),

$$(\mathbf{E}_2 - \mathbf{E}_1)\cdot\mathbf{n} = 4\pi k\sigma, \tag{3.80}$$

i.e. in the present case

$$E_{2x} = -\left(\frac{\partial\phi}{\partial x}\right)_{x\to 0} = -\frac{2qdk}{(d^2+Y^2+Z^2)^{3/2}} \quad \text{newtons/coulomb}. \tag{3.81}$$

Hence

$$\sigma = -\frac{qd}{2\pi(d^2+Y^2+Z^2)^{3/2}} \quad \text{coulombs/meter}^2. \tag{3.82}$$

The desired ratio is therefore (with $r^2 = Y^2 + Z^2$)

$$\begin{aligned}\frac{Q_0}{q} &= -d\int\frac{dF}{2\pi(d^2+Y^2+Z^2)^{3/2}} = -d\int_{\theta=0}^{2\pi}\int_{r=0}^{R_0}\frac{rd\theta dr}{2\pi(d^2+r^2)^{3/2}}\\ &= -d\int_0^{R_0}\frac{rdr}{(d^2+r^2)^{3/2}} = d\left[\frac{1}{(r^2+d^2)^{1/2}}\right]_0^{R_0}\\ &= d\left[\frac{1}{\sqrt{R_0^2+d^2}} - \frac{1}{d}\right]\end{aligned} \tag{3.83a}$$

and

$$Q_0 = -q(1 - \cos\alpha), \qquad \cos\alpha = \frac{d}{\sqrt{R_0^2 + d^2}}. \tag{3.83b}$$

## 3.7 Method V: Conjugate Functions

This method, also called the method of conformal transformations,[10] is applicable only in two-dimensional problems and depends on 5 main points which we therefore consider first.[11]

Let $W = U + iV$ be a function of the complex variable $z = x + iy$ with $U(x,y), V(x,y)$ real functions of the variables $x$ and $y$. The function $W$ is called a "*complex potential*". Assume also that $\lim_{\delta z \to 0} \delta W/\delta z = dW/dz$. Then one verifies readily one after the other the following relations:

1. The relation between partial derivatives:

$$\frac{\partial}{\partial y} = i\frac{\partial}{\partial x} \quad \left(\text{from } \frac{\partial}{\partial y} = \frac{\partial}{\partial z}\frac{\partial z}{\partial y} = i\frac{\partial}{\partial z}, \ \frac{\partial}{\partial x} = \frac{\partial}{\partial z}\frac{\partial z}{\partial x} = \frac{\partial}{\partial z}\right). \tag{3.84}$$

2. Two relations between partial derivatives of $U$ and $V$:

$$\frac{\partial V}{\partial y} = \frac{\partial U}{\partial x}, \quad \frac{\partial V}{\partial x} = -\frac{\partial U}{\partial y}. \tag{3.85}$$

   These equations follow from 1. applied to $W = U + iV$. The equations are called *Cauchy–Riemann equations*. They express that $U + iV = W(x + iy)$, where $W(z)$ is a regular function of $z$. The only conformal transformations of a domain in the $z$-plane into a domain of the $W$-plane are of the form $W = F(z)$, where $F(z)$ is a regular function of $z$.
3. At points where they intersect, the lines $U =$ const. are perpendicular to lines $V =$ const. With the help of the previous relations, one can verify that

$$\boldsymbol{\nabla} V \cdot \boldsymbol{\nabla} U = 0, \quad \text{i.e.} \quad \boldsymbol{\nabla} V \perp \boldsymbol{\nabla} U. \tag{3.86}$$

   It follows that lines of constant $U$ are always perpendicular to lines of constant $V$. In other words, lines of constant force are perpendicular to lines of constant potential (i.e. the equipotentials).

---
[10]See e.g. Phillips [115], p. 36.
[11]See e.g. Schelkunoff [129], p. 179. For more mathematical aspects of this method, in particular for the calculation of equipotentials, see Cook [33], pp. 228–232.

4. The functions $U$ and $V$ satisfy the two-dimensional Laplace equation, i.e.

$$\nabla^2 U = 0, \quad \nabla^2 V = 0, \tag{3.87}$$

as follows from 2. One of the reasons for the importance of the Cauchy–Riemann equations in physics lies in the fact that any of their solutions are automatically harmonic functions, i.e. solutions of Laplace's equation.

5. Since

$$\begin{aligned}(\boldsymbol{\nabla} U)\cdot(\boldsymbol{\nabla} U) &= \left(\frac{\partial U}{\partial x}, \frac{\partial U}{\partial y}\right)\cdot\left(\frac{\partial U}{\partial x}, \frac{\partial U}{\partial y}\right) \\ &\overset{(3.85)}{=} \left(\frac{\partial V}{\partial y}, -\frac{\partial V}{\partial x}\right)\cdot\left(\frac{\partial V}{\partial y}, -\frac{\partial V}{\partial x}\right),\end{aligned}$$

we have

$$\left(\frac{\partial U}{\partial x}\right)^2 + \left(\frac{\partial U}{\partial y}\right)^2 = \left(\frac{\partial V}{\partial x}\right)^2 + \left(\frac{\partial V}{\partial y}\right)^2,$$

and hence

$$|\boldsymbol{\nabla} V| = |\boldsymbol{\nabla} U|. \tag{3.88}$$

Therefore the magnitudes of these gradients are equal, but their directions perpendicular (cf. Eq. (3.86)).

We now calculate the capacity of a 2-plate condenser or capacitor in two dimensions with charge $q$ on plate $A$. To do this, we assume the two plates $A, A'$, of the condenser have unit length in the additional third dimension. The plates therefore lie along lines of constant potential $V$ as in Fig. 3.11. One may note that the plates are not necessarily parallel (see examples below). Let $\delta r$ be an element of length of plate $A$.

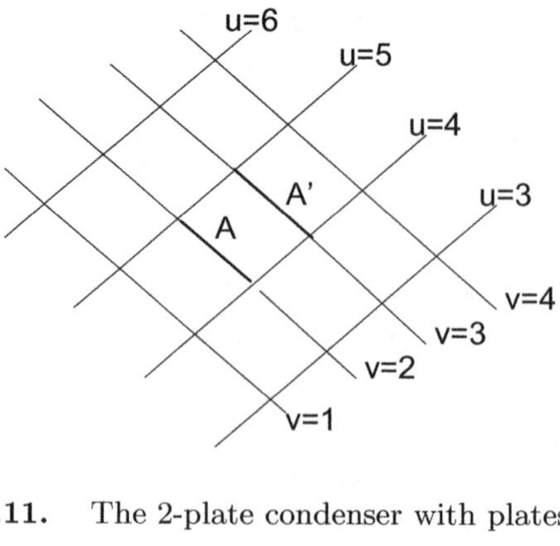


**Fig. 3.11.** The 2-plate condenser with plates $A, A'$.

Then, (always of unit length in the (here) irrelevant third dimension)

$$q = \int \sigma dl \quad \text{coulombs}, \tag{3.89}$$

where $\sigma$ is again the charge per unit area. Since the electric field strength $E$ is again approximately zero on the backside of a plate, $E$ is given by $E = 4\pi k\sigma$ (cf. Eq. (2.90)). Hence for the charge $q$ on one plate with $dl$ along $V = \text{const.}$

$$q = \frac{1}{4\pi k}\int |E|dl = \frac{1}{4\pi k}\int |\boldsymbol{\nabla} U|dl = \frac{1}{4\pi k}\int \frac{\partial U}{\partial l}dl. \tag{3.90}$$

It follows that with $|q|$ the same for the other plate,

$$q = \frac{1}{4\pi k}[U] \quad \text{coulombs}, \tag{3.91}$$

where [...] stands for the corresponding difference. For the capacity of the condenser we then obtain the formula:

$$C = \frac{q}{[V]} = \frac{1}{4\pi k}\frac{[U]}{[V]} \quad \text{farads}. \tag{3.92}$$

Thus $V$ need only be proportional to the potential.

**Example 3.9 (The simple parallel plate condenser).** Calculate the capacity of the parallel plate condenser with the help of the preceding considerations.

**Solution:** We put the plates $A, A'$ along the $x$-direction as in Fig. 3.11 and assume they are a distance $y = d$ apart. We set $W = z$, so that $U = x, V = y$. Then the capacity is

$$C = \frac{1}{4\pi k}\frac{[U]}{[V]} = \frac{1}{4\pi k}\frac{[x]}{[y]} = \frac{1}{4\pi kd} \quad \text{farads} \tag{3.93}$$

per unit area (with $x = 1$ and again with unit length in the third direction which is normally called $z$, but here $z = x + iy$). The result agrees with that of Eq. (2.101).

**Example 3.10 (The condenser with radially assembled plates).** Calculate the capacity of the condenser with angle $\triangle\theta$ between the plates.

**Solution:** In this case we choose $W = \ln z = \ln r + i\theta$, so that $U = \ln r$, $V = \theta$. Let the angle between the plates with radial lengths $b - a$ be the

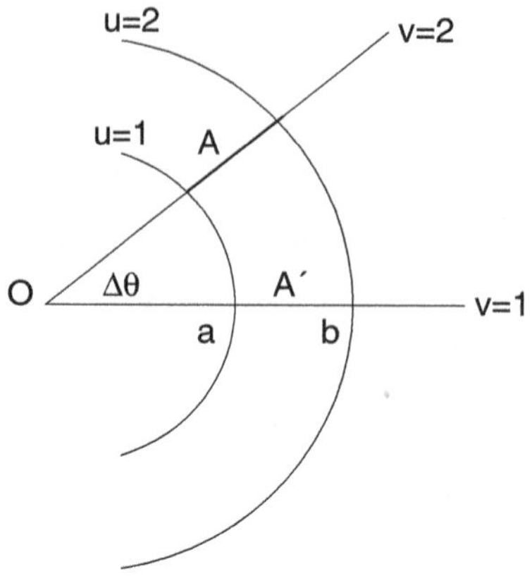


**Fig. 3.12.** Condenser plates $A, A'$.

polar angle $\triangle\theta$ as in Fig. 3.12. In this case we have

$$C = \frac{1}{4\pi k}\frac{[U]}{[V]} = \frac{1}{4\pi k}\frac{[\ln r]}{[\theta]} = \frac{\ln(b/a)}{4\pi k\triangle\theta} \quad \text{farads.} \tag{3.94}$$

**Example 3.11 (The cylindrical condenser).** Calculate The capacity of a cylindrical condenser.

**Solution:** In the preceding example we reverse the roles of $U$ and $V$, so that $U = \theta, V = \ln r$. We then obtain the condenser consisting of two coaxial cylinders of radii $a$ and $b > a$ (see the arcs of the cylinders in Fig. 3.12). The capacity is then obtained as

$$C = \frac{1}{4\pi k}\frac{[U]}{[V]} = \frac{1}{4\pi k}\frac{[\theta]}{[\ln r]} = \frac{2\pi}{4\pi k\ln(b/a)} = \frac{1}{2k\ln(b/a)} \quad \text{farads.} \tag{3.95}$$

This result (per unit length of the cylinders) agrees with the result (2.118). Obviously spheres cannot be handled with the present method.

With this 2-dimensional method we can consider very different cases — e.g. the case of non-coaxial cylinders. To do this, we consider a new function $W(z)$. We let this be $W(z) = V + iU$ ($V$ is later identified with the potential) and $z = x + iy$. Let $P(x, y)$ be the point in the complex plane as indicated in Fig. 3.13, and let $r_+, r_-$ be the distances from $P(x, y)$ to two points, which are along the $x$-axis a distance $2d$ apart. We choose the following function $W$:

$$W = \ln\frac{z_+}{z_-} = \ln\frac{r_+}{r_-} + i(\theta_+ - \theta_-) \equiv V + iU. \tag{3.96}$$

We have therefore

$$V = \ln \frac{r_+}{r_-}, \quad U = \theta_+ - \theta_-. \tag{3.97}$$

We can rewrite the first equation for $V$ with the help of Fig. 3.13 as

$$e^{2V} = \frac{r_+^2}{r_-^2} = \frac{(x+d)^2 + y^2}{(x-d)^2 + y^2}. \tag{3.98}$$

With further algebraic manipulation of this equation we have

$$x^2(1 - e^{2V}) + d^2(1 - e^{2V}) + y^2(1 - e^{2V}) + 2xd(1 + e^{2V}) = 0, \tag{3.99}$$

from which with

$$\coth^2 V - \operatorname{cosech}^2 V = 1 \quad \text{and} \quad \coth V = \frac{\cosh V}{\sinh V},$$

the following equation of a circle results

$$y^2 + (x - d\coth V)^2 = \left(\frac{d}{\sinh V}\right)^2. \tag{3.100}$$

The equipotentials $\pm V$ = const. are circles (with change of sign of $V$ also coth $V$ and sinh $V$ change signs). Let $V = V_+$ be the potential on one (cylindrical) plate of the condenser with radius $a_r = d/\sinh V_+$ and with its

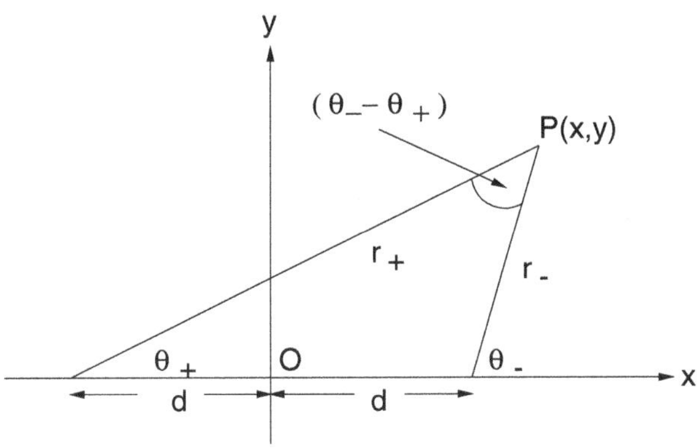


**Fig. 3.13.** Point $P$ in the complex plane.

center at $b_+ = d \coth V_+$ on the $x$-axis. Then $b_+/a_r = \cosh V_+$ and so

$$V_+ = \cosh^{-1} \frac{b_+}{a_+}. \tag{3.101}$$

Let $V = -V_-$ on the other plate of the condenser. In this case we have

$$V_- = \cosh^{-1} \frac{b_-}{a_-}, \tag{3.102}$$

where $b_- = d \coth V_-$ and $a_- = d/\sinh V_-$. The capacity between the cylinders is now ($q$ being the charge on one)

$$C = \frac{q}{[V]} = \frac{1}{4\pi k}\frac{[U]}{[V]} = \frac{1}{4\pi k}\frac{[\theta_+ - \theta_-]}{[V_+ + V_-]} \quad \text{farads.} \tag{3.103}$$

The entire domain of $(\theta_+ - \theta_-)$ is $2\pi$. It therefore follows that

$$C = \frac{1}{4\pi k}\frac{2\pi}{(\cosh^{-1}(b_+/a_+) + \cosh^{-1}(b_-/a_-))} \quad \text{farads.} \tag{3.104}$$

We now consider some applications.

**Example 3.12 (Parallel wires a distance $D$ apart and with radii $a$[12]).** Again the problem is to calculate the capacity.

**Solution:** In this case $a_+ = a_- = a$ and $b_+ = b_- = D/2$, so that the capacity is given by

$$C = \frac{1}{4k \cosh^{-1}(D/2a)} = \frac{\pi\epsilon_0}{\ln[D/2a + \sqrt{(D/2a)^2 - 1}]} \quad \text{farads,} \tag{3.105}$$

where we used the relation[13]

$$\cosh^{-1} A = \pm \ln(A + \sqrt{A^2 - 1}) + 2in\pi.$$

**Example 3.13 (The wire in air (equivalent to a cylinder and a plane)).** Calculate its capacity.

**Solution:** In this case one of the cylinders is the surface of the earth with $a_- = b_-$ and $b_- \to \infty$ (the circle with center $b_- = d \coth V_-$ to $-\infty$ and

[12] For a more elementary derivation see Bolton [14], pp. 45–46.
[13] Dwight [42], formula 721.5a, p. 159.

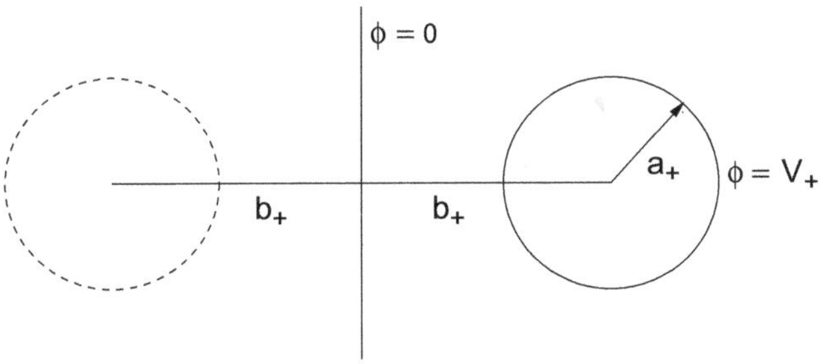


**Fig. 3.14.** The fictitious cylinder in Kelvin's method.

radius $d/\sinh V_-$ to $+\infty$, becomes a straight line representing the surface of the earth). This means $V_- = 0$. It then follows from Eq. (3.104) that

$$C = \frac{1}{2k\cosh^{-1}(b_+/a_+)} \quad \text{farads.} \tag{3.106}$$

This result can also be obtained with Kelvin's method of image charges; cf. Fig. 3.14. The capacity of the system consisting of one cylinder and the parallel image cylinder with opposite charge is given by the expression (3.105) as

$$C_{\text{parallel cylinders}} = \frac{1}{4k\cosh^{-1}(b_+/a_+)} \quad \text{farads,} \tag{3.107}$$

where $b_+$ is the height of the actual cylinder above the ground (the earth's surface where $V = 0$) and $a_+$ its radius. The difference of potentials of the two parallel cylinders is $[V] = V_+ - (-V_+) = 2V_+$. Since $C \propto 1/[V]$, and in the case of one cylinder above the surface at which $V = 0$, we have in the latter case $[V] = V_+ - 0 = V_+$, so that

$$C_{\text{cylinder to earth}} = 2C_{\text{parallel cylinders}} = \frac{1}{2k\cosh^{-1}(b_+/a_+)} \quad \text{farads.} \tag{3.108}$$

**Example 3.14 (Electrolytic determination of the capacity).** Use the cylindrical condenser to suggest an experiment for the electrolytic determination of its capacity with the help of Eq. (2.90), i.e. the relation $\mathbf{E}\cdot\mathbf{n} = 4\pi k\sigma_{\text{charge}}$, where $\sigma_{\text{charge}}$ is the charge per unit area on one cylindrical electrode, and $\mathbf{E} = 0$ inside the conducting material of the electrode. (Note: This Example assumes familiarity with Sections 5.2 and 6.1).

**Solution:** As a simple application of the cylindrical condenser one can imagine — as indicated in Fig. 3.15 — two cylindrical electrodes in a

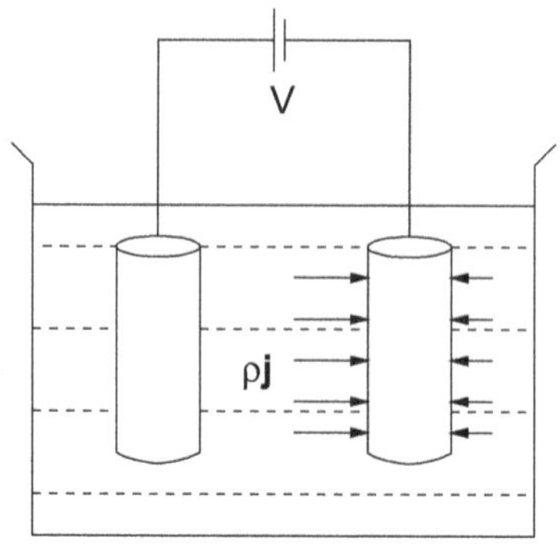


**Fig. 3.15.** The electrolytic tank containing two electrodes.

container with a salt solution (cf. Section 6.1 where Ohm's law $V = IR$, current $I = \int \mathbf{j} \cdot d\mathbf{F}$, is written $j = \sigma E$ with conductivity $\sigma = 1/\rho$, where $\rho$ is the resistivity; this is unfortunately standard notation, so that this $\rho$ must not be confused with the charge density, nor conductivity $\sigma$ (in units of siemens/meter or ohm$^{-1}$/meter with charge per unit area). Consider the salt solution with resistivity $\rho$. Macroscopically positive and negative (ion) charges in this solution neutralize each other so that the macroscopic charge density $\rho_{\text{charge}}$ in the solution is nought (hence the solution can be looked at as a dielectric medium). Thus the potential $\phi$ and hence the voltage $V$ between the electrodes is given by the Laplace equation $\nabla^2\phi = 0$, i.e. with source term zero. One can use a probe connected to a voltmeter to determine the potential distribution which would have existed with the same conducting plates in a homogeneous dielectric. Let $\mathbf{j}$ be the current density in the solution (cf. Section 5.2) and $dF$ an element of area of one of the electrodes. Since (cf. Section 6.1) $\mathbf{j} = \mathbf{E}/\rho$, the resistance $R$ of the solution is given by Ohm's law as $R = V/I$ with the current $I$ entering each electrode given by $I = \int \mathbf{j} \cdot d\mathbf{F}$, so that

$$R = \frac{V}{\int \mathbf{j} \cdot d\mathbf{F}} = \frac{\rho V}{\int \mathbf{E} \cdot d\mathbf{F}} = \frac{\rho V}{4\pi k \int \sigma_{\text{charge}} dF} \quad \text{ohms.} \tag{3.109}$$

The capacity between the electrodes is therefore, if $\sigma_{\text{charge}}$ denotes the charge $q$ of an electrode per unit area,

$$C = \frac{q}{V} = \frac{\int \sigma_{\text{charge}} dF}{V} = \frac{\rho}{4\pi k R} \quad \text{farads,} \tag{3.110}$$

as one would have written in the case of a dielectric. Thus the capacity of the system of cylinders can be determined experimentally from the known resistivity $\rho$ of the solution into which the cylinders are immersed. $R$ is the resistance of the electrolyte.

## 3.8 Orthogonal Functions

### 3.8.1 *Orthogonal functions in general*

We now have to develop some mathematics so that we can consider the generalization of a dipole (referred to in Example 2.13) to a multiple configuration of charges. We begin with the expansion of arbitrary functions in terms of a complete set of basis functions and consider first the case of one dimensional variables.

Functions $U_n(x), n = 1, 2, \ldots, N$ (with $N$ finite or infinite), are said to be *orthogonal* in $[a, b]$, if

$$\int_a^b U_n^\star(x) U_m(x) dx = s_n \delta_{nm}. \tag{3.111}$$

The nonnegative number

$$||U_n|| = \sqrt{\int_a^b U_n^\star(x) U_n(x) dx} \tag{3.112}$$

is called *norm* of $U_n(x)$. The function $U_n$ is said to be *normalized*, if

$$\int_a^b U_n^\star(x) U_n(x) dx = 1. \tag{3.113}$$

The *analogy with orthogonal vectors* is obvious. The *expansion of an arbitrary function* $f(x)$ *in terms of* a set of basis functions $\{U_n(x)\}$ is given by the sum

$$\sum_{n=1}^{N} a_n U_n(x). \tag{3.114}$$

It is necessary to establish whether the sum converges for $N \to \infty$ towards $f(x)$, in the sense of *convergence in the mean*, i.e. (definition) one demands

$$\lim_{N\to\infty} \int_a^b \left| f(x) - \sum_{n=1}^{N} a_n U_n(x) \right|^2 dx = 0, \tag{3.115}$$

so that

$$f(x) = \sum_{n=1}^{\infty} a_n U_n(x) \tag{3.116}$$

(implying uniform convergence), and what the coefficients are. We construct the *mean squared error*

$$M_N = \int_a^b \left| f(x) - \sum_{n=1}^{N} a_n U_n(x) \right|^2 dx \tag{3.117}$$

and determine $a_n$ such that $M_N$ is minimal. $M_N$ is minimized for

$$\frac{\partial M_N}{\partial a_j} = 0, \quad \frac{\partial M_N}{\partial a_j^\star} = 0, \tag{3.118}$$

i.e.

$$\int_a^b U_j \left( f^\star(x) - \sum_{n=1}^{N} a_n^\star U_n^\star \right) dx = 0 \tag{3.119}$$

and the complex conjugate. This implies for normalized $U_n(x)$ the relations

$$a_j^\star = \int_a^b f^\star(x) U_j dx, \quad a_j = \int_a^b f(x) U_j^\star dx. \tag{3.120}$$

Since

$$\begin{aligned} M_N &= \int_a^b \Bigg\{ f^\star(x) f(x) + \sum_{n,n'} a_n^\star a_{n'} U_n^\star(x) U_{n'}(x) \\ &\quad - f^\star(x) \sum_n a_n U_n(x) - f(x) \sum_n a_n^\star U_n^\star(x) \Bigg\} dx \\ &\overset{(3.116)}{=} \int_a^b dx f^\star(x) f(x) - \sum_n |a_n|^2, \end{aligned} \tag{3.121}$$

it follows immediately from $M_N \geq 0$ that

$$\int_a^b |f(x)|^2 dx \geq \sum_{n=1}^{N} |a_n|^2. \tag{3.122}$$

This relation is known as *Parseval relation.* When the equality applies, the quantities $a_n$ describe functions $f(x)$ completely — the Parseval relation is then called *completeness relation.* Analogous considerations apply to functions in higher dimensions.

### 3.8.2 *Fourier series, Fourier expansions*

The functions

$$\varphi_n(x) : \frac{1}{\sqrt{2\pi}}, \quad \frac{\sin x}{\sqrt{\pi}}, \quad \frac{\cos x}{\sqrt{\pi}}, \quad \frac{\sin 2x}{\sqrt{\pi}}, \quad \frac{\cos 2x}{\sqrt{\pi}}, \ldots \tag{3.123}$$

constitute in the interval $-\pi \leq x \leq \pi$ a complete orthonormal system. This follows from the integrals

$$\begin{aligned} \frac{1}{\pi}\int_{-\pi}^{\pi} \cos mx \cos nx dx &= \delta_{mn}, \\ \frac{1}{\pi}\int_{-\pi}^{\pi} \sin mx \sin nx dx &= \delta_{mn}, \\ \int_{-\pi}^{\pi} \cos mx \sin nx dx &= 0. \end{aligned} \tag{3.124}$$

Then in the interval $-\pi \leq x \leq \pi$ we can set

$$f(x) = \sum_{n=0}^{\infty} c_n \varphi_n(x), \quad c_n = \int_{-\pi}^{\pi} f(x)\varphi_n(x)dx. \tag{3.125}$$

In addition $f(x + 2\pi) = f(x)$, i.e. the function is periodic.

**Complex Fourier series:**
The functions

$$\varphi_n(x) = \frac{1}{\sqrt{2l}} e^{in\pi x/l}, \quad n = 0, \pm 1, \pm 2, \ldots \tag{3.126}$$

constitute in the interval $-l \leq x \leq l$ a complete orthonormal system with

$$\int_{-l}^{l} dx \varphi_n(x)\varphi_m^{\star}(x) = \delta_{mn}, \tag{3.127}$$

so that the periodic function $f(x) = f(x + 2l)$ can be expanded as

$$f(x) = \sqrt{2l} \sum_{n=-\infty}^{\infty} c_n \varphi_n(x), \quad c_n = \frac{1}{2l}\int_{-l}^{l} f(x) e^{-in\pi x/l} dx. \tag{3.128}$$

**Fourier integrals:**
We now want to explore the *expansion of nonperiodic functions* by considering in the relations (3.128) the limit $l \to \infty$. For $l \to \infty$ the interval $i\pi x/l$

between the exponents of terms in the series

$$f(x) = \sum_{n=-\infty}^{\infty} c_n e^{in\pi x/l} \tag{3.129}$$

becomes increasingly smaller; in the limit, one has the continuum. With

$$k_n = \frac{n\pi}{l} \quad (\text{so that} \quad k_{n+1} - k_n = \pi/l), \quad \frac{lc_n}{\pi} = g(k_n), \tag{3.130}$$

we have

$$f(x) = \pi \sum_{n=-\infty}^{\infty} e^{ik_n x} \frac{g(k_n)}{l} = \sum_{n=-\infty}^{\infty} e^{ik_n x} g(k_n)(k_{n+1} - k_n), \tag{3.131}$$

which in the limit $l \to \infty$ yields the representation

$$f(x) = \int_{-\infty}^{\infty} dk g(k) e^{ikx} \tag{3.132}$$

($k_{n+1} - k_n \sim dk = \pi/l$, and for $l$ very large: $\pi/l \sim \pi\epsilon = \delta k$; for $n$ finite we have $k_n = n\pi/l \sim n\pi\epsilon$, and for $n \to \pm\infty : k_n \to k_{\pm\infty} = \pm\infty$). Using the second of Eq. (3.130), we have (cf. Eq. (3.128))

$$c_n = \frac{1}{2l} \int_{-l}^{l} f(x) e^{-in\pi x/l} dx \overset{(3.127)}{=} \frac{\pi g(k_n)}{l}, \tag{3.133}$$

and so obtain correspondingly with Eq. (3.128)

$$g(k_n) = \frac{1}{2\pi} \int_{-\infty}^{\infty} f(x) e^{-ik_n x} dx,$$

i.e.

$$g(k) = \frac{1}{2\pi} \int_{-\infty}^{\infty} f(x) e^{-ikx} dx. \tag{3.134}$$

This expression with wave number k dependence is called *Fourier transform* or *spectral function* of $f(x)$. We can readily obtain the inverse of Eq. (3.134) using

$$\delta(x) = \frac{1}{2\pi} \int_{-\infty}^{\infty} e^{ikx} dk, \tag{3.135}$$

because then

$$\begin{aligned}\int_{-\infty}^{\infty} g(k)e^{iky}dk &= \frac{1}{2\pi}\int_{-\infty}^{\infty} e^{iky}dk \int_{-\infty}^{\infty} f(x)e^{-ikx}dx \\ &= \int_{-\infty}^{\infty} dx f(x)\frac{1}{2\pi}\int_{-\infty}^{\infty} e^{ik(y-x)}dk \\ &= \int_{-\infty}^{\infty} f(x)\delta(y-x)dx = f(y). \end{aligned} \tag{3.136}$$

A necessary condition for the possibility to represent a function $f(x)$ as a Fourier integral is the convergence of the integral

$$\int_{-\infty}^{\infty} |f(x)|dx, \quad \text{i.e.} \quad \lim_{x\to\pm\infty} f(x) = 0. \tag{3.137}$$

### 3.8.3 *Spherical functions (Legendre, associated Legendre and hyperspherical functions)*

We encountered the following factor earlier (in the integral representation of the solution of Poisson's equation, see e.g. Eqs. (2.38) and (3.33)), and now consider it in more detail. Written as an expansion the factor is

$$\begin{aligned}\frac{1}{|\mathbf{r}-\mathbf{r}'|} &= \frac{1}{\sqrt{r^2+r'^2-2rr'\cos\alpha}} \\ &= \frac{1}{r}\left\{1+P_1(x)\frac{r'}{r}+P_2(x)\left(\frac{r'}{r}\right)^2+\cdots\right\}\theta(r-r')+(r\leftrightarrow r') \\ &= \frac{1}{r}\sum_{l=0}^{\infty} P_l(x)\left(\frac{r'}{r}\right)^l \theta(r-r')+(r\leftrightarrow r'), \end{aligned} \tag{3.138}$$

where with $x=\cos\alpha$

$$P_0(x)=1, \quad P_1(x)=x, \quad P_2(x)=\frac{1}{2}(3x^2-1), \ \ldots. \tag{3.139}$$

These polynomials are called *Legendre polynomials.* These are in general given by the formula[14]

$$P_l(x) = \frac{1}{2^l l!}\frac{d^l}{dx^l}(x^2-1)^l \tag{3.140}$$

[14]See e.g. Magnus and Oberhettinger [90], also for some other relations cited below.

and define a complete orthonormal system in the interval $x \in [1, -1]$, $\alpha \in [0, \pi]$:

$$\int_{-1}^{1} P_l(x) P_m(x) dx = \frac{2}{2l+1} \delta_{lm}. \tag{3.141}$$

A complete orthonormal system of functions which is defined on the *unit sphere* with $\theta \in [0, \pi], \varphi \in [0, 2\pi]$ is given by the *hyperspherical functions* $Y_{lm}$ which are defined as

$$Y_{lm}(\theta, \varphi) = \left\{ \frac{(2l+1)(l-m)!}{4\pi(l+m)!} \right\}^{1/2} P_l^m(\cos\theta) e^{im\varphi}, \tag{3.142}$$

where $-l \leq m \leq l$ and integral, and the functions

$$P_l^m(x) = (-1)^m (1 - x^2)^{m/2} \frac{d^m}{dx^m} P_l(x) \tag{3.143}$$

are the *associated Legendre polynomials.* For instance

$$Y_{0,0} = \frac{1}{\sqrt{4\pi}}, \quad Y_{10} = \sqrt{\frac{3}{4\pi}} \cos\theta, \quad Y_{11} = -\sqrt{\frac{3}{8\pi}} \sin\theta e^{i\varphi}. \tag{3.144}$$

The associated Legendre polynomials with negative indices $m$ are defined by

$$P_l^{-m}(x) = (-1)^m \frac{(l-m)!}{(l+m)!} P_l^m(x). \tag{3.145}$$

General relations of these are

$$P_l^m(x) = \frac{(-1)^m}{2^l l!} (1 - x^2)^{m/2} \frac{d^{l+m}}{dx^{l+m}} (x^2 - 1)^l, \tag{3.146}$$

$$\int_{-1}^{1} P_{l'}^m(x) P_l^m(x) dx = \frac{2}{2l+1} \frac{(l+m)!}{(l-m)!} \delta_{ll'} \tag{3.147}$$

or

$$\int_0^{2\pi} d\varphi \int_0^{\pi} Y_{l'm'}^\star(\theta, \varphi) Y_{lm}(\theta, \varphi) \sin\theta d\theta = \delta_{ll'} \delta_{mm'}. \tag{3.148}$$

The *completeness relation* follows in analogy to the basis change of vectors, $\{\mathbf{e}_i\} \to \{\mathbf{E}_k\}$, from $\mathbf{e}_i \cdot \mathbf{e}_j = \delta_{ij}$ to $\sum_k (\mathbf{e}_i \cdot \mathbf{E}_k)(\mathbf{E}_k \cdot \mathbf{e}_j) = \delta_{ij}$, as the relation

$$\sum_{l=0}^{\infty} \sum_{m=-l}^{l} Y_{lm}^\star(\theta', \varphi') Y_{lm}(\theta, \varphi) = \delta(\varphi - \varphi') \delta(\cos\theta - \cos\theta') \tag{3.149}$$

(in the above analogy $k$ corresponds to $m, l$, and $i, j$ correspond to $\theta, \varphi$). In books on Special Functions also the following relation called *addition theorem* can be found:

$$P_l(\cos\alpha) = \frac{4\pi}{2l+1}\sum_{m=-l}^{l} Y^\star_{lm}(\theta', \varphi')Y_{l,m}(\theta, \varphi), \tag{3.150}$$

where (see below)

$$\cos\alpha = \sin\theta\sin\theta'\cos(\varphi - \varphi') + \cos\theta\cos\theta' = \cos(\theta - \theta'), \quad \text{if} \quad \varphi = \varphi' \tag{3.151}$$

(hence its name "addition theorem"), where $\theta$ and $\varphi$ are the spherical coordinates of a vector $\mathbf{r}$, i.e.

$$\mathbf{r} = (r\sin\theta\cos\varphi, r\sin\theta\sin\varphi, r\cos\theta), \tag{3.152}$$

$$\mathbf{r}' = (r'\sin\theta'\cos\varphi', r'\sin\theta'\sin\varphi', r'\cos\theta'), \tag{3.153}$$

so that

$$\mathbf{r}\cdot\mathbf{r}' = rr'(\sin\theta\sin\theta'\cos(\varphi - \varphi') + \cos\theta\cos\theta'). \tag{3.154}$$

We use the addition theorem in order to rewrite Eq. (3.138):

$$\begin{aligned}\frac{1}{|\mathbf{r}-\mathbf{r}'|} &= \frac{1}{r}\sum_{l=0}^{\infty} P_l(x)\left(\frac{r'}{r}\right)^l \theta(r-r') + (r \leftrightarrow r')\\ &= \sum_{l=0}^{\infty}\sum_{m=-l}^{l} \frac{4\pi}{2l+1}\frac{r'^l}{r^{l+1}} Y^\star_{lm}(\theta', \varphi')Y_{lm}(\theta, \varphi)\theta(r-r') + (r \leftrightarrow r').\end{aligned} \tag{3.155}$$

The hyperspherical functions are also obtained as the solutions of Laplace's equation in spherical polar coordinates.

## 3.9 The Multipole Expansion

We consider a *charge distribution* (density $\rho$) in a domain $|\mathbf{r}| < a$. Then the potential (cf. Eq. (2.38)) is for

$$r > a: \quad \phi(\mathbf{r}) = k\int \frac{\rho(\mathbf{r}')}{|\mathbf{r}-\mathbf{r}'|} dV' \tag{3.156}$$

(the origin of coordinates for instance in the charge center of mass), provided the volume of integration extended to infinity does not contain conducting surfaces (equipotentials specified by boundary conditions). Thus, only the expression (3.156) remains of the general relation (3.46). This remaining part (3.156) collects contributions only from points $\mathbf{r}'$ in space at which $\rho(\mathbf{r}') \neq 0$. Since $r' < a < r$, this means, that we can expand $1/|\mathbf{r}-\mathbf{r}'|$ in powers of $r'/r$. This allows us to substitute the expression (3.155) in Eq. (3.156):

$$\phi(\mathbf{r}) = k\sum_{l=0}^{\infty}\sum_{m=-l}^{l}\frac{4\pi}{2l+1}\frac{1}{r^{l+1}}Y_{lm}(\theta,\varphi)\int\rho(\mathbf{r}')r'^{l}Y^{\star}_{lm}(\theta',\varphi')dV'. \tag{3.157}$$

The expressions

$$q^{\star}_{lm} = k\int Y^{\star}_{lm}(\theta',\varphi')r'^{l}\rho(\mathbf{r}')dV' \tag{3.158}$$

are called *multipole moments of the charge distribution* $\rho(\mathbf{r}')$: Specifically the case $l = 0$ is called *monopole moment*, the case $l = 1$ *dipole moment*, the case $l = 2$ *quadrupole moment*, and so on. Since $l = 0, 1, 2, \ldots$, and $m = -l, -l+1, \ldots, l$, the $2l+1$ components of $q_{lm}$ form a quantity known as a *spherical tensor of the lth rank*. With the relation

$$Y_{l,-m} = (-1)^m Y^{\star}_{lm} \tag{3.159}$$

it follows that

$$q^{\star}_{l,-m} = (-1)^m q_{lm}. \tag{3.160}$$

The multipole moments describe how the charge is distributed:

$$q_{00} = q^{*}_{00} = k\int Y^{\star}_{00}(\theta',\varphi')\rho(\mathbf{r}')dV' = \frac{k}{\sqrt{4\pi}}\int\rho(\mathbf{r}')dV' = \frac{kQ}{\sqrt{4\pi}}, \tag{3.161}$$

and using Eq. (3.144),

$$\begin{aligned} q^{\star}_{11} &= k\int\rho(\mathbf{r}')r'Y^{\star}_{11}(\theta',\varphi')dV' = -k\sqrt{\frac{3}{8\pi}}\int\rho(\mathbf{r}')r'\sin\theta' e^{-i\varphi'}dV' \\ &= -k\sqrt{\frac{3}{8\pi}}\int\rho(\mathbf{r}')r'(\sin\theta'\cos\varphi' - i\sin\theta'\sin\varphi')dV', \end{aligned} \tag{3.162}$$

and with Eq. (3.153)

$$q_{11}^{\star} = -k\sqrt{\frac{3}{8\pi}}\int \rho(\mathbf{r}')(x' - iy')dV', \tag{3.163}$$

or, with the definitions

$$p_{x,y,z} = \int \rho(\mathbf{r}')(x', y', z')dV', \tag{3.164}$$

we obtain

$$q_{11}^{\star} = -k\sqrt{\frac{3}{8\pi}}(p_x - ip_y), \tag{3.165}$$

$$q_{1,-1} = -q_{11}^{\star} = k\sqrt{\frac{3}{8\pi}}(p_x - ip_y), \tag{3.166}$$

and

$$\begin{aligned} q_{10} &= k\int \rho(\mathbf{r}')r'Y_{10}(\theta', \varphi')dV' \\ &= k\sqrt{\frac{3}{4\pi}}\int \rho(\mathbf{r}')r'\cos\theta' dV' = k\sqrt{\frac{3}{4\pi}}p_z. \end{aligned} \tag{3.167}$$

Here $p_x, p_y, p_z$ are the Cartesian components of the *electric dipole moment* $\mathbf{p}$.

We thus obtain (to be verified below) for $\phi(\mathbf{r})$ the following series expansion (we leave the verification of the electric quadrupole term as an exercise):

$$\phi(\mathbf{r}) = \frac{kQ}{r} + \frac{k\mathbf{p}\cdot\mathbf{r}}{r^3} + \frac{k}{6}\sum_{i,j}\frac{Q_{ij}(3x_ix_j - r^2\delta_{ij})}{r^5} + \cdots \quad \text{volts.} \tag{3.168}$$

This expansion is obtained since from Eq. (3.157):

$$\begin{aligned} \phi(\mathbf{r}) &= \frac{4k\pi}{r}Y_{00}(\theta, \varphi)\int Y_{00}(\theta', \varphi')\rho(\mathbf{r}')dV' \\ &+ \frac{4k\pi}{3r^2}Y_{11}(\theta, \varphi)\int Y_{11}^{\star}(\theta', \varphi')r'\rho(\mathbf{r}')dV' \\ &+ \frac{4k\pi}{3r^2}Y_{10}(\theta, \varphi)\int Y_{10}(\theta', \varphi')r'\rho(\mathbf{r}')dV' \\ &+ \frac{4k\pi}{3r^2}Y_{1,-1}(\theta, \varphi)\int Y_{1,-1}(\theta', \varphi')r'\rho(\mathbf{r}')dV' + \cdots, \end{aligned} \tag{3.169}$$

which reduces with Eqs. (3.144) and (3.158) to

$$\begin{aligned}\phi(\mathbf{r}) &= \frac{kQ}{r} + \frac{4k\pi}{3r^2}\left(-\sqrt{\frac{3}{8\pi}}\sin\theta e^{i\varphi}\right)\left(-\sqrt{\frac{3}{8\pi}}(p_x - ip_y)\right)\\ &\quad + \frac{4k\pi}{3r^2}\left(\sqrt{\frac{3}{4\pi}}\cos\theta\right)\left(\sqrt{\frac{3}{4\pi}}p_z\right)\\ &\quad + \frac{4k\pi}{3r^2}\left(\sqrt{\frac{3}{8\pi}}\sin\theta e^{-i\varphi}\right)\left(\sqrt{\frac{3}{8\pi}}(p_x + ip_y)\right) + \cdots\\ &= \frac{kQ}{r} + \frac{k}{2r^2}\sin\theta e^{i\varphi}(p_x - ip_y) + \frac{k\cos\theta}{r^2}p_z\\ &\quad + \frac{k\sin\theta}{2r^2}e^{-i\varphi}(p_x + ip_y) + \cdots, \end{aligned} \tag{3.170}$$

and so with $\mathbf{r}$ of Eq. (3.152) to

$$\begin{aligned}\phi(\mathbf{r}) &= \frac{kQ}{r} + \frac{kr\cos\theta}{r^3}p_z + \frac{kr\sin\theta\cos\varphi}{r^3}p_x + \frac{kr\sin\theta\sin\varphi}{r^3}p_y + \cdots\\ &= \frac{kQ}{r} + \frac{k\mathbf{p}\cdot\mathbf{r}}{r^3} + \cdots \quad \text{volts.}\end{aligned} \tag{3.171}$$

The first term $kQ/r$ is the potential of a point charge $Q$ at the origin, seen at $\mathbf{r}$. The potential of a dipole $\mathbf{p}$ at the origin and observed at $\mathbf{r}$ is $k\mathbf{p}\cdot\mathbf{r}/r^3$, and so on. For the electric field strength we then obtain (cf. Eq. (2.53))

$$\mathbf{E}(\mathbf{r}) = -\boldsymbol{\nabla}\phi(\mathbf{r}) = -\boldsymbol{\nabla}\left[k\frac{Q}{|\mathbf{r}|} - k\mathbf{p}\cdot\boldsymbol{\nabla}\frac{1}{|\mathbf{r}|} + \cdots\right] \quad \text{newtons/coulomb.} \tag{3.172}$$

This expression will be needed later. The potential of a dipole $\mathbf{p}$ at the origin, but observed at $\mathbf{r}$ is therefore

$$\phi_d(\mathbf{r}) = k\frac{\mathbf{p}\cdot\mathbf{r}}{r^3}. \tag{3.173}$$

Its field strength $\mathbf{E}_d$ is

$$\mathbf{E}_d = -\boldsymbol{\nabla}\phi_d(\mathbf{r}) = -k\frac{\mathbf{p} - 3\mathbf{n}(\mathbf{n}\cdot\mathbf{p})}{r^3} \quad \text{newtons/coulomb,} \tag{3.174}$$

where $\mathbf{n} = \mathbf{r}/r$. The expansion (3.171) tells us that, as we approach the individual charges by going to smaller values of $r$, we see more and more of the structure of the charge distribution.

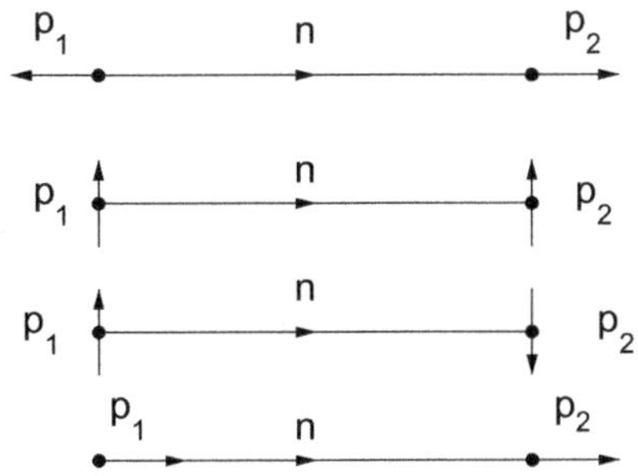


**Fig. 3.16.** Relative orientations of dipole pairs.

We close this consideration with a remark on the *interaction of 2 dipoles.* What we mean by this interaction is the energy of one dipole (say $\mathbf{p}_2$) in the field of the other at $\mathbf{r}$. This interaction is defined by the expression

$$\begin{aligned} W_{12} = -\mathbf{p}_2 \cdot \mathbf{E}_1 &= -\mathbf{p}_2 k \left[ -\frac{\mathbf{p}_1 - 3\mathbf{n}(\mathbf{n}\cdot\mathbf{p}_1)}{r^3} \right] \\ &= \frac{k}{r^3}\left[\mathbf{p}_1\cdot\mathbf{p}_2 - 3(\mathbf{n}\cdot\mathbf{p}_1)(\mathbf{n}\cdot\mathbf{p}_2)\right] \quad \text{joules.} \end{aligned} \tag{3.175}$$

We see that this energy and hence force depends also on the orientation of the dipoles to the straight line connecting them. We set

$$W \equiv k\frac{p_1 p_2}{r^3} \quad \text{joules.} \tag{3.176}$$

In the case of the orientations depicted in Fig. 3.16 we then have the following energy: $(a)\, W_{12} = 2W$, $(b)\, W_{12} = W$, $(c)\, W_{12} = -W$, $(d)\, W_{12} = -2W$ (minimum of $W_{12}$). We make an important and fundamental observation: The minimum of the energy ensues for the *same orientation* of the dipoles. This is a fundamental phenomenon, which is also seen to hold for instance for atomic spin orientations. In quantum statistics it is shown that this behavior is reached at the zero point of the absolute temperature scale.

**Example 3.15 (Dipole and quadrupole distributions).** For the dipole and quadrupole distributions shown in Fig. 3.17 calculate the electric field at the point $P$.

**Solution:** In the first case the electric field at the point $P$ in Fig. 3.17 is the sum of a repulsive contribution and an attractive contribution with the resultant as shown antiparallel to the dipole (easiest to see with

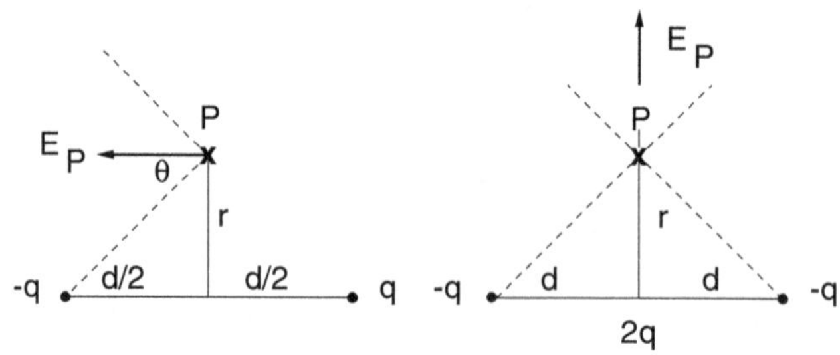


**Fig. 3.17.** Dipole and quadrupole distributions.

e.g. Eq. (2.3)):

$$E_P = \frac{(+q)\cos\theta}{4\pi\epsilon_0[r^2 + d^2/4]} - \frac{(-q)\cos\theta}{4\pi\epsilon_0[r^2 + d^2/4]} \quad \text{newtons/coulomb}, \tag{3.177}$$

where $\cos\theta = (d/2)/\sqrt{r^2 + d^2/4}$, so that

$$E_P = \frac{qd}{4\pi\epsilon_0[r^2 + d^2/4]^{3/2}} \simeq \frac{p}{4\pi\epsilon_0 r^3} \quad \text{newtons/coulomb}, \tag{3.178}$$

for $r \gg d$, where $p = qd$ is the dipole moment.

In the second case the resultant field is directed along $r$ and away from the dipole:

$$\begin{aligned} E_P &= \frac{2q}{4\pi\epsilon_0 r^2} - 2\frac{qr}{4\pi\epsilon_0(r^2 + d^2)^{3/2}} \\ &= \frac{2q}{4\pi\epsilon_0}\left[\frac{1}{r^2} - \frac{r}{(r^2 + d^2)^{3/2}}\right] \\ &\simeq \frac{3pd}{4\pi\epsilon_0 r^4} \quad \text{newtons/coulomb} \quad \text{for} \quad \mathrm{r} \gg \mathrm{d}, \end{aligned} \tag{3.179}$$

again with $p = qd$.

**Example 3.16 (Dipole-dipole interaction).** Consider two atoms as harmonic oscillators (spring constant $k_0 = \omega_0^2$) and add to these the dipole-dipole interaction (separation of the atoms: $R$). What is then the energy of the system? In a similar way one can consider two electric $LC$ circuits (a distance $R$ apart) with eigenfrequency $\omega_0 = 1/\sqrt{LC}$, in which the condensers are treated like dipoles.

**Solution:** Without the dipole-dipole interaction the energy of the system is

$$E_0 = \frac{1}{2}m(\dot{x}_1^2 + \dot{x}_2^2) + \frac{1}{2}\omega_0^2(x_1^2 + x_2^2) \quad \text{joules}, \tag{3.180}$$

and quantum mechanically

$$E_0 = \left(N_1 + \frac{1}{2}\right)\hbar\omega_0 + \left(N_2 + \frac{1}{2}\right)\hbar\omega_0 \quad \text{joules}, \tag{3.181}$$

where $N_1, N_2 = 0, 1, 2, \ldots$. With attractive dipole-dipole interaction (cf. Eq. (2.176))

$$-ax_1x_2, \quad a \propto \frac{1}{R^3} \tag{3.182}$$

and $k_0 \equiv \omega_0^2$, the energy becomes

$$\begin{aligned} E &= \frac{1}{2}m(\dot{x}_1^2 + \dot{x}_2^2) + \frac{1}{2}k_0(x_1^2 + x_2^2) - ax_1x_2 \\ &= \frac{1}{2}m(\dot{x}_1^2 + \dot{x}_2^2) + \frac{1}{2}(k_0 - a)(x_1^2 + x_2^2) + \frac{1}{2}a(x_1 - x_2)^2. \end{aligned} \tag{3.183}$$

We set

$$y_1 = x_1 + x_2, \quad y_2 = x_1 - x_2, \tag{3.184}$$

so that

$$x_1 = \frac{1}{2}(y_1 + y_2), \quad x_2 = \frac{1}{2}(y_1 - y_2), \tag{3.185}$$

and $E$ is then diagonalized, i.e.

$$\begin{aligned} E &= \frac{1}{2}m\left(\frac{\dot{y}_1}{\sqrt{2}}\right)^2 + \frac{1}{2}m\left(\frac{\dot{y}_2}{\sqrt{2}}\right)^2 + \frac{1}{2}(k_0 - a)\frac{1}{2}m\left(\frac{y_1}{\sqrt{2}}\right)^2 \\ &\quad + \frac{1}{2}(k_0 + a)\frac{1}{2}m\left(\frac{y_2}{\sqrt{2}}\right)^2 \quad \text{joules}. \end{aligned} \tag{3.186}$$

Quantum mechanically this is

$$E = \left(N_1 + \frac{1}{2}\right)\hbar\sqrt{k_0 - a} + \left(N_2 + \frac{1}{2}\right)\hbar\sqrt{k_0 + a} \quad \text{joules}. \tag{3.187}$$

For the difference with $E_0$, i.e. with no dipole-dipole interaction, we obtain for the ground state of the system $(N_1 = 0, N_2 = 0)$

$$\begin{aligned}\triangle E = E - E_0\Big|_{N_1,N_2=0} &= \frac{1}{2}\hbar(\sqrt{k_0 - a} - \sqrt{k_0}) + \frac{1}{2}\hbar(\sqrt{k_0 + a} - \sqrt{k_0}) \\ &= \frac{1}{2}\hbar\omega_0\left(\sqrt{1 - \frac{a}{\omega_0^2}} + \sqrt{1 + \frac{a}{\omega_0^2}} - 2\right) \\ &\simeq \frac{1}{2}\hbar\omega_0\left(-\frac{a^2}{4\omega_0^4}\right) \propto \frac{1}{R^6}. \end{aligned} \tag{3.188}$$

This is an energy similar to that of a van der Waals-force. One can look at this calculation as a rough physical model of the Casimir effect in producing the potential $\propto 1/R^6$, at least this was suggested in the literature.[15] Actually the Casimir effect is a boundary effect, as we shall see in Chapter 20.

## 3.10 Exercises

**Example 3.17 (A charge in a conducting corner).** A point charge $q$ is placed at the point $(x = a, y = a)$ with respect to a corner of conducting plates at planes $x = 0$ and $y = 0$. Show that the components $F_x, F_y$ of the force acting on the charge are given by

$$F_a = -\frac{kq^2}{16a^2}(4 - \sqrt{2}) \quad \text{newtons} \tag{3.189}$$

$(k = 1/4\pi\epsilon_0)$. Explain the sign of the force.

**Example 3.18 (The reciprocity theorem).** Show with the help of the second of Green's theorems that if

$$\{q_i, \phi_i\} \quad \text{and} \quad \{q_i', \phi_i'\}, \quad i = 1, \ldots, n, \tag{3.190}$$

are two possible sets of charges and potentials for surfaces $S_1, \ldots, S_n$, then the *reciprocity theorem* results:

$$\sum_{i=1}^{n} q_i\phi_i' = \sum_{i=1}^{n} q_i'\phi_i. \tag{3.191}$$

---

[15] See Kleppner [78]; cf. also the Casimir effect in Chapter 20 and footnote before Eq. (20.85).

**Example 3.19 (Constancy of capacitance).** Deduce from the linearity of the relation between charge $q$ and potential $\phi$ that the capacitance $C$ of a condenser is a constant. Hence show that if $q, \phi$ and $q', \phi'$ are possible sets, and $v$ and $V'$ are the potential differences between the two condensers in these cases, then

$$V' = \frac{q'}{q} V. \tag{3.192}$$

**Example 3.20 (Capacity of a conducting disc[16]).** Show that the capacity $C$ of a conducting disc of radius $a$ is given by

$$C = 8\epsilon_0 a \quad \text{farads.} \tag{3.193}$$

[Hint: The capacity $C$ is obtained from the relation: potential $\phi = q/4\pi\epsilon_0 r$ for $r$ large and $q$ the total charge. Consider $\triangle\phi$ expressed in oblate spheroidal coordinates and finally the limit of one axis vanishing].

**Example 3.21 (Infinite number of images in case of parallel planes).** Consider two parallel planes at $z = 0$ and $z = c$ as in Fig. 3.4 and a point charge $q$ at $z = h$. Use the method of image charges to obtain the potential $V$ of the charge $q$. (Answer:

$$V = \frac{q}{4\pi\epsilon_0} \sum_{n=-\infty}^{\infty} \frac{1}{\sqrt{y^2 + (z - h + 2nc)^2}} - \frac{q}{4\pi\epsilon_0} \sum_{n=-\infty}^{\infty} \frac{1}{\sqrt{y^2 + (z + h + 2nc)^2}}, \tag{3.194}$$

where $y$ is the distance from the line of the charges).

**Example 3.22 (Capacity of two conducting spheres).** The centers of two conducting spheres of radii $a$ and $b$ are a distance $D$ apart, $a, b < D$. Show that the capacitance $C$ is approximately given by

$$C = \frac{4\pi\epsilon_0}{(1/a) + (1/b) - (2/D)} \quad \text{farads.} \tag{3.195}$$

**Example 3.23 (Charge q between two conducting plates).** A charge $q$ is put midway between two infinite, parallel conducting plates which are

[16]Bradley [22], pp. 57–59. See comments to Example 2.24. The oblate coordinates as well as the Laplacian expressed in terms of these can conveniently be looked up in Magnus and Oberhettinger [90], p. 150. For an alternative derivation see Jackson [73], Section 3.12.

earthed. Use Kelvin's method of image charges to determine the potential in the region between the plates.[17]

**Example 3.24 (The hyperbolic capacitor).** A capacitor is bounded by the surfaces $xy = 1, xy = 0$. Verify that the potential satisfies Laplace's equation and sketch the direction of the electric field.

**Example 3.25 (Electric field at a field point P).** Show that the electric field at a point $P$ above a uniformly charged surface $S$ is given by $k\sigma\Omega$, where $k = 1/4\pi\epsilon_0$, $\sigma$ is the surface charge density and $\Omega$ the solid angle subtended by $S$ at the field point $P$.

**Example 3.26 (Potential at the rim of a charged disc).** Consider a disc of radius $a$ which is uniformly charged with charge $\sigma$ per unit area. Show that the potential at the rim is proportional to the radius $a$. The disc is not said to be a conducting surface. Argue — without explicit evaluation of integrals — how the potential of the disc at a radius $r_0 < a$ differs from that at the rim (cf. Example 2.24).

**Example 3.27 (Potential at a point within a charged surface).** Using Eq. (3.9) show that if $P$ is a point within a charged surface $S$ the potential $\phi$ can be uniquely determined at $P$.
[Hint: Use plane polar coordinates].

**Example 3.28 (Equipotentials of a finite charged line).** Consider a uniformly charged line of finite length as in the case of the rod of Example 3.3 and calculate the electrostatic potential $\phi(x, y)$ at an arbitrary point $(x, y)$. Then consider $\phi(x, y) = \text{const.}$ and show that the equipotential curves are ellipses with foci at the ends of the line.

**Example 3.29 (Capacitance of parallel cylinders).** Consider two parallel cylinders of radius $a$ with their axes a distance $D$ apart. Use the method of image charges (one in each cylinder but of opposite sign) to derive the capacitance per unit length given by Eq. (3.104).
[Hint: Use the potential of a cylinder given by Eq. (2.117)].[18]

**Example 3.30 (Charge in right angular conducting corner).** Show that the $x$-component $F_x$ of the force acting on a charge $q$ at point $(x, y)$ in

---

[17]Problem P8-26 in Cook [33], p. 236. For discussion see also Scanio [128] and Dick [36].
[18]See, if you like, Ferrari [54], p. 183.

a conducting corner is given by

$$F_x = \frac{kq^2}{4}\left[\frac{x}{(x^2+y^2)^{3/2}} - \frac{1}{x^2}\right] \quad \text{newtons.} \tag{3.196}$$

Using this result find the charge distribution $\sigma$ on each plane. What is the value of $\sigma$ at the point of intersection of the two planes? How far from there are the maxima of the charge distributions on the two planes?

**Example 3.31 (Charge in front of a conducting plane).** A charge $q$ is placed at a shortest distance $d$ in front of a conducting plane. What is the force acting on the charge?

**Example 3.32 (The conducting sphere).** If the conducting sphere of Section 3.6.1 with radius $a$ and charge $q$ a distance $d$ away from its center is required to have charge 1, what is the potential of the sphere?

(Answer: $k(1/a + q/d)$).

**Example 3.33 (Dipole moment induced on a sphere).** Consider two conducting spheres of radius $a$ with centers a distance $l > 2a$ apart. One sphere carries charge $q$. Calculate to lowest order of approximation the dipole moment induced on the other sphere.

**Example 3.34 (Kelvin's generalization of Green's theorem).** Show that a generalization of Green's theorem (3.34) due to Kelvin is

$$\int dV[\varphi\nabla\cdot(\chi\nabla\psi) + \chi(\nabla\psi)\cdot(\nabla\varphi)] = \int \varphi\chi(\nabla\psi)\cdot d\mathbf{F}. \tag{3.197}$$

**Example 3.35 (Green's theorem for two space dimensions).** Show that in two space dimensions Green's theorem (3.51) is[19]

$$\int_A [\phi\Delta_2\psi + (\nabla_2\phi)\cdot(\nabla_2\psi)]ds = \oint_C \phi(\nabla_2\psi)\cdot d\mathbf{l}, \tag{3.198}$$

where $C$ is the path around the area $A$.

**Example 3.36 (Laplace transform derived from Fourier transform).** Starting from the following Fourier transform $a(t)$ and its inverse $s(\omega)$,

$$a(t) = \frac{1}{2\pi}\int_{-\infty}^{\infty} s(\omega)e^{i\omega t}d\omega, \quad s(\omega) = \int_{-\infty}^{\infty} a(t)e^{-i\omega t}dt, \tag{3.199}$$

assuming that $a(t) = 0$ for $t < 0$, setting $\omega = -i(p-c)$, $c =$ constant, $A(t) = a(t)\exp(ct)$, obtain the *Laplace transform* $A(t)$ and its

[19]See Jackson [73], p. 347.

inverse $S(p)$, i.e.

$$A(t) = \frac{1}{2\pi i}\int_{p=c-i\infty}^{c+i\infty} e^{pt}S(p)dp, \quad S(p) = \int_{t'=0}^{\infty} e^{-pt'}A(t')dt', \tag{3.200}$$

where $c$ must be greater than the value of $p$ at which the singularity occurs in $S(p)$ (the singularity must be within the contour in the plane of complex $p$ to give a non-zero contribution).

**Example 3.37 (Delta function as limit of a rectangular pulse).** Consider a rectangular pulse defined by $a(t) = Y$ for $-\delta/2 \leq t \leq \delta/2$, and zero beyond. Evaluate the Fourier transform $s(\omega)$ of $a(t)$, i.e.

$$s(\omega) = \int_{-\infty}^{\infty} a(t)e^{-i\omega t}dt, \tag{3.201}$$

and hence the limit $Y \to \infty, \delta \to 0, Y\delta \to 1$.

**Example 3.38 (Laplace transform of the unit step function).** The Laplace transform $S(p)$ of a function $A(t)$ and the inverse transform are defined in Example 3.36. Determine the Laplace transform $S(p)$ of the unit step function $A(t) = \theta(t)$, and check the result by evaluation of the inverse Laplace transform. Check also the Laplace transform of the delta function.

**Example 3.39 (Electric field of a semicircular line charge).** Show that the electric field $\mathbf{E}$ of a semicircular uniform distribution of line charge of linear charge density $\lambda$ and of radius $R$ situated in the $z = 0$ plane in free space is given by ($k = 1/4\pi\epsilon_0$)

$$\mathbf{E} = k\frac{\pi R\lambda}{(R^2+z^2)^{3/2}}\left(-\frac{2R}{\pi}\mathbf{e}_x + z\mathbf{e}_z\right) \quad \text{newtons/coulomb.} \tag{3.202}$$

For the solution and discussion of possible pitfalls see Magid [91], pp. 165–167.

**Example 3.40 (A Fourier transform).** Show that the Fourier transform of

$$a(t) = 1 - 2\frac{|t|}{\delta} \tag{3.203}$$

between $t = -\delta/2$ and $t = \delta/2$ is

$$S(\omega) = \int a(t)e^{-i\omega t}dt = \frac{1}{2}\delta\left(\frac{\sin(\omega\delta/4)}{\omega\delta/4}\right)^2. \tag{3.204}$$

# Chapter 4

# Macroscopic Electrostatics

## 4.1 Introductory Remarks

In this chapter we extend the previous considerations to *dielectric media* (generally with negligible conductivity) which are also known as *insulators*. These differ from the conductors considered in the preceding chapter in not being characterized by charged particles which can be set into motion by a potential gradient. Thus if the space between the plates of a condenser is filled with such a medium the charges on the condenser plates remain unaffected. This leads to the consideration of the polarization of these media and their macroscopic effects, the most prominent concept introduced here being that of the *dielectric displacement* **D**.[1]

## 4.2 Dielectrics and Dielectric Displacement

So far we considered point charges, for which the following equations hold:

$$\nabla \cdot \mathbf{E} = 4k\pi\rho, \quad \nabla \times \mathbf{E} = 0, \quad k = \frac{1}{4\pi\epsilon}. \tag{4.1}$$

Here, as in our previous considerations, **E** and $\rho$ are constant in time (in electrostatics). In macroscopic dimensions, yet in a small volume, the number of charge-carrying particles is of the order of (say) $10^{18}$, most of which are in microscopic motion. In the context of macroscopic electrostatics we consider only such cases which are macroscopically static, i.e. without an observable change in time. This means we assume that all fluctuations in space and time taking place microscopically average out such that they leave no effect macroscopically. Thus we again assume the charge distributions and fields

[1]For a very readable and elementary introduction to this topic see Rossiter [126].

to be time independent. Later it will be a decisive step to assume that the equations of electrostatics hold also for time dependent cases (in the absence of magnetic fields). We define the following mean values

$$\mathbf{r} \notin V' : \langle \mathbf{E}(\mathbf{r}) \rangle := \frac{1}{V'} \int_{V'} \mathbf{E}(\mathbf{r}+\mathbf{r}')dV' \quad \text{newtons/coulomb,}$$

$$\mathbf{r}'' \notin V' : \langle \rho(\mathbf{r}'') \rangle := \frac{1}{V'} \int_{V'} \rho(\mathbf{r}''+\mathbf{r}')dV' \quad \text{coulombs/meter}^3. \qquad (4.2)$$

Here $V'$ is as shown in Fig. 4.1 the volume containing the molecules.

We imagine the electric field $\mathbf{E}$ with definite direction directed from a charge $+q$ to a charge $-q$. If we place charges or charged particles into this field the positive charges will be attracted by $-q$ and repelled by $q$, and corresponding effects apply to the negative charges. In this way a separation of the charges, contained in the medium or material arises and is described as its *polarization.* This is indicated in Fig. 4.2.

Thus, in an externally applied electric field the dielectric substance is polarized with the face at one end exhibiting a positive induced charge, and that at the other the equal and opposite charge. The object can then be considered as an arrangement with alternately positively charged and

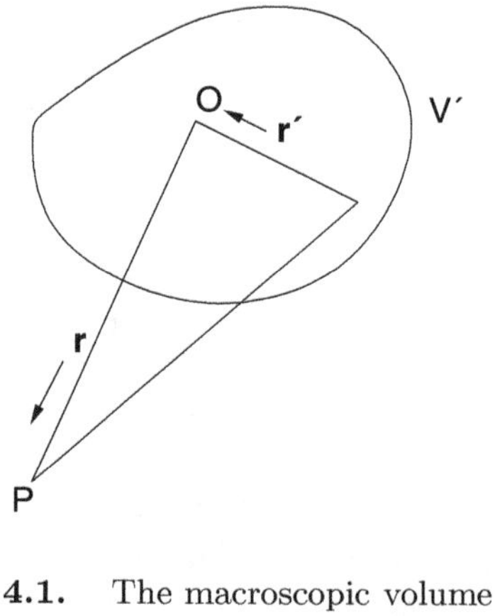


**Fig. 4.1.** The macroscopic volume $V'$.

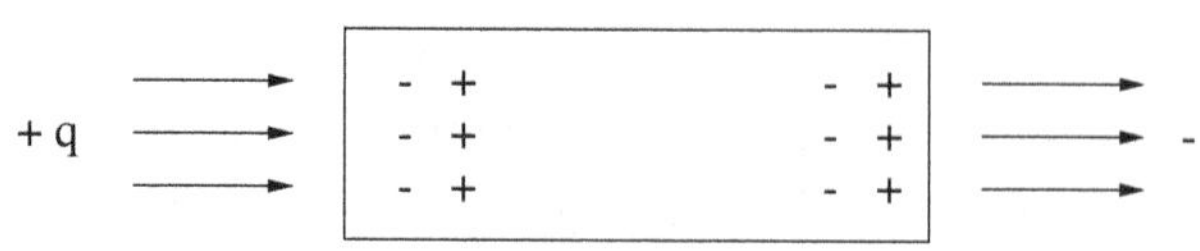


**Fig. 4.2.** Polarization of a dielectric in an external field.

negatively charged cross sections resulting from a corresponding alignment of its constituent atoms or molecules as dipoles. Thus every such constituent particle behaves like an oriented dipole. The induced charges are described as *polarization charges*. The quadrupole moments of molecules are usually negligibly small. We define as *(charge) center of mass* of a molecule $j$ with charge $q_j = \int \rho_j(\mathbf{r}')dV'$ the expression

$$\mathbf{r}_j = \frac{\int \mathbf{r}'\rho_j(\mathbf{r}')dV'}{\int \rho_j(\mathbf{r}')dV'} = \frac{\int \mathbf{r}'\rho_j(\mathbf{r}')dV'}{q_j} \quad \text{meters} \tag{4.3}$$

(integration over the volume of the molecule). Then the electric field strength $\mathbf{E}$ at a point $\mathbf{r}$ outside the volume containing $n$ molecules is, with

$$\frac{\mathbf{r}-\mathbf{r}'}{|\mathbf{r}-\mathbf{r}'|^3} = -\nabla_{\mathbf{r}}\frac{1}{|\mathbf{r}-\mathbf{r}'|} = \nabla_{\mathbf{r}'}\frac{1}{|\mathbf{r}'-\mathbf{r}|}, \tag{4.4}$$

given by (recall Eq. (3.172))

$$\mathbf{E}(\mathbf{r}) = -\nabla\sum_{j=1}^{n}\left[\frac{kq_j}{|\mathbf{r}-\mathbf{r}_j|} + k\mathbf{p}_j\cdot\nabla_j\frac{1}{|\mathbf{r}-\mathbf{r}_j|}\right] \quad \text{newtons/coulomb.} \tag{4.5}$$

We have two ways to interpret this equation:

(a) Given the charge distribution with corresponding dipole moments (e.g. polar molecules), then $\mathbf{E}(\mathbf{r})$ is the electric field strength produced by these.
(b) Given electrically neutral atoms, which are placed in an externally applied field $\mathbf{E}$ (this is frequently the case, since dipole moments can average out as a result of their motion), then this external field produces the polarization with induced charges and dipole moments $\mathbf{p}_j$ of the molecules.

With the help of delta distributions which are integrated over we can rewrite the expression (4.5) as

$$\mathbf{E}(\mathbf{r}) = -\nabla\sum_j\int dV''\left[\frac{kq_j\delta(\mathbf{r}_j-\mathbf{r}'')}{|\mathbf{r}-\mathbf{r}''|} + k\delta(\mathbf{r}_j-\mathbf{r}'')\mathbf{p}_j\cdot\nabla_{\mathbf{r}''}\frac{1}{|\mathbf{r}-\mathbf{r}''|}\right]. \tag{4.6}$$

It follows that the mean value of the electric field strength at $\mathbf{r}$ is ($V = V'$ being the volume containing the molecules)

$$\begin{aligned}
\langle \mathbf{E}(\mathbf{r}) \rangle &= \frac{1}{V'}\int_{V'} E(\mathbf{r}+\mathbf{r}')dV' \\
&= -\frac{1}{V}\sum_j \int_V dV' \boldsymbol{\nabla} \int_V dV'' \left[\frac{kq_j\delta(\mathbf{r}_j - \mathbf{r}'')}{|\mathbf{r}+\mathbf{r}'-\mathbf{r}''|}\right. \\
&\quad \left. + k\delta(\mathbf{r}_j - \mathbf{r}'')\mathbf{p}_j \cdot \boldsymbol{\nabla}'' \frac{1}{|\mathbf{r}+\mathbf{r}'-\mathbf{r}''|}\right] \\
&\overset{\mathbf{s}=\mathbf{r}''-\mathbf{r}'}{=} -\frac{1}{V}\sum_j \int_V dV' \boldsymbol{\nabla} \int_V d\mathbf{s} \left[\frac{kq_j\delta(\mathbf{r}_j - \mathbf{s} - \mathbf{r}')}{|\mathbf{r}-\mathbf{s}|}\right. \\
&\quad \left. + k\delta(\mathbf{r}_j - \mathbf{s} - \mathbf{r}')\mathbf{p}_j \cdot \boldsymbol{\nabla}_s \frac{1}{|\mathbf{r}-\mathbf{s}|}\right], \qquad (4.7)
\end{aligned}$$

where e.g. $d\mathbf{s} \equiv d\mathbf{r}'' \equiv dV''$. Integrating over $\mathbf{s}$ reproduces — apart from the averaging — Eq. (4.5) (with $\mathbf{r} \rightarrow \mathbf{r}+\mathbf{r}'$); the variables $\mathbf{s}$ and $\mathbf{r}'$ must therefore be considered as independent variables of integration, and we obtain

$$\begin{aligned}
\langle \mathbf{E}(\mathbf{r}) \rangle = -\boldsymbol{\nabla} \int_V d\mathbf{s} &\left[\frac{1}{|\mathbf{r}-\mathbf{s}|}\frac{1}{V}\int_V dV' \sum_j q_j k\delta(\mathbf{r}_j - \mathbf{s} - \mathbf{r}')\right. \\
&\left. + \left(\boldsymbol{\nabla}_s \frac{1}{|\mathbf{r}-\mathbf{s}|}\right)\frac{k}{V}\int_V dV' \sum_j \mathbf{p}_j\delta(\mathbf{r}_j - \mathbf{s} - \mathbf{r}')\right] \qquad (4.8)
\end{aligned}$$

(interchanging averaging and taking of gradient). We now set

$$\begin{aligned}
\frac{1}{V}\int_V dV' \sum_j q_j\delta(\mathbf{r}_j - \mathbf{s} - \mathbf{r}') &\equiv N(\mathbf{s})\langle q_{\rm mol}(\mathbf{s})\rangle \equiv \rho(\mathbf{s}), \\
\frac{1}{V}\int_V dV' \sum_j \mathbf{p}_j\delta(\mathbf{r}_j - \mathbf{s} - \mathbf{r}') &\equiv N(\mathbf{s})\langle \mathbf{p}_{\rm mol}(\mathbf{s})\rangle \equiv \mathbf{P}(\mathbf{s}), \qquad (4.9)
\end{aligned}$$

where we define as

$N(\mathbf{s})$: *the number of molecules per unit volume* at $\mathbf{s}$,
$\langle q_{\rm mol}(\mathbf{s})\rangle$: *average charge per molecule* at $\mathbf{s}$,
$\rho(\mathbf{s})$: *macroscopic charge density* at $\mathbf{s}$,
$\mathbf{P}(\mathbf{s})$: *polarization vector* (*=dipole moment per unit volume*) at $\mathbf{s}$,
$\langle \mathbf{p}_{\rm mol}(\mathbf{s})\rangle$: *average dipole moment per molecule* at $\mathbf{s}$.

We then have (with $\mathbf{s}$ replaced by $\mathbf{r}'$)

$$\begin{aligned}\mathbf{E}^*(\mathbf{r}) \equiv \langle \mathbf{E}(\mathbf{r})\rangle &= -\boldsymbol{\nabla}_{\mathbf{r}} \int_V dV' \left[\frac{k\rho(\mathbf{r}')}{|\mathbf{r}-\mathbf{r}'|} + k\mathbf{P}(\mathbf{r}')\cdot\boldsymbol{\nabla}'\frac{1}{|\mathbf{r}-\mathbf{r}'|}\right] \\ &\equiv -\boldsymbol{\nabla}\phi_\rho - \boldsymbol{\nabla}\phi^P. \end{aligned} \tag{4.10}$$

In view of continuity for $\mathbf{r} \neq \mathbf{r}'$ it is permissible to write the divergence (see explanations below)

$$\begin{aligned}\boldsymbol{\nabla}\cdot\mathbf{E}^*(\mathbf{r}) &= -\int_V dV' \left[k\rho(\mathbf{r}')\boldsymbol{\nabla}^2\frac{1}{|\mathbf{r}-\mathbf{r}'|} + k\mathbf{P}(\mathbf{r}')\cdot\boldsymbol{\nabla}'\left\{\boldsymbol{\nabla}^2\frac{1}{|\mathbf{r}-\mathbf{r}'|}\right\}\right] \\ &= 4\pi\int_V dV' \left[k\rho(\mathbf{r}')\delta(\mathbf{r}-\mathbf{r}') + k\mathbf{P}(\mathbf{r}')\cdot\boldsymbol{\nabla}'\delta(\mathbf{r}-\mathbf{r}')\right] \\ &= 4\pi k\rho(\mathbf{r}) - 4\pi k\boldsymbol{\nabla}\cdot\mathbf{P}(\mathbf{r}) \\ &= \frac{1}{\epsilon_0}\left[\rho(\mathbf{r}) - \boldsymbol{\nabla}\cdot\mathbf{P}(\mathbf{r})\right], \end{aligned} \tag{4.11}$$

where in the step from the first line to the second we used the replacement of Eq. (2.51),

$$\nabla^2_{\mathbf{r}}\frac{1}{|\mathbf{r}-\mathbf{r}'|} \to -4\pi\delta(\mathbf{r}-\mathbf{r}'), \tag{4.12}$$

and in the step from the second line to the third $\boldsymbol{\nabla}' \to -\boldsymbol{\nabla}$, and then $\boldsymbol{\nabla}$ was put in front of the integral. We now set

$$\mathbf{D} \equiv \epsilon_0\mathbf{E}^* + \mathbf{P}. \tag{4.13}$$

The vector $\mathbf{D}$ is called *dielectric displacement.* As a consequence we have the important relation

$$\boldsymbol{\nabla}\cdot\mathbf{D} = \rho. \tag{4.14}$$

We observe here: The polarization charges (cf. $\rho_P$ below) are not sources of $\mathbf{D}$. Since $\mathbf{E}^* = -\boldsymbol{\nabla}$ of something, but always *curl grad* $= 0$, it follows that

$$\boldsymbol{\nabla}\times\mathbf{E}^* = 0. \tag{4.15}$$

Equations (4.14) and (4.15) are the macroscopic equivalents of the microscopic equations $\boldsymbol{\nabla}\cdot\mathbf{E} = 4\pi k\rho$ and $\boldsymbol{\nabla}\times\mathbf{E} = 0$.

We observed at the beginning, that in the MKSA-system of units the factor $k = 1/4\pi\epsilon_0$ is not dimensionless. As a consequence $\mathbf{D}$ has a dimension different from that of **E**. In the MKSA-system the unit of dielectric displacement **D** is *one coulomb* (C) *per square meter*; in Gaussian units[2]

$$1\ \mathrm{C/m^2} = 12\pi \times 10^5\ \mathrm{statvolts/cm}\ \ (\mathrm{statcoulombs/cm^2}).$$

We deduce from the above relation (4.13) that the polarization **P** is measured in the MKSA-system in the same units as **D**.

We distinguish between the two ways to see this:

(a) Given a charge distribution in space (like, for example, of molecules, also polar molecules) whose electrostatic potential contains contributions of dipole moments (higher multipole moments being negligible), then $\mathbf{E}(\mathbf{r})$ is the strength of the electric field of this charge distribution.
(b) Given electrically neutral atoms which macroscopically form a neutral dielectric, and that these are placed in an externally applied electric field $\mathbf{E}(\mathbf{r})$. Then this external field generates the polarization of the dielectric with induced charges or induced dipole moments which are then defined by the contributions in the above expression for $\mathbf{E}(\mathbf{r})$. (In the case of a neutral atom, which is considered classically, we can have a dipole moment; quantum mechanically, however, where, for instance, the electron of the hydrogen atom would be found only with a certain probability somewhere near the proton, hence also its charge, the dipole moment of such an atom averages out).

Consider an element of length $l$ of a dielectric as illustrated in Fig. 4.3. Here $\triangle F$ is the cross sectional area with charge $Q$ or charge per unit area $\sigma_P = Q/\triangle F$. The dipole moment is charge $\times$ distance between the charges, i.e. if **P** is the density of dipole moments,

$$V|\mathbf{P}| = Ql = \sigma_P \triangle F l = \sigma_P V. \tag{4.16}$$

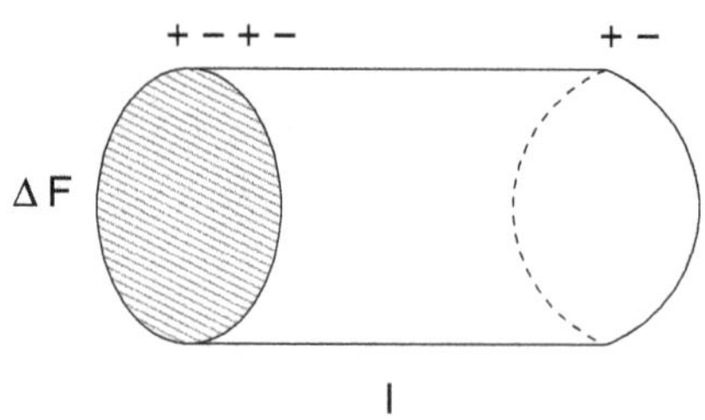


**Fig. 4.3.** An element of a dielectric.

[2]See Appendix B.

It follows that

$$|\mathbf{P}| = \sigma_P \equiv \mathbf{P} \cdot \mathbf{n} \quad \text{coulombs/meter}^2. \tag{4.17}$$

We now look in more detail at the expression $\phi^P$ introduced in Eq. (4.10) and contained in $\mathbf{E}^*(\mathbf{r})$. We have[3]

$$\begin{aligned}
\phi^P(\mathbf{r}) &= k\int_V dV'\mathbf{P}(\mathbf{r}') \cdot \boldsymbol{\nabla}' \frac{1}{|\mathbf{r}-\mathbf{r}'|} \\
&= k\int_V dV' \left[\boldsymbol{\nabla}' \cdot \left(\mathbf{P}(\mathbf{r}')\frac{1}{|\mathbf{r}-\mathbf{r}'|}\right) - \frac{1}{|\mathbf{r}-\mathbf{r}'|}\boldsymbol{\nabla}' \cdot \mathbf{P}(\mathbf{r}')\right] \\
&= k\int_{F'(V)} d\mathbf{F}' \cdot \mathbf{P}(\mathbf{r}')\frac{1}{|\mathbf{r}-\mathbf{r}'|} - k\int dV' \frac{\boldsymbol{\nabla}' \cdot \mathbf{P}(\mathbf{r}')}{|\mathbf{r}-\mathbf{r}'|} \\
&\equiv \phi^P_\sigma + \phi^P_\rho \quad \text{volts}.
\end{aligned} \tag{4.18}$$

This equation defines the quantities $\phi^P_\sigma, \phi^P_\rho$. We let $\sigma_P$ be the induced charge density per unit area (induced through polarization by the external field) defined by

$$\sigma_P = \mathbf{P}(\mathbf{r}) \cdot \mathbf{n} \quad \text{coulombs/meter}^2, \tag{4.19}$$

and we let $\rho_P$ be the similarly induced charge density per unit volume defined by

$$\boldsymbol{\nabla} \cdot \mathbf{P} = -\rho_P \quad \text{coulombs/meter}^3. \tag{4.20}$$

Without the external field $\mathbf{E}$, the polarization $\mathbf{P}$ would be zero (because the dipoles would average out). For $|\mathbf{E}|$ not too large (and not for ferroelectric substances, i.e. not universally) we have — from here on we replace $\mathbf{E}^*$ by $\mathbf{E}$ — the linear relation

$$\mathbf{P} = \chi_e\epsilon_0\mathbf{E} \quad \text{coulombs/meter}^2, \tag{4.21}$$

in which the constant $\chi_e$ is the so-called *electric susceptibility*. For *isotropic dielectrics* $\mathbf{P} \propto \mathbf{E}$:

$$P_i(\mathbf{E}) = P_i(0) + \sum_j a_{ij}E_j + \cdots, \quad P_i(0) = 0, \quad a_{ij} \propto \delta_{ij}. \tag{4.22}$$

---

[3] A corresponding result can be derived for the scalar magnetic potential; see Example 5.19.

Since (cf. Eq. (4.13)) $\mathbf{D} = \epsilon_0 \mathbf{E} + \mathbf{P}$, it follows that

$$\mathbf{D} = (\epsilon_0 + \epsilon_0 \chi_e)\mathbf{E} \equiv \epsilon \mathbf{E}, \quad \epsilon \equiv \epsilon_0 + \epsilon_0 \chi_e, \quad \chi_e = \frac{1}{\epsilon_0}(\epsilon - \epsilon_0) \geq 1. \tag{4.23}$$

The quantity $\epsilon$ is called *permittivity* or *dielectric constant* and is measured like $\epsilon_0$ in farads per meter.[4] The ratio $\epsilon/\epsilon_0$ for the case of the vacuum is 1 and for example for air 1.0005, for glass 5 up to 8 and for water 81. Above we had (see Eq. (4.14)) $\boldsymbol{\nabla} \cdot \mathbf{D} = \rho$, so that, if $\epsilon$ is constant,

$$\boldsymbol{\nabla} \cdot \mathbf{E} = \frac{\rho}{\epsilon}. \tag{4.24}$$

This means that our vacuum relation (2.19a) with $\boldsymbol{\nabla} \cdot \mathbf{E} = 4\pi k \rho = \rho/\epsilon_0$ becomes for a dielectric $\boldsymbol{\nabla} \cdot \mathbf{E} = 4\pi k \rho$ with $k = 1/4\pi\epsilon$, and this factor $k$ now replaces the vacuum factor in $\mathbf{E}$ and $\phi$ of Eqs. (2.7) and (2.9). One defines as *relative permittivity* the expression

$$\kappa = \frac{\epsilon}{\epsilon_0} = 1 + \chi_e, \quad E = \frac{D}{\kappa \epsilon_0}. \tag{4.25}$$

We imagine the formation of polarization charges as explained above: The electrically neutral atom when placed in an electric field assumes the shape of a dipole as a result of the deformation polarization (classically the shifting of positive charge to one side, and negative charge to the other). *Polar molecules* are those which possess their own dipole moment. In many other cases rotation, vibration etc. average the dipole moments out to zero in the laboratory frame of reference.

It remains to consider the modification of the Gauss law. We had in Eq. (4.13)

$$\mathbf{D} = \epsilon_0 \mathbf{E} + \mathbf{P} \quad \text{coulombs/m}^2. \tag{4.26}$$

It follows that

$$\int dV' \boldsymbol{\nabla} \cdot \mathbf{D} = \epsilon_0 \int dV' \boldsymbol{\nabla} \cdot \mathbf{E} + \int dV' \boldsymbol{\nabla} \cdot \mathbf{P}, \tag{4.27}$$

i.e. with Eqs. (4.14) and (4.20),

$$\int dV' \rho = \epsilon_0 \int \mathbf{E} \cdot d\mathbf{F} + \int dV'(-\rho_P), \tag{4.28}$$

[4] Clemmow [32], p. 104, calls $\epsilon/\epsilon_0$ dielectric constant which is also called relative permittivity, e.g. The Electromagnetic Problem Solver [47], p. 213.

or

$$Q = \epsilon_0 \int \mathbf{E} \cdot d\mathbf{F} - Q_P$$

and hence

$$\epsilon_0 \int \mathbf{E} \cdot d\mathbf{F} = Q + Q_P. \tag{4.29}$$

This is the *modified Gauss law*, in which $Q$ is the *true macroscopic charge* and $Q_P$ the *induced charge*. One should note that $Q_P$ does not have to be zero; in the case of unpolarized dielectrics the sum of the charges vanishes for every volume element, i.e. $\rho = 0$. In the case of polarization in general $\rho_P dV \neq 0$, in view of a different distribution (cf. a volume element of the box in Fig. 4.2). See, however, Section 4.5, where $\rho_P = 0$.

**Example 4.1 (Relative permittivity[5]).** An insulating medium fills the space between two parallel plates with charge densities $\sigma$ and $-\sigma$ coulomb/meter$^2$. Determine the relative permittivity of the dielectric, i.e. $\kappa$ (with $E = D/\kappa\epsilon_0$), expressed in terms of the induced charges ($\pm\sigma_P$ coulombs/meter$^2$) at the end-faces.

**Solution:** We start from the standard equations, i.e. Eq. (4.26),

$$\epsilon_0 \mathbf{E} = \mathbf{D} - \mathbf{P}, \quad \text{with} \quad \boldsymbol{\nabla} \cdot \mathbf{D} = \rho, \quad \boldsymbol{\nabla} \cdot \mathbf{P} = -\rho_P. \tag{4.30}$$

Polarization charges are not sources of $\mathbf{D}$. The decisive point here is to distinguish clearly between the real applied charges on the plates at either end ($\sigma, -\sigma$ per unit area), and the charges induced by them in the medium, i.e. $-\sigma_P$ at the medium end next to $+\sigma$ and $+\sigma_P$ at the medium end next to $-\sigma$. Hence we have

$$\int \boldsymbol{\nabla} \cdot \mathbf{D} dV = \int \mathbf{D} \cdot d\mathbf{F} = \int \rho dV = Q, \tag{4.31}$$

so that for an element $\triangle F$ of area of a plate

$$\mathbf{D} \cdot \triangle \mathbf{F} = Q, \quad D = \frac{Q}{\triangle F} = \sigma, \tag{4.32}$$

with $\mathbf{D}$ parallel to $\triangle \mathbf{F}$. However, in the relation

$$\int \boldsymbol{\nabla} \cdot \mathbf{P} dV = \int \mathbf{P} \cdot d\mathbf{F} = -\int \rho_P dV = -Q_P, \tag{4.33}$$

[5]See also *The Electromagnetic Problem Solver* [47], p. 213.

$\mathbf{P}$ is parallel to $-\mathbf{D}$ and hence

$$-\mathbf{P}\cdot\triangle\mathbf{F} = -Q_P, \quad P = \frac{Q_P}{\triangle F} = \sigma_P. \tag{4.34}$$

Thus

$$E = \frac{D-P}{\epsilon_0} = \frac{\sigma - \sigma_P}{\epsilon_0} = \frac{\sigma(\sigma-\sigma_P)}{\epsilon_o\sigma} = \frac{D}{\epsilon_0}\frac{(\sigma-\sigma_P)}{\sigma} \equiv \frac{D}{\epsilon_0\kappa}, \tag{4.35}$$

where

$$\kappa = \frac{\sigma}{\sigma - \sigma_P}. \tag{4.36}$$

## 4.3 The Behavior of D at an Interface

We are now interested in the behavior of the electric flux density $\mathbf{D}$ on both sides of a macroscopically thin, charged surface, in much the same way as in the case of $\mathbf{E}$. We have

$$\nabla\cdot\mathbf{D} = \rho \quad \text{coulombs/m}^3 \tag{4.37}$$

(the polarization charges are not sources of $\mathbf{D}$). Thus

$$\int\nabla\cdot\mathbf{D} = \int\mathbf{D}\cdot d\mathbf{F} = \int dV\rho = Q, \tag{4.38}$$

and hence for an infinitesimal surface element as indicated in Fig. 4.4 we have

$$\mathbf{D}_1\cdot\mathbf{n}_1 + \mathbf{D}_2\cdot\mathbf{n}_2 = \frac{Q}{\triangle F} = \sigma, \tag{4.39}$$

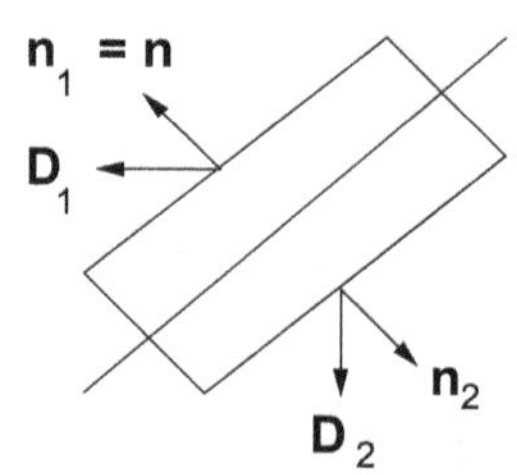


**Fig. 4.4.** Element of a charged boundary surface.

or with $\mathbf{n}$ as a unit vector in the direction of $\mathbf{n}_1$[6]:

$$(\mathbf{D}_1 - \mathbf{D}_2) \cdot \mathbf{n} = \sigma. \tag{4.40}$$

But since (cf. Eq. (4.29))

$$\epsilon_0 \int \mathbf{E} \cdot d\mathbf{F} = (Q + Q_P), \tag{4.41}$$

we obtain for $\mathbf{E}$ instead of $\mathbf{D}$:

$$\epsilon_0(\mathbf{E}_1 - \mathbf{E}_2) \cdot \mathbf{n} = (\sigma + \sigma_P). \tag{4.42}$$

On the other hand, since $\boldsymbol{\nabla} \times \mathbf{E} = 0$, one obtains for a closed contour from just above to just below the interface that

$$\oint \mathbf{E} \cdot d\mathbf{s} = \int \boldsymbol{\nabla} \times \mathbf{E} \cdot d\mathbf{F} \overset{(4.15)}{=} 0, \tag{4.43}$$

so that the tangential components of $\mathbf{E}$, i.e. those in the plane of the interface surface, are equal, i.e.

$$E_{1t} - E_{2t} = 0. \tag{4.44}$$

Since $\boldsymbol{\nabla} \times \mathbf{D} \neq 0$ for $\mathbf{P} \neq 0$, there is no corresponding relation for $\mathbf{D}$.

**Example 4.2 (Lines of E at interfaces).** A charge $q$ is placed at $z = 1$ along the $z$-axis in a medium with dielectric constant $\epsilon_1$ in the region $z > 0$. The medium in the region $z < 0$ has dielectric constant $\epsilon_2$. Sketch the paths of lines of $\mathbf{E}$ in the cases $\epsilon_1 > \epsilon_2$ and $\epsilon_2 > \epsilon_1$.

**Solution:** With $\rho = \sqrt{x^2 + y^2}$ the potential at a point $(\rho, z)$ in the first medium is given by (observe the factor $1/4\pi\epsilon_1$)

$$\phi = \frac{q}{4\pi\epsilon_1\sqrt{\rho^2 + (z-1)^2}} \quad \text{volts}, \tag{4.45}$$

[6]Note that in the case of a parallel plate capacitor with $\mathbf{E}$ zero on one side, one has $D = \sigma$ coulombs/m$^2$ inside the capacitor.

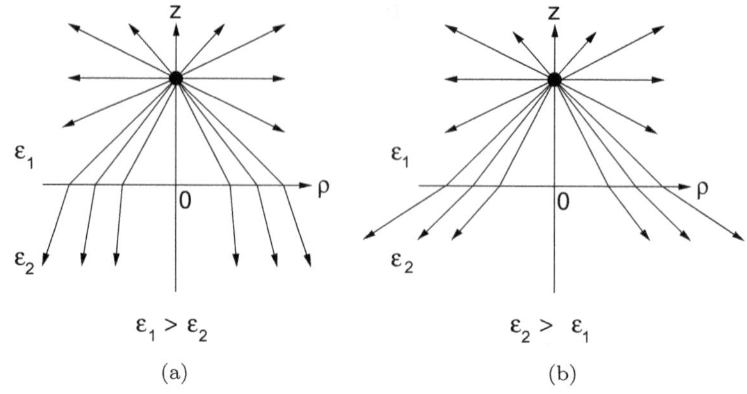


**Fig. 4.5.** Lines of constant **E** for (a) $\epsilon_1 > \epsilon_2$ and (b) $\epsilon_2 > \epsilon_1$.

so that

$$\text{for} \quad z > 0: \quad \mathbf{E}^{(1)} = -\boldsymbol{\nabla}\phi = \frac{q}{4\pi\epsilon_1}\left(\frac{\rho\mathbf{e}_\rho + (z-1)\mathbf{e}_z}{[\rho^2 + (z-1)^2]^{3/2}}\right) \quad \text{volts/meter.} \tag{4.46}$$

At the interface to the second medium along the interface normals we have ("$\pm 0$" meaning $\pm\epsilon, \epsilon > 0$ and small) from Eq. (4.40):

$$(\epsilon_1 E_z^{(1)})_{z=+0} = (\epsilon_2 E_z^{(2)})_{z=-0} \tag{4.47}$$

(no charge given on the interface). Moreover tangentially from Eq. (4.44):

$$[E_\rho^{(1)}]_{z=+0} = [E_\rho^{(2)}]_{z=-0}. \tag{4.48}$$

Hence (see Eq. (4.46))

$$(E_z^{(2)})_{z=-0} = \frac{\epsilon_1}{\epsilon_2}(E_z^{(1)})_{z=+0} = -\frac{q}{4\pi\epsilon_2[\rho^2+1]^{3/2}}, \tag{4.49}$$

and (see Eq. (4.44))

$$(E_\rho^{(2)})_{z=-0} = (E_\rho^{(1)})_{z=+0} = \frac{q\rho}{4\pi\epsilon_1[\rho^2+1]^{3/2}}. \tag{4.50}$$

The trajectories of lines of **E** are shown in Figs. 4.5(a) and 4.5(b). For $\epsilon_2 \gg \epsilon_1$, i.e. $\epsilon_2 \to \infty$, the second medium behaves approximately like a conductor (no field).

**Example 4.3 (The spherical condenser).** A condenser consists of two concentric, conducting, spherical shells with inner charge $q$ and radii $a$ and $b$, $a < b$. The space between the spheres from $a$ to $r_0$ is filled with a dielectric with dielectric constant $\epsilon_1$, and from $r_0$ to $b$ with a different medium of dielectric constant $\epsilon_2$ (both linear). The outer sphere is earthed. Calculate the capacity of the condenser.

**Solution:** We have spherical symmetry with $\boldsymbol{\nabla} \cdot \mathbf{D} = \rho$ and hence

$$\int \mathbf{D} \cdot d\mathbf{F} = D_r \cdot 4\pi r^2 = q, \quad D_r = \frac{q}{4\pi r^2} \tag{4.51}$$

for all $r$. Moreover, $E_r = -\partial\phi/\partial r$ for potential $\phi$. We assume the linearity $\mathbf{D} = \epsilon\mathbf{E}$. Then:

for $a \leq r < r_0$: $E_r = D_r/\epsilon_1 = q/4\pi\epsilon_1 r^2$, so that $\phi = -\int E_r dr = q/4\pi\epsilon_1 r + C_1$;

for $r_0 < r \leq b$: $E_r = D_r/\epsilon_2 = q/4\pi\epsilon_2 r^2$, so that $\phi = q/4\pi\epsilon_2 r + C_2$.

At $r = a$: $\phi(r = a) = q/4\pi\epsilon_1 a + C_1$.

Continuity of $\phi$ at $r = r_0$:

$$\frac{q}{4\pi\epsilon_1 r_0} + C_1 = \frac{q}{4\pi\epsilon_2 r_0} + C_2.$$

Earthing at $r = b$:

$$\phi(r = b) = 0, \quad \frac{q}{4\pi\epsilon_2 b} + C_2 = 0.$$

Thus

$$\phi(r = a) = \frac{q}{4\pi\epsilon_1 a} + \frac{q}{4\pi\epsilon_2 r_0} - \frac{q}{4\pi\epsilon_1 r_0} - \frac{q}{4\pi\epsilon_2 b} \quad \text{volts.} \tag{4.52}$$

Hence the capacity $C$ is

$$C = \frac{q}{\phi(r = a)} = \frac{4\pi}{[1/\epsilon_1 a - 1/\epsilon_2 b + (1/r_0)(1/\epsilon_2 - 1/\epsilon_1)]} \quad \text{farads.} \tag{4.53}$$

**Example 4.4 (The coaxial cable).** A coaxial cable consists of a conducting wire with circular cross section of radius $a$ and charge $e$ per unit length, which is surrounded by an insulating layer of thickness $d$ and a thin, hollow, conducting and earthed cylinder of radius $c \gg a + d$ as indicated in Fig. 4.6. Determine the capacity of the cable per unit length.

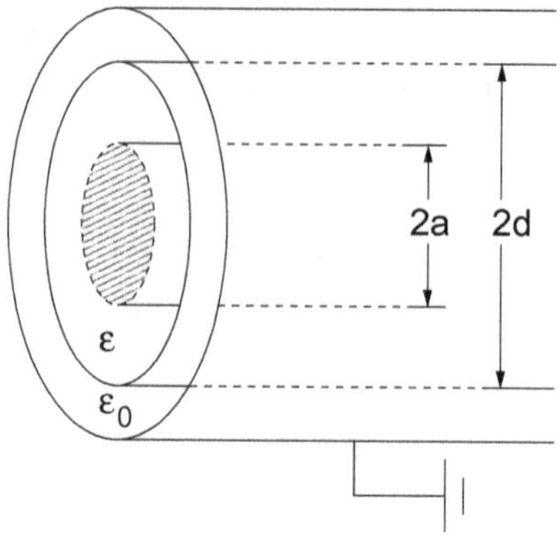


**Fig. 4.6.** The coaxial cable.

**Solution:** We have cylindrical symmetry. The outer cylinder is earthed, and the dielectric is assumed to be linear, i.e. $D \propto E, D = \epsilon E$. We use

$$\int_F \mathbf{D} \cdot d\mathbf{F} = \text{charge enclosed} \tag{4.54}$$

(i.e. $e$ per unit length with the integral over the cylindrical surface). We then have in the domain

$$0 < r < a: \quad E = 0 \quad \text{(conductor)}.$$

According to the treatment of the cylindrical condenser in Example 2.12 (Eq. (2.116)) we have for

$$a < r < a + d: \quad E = \frac{e}{2\pi\epsilon r} \quad \text{volts/meter,}$$

and for

$$a + d < r < c: \quad E = \frac{e}{2\pi\epsilon_0 r} \quad \text{volts/meter.}$$

Integrating with the help of $E = -d\phi/dr$:

$$\begin{aligned} 0 < r < a: &\quad \phi = \phi_1 = \text{const.,} \\ a < r < a + d: &\quad \phi = -\frac{e}{2\pi\epsilon}\ln r + C_1, \\ a + d < r < c: &\quad \phi = -\frac{e}{2\pi\epsilon_0}\ln r + C_2. \end{aligned} \tag{4.55}$$

The boundary conditions are:

$$r = a: \quad \phi = \phi_1 = -\frac{e}{2\pi\epsilon}\ln a + C_1,$$

$$r = a + d \text{ and } \phi \text{ continuous:} \quad -\frac{e}{2\pi\epsilon}\ln(a+d) + C_1 = -\frac{e}{2\pi\epsilon_0}\ln(a+d) + C_2,$$

$$r = c: \quad \phi_2 = 0 \quad \text{(earthed)}, \quad -\frac{e}{2\pi\epsilon_0}\ln c + C_2 = 0.$$

Inserting the latter two expressions into the first for the constants, we obtain

$$\begin{aligned}
\phi_1 &= -\frac{e}{2\pi\epsilon}\ln a + C_1 \\
&= -\frac{e}{2\pi\epsilon}\ln a - \frac{e}{2\pi\epsilon_0}\ln(a+d) + \frac{e}{2\pi\epsilon_0}\ln c + \frac{e}{2\pi\epsilon}\ln(a+d) \\
&= \frac{e}{2\pi\epsilon_0}\left[\frac{\epsilon_0}{\epsilon}\ln\left(\frac{a+d}{a}\right) + \ln\left(\frac{c}{a+d}\right)\right] \quad \text{volts.}
\end{aligned} \tag{4.56}$$

Thus, for the capacity $C$ follows

$$C = \frac{e}{\phi_1 - \phi_2} = \frac{2\pi\epsilon_0}{[(\epsilon_0/\epsilon)\ln\{(a+d)/a\} + \ln\{c/(a+d)\}]} \quad \text{farads/meter.} \tag{4.57}$$

## 4.4 Polarization of a Sphere in an Electric Field $\mathbf{E}_0$

The sphere[7] is electrically neutral (i.e. originally uncharged, $\rho = 0$) and is assumed to be made of a material with dielectric constant $\epsilon_2$. The externally applied electric field $\mathbf{E}_0$ is taken to be homogeneous and parallel to the unit vector $\mathbf{e}_z$ in a medium with dielectric constant $\epsilon_1$. We take the center of the sphere as the origin of the frame of coordinates as shown in Fig. 4.7.

The polarization charges on the surface give the sphere the appearance of a dipole placed at the origin and pointing in the direction of $z$ (we shall see that the induced volume charge density in the sphere, i.e. $\rho_P$, vanishes). Let $\mathbf{E}_i$ be the field $\mathbf{E}$ inside the sphere and $\mathbf{E}_a$ the field outside.

We begin by writing down the general equations of (now macroscopic) electrostatics for field $\mathbf{E}$ in a linearly isotropic dielectric ($\rho = 0$ implies no

[7] The following considerations apply to a sphere. In the case of a cylinder treated in Example 4.5 some differences arise as can be seen e.g. in the steps from Eq. (4.84) to Eq. (4.87).

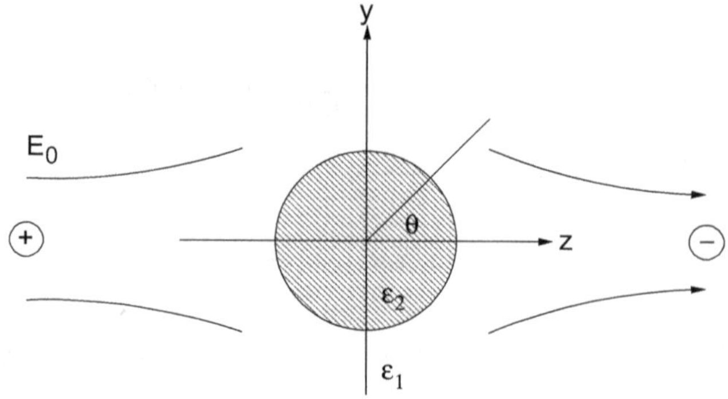


**Fig. 4.7.** A spherical dielectric in the field $\mathbf{E}_0$.

"*true*" charges as compared to polarization charges):

$$\nabla \cdot \mathbf{D} = 0, \quad \mathbf{D} = \epsilon_0 \mathbf{E} + \mathbf{P} = \epsilon \mathbf{E},$$
$$\mathbf{P} \cdot \mathbf{n} = \sigma_P, \quad \nabla \cdot \mathbf{P} = -\rho_P, \tag{4.58}$$

and

$$\nabla \times \mathbf{E} = 0, \quad \mathbf{E} = -\nabla \varphi. \tag{4.59}$$

From these equations we obtain

$$\epsilon_0 \nabla \cdot \mathbf{E} + \nabla \cdot \mathbf{P} = 0, \tag{4.60}$$

and so

$$\epsilon_0 \triangle \varphi = -\rho_P. \tag{4.61}$$

On the other hand

$$0 = \nabla \cdot \mathbf{D} = \epsilon \nabla \cdot \mathbf{E} = -\epsilon \triangle \varphi. \tag{4.62}$$

Hence, (for $\epsilon \neq 0$) we obtain the Poisson (or here Laplace) equation $\triangle \varphi = 0$ and so $\rho_P = 0$ for the assumed linear dielectric, i.e. for $\mathbf{P}$ parallel to $\mathbf{E}$ (not for other cases). This means, there is no volume distribution $\rho_P$, as we see from Eq. (4.20).

At the interface between two media the boundary conditions (4.44) and (4.40) have to be satisfied. This means,

1. $E_{1t} = E_{2t}$, and so

$$E_{1\theta}\Big|_{r=a} = E_{2\theta}\Big|_{r=a}, \quad \text{or} \quad \frac{\partial \varphi_{\rm i}}{\partial \theta}\Big|_{\rm r=a} = \frac{\partial \varphi_{\rm a}}{\partial \theta}\Big|_{\rm r=a}; \tag{4.63}$$

2.

$$(\mathbf{D}_1 - \mathbf{D}_2) \cdot \mathbf{n} = \sigma = 0, \quad \text{or} \quad (\mathbf{D}_1 \cdot \mathbf{n}) = (\mathbf{D}_2 \cdot \mathbf{n}), \tag{4.64}$$

(where $\sigma$, the "*true*" surface charge density, is zero). Thus (since $\mathbf{D} = \epsilon\mathbf{E} = -\epsilon\boldsymbol{\nabla}\varphi$)

$$-\epsilon_1 \frac{\partial \varphi_a}{\partial r}\bigg|_{r=a} = -\epsilon_2 \frac{\partial \varphi_i}{\partial r}\bigg|_{r=a}. \tag{4.65}$$

In order to be able to solve these two equations, one possibility is to make a plausible ansatz for $E_i$, taking into account $\boldsymbol{\nabla} \times \mathbf{E} = 0$ and the assumed isotropy (4.21) of the medium. We make the *ansatz* of a homogeneous field $E_i\mathbf{e}_z$ in the interior of the sphere (and demonstrate that this ansatz together with $\varphi_a$ satisfies the above boundary conditions). The ansatz $E_i$ implies

$$\varphi_i = -E_i z, \quad z = r\cos\theta. \tag{4.66}$$

From outside the sphere appears as a dipole along $\mathbf{e}_z$ with appropriate dipole potential of Eq. (3.171) (taking into account, that $\mathbf{p}_z \cdot \mathbf{r} = p_z r\cos\theta = p_z z$)

$$\varphi_a = -E_0 z + k\frac{\mathbf{p}_z \cdot \mathbf{r}}{r^3} = -E_0 z + k\frac{p_z z}{(x^2+y^2+z^2)^{3/2}} \tag{4.67}$$

(so that $E_\infty = -\partial\varphi_a/\partial z = E_0$). The second term on the right really results from solution of the Poisson equation $\triangle\varphi_a = 0$ with *Kelvin's theorem*, which says that if $\varphi_a \propto r^n$ is a solution, then there is another solution $\propto 1/r^{n+1}$ for the same angular dependence (two independent solutions of the second order Laplace equation in spherical coordinates, see Examples 4.10 and 4.11). In the present case of a sphere

$$r = (x^2+y^2+z^2)^{1/2}, \quad dr/dy = y/r, \tag{4.68}$$

so that (note the use of Cartesian coordinates)

$$\frac{d\varphi_a}{dy} = -k\frac{3p_z zy}{r^5}, \quad E_{ay} = k\frac{3p_z zy}{r^5}. \tag{4.69}$$

Analogously, we obtain

$$E_{az} = E_0 - k\frac{p_z}{r^3} + k\frac{3p_z z^2}{r^5}. \tag{4.70}$$

More generally this is

$$\mathbf{E} = \mathbf{E}_0 + k\left[\frac{3(\mathbf{p}\cdot\mathbf{r})\mathbf{r}}{r^5} - \frac{\mathbf{p}}{r^3}\right] \quad \text{newtons/coulomb}. \tag{4.71}$$

Thus outside the sphere and for $r$ near $a$ the field components $E_{ay}, E_{ax}, E_z$, are different from zero and lead to a curving of the lines of constant $E$ or $D$ (see below). The boundary condition 1., in which we can replace $\partial/\partial\theta$ by $\partial/\partial\cos\theta$, yields with Eq. (4.66) when divided by $-a$:

$$-\frac{1}{a}\frac{\partial\varphi_i}{\partial\cos\theta} = E_i = -\frac{1}{a}\frac{\partial\varphi_a}{\partial\cos\theta} = E_0 - k\frac{p_z}{a^3}. \tag{4.72}$$

We note in passing, that this equation follows also from the continuity of the potential at the boundary of the sphere, i.e. at $r = a$. Analogously, boundary condition 2. yields with Eq. (4.67) the equation (apart from a factor $\cos\theta$)

$$\epsilon_1\left(E_0 + k\frac{2p_z}{a^3}\right) = \epsilon_2 E_i. \tag{4.73}$$

From Eqs. (4.72) and (4.73) we obtain

$$E_i - E_0 = -k\frac{p_z}{a^3} = \frac{\epsilon_1 E_0 - \epsilon_2 E_i}{2\epsilon_1}, \tag{4.74}$$

i.e.

$$E_i = \frac{3\epsilon_1}{\epsilon_2 + 2\epsilon_1}E_0 \quad \text{volts/meter}. \tag{4.75}$$

In Example 4.5 the corresponding result is derived for a cylinder. From Eq. (4.72) we also obtain

$$kp_z = (E_0 - E_i)a^3 = E_0\left(1 - \frac{3\epsilon_1}{\epsilon_2 + 2\epsilon_1}\right)a^3,$$

and hence

$$kp_z = \frac{\epsilon_2 - \epsilon_1}{\epsilon_2 + 2\epsilon_1}a^3 E_0. \tag{4.76}$$

This expression inserted into Eqs. (4.67) and (4.66) gives the potentials outside and inside the sphere respectively as

$$\varphi_a = -E_0 r\cos\theta + \frac{a^3}{r^2}\left(\frac{\epsilon_2 - \epsilon_1}{\epsilon_2 + 2\epsilon_1}\right)E_0\cos\theta, \quad r > a,$$

$$\varphi_i = -\frac{3\epsilon_1}{\epsilon_2 + 2\epsilon_1}E_0 r\cos\theta, \quad r < a. \tag{4.77}$$

It follows from Eq. (4.76), that $p_z >$ (or $<$) 0, if $\epsilon_2 >$ (or $<$)$\epsilon_1$. Substituting the expression (4.76) for $kp_z$ into $E_{ay}$ of Eq. (4.69) allows us to determine the distortion of the field lines in the neighborhood of the sphere. Let $\tan\theta$

be the gradient of the **E** field line of force at a point. Then, there for $z > 0, y > 0$:

$$\tan\theta := \frac{E_{ay}}{E_{az}} \overset{(4.70)}{\sim} \frac{E_{ay}}{E_0} \overset{(4.69),(4.76)}{=} \frac{3zya^3}{r^5}\left(\frac{\epsilon_2 - \epsilon_1}{\epsilon_2 + 2\epsilon_1}\right) \begin{cases} > 0 & \text{for} \quad \epsilon_2 > \epsilon_1, \\ < 0 & \text{for} \quad \epsilon_2 < \epsilon_1. \end{cases} \tag{4.78}$$

For lines of **D** this ratio has to be multiplied by $\epsilon_1/\epsilon_2$ which affects the magnitude but not the sign of $\tan\theta$. For $\epsilon_2 = \epsilon_1$ the ratio vanishes implying $\theta = 0$, and the lines pass through without deflection. The correspondingly distorted lines of **E** are shown in Fig. 4.8. Figure 4.8(a) shows $\tan\theta > 0$ for $z > 0$ and Fig. 4.8(b) $\tan\theta < 0$ for $z > 0$. The next figure, Fig. 4.9, shows the paths for (a) $\epsilon_2 > \epsilon_1$ and (b) $\epsilon_2 < \epsilon_1$.

*The limiting case* of $\epsilon_2 \to \infty$: This is the case of the conducting sphere: The main observation is (in view of Eq. (4.75)) $E_i = 0$ (i.e. potential $\varphi = \text{const.}$) in the interior. This case is depicted in Fig. 4.10.

One should note in Fig. 4.9(a) the behavior of the $D$-lines in the interior: $\epsilon_2 \to \infty$, $E_i \to 0$, but $D = \epsilon_2 E_i = \text{finite}$.

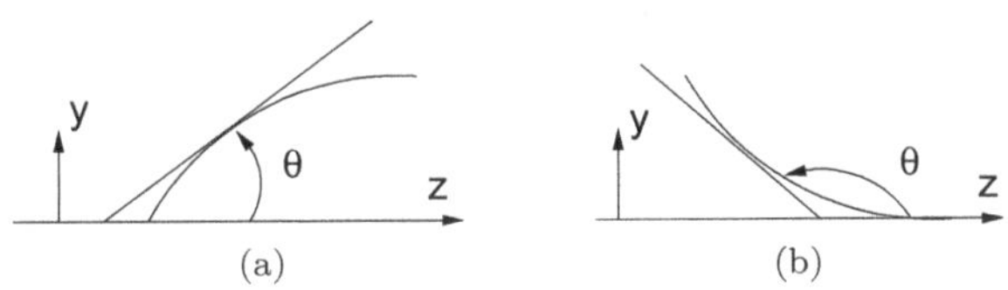


**Fig. 4.8.** Bending of lines of **E** for $z > 0$, to right of sphere.

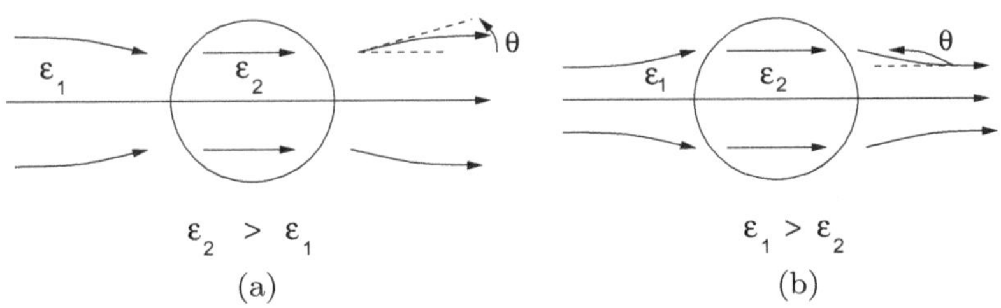


**Fig. 4.9.** Lines of $\mathbf{E}_a$ with deflection $\theta$ in region $z > 0$.

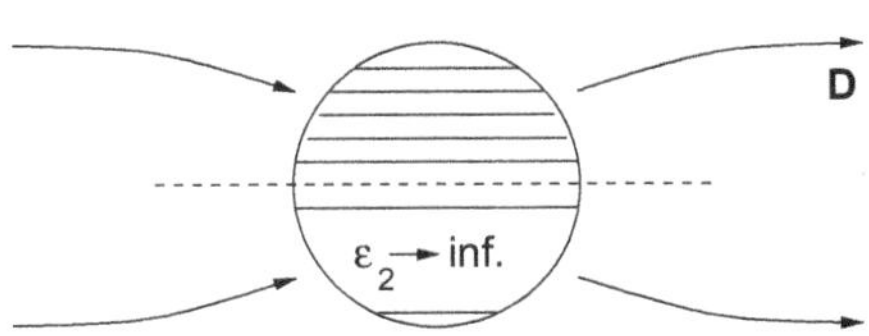


**Fig. 4.10.** Lines of $\mathbf{D} = \epsilon\mathbf{E}$.

*The limiting case* $\epsilon_2 \lll \epsilon_1$: In this case we have from Eq. (4.75)

$$E_i = \frac{3\epsilon_1}{\epsilon_2 + 2\epsilon_1} E_0 \sim \frac{3}{2} E_0, \tag{4.79}$$

but $D_i = \epsilon_2 E_i$, so that $D_i = (3/2)\epsilon_2 E_0$ is small for $\epsilon_2 \to 0$. This means, in this case the $D$-field is expelled from the sphere, as indicated in Fig. 4.11. This is the electric analogy to expulsion of the magnetic field from a superconductor[8] in magnetism.

Finally we ask: *How do lines of* $\mathbf{E}$ *differ from lines of* $\mathbf{D}$? $D$-lines cannot be created at the boundary since $\boldsymbol{\nabla} \cdot \mathbf{D} = 0$ (no external charges). In the interior as well as the exterior of the sphere, $\mathbf{P}$ is parallel to $\mathbf{E}$, if the dielectric is isotropic, i.e. $\mathbf{P} = \chi \mathbf{E}$ as assumed above. In the *nonisotropic case* this does not apply. Then the polarization charges $\rho_P$ are additional sources for lines of $\mathbf{E}$ as indicated in Fig. 4.12, because then

$$0 = \boldsymbol{\nabla} \cdot \mathbf{D} = \epsilon_0 \boldsymbol{\nabla} \cdot \mathbf{E} + \boldsymbol{\nabla} \cdot \mathbf{P} = \epsilon_0 \boldsymbol{\nabla} \cdot \mathbf{E} - \rho_P,$$

i.e.

$$\boldsymbol{\nabla} \cdot \mathbf{E} = \frac{1}{\epsilon_0} \rho_P. \tag{4.80}$$

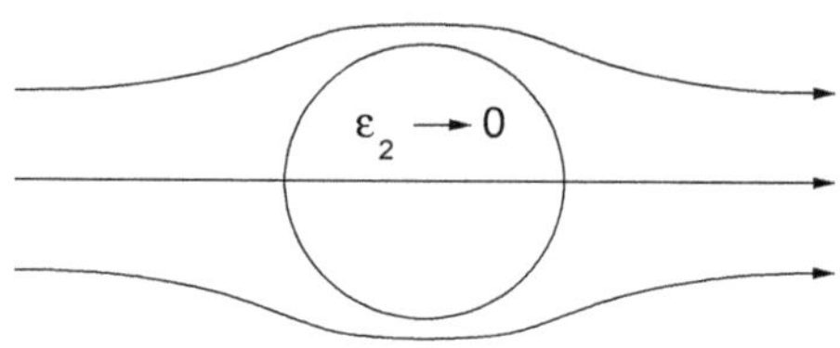


**Fig. 4.11.** Expulsion of lines of $\mathbf{D}$ for $\epsilon_2 \to 0$.

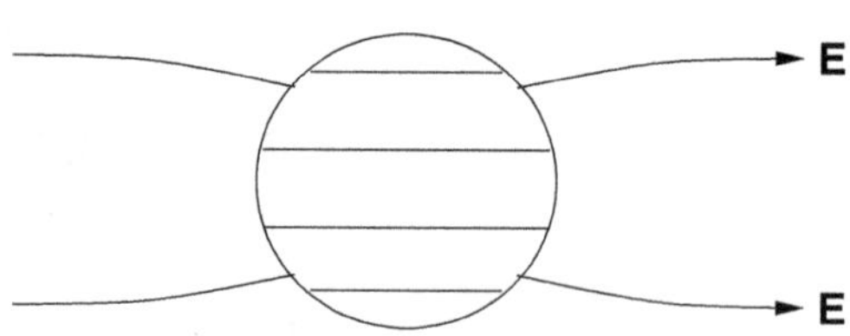


**Fig. 4.12.** *Additional* lines of $\mathbf{E}$ in the presence of polarization charges.

[8]Knowledge of elementary classical physics does not suffice to understand the phenomenon of superconductivity; knowledge of quantum statistics is required. A paper which explains the phenomenon with this but without heavy mathematical machinery is that of Weisskopf [161], who is wellknown for his deep insight into physics.

**Example 4.5 (Polarization of a cylinder in a homogeneous field $\mathbf{E}_0$).** Determine the field inside a cylinder of radius $r = a$ and with axis along the $x$-axis in analogy to the case of the sphere above but in terms of cylindrical coordinates.

**Solution:** The potential $\varphi$ outside and far away ($r \gg a$) is

$$\varphi_a \to -E_0 z, \quad z = r\cos\theta, \tag{4.81}$$

and we have $\triangle\varphi = 0$, except where there are charges. For the potentials outside and inside we set respectively (see the comment on *Kelvin's theorem* after Eq. (4.67))

$$\varphi_a = -\left(E_0 r + \frac{B}{r}\right)\cos\theta, \tag{4.82}$$

and

$$\varphi_i = -E_i z. \tag{4.83}$$

At $r = a$ we must have $\varphi_a = \varphi_i$, i.e.

$$E_0 a + \frac{B}{a} = E_i a, \tag{4.84}$$

and continuity of $\mathbf{D}\cdot\mathbf{n} = \epsilon E_r = -\epsilon\partial\varphi/\partial r$. Hence

$$\epsilon_1 \frac{\partial\varphi_a}{\partial r}\bigg|_{r=a} = \epsilon_2 \frac{\partial\varphi_i}{\partial r}\bigg|_{r=a}, \quad \epsilon_1\left(E_0 - \frac{B}{a^2}\right) = \epsilon_2 E_i. \tag{4.85}$$

The two Eqs. (4.84) and (4.85) can be solved for $B$ and $E_i$ giving

$$B = \left(\frac{\epsilon_1 - \epsilon_2}{\epsilon_1 + \epsilon_2}\right) a^2 E_0, \quad E_i = \left(\frac{2\epsilon_1}{\epsilon_1 + \epsilon_2}\right) E_0. \tag{4.86}$$

With insertion of these expressions into Eqs. (4.82) and (4.83) the potentials outside and inside are respectively given by

$$\begin{aligned} \varphi_a &= -E_0 r\cos\theta + \left(\frac{\epsilon_2 - \epsilon_1}{\epsilon_2 + \epsilon_1}\right) a^2 \frac{\cos\theta}{r} E_0, \quad r > a, \\ \varphi_i &= -\frac{2\epsilon_1}{\epsilon_1 + \epsilon_2} E_0 r\cos\theta, \quad r < a. \end{aligned} \tag{4.87}$$

An application of this result for $E_i$ can be found in Example 4.6.

## 4.5 Energy of the Electric Field

### 4.5.1 *The energy density formula*

In accordance with the definition of the electrostatic potential (2.9) or (3.3) in the whole of space, the potential *energy* of a charge of $q$ coulombs in the potential $\phi$ volts is $W = q\phi$ joules. In the case of a number of discrete charges the energy is, as we saw in Section 2.10,

$$W = \frac{1}{2}\sum_i q_i\phi_i, \quad \phi_i = \sum_{q_k \neq q_i} k\frac{q_k}{|\mathbf{r}_i - \mathbf{r}_k|} \quad \text{joules/coulomb.} \tag{4.88}$$

In the case of continuous charge distributions we have

$$W = \frac{1}{2}\int \phi(\mathbf{r})\rho(\mathbf{r})d\mathbf{r} \quad \text{joules.} \tag{4.89}$$

Since $\boldsymbol{\nabla}\cdot\mathbf{D} = \rho$, we have

$$\begin{aligned} W &\stackrel{(2.178)}{=} \frac{1}{2}\int \phi(\mathbf{r})\boldsymbol{\nabla}\cdot\mathbf{D}d\mathbf{r} \\ &= \frac{1}{2}\left\{\int \boldsymbol{\nabla}\cdot(\phi\mathbf{D})d\mathbf{r} - \int \mathbf{D}\cdot\boldsymbol{\nabla}\phi d\mathbf{r}\right\} \\ &= \frac{1}{2}\left\{\int_F \phi\mathbf{D}\cdot d\mathbf{F} + \int \mathbf{D}\cdot\mathbf{E}d\mathbf{r}\right\} \\ &= \frac{1}{2}\int \mathbf{D}\cdot\mathbf{E}d\mathbf{r} \quad \text{joules.} \end{aligned} \tag{4.90}$$

To obtain the total energy, and in view of the concept of a field, we have to integrate over all space. Then the surface integral yields a contribution which tends to zero.

Recall now that in Section 3.6.2 we considered the situation in which boundary conditions are also given, such as those provided by conducting surfaces (recall e.g. Eq. (3.46)). We want to show now that the total energy of the electrostatic field is

$$\frac{1}{2}\int \mathbf{D}\cdot\mathbf{E}d\mathbf{r},$$

*even if the space encloses conductors*, provided the volume of integration is only that of the dielectric. This result can be deduced as follows. We consider the potential energy of a dielectric with enclosed conductors of potentials $\phi_n$

which enclose charges $q_n$. We have with

$$q_n = \int_{V_n} \nabla \cdot \mathbf{D} d\mathbf{r} = \int_{\mathbf{F}_n^*} \mathbf{D}_n \cdot d\mathbf{F}^*, \quad d\mathbf{r} \equiv dV, \tag{4.91}$$

and $d\mathbf{F}^*{}_n = -d\mathbf{F}_n$, as can be seen from Fig. 4.13. Then the potential energy contained in the dielectric is given by

$$\begin{aligned} W &= \frac{1}{2}\int_{V_{\text{diel}}} \phi\rho dV + \frac{1}{2}\sum_n \phi_n q_n \\ &= \frac{1}{2}\int_{V_{\text{diel}}} \phi\nabla \cdot \mathbf{D} dV + \frac{1}{2}\sum_n \phi_n \int_{\mathbf{F}_n^*} \mathbf{D}_n \cdot d\mathbf{F}_{\mathbf{n}}^* \\ &= \frac{1}{2}\left[\int_{V_{\text{diel}}} \nabla \cdot (\phi\mathbf{D}) dV - \int_{V_{\text{diel}}} \mathbf{D} \cdot \nabla\phi dV \right. \\ &\quad \left. + \sum_n \phi_n \int_{\mathbf{F}_n^*} \mathbf{D}_n \cdot d\mathbf{F}^*{}_n \right] \quad \text{joules.} \end{aligned} \tag{4.92}$$

This relation can now be rewritten as (one may note in the second step the contributions of $F_{\text{diel}}$ from inside and outside boundaries of the dielectric)

$$\begin{aligned} W &= \frac{1}{2}\left[\int_{F_{\text{diel}}} \phi\mathbf{D} \cdot d\mathbf{F} + \int_{V_{\text{diel}}} \mathbf{D} \cdot \mathbf{E} dV - \sum_n \phi_n \int_{F_n} \mathbf{D}_n \cdot d\mathbf{F}_n \right] \\ &= \frac{1}{2}\left[\int_{F_\infty} \phi\mathbf{D} \cdot d\mathbf{F} + \sum_n \int_{F_n} \phi_n \mathbf{D}_n \cdot d\mathbf{F}_n \right. \\ &\quad \left. + \int_{V_{\text{diel}}} \mathbf{D} \cdot \mathbf{E} dV - \sum_n \phi_n \int_{F_n} \mathbf{D}_n \cdot d\mathbf{F}_n \right] \\ &= \frac{1}{2}\int_{V_{\text{diel}}} \mathbf{D} \cdot \mathbf{E} dV \quad \text{joules.} \end{aligned} \tag{4.93}$$

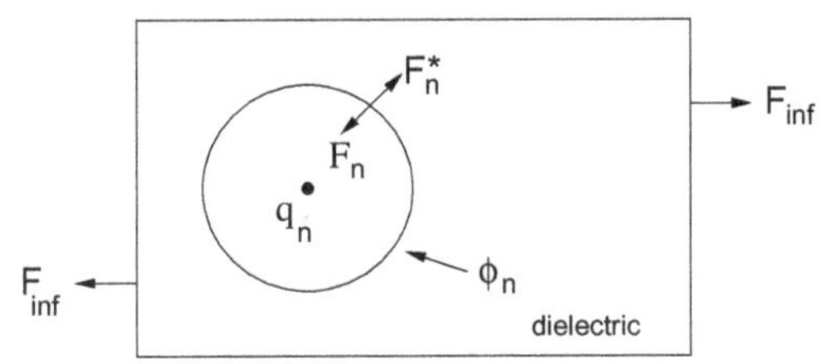


**Fig. 4.13.** A dielectric with conductor cavities.

Thus the energy contained in the dielectric is the integral of the density $\mathbf{D} \cdot \mathbf{E}/2$ over the volume of the dielectric.

### 4.5.2 *Polarization energy*

Finally we consider the change in energy when a vacuum is replaced by a medium. Let $\triangle_0 W$ be this change from vacuum values $\mathbf{D}_0 = \epsilon_0 \mathbf{E}_0$ and $\mathbf{E}_0$ to medium values $\mathbf{D} = \epsilon_0 \mathbf{E} + \mathbf{P}$ and $\mathbf{E}$. Then

$$\triangle_0 W = \frac{1}{2} \int d\mathbf{r}[\mathbf{D} \cdot \mathbf{E} - \mathbf{D}_0 \cdot \mathbf{E}_0]. \tag{4.94}$$

We set $\mathbf{D} = \mathbf{D}_0 + \delta \mathbf{D}$, so that

$$\triangle_0 W = \frac{1}{2} \int d\mathbf{r}[\mathbf{D}_0 \cdot \mathbf{E} - \mathbf{D} \cdot \mathbf{E}_0] + \frac{1}{2} \int d\mathbf{r}(\mathbf{E} + \mathbf{E}_0) \cdot \delta \mathbf{D}. \tag{4.95}$$

Here $(\mathbf{E} + \mathbf{E}_0) = -\boldsymbol{\nabla} \phi'$, where $\phi'$ is a potential. Using the relation

"div(scalar × vector) = grad(scalar)·(vector) + (scalar) div(vector)",

we have (with $\boldsymbol{\nabla} \cdot \delta \mathbf{D} = \delta(\boldsymbol{\nabla} \cdot \mathbf{D}) = \delta \rho$)

$$-\boldsymbol{\nabla} \phi' \cdot \delta \mathbf{D} = -\boldsymbol{\nabla} \cdot (\phi' \delta \mathbf{D}) + \phi' \boldsymbol{\nabla} \cdot \delta \mathbf{D}. \tag{4.96}$$

With Gauss' divergence theorem the first term integrated over the volume becomes a surface integral which vanishes. The second contribution vanishes if no new charge $\delta \rho$ is introduced. Hence the second integral in Eq. (4.95) is zero and we obtain the following result (which will be used in the next example):

$$\begin{aligned} \triangle_0 W &= \frac{1}{2} \int d\mathbf{r}[\mathbf{D}_0 \cdot \mathbf{E} - \mathbf{D} \cdot \mathbf{E}_0] \\ &= \frac{1}{2} \int d\mathbf{r}[\epsilon_0 \mathbf{E}_0 \cdot \mathbf{E} - (\epsilon_0 \mathbf{E} + \mathbf{P}) \cdot \mathbf{E}_0] \\ &= -\frac{1}{2} \int d\mathbf{r}[\mathbf{P} \cdot \mathbf{E}_0] \quad \text{joules}. \end{aligned} \tag{4.97}$$

**Example 4.6 (Measurement of electric susceptibility[9]).** Consider a cylindrical U-tube filled with some liquid ($l$) to the same level in both arms in the original equilibrium position with vapour ($v$) above. A homogeneous electric field $E_0$ is applied to one arm of the U-tube. In the presence of the field the liquid rises to a height $h$ above the equilibrium position.

[9] This is also problem 15, p. 100, of Robinson [124] (there unsolved).

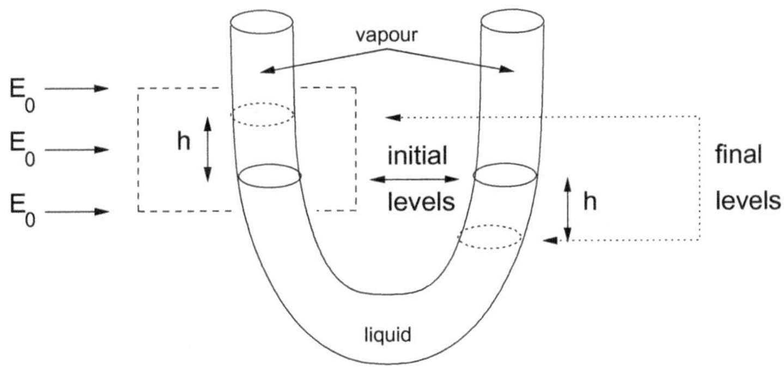


**Fig. 4.14.** The U-tube with one arm in field $E_0$.

Considering this arrangement as a cylindrical dielectric exposed to an external field $E_0$, find an expression from which the electric susceptibility of the liquid ($\chi_l = \epsilon_l/\epsilon_0 - 1$) can be obtained if that of the vapour ($\chi_v$) is known.

**Solution:** Let $F$ be the constant cross section of the tube and $\rho$ the density of the liquid, and $g$ the acceleration due to gravity. The hydrostatic force producing the rise of the liquid to height $h$ above the original equilibrium position against the vapour is, see Fig. 4.14,

$$\rho F(2h)g \quad \text{newtons},$$

and the corresponding energy is

$$\int_0^h \rho F(2h)g dh = \rho g F h^2 \quad \text{joules}. \tag{4.98}$$

The change in electrostatic energy is — using Eq. (4.97) — the change in polarization energy

$$\begin{aligned}\triangle_0 W_l - \triangle_0 W_v &= -\frac{1}{2}\int d\mathbf{r}(P_l - P_v)E_0 \\ &= -\frac{1}{2}\int_0^h (F dh)(P_l - P_v)E_0 \\ &= -\frac{1}{2}Fh(P_l - P_v)E_0 \quad \text{joules}.\end{aligned} \tag{4.99}$$

The total change in energy is therefore

$$W_{\text{total}} = \rho g F h^2 - \frac{1}{2}Fh(P_l - P_v)E_0 \quad \text{joules}. \tag{4.100}$$

The condition for equilibrium is $\partial W_{\text{total}}/\partial h = 0$, i.e.

$$h = \frac{(P_l - P_v)E_0}{4\rho g} \quad \text{meters.} \tag{4.101}$$

Now, for $i = l, v$ we have $P_i = \epsilon_0 \chi_i E_i$, $\chi_i = \epsilon_i/\epsilon_0 - 1$, where $E_i$ is the field in the medium $i = l, v$. From Example 4.5 we obtain for the cylindrical media $l, v$, cf. Eq. (4.86),

$$E_l = \frac{2\epsilon_0}{\epsilon_0 + \epsilon_l} E_0, \quad E_v = \frac{2\epsilon_0}{\epsilon_0 + \epsilon_v} E_0. \tag{4.102}$$

Then

$$P_i = \frac{2\epsilon_0(\epsilon_i - \epsilon_0)}{\epsilon_0 + \epsilon_i} E_0 = 2\epsilon_0 \left(1 - \frac{2\epsilon_0}{\epsilon_0 + \epsilon_i}\right) E_0. \tag{4.103}$$

Thus

$$h = \frac{2\epsilon_0 E_0^2}{4\rho g}\left[-\frac{2\epsilon_0}{\epsilon_0 + \epsilon_l} + \frac{2\epsilon_0}{\epsilon_0 + \epsilon_v}\right] = \frac{(\epsilon_0 E_0)^2}{\rho g} \frac{\epsilon_l - \epsilon_v}{(\epsilon_0 + \epsilon_l)(\epsilon_0 + \epsilon_v)} \quad \text{meters,} \tag{4.104}$$

which can be re-expressed as

$$h = \frac{\epsilon_0(\chi_l - \chi_v)}{\rho g(\chi_l + 2)(\chi_v + 2)} E_0^2 \quad \text{meters.} \tag{4.105}$$

From this we obtain

$$\chi_l = \frac{\epsilon_0 E_0^2 \chi_v + 2h\rho g(\chi_v + 2)}{\epsilon_0 E_0^2 - h\rho g(\chi_v + 2)}. \tag{4.106}$$

Hence by measuring $h$, and since the other quantities appearing on the right are assumed to be known, the susceptibility $\chi_l$ of the liquid can be determined.

## 4.6 Summary of Formulas of Macroelectrostatics

We summarize here the most important formulas of macroelectrostatics.

1. **Multipole expansion** (cf. Eqs. (3.171) and (3.172)):

$$\phi(\mathbf{r}) = k\frac{Q}{r} + k\frac{\mathbf{p}\cdot\mathbf{r}}{r^3} + \cdots \quad \text{volts,} \tag{4.107}$$

$$\mathbf{E} = -\boldsymbol{\nabla}\phi = -k\boldsymbol{\nabla}\left[\frac{Q}{|\mathbf{r}|} - \mathbf{p}\cdot\boldsymbol{\nabla}\frac{1}{|\mathbf{r}|} + \cdots\right] \quad \text{newtons/coulomb,} \tag{4.108}$$

but (note the signs of the dipole contributions)

$$\mathbf{E}(\mathbf{r}) = -k\boldsymbol{\nabla}_{\mathbf{r}}\sum_j \left[\frac{q_j}{|\mathbf{r}-\mathbf{r}_j|} + \mathbf{p_j}\cdot\boldsymbol{\nabla}_j\frac{1}{|\mathbf{r}|} + \cdots\right] \quad \text{newtons/coulomb,} \tag{4.109}$$

since (cf. Eqs. (2.6) and (2.53))

$$\boldsymbol{\nabla}_{\mathbf{r}}\frac{1}{|\mathbf{r}-\mathbf{r}'|} = -\frac{\mathbf{r}-\mathbf{r}'}{|\mathbf{r}-\mathbf{r}'|^3}, \quad \frac{\mathbf{r}}{r^3} = -\boldsymbol{\nabla}\frac{1}{|\mathbf{r}|} \tag{4.110}$$

and

$$\boldsymbol{\nabla}_{\mathbf{r}'}\frac{1}{|\mathbf{r}'-\mathbf{r}|} = -\frac{\mathbf{r}'-\mathbf{r}}{|\mathbf{r}-\mathbf{r}'|^3} = -\boldsymbol{\nabla}_{\mathbf{r}}\frac{1}{|\mathbf{r}-\mathbf{r}'|}. \tag{4.111}$$

2. **Macroscopic electrostatics** (cf. Eqs. (4.13), (4.14), (4.17) and (4.20)):

$$\boldsymbol{\nabla}\cdot\mathbf{D} = \rho \quad \text{coulombs/m}^3, \quad \mathbf{D} = \epsilon_0\mathbf{E} + \mathbf{P} \quad \text{coulombs/m}^2, \tag{4.112}$$

$$\mathbf{P}\cdot\mathbf{n} = \sigma_P, \quad \boldsymbol{\nabla}\cdot\mathbf{P} = -\rho_P, \tag{4.113}$$

$$\boldsymbol{\nabla}\times\mathbf{E} = 0, \quad (\mathbf{E} = -\boldsymbol{\nabla}\phi). \tag{4.114}$$

Note: We have to distinguish between $\rho_P, \sigma_P$ und $\rho, \sigma$. For homogeneous, linearly isotropic dielectrics the following **dielectric material equations** can be used:

$$\mathbf{D} = \epsilon\mathbf{E} \quad \text{coulombs/m}^2, \quad \mathbf{P} = \epsilon_0\chi\mathbf{E} \quad \text{coulombs/m}^2. \tag{4.115}$$

3. **The behavior at boundary surfaces** (cf. Eqs. (4.40) and (4.42)):

$$\epsilon_0(\mathbf{E}_1 - \mathbf{E}_2)\cdot\mathbf{n} = \sigma + \sigma_P, \quad (\mathbf{D}_1 - \mathbf{D}_2)\cdot\mathbf{n} = \sigma. \tag{4.116}$$

But always for tangential components of $\mathbf{E}$ (cf. Eq. (4.44)):

$$E_{1t} - E_{2t} = 0. \tag{4.117}$$

For $\mathbf{P} \neq 0$ there is no corresponding relation for $\mathbf{D}$.

4. **Energy of the electric field** (cf. Eqs. (4.90) and (4.93)):
The total energy of the electrostatic field is

$$W = \frac{1}{2}\int \mathbf{D}\cdot\mathbf{E}dV \quad \text{joules.} \tag{4.118}$$

## 4.7 Exercises

**Example 4.7 (Classical radius of the electron).** Consider the electron as having charge $e$ distributed over a sphere of radius $R$, the field $E$ at a point $r$ outside the sphere will be $e/4\pi\epsilon_0 r^2$. From this the total electrical energy is the volume integral over all space outside the sphere of radius $R$. Evaluate this. Then neglecting nonelectromagnetic corrections equate the result to $m_e c^2$ and solve for $R$. Evaluate the result to obtain a numerical value.

(Answer: $R = e^2/4\pi\epsilon_0 m_e c^2 = 2.84 \times 10^{-15}$ meters).[10]

**Example 4.8 (Total field of induced dipoles at a point).** Taking from the expression (4.26) the contribution of a single dipole, consider the total field of the induced molecular dipoles at the center of a sphere in the direction of the electric field (say the $z$-direction) by evaluating the sum over the field contributions of the large number of molecules within the sphere as averages (e.g. take the average of $z^2$ as $r^2/3, r^2 = x^2 + y^2 + z^2$).

(Answer: Zero).

**Example 4.9 (The Lorentz local field).** The electric field $E$ in a parallel plate condenser was shown to be $\sigma/\epsilon_0$ (cf. Eq. (2.99)). This is the contribution resulting from the charge on the external surfaces and the polarization of the material. However, there is another contribution from the molecular dipoles around the point under consideration. Show that the average of this other local contribution is[11] $P/3\epsilon_0$, where $\mathbf{P}$ is the polarization vector defined in Section 4.2. Thus the Lorentz local field is

$$E = \frac{\sigma}{\epsilon_0} + \frac{P}{3\epsilon_0} \quad \text{newtons/coulomb.} \tag{4.119}$$

**Example 4.10 (Kelvin's theorem in cylindrical coordinates).** Consider Laplace's equation in cylindrical coordinates. Show that for no change in the z-direction ($z = z_0$ fixed) but certain angular dependences of the solution $\phi(z, r, \theta)$, e.g. $\phi(z, r, \theta) \propto \cos\theta$, for which the right hand side of the equation

$$\frac{r}{\phi}\frac{\partial}{\partial r}\left(r\frac{\partial \phi}{\partial r}\right) = -\frac{1}{\phi}\frac{\partial^2 \phi}{\partial \theta^2} \tag{4.120}$$

[10]See Example 10.9 and Jackson [73], p. 681 (there in Gaussian units).
[11]For an elementary calculation see e.g. Rossiter [126], pp. 35–38.

is fixed, being independent of any other variations in $\phi(z,r,\theta)$, that if one known solution has $\phi(z,r,\theta) \propto r^n$, there is another solution $\phi(z,r,\theta) \propto r^{-n}$.

**Example 4.11 (Kelvin's theorem in spherical polar coordinates).** Consider Laplace's equation in spherical polar coordinates $r, \theta, \varphi$. Show that for certain angular dependences (e.g. $\phi \propto \cos\theta \sin\varphi$) for which the right hand side of the equation

$$\frac{1}{\phi}\frac{\partial}{\partial r}\left(r^2\frac{\partial\phi}{\partial r}\right) = -\frac{1}{\phi\sin\theta}\frac{\partial}{\partial\theta}\left(\sin\theta\frac{\partial\phi}{\partial\theta}\right) - \frac{1}{\phi\sin^2\theta}\frac{\partial^2\phi}{\partial\varphi^2} \tag{4.121}$$

is fixed, so that the left hand side remains fixed, that if one known solution has $\phi \propto r^n$, there is another solution $\phi$ with the same angular dependence but with $\phi \propto r^{-(n+1)}$.

**Example 4.12 (Dielectric filling of condenser).** Recall from Example 2.9 the relation

$$V = \phi_1 - \phi_2 = \int_{l_1}^{l_2} \mathbf{E}\cdot d\mathbf{l} \quad \text{volts} \tag{4.122}$$

for the potential difference between condenser plates 1 and 2, and that $\mathbf{E}$ is normal to the plates and of magnitude $\sigma/\epsilon_0$ in a vacuum. Now assume that a parallel slab of a dielectric with permittivity $\epsilon'$ and thickness $d < D$ is inserted into a parallel plate condenser with plate separation $D$ and plate charge $\sigma A$, $A$ being the area of each of the two plates. What is the capacity of this condenser? Examine the limits $d \to 0, d \to D$.

**Example 4.13 (Storage of electrical energy in a parallel plate condenser).** The conducting plates of a parallel plate condenser are a distance 0.1 meter apart. A voltage of 200 volts is applied. Obtain the electric field intensity $\mathbf{E}$ and the electric flux density $\mathbf{D}$. What are the values of these quantities if a material of permittivity $\epsilon = 5\epsilon_0$ is inserted between the plates (a) before and (b) after the voltage is disconnected?

(Answers: 2000 V/m, $17.7 \times 10^{-3}$ μC/m$^2$,
(a) 2000 V/m, $88.5 \times 10^{-3}$ C/m$^2$,
(b) 400 V/m, $17.7 \times 10^{-3}$ μC/m$^2$).

**Example 4.14 (Electric field inside real spherical hole).** Let the electric field at a large distance $r$ from a spherical hole of radius $a$ be $E_0$ in the $z$-direction (as in Section 4.4). Assume as solutions of Laplace's equation (with the same angular dependence)

$$\varphi_a = -E_0 r\cos\theta - \frac{B}{r^2}\cos\theta, \quad \varphi_i = -Ar\cos\theta. \tag{4.123}$$

Determine the constants $A$ and $B$ from the conditions (i) $\varphi$ continuous at $r = a$, and (ii) $D_{\rm normal} = -\epsilon(\partial\varphi/\partial r)$ continuous at $r = a$. Show that if a *real* spherical hole (inside $\epsilon_2 = \epsilon_0$) is cut in a medium of dielectric constant $\epsilon_1 = \epsilon$, then the electric field within the hole is

$$E_i = E_0\left(1 + \frac{\epsilon - 1}{2\epsilon + 1}\right) \quad \text{N/C}. \tag{4.124}$$

**Example 4.15 (Electric field inside an imaginary spherical hole).** Consider the electric field acting on a molecule in a medium of dielectric constant $\epsilon$. In this case it is necessary to find the contributions to the electric field arising from the region exterior to a spherical surface. The spherical hole is in this consideration purely a figment of the imagination, and the lines of force and induction are undeviated, and the field is everywhere $E_0$. What is required is the internal field of the *imaginary* spherical hole arising from the region exterior to it. Proceed as follows. Since with field $E_0$ we have $D_0 := \epsilon E_0 = \epsilon_0 E_0 + P$, the electric dipole moment per unit volume is $P = (\epsilon - \epsilon_0)E_0$. This polarization of the medium leaves unbalanced the charges on the surface of the imaginary sphere. These charges are of magnitude $\sigma = P\cos\theta$ per unit area of the sphere. Calculate, by integrating over the surface of the imaginary sphere, the extra field $E'$ due to these charges (resolved along the direction of $E_0$ as the rest cancels).

(Answer: $E_i = E_0 + E' = [1 + (\epsilon - \epsilon_0)/3\epsilon_0]E_0$ N/C).

Comment: The field acting on a molecule is the sum of (i) the above from the region exterior to the imaginary hole of sufficient radius so that the molecular structure can be treated as a continuum, and (ii) the field from individual molecules within the imaginary hole. In the case of molecules arranged in a cubic lattice, or randomly, this contribution vanishes by symmetry and the total field is that of (i). However, as argued by Onsager,[12] the above reasoning (due to Lorentz and Debye) is usually not valid as one normally requires to know the field acting on a molecule when this has a given orientation which

[12]Onsager [107]. For discussion see Bleaney and Bleaney [12], p. 312.

conflicts with the above. Thus it is usually better to take the field inside as that of Example 4.14.

**Example 4.16 (Field of uniformly polarized sphere).** A sphere of radius $a$ and made of dielectric material is polarized along the $z$-axis, i.e. $\mathbf{P} = P\mathbf{e}_z$. Show that the internal electric field is uniform and given by $-P/3\epsilon_0$ newtons/coulomb, and that the field outside is that of a dipole of moment $4\pi Pa^3/3$ coulomb · meters placed at the center of the sphere.[13]

[13]See Lawden [82], problem 2.1.

## Chapter 5

# Magnetostatics

## 5.1 Introductory Remarks

Magnetostatics deals with *stationary currents.* So far, we considered only static charges. A current, however, consists of moving charges. The fundamental observation, that magnetic fields exist in the neighborhood of currents, and hence of moving charges, was made by Oersted in 1819. Oersted had observed that a magnetic needle aligns itself in the vicinity of a current carrying wire perpendicular to this wire. Within a very short time after this discovery (i.e. within a few years), Ampère published his results of a series of experiments which established the law of the force today named after him, and Biot and Savart observed the corresponding law (for a current element) that carries their names today. We are concerned with these laws in this chapter.[1]

## 5.2 Continuity Equation and Stationary Current

Here, by *current density* we mean only that of the conduction electrons; we do not mean the current density of (classically considered) circular currents of molecular electrons (which, by the way, are not even sufficiently well known). First, we consider the case without matter effects. A precisely known current density **j** is defined by the charge density of positive charges multiplied by their velocity, i.e.

$$\mathbf{j} = \rho \frac{d\mathbf{s}}{dt} \quad \mathrm{A/m^2} \tag{5.1}$$

[1]For a brief and lucid synopsis of the historical development, see Attwood [5], pp. 455–463.

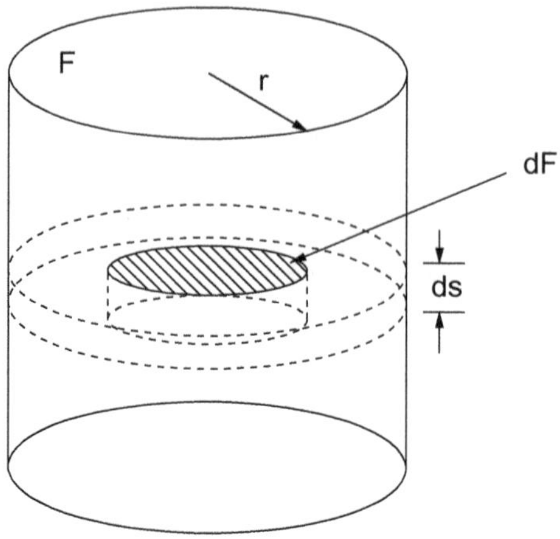


**Fig. 5.1.** The current element.

in ampere/m$^2$ or C/s · m$^2$. We use $d\mathbf{s}$ for an element of the trajectory or path as indicated in Fig. 5.1, in order not to confuse this with the element of some distance, $dr$, or the volume element $d\mathbf{r}$. As usual, $t$ represents the time variable.

The *current* $I$ is defined as that amount of positive[2] charge which passes through an area $F$ per unit time, i.e. ($V$ meaning volume)

$$\begin{aligned} I &= \int_F \mathbf{j} \cdot d\mathbf{F} = \int_F \rho \frac{d\mathbf{s} \cdot d\mathbf{F}}{dt} \\ &= \rho \frac{d\mathbf{s}}{dt} \cdot \mathbf{F} = \rho \frac{dV}{dt} = \frac{dq}{dt} \qquad \text{A or C/s}, \end{aligned} \tag{5.2}$$

if

$$\frac{d\rho}{dt} = 0 \text{ and } q = \rho V \tag{5.3}$$

(the latter is, in fact, the integrated form of the equation of continuity to be discussed in the following and implies that the number of charges per unit volume and per unit time is constant — as much charge as flows in also flows out). With the help of the previous equation, we also have

$$I d\mathbf{s} = \rho \frac{dV}{dt} d\mathbf{s} = \mathbf{j} dV = (\rho dV) \frac{d\mathbf{s}}{dt} = dq \frac{d\mathbf{s}}{dt}. \tag{5.4}$$

In the following, we also write $\dot{\mathbf{s}} = d\mathbf{s}/dt$. In addition, we have — these manipulations serve the purpose of making the definition of a *surface current*

[2]For an explanation, see, for instance, Jackson [73], p. 169.

plausible,

$$dq\frac{d\mathbf{s}}{dt} = \frac{dq/dt}{2\pi r}2\pi r d\mathbf{s} = \frac{\mathbf{I}(2\pi r ds)}{2\pi r} = \frac{\mathbf{I}}{2\pi r}dF_{\text{cyl}}, \tag{5.5}$$

where[3]

$$\mathbf{K} \equiv \frac{\mathbf{I}}{2\pi r} = \frac{dq}{dF_{\text{cyl}}}\dot{\mathbf{s}} = \sigma\mathbf{v} \quad \text{A/m} \tag{5.6}$$

is a *surface current*, also called layer of current, and $\sigma$ charge per unit area. Hence, we can write

$$\mathbf{I}ds = \mathbf{j}dV = \mathbf{K}dF \quad \text{A} \cdot \text{m}, \tag{5.7}$$

where $\mathbf{I}$ represents the *line or linear current*, $\mathbf{j}$ the *current density*, and $\mathbf{K}$ the *surface current* (density). We now construct the divergence:

$$\boldsymbol{\nabla} \cdot \mathbf{j} = \boldsymbol{\nabla} \cdot (\rho\dot{\mathbf{s}}) = \rho\boldsymbol{\nabla} \cdot \dot{\mathbf{s}} + (\boldsymbol{\nabla}\rho) \cdot \dot{\mathbf{s}}. \tag{5.8}$$

But

$$\boldsymbol{\nabla} \cdot \dot{\mathbf{s}}\Big|_{\mathbf{s}\equiv\mathbf{r}} = \boldsymbol{\nabla} \cdot \dot{\mathbf{r}} = \frac{d}{dt}\boldsymbol{\nabla} \cdot \mathbf{r} = \frac{d}{dt}3 = 0. \tag{5.9}$$

This condition, the vanishing of the divergence of a velocity, is known as the condition of *incompressibility* (as in hydrodynamics). We obtain with this

$$\boldsymbol{\nabla} \cdot \mathbf{j} = (\boldsymbol{\nabla}\rho) \cdot \dot{\mathbf{s}}. \tag{5.10}$$

In general, one has

$$\frac{d\rho(\mathbf{s},t)}{dt} = \frac{\partial\rho}{\partial t} + (\boldsymbol{\nabla}\rho) \cdot \dot{\mathbf{s}} = \frac{\partial\rho}{\partial t} + \boldsymbol{\nabla} \cdot \mathbf{j}. \tag{5.11}$$

Since (see Eq. (5.3)) $d\rho/dt = 0$, it follows that

$$0 = \frac{\partial\rho}{\partial t} + \boldsymbol{\nabla} \cdot \mathbf{j}. \tag{5.12}$$

This important equation is known as *equation of continuity*. The equation plays an important role in many branches of physics, including quantum mechanics and statistical mechanics. The condition $d\rho/dt = 0$ implies that in every time interval the number of charges in every unit volume remains constant. This again implies, since charges move in and out, *that no charges are created or annihilated*. The significance of the equation of continuity is therefore that as much charge as flows into a volume element per unit time

[3]Surface currents will be discussed in Section 11.2.

also flows out of this volume element again per unit time. Thus, the amount of charge in every volume element remains constant, or — as one says — is conserved. Assume now that in addition $\partial\rho/\partial t = 0$. This condition says that the density does not change explicitly with time. This condition is described as the condition of *stationarity* or (specifically in statistical mechanics) of *equilibrium*. We thus have as condition for a *stationary* or *steady current*

$$\boldsymbol{\nabla}\cdot\mathbf{j} = 0. \tag{5.13}$$

Stationarity is attained only after some time interval, e.g. after a condenser has been charged. The non-stationarity in between is described as a *transient* effect. Thus, $\partial\rho/\partial t = 0$ does not apply immediately at the time of charging or discharging of a condenser.

**Example 5.1 (The relaxation time of a dielectric).** The homogeneous dielectric of a condenser has a weak conductivity $\sigma$ ohm$^{-1}$meter$^{-1}$. Calculate with the help of the equation of continuity and Ohm's law in the form $\mathbf{j} = \sigma\mathbf{E}$ amperes/m$^2$ (cf. Section 6.1) the relaxation time of the dielectric (dielectric constant $\epsilon$), i.e. the time interval in the course of a discharging of the condenser, in which the charge density drops to the fraction $1/e$ of its original value.

**Solution:** We have from Eq. (5.12)

$$\boldsymbol{\nabla}\cdot\mathbf{j} + \frac{\partial\rho}{\partial t} = 0, \quad \mathbf{j} = \sigma\mathbf{E} \quad \text{amperes/m}^2,$$

from which we obtain

$$\sigma\boldsymbol{\nabla}\cdot\mathbf{E} + \frac{\partial\rho}{\partial t} = 0. \tag{5.14}$$

Since the dielectric is homogeneous, i.e. isotropic, we have $\mathbf{D} = \epsilon\mathbf{E}$, $\epsilon =$ const., and from Eq. (4.37), i.e. $\boldsymbol{\nabla}\cdot\mathbf{D} = \rho$, we obtain $\boldsymbol{\nabla}\cdot\mathbf{E} = \rho/\epsilon$. With this relation and the preceding equation, we obtain

$$\frac{\partial\rho}{\partial t} = -\frac{\sigma\rho}{\epsilon}, \quad \text{i.e.} \quad \rho = \rho_0 e^{-\sigma t/\epsilon} \quad \text{coulombs/m}^3, \tag{5.15}$$

where at time $t = 0$ we have $\rho = \rho_0$. The relaxation time is therefore $\epsilon/\sigma$. This expression is large for small $\sigma$. We observe here incidentally that in the Gaussian system of units with $\epsilon_0 = 1$ and dimensionless the conductivity is given in seconds$^{-1}$.

In much the same way as one can produce a steady flow of water with the help of a pump, one can use a battery in a closed circuit to produce a

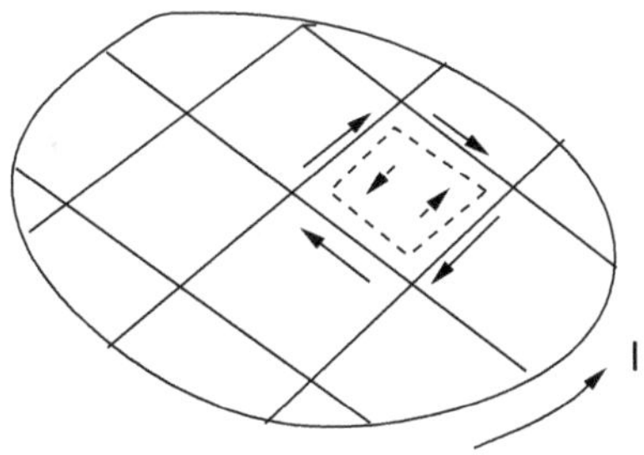


**Fig. 5.2.** The net of conducting wires.

steady or stationary current. The area enclosed by this circuit is of particular significance for its magnetic properties. We can imagine the addition of further conducting wires as illustrated in Fig. 5.2 until the entire area is filled with a net of these covering the entire area of the original circuit which then appears as the boundary of the net. The stationary current is flowing in this boundary; whereas, currents in the net segments inside all cancel each other. It is because the internal currents all cancel each other (no matter what the size of an element of the net may be) that the net has the same magnetic properties as the original circuit.

## 5.3 Ampère's Experiments and the Law of Biot and Savart

In electrostatics, we saw that Coulomb and Cavendish had discovered the law of force $\propto 1/r^2$ experimentally, but that the proportionality of the force to the charges was more or less guessed. In the case of magnetostatics, Ampère discovered that the force acting between current carrying circuits which had been arranged in a particularly symmetric fashion is similarly proportional to $1/r^2$, where $r$ is the distance between the circuits, and he deduced from this that this behavior is of a general kind.

Ampère was largely concerned with the study of *forces* due to current-carrying wires. In *one experiment*, he demonstrated that two wires close to each other and carrying equal but opposite currents do not exert a force on some distant conducting wire. Today, we know that the magnetic fields annul each other in this case, as illustrated in Fig. 5.3. In a *second experiment*, Ampère demonstrated, as far as the effect on other current-carrying conductors was concerned, that arrangements of conductors as in Fig. 5.4 have the same effect (i.e. the same field). In a *third experiment*, Ampère showed that the forces stemming from magnets or currents which act on a wire having the shape of a circular arc carrying the current $I$

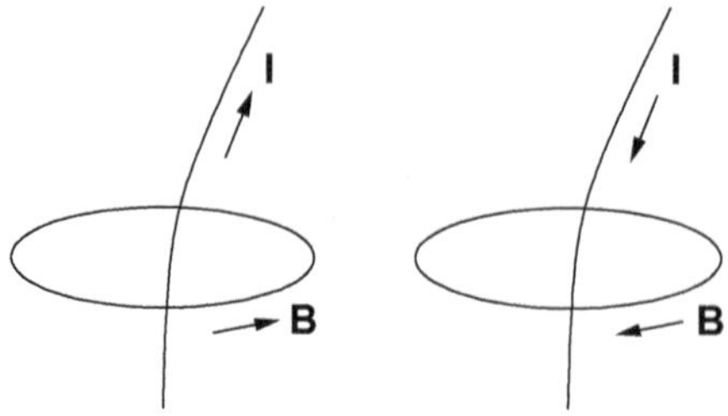


**Fig. 5.3.** The first experiment of Ampère.

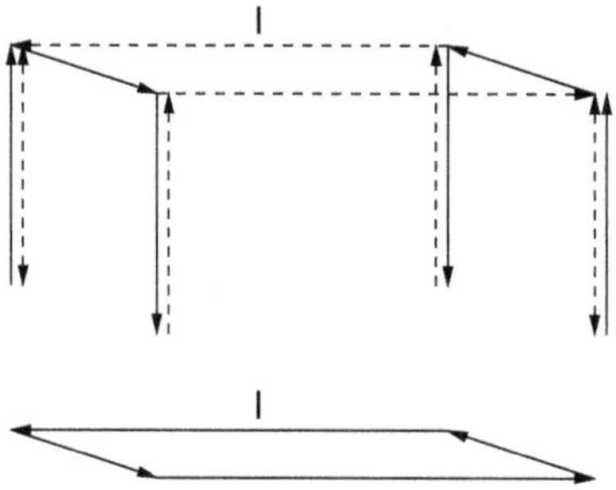


**Fig. 5.4.** The second experiment of Ampère.

do not move this wire and hence must be vertical to this wire ($\mathbf{F} = \int I d\mathbf{s} \times \mathbf{B}(\mathbf{r})$). In this experiment, Ampère used the subtle arrangement illustrated schematically in Fig. 5.5. The two mercury columns provided the necessary electric contact with the element of wire, which could move freely in its plane around a vertical.

Finally, in a *fourth experiment*, Ampère showed that the force acting between current-carrying loops is again proportional to $1/r^2$, $r$ being their separation.

To this end, he constructed a symmetric arrangement of three circular coils carrying current $I$, whose radii were in the ratio $n^0(=1) : n^1 : n^2$. Let the distances of the first and the last coils from the second coil be $a$ and $b$ as indicated in Fig. 5.6. The central coil was then adjusted in such a way that the forces due to the currents balance each other out. If, as stated, the radii of the coils are respectively $1, n, n^2$, then equilibrium (no motion) is obtained for (this was Ampère's experimental observation)

$$b = na.$$

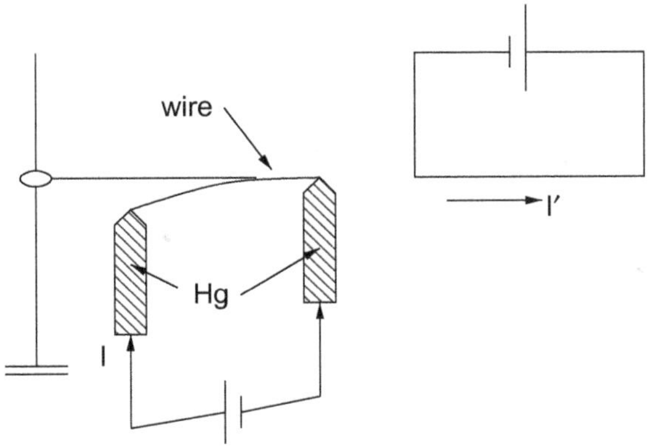


**Fig. 5.5.** The third experiment of Ampère.

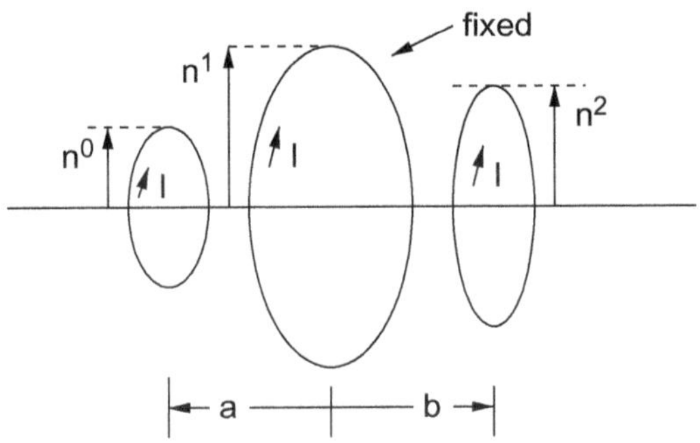


**Fig. 5.6.** The fourth experiment of Ampère.

If the forces are proportional to the circumferences or equi-angular arcs, then this means (with $f(x)$ for the dependence on the distance) for equilibrium at $b = na$,

$$2\pi \cdot 2\pi n \cdot f(a) = 2\pi n \cdot 2\pi n^2 \cdot f(b)\Big|_{b=na}$$

(ignoring the effects of the outer coils on each other), i.e.

$$f(a) = n^2 f(na),$$

i.e.

$$f(a) = \frac{\text{const.}}{a^2}$$

(so that const.$/a^2 = n^2$const.$/n^2a^2$). This is the $1/r^2$ law for currents.

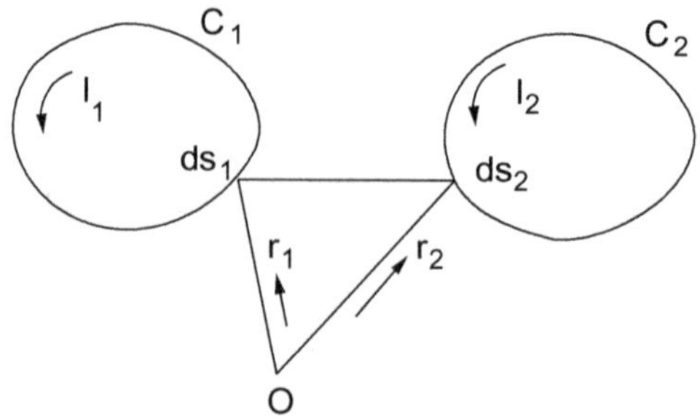


**Fig. 5.7.** Two conductor elements.

The fundamental formula for the field outside a current carrying wire is today named after Biot and Savart, who had also discovered the law $\propto 1/r^2$ shortly before Ampère (1820).

The experiments of Ampère (1820–1825), however, were considerably more rigorous and were praised by Maxwell[4] as "*one of the most brilliant achievements in science*".

The actual postulate or the law of magnetostatics derived from observation is *Ampère's law*, which (like that of the Coulomb force between charges in electrostatics) gives the force **F** between two string-like current-carrying conductors as illustrated in Fig. 5.7, and can be written (with a constant $k' = \mu_0/4\pi$ in front) as

$$\mathbf{F} = \frac{\mu_0}{4\pi} \oint_{C_1} \oint_{C_2} \frac{I_1 d\mathbf{s}_1 \times \{I_2 d\mathbf{s}_2 \times (\mathbf{r}_1 - \mathbf{r}_2)\}}{|\mathbf{r}_1 - \mathbf{r}_2|^3} \quad \text{newtons.} \tag{5.16}$$

Using the vector triple product relation

$$d\mathbf{s}_1 \times \{d\mathbf{s}_2 \times (\mathbf{r}_1 - \mathbf{r}_2)\} = \{d\mathbf{s}_1 \cdot (\mathbf{r}_1 - \mathbf{r}_2)\} d\mathbf{s}_2 - (d\mathbf{s}_1 \cdot d\mathbf{s}_2)(\mathbf{r}_1 - \mathbf{r}_2) \tag{5.17}$$

and Stokes' theorem, with which (see e.g. Eq. (2.6))

$$\begin{aligned} \oint_{C_1} d\mathbf{s}_1 \cdot \frac{\mathbf{r}_1 - \mathbf{r}_2}{|\mathbf{r}_1 - \mathbf{r}_2|^3} &= -\oint_{C_1} d\mathbf{s}_1 \cdot \boldsymbol{\nabla}_{\mathbf{r}_1} \frac{1}{|\mathbf{r}_1 - \mathbf{r}_2|} \\ &= -\int_{F_{C_1}} d\mathbf{F} \cdot \boldsymbol{\nabla} \times \left\{ \boldsymbol{\nabla} \frac{1}{|\mathbf{r}_1 - \mathbf{r}_2|} \right\}, \\ &= 0, \end{aligned} \tag{5.18}$$

[4]See e.g. Harnwell [69], p. 298.

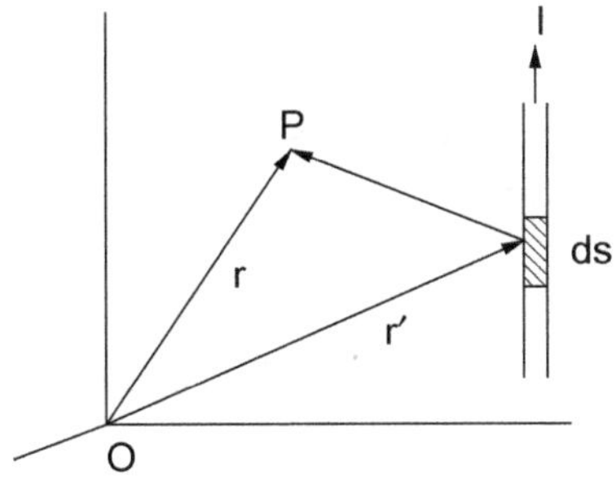


**Fig. 5.8.** The conductor element.

(i.e. *curl grad* = 0), the force **F** is also given by

$$\mathbf{F} = -\frac{\mu_0}{4\pi}\oint_{C_1}\oint_{C_2} I_1 d\mathbf{s}_1 \cdot I_2 d\mathbf{s}_2 \frac{(\mathbf{r}_1 - \mathbf{r}_2)}{|\mathbf{r}_1 - \mathbf{r}_2|^3} \quad \text{newtons.} \tag{5.19}$$

The *law of Biot–Savart* thus abstracts the observations of Biot, Savart, and Ampère by specifying an expression for the magnetic field **B**, called *magnetic induction* or *magnetic flux density*,[5] but frequently also simply magnetic field, which a conductor with current $I$ or the conductor element of length $ds$ — see Fig. 5.8 — generates at a point **r** outside the conductor, and is defined by

$$d\mathbf{B}(\mathbf{r}) := k' \frac{I(\mathbf{r}')d\mathbf{s}(\mathbf{r}') \times (\mathbf{r} - \mathbf{r}')}{|\mathbf{r} - \mathbf{r}'|^3}, \quad k' = \frac{\mu_0}{4\pi}, \tag{5.20}$$

or

$$\begin{aligned} \mathbf{B}(\mathbf{r}) &= k' \int_{\text{conductor}} \frac{I(\mathbf{r}')d\mathbf{s}(\mathbf{r}') \times (\mathbf{r} - \mathbf{r}')}{|\mathbf{r} - \mathbf{r}'|^3} \\ &\overset{(5.7)}{=} k' \int_{\text{conductor}} d\mathbf{r}' \frac{\mathbf{j}(\mathbf{r}') \times (\mathbf{r} - \mathbf{r}')}{|\mathbf{r} - \mathbf{r}'|^3} \quad \text{teslas.} \end{aligned} \tag{5.21}$$

In MKSA units, the constant $k'$ is given in newtons per ampere$^2$ by

$$k' = \frac{\mu_0}{4\pi}, \quad \mu_0 = 4\pi \times 10^{-7} = 12.566 \times 10^{-7}\ \text{N/A}^2. \tag{5.22}$$

The magnetic induction **B** is then given in units of weber/meter$^2$ or in units called tesla (T) (named after the engineer of this name); 1 tesla = $10^4$ gauss.

---

[5]We begin here with the induction, although we have not yet considered media. Some authors therefore start with the magnetic field strength **H** (in this chapter, $\mathbf{H} = \mathbf{B}/\mu_0$), which we introduce later.

In Gaussian units, the constant is $k' = 1/c$, $c$ being the velocity of light in vacuum.

As mentioned above, Ampère was mainly concerned with the investigation of the force that one current exerts on another current carrying conductor. Consequently, it became customary to describe as *Ampère's law* the expression for the force that a field $\mathbf{B}$ exerts on a conductor with current $I$. This means the expression obtained by inserting the above expression for $d\mathbf{B}$ into that for $\mathbf{F}$, i.e.

$$\mathbf{F} = \int I(\mathbf{r}')d\mathbf{s}(\mathbf{r}') \times \mathbf{B}(\mathbf{r}') \quad \text{newtons.} \tag{5.23}$$

In the MKSA system of units, the proportionality constant appearing in this expression is 1. This constant follows by demanding that force equals time derivative of momentum; indeed, the logical derivation of this expression is to go to the relativistic theory, obtain there the four-momentum of a charge in the gauge field, and then differentiate the latter's space-like components with respect to time. Some authors, e.g. Greiner [61], say simply, the constant $k'$ "*is obtained experimentally as 1/c*" in the Gaussian system.[6] One may also observe here that the force $\mathbf{F}$ of Eq. (5.23) involves $\mathbf{B} = \mu_0\mathbf{H}$, where $\mathbf{H}$ is called the *magnetic field strength.*

With the relation $Id\mathbf{s} = \mathbf{j}dV$, it follows that

$$\mathbf{F} = \int_{\text{conductor}} \mathbf{j}(\mathbf{r}') \times \mathbf{B}(\mathbf{r}')d\mathbf{r}' \quad \text{newtons.} \tag{5.24}$$

For a point charge $q$ moving with velocity $\mathbf{v}$, we can write

$$\mathbf{j} = q\mathbf{v}\delta(\mathbf{r} - \mathbf{r}') \quad \text{coulombs/s} \cdot \text{m}^2 \tag{5.25}$$

(a precisely known current density) so that inserted into Eq. (5.24) we obtain

$$\mathbf{F} = q\mathbf{v}(\mathbf{r}) \times \mathbf{B}(\mathbf{r}) \quad \text{newtons.} \tag{5.26}$$

This expression is known as *Lorentz force* (cf. the work of Jackson [73], p. 238, where this expression is initially written down with little justification). A version of this equation which is of practical importance, and is also

[6]One way of determining experimentally the constant relating electric units to magnetic units (which turns out to be $c$) is discussed briefly in Appendix B in relation to the experiment of Rosa and Dorsey.

described as *Ampère's law*, is obtained by writing it, e.g. in the simple perpendicular case, as

$$dF = IBds, \quad I = dq/dt. \tag{5.27}$$

In the case of a coil with N turns in the field $B$, the torque (moment) of this force is then $dFdl = INBdA$, where $dA$ is the area $dsdl$.

The expression (5.21) for $\mathbf{B}$ should be compared with that for $\mathbf{E}$, i.e. the formula (2.7):

$$\mathbf{E}(\mathbf{r}) = k\int d\mathbf{r}'\frac{\rho(\mathbf{r}')(\mathbf{r}-\mathbf{r}')}{|\mathbf{r}-\mathbf{r}'|^3} \quad \text{newtons/coulomb}, \quad k = \frac{1}{4\pi\epsilon_0}. \tag{5.28}$$

The field $\mathbf{B}$ follows for moving charges (current density $\mathbf{j}$) in a similar way as the field $\mathbf{E}$ for static charges (charge density $\rho$).

## 5.4 Examples

In the following, we consider examples which are important in later contexts and therefore do not simply serve the purpose of illustration or exercise.

**Example 5.2 (The field B of a long, straight conducting wire).** Calculate the field $\mathbf{B}$ of an infinitely long, straight conductor carrying current $I$ (see Fig. 5.9).

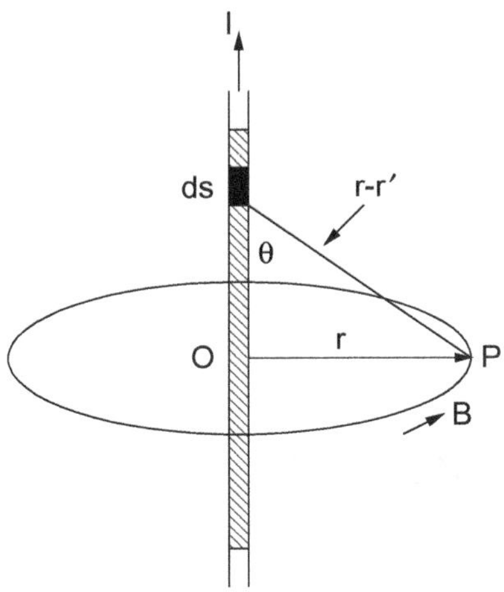


**Fig. 5.9.** The long, straight conducting wire.

**Solution:** From the Biot–Savart law (5.21), we obtain (with $k' = \mu_0/4\pi$)

$$\mathbf{B}(\mathbf{r}) = Ik' \int_{-\infty}^{\infty} \frac{d\mathbf{s} \times (\mathbf{r} - \mathbf{r}')}{|\mathbf{r} - \mathbf{r}'|^3} \quad \text{teslas.} \tag{5.29}$$

With the geometry of Fig. 5.9, this can be written in magnitude

$$B(\mathbf{r}) = Ik' \int_{-\infty}^{\infty} \frac{ds|\mathbf{r} - \mathbf{r}'| \sin\theta}{|\mathbf{r} - \mathbf{r}'|^3} \quad \text{teslas} \tag{5.30}$$

so that[7] ($k' = \mu_0/4\pi$) since $\mathbf{r}$ is perpendicular to $\mathbf{r}'$:

$$\begin{aligned} B(\mathbf{r}) &= Ik' \int_{-\infty}^{\infty} \frac{r\,ds}{|\mathbf{r} - \mathbf{r}'|^3} = Irk' \int_{-\infty}^{\infty} \frac{ds}{(r^2 + s^2)^{3/2}} \\ &= \frac{Irk'}{r^2} \left[ \frac{s}{\sqrt{r^2 + s^2}} \right]_{-\infty}^{\infty} = \frac{2Irk'}{r^2} = \frac{2I}{r} k' \\ &= \frac{\mu_0}{2\pi} \frac{I}{r} \quad \text{teslas.} \end{aligned} \tag{5.31}$$

The corresponding magnetic field strength $\mathbf{H}$ is accordingly $\mathbf{B}/\mu_0$, i.e. $\mathbf{H} = (I/2\pi r)\mathbf{e}_\theta$ A/m.

**Example 5.3 (The force between two long, thin, parallel conducting wires).** Calculate the force which two infinitely long, parallel wires a distance $d$ apart exert on each other.

**Solution:** We consider the case as depicted in Fig. 5.10. Wire 2 generates the field $\mathbf{B}_2$ at wire 1 which follows from Eq. (5.31) and is in magnitude

$$B_2 = \frac{2I_2}{d} k' \quad \text{teslas,} \tag{5.32}$$

where $d$ is the distance between the conductors. The field $\mathbf{B}_2$ is perpendicular to the currents $\mathbf{I}_1$, $\mathbf{I}_2$. Hence, the force, that $\mathbf{B}_2$ exerts on wire 1, is

[7] Using Dwight [42], formula 200.03, p. 45.

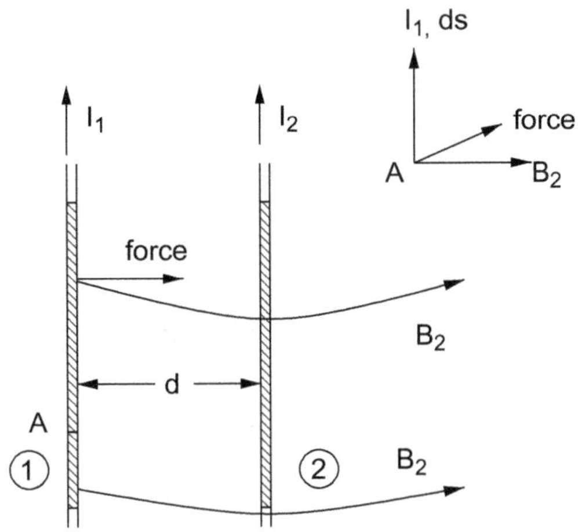


**Fig. 5.10.** Two long wires carrying parallel currents.

(cf. Eq. (5.23))

$$F = k' \int I_1 ds(\mathbf{r}') \frac{2I_2}{d} = k' \frac{2I_1 I_2}{d} \int ds(\mathbf{r}') \quad \text{newtons.} \tag{5.33}$$

The force per unit length of the conductors is therefore in magnitude

$$k' \frac{2I_1 I_2}{d} = \mu_0 \frac{I_1 I_2}{2\pi d} \quad \text{newtons/meter.} \tag{5.34}$$

For $\mathbf{I}_1$, $\mathbf{I}_2$ parallel, we have the situation illustrated in Fig. 5.10. The force is directed from wire 1 to wire 2 and is therefore attractive. Thus, currents in the same direction attract, currents in opposite directions repel.[8]

**Example 5.4 (The circular or ring conductor).** As another important example for the calculation of the magnetic field, we consider the circular or ring-shaped conductor shown in Fig. 5.11 with current $I$ and radius $a$.

**Solution:** The components of the field $\mathbf{B}$ perpendicular to the $z$-axis cancel each other along this axis. The field $d\mathbf{B}$ at $\mathbf{r}$ along the axis due to the element

[8]See beautiful sketches in the book of Griffiths [63], p. 203.

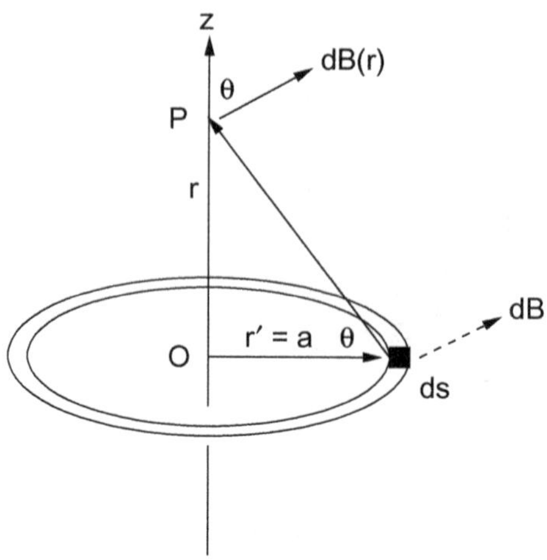


**Fig. 5.11.** The ring-shaped conductor.

$d\mathbf{s}$ of the conductor is according to the Biot–Savart law (5.20) given by

$$d\mathbf{B}(\mathbf{r}) = Ik' \frac{d\mathbf{s}(\mathbf{r}') \times (\mathbf{r} - \mathbf{r}')}{|\mathbf{r} - \mathbf{r}'|^3} \quad \text{teslas}. \tag{5.35}$$

Along the $z$-axis, $d\mathbf{B}$ is

$$dB(\mathbf{r}) \cos\theta = \frac{dB(\mathbf{r})a}{|\mathbf{r} - \mathbf{r}'|} \quad \text{teslas}. \tag{5.36}$$

Hence, the magnetic induction or magnetic flux density $B$ at point P along the $z$-axis is

$$\begin{aligned} B &= \int_{\text{wire}} \frac{a dB(\mathbf{r})}{|\mathbf{r} - \mathbf{r}'|} \overset{(5.35)}{=} k' \int \frac{I|d\mathbf{s} \times (\mathbf{r} - \mathbf{r}')|}{|\mathbf{r} - \mathbf{r}'|^3} \frac{a}{|\mathbf{r} - \mathbf{r}'|} \\ &= Iak' \int \frac{a d\theta'}{|\mathbf{r} - \mathbf{r}'|^3} \quad (\text{since } d\mathbf{s} \perp \mathbf{r} - \mathbf{r}',\ ds = a d\theta') \\ &= Iak' \frac{2\pi a}{(r^2 + a^2)^{3/2}} \quad \text{teslas} \quad (\text{since } \mathbf{r}' \perp \mathbf{r}), \end{aligned} \tag{5.37}$$

i.e. for $r \gg a$,

$$B \simeq 2I\pi a^2 \frac{k'}{r^3} = \frac{\mu_0}{2} \frac{Ia^2}{r^3} \quad \text{teslas} \quad \left(k' = \frac{\mu_0}{4\pi}\right). \tag{5.38}$$

We will deal with this problem again later (see Example 5.7) because one expects the field $B$ at a point $\mathbf{r}$ to be related to the solid angle which the ring-shaped wire subtends at this point.

**Example 5.5 (The magnetic moment of a closed current loop).** The magnetic moment **m** of a circuit is defined as 1/2 the volume integral of the moment of the current density, i.e. as[9]

$$\mathbf{m} := \frac{1}{2}\int d\mathbf{r}[\mathbf{r}\times\mathbf{j}(\mathbf{r})] \stackrel{(5.7)}{=} \frac{1}{2}\int \mathbf{r}\times I d\mathbf{s}(\mathbf{r}) = I\int d\mathbf{F} \quad \text{amperes}\cdot\text{m}^2, \quad (5.39)$$

where, see Fig. 5.12(a), $d\mathbf{F} = \frac{1}{2}\mathbf{r}\times d\mathbf{s}$ is an element of area. Calculate the field **B**. In other words, show that the magnetic field due to a closed current loop is the same as that from a distribution of *magnetic dipoles*, moment $m = IF$ or $I$ per unit area covering any surface which has the current loop as its perimeter (See also Example 5.7(a)).

**Solution:** In the special case of the ring-shaped wire considered in Example 5.4, we have

$$m = I\pi a^2 \quad \text{A}\cdot\text{m}^2, \quad (5.40)$$

a relation which is also known as *Ampère's dipole law.* By definition, the current $I$ has the direction of positive charge; hence, **m** is opposite to the direction of negative charge. In terms of the expression for $m$, we can rewrite $B$ of Eq. (5.38) as

$$B = \frac{\mu_0}{2}\frac{m}{\pi r^3}, \quad B = k'\frac{2m}{r^3} \quad \text{teslas}. \quad (5.41)$$

It is instructive to compare this result with the field of an *electric dipole* with moment **p**, as illustrated in Fig. 5.12(b). The electric field **E** at point $P$

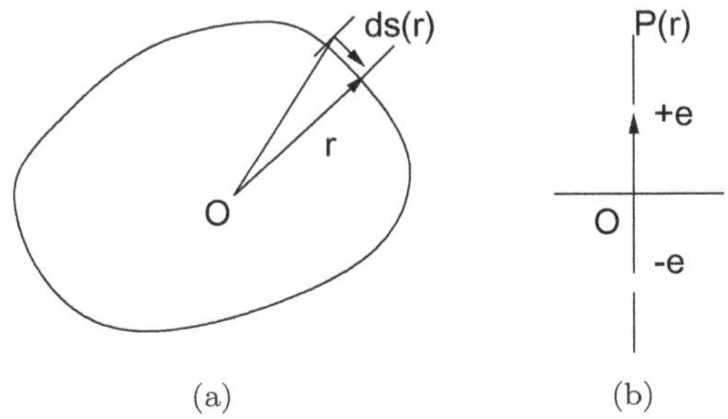


**Fig. 5.12.** (a) Circuit element, (b) dipole.

[9]Thus, in the simple case with $I = dq/dt$, we have the magnetic moment $\mathbf{m} = (q/2)(\mathbf{r}\times\mathbf{v})$, which is not to be confused with the electric dipole moment $\mathbf{p} = q\mathbf{r}$; see Eqs. (2.127) and (3.164).

follows, as we saw with Eqs. (3.173) and (3.174), from

$$\mathbf{E} = -\boldsymbol{\nabla}\phi_{\rm dipole}, \quad \phi_{\rm dipole} = k\frac{\mathbf{p}\cdot\mathbf{r}}{r^3} \overset{\text{here}}{=} k\frac{pr}{r^3} = k\frac{p}{r^2} \quad \text{volts} \tag{5.42}$$

so that

$$E = k\frac{2p}{r^3} \quad \text{newtons/coulomb}. \tag{5.43}$$

The general expression for $\mathbf{E}$, i.e. at a point $P$ in the direction of the polar angle $\theta$ as seen from the origin, can immediately be written down by taking scalar and vector products of Eq. (3.174) with $\mathbf{n} = \mathbf{r}/r$ (or see e.g. the work of Greiner [61], p. 36):

$$\mathbf{E} = k\frac{2p\cos\theta}{r^3}\mathbf{e}_r + k\frac{p\sin\theta}{r^3}\mathbf{e}_\theta, \quad k = \frac{1}{4\pi\epsilon_0}, \tag{5.44}$$

in agreement with the above expression for polar angle $\theta = 0$. Considering the expressions (5.41) and (5.43) and taking into account that (see Chapter 1)

$$k'/k = 1/c^2, \quad B = \frac{k}{c^2}\frac{2m}{r^3} = \frac{E}{p}\frac{m}{c^2}, \tag{5.45}$$

we obtain the correspondence

$$B \leftrightarrow \frac{1}{c}E, \quad m \leftrightarrow cp. \tag{5.46}$$

This means that the magnetic moment $\mathbf{m}$ generates the field $\mathbf{B}$ in complete analogy to the generation of the field $\mathbf{E}$ by an electric dipole, and corresponding to Eq. (5.44), we have[10]

$$\begin{aligned}\mathbf{B}(\mathbf{r}) &= k'\frac{2m\cos\theta}{r^3}\mathbf{e}_r + k'\frac{m\sin\theta}{r^3}\mathbf{e}_\theta \\ &= k'\frac{3\mathbf{r}(\mathbf{m}\cdot\mathbf{r}) - r^2\mathbf{m}}{r^5} \\ &= -k'\boldsymbol{\nabla}\frac{\mathbf{m}\cdot\mathbf{r}}{r^3} \quad \text{teslas} \quad (\text{or Wb/m}^2).\end{aligned} \tag{5.47}$$

More generally, we can write

$$\mathbf{B}(\mathbf{r}) = k'\left[\frac{-\mathbf{m}}{|\mathbf{r}-\mathbf{r}'|^3} + \frac{3\{\mathbf{m}\cdot(\mathbf{r}-\mathbf{r}')\}(\mathbf{r}-\mathbf{r}')}{|\mathbf{r}-\mathbf{r}'|^5}\right] \quad \text{teslas} \tag{5.48}$$

[10]See the steps from Eq. (3.173) to Eq. (3.174).

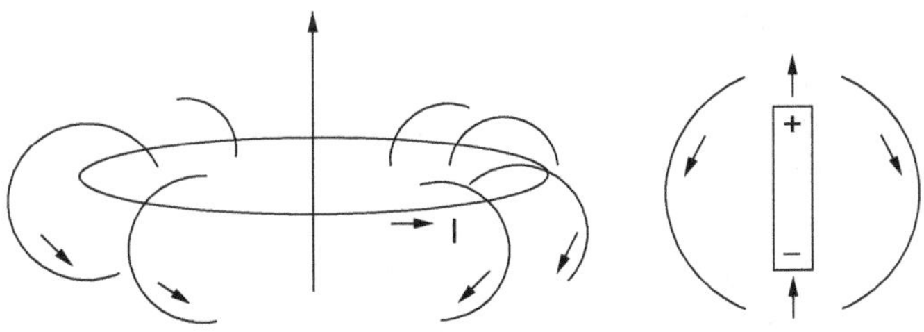


**Fig. 5.13.** Analogy between the fields of a circular current and a magnet.

with, as will be shown in Example 6.1, see Eq. (6.16),

$$\mathbf{B}(\mathbf{r}) = \boldsymbol{\nabla} \times \mathbf{A}, \quad \mathbf{A}(\mathbf{r}) = k' \frac{\mathbf{m} \times (\mathbf{r} - \mathbf{r}')}{|\mathbf{r} - \mathbf{r}'|^3} \quad \text{tesla} \cdot \text{meters}. \tag{5.49}$$

We see therefore that a circular current generates an induction field **B**, which sufficiently far away looks like that of a dipole, as indicated in Fig. 5.13. The dipole field is seen to be similar to the electric field due to an electric dipole. The relation (5.49) will be verified in Example 6.1.

## 5.5 The Electromagnetic Vector Potential A

Our next objective is to obtain the vector potential **A** from which the field **B** can be derived. We return to the law of Biot–Savart. With (cf. Eq. (4.4))

$$\frac{\mathbf{r} - \mathbf{r}'}{|\mathbf{r} - \mathbf{r}'|^3} = -\boldsymbol{\nabla}_{\mathbf{r}} \frac{1}{|\mathbf{r} - \mathbf{r}'|} = \boldsymbol{\nabla}' \frac{1}{|\mathbf{r} - \mathbf{r}'|} \equiv \boldsymbol{\nabla}_{\mathbf{r}'} \frac{1}{|\mathbf{r} - \mathbf{r}'|} \tag{5.50}$$

we obtain from Eq. (5.21)

$$\begin{aligned} \mathbf{B}(\mathbf{r}) &= -k' \int d\mathbf{r}' \mathbf{j}(\mathbf{r}') \times \boldsymbol{\nabla}_{\mathbf{r}} \frac{1}{|\mathbf{r} - \mathbf{r}'|} \\ &= \boldsymbol{\nabla}_{\mathbf{r}} \times k' \int \frac{\mathbf{j}(\mathbf{r}') d\mathbf{r}'}{|\mathbf{r} - \mathbf{r}'|} \quad \text{teslas} \end{aligned} \tag{5.51}$$

(where $k' = \mu_0/4\pi$). But "*div curl*" is always zero. Hence,

$$\boldsymbol{\nabla} \cdot \mathbf{B}(\mathbf{r}) = 0. \tag{5.52}$$

This equation has *no source term* on the right-hand side. This means in physical terms the following: There are *no single magnetic poles*, i.e. *monopoles*. How did we arrive at the following result? The empirically obtained law of Biot–Savart gives the field **B** which surrounds a conductor with current $I$; we saw that this field corresponds to that of a magnetic moment; a magnetic moment, however, corresponds to a pairing of magnetic poles — hence, no single magnetic poles enter the consideration. It should be noted that we are (apparently) dealing with macroscopic considerations (the Biot–Savart law applies to macroscopic conductors); these can, however, also be interpreted microscopically (the classically considered electron encircling a nucleus represents a current). The effect of the magnetic induction **B** on magnetizable matter will later be considered separately, in analogy with macroscopic electrostatics. However, we mention already here the further meaning of Eq. (5.52), which is Gauss' law for magnetostatistics. Applying Gauss' divergence theorem, it implies that the flux $\propto \mathbf{B}\cdot\mathbf{F}$, where **F** is a surface area, out of any closed surface always vanishes, even through magnetized material.

We now define the *electromagnetic vector potential* **A** by

$$\mathbf{A}(\mathbf{r}) = k'\int\frac{\mathbf{j}(\mathbf{r}')d\mathbf{r}'}{|\mathbf{r}-\mathbf{r}'|} \overset{(5.7)}{=} k'I\int\frac{d\mathbf{s}'}{|\mathbf{r}-\mathbf{r}'|} \quad \text{tesla}\cdot\text{meters} \tag{5.53}$$

so that

$$\mathbf{B} = \boldsymbol{\nabla}\times\mathbf{A} \tag{5.54}$$

(**A** is not unique since $\mathbf{A}+\boldsymbol{\nabla}\phi$, for $\phi$ a scalar function, is also possible; see gauge transformations in the following). We next evaluate the expression $\boldsymbol{\nabla}\times\mathbf{B}$, in which we use first the formula

$$\boldsymbol{\nabla}\times(\boldsymbol{\nabla}\times\mathbf{A}) = \boldsymbol{\nabla}(\boldsymbol{\nabla}\cdot\mathbf{A}) - \triangle\mathbf{A} \tag{5.55}$$

("*curl curl = grad div minus div grad*"), then we use (cf. Eq. (2.51))

$$\triangle_{\mathbf{r}}\frac{1}{|\mathbf{r}-\mathbf{r}'|} = -4\pi\delta(\mathbf{r}-\mathbf{r}') \tag{5.56}$$

(and integrate over the delta function), and finally we use Gauss' divergence theorem. We then have

$$\begin{aligned}\boldsymbol{\nabla}\times\mathbf{B} &= \boldsymbol{\nabla}\times\left(\boldsymbol{\nabla}_{\mathbf{r}}\times k'\int\frac{\mathbf{j}(\mathbf{r}')d\mathbf{r}'}{|\mathbf{r}-\mathbf{r}'|}\right)\\ &= k'\boldsymbol{\nabla}\int\boldsymbol{\nabla}_{\mathbf{r}}\cdot\frac{\mathbf{j}(\mathbf{r}')d\mathbf{r}'}{|\mathbf{r}-\mathbf{r}'|} - k'\int\triangle_{\mathbf{r}}\frac{\mathbf{j}(\mathbf{r}')d\mathbf{r}'}{|\mathbf{r}-\mathbf{r}'|}\\ &= k'\boldsymbol{\nabla}\int\left(\boldsymbol{\nabla}_{\mathbf{r}}\cdot\frac{\mathbf{j}(\mathbf{r}')}{|\mathbf{r}-\mathbf{r}'|}\right)d\mathbf{r}' + 4\pi k'\mathbf{j}(\mathbf{r}).\end{aligned} \tag{5.57}$$

Since $\boldsymbol{\nabla}_{\mathbf{r}}$ acts only on the denominator and there as $-\boldsymbol{\nabla}_{\mathbf{r}'}$, this can be rewritten as

$$\begin{aligned}\boldsymbol{\nabla}\times\mathbf{B} &= -k'\boldsymbol{\nabla}\int \mathbf{j}(\mathbf{r}')\cdot\left(\boldsymbol{\nabla}_{\mathbf{r}'}\frac{1}{|\mathbf{r}-\mathbf{r}'|}\right)d\mathbf{r}' + 4\pi k'\mathbf{j}(\mathbf{r})\\ &= -k'\boldsymbol{\nabla}\int d\mathbf{r}'\left[\boldsymbol{\nabla}'\cdot\left(\frac{\mathbf{j}(\mathbf{r}')}{|\mathbf{r}-\mathbf{r}'|}\right) - \frac{1}{|\mathbf{r}-\mathbf{r}'|}\boldsymbol{\nabla}'\cdot\mathbf{j}(\mathbf{r}')\right] + 4\pi k'\mathbf{j}(\mathbf{r})\\ &= -k'\boldsymbol{\nabla}\int_{F_\infty} d\mathbf{F}\cdot\frac{\mathbf{j}(\mathbf{r}')}{|\mathbf{r}-\mathbf{r}'|} + k'\boldsymbol{\nabla}\int\frac{\boldsymbol{\nabla}'\cdot\mathbf{j}(\mathbf{r}')}{|\mathbf{r}-\mathbf{r}'|}d\mathbf{r}' + 4\pi k'\mathbf{j}(\mathbf{r}).\end{aligned} \tag{5.58}$$

Here, the first term vanishes, if we extend the area of integration $F$ to infinity and remember, that $\mathbf{j} = 0$ at $r = \infty$, since the conductor and hence $\mathbf{j}$ are located in a finite part of space. The second term also vanishes, since in magnetostatics $\boldsymbol{\nabla}\cdot\mathbf{j} = 0$, which is the condition for stationary currents. Thus, there remains *in magnetostatics* (with $k' = \mu_0/4\pi$)

$$\boldsymbol{\nabla}\times\mathbf{B} = \mu_0\mathbf{j}(\mathbf{r}). \tag{5.59}$$

As a *subsidiary result*, we obtain

$$\boldsymbol{\nabla}\cdot\mathbf{A}(\mathbf{r}) = k'\boldsymbol{\nabla}_{\mathbf{r}}\cdot\int\left(\frac{\mathbf{j}(\mathbf{r}')}{|\mathbf{r}-\mathbf{r}'|}\right)d\mathbf{r}' = -k'\int\mathbf{j}(\mathbf{r}')\cdot\left(\boldsymbol{\nabla}_{\mathbf{r}'}\frac{1}{|\mathbf{r}-\mathbf{r}'|}\right)d\mathbf{r}',$$

which can be rewritten as

$$\begin{aligned}\boldsymbol{\nabla}\cdot\mathbf{A}(\mathbf{r}) &= -k'\int d\mathbf{r}'\left[\boldsymbol{\nabla}'\cdot\left(\frac{\mathbf{j}(\mathbf{r}')}{|\mathbf{r}-\mathbf{r}'|}\right) - \frac{1}{|\mathbf{r}-\mathbf{r}'|}\boldsymbol{\nabla}'\cdot\mathbf{j}(\mathbf{r}')\right]\\ &= -k'\int_{F_\infty} d\mathbf{F}\cdot\frac{\mathbf{j}(\mathbf{r}')}{|\mathbf{r}-\mathbf{r}'|} + k'\int d\mathbf{r}'\frac{\boldsymbol{\nabla}'\cdot\mathbf{j}(\mathbf{r}')}{|\mathbf{r}-\mathbf{r}'|}\\ &= 0\end{aligned} \tag{5.60}$$

for the same reasons as those used in deriving Eq. (5.59). Thus,

$$\boldsymbol{\nabla}\cdot\mathbf{A}(\mathbf{r}) = 0. \tag{5.61}$$

This relation is known as *Coulomb gauge.* It can be shown[11] that this condition has the physical meaning that only two components of the vector potential $\mathbf{A}$ are independent, and that the vector $\mathbf{A}$ is orthogonal to the propagation vector $\mathbf{k}$ of the free electromagnetic wave, and hence implies

[11]See Chapter 18.

the *transversality* of the wave (fields $\mathbf{E}, \mathbf{H}$ perpendicular to the direction of propagation).

One should note that here we arrived at the relation $\nabla \cdot \mathbf{A} = 0$ only as a consequence of $\nabla \cdot \mathbf{j} = 0$, i.e. in the case of stationary currents. A condition of this kind, however, is always required for the vector potential, as a consequence of the so-called *gauge invariance of the theory.* By this, one means (put simply) the *invariance* of the field equations or equations of motion, here the Maxwell equations, under transformations (here for magnetostatics) called *gauge transformations*, i.e.

$$\mathbf{A} \rightarrow \mathbf{A}' = \mathbf{A} + \nabla\chi, \quad \phi \rightarrow \phi' = \phi, \tag{5.62}$$

(more generally $\phi' = \phi - \partial\chi/\partial t$), where $\triangle\chi = 0$. The invariance of the equations obtained so far follows immediately:

$$\begin{aligned} \mathbf{E} \rightarrow \mathbf{E}' &= -\nabla\phi' = -\nabla\phi = \mathbf{E}, \\ \mathbf{B} \rightarrow \mathbf{B}' = \nabla \times \mathbf{A}' &= \nabla \times (\mathbf{A} + \nabla\chi) = \nabla \times \mathbf{A} = \mathbf{B}, \\ \nabla \cdot \mathbf{A} = 0 \rightarrow \nabla \cdot \mathbf{A}' &= \nabla \cdot (\mathbf{A} + \nabla\chi) = \nabla \cdot \mathbf{A} = 0, \end{aligned} \tag{5.63}$$

since $\triangle\chi = 0$. Again one should note that the invariant fields $\mathbf{E}$ and $\mathbf{B}$ are macroscopically observable quantities, the non-unique gauge potential, of course, is not.

## 5.6 Integral Form of Ampère's Law

We return to our earlier considerations. We know that

$$\triangle \frac{1}{|\mathbf{r} - \mathbf{r}'|} = -4\pi\delta(\mathbf{r} - \mathbf{r}'). \tag{5.64}$$

Hence, we have with Eq. (5.53)

$$\triangle\mathbf{A}(\mathbf{r}) = -4\pi k' \int \delta(\mathbf{r} - \mathbf{r}')\mathbf{j}(\mathbf{r}')d\mathbf{r}',$$

i.e.

$$\triangle\mathbf{A}(\mathbf{r}) = -\mu_0\mathbf{j}(\mathbf{r}). \tag{5.65}$$

Thus, the *current density* $\mathbf{j}$ *is the source of the vector potential* $\mathbf{A}$. This vector equation should be compared with the scalar Poisson equation $\triangle\phi = -4\pi k\rho = \rho/\epsilon_0$ in electrostatics, in which the charge density $\rho$ appears as the source function of the potential $\phi$. Equation (5.65) thus is the magnetic analogue of the Poisson equation in electrostatics.

We now take a closer look at Eq. (5.59). This equation is also described as *Ampère's law*; in fact, it is the differential form of the Biot–Savart law. (Since for a gauge transformation $\boldsymbol{\nabla} \times \mathbf{B} = \boldsymbol{\nabla} \times \mathbf{B}'$, the invariance of this equation requires that $\mathbf{j} = \mathbf{j}'$). We apply to Eq. (5.59) the *theorem of Stokes* and obtain

$$\int \boldsymbol{\nabla} \times \mathbf{B} \cdot d\mathbf{F} = \oint_C \mathbf{B} \cdot d\mathbf{s} = \mu_0 \int_{F(C)} \mathbf{j} \cdot d\mathbf{F}, \tag{5.66}$$

where $d\mathbf{s}$ is an element of the path $C$ enclosing the area $F(C)$. Let us assume that $n$ conductors are enclosed by $C$ as indicated in Fig. 5.14 (or one with several turns or wiggles passing through the area enclosed by $C$, some also in opposite directions). Let $I$ be the current flowing $n$ times through the surface enclosed by $C$ in one direction (or the net number, if currents are also going in the opposite direction). Then,

$$\mu_0 \int_{F(C)} \mathbf{j} \cdot d\mathbf{F} = \mu_0 n I, \tag{5.67}$$

and hence,

$$\oint_C \mathbf{B} \cdot d\mathbf{s} = \mu_0 n I. \tag{5.68}$$

This important formula is the *integral form of Ampère's law.* In view of Eqs. (5.59) and (5.65), we have

$$\boldsymbol{\nabla} \times \mathbf{B} + \triangle \mathbf{A} = 0. \tag{5.69}$$

But

$$\boldsymbol{\nabla} \times (\boldsymbol{\nabla} \times \mathbf{A}) = \boldsymbol{\nabla}(\boldsymbol{\nabla} \cdot \mathbf{A}) - \triangle \mathbf{A} = 0 - \triangle \mathbf{A} \tag{5.70}$$

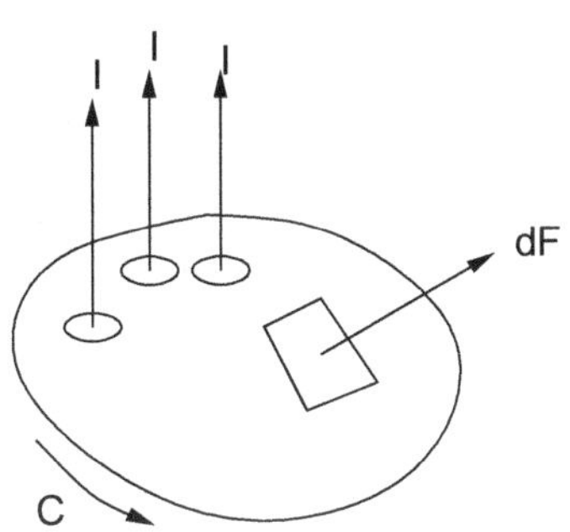


**Fig. 5.14.** Currents through a surface.

and always $\nabla \times \nabla = 0$. Hence, for some *scalar* $\phi_0$,

$$\mathbf{B} = \nabla \times \mathbf{A} + \nabla\phi_0. \tag{5.71}$$

Since, as we saw, $\nabla \cdot \mathbf{B} = 0$ and always "*div curl* = 0", it follows that $\phi_0$ is the solution of $\triangle\phi_0 = 0$. If $\mathbf{A} = 0$, then it follows that $\mathbf{B}$ can always be expressed as a gradient (this is the so-called Coulomb law of magnetostatics for hypothetical monopoles). The possibility of deriving the electric field strength $\mathbf{E}$ in electrostatics from a scalar potential $\phi$ depended on the relation $\nabla \times \mathbf{E} = 0$. Since $\nabla \times \mathbf{B} \neq 0$ (Eq. (5.59)), such a representation is not possible unless the current is confined to thin wires. The scalar representation fails in magnetostatics for currents in continuous media (the subject of Chapter 6), which means (compared with electrostatics) for a distributed magnetomotive force.

## 5.7 Further Examples

In this section, we consider examples to illustrate various applications of magnetostatics.

**Example 5.6 (The magnetic field of a long solenoid with current $I$).** Since the solenoid is stated to be *long*, the magnetic field in its interior is practically homogeneous and outside practically zero, as we can see immediately by drawing lines of constant $\mathbf{B}$ into the diagram as in Fig. 5.15. One observes that the lines crowd together in the inside region and thin out outside. The field inside is to be calculated.

**Solution:** In Fig. 5.15, a path $C$ is shown, which encloses $N$ turns of the solenoid (cross-section of turns indicated by dots). Hence, from Eq. (5.68),

$$\oint_C \mathbf{B} \cdot d\mathbf{s} = IN\mu_0, \tag{5.72}$$

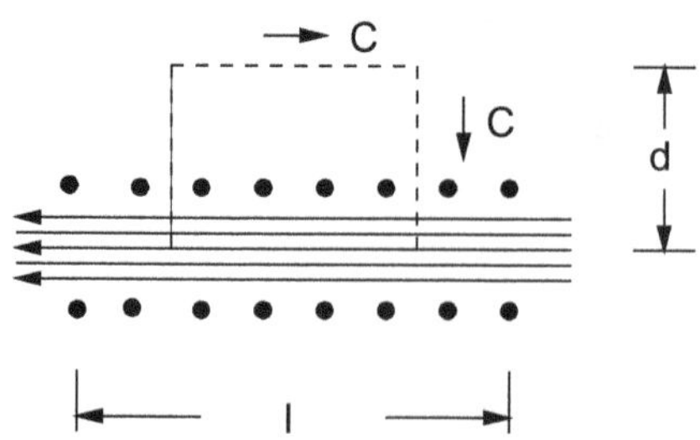


**Fig. 5.15.** Longitudinal cross-section of the solenoid.

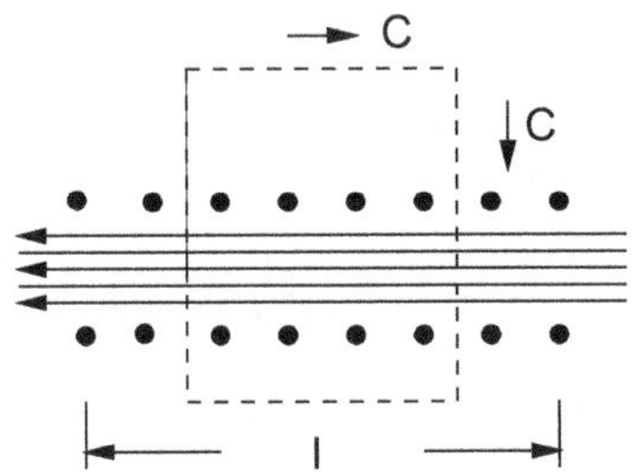


**Fig. 5.16.** A different closed path $C$.

where $IN$ amperes is the current flowing through the surface enclosed by $C$. Such a closed path with currents passing through the surface it encloses is frequently called an *amperian loop.*[12] Since the field $\mathbf{B}$ outside the solenoid is practically zero, and $\mathbf{B}$ along the two transverse sections of lengths $d$ in Fig. 5.15 is also practically zero, and inside the solenoid the field is practically homogeneous, it follows that, with $N = nl$,

$$Bl = IN\mu_0, \quad \text{i.e.} \quad B = In\mu_0 = \frac{\mu_0 IN^2}{l^2} \quad \text{weber/m}^2, \tag{5.73}$$

where $n = N/l$ is the number of turns per unit length of the solenoid.

For a better understanding, we choose another closed path $C$ as indicated in Fig. 5.16. In this case, the field along the path is everywhere zero, i.e. $\oint_C \mathbf{B} \cdot d\mathbf{s} = 0$. On the other hand (since the path this time encloses ingoing as well as outgoing currents — see Eq. (5.67)),

$$\int_C \mathbf{j} \cdot d\mathbf{F} = 2I - 2I = 0. \tag{5.74}$$

Thus, here the path was selected in such a way that one is unable to determine the field $B$.

**Example 5.7 (Significance of solid angle subtended by a current loop).**

(a) Show that the field $\mathbf{B}(\mathbf{r})$ at point $\mathbf{r}$ due to a closed current-carrying loop (current $I$) is given by

$$\mathbf{B}(\mathbf{r}) = -\frac{\mu_0}{4\pi}\boldsymbol{\nabla}(I\omega) \quad \text{teslas}, \tag{5.75}$$

where $\omega$ is the solid angle subtended by the current loop at the point $\mathbf{r}$.

[12]See e.g. Griffiths [63], p. 225.

(b) With the help of the expression obtained under (a) for the field $\mathbf{B}(\mathbf{r})$, calculate the field $B(z)$ of a ring-shaped conductor of radius $a$ at a point $P(z)$, distance $z$ above the center $O$ of the ring-shaped conductor.

(c) A solenoid of radius $a$, length $l$, with current $I$ consists of $n$ uniform turns per unit length. Determine the component of the field $\mathbf{B}$ in axial direction at a point a distance $x$ away from one end of the solenoid (e.g. from the origin $O$ there on the left side of the axis).

**Solution:** Note that in the following $d\mathbf{r}$ is the displacement of $\mathbf{r}$ and not the volume element!

(a) We consider, as indicated in Fig. 5.17(a), a displacement of the field point $P$ by $d\mathbf{r}$, which is equivalent to a displacement $-d\mathbf{r}'$ of the coordinate $\mathbf{r}'$ of the current element $Id\mathbf{s}(\mathbf{r}')$, the latter thereby sweeping out the area $d\mathbf{F}' = d\mathbf{r}' \times d\mathbf{s}(\mathbf{r}')$, i.e. $d\mathbf{r} = -d\mathbf{r}'$, which means shifting the field point while keeping the circuit stationary is equivalent to shifting the circuit in the opposite direction while keeping the field point stationary. Thus, $dF'$ is the element of area swept out by the current element $Id\mathbf{s}$ in moving through a distance $d\mathbf{r}'$.

Starting from the formula (5.21), i.e. with vector integration around a closed loop,

$$\mathbf{B}(\mathbf{r}) = \frac{\mu_0 I}{4\pi}\oint \frac{d\mathbf{s}(\mathbf{r}') \times (\mathbf{r} - \mathbf{r}')}{|\mathbf{r} - \mathbf{r}'|^3} \quad \text{teslas,} \tag{5.76}$$

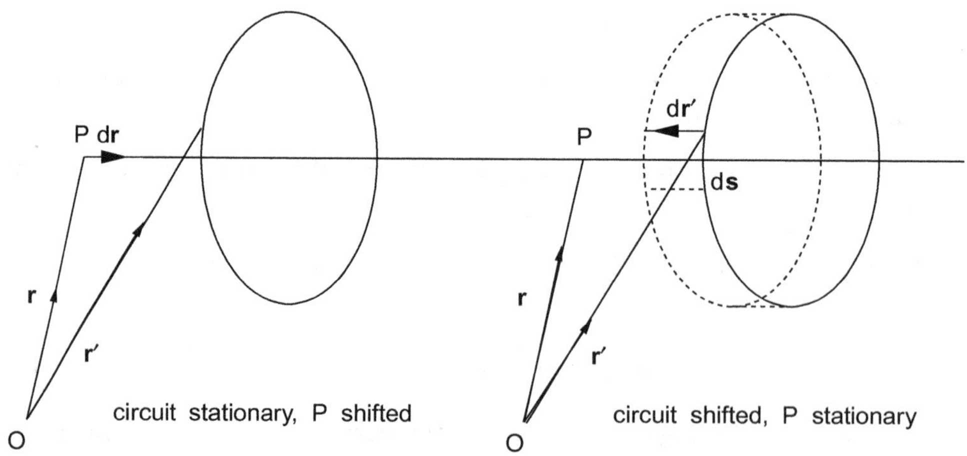


**Fig. 5.17(a).** Equivalent shifts with $d\mathbf{r} = -d\mathbf{r}'$.

we then have, recalling that in a scalar triple product scalar multiplication and vector multiplication may be exchanged, and recalling Eq. (2.13),

$$\mathbf{B}(\mathbf{r}) \cdot d\mathbf{r} = -\frac{\mu_0 I}{4\pi} \oint d\mathbf{r}' \cdot \frac{d\mathbf{s}(\mathbf{r}') \times (\mathbf{r} - \mathbf{r}')}{|\mathbf{r} - \mathbf{r}'|^3} = -\frac{\mu_0 I}{4\pi} \oint \frac{d\mathbf{r}' \times d\mathbf{s}(\mathbf{r}') \cdot (\mathbf{r} - \mathbf{r}')}{|\mathbf{r} - \mathbf{r}'|^3}$$
$$= -\frac{\mu_0 I}{4\pi} \oint \frac{d\mathbf{F}' \cdot (\mathbf{r} - \mathbf{r}')}{|\mathbf{r} - \mathbf{r}'|^3} \equiv -k' I d\omega', \quad k' = \frac{\mu_0}{4\pi}, \tag{5.77}$$

where $d\omega'$ is the solid angle subtended at $P$ by the entire ring-shaped area that arises in the shift of the loop. This area swept out by the current loop subtends the solid angle $d\omega'$ at the field point $P$. This $d\omega'$ is the same as the *change* $d\omega$ thereby in the solid angle $\omega$ subtended by the current loop at $P$. Thus (with $\nabla = d/d\mathbf{r}$)

$$\mathbf{B}(\mathbf{r}) \cdot d\mathbf{r} = -k' d(I\omega),$$
$$\omega = \int_{F'} \frac{d\mathbf{F}' \cdot (\mathbf{r} - \mathbf{r}')}{|\mathbf{r} - \mathbf{r}'|^3} \overset{(5.50)}{=} \int_{F'} d\mathbf{F}' \cdot \boldsymbol{\nabla}_{\mathbf{r}'} \frac{1}{|\mathbf{r} - \mathbf{r}'|}, \tag{5.78}$$

where $\omega$ is the solid angle subtended by the conductor loop at $P$. The solution of the last equation is therefore

$$\mathbf{B}(\mathbf{r}) = -k' \boldsymbol{\nabla}(I\omega) \quad \text{teslas}. \tag{5.79}$$

We observe that here $\mathbf{H} = \mathbf{B}/\mu_0$ can be derived from the *scalar magnetic potential*[13]

$$V_m = \frac{I\omega}{4\pi}, \quad \mathbf{H} = -\boldsymbol{\nabla} V_m, \quad V_m = \frac{I}{4\pi} \int_{F'} d\mathbf{F}' \cdot \boldsymbol{\nabla}_{\mathbf{r}'} \frac{1}{|\mathbf{r} - \mathbf{r}'|}, \tag{5.80}$$

in the same way as the electric field $\mathbf{E}$ from $V \equiv \phi$ in electrostatics. Considering microscopic currents with magnetic moments $\mathbf{m}(\mathbf{r}') = I d\mathbf{F}'$, we can jump to the macroscopic scalar potential $V_m$ over a *distribution of magnetic dipole moments* of density $\mathbf{M}$ (see Example 5.19).

(b) We now consider a circular conductor of radius $a$. The problem is to determine the solid angle subtended by this circular conductor at the point $P$

[13] In most of the British literature, the symbol $\Omega$ is used for $V_m$. For considerable discussion of this potential, see Bradley [22], pp. 65–80. Observe that $V_m$ is measured in amperes and $H$ in amperes/meter. In Eq. (5.80), the symbol $\omega$ stands for an angle, not for an angular frequency. If the dimensions of the loop are small compared with the distance to the field point, the integral can be evaluated approximately in the form of a multipole expansion akin to the expansion in electrostatics. See Eyges [49], p. 129.

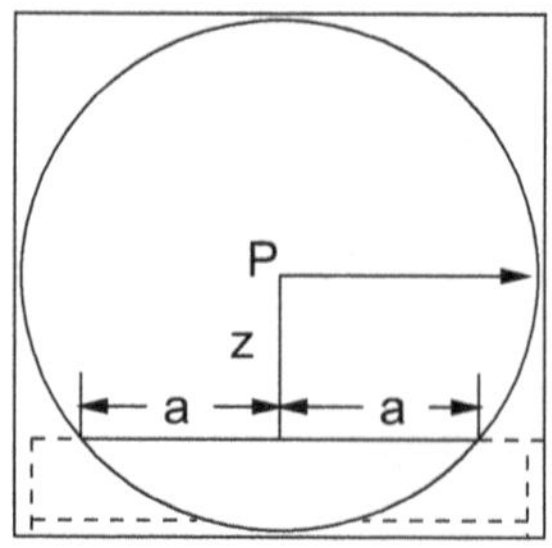


**Fig. 5.17(b).** The spherical shell and its cylindrical envelope (dashed).

in Fig. 5.17(b). We determine this as follows. As indicated in Fig. 5.17(b), we enclose the sphere (shown as a circle with center $P$) in a cylinder (indicated by the square) with open ends at the top and bottom of the $z$ direction. This cylindrical envelope of the entire sphere has surface area $2\pi$ times radius times height, i.e. $2\pi\sqrt{a^2+z^2} \times 2\sqrt{a^2+z^2} = 4\pi(a^2+z^2)$. We recognize this as the surface area of the sphere. One can prove that the surface area of that fraction $F_S$ of surface of the sphere which has the conductor ring of radius $a$ as boundary can be obtained analogously, i.e. this surface area is again given by that of its cylindrical envelope (shown as dashed in Fig. 5.17(b)), so that[14] i.e.

$$F_S = (2\pi \times \text{cylindrical radius}) \times \text{cylinder height},$$

implying

$$F_S = 2\pi\sqrt{z^2+a^2}\left(\sqrt{z^2+a^2} - z\right). \tag{5.81a}$$

It follows that

$$\omega = \frac{\text{area}}{(\text{radius})^2} = \frac{F_S}{(\sqrt{z^2+a^2})^2} = 2\pi\left(1 - \frac{z}{\sqrt{z^2+a^2}}\right). \tag{5.81b}$$

Hence, with Eq. (5.79) for the component $B_z$,

$$B_z = -k'I\frac{\partial}{\partial z}\omega = k'I2\pi\left\{\frac{1}{\sqrt{z^2+a^2}} - \frac{z^2}{(z^2+a^2)^{3/2}}\right\} = \frac{2\pi k'Ia^2}{(z^2+a^2)^{3/2}} \quad \text{teslas} \tag{5.82}$$

in agreement with Eq. (5.37).

(c) Next, we consider a cylindrical solenoid of radius $a$ and length $l$ as indicated in Fig. 5.17(c). The problem is to determine $B$ in axial direction

[14]This (relationship) is known as a theorem of Pappus, see e.g. Müller–Kirsten [101], p. 284 or Bullen [23], p. 171.

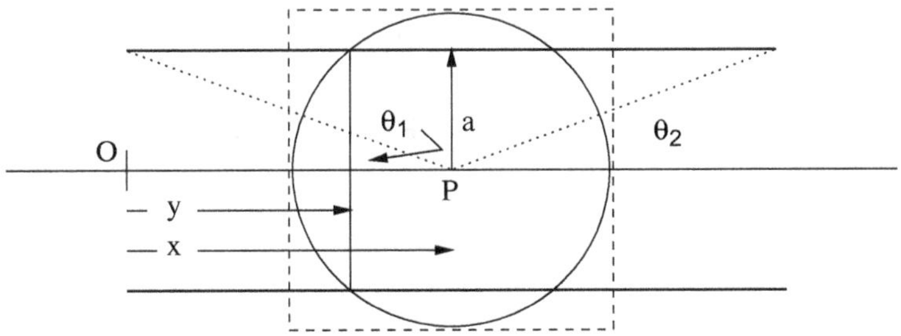


**Fig. 5.17(c).** The solenoid.

at point $P$, distance $x$ from $O$. First, we consider only one turn at $y$ and so $ndy$ turns in the immediate neighborhood. Hence, we calculate the solid angle subtended by the turns at $y$ at the point $P$. This solid angle is again calculated as above with the help of the enveloping cylinder (observe $\sqrt{a^2+(x-y)^2}$ is the radius of the sphere which is equal to the radius of the enveloping cylinder):

$$\omega = \frac{\text{area}}{(\text{radius})^2} = \frac{2\pi\sqrt{a^2+(x-y)^2}\{\sqrt{a^2+(x-y)^2}-(x-y)\}}{(\sqrt{a^2+(x-y)^2})^2}$$

$$= 2\pi\left\{1-\frac{x-y}{\sqrt{a^2+(x-y)^2}}\right\}. \tag{5.83}$$

Hence, $B_x$ with Eq. (5.79),

$$B_x^{(1)} = -k'I\frac{\partial}{\partial x}\left\{2\pi\left(1-\frac{x-y}{\sqrt{a^2+(x-y)^2}}\right)\right\}. \tag{5.84}$$

We have to sum over the field contributions due to all turns. The interval $dy$ at $y$ contains $ndy$ turns so that its field contribution is

$$dB_x = -k'I\frac{\partial}{\partial x}\left\{2\pi\left(1+\frac{y-x}{\sqrt{a^2+(x-y)^2}}\right)\right\}ndy \quad \text{teslas}, \tag{5.85}$$

and altogether over the length $l$,

$$\begin{aligned} B_x &= -k'I\frac{\partial}{\partial x}2\pi n\int_0^l dy\left(1+\frac{y-x}{\sqrt{a^2+(y-x)^2}}\right) \\ &= -k'I2\pi n\frac{\partial}{\partial x}\left\{l+\left[\sqrt{a^2+(y-x)^2}\right]_0^l\right\} \\ &= -k'I2\pi n\frac{\partial}{\partial x}\left\{l+\sqrt{a^2+(l-x)^2}-\sqrt{a^2+x^2}\right\} \\ &= k'In2\pi\left[\frac{l-x}{\sqrt{a^2+(l-x)^2}}+\frac{x}{\sqrt{a^2+x^2}}\right] \quad \text{teslas.} \end{aligned} \tag{5.86}$$

Hence, in terms of the angles in Fig 5.17(c),

$$B_x = k'In2\pi[\cos\theta_2+\cos\theta_1] \overset{l\to\infty}{\to} 4\pi k'In = \mu_0 In \quad \text{teslas} \tag{5.87}$$

in agreement with Eq. (5.73).[15]

**Example 5.8 (The quadratic current loop).** A current $I$ flows in a planar, quadratic loop of wire with sides of length $l$. Determine the field $\mathbf{B}$ at the center of the square.

**Solution:** We consider first the field at a point $P$, originating from one side of the quadratic loop. According to the Biot–Savart law (5.20), we have, see Fig. 5.18(a),

$$\mathbf{dB} = \frac{\mu_0}{4\pi}\frac{Id\mathbf{s}(\mathbf{r}')\times(\mathbf{r}-\mathbf{r}')}{|\mathbf{r}-\mathbf{r}'|^3} \quad \text{teslas,} \tag{5.88}$$

i.e. (see Fig. 5.18(a))

$$dB = \frac{\mu_0}{4\pi}\frac{Idx\sin\alpha PQ}{PQ^3} \quad \text{teslas.} \tag{5.89}$$

We have

$$PQ = \frac{r}{\sin\alpha}, \quad \tan\alpha = -\frac{r}{x}, \quad -x = r\cot\alpha, \quad dx = \frac{rd\alpha}{\sin^2\alpha} \tag{5.90}$$

so that

$$dB = \frac{\mu_0}{4\pi}\frac{Ird\alpha(\sin\alpha)^3}{(\sin\alpha)^2 r^2} = \frac{\mu_0}{4\pi}\frac{I}{r}\sin\alpha d\alpha, \tag{5.91}$$

[15]For a somewhat different derivation, see Attwood [5], pp. 262–264.

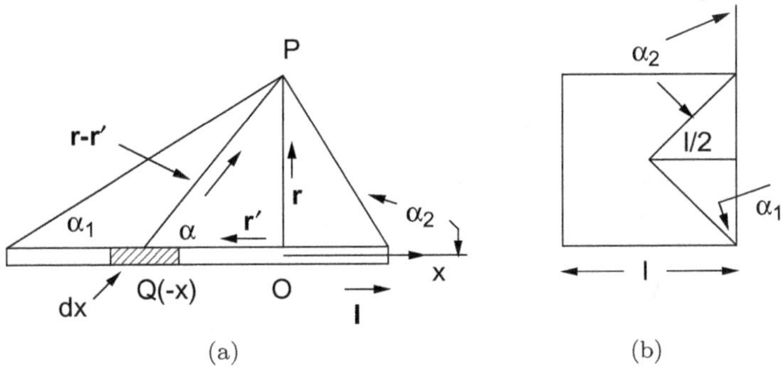


**Fig. 5.18.** (a) The field point $P$, (b) the quadratic loop.

and hence by integration (see Fig. 5.18(b)),

$$B = \frac{\mu_0}{4\pi}\frac{I}{r}(\cos\alpha_1 - \cos\alpha_2) \quad \text{teslas}. \tag{5.92}$$

The requested total field at $P$ is therefore the sum of four such contributions:

$$B = 4\frac{\mu_0}{4\pi}\frac{I}{l/2}(\cos 45^\circ - \cos 135^\circ) = \frac{\mu_0}{4\pi}\frac{4I\cdot 2}{l}\frac{2}{\sqrt{2}} = \frac{\mu_0}{4\pi}\frac{8\sqrt{2}I}{l} \quad \text{teslas}. \tag{5.93}$$

**Example 5.9 (An electron in a vector potential).** An electron at rest (initial velocity zero!) is exposed to electromagnetic radiation given by the vector potential

$$\mathbf{A} = (0, A(x - ct), 0). \tag{5.94}$$

Formulate the (non-relativistic) equation of motion of the electron and derive, one after the other, the $z$-, $y$-, $x$-components $(\dot{z}, \dot{y}, \dot{x})$ of the velocity of the electron. Assume that $A(u)$ tends to zero for $u \to \infty$.

**Solution:** The force acting on the electron is the *Lorentz force* (5.26). Hence, the equation of motion is with inclusion of an electric field $\mathbf{E}$

$$\dot{\mathbf{p}} = e(\mathbf{E} + \mathbf{v} \times \mathbf{B}) \quad \text{newtons}, \tag{5.95}$$

where $\mathbf{v} = (\dot{x}, \dot{y}, \dot{z})$ (with the electron initially at the origin). Here,[16] $\mathbf{E} = -\partial\mathbf{A}/\partial t$ (the contribution $-\boldsymbol{\nabla}\phi$ drops out with the given condition), and

[16]This is more fully explained after the introduction of time. See Section 7.5. Time variations of the magnetic field (which create a distributed electromotive force) thus imply in the electrostatic case a failure of a purely scalar representation of the electric field.

$\mathbf{B} = \boldsymbol{\nabla} \times \mathbf{A}$. Hence, in our case,

$$\mathbf{E} = -\left(0, \frac{\partial A(x-ct)}{\partial t}, 0\right) = -(0, -cA'(x-ct), 0), \tag{5.96}$$

where $A' = \partial A(x-ct)/\partial x$. Moreover,

$$\mathbf{B} = \boldsymbol{\nabla} \times \mathbf{A} = \begin{vmatrix} \mathbf{e}_x & \mathbf{e}_y & \mathbf{e}_z \\ \partial/\partial x & \partial/\partial y & \partial/\partial z \\ 0 & A(x-ct) & 0 \end{vmatrix} = (0, 0, A'), \tag{5.97}$$

i.e. $\mathbf{B} = (0, 0, A')$. Hence,

$$\mathbf{v} \times \mathbf{B} = (\dot{x}, \dot{y}, \dot{z}) \times (0, 0, A') = (\dot{y}A', -\dot{x}A', 0). \tag{5.98}$$

The expressions (5.96) and (5.98) now have to be inserted into (5.95) and yield

$$\dot{\mathbf{p}} = e(\dot{y}A', (c-\dot{x})A', 0). \tag{5.99}$$

Thus, $\dot{p}_z = 0$, i.e. $\dot{z} = 0$. For the $y$-component, we have

$$m\ddot{y} = e(c-\dot{x})A',\ c - \dot{x} = -\frac{d}{dt}(x-ct),$$

$$\therefore (c-\dot{x})A' = -A'\frac{d}{dt}(x-ct) = -\dot{A},$$

$$\therefore m\ddot{y} = -e\dot{A},$$

$$\therefore \dot{y} = -\frac{e}{m}A + \text{const.}, \quad \text{const.} = 0 \quad \text{with} \quad \dot{y} = 0, A = 0 \quad \text{at} \quad t = 0. \tag{5.100}$$

Finally, the $x$-component is given by

$$m\ddot{x} = e\dot{y}A', \quad \therefore \ddot{x} = -\frac{e^2}{m^2}AA' = -\frac{e^2}{m^2}\frac{1}{2(\dot{x}-c)}\frac{d}{dt}A^2. \tag{5.101}$$

Thus,

$$2\ddot{x}(\dot{x}-c) = -\frac{e^2}{m^2}\frac{d}{dt}A^2 \tag{5.102}$$

so that

$$\dot{x}^2 - 2c\dot{x} = -\frac{e^2}{m^2}A^2 + \underbrace{\text{const.}}_{0}, \quad \dot{x} = c \pm \sqrt{c^2 - \frac{e^2}{m^2}A^2}, \tag{5.103}$$

i.e.

$$\dot{x} = c - \sqrt{c^2 - \frac{e^2}{m^2}A^2} \simeq \frac{1}{2c}\left(\frac{eA}{m}\right)^2 \quad \text{m/s}. \tag{5.104}$$

## 5.8 Exercises

**Example 5.10 (Force of field on magnetic moment).** Show that the force $\mathbf{F}$ exerted by a static field $\mathbf{B}(\mathbf{r})$ on a magnetic dipole moment $\mathbf{m}$ at the point $\mathbf{r}$ is

$$\mathbf{F} = (\mathbf{m} \cdot \boldsymbol{\nabla})\mathbf{B}(\mathbf{r}) \quad \text{newtons}. \tag{5.105}$$

**Example 5.11 (B does not contribute to kinetic energy).** Show by forming the scalar product of the Lorentz force with the velocity $\mathbf{v}$ of a charge that the magnetic field $\mathbf{B}(\mathbf{r})$ does not contribute to the kinetic energy of the charge, i.e. that in a magnetic field the energy of a charged particle is constant.

**Example 5.12 (Vector potential for a uniform field B).** In Eq. (5.97), the magnetic flux density $\mathbf{B}$ was written in matrix form. What are three possible vector potentials $\mathbf{A}$ for a uniform magnetic field $B$ in the $z$-direction (e.g. one is $\mathbf{A} = (0, Bx, 0)$)? Verify in each case that $\boldsymbol{\nabla} \cdot \mathbf{A} = 0$.

**Example 5.13 (Non-uniqueness of vector potential A).** Show that $\mathbf{B} = \boldsymbol{\nabla} \times \mathbf{A}$ is unchanged if $f(x)$ is added to $A_x$, $g(y)$ to $A_y$, and $h(z)$ to $A_z$.

**Example 5.14 (Maximum orbital radius of electron in magnetic field).** What would be the maximum orbital radius for electrons moving with speed $v$ in a constant uniform magnetic field?

**Example 5.15 (Distribution of magnetic dipoles).** By linking the magnetic moment $\mathbf{m}$ of Eq. (5.39) with the scalar magnetic potential $V_m$ of Eq. (5.80) show that the magnetic field due to a closed current loop is the same as that from a distribution of magnetic dipoles of moment $I$ per unit area covering any surface which has the current loop as perimeter.

**Example 5.16 (The solenoid of finite length).** A solenoid of $n$ turns per unit length has length $l$ and radius $l/2$. What is the magnetic flux density $B$ at the center?

**Example 5.17 (Magnetic scalar potential of a straight wire).** The field $\mathbf{B}$ around a long straight wire carrying current $I$ amperes was obtained in Example 5.2. Use this result and cylindrical polar coordinates $(\rho, \theta)$ in order to derive the scalar magnetic scalar potential $V_m$.

(Answer: $-I\theta/2\pi$ amperes).[17]

**Example 5.18 (Magnetic dipole above circular wire).** A magnetic dipole of moment $\mathbf{m}$ is placed on and parallel to the axis of a circular wire of radius $a$ which carries the current $I$. The dipole is placed at a point a distance $D$ from the center of the wire. Show that for $D \gg a$ the force $\mathbf{F}$ acting on the circuit is given by

$$\mathbf{F} = \frac{3\mu_0 I \mathbf{m} a^2}{2D^4} \quad \text{newtons.} \tag{5.106}$$

**Example 5.19 (The macroscopic scalar potential).** Proceeding as explained after Eq. (5.80) show that the macroscopic scalar potential is given by

$$V_m(\mathbf{r}) = \frac{1}{4\pi}\int dV \mathbf{M}(\mathbf{r}') \cdot \boldsymbol{\nabla}_{\mathbf{r}'} \frac{1}{|\mathbf{r} - \mathbf{r}'|} \quad \text{amperes,} \tag{5.107}$$

and that this can be re-expressed in the form ($\mathbf{M}$ being defined in Chapter 6 after Eq. (6.18))

$$V_m(\mathbf{r}) = \frac{1}{4\pi}\left[\int_{F'} \frac{\mathbf{M}(\mathbf{r}') \cdot d\mathbf{F}'}{|\mathbf{r} - \mathbf{r}'|} - \int_{V'} \frac{\boldsymbol{\nabla}_{\mathbf{r}'} \cdot \mathbf{M}(\mathbf{r}')}{|\mathbf{r} - \mathbf{r}'|} dV'\right]. \tag{5.108}$$

Compare this result with the corresponding result in electrostatics (cf. Eq. (4.18)).

**Example 5.20 (Magnetic flux of a solenoid).** A solenoid consists of $N$ turns of wire wound round a hollow pipe of length $L$ and having a cross-sectional area of $A$ meters squared. The resistance of the wire is $R$ ohms. A voltage of $V$ volts is applied across the wire. What is the magnetic flux $\Phi$?

(Answer: $VN\mu_0 A/RL$ Wb).

**Example 5.21 (The magnetic field inside a solenoid).** In Example 5.4, the magnetic field $B$ at $z$ on the axis of a circular wire loop of radius $a$ centered at the origin and carrying the current of $I$ amperes was shown to

[17] For discussion of the non-uniqueness of the potential (the field not being conservative), see Lawden [82], pp. 66–67.

be (cf. Eq. (5.37))

$$B = \frac{\mu_0 I}{2} \frac{a^2}{(z^2 + a^2)^{3/2}} \quad \text{teslas}. \tag{5.109}$$

Integrate this result to show that the magnetic field inside an infinitely long solenoid with $n$ loops and carrying current $I$ is $\mu_0 I n$.

**Example 5.22 (Magnetic moment of oscillating charge).** A charge $q$ is moving with simple harmonic motion $\mathbf{r}(t) = \mathbf{r}_0 \sin(\omega t)$ about the origin. Show that the magnetic moment $\mathbf{m} = (q/2)(\mathbf{r} \times \mathbf{v})$, $\mathbf{v}$ the velocity of the charge, is constant (and therefore does not radiate, as can be deduced from the results of Chapter 10).

**Example 5.23 (Trajectory of a charged particle in a magnetic field B).** The equation of motion of a particle of mass $m$ and charge $e$ in a uniform magnetic field $\mathbf{B}$ is given by

$$m\ddot{\mathbf{r}} = e\dot{\mathbf{r}} \times \mathbf{B}, \quad \mathbf{B} = B_0 \mathbf{e}_z. \tag{5.110}$$

Show that the trajectory of the particle is a helix:

$$\begin{aligned} x(t) = A\sin(\Omega t + \alpha) + x_0, \quad y(t) = A\cos(\Omega t + \alpha), \\ z(t) = at + b, \quad \Omega = |e|B_0|m. \end{aligned} \tag{5.111}$$

See e.g. the work of Coulson and Boyd [35] or Cheston [30], pp. 323–325.

# Chapter 6

# Macroscopic Magnetostatics

## 6.1 Introductory Remarks

Up to this point we have only considered currents $I$ or precisely known current densities $\mathbf{j}$ which describe a stationary microscopic flow of charge. In macroscopic problems — and this means in continuous media — the total current density is not known precisely. The atomic or molecular ring-like currents (and spins) of the electrons in matter determine the latter's magnetic properties, whose effective magnetic moments contribute to the vector potential $\mathbf{A}(\mathbf{r})$, as also the conduction electrons of the macroscopic transport of charge. For macroscopic effects again only an averaging over a macroscopic volume is meaningful. We can keep the microscopic current density $\mathbf{j}$ considered thus far as the density of macroscopic charge transport; this has to be supplemented, however, by the contributions of microscopic molecular currents. Hence we write

$$\text{total current density} = \mathbf{j} + \mathbf{j}_{\text{mol}} \quad \text{amperes/meter}^2. \tag{6.1}$$

We are interested here in *currents in conductors*. We defined the ideal conductor electrostatically as an *equipotential domain* in which the electrons can move about without doing work. If a *potential difference* $V$ is applied, and hence a field $\mathbf{E}$,

$$\phi \equiv V = \int \mathbf{E} \cdot d\mathbf{s} \quad \text{volts}, \tag{6.2}$$

e.g. $V = V_1 - V_2 = EL$ ($L$ = length of the conductor), then the conduction electrons move in the direction determined by $\mathbf{E}$. One thus has a current $I$ and a current density $\mathbf{j}$. However, in general these conduction electrons can not move about completely freely. The nuclei of the atoms of the conductor occupy points in the lattice of the rigid body structure of the metal, and this lattice interacts with the conduction electrons, although relatively weakly. Something similar applies in the case of an externally applied magnetic field. The *resistance of the lattice* is evident in *Ohm's law*, i.e. in the linear relation

$$\mathbf{j} = \sigma \mathbf{E} \quad \text{amperes/m}^2, \tag{6.3}$$

where $\sigma = 1/\rho$ ohm$^{-1}$meter$^{-1}$ is the *conductivity* (gasses, for instance, are not linear conductors, i.e. in their case Eq. (6.3) does not apply, instead $\mathbf{j} = f(\mathbf{E})$). Hence, with $F$ the appropriate cross sectional area ($I = \int \mathbf{j} \cdot d\mathbf{F}$),

$$I = Fj = F\sigma E \quad \text{amperes}, \tag{6.4}$$

and so (see above)[1]

$$\frac{V}{I} = \frac{EL}{F\sigma E} = \frac{L}{F\sigma} \equiv \rho\frac{L}{F} \equiv R \quad \text{ohms}. \tag{6.5}$$

Here $\rho$ is the *resistivity*, also called *specific resistance*, and $R$ the *resistance*, and we see that this depends on the geometry of the conductor. Here we usually avoid the use of $\rho$ for resistivity since this can be confused with $\rho$ for charge density. Thus the charge density appears in the product $\mathbf{E} \cdot \mathbf{j} = \rho \mathbf{E} \cdot d\mathbf{s}/dt$ which (since it represents work per volume and time) is the energy transfered per unit volume and per unit time from the conduction electrons to the lattice. Note the difference between resistance and resistivity, and correspondingly the difference between conductance (measured in siemens (S) or ohms$^{-1}$ ($\Omega^{-1}$)) and conductivity (measured in (ohm $\cdot$ meter)$^{-1}$).

## 6.2 Macroscopic Magnetization

We return to the original considerations of this chapter, i.e. to those pertaining to continuous media. In keeping with the subdivision of the entire current density we write the total vector potential correspondingly

$$\mathbf{A} = \mathbf{A}_{\text{macr}} + \mathbf{A}_{\text{mol}} \quad \text{tesla} \cdot \text{meters} \tag{6.6}$$

or, using Eq. (5.53),

$$\mathbf{A} = k' \int d\mathbf{r}' \frac{\mathbf{j}(\mathbf{r}')}{|\mathbf{r} - \mathbf{r}'|} + k' \int d\mathbf{r}' \frac{\mathbf{j}_{\text{mol}}(\mathbf{r}')}{|\mathbf{r} - \mathbf{r}'|} \quad \text{tesla} \cdot \text{meters}, \tag{6.7}$$

[1]Observe that in Gaussian units, in which (cf. Eq. (5.15)) the conductivity is given in seconds$^{-1}$, the resistance is given in seconds/cm.

where $\mathbf{j}$ is the conduction current density, $\mathbf{j}_{\rm mol}$ the contribution of the molecular ring currents and $k' = \mu_0/4\pi$. On the other hand we saw in the preceding chapter, Eq. (5.48), that the field $\mathbf{B}$ of a magnetic dipole moment $\mathbf{m}$ is given by

$$\mathbf{B} = k'\frac{3\mathbf{n}(\mathbf{n}\cdot\mathbf{m}) - \mathbf{m}}{r^3} \quad \text{with} \quad \mathbf{n} = \frac{\mathbf{r}}{r} \quad \text{and} \quad \mathbf{B} = \boldsymbol{\nabla}\times\mathbf{A}. \tag{6.8}$$

We verify first that the vector potential of this expression for $\mathbf{B}$ is given by

$$\mathbf{A} = k'\frac{\mathbf{m}\times\mathbf{r}}{r^3} \quad \text{tesla}\cdot\text{meters}, \tag{6.9}$$

as stated earlier (cf. Eq. (5.49)), and our intention is to express $\mathbf{A}_{\rm mol}$ in this form.

**Example 6.1 (Vector potential A of a magnetic dipole moment m).** Verify Eq. (6.9).

**Solution:** We use the formula

$$\boldsymbol{\nabla}\times(\mathbf{a}\times\mathbf{b}) = \mathbf{a}(\boldsymbol{\nabla}\cdot\mathbf{b}) - \mathbf{b}(\boldsymbol{\nabla}\cdot\mathbf{a}) + (\mathbf{b}\cdot\boldsymbol{\nabla})\mathbf{a} - (\mathbf{a}\cdot\boldsymbol{\nabla})\mathbf{b}, \tag{6.10}$$

which implies

$$\begin{aligned}\boldsymbol{\nabla}\times\left(\frac{\mathbf{m}\times\mathbf{r}}{r^3}\right) &= \boldsymbol{\nabla}\times\left(\mathbf{m}\times\frac{\mathbf{r}}{r^3}\right)\\ &= \mathbf{m}\left(\boldsymbol{\nabla}\cdot\frac{\mathbf{r}}{r^3}\right) - \frac{\mathbf{r}}{r^3}(\boldsymbol{\nabla}\cdot\mathbf{m}) + \left(\frac{\mathbf{r}}{r^3}\cdot\boldsymbol{\nabla}\right)\mathbf{m} - (\mathbf{m}\cdot\boldsymbol{\nabla})\frac{\mathbf{r}}{r^3}\\ &= \mathbf{m}\left(\boldsymbol{\nabla}\cdot\frac{\mathbf{r}}{r^3}\right) - (\mathbf{m}\cdot\boldsymbol{\nabla})\frac{\mathbf{r}}{r^3},\end{aligned} \tag{6.11}$$

since $\mathbf{m} = \text{const}$. Since, moreover,

$$\frac{\partial r}{\partial x} = \frac{\partial\sqrt{x^2+y^2+z^2}}{\partial x} = \frac{x}{r}, \quad \text{or} \quad \boldsymbol{\nabla}r = \frac{\mathbf{r}}{r}, \tag{6.12}$$

and

$$\frac{\partial}{\partial x}\left(\frac{x}{r^3}\right) = \frac{1}{r^3} - \frac{3x}{r^4}\frac{\partial r}{\partial x} = \frac{1}{r^3} - \frac{3x}{r^4}\frac{x}{r}, \tag{6.13}$$

so that

$$\left(\boldsymbol{\nabla}\cdot\frac{\mathbf{r}}{r^3}\right) = \frac{3}{r^3} - \frac{3}{r^5}r^2 = 0, \tag{6.14}$$

it follows that

$$\nabla \times \left(\frac{\mathbf{m} \times \mathbf{r}}{r^3}\right) = -(\mathbf{m} \cdot \nabla)\frac{\mathbf{r}}{r^3}. \tag{6.15}$$

However,

$$\begin{aligned}(\mathbf{m} \cdot \nabla)\frac{\mathbf{r}}{r^3} &= m_x \frac{\partial}{\partial x}\left(\frac{\mathbf{r}}{r^3}\right) + \cdots = m_x\left(\frac{\mathbf{e}_x}{r^3} - \frac{3}{r^4}\mathbf{r}\frac{x}{r}\right) + \cdots \\ &= \frac{\mathbf{m}r^2}{r^5} - \frac{3\mathbf{r}(\mathbf{m} \cdot \mathbf{r})}{r^5}.\end{aligned}$$

It follows that

$$\nabla \times \left(\frac{\mathbf{m} \times \mathbf{r}}{r^3}\right) k' = k' \frac{3\mathbf{r}(\mathbf{m} \cdot \mathbf{r}) - \mathbf{m}r^2}{r^5} \tag{6.16}$$

in agreement with the above claim.

Thus we can write for the vector potential at the point $\mathbf{r}$, which is due to molecule "$i$" ($\mathbf{m}_i$ the latter's total magnetic moment) as in Eq. (6.9),

$$\mathbf{A}_{\text{mol,i}}(\mathbf{r}) = k' \frac{\mathbf{m}_i \times (\mathbf{r} - \mathbf{r}_i)}{|\mathbf{r} - \mathbf{r}_i|^3} \quad \text{tesla} \cdot \text{meters}. \tag{6.17}$$

Hence the vector potential of Eq. (6.7) becomes

$$\mathbf{A}(\mathbf{r}) = k' \int d\mathbf{r}' \frac{\mathbf{j}(\mathbf{r}')}{|\mathbf{r} - \mathbf{r}'|} + k' \sum_i \frac{\mathbf{m}_i \times (\mathbf{r} - \mathbf{r}_i)}{|\mathbf{r} - \mathbf{r}_i|^3} \quad \text{tesla} \cdot \text{meters}. \tag{6.18}$$

We now define as

$N(\mathbf{r})$: the *average number of molecules per unit volume* at $\mathbf{r}$,
$\langle \mathbf{m}(\mathbf{r}) \rangle$: the *average molecular magnetic dipole moment* at $\mathbf{r}$,
$\mathbf{M}:=\mathrm{N}\langle \mathbf{m} \rangle$: the *macroscopic magnetization*, i.e. *magnetic dipole density.*

Hence we can write

$$\Sigma_i(\text{contribution})_i = \int (\text{density of contributions}) d\mathbf{r}', \tag{6.19}$$

and thus Eq. (6.18) becomes

$$\begin{aligned}\mathbf{A}(\mathbf{r}) &= k' \int d\mathbf{r}' \frac{\mathbf{j}(\mathbf{r}')}{|\mathbf{r} - \mathbf{r}'|} + k' \int d\mathbf{r}' \frac{\mathbf{M}(\mathbf{r}') \times (\mathbf{r} - \mathbf{r}')}{|\mathbf{r} - \mathbf{r}'|^3} \\ &\overset{(5.50)}{=} k' \int d\mathbf{r}' \frac{\mathbf{j}(\mathbf{r}')}{|\mathbf{r} - \mathbf{r}'|} + k' \int d\mathbf{r}' \mathbf{M}(\mathbf{r}') \times \nabla_{\mathbf{r}'} \frac{1}{|\mathbf{r} - \mathbf{r}'|}.\end{aligned} \tag{6.20}$$

For the curl of the product of a scalar function $\phi$ multiplied by a vector function $\mathbf{M}$ the following formula can be found in Appendix C or in Tables of Formulas:

$$\boldsymbol{\nabla} \times (\phi\mathbf{M}) = (\boldsymbol{\nabla}\phi) \times \mathbf{M} + \phi\boldsymbol{\nabla} \times \mathbf{M}, \tag{6.21}$$

or for minus the first term on the right side of this equation

$$-(\boldsymbol{\nabla}\phi) \times \mathbf{M} = \mathbf{M} \times (\nabla\phi) = -\boldsymbol{\nabla} \times (\phi\mathbf{M}) + \phi\boldsymbol{\nabla} \times \mathbf{M}. \tag{6.22}$$

With $\phi = 1/|\mathbf{r} - \mathbf{r}'|$ we obtain for the integral in the second term on the right side of Eq. (6.20):

$$\begin{aligned}&\int d\mathbf{r}'\mathbf{M}(\mathbf{r}') \times \boldsymbol{\nabla}_{\mathbf{r}'}\frac{1}{|\mathbf{r} - \mathbf{r}'|}\\ &\quad = -\int d\mathbf{r}'\boldsymbol{\nabla}' \times \left(\frac{\mathbf{M}}{|\mathbf{r} - \mathbf{r}'|}\right) + \int d\mathbf{r}'\frac{1}{|\mathbf{r} - \mathbf{r}'|}\boldsymbol{\nabla}' \times \mathbf{M}.\end{aligned} \tag{6.23}$$

But, as we shall verify in a moment,

$$\int d\mathbf{r}\boldsymbol{\nabla} \times \mathbf{B} = \int d\mathbf{F} \times \mathbf{B}. \tag{6.24}$$

**Example 6.2 (Verification of Eq. (6.24)).** Verify Eq. (6.24) by scalar multiplication by a constant vector $C$.

**Solution:** For the verification we multiply the right hand side of Eq. (6.24) by an arbitrary vector $\mathbf{C} =$ const. and consider

$$\mathbf{C} \cdot \int d\mathbf{F} \times \mathbf{B} \equiv \int \mathbf{C} \cdot (\mathbf{n} \times \mathbf{B})dF. \tag{6.25}$$

For the *scalar triple product* we have

$$\mathbf{C} \cdot (\mathbf{n} \times \mathbf{B}) = (\mathbf{C} \times \mathbf{n}) \cdot \mathbf{B} = \mathbf{B} \cdot (\mathbf{C} \times \mathbf{n}) = (\mathbf{B} \times \mathbf{C}) \cdot \mathbf{n}. \tag{6.26}$$

Hence

$$\begin{aligned}\mathbf{C} \cdot \int d\mathbf{F} \times \mathbf{B} &= \int (\mathbf{B} \times \mathbf{C}) \cdot d\mathbf{F}\\ &\overset{\text{Gauss}}{=} \int d\mathbf{r}\boldsymbol{\nabla} \cdot (\mathbf{B} \times \mathbf{C}) \overset{\text{see below}}{=} \int d\mathbf{r}\mathbf{C} \cdot (\boldsymbol{\nabla} \times \mathbf{B}),\end{aligned} \tag{6.27}$$

where in the last step we used for $\mathbf{u} = \mathbf{B}$ and $\mathbf{v} = \mathbf{C}$ and $\mathbf{C} =$ const. the formula (see Appendix C):

$$\begin{aligned} \boldsymbol{\nabla} \cdot (\mathbf{u} \times \mathbf{v}) &= \mathbf{v} \cdot (\boldsymbol{\nabla} \times \mathbf{u}) - \mathbf{u} \cdot (\boldsymbol{\nabla} \times \mathbf{v}), \\ \boldsymbol{\nabla} \cdot (\mathbf{B} \times \mathbf{C}) &= \mathbf{C} \cdot (\boldsymbol{\nabla} \times \mathbf{B}). \end{aligned} \tag{6.28}$$

Hence Eq. (6.24) applied to the first integral on the right hand side of Eq. (6.23) yields

$$-\int d\mathbf{r}' \boldsymbol{\nabla}' \times \left( \frac{\mathbf{M}}{|\mathbf{r} - \mathbf{r}'|} \right) = -\int d\mathbf{F}' \times \frac{\mathbf{M}}{|\mathbf{r} - \mathbf{r}'|}. \tag{6.29}$$

Here (see definition after Eq. (6.18)) $\mathbf{M}(\mathbf{r}') = N(\mathbf{r}')\langle \mathbf{m} \rangle$. The integration extends over the entire volume of the macroscopic body. Assuming localization, there are no molecules outside and hence the surface integral there vanishes. Then

$$\int_{F'} d\mathbf{F}' \times \frac{\mathbf{M}(\mathbf{r}')}{|\mathbf{r} - \mathbf{r}'|} = 0, \tag{6.30}$$

and with Eqs. (6.20) and (6.23) we are left with

$$\mathbf{A}(\mathbf{r}) = k' \int d\mathbf{r}' \frac{\mathbf{j}(\mathbf{r}') + \boldsymbol{\nabla}' \times \mathbf{M}(\mathbf{r}')}{|\mathbf{r} - \mathbf{r}'|} \quad \text{tesla} \cdot \text{meters} \tag{6.31}$$

for the vector potential at a point $\mathbf{r}$ outside the matter of the macroscopic body. Thus the macroscopic magnetization $\mathbf{M}$ implies a *magnetization current*

$$\mathbf{j}_M = \boldsymbol{\nabla} \times \mathbf{M} \quad \text{amperes/m}^2. \tag{6.32}$$

We therefore obtain for the macroscopic flux density $\mathbf{B}$ resulting from the total current density (6.1):

$$\boldsymbol{\nabla} \times \mathbf{B} = \mu_0(\mathbf{j}(\mathbf{r}) + \mathbf{j}_M(\mathbf{r})) = \mu_0 \mathbf{j}_{\text{total}} \quad \text{teslas/m}, \tag{6.33}$$

where $\mathbf{j}_{\text{total}}$ is the total current density. With Eq. (6.32) we then have

$$\boldsymbol{\nabla} \times \mathbf{B} = \mu_0(\mathbf{j} + \boldsymbol{\nabla} \times \mathbf{M}), \tag{6.34}$$

or

$$\boldsymbol{\nabla} \times (\mathbf{B} - \mu_0 \mathbf{M}) = \mu_0 \mathbf{j}, \tag{6.35}$$

or, introducing the vector $\mathbf{H} = (\mathbf{B} - \mu_0\mathbf{M})/\mu_0$,

$$\nabla \times \mathbf{H} = \mathbf{j} \quad \text{amperes/meter}^2 \tag{6.36}$$

with (units are discussed below)

$$\mathbf{H} = \frac{1}{\mu_0}\mathbf{B} - \mathbf{M} \quad \text{amperes/meter}, \quad \mathbf{B} = \mu_0(\mathbf{H} + \mathbf{M}) \quad \text{teslas}. \tag{6.37}$$

The vector $\mathbf{H}$ is called *magnetic field strength.* This field strength $\mathbf{H}$ does not depend on the molecular magnetic moments. (Note the analogy with electrostatics: There, since $\nabla \cdot \mathbf{D} = \rho$, the polarization charges with density $\rho_P$ are not sources of $\mathbf{D}$). This is the reason why in many texts $\mathbf{H}$ is introduced first, and not $\mathbf{B}$, since the current is first considered without the effect of magnetization. This means the $\mathbf{B}$ we introduced originally is $\mu_0\mathbf{H}$. In vacuum $\mathbf{B} = \mu_0\mathbf{H}$, but within a magnetic material $\mathbf{B} = \mu_0(\mathbf{H}+\mathbf{M})$.

The integral form of Eq. (6.36) is very important in applications, that is

$$\int_F (\nabla \times \mathbf{H}) \cdot d\mathbf{F} \overset{\text{Stokes}}{=} \oint_{C(F)} \mathbf{H} \cdot d\mathbf{l} = \int_F \mathbf{j} \cdot d\mathbf{F} \quad \text{amperes}. \tag{6.38}$$

On the right hand side we do not write $I$, because in general (i.e. depending on the problem) the current can cross the area $F$ several times and in different directions, and then the right hand side is an integral multiple of $I$. The formula (6.38) will be an important starting point in problems to be discussed later.

For many cases, i.e. types of matter, one observes the linearity

$$\mathbf{M} = \kappa\mathbf{H} \quad \text{amperes/meter}, \quad \kappa = \text{const}. \tag{6.39}$$

In these *linear* cases

$$\mathbf{B} = \mu_0(\mathbf{H} + \mathbf{M}) = \mu\mathbf{H} \quad \text{teslas}, \quad \mu = \mu_0(1 + \kappa) \quad \text{henries/meter}. \tag{6.40}$$

Today the unit of *magnetic field strength* is amperes/meter, written A/m, and sometimes ampere-revolutions/meter (the obsolete unit is oersted (Oe)). The magnetic permeability $\mu$ is measured in units of newtons/ampere$^2$, and $\mu_0 = 1.257 \times 10^{-6}$ newtons/ampere$^2$ is called the *magnetic constant* or the *permeability of free space.* Thus here in *macroscopic* magnetostatics we have

$$\mathbf{A}(\mathbf{r}) = k' \int d\mathbf{r}' \frac{\mathbf{j}_{\text{total}}(\mathbf{r}')}{|\mathbf{r} - \mathbf{r}'|} \quad \text{tesla} \cdot \text{meters}, \tag{6.41}$$

where

$$\begin{aligned}\mathbf{j}_{\text{total}} &= \mathbf{j} + \mathbf{j}_M = \mathbf{j} + \nabla \times \mathbf{M} \\ &= \mathbf{j} + \nabla \times (\kappa \mathbf{H}) = \mathbf{j} + \kappa \nabla \times \mathbf{H} \\ &= \mathbf{j} + \kappa \mathbf{j} \equiv \frac{\mu}{\mu_0}\mathbf{j} \quad \text{amperes/m}^2, \end{aligned} \tag{6.42}$$

and $\mu = \mu_0(1+\kappa)$. Thus we have here in macroscopic magnetostatics

$$\mathbf{A}(\mathbf{r}) = k' \frac{\mu}{\mu_0} \int d\mathbf{r}' \frac{\mathbf{j}(\mathbf{r}')}{|\mathbf{r} - \mathbf{r}'|} \quad \text{tesla} \cdot \text{meters}, \tag{6.43}$$

where

$$k' \frac{\mu}{\mu_0} = \frac{\mu_0}{4\pi} \frac{\mu}{\mu_0} = \frac{\mu}{4\pi}. \tag{6.44}$$

Thus one has almost the same formula (in these linear cases) as in the microscopic case, except that $\mu_0 \to \mu$, with $\mu$ incorporating matter effects.

## 6.3 Magnetic Properties of Matter

We saw above that

$$\frac{\mu}{\mu_0} = 1 + \kappa. \tag{6.45}$$

Here $\mu/\mu_0$ is the *relative permeability of the medium* and $\kappa$ the *magnetic susceptibility*. One distinguishes between diamagnetic, paramagnetic and ferromagnetic materials.

*Diamagnetic materials* are those consisting of atoms, for which the sum of the individual orbital angular momenta of the electrons is zero. Such atoms therefore do not possess a permanent total magnetic moment. However, when placed in an external field **B**, they develop an induced magnetic moment on account of their motion in this field **B** (in accordance with the law of induction to be discussed in Chapter 7 which implies an induced electromotive force which accelerates the electrons). According to the law of induction (cf. Section 7.2: the Lenz rule) the induced magnetic moment or rather its field has the direction opposite to that of the applied field **B**, so that

$$\kappa < 0, \quad \mu < 1.$$

The effect is independent of temperature (as the explanation implies).

*Paramagnetic materials* are those whose atoms possess permanent magnetic moments (in fact, their electrons outside closed electronic shells), which favour alignment with the applied field, so that for these materials

$$\kappa > 0, \quad \mu > 1.$$

(In the case of these the effect of induced magnetic moments is much smaller).

*Ferromagnetic materials* are those, for which

$$\kappa > 0, \quad \text{and} \quad \mu = \mu(\mathbf{H}) \gg 1$$

(see discussion of *hysteresis curves* ($B$ versus $H$) e.g. in literature such as Jackson [73]). In the case of these materials the connection between **B** and **H** depends on the past of the system; the connection is therefore not unique. In addition the connection depends on the temperature (above the so-called Curie-temperature ferromagnetic materials become paramagnetic).

## 6.4 Energy of the Magnetic Field

For the *magnetic energy* one obtains, as we shall show in Section 7.3, an expression very similar to that for electric energy. We cite the result at this point for reasons of completeness:

$$W = \frac{1}{2}\int d\mathbf{r}\, \mathbf{B}\cdot\mathbf{H} \quad \text{joules.} \tag{6.46}$$

## 6.5 Behavior of B and H at Boundary Surfaces

It is instructive to keep in mind the corresponding results in electrostatics. In the case of the field **E** we saw that *without considerations of properties of matter* at a thin charged surface (see Eq. (2.90))

$$(\mathbf{E}_1 - \mathbf{E}_2)\cdot\mathbf{n} = 4\pi k\sigma, \quad k = \frac{1}{4\pi\epsilon_0}, \tag{6.47}$$

or *in the case of macroscopic electrostatics*, which takes the polarization of the dielectric into account,

$$(\mathbf{E}_1 - \mathbf{E}_2)\cdot\mathbf{n} = 4\pi k(\sigma + \sigma_P), \tag{6.48}$$

and (cf. Eq. (4.40))

$$(\mathbf{D}_1 - \mathbf{D}_2)\cdot\mathbf{n} = 4\pi k\sigma. \tag{6.49}$$

In the case of the field **B** we have $\boldsymbol{\nabla}\cdot\mathbf{B}=0$, i.e.

$$0=\int_V dV\boldsymbol{\nabla}\cdot\mathbf{B}=\int_{F(V)}\mathbf{B}\cdot d\mathbf{F}, \tag{6.50}$$

so that

$$(\mathbf{B}_2-\mathbf{B}_1)\cdot\mathbf{n}\triangle F=0. \tag{6.51}$$

The zero on the right side implies that no single magnetic poles exist. Hence the normal component of the field at the surface is continuous, i.e.

$$B_{2n}=B_{1n}. \tag{6.52}$$

In the case of the field **E** we obtained from $\boldsymbol{\nabla}\times\mathbf{E}=0$ the relation (4.44),

$$E_{2t}=E_{1t}. \tag{6.53}$$

In the case of **H** we have

$$\boldsymbol{\nabla}\times\mathbf{H}=\mathbf{j}, \tag{6.54}$$

i.e.

$$\int_F\boldsymbol{\nabla}\times\mathbf{H}\cdot d\mathbf{F}=\oint_C\mathbf{H}\cdot d\mathbf{s}=\int_F\mathbf{j}\cdot d\mathbf{F}=I. \tag{6.55}$$

We now consider the contribution on the right side in relation to an element of area $d\mathbf{F}$ perpendicular to the interface between two media as in Fig. 6.1. Then $d\mathbf{F}$ is a vector in the plane of the interface. Applied to this element of area we can write Eq. (6.55):

$$\int_{\triangle F}\mathbf{j}\cdot d\mathbf{F}=\mathbf{j}\cdot\triangle\mathbf{F}=jld,$$

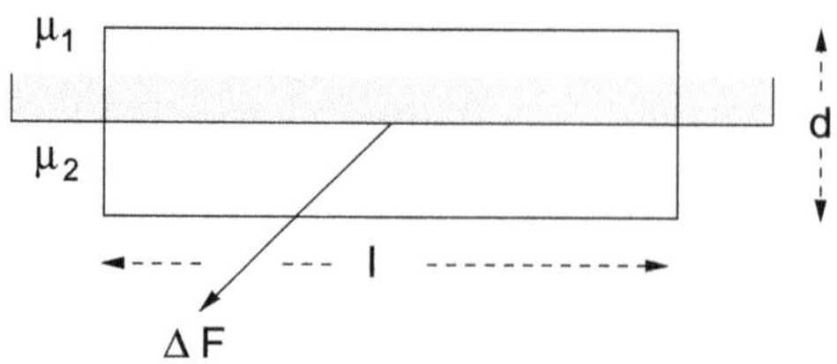


**Fig. 6.1.** Element of area $\triangle F$.

where $l$ is the length and $d$ the height of the element of area. For conductors we have Ohm's law[2] $\mathbf{j} = \sigma \mathbf{E}$ or $j = \sigma E$. In ideal conductors $\sigma = \infty$ with $\mathbf{E} = -\boldsymbol{\nabla}\phi = 0$. It is then possible that

$$jld = \sigma Eld \neq 0 \quad \text{and} \quad \text{finite}, \tag{6.56}$$

and one defines (cf. Eqs. (5.6) and (5.7))

$$\lim_{\sigma\to\infty, d\to 0} \sigma Ed := K, \quad \int \mathbf{j}\cdot d\mathbf{F} = Kl, \tag{6.57}$$

where $K$ is a surface current density,[3] $\mathbf{j}$ is the volume current density. $K$ is also called *linear current density*.[4] If the conductivity $\sigma$ is finite, we have $K = 0$; i.e. in this case there is actually no surface current.[5]

The continuity condition for $\mathbf{H}$ can now be written (using $\oint_C \mathbf{H}\cdot d\mathbf{s}$ as in the case of $\mathbf{E}$)

$$H_{2t} - H_{1t} = K \quad (= 0 \text{ for conductivity } \sigma \text{ finite}), \tag{6.58}$$

i.e. (in different formulation required in Chapter 11)

$$\mathbf{n} \times (\mathbf{H}^{(2)} - \mathbf{H}^{(1)}) = \mathbf{K}. \tag{6.59}$$

The surface current introduced here is the same as the surface current $K$ referred to earlier as in Eq. (5.6). The name "linear current density" has its origin in the fact that in Eq. (6.55),

$$H_{2t} - H_{1t} = \frac{I}{2l} \quad \text{amperes/meter}, \tag{6.60}$$

the current appears linearly. This result is of considerable importance and, like its counterpart (2.90) in electrostatics, will be used in examples, as in Example 14.2.

---

[2]One should not confuse $\sigma$ here, the conductivity, with $\sigma$ in Eq. (5.6), the charge per unit area. This is, unfortunately, standard notation.
[3]See Jackson [73], p. 19.
[4]See Lim [88], p. 42.
[5]See Jackson [73], p. 336. Nonetheless an effective surface current will be defined later in the treatment of wave guides, in fact with the help of $\mathbf{j}$.

## 6.6 Current Circuits Compared with Flux Circuits

It is interesting and instructive to enquire about analogies between electric and magnetic properties. In an electric circuit, i.e. *current carrying circuit*, with stationary current $I = jF$ defined by the condition $\nabla \cdot \mathbf{j} = 0$, the current $I$ is the constant quantity. The potential $V$, or the corresponding *electromotive force* of the circuit, is with $\mathbf{j} = \sigma\mathbf{E}, \mathbf{I} = \mathbf{j}F$ (see Eq. (6.4)) given by

$$V = \oint \mathbf{E} \cdot d\mathbf{l} = \oint \frac{\mathbf{j} \cdot d\mathbf{l}}{\sigma} = \oint \frac{\mathbf{I} \cdot d\mathbf{l}}{\sigma F} = I \oint \frac{dl}{\sigma F} = IR \quad \text{volts}, \tag{6.61}$$

where $R$ is the *electric resistance* of the circuit.

The magnetic analogy is (see Eq. (6.38)) the *magnetomotive force*

$$\oint \mathbf{H} \cdot d\mathbf{l} = \int_F \mathbf{j} \cdot d\mathbf{F} = IN \quad \text{ampere-turns}, \tag{6.62}$$

where now $\nabla \cdot \mathbf{B} = 0$ with constant *magnetic flux* $\Phi = \int \mathbf{B} \cdot d\mathbf{F_0}$ webers (Wb). Here $N$ is the effective number of currents $I$ passing through the surface $C(F)$ (i.e. the number of currents $I$ coming out minus the number going in, i.e. in the opposite direction). The relation $B_{1n} = B_{2n} = \cdots$ tells us that the magnetic flux is conserved, i.e. the lines of flux are continuous at a boundary surface $F_0$ (note the difference between the areas $F(C)$ and $F_0$). It follows that

$$\int_F \mathbf{j} \cdot d\mathbf{F} = \oint \frac{\mathbf{B} \cdot d\mathbf{l}}{\mu} = \oint \frac{F_0 \mathbf{B} \cdot d\mathbf{l}}{F_0 \mu} = \oint \frac{(B_n F_0) dl}{F_0 \mu} = \Phi \oint \frac{dl}{\mu F_0}. \tag{6.63}$$

If the magnetic flux $\Phi$ corresponds in the electric case to the current $I$, then the expression

$$\oint \frac{dl}{\mu F_0} \tag{6.64}$$

may be interpreted as *magnetic resistance*. It follows that we can calculate the magnetic flux with the relation (6.64), i.e. with

$$\Phi = \frac{\int_F \mathbf{j} \cdot d\mathbf{F}}{\oint (dl/\mu F_0)} = \frac{IN}{\oint (dl/\mu F_0)} \quad \text{webers}. \tag{6.65}$$

This result will be used in Example 6.4 below.

## 6.7 Examples

The first two examples below deal with flux or magnetic circuits in electromagnets. This topic is particularly dealt with in detail in the book of Ferrari [53]. Further examples deal with the calculation of the vector potential $\mathbf{A}$, the calculation of a magnetic field strength $\mathbf{H}$, and the calculation of energy.

**Example 6.3 (The electromagnet).** Determine the magnetic field strength $H_2$ in the gap of the circular electromagnet of permeability $\mu$ shown in Fig. 6.2 (assuming constant and homogeneous fields).

**Solution:** The ring-shaped magnetic material with circular cross-sectional area $F_0$, permeability $\mu$ and mean circumference $L - l$, where $l$ is the length of the gap (of air or vacuum of permeability $\mu_0$), has $N$ turns of a wire-like conductor carrying current $I$ wound around it. We use the integral form of Eq. (6.36), i.e.

$$\oint_C \mathbf{H} \cdot d\mathbf{l} = \int_{F(C)} \mathbf{j} \cdot d\mathbf{F} \quad \text{amperes,} \tag{6.66}$$

where $C$ is the path shown in Fig. 6.2 and $F(C)$ the area enclosed by it. Now $\mathbf{B} = \mu\mathbf{H}$, and at a boundary surface, which means at the two transverse areas, where the gap begins, as we saw, $B_{1n} = B_{2n}$. We set $B = B_1$ in the magnetic material. Then

$$H_1\mu = H_2\mu_0. \tag{6.67}$$

On the other hand, from Eq. (6.38) (since the wire crosses the area $F(C)$ a total of $N$ times in the same direction)

$$H_1 L + H_2 l = IN. \tag{6.68}$$

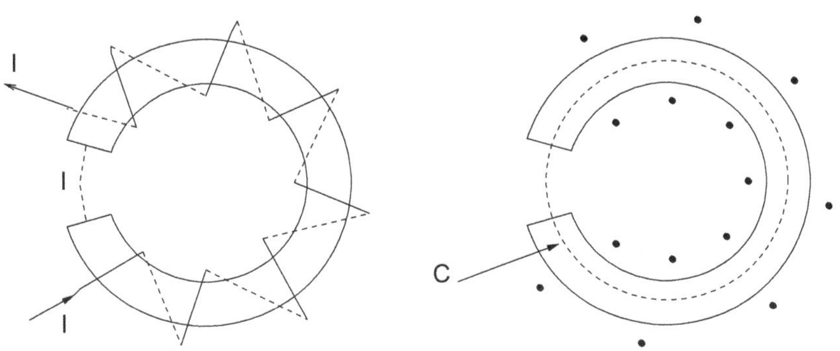


**Fig. 6.2.** The electromagnet.

From Eqs. (6.67) and (6.69) we eliminate $H_1$ and obtain

$$H_2 = \frac{IN}{l + \mu_0 L/\mu} \tag{6.69}$$

in ampere-turns/meter. The same result follows immediately from application of the flux formula (6.65), i.e. with

$$\Phi = \int_F \frac{\mathbf{j} \cdot d\mathbf{F}}{\oint dl/\mu F_0}. \tag{6.70}$$

Inserting the appropriate quantities, we obtain (Wb meaning webers)

$$\mu_0 H_2 F_0 = \frac{IN}{(1/F_0)[L/\mu + l/\mu_0]} \quad \text{Wb}, \tag{6.71}$$

in agreement with the result above. The next example is of a similar type but requires a little more geometry.

**Example 6.4 (The electromagnet with tapered poles towards the gap).** We consider a toroidal electromagnet of magnetic permeability $\mu_1$ with circular cross section and radius $R$ (of the cross-sectional area) along the torus part of length $L$ (as in Example 6.3). Towards the gap (of length $h$), at both ends along a length $l$ (always along the circle which coincides with the axis of the torus), the radius decreases from $R$ to $r$ at the poles. Determine the magnetic field strength $H_2$ in the gap.

**Solution:** We use Eq. (6.65). For this we have to evaluate

$$\underbrace{H_2 \mu_0 \pi r^2}_{\text{flux in the gap}} = \frac{IN}{\oint dl/\mu F_0} \quad \text{Wb}. \tag{6.72}$$

Here the line integrals along the gap and the main part of the torus follow as in Example 6.3, the respective cross-sectional areas being $\pi r^2$ and $\pi R^2$. Along each of the two tapered ends the radius $y$ of the cross section of the electromagnet at a distance $x$ measured from the small end is given by — as one can calculate with the method of congruent triangles (see Fig. 6.3) —

$$\frac{x}{l} = \frac{y - r}{R - r}, \quad \text{i.e.} \quad y = r + \frac{x}{l}(R - r) \tag{6.73}$$

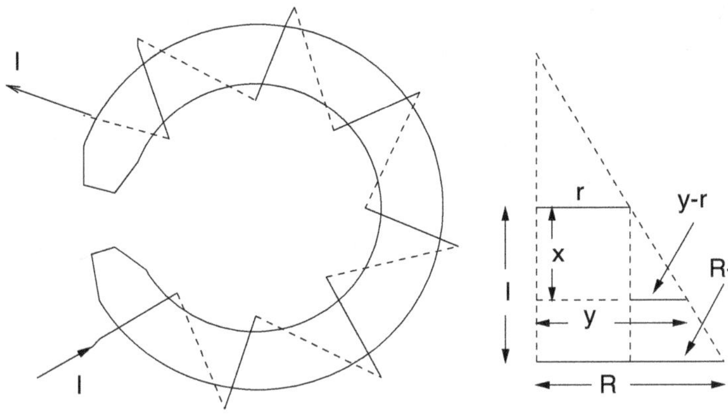


**Fig. 6.3.** The electromagnet with tapered poles.

(verification: for $x = 0$ this is $r$ and for $x = l$ this is $R$), so that for each of the two pole pieces

$$\begin{aligned}\int \frac{dl}{\mu_1 F_0} &= \frac{1}{\mu_1}\int_0^l dx \frac{1}{\pi[r + x(R-r)/l]^2} = -\frac{l}{\pi\mu_1(R-r)}\left[\frac{1}{r + x(R-r)/l}\right]_0^l \\ &= -\frac{l}{\pi\mu_1(R-r)}\left[\frac{1}{R} - \frac{1}{r}\right] = \frac{l}{\pi\mu_1 r R}. \end{aligned} \tag{6.74}$$

Altogether we have

$$H_2\mu_0\pi r^2 = \frac{IN}{h/\mu_0\pi r^2 + L/\mu_1\pi R^2 + 2l/\pi\mu_1 r R} \quad \text{Wb},$$

i.e.

$$H_2 = \frac{IN}{h + (\mu_0 L r^2)/(\mu_1 R^2) + (\mu_0 2lr)/(\mu_1 R)} \quad \text{ampere-revolutions/meter.} \tag{6.75}$$

We see that for $l \to 0, r \to R$ the result reduces to that of Example 6.3, i.e. Eq. (6.69).

**Example 6.5 (The vector potential along a thin, circular wire[6]).** Calculate the vector potential along a current carrying wire of tiny cross section and circular shape.

[6]See also Witten [166].

**Solution:** We use Eq. (6.43) along with $\mathbf{j}d\mathbf{r} = Id\mathbf{s}$ of Eq. (5.7). With $ds = rd\phi$, where $r$ is the radius formed by the wire, we have

$$\begin{aligned}
\mathbf{A}(\mathbf{r}) &= \frac{\mu}{4\pi}\int \frac{Id\mathbf{s}(\mathbf{r}')}{|\mathbf{r}-\mathbf{r}'|} \\
&= \frac{\mu}{4\pi}\int_{\phi=0}^{2\pi} \frac{\mathbf{I}(r'd\phi)}{r'\sqrt{2-2\cos\phi}} \quad (\mathbf{r} \quad \text{along} \quad \text{wire}) \\
&= \mathbf{I}\frac{\mu}{4\pi\sqrt{2}}\int_0^{2\pi} \frac{d\phi}{\sqrt{2}\sin(\phi/2)} = \mathbf{I}\frac{\mu}{4\pi}\int_0^{\pi}\frac{dx}{\sin x} \\
&= \mathbf{I}\frac{\mu}{4\pi}\left[\ln\left|\tan\frac{x}{2}\right|\right]_0^{\pi} \\
&= \mathbf{I}\frac{\mu}{4\pi}\left[\ln\left|\frac{x}{2}\right| + \frac{1}{3}\left(\frac{x}{2}\right)^2 + \cdots\right]_{\delta\phi=2/\Lambda r\to 0}^{\pi-\delta\phi} \\
&\sim \mathbf{I}\frac{\mu}{4\pi}\ln(\Lambda r) + \text{const.} \quad \text{tesla}\cdot\text{meters},
\end{aligned} \tag{6.76}$$

where we used formula 432.10, p. 90 of Dwight [42] and a not so familiar expansion from a Table of Formulas.[7] The divergence that we encounter here (for $\Lambda \to 0$) appears because we assumed that the wire has (effectively) vanishing cross section ($r = r'$). For small thickness $1/\Lambda$ the dominant contribution is proportional to $\ln(\Lambda r)$, $\Lambda$ having the dimension of a reciprocal length. Since the current is constant along the wire, the same is true for $\mathbf{A}(\mathbf{r})$.

**Example 6.6 (Vector potential of a long, straight wire with current I).** Determine the vector potential of the wire.

**Solution:** We start with the same relation for the vector potential as in Example 6.5, i.e. with

$$\mathbf{A}(\mathbf{r}) = \frac{\mu}{4\pi}\int \frac{Id\mathbf{s}(\mathbf{r}')}{|\mathbf{r}-\mathbf{r}'|} \quad \text{tesla}\cdot\text{meters}. \tag{6.77}$$

We take the wire as lying along the $y$-direction. Then the formula implies that the components $A_x, A_z$ are zero, and for the $y$-component we obtain for the field at a point $x$ along the $x$-axis

$$A_y(x) = \frac{\mu I}{4\pi}\int_{y_1}^{y_2} \frac{dy}{\sqrt{x^2+y^2}} = \frac{\mu I}{4\pi}\left\{\sinh^{-1}\frac{y_2}{x} - \sinh^{-1}\frac{y_1}{x}\right\} \quad \text{tesla}\cdot\text{meters}. \tag{6.78}$$

[7] Gradshteyn and Ryzhik [60], formula 1.518(3), p. 46.

The last expression follows in a similar way as in the case of the electrostatic potential we encountered in Example 3.3, Eq. (3.10).

**Example 6.7 (Energy of the magnetic moment m in the field B).** Determine the energy of the magnetic moment $\mathbf{m}$.

**Solution:** We consider first the case of an electric dipole $\mathbf{p}$ in an applied field $\mathbf{E}$. The energy $U$ of the dipole is the sum of the energies of the two charges $q$ and $-q$ at points $\mathbf{r}_+$ and $\mathbf{r}_-$, a small distance $d$ apart, i.e.

$$U = q\phi(\mathbf{r}_+) - q\phi(\mathbf{r}_-) = q\left[\phi(\mathbf{r}_- + \mathbf{d}) - \phi(\mathbf{r}_-)\right] = q\mathbf{d}\cdot\boldsymbol{\nabla}\phi = -\mathbf{p}\cdot\mathbf{E}, \quad (6.79)$$

with $\mathbf{p} = q\mathbf{d}$. We define the force $\mathbf{F}$ by the expression

$$U = -\int \mathbf{F}\cdot d\mathbf{s} = -\mathbf{p}\cdot\mathbf{E} \quad \text{joules.} \quad (6.80)$$

Then

$$\mathbf{F} = -\boldsymbol{\nabla}U = \boldsymbol{\nabla}(\mathbf{p}\cdot\mathbf{E}) \quad \text{newtons.} \quad (6.81)$$

In the magnetic case we have to remember first of all that $\mathbf{B} = \boldsymbol{\nabla}\times\mathbf{A}$, i.e. $\mathbf{B}$ is in general not expressible as a gradient. However, we now consider an *external field*, i.e. a field $\mathbf{B}$, where the current density is zero. Then $\boldsymbol{\nabla}\times\mathbf{B} = 0$, and $\mathbf{B}$ can be written as a gradient, i.e. derivable from a potential $\phi$. It follows that we can use the relations of above and we can write

$$U = -\mathbf{m}\cdot\mathbf{B} \quad \text{joules,} \quad \mathbf{F} = \boldsymbol{\nabla}(\mathbf{m}\cdot\mathbf{B}) \quad \text{newtons.} \quad (6.82)$$

**Example 6.8 (Magnetic field of a solid cylinder with current I).** An infinitely long solid cylinder of radius $R$ (made of material with relative permeability $\mu/\mu_0 > 1$) carries the constant current density $\mathbf{j}_0$ in the direction of its axis. Calculate (a) the vector potential $\mathbf{A}$, and (b) the magnetic field strength $\mathbf{H}$ within and without the conductor by solving the Poisson equation analogue for the vector potential.

**Solution:**
(a) Since from Eq. (6.43)

$$\mathbf{A}(\mathbf{r}) = \frac{\mu}{4\pi}\int d\mathbf{r}'\frac{\mathbf{j}(\mathbf{r}')}{|\mathbf{r}-\mathbf{r}'|} \quad \text{tesla}\cdot\text{meters,} \quad (6.83)$$

the vector $\mathbf{A}$ is parallel to $\mathbf{j}$ and parallel to $\mathbf{e}_z$. The Poisson equation analogue (cf. Eq. (5.65) with $\mu_0 \to \mu$) is $\triangle\mathbf{A} = -\mu\mathbf{j}, \mathbf{j} = j_0\mathbf{e}_z\theta(R-\rho), j_0 = I/\pi R^2$, with cylindrical coordinates $\rho, \varphi, z$. In view of the cylindrical symmetry we have $A = A_z(\rho)$ and the Poisson equation analogue is (cf. Eq. (2.21b))

$$\frac{1}{\rho}\frac{\partial}{\partial \rho}\left(\rho\frac{\partial}{\partial \rho}A_z(\rho)\right) = -\mu j_0 \theta(R-\rho). \tag{6.84}$$

Here $\theta$ is the theta function defined in Eq. (2.45). We have therefore

$$\begin{aligned} &(i)\ \ \rho > R: \ \ \rho\frac{\partial}{\partial \rho}A_z(\rho) = c \text{ (const.)};\\ &(ii)\ \ \rho < R: \ \ \frac{1}{\rho}\frac{\partial}{\partial \rho}\left(\rho\frac{\partial}{\partial \rho}A_z(\rho)\right) = -\mu j_0. \end{aligned} \tag{6.85}$$

In case (i) we have

$$A_z(\rho) = c\ln\rho + A_{z0}. \tag{6.86}$$

In case (ii) we have

$$\frac{\partial}{\partial \rho}\left(\rho\frac{\partial}{\partial \rho}A_z(\rho)\right) = -\mu j_0 \rho, \tag{6.87}$$

and hence

$$A_z(\rho) = -\frac{\mu j_0}{4}\rho^2 + c_1\ln\rho + c_2. \tag{6.88}$$

We choose $c_2 = 0$. Regularity at $\rho = 0$ requires $c_1 = 0$. Continuity at $\rho = R$ implies:

$$c\ln R + A_{z0} = -\frac{1}{4}\mu j_0 R^2, \quad A_{z0} = -c\ln R - \frac{1}{4}\mu j_0 R^2. \tag{6.89}$$

We obtain therefore with $A_{z0}$ inserted into Eq. (6.86):

$$A_z(\rho) = \begin{cases} -\frac{1}{4}\mu j_0 \rho^2 & \text{for} \quad \rho \leq R, \\ c\ln(\rho/R) - \frac{1}{4}\mu j_0 R^2 & \text{for} \quad \rho \geq R. \end{cases} \tag{6.90}$$

(b) Considering the magnetic field strength $\mathbf{H}$, we have

$$\mathbf{B} = \mu\mathbf{H} = \nabla\times\mathbf{A} = \mu H_\varphi(\rho)\mathbf{e}_\varphi = \left(\frac{\partial}{\partial z}A_\rho - \frac{\partial}{\partial \rho}A_z\right)\mathbf{e}_\varphi = -\frac{\partial}{\partial \rho}A_z\mathbf{e}_\varphi \quad \text{teslas}. \tag{6.91}$$

With Eq. (6.90) it follows that

$$H_\varphi(\rho) = -\frac{1}{\mu}\frac{\partial}{\partial \rho}A_z(\rho) = \begin{cases} \frac{1}{2}j_0\rho & \text{for} \quad \rho \leq R, \\ -c/\mu\rho & \text{for} \quad \rho \geq R. \end{cases} \tag{6.92}$$

Continuity at $\rho = R$ implies:

$$-\frac{c}{\mu R} = \frac{1}{2} j_0 R, \quad -c = \frac{1}{2} j_0 \mu R^2,$$

and therefore

$$H_\varphi(\rho) = \begin{cases} j_0 \rho/2 \ \text{ amperes/meter} & \text{for} \quad \rho \leq R, \\ j_0 R^2/2\rho \ \text{ amperes/meter} & \text{for} \quad \rho \geq R. \end{cases} \tag{6.93}$$

## 6.8 Summary of Formulas of Magnetostatics

We collect here for reference purposes the most important results of magnetostatics:

1. **Current, equation of continuity and stationarity** (cf. Eqs. (5.2), (5.12) and (5.13)):

$$I = \frac{dq}{dt}, \quad dq = \rho dV, \quad \frac{\partial \rho}{\partial t} + \nabla \cdot \mathbf{j} = 0, \quad \text{stationarity} \quad \nabla \cdot \mathbf{j} = 0. \tag{6.94}$$

2. **Magnetic field strength** (cf. Eqs. (6.36), (6.38) and (5.1)):

$$\nabla \times \mathbf{H} = \mathbf{j}, \quad \text{or} \quad \oint_C \mathbf{H} \cdot dl = \int_{F(C)} \mathbf{j} \cdot d\mathbf{F}, \tag{6.95}$$

(valid only if the current does not vary with time too rapidly),

$$\mathbf{j} = \rho \frac{d\mathbf{s}}{dt} \ \text{ amperes/meter}^2, \quad \mathrm{I} = \int \mathbf{j} \cdot d\mathbf{F} \ \text{ amperes.} \tag{6.96}$$

3. **Macroscopic magnetostatics** (cf. Eqs. (6.42), (6.32) and (6.40))[8]:

$$\mathbf{j}_{\text{total}} = \mathbf{j} + \mathbf{j}_M \ \text{ amperes/meter}^2, \tag{6.97}$$

magnetization current density

$$\mathbf{j}_M = \nabla \times \mathbf{M}, \tag{6.98}$$

$$\mathbf{B} = \mu_0(\mathbf{H} + \mathbf{M}) = \mu \mathbf{H} \quad \text{teslas.} \tag{6.99}$$

[8]Some authors relate **B** to **H** by $\mathbf{B} = \mu_0 \mathbf{H} + \mathbf{M}'$, in which case $\mathbf{M}'$ has units of **B** (webers/meter$^2$) instead of **H** (amperes/meter). See e.g. *The Electromagnetic Problem Solver* [47], Section II.

4. **Magnetic material equations** (cf. Eqs. (6.39) and (6.40)):

$$\mathbf{M} = \kappa\mathbf{H} \quad \text{amperes/meter}, \quad \mathbf{B} = \mu\mathbf{H}, \quad \frac{\mu}{\mu_0} = 1 + \kappa. \tag{6.100}$$

5. **Biot–Savart law** (cf. Eq. (5.21)):

$$\mathbf{B}(\mathbf{r}) = \frac{\mu}{4\pi}\int \frac{I(\mathbf{r}')d\mathbf{s}(\mathbf{r}') \times (\mathbf{r} - \mathbf{r}')}{|\mathbf{r} - \mathbf{r}'|^3} \quad \text{teslas} \quad (\text{with} \quad Id\mathbf{s} = \mathbf{j}dV). \tag{6.101}$$

6. **Ampère's law and Lorentz-force** (cf. Eqs. (5.23) and (5.26)):

$$\mathbf{F} = \int I(\mathbf{r}')d\mathbf{s}(\mathbf{r}') \times \mathbf{B}(\mathbf{r}'), \quad \mathbf{F} = q\mathbf{v}(\mathbf{r}) \times \mathbf{B}(\mathbf{r}) \quad \text{newtons}. \tag{6.102}$$

7. **No magnetic monopoles** (cf. Eq. (5.52)):

$$\boldsymbol{\nabla} \cdot \mathbf{B} = 0. \tag{6.103}$$

8. **Behavior at boundary surfaces** (cf. Eqs. (6.52) and (6.59)):

$$B_{2n} = B_{1n}, \quad H_{2t} - H_{1t} = K \tag{6.104}$$

(surface current $K = 0$ for finite conductivity $\sigma$).

9. **B-field energy** (cf. Eq. (6.46)):

$$W = \frac{1}{2}\int \mathbf{B} \cdot \mathbf{H}dV \quad \text{joules}. \tag{6.105}$$

## 6.9 Exercises

**Example 6.9 (Calculation of a resistance R).** The resistivity of copper is $1.7 \times 10^{-8}$ ohm · m. What is the resistance between two opposing edges of a square sheet of copper $10^{-3}$ m thick?

**Example 6.10 (Couple acting on a current loop).** A *couple* $C$ is defined as

$$C = -\frac{\partial U}{\partial \theta} \quad \text{joules/radian}, \tag{6.106}$$

where $U$ is a potential energy. Show that the couple acting on a current loop of area $F$ and current $I$ in a magnetic field $\mathbf{B}$ is given by

$$\mathbf{C} = -I\mathbf{F} \times \mathbf{B} \quad \text{joules/radian}. \tag{6.107}$$

**Example 6.11 (Measurement of paramagnetic susceptibility).** Calculate an expression for the energy contained in a medium of given magnetic permeability $\mu$ when this is placed in a magnetic field. Obtain further the change in magnetic energy which would occur if, over a certain region, the first medium were replaced by a second one of different permeability, allowing for the resultant change in the magnetic field in the vicinity. Use the result to set up the theory of a U-tube method analogous to that of Example 4.6 for the determination of the paramagnetic susceptibility of a liquid.

**Example 6.12 (Law of induction).** Recall that the magnetic effects of a closed current-carrying loop are equivalent to those of a certain distribution of magnetic dipoles. Use this result to obtain an expression for the energy of a current-carrying loop in a magnetic field. By equating the rate at which magnetic energy is fed in, to the rate it is converted into thermal energy for constant current, establish the law of induction:

$$\oint \mathbf{E} \cdot d\mathbf{l} = V = -\frac{\partial}{\partial t} \int \mathbf{B} \cdot d\mathbf{F}, \tag{6.108}$$

and thence the Maxwell equation

$$\boldsymbol{\nabla} \times \mathbf{E} = -\frac{\partial \mathbf{B}}{\partial t}. \tag{6.109}$$

**Example 6.13 (Energy of m in a field B).** A magnetic dipole of moment $\mathbf{m}'$ is placed in the field of induction $\mathbf{B}$ of a dipole of moment $\mathbf{m}$, both dipoles being in the same plane. Calculate the energy $W$ of $\mathbf{m}'$ in the field $\mathbf{B}$ and thence its orientation with respect to the other dipole at a minimum of energy.

**Example 6.14 (Fundamental equations when $\mathbf{M} \neq \kappa\mathbf{H}$).** Show that for constant permeability $\mu$ the scalar magnetic potential $V_m$ given by Eq. (5.80), i.e. $\mathbf{H} = -\boldsymbol{\nabla} V_m$, satisfies Laplace's equation $\triangle V_m = 0$. Show also that when $\mathbf{M} = \kappa\mathbf{H}$ amperes/meter is not valid, the scalar magnetic potential satisfies the Poisson analogue equation $\triangle V_m = \boldsymbol{\nabla} \cdot \mathbf{M}$, where $\mathbf{B} = \mu_0(\mathbf{H} + \mathbf{M})$ teslas.

**Example 6.15 (The hollow cylinder with current I).** A solid cylinder of permeability $\mu$ has a circular cross section of radius $a$ and is coaxially hollow with radius $b < a$. A current $I$ flows along the solid part of the cylinder. Calculate the magnetic field within and without the hollow cylinder.

**Example 6.16 (Potential of a magnetized sphere).** A sphere of radius $a$ is magnetized so that at a point $\mathbf{r}$ away from the origin at the center of the sphere the intensity $\mathbf{H} = \alpha z \mathbf{e}_z$, $\alpha$ a constant. Derive the scalar magnetic potential $V_m$ outside the sphere.

(Answer: $V_m = 2\alpha a^5 P_2(\cos\theta)/15r^3$ ampere · radian).

**Example 6.17 (Proof of Eq. (6.24)).** Prove Eq. (6.24), i.e.

$$\int dV(\boldsymbol{\nabla} \times \mathbf{B}) = \int d\mathbf{F} \times \mathbf{B}, \tag{6.110}$$

by replacing $\mathbf{B}$ by $\mathbf{B} \times \mathbf{C}$ in the divergence theorem.

**Example 6.18 (Another integral relation).** Show — by a suitable substitution in Stokes's theorem — that

$$\int_{\text{surface}} d\mathbf{F} \times \boldsymbol{\nabla} G = \oint_{\text{path}} G d\mathbf{s}. \tag{6.111}$$

**Example 6.19 (Field B of a long, straight conducting wire).** Rederive the field $\mathbf{B}$ of a long, straight conducting wire, i.e. Eq. (5.31), from the vector potential $\mathbf{A}$, i.e. Eq. (6.77). Check that in Gaussian units the result is $B = 2I/cr$.

# Chapter 7

# The Maxwell Equations

## 7.1 Preliminary Remarks

Now that we have dealt with electrostatics and magnetostatics, also for macroscopic objects, the next step is to introduce time dependences. Proceeding in our phenomenological and historical approach we are led to consider next Faraday's law of induction. With this we can complete the equations of macroscopic electrodynamics with the addition of *Maxwell's displacement current*. The result is the full set of *Maxwell's equations*.

## 7.2 Time-Dependent Fields and Faraday's Law of Induction

Faraday discovered in 1831 *that an electric current arises in a closed wire loop if the wire is moved through a magnetic field*, in other words when the position or orientation of the wire with respect to the magnetic field is changed, or if the magnetic field varies with time. We consider two situations in which Faraday's observation applies. In the first case the field **B** is maintained constant in time.

(a) In an electric field **E** the charge $q$ experiences the force

$$\mathbf{F} = q\mathbf{E} \quad \text{newtons}, \quad \frac{\mathrm{d}\mathbf{F}}{\mathrm{dq}} = \mathbf{E} \quad \text{newtons/coulomb}. \tag{7.1}$$

This force **F** results from a nonvanishing potential difference in the conductor. On the other hand (cf. Lorentz-force, Eq. (5.26)), the field **B**, cf. Fig. 7.1(a), acting on a charge $dq$ moving with velocity $\mathbf{v} = d\mathbf{l}/dt$ implies that the latter experiences the force $d\mathbf{F}$ given by

$$d\mathbf{F} = dq\frac{d\mathbf{l}}{dt} \times \mathbf{B} \quad \text{newtons},$$

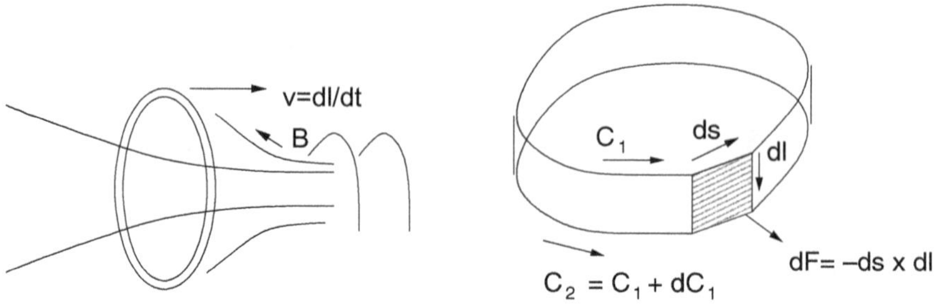


(a) (b)

**Fig. 7.1.** (a), (b) The moving current loop.

so that

$$\frac{d\mathbf{F}}{dq} = \frac{d\mathbf{l}}{dt} \times \mathbf{B} \quad \text{newtons/coulomb.} \tag{7.2}$$

Identifying the forces of Eqs. (7.1) and (7.2) in order to arrive at an explanation of Faraday's observation, we obtain

$$dq\mathbf{E} = \frac{d\mathbf{l}}{dt} \times dq\mathbf{B}, \tag{7.3}$$

i.e.

$$\mathbf{E} = \frac{d\mathbf{l}}{dt} \times \mathbf{B} \quad \text{newtons/coulomb,} \tag{7.4}$$

provided the right side (or one component) is parallel to **E**. This means, the electric force acting on the charge $dq$ is equal to the force which the field **B** exerts on the charge $dq$ moving with velocity **v**. Put differently: The right side of Eq. (7.4), i.e. the force that **B** exerts on $dq$, *induces* the electric force $\mathbf{E}dq$, i.e. the Lorentz-force, acts on the electrons in the conductor, and hence a current in the conductor is observed, and so a potential difference or induced voltage. We assume this here as an *empirical finding.*[1]

We let $C_1$ be the initial position of the conducting loop and $C_2 = C_1 + \delta C_1$ its position a time interval $\delta t$ later (as indicated in Fig. 7.1(b)). The quantity $\delta C_1$ represents the displacement of the coordinates of the loop or, alternatively, the deformation of the loop. The potential difference $V$ induced

[1]The induced voltage or potential difference $V$ is also called *induced electromotive force.* The potential difference $V$ has the dimension of energy. The term "*force*" has a historical origin; in the 18th century, as also at the beginning of the 19th century, various quantities which today represent energy were described as "force", e.g. also in writings of Kant.

in the loop $C_1$ is according to its definition and after inserting the expression for $\mathbf{E}$ from above given by

$$\begin{aligned} V &= \oint_{C_1} \mathbf{E}\cdot d\mathbf{s} = \oint_{C_1} \left(\frac{d\mathbf{l}}{dt}\times\mathbf{B}\right)\cdot d\mathbf{s} \\ &= \int_{C_1} \left(d\mathbf{s}\times\frac{d\mathbf{l}}{dt}\right)\cdot\mathbf{B} = -\int_{\text{displacement area}} \frac{d\mathbf{F}\cdot\mathbf{B}}{dt} \quad \text{volts}, \end{aligned} \tag{7.5}$$

where the integration is over the area swept out by $C_1$ in its displacement to $C_2$. However, we always have $\boldsymbol{\nabla}\cdot\mathbf{B} = 0$, i.e. integrating over the displacement-volume between $C_1$ and $C_2$ (here $d\mathbf{r} \equiv dV$ is the volume element)

$$\begin{aligned} 0 &= \int_{V(\text{displacement})} d\mathbf{r}\boldsymbol{\nabla}\cdot\mathbf{B} \\ &= \int_{F(V)} \mathbf{B}\cdot d\mathbf{F} \\ &= \left\{\int_{\text{displacement}} + \int_{F(C_1)} + \int_{F(C_1+\delta C_1)}\right\}\mathbf{B}\cdot d\mathbf{F}. \end{aligned} \tag{7.6}$$

Here $F(C_1)$ is the area enclosed by $C_1$, and correspondingly $F(C_2)$ the area enclosed by $C_2$, and $F$(displacement) is as before the area swept out by the boundary. Then

$$\begin{aligned} -\int_{\text{boundary}} \mathbf{B}\cdot d\mathbf{F} &= \int_{F(C_1)} \mathbf{B}\cdot d\mathbf{F}_1 + \int_{F(C_1+\delta C_1)} \mathbf{B}\cdot d\mathbf{F}_2 \\ &= -\left[\int_{F(C_1+\delta C_1)} \mathbf{B}\cdot d\mathbf{F}_1 - \int_{F(C_1)} \mathbf{B}\cdot d\mathbf{F}_1\right] \\ &= -\delta\int_{F(C_1)} \mathbf{B}\cdot d\mathbf{F}_1. \end{aligned} \tag{7.7}$$

It then follows with Eq. (7.5) that

$$V = \oint_{C_1} \mathbf{E}\cdot d\mathbf{s} \overset{\text{Stokes}}{\equiv} \int_{F(C_1)} \boldsymbol{\nabla}\times\mathbf{E}\cdot d\mathbf{F} = -\frac{d}{dt}\int_{F(C_1)} \mathbf{B}\cdot d\mathbf{F}_1. \tag{7.8}$$

Thus in this case with the field $\mathbf{B}$ constant in time:

$$\int_{F(C_1)} \boldsymbol{\nabla}\times\mathbf{E}\cdot d\mathbf{F} = -\frac{d}{dt}\int_{F(C_1(t))} \mathbf{B}\cdot d\mathbf{F}_1. \tag{7.9}$$

In the next case we consider $C_1$ as fixed and $\mathbf{B}$ as varying with time.

(b) We have seen in Chapter 5 that at a sufficient distance away from a current carrying loop the *magnetic properties of the loop* can be considered as those of a magnetic dipole with moment $\mathbf{m}$ given by (cf. Eq. (5.39))

$$\mathbf{m} = I\int_{C_1 \text{ fixed}} d\mathbf{F} \quad \text{ampere} \cdot \text{meters}^2. \tag{7.10}$$

The energy of this dipole in an external field $\mathbf{B}$ is (see Eq. (6.82))

$$U = -\mathbf{m}\cdot\mathbf{B} = -I\int_{C_1 \text{ fixed}} \mathbf{B}\cdot d\mathbf{F} \quad \text{joules}. \tag{7.11}$$

We now assume: In the time interval $\delta t$ the field $\mathbf{B}$ changes by an amount $\delta\mathbf{B}$ in such a way, that the amount of magnetic energy thereby gained per second is exactly compensated by the loss of a corresponding amount of energy as heat whilst *the current $I$ is kept constant.* Then the energy lost as heat per second (or power) is

$$-\frac{dqV}{dt} = -IV = -I\oint \mathbf{E}\cdot d\mathbf{l} \quad \text{joules/second}. \tag{7.12}$$

On the other hand the magnetic energy gained per second is

$$+\frac{\partial U}{\partial t} = -I\int_{C_1 \text{ fixed}} \frac{\partial \mathbf{B}}{\partial t}\cdot d\mathbf{F} \quad \text{joules/second}. \tag{7.13}$$

Hence the sum "loss + gain = 0" yields the relation

$$\oint \mathbf{E}\cdot d\mathbf{l} = -\int \frac{\partial \mathbf{B}}{\partial t}\cdot d\mathbf{F} \quad \text{volts}. \tag{7.14}$$

Let $\Phi$ be the *magnetic flux* defined by

$$\Phi = \int_F \mathbf{B}(t)\cdot d\mathbf{F} \quad \text{webers}. \tag{7.15}$$

The MKSA-unit of magnetic flux is the weber (Wb) (for conventions of notation see Appendix B). Equation (7.12) can then be written

$$IV = -I\frac{d\Phi}{dt} \quad \text{joules/second}. \tag{7.16}$$

In this formulation Faraday's law is also known as *Neumann's law.* The minus sign in Eq. (7.16) expresses effectively what the so-called *Lenz rule* says (see below). The quantity

$$P := IV$$

is also described as *power* and is also given in watts (W). Linked with Ohm's law (7.31) $P = I^2R$ W is the amount of power dissipated in a transmission line. The amount of power supplied by a generating system at any given time is called its *load*.

Electrical energy is distributed in towns as alternating current (AC) in view of the latter's advantages. Alternating current reverses periodically direction and changes its magnitude in time. The voltage $V$ may be increased or decreased by means called *transformers*. This permits the power to be transmitted through power transmission lines at high voltage and transformed to a lower, safer voltage for use e.g. in households. The power $P = IV$ transmitted requires less loss–producing current than for the same power at a lower voltage. For domestic use power is transmitted at high kilovolts on pylons (seen in the landscape) and then transmitted on lower level lines and transformed down to e.g. 100 V–240 V for domestic use.

Recalling — as an instructive analogy — the following result from calculus[2]

$$\frac{d}{dt}\int_{a(t)}^{b(t)} f(x,t)dx = f(b(t),t)\frac{db}{dt} - f(a(t),t)\frac{da}{dt} + \int_{a(t)}^{b(t)} \frac{\partial f(b(t),t)}{\partial t}dx, \tag{7.17}$$

we see that the results (7.9), (7.14) can be combined in the following form (where instead of boundary values we have the path of integration $C_1(t)$):

$$\oint_{C_1(t)} \mathbf{E}\cdot d\mathbf{l} = \int_{F(C_1(t))} \boldsymbol{\nabla}\times\mathbf{E}\cdot d\mathbf{F} = -\frac{d}{dt}\int_{F(C_1(t))} \mathbf{B}(\mathbf{r},t)\cdot d\mathbf{F}. \tag{7.18}$$

Here we are at present not so much interested in the variation of the positions of current circuits with time (but see Example 7.3), as in time-dependent fields $\mathbf{B}$. We assume therefore, that the form of $C_1$ remains unchanged and that $C_1$ is fixed in space. Then we have the case of Eq. (7.9), i.e.

$$\int_{F(C_1)} \boldsymbol{\nabla}\times\mathbf{E}\cdot d\mathbf{F} = -\int_{F(C_1)} \frac{\partial\mathbf{B}}{\partial t}\cdot d\mathbf{F}. \tag{7.19}$$

[2]See e.g. Whittaker and Watson [162], p. 67.

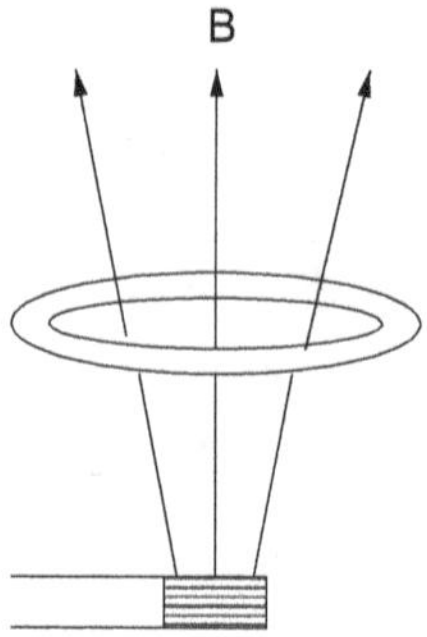


**Fig. 7.2.** The Lenz rule.

Since $C_1$ is arbitrary, we can equate the integrands and obtain

$$\nabla \times \mathbf{E} = -\frac{\partial \mathbf{B}}{\partial t} \quad \text{teslas/second.} \tag{7.20}$$

This is the differential form of *Faraday's law.*

Finally we comment briefly on the so-called *Lenz rule.* This rule is useful in practical applications. In other respects there is no necessity to use it. What the Lenz rule says is contained in Maxwell's equations. In the literature the Lenz rule is frequently — at least on a first reading — difficult to follow, e.g. with a statement like: "The Lenz rule says that the induced current and the associated magnetic flux have directions such that they act against the change of the external flux." Let us consider a conducting ring which falls in a field $\mathbf{B}$ as depicted in Fig. 7.2 (i.e. the ring maintains its shape but the flux through it changes). The change of flux $d\Phi$ or change in the number of lines of force cut by the falling ring in time $dt$ is given by $d\Phi/dt$. According to Eq. (7.16) the magnitude of this rate gives the voltage $V$ induced in the ring. Lenz's rule determines the sign in Eq. (7.16) by calling $Vdt$ now induced flux (i.e. magnetic flux associated with the induced current) and saying, its sign is such that it cancels the corresponding external flux $d\Phi$.

At this point the natural question arises: Is there also a magnetic field strength $\mathbf{H}$ which is induced by time-variation of $\mathbf{E}$? The answer is yes, and is provided by Maxwell's displacement current (see Example 7.1). Faraday himself had asked himself this question and performed experiments, but with no success.

## 7.3 Energy of the Magnetic Field

Starting from the expression (cf. Eq. (5.39))

$$\mathbf{m} = I\mathbf{F} \quad \text{ampere} \cdot \text{meters}^2 \tag{7.21}$$

for the magnetic moment of a circuit or network with current $I$ and area $F$, and also from the expression for the potential energy $U$ of the moment $\mathbf{m}$ of a magnetic dipole in the field $\mathbf{B}$, i.e.

$$U = -\mathbf{m} \cdot \mathbf{B} = -I\mathbf{B} \cdot \mathbf{F} \equiv -I \int \mathbf{B} \cdot d\mathbf{F} \quad \text{joules}, \tag{7.22}$$

we have for the variation of the energy of the field (where we have to re-express the current $I$ in terms of the field)

$$\begin{aligned} -\delta U \equiv \delta W &= I\delta \int_{F \text{ fixed}} \mathbf{B} \cdot d\mathbf{F} = I \int_F (\nabla \times \delta\mathbf{A}) \cdot d\mathbf{F} \\ &\overset{\text{Stokes}}{=} I \oint \delta\mathbf{A} \cdot d\mathbf{s} \overset{(5.7)}{=} \int d\mathbf{r}\delta\mathbf{A} \cdot \mathbf{j}, \end{aligned} \tag{7.23}$$

i.e.

$$\delta W = \int d\mathbf{r}\, \delta\mathbf{A} \cdot \mathbf{j}. \tag{7.24}$$

This expression has to be compared with the corresponding expression in electrostatics, i.e.

$$\delta W = \int d\mathbf{r}\delta\phi(\mathbf{r})\rho(\mathbf{r}) \tag{7.25}$$

(observe that one can define an expression $\delta W' = \int d\mathbf{r}\delta\rho(\mathbf{r})\phi(\mathbf{r})$). Inserting into Eq. (7.10) the expression for $\mathbf{j}$ which we obtain from the equation $\nabla \times \mathbf{H} = \mathbf{j}$ of Chapter 6, and using the relation

$$\nabla \cdot (\mathbf{a} \times \mathbf{b}) = \mathbf{b} \cdot (\nabla \times \mathbf{a}) - \mathbf{a} \cdot (\nabla \times \mathbf{b}), \tag{7.26}$$

it follows that

$$\begin{aligned} \delta W &= \int_{V_\infty} d\mathbf{r}\delta\mathbf{A} \cdot (\nabla \times \mathbf{H}) \\ &= \int d\mathbf{r}[\nabla \cdot (\mathbf{H} \times \delta\mathbf{A}) + \mathbf{H} \cdot (\nabla \times \delta\mathbf{A})] \\ &= 0 + \int d\mathbf{r}\mathbf{H} \cdot \delta\mathbf{B}, \end{aligned} \tag{7.27}$$

where the first term was converted into a surface integral at infinity with the help of Gauss' divergence theorem and therefore vanishes. For $\mathbf{B} = \mu\mathbf{H}$ we have $\delta(\mathbf{H} \cdot \mathbf{B}) = \mathbf{B} \cdot \delta\mathbf{H} + \mathbf{H} \cdot \delta\mathbf{B} = 2\mathbf{H} \cdot \delta\mathbf{B}$ and hence

$$W = \frac{1}{2} \int d\mathbf{r}\, \mathbf{H} \cdot \mathbf{B} \quad \text{joules}. \tag{7.28}$$

This result should be compared with the corresponding energy in macroscopic electrostatics, i.e. Eq. (4.93).

## 7.4 The Generalized Ohm's Law

In the case of a simple circuit with resistance $R$, generator voltage $V_e$, Ohm's law is $V_e = IR$, where $I$ is the conduction current. Next we consider the case of a coil with $n$ turns per unit length. In this case one has a magnetic field along the axis of the coil (as we saw e.g. in the case of a solenoid), and so

$$H = nI, \quad B = \mu H. \tag{7.29}$$

If $H$ is changed by changing $I$, also the flux $\Phi = n\times$ area of one turn $\times B$ changes and this induces in the conductor of the circuit an *induced potential*

$$V_i = -\frac{d\Phi}{dt}. \tag{7.30}$$

Hence the *generalized Ohm's law* is

$$IR = V = V_e + V_i = V_e - \frac{d\Phi}{dt}, \tag{7.31}$$

where $V_e$ is the potential or e.m.f. provided by an external source.

## 7.5 E and B with Time-Dependent Potentials

We saw: Since $\nabla \cdot \mathbf{B} = 0$ and "*div curl = 0*", the field $\mathbf{B}$ can always be written as $\mathbf{B} = \nabla \times \mathbf{A}$. According to Faraday's law (7.20) we then have

$$\nabla \times \mathbf{E} = -\frac{\partial \mathbf{B}}{\partial t} = -\frac{\partial}{\partial t}\nabla \times \mathbf{A} = -\nabla \times \frac{\partial \mathbf{A}}{\partial t}, \tag{7.32}$$

so that $\mathbf{E} = -\partial \mathbf{A}/\partial t$ + something, whose curl is zero, i.e.

$$\mathbf{E} = -\frac{\partial \mathbf{A}}{\partial t} - \nabla\phi \quad \text{newtons/coulomb}. \tag{7.33}$$

This equation differs in the first term from the equation obtained in electrostatics (recall Example 5.9).[3]

[3]Note that in Gaussian units $\mathbf{E}$ is $\mathbf{E} = -\nabla\phi - \partial\mathbf{A}/c\partial t$.

## 7.6 Displacement Current and Maxwell Equations

We obtained in Chapter 6 the following equation (cf. Eq. (6.36))

$$\nabla \times \mathbf{H} = \mathbf{j} \quad \text{amperes/m}^2. \tag{7.34}$$

Taking the divergence of both sides, we have

$$\nabla \cdot (\nabla \times \mathbf{H}) = \nabla \cdot \mathbf{j}. \tag{7.35}$$

The left side is always zero, since always "*div curl*" equals zero. But $\nabla \cdot \mathbf{j}$ is not zero except in magnetostatics where $\nabla \cdot \mathbf{j} = 0$ is the stationarity condition (see e.g. the continuity equation (6.94)). Hence, here (this means not in magnetostatics, where $\partial\rho/\partial t = 0$) we have

$$\nabla \cdot \mathbf{j} + \frac{\partial \rho}{\partial t} = 0, \tag{7.36}$$

or with the Gauss law

$$\nabla \cdot \mathbf{D} = \rho, \tag{7.37}$$

where $\mathbf{D}$ is the dielectric displacement,

$$\nabla \cdot \mathbf{j} + \nabla \cdot \frac{\partial \mathbf{D}}{\partial t} = 0. \tag{7.38}$$

Thus, Eq. (7.34) would have to be (with inclusion of time dependence)

$$\nabla \times \mathbf{H} = \mathbf{j} + \frac{\partial \mathbf{D}}{\partial t}. \tag{7.39}$$

The expression on the right of this equation is called *Maxwell current density* and the term proportional to $\partial\mathbf{D}/\partial t$ is called the density of *Maxwell's displacement current.* This displacement current puzzled people considerably in the early days of Maxwell theory. A detailed discussion can be found in the book of O'Rahilly.[4] On p. 89, O'Rahilly remarks: "*In a flash of mathematical insight he (Maxwell) saw, that the addition of an extra term to one of the electromagnetic equations would make an immense difference.*" What irritated people at that time was that $\mathbf{j}$ can be clearly visualized as a current (density), but $\partial\mathbf{D}/\partial t$? Pohl (cf. O'Rahilly, p. 100) writes in his book: "*By the term 'displacement current' we denote an alteration of the electric field in time, i.e. the appearance and disappearance of lines of force. The term 'current' has no doubt been historically adopted from the analogy with water. In the conduction current atoms of electricity really do move*

[4]O'Rahilly [122], Vol. 1, see pp. 95–101 and pp. 81–101, see also p. 232.

*or flow. In the term 'displacement current' there remains no trace of the original meaning of the word 'current' .... We shall see later, however, that the idea of the displacement current reveals unsuspected relationships to us and vastly extends our physical conceptions of the world. Here we shall content ourselves with mentioning that the light which reaches our eyes from any source is, from the point of view of physics, nothing but a displacement current.*" Citing Thomson, O'Rahilly (p. 101) says: "*We may cheerfully confess that it (i.e. the displacement current) was his (Maxwell's) 'greatest contribution to physics', i.e. as an analytical formula.*"

Today we are accustomed to the concept of fields already intuitively, so that the puzzles of long-ago lose their significance. In the book of Sommerfeld[5] one reads correspondingly: "*The vector field strength* $\mathbf{E}$ *is associated with a second electric vector* $\mathbf{D}$. *We call this electric 'excitation', but will frequently ... adhere to the customary term of 'dielectric displacement' (Maxwell's 'displacement') .... We assumed here the difference between conductor and nonconductor (dielectric) as being self-evident. In reality there is no perfect dielectric, since even the best nonconducting material can become weakly conducting as e.g. under the influence of cosmic radiation. Maxwell therefore augments the displacement current to the total current* $\mathbf{C} = \dot{\mathbf{D}} + \mathbf{j}$ *.... This step of attributing equal importance to* $\dot{\mathbf{D}}$ *and* $\mathbf{j}$ *was a fundamentally new idea of Maxwell's and was a crucial prerequisite in enabling a complete formulation of electrodynamical phenomena. Similarly for a metallic conductor he supplements the conduction current* $\mathbf{j}$ *by the addition of a hypothetical displacement current* $\dot{\mathbf{D}}$, *although the first is much more important than the second except in the case of rapidly changing fields.*"

The displacement current $\partial\mathbf{D}/\partial t$ is in general of importance only,

(a) if $\mathbf{j} = 0$, thus, for instance, in the case of propagation of light through a dielectric, or
(b) in the case of high frequencies, when $\partial\mathbf{D}/\partial t \propto$ frequency, e.g. in the case of propagation of light through a metal.

We have thus completed *Maxwell's equations*. We summarize these now. First in *differential form*, thereafter in *integral form*.

**Maxwell's equations in differential form are:**

$$\nabla \cdot \mathbf{D} = \rho, \tag{7.40}$$

$$\nabla \times \mathbf{E} = -\frac{\partial \mathbf{B}}{\partial t}, \tag{7.41}$$

[5] Sommerfeld [139], pp. 8–9.

$$\nabla \cdot \mathbf{B} = 0, \tag{7.42}$$

$$\nabla \times \mathbf{H} = \mathbf{j} + \frac{\partial \mathbf{D}}{\partial t}. \tag{7.43}$$

These are supplemented by the *equation of continuity*:

$$\nabla \cdot \mathbf{j} + \frac{\partial \rho}{\partial t} = 0, \tag{7.44}$$

and the *generalized Lorentz-force* per unit volume (now combining the electric force with the Lorentz force (5.26) and using Eq. (5.1), $\mathbf{j} = \rho \mathbf{v}$):

$$\mathbf{F} = \rho \mathbf{E} + \mathbf{j} \times \mathbf{B}. \tag{7.45}$$

Maxwell's equations are completed by the *connecting relations*:

$$\mathbf{D} = \epsilon_0 \mathbf{E} + \mathbf{P}, \qquad \mathbf{B} = \mu_0(\mathbf{H} + \mathbf{M}). \tag{7.46}$$

The *matter equations* apply to special cases

$$\mathbf{D} = \epsilon \mathbf{E}, \quad \mathbf{B} = \mu \mathbf{H}. \tag{7.47}$$

The *generalized Ohm's law* is the relation

$$\mathbf{j} = \sigma \mathbf{E}. \tag{7.48}$$

**Maxwell's equations in integral form are:**

$$\oint \mathbf{D} \cdot d\mathbf{F} = Q, \tag{7.49}$$

$$V = \oint_C \mathbf{E} \cdot d\mathbf{l} = \int_{F(C)} (\nabla \times \mathbf{E}) \cdot d\mathbf{F} = -\frac{d\Phi}{dt}, \quad \Phi = \int_{F(C)} \mathbf{B} \cdot d\mathbf{F}, \tag{7.50}$$

$$\int_{F(\text{closed})} \mathbf{B} \cdot d\mathbf{F} = 0, \tag{7.51}$$

$$\oint \mathbf{H} \cdot d\mathbf{l} = \int_{F \text{ fixed}} \mathbf{j} \cdot d\mathbf{F} + \frac{d}{dt} \int_{F \text{ fixed}} \mathbf{D} \cdot d\mathbf{F}. \tag{7.52}$$

The *equation of continuity* follows from

$$\frac{d}{dt} \int_V \rho dV = -\int_F \mathbf{j} \cdot d\mathbf{F}. \tag{7.53}$$

The *main assumptions*, we made in our historical and phenomenological approach in order to arrive at Maxwell's equations were the empirically discovered laws of Coulomb, Ampère/Biot–Savart and Faraday (Coulomb in the case of electric charges, Ampère and Biot and Savart for magnetic fields of currents, and Faraday for the induced electromotive force). We also assumed, without saying so, that the equations obtained originally for static systems are also valid in the case of dynamic (i.e. time-dependent) systems. The following example is meant to illustrate how the displacement current occurs.

**Example 7.1 (The displacement current in a condenser).** A parallel plate condenser consists of two circularly shaped parallel plates of radius $R = 0.1$ meter. The plates are charged such that the electric field $E$ between them is subjected to a constant time variation $dE/dt = 10^{13}$ volts/meter · second. The following relations apply: $B = \mu H, D = \epsilon E$.

(a) Calculate the displacement current between the plates of the condenser.
(b) Determine the induced magnetic field $B$ between the plates as a function of the distance from the center of the cylindrically symmetric condenser in a direction parallel to the plates. What is the value of $B$ at $r = R$? (Note: $\epsilon_0 = 8.9 \times 10^{-12}$ ampere·second/volt·meter, $\mu_0 = 4\pi \times 10^{-7}$ volt·second/ampere · meter).

**Solution:**

(a) The current $I$ is given by $I = \int \mathbf{j} \cdot d\mathbf{F}$ amperes, where $\mathbf{j}$ is the current density. The density of the displacement current is $\partial\mathbf{D}/\partial t$, so that the displacement current itself is given by

$$\begin{aligned} I_D &= \int_{F=\pi R^2} \frac{\partial \mathbf{D}}{\partial t} \cdot d\mathbf{F} = \epsilon_0 \frac{dE}{dt} \pi R^2 \\ &= 8.9 \times 10^{-12} \times 10^{13} \times 3.14 \times (0.1)^2 \text{ amperes} = 2.8 \text{ amperes}. \end{aligned} \tag{7.54}$$

(b) We start with the Ampère–Maxwell-law $\boldsymbol{\nabla} \times \mathbf{H} = \mathbf{j} + \partial\mathbf{D}/\partial t$ with conduction current density $\mathbf{j} = 0$. Multiplying by $\mu_0$ we obtain

$$\mu_0 \boldsymbol{\nabla} \times \mathbf{H} = \boldsymbol{\nabla} \times \mathbf{B} = \mu_0 \epsilon_0 \frac{\partial \mathbf{E}}{\partial t}. \tag{7.55}$$

We integrate this equation over a cross-sectional area and use Stokes's theorem. Then

$$\int_{F(C)} \boldsymbol{\nabla} \times \mathbf{B} \cdot d\mathbf{F} = \oint_{C(F)} \mathbf{B} \cdot d\mathbf{l} = \mu_0 \epsilon_0 \int_{F(C)} \frac{\partial \mathbf{E}}{\partial t} \cdot d\mathbf{F}. \tag{7.56}$$

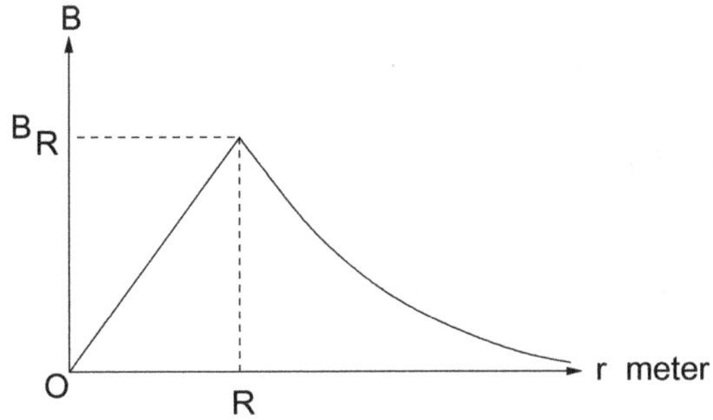


**Fig. 7.3.** Variation of the induced $B$ with radius $r$.

For $r \leq R$, and $C$ a circular path with radius $r$ we obtain

$$B(2\pi r) = \mu_0 \epsilon_0 \frac{dE}{dt} \pi r^2, \quad \text{or} \quad B = \frac{1}{2} \mu_0 \epsilon_0 r \frac{dE}{dt}. \tag{7.57}$$

For $r \geq R$, on the other hand, we obtain, since the electric field is nonvanishing only between the plates (ignoring boundary effects),

$$B(2\pi r) = \mu_0 \epsilon_0 \frac{dE}{dt} \pi R^2, \quad \text{i.e.} \quad B = \frac{1}{2} \mu_0 \epsilon_0 \frac{R^2}{r} \frac{dE}{dt}. \tag{7.58}$$

The variation of this induced magnetic field with $r$ is shown in Fig. 7.3.

$$\begin{aligned} B_R &= \frac{1}{2} \mu_0 \epsilon_0 R \frac{dE}{dt} = \frac{1}{2} \times 4\pi \times 10^{-7} \times 8.9 \times 10^{-12} \times 0.1 \times 10^{13} \\ &= 5.6 \times 10^{-6} \text{ weber} \cdot \text{meters}^{-2}. \end{aligned} \tag{7.59}$$

Although the displacement current is comparatively large, the magnetic field is rather small.

## 7.7 Poynting Vector and Conservation of Energy

The work $W_0$ done by charges per unit time in the presence of an electromagnetic field is the *power* (in units of watts, $W$ or $J/s$)

$$\begin{aligned} \frac{dW_0}{dt} &\equiv \frac{d \text{ work}}{dt} \\ &= \sum_i (\text{force acting on charge } i) \cdot (\text{velocity of charge } i) \\ &= \int \mathbf{v} \cdot (\rho \mathbf{E} + \mathbf{j} \times \mathbf{B}) dV \quad \text{joules/second}, \end{aligned} \tag{7.60}$$

where we inserted Eq. (7.45). Since $\mathbf{j} = \rho\mathbf{v}$, we have $\mathbf{v}\cdot\mathbf{j}\times\mathbf{B} = \rho\mathbf{v}\cdot\mathbf{v}\times\mathbf{B} = \rho\mathbf{v}\times\mathbf{v}\cdot\mathbf{B} = 0$. This implies (using $\mathbf{D} = \epsilon\mathbf{E},\ \mathbf{B} = \mu\mathbf{H}$ later)

$$\begin{aligned}\frac{dW_0}{dt} &= \int \rho\mathbf{v}\cdot\mathbf{E}dV \equiv \int \mathbf{j}\cdot\mathbf{E}dV \overset{(7.43)}{=} \int dV\mathbf{E}\cdot\left(\boldsymbol{\nabla}\times\mathbf{H} - \frac{\partial\mathbf{D}}{\partial t}\right)\\ &= \int dV(\mathbf{E}\cdot\boldsymbol{\nabla}\times\mathbf{H} - \mathbf{H}\cdot\boldsymbol{\nabla}\times\mathbf{E})\\ &\quad - \int dV\left(\mathbf{E}\cdot\frac{\partial\mathbf{D}}{\partial t} - \mathbf{H}\cdot\underbrace{\boldsymbol{\nabla}\times\mathbf{E}}_{-\partial\mathbf{B}/\partial t}\right) \quad \text{watts.}\end{aligned} \tag{7.61}$$

Now using the relation (cf. Appendix C)

$$\boldsymbol{\nabla}\cdot(\mathbf{E}\times\mathbf{H}) = \mathbf{H}\cdot(\boldsymbol{\nabla}\times\mathbf{E}) - \mathbf{E}\cdot(\boldsymbol{\nabla}\times\mathbf{H}), \tag{7.62}$$

we obtain

$$\begin{aligned}\frac{dW_0}{dt} &= -\int dV\boldsymbol{\nabla}\cdot(\mathbf{E}\times\mathbf{H}) - \frac{1}{2}\frac{\partial}{\partial t}\int dV(\mathbf{E}\cdot\mathbf{D} + \mathbf{H}\cdot\mathbf{B})\\ &= -\int(\mathbf{E}\times\mathbf{H})\cdot d\mathbf{F} - \frac{\partial W}{\partial t},\end{aligned} \tag{7.63}$$

where we assumed that $\mathbf{D} \propto \mathbf{E}$ and $\mathbf{B} \propto \mathbf{H}$ in time dependence. The vector

$$\mathbf{S} = \mathbf{E}\times\mathbf{H} \quad \text{watt}\cdot\text{meter}^{-2} \tag{7.64}$$

is called *Poynting vector*. The expression for $W$ is the *electromagnetic energy*,

$$W = \frac{1}{2}\int_V dV(\mathbf{E}\cdot\mathbf{D} + \mathbf{H}\cdot\mathbf{B}) \quad \text{joules.} \tag{7.65}$$

We have therefore:

$$\frac{d\,\text{work}}{dt} = \underbrace{\int dV\mathbf{j}\cdot\mathbf{E}}_{\text{Ohmic power}} = -\underbrace{\int_{F(V)} \mathbf{S}\cdot d\mathbf{F}}_{d\mathbf{F}\text{ directed to outside}} - \frac{\partial W}{\partial t} \quad \text{watts.} \tag{7.66}$$

Thus *the work done per unit time is equal to the loss of electromagnetic energy per unit time* ($\partial W/\partial t$) *plus gain or loss of energy through the walls of the system* ($\int \mathbf{S}\cdot d\mathbf{F}$). It should be noted that only the integral over the Poynting vector is a measurable quantity, i.e. the integral over a closed surface, not the transport of energy through some part of the volume, since only the total energy is a conserved quantity. Furthermore one should note that the definition of the Poynting vector involves some arbitrariness because we can always add to $\mathbf{S}$ a vector $\mathbf{T}$, whose divergence is zero, so that

$$\int_F \mathbf{T} \cdot d\mathbf{F} = \int_{V(F)} dV \boldsymbol{\nabla} \cdot \mathbf{T} = 0. \tag{7.67}$$

In the case of the "ideal" or "perfect conductor" the conduction electrons do no work, i.e. conductivity $\sigma = \infty$ and so resistance $R = 0$, and $\mathbf{E} = -\boldsymbol{\nabla}\phi = 0$, so that $\mathbf{j} = \sigma\mathbf{E} =$ finite, but $\int \mathbf{j}\cdot\mathbf{E}dV = 0$. (This is the reverse case to that of insulators or dielectrics, for which in the ideal case $\sigma = 0$).

Thus in the case *with* resistance $dW_0/dt \neq 0$ but in the "ideal conductor" we have $\int \mathbf{j} \cdot \mathbf{E}dV = 0$, since there $\mathbf{E} = 0$, i.e.

$$\int_{F(V)} \mathbf{S} \cdot d\mathbf{F} = -\frac{\partial W}{\partial t} \quad \text{watts,} \tag{7.68}$$

which implies: energy transmitted through the walls per unit time = loss of electromagnetic energy per unit time. The (somewhat paradox) example usually given to the above is that of *transport of energy in a conduction wire.*

### 7.7.1 *Application: The conduction wire*

The conduction wire we consider now is not made of ideal conductor material, since otherwise we would have to have $V = IR = 0$. Thus the conduction electrons suffer resistance and hence perform work $\neq 0$. The diagram in Fig. 7.4 shows, that the electric field $\mathbf{E}$ here points in the direction of the current and the vector $\mathbf{H}$ to its right into the paper.

We can obtain the direction of the Poynting vector with the right hand "thumb rule" ($\mathbf{E}$ along the index finger, $\mathbf{H}$ along the middle finger and so $\mathbf{E}\times\mathbf{H}$ parallel to the thumb), as indicated in the figure. We have *in the wire*:

$$\mathbf{E} = -\boldsymbol{\nabla}\phi, \quad E_{\text{tang}} \text{ continuous at surface,} \tag{7.69}$$

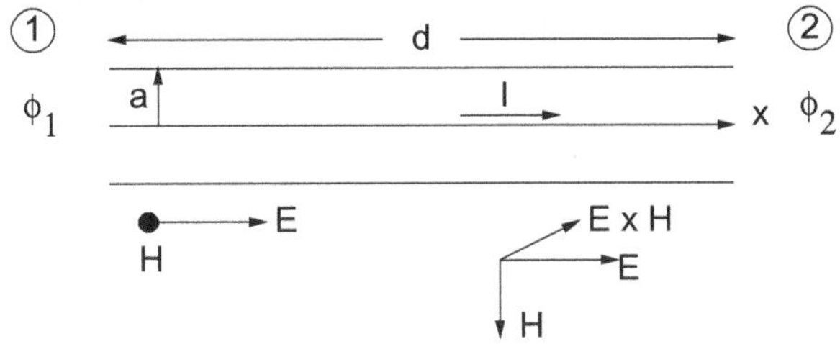


**Fig. 7.4.** The conduction wire.

and, with the wire over a length $d$,

$$\int_1^2 \mathbf{E}\cdot d\mathbf{l} = -\int_1^2 \nabla\phi\cdot d\mathbf{l} = -\int_1^2 d\phi = -(\phi_2-\phi_1) = \phi_1-\phi_2 \equiv V. \quad (7.70)$$

But $E_{\text{tangential}}$ is continuous at an interface, as we saw (cf. Eq. (4.44)). Hence also on the surface from outside: $E = V/d$. On the other hand, from the Maxwell equation (7.43),

$$\nabla\times\mathbf{H} = \mathbf{j} + \frac{\partial\mathbf{D}}{\partial t}, \quad (7.71)$$

follows or (neglecting the displacement current $\partial\mathbf{D}/\partial t$) from the integral once around the wire:

$$\oint \mathbf{H}\cdot d\mathbf{l} = \int \mathbf{j}\cdot d\mathbf{F} = I, \quad (7.72)$$

that

$$H = \frac{I}{2\pi a} \quad \text{amperes/meter}. \quad (7.73)$$

Since $\mathbf{E}\perp\mathbf{H}$,

$$\mathbf{E}\times\mathbf{H} = \frac{V}{d}\frac{I}{2\pi a} = \frac{VI}{2\pi a d} \quad \text{watt}\cdot\text{meters}^{-2}, \quad (7.74)$$

or $\mathbf{S} = -(EI/2\pi a)\mathbf{e}_r$. Thus the amount of energy transported *into* the wire per unit time is:

$$-\int(\mathbf{E}\times\mathbf{H})\cdot d\mathbf{F} = \frac{IV}{2\pi a d}2\pi a d = IV \quad \text{watts} \quad (7.75)$$

and is equal to the amount of energy, that the wire loses per unit time in the form of heat. We see therefore: Energy is radiated away, but at the same time absorbed, *in such a way that the current is kept constant* (this latter point is not always given sufficient emphasis, and then makes the problem appear paradoxical).

The fields $\mathbf{E}$ and $\mathbf{H}$ are a consequence of the current $I$ or the potential difference $V$; it is *not* possible to have $I$ and $V$ without $\mathbf{E}$ and $\mathbf{H}$, or the converse. It is thus a question of one's point of view whether one says, the power $-IV$ or the energy transported per unit time $IV$ are a consequence of $I$ and $V$ or a consequence of the fields $\mathbf{E}$ and $\mathbf{H}$. The easiest way to understand this problem is probably as an example of the conservation of energy. For a fixed current $I$ and thus in a static case ($\partial W/\partial t = 0$) the loss of energy as a result of the resistance $R$ must be compensated by radiation *into* the conductor, i.e.

$$\int \mathbf{S}\cdot(-d\mathbf{F}) = IV = I(\phi_1-\phi_2). \tag{7.76}$$

We can see this equivalence as follows: The Ohmic power is

$$\begin{aligned}\int_1^2 \mathbf{E}\cdot\mathbf{j}dV &= -\int \mathbf{j}\cdot\boldsymbol{\nabla}\phi dV = -\int_1^2 \mathbf{j}\cdot\left(\mathbf{e}_x\frac{d\phi}{dx}\underbrace{dx\,dF}_{dV}\right)\\ &= -\int_1^2\int(\mathbf{j}\cdot d\mathbf{F})d\phi = -\int_1^2 I d\phi = -I(\phi_2-\phi_1)\\ &= I(\phi_1-\phi_2) = IV \quad \text{watts.}\end{aligned} \tag{7.77}$$

### 7.7.2 *The field momentum density*

Let a volume $V$ be given with charge distribution (i.e. density) $\rho$. The *force that the field exerts on the charges*, is the external force in Newton's equation for the charged particle, i.e.

$$\begin{aligned}&\frac{d\,(\text{mechanical momentum})}{dt} = \text{external force}\\ &\quad = \int(\mathbf{E}+\mathbf{v}\times\mathbf{B})\rho dV = \int\left(\underbrace{\rho}_{\boldsymbol{\nabla}\cdot\mathbf{D}}\mathbf{E}+\underbrace{\mathbf{j}}_{\boldsymbol{\nabla}\times\mathbf{H}-\partial\mathbf{D}/\partial t}\times\mathbf{B}\right)dV\\ &\quad = \int\left[\mathbf{E}\boldsymbol{\nabla}\cdot\mathbf{D}+\left(\boldsymbol{\nabla}\times\mathbf{H}-\frac{\partial\mathbf{D}}{\partial t}\right)\times\mathbf{B}\right]dV.\end{aligned} \tag{7.78}$$

Adding and subtracting a vector product contribution this expression becomes:

$$\begin{aligned}&\frac{d\,(\text{mechanical momentum})}{dt}\\ &\quad = -\int\left[\frac{\partial\mathbf{D}}{\partial t}\times\mathbf{B}+\mathbf{D}\times\frac{\partial\mathbf{B}}{\partial t}\right]dV\\ &\qquad +\int\left[\mathbf{E}\boldsymbol{\nabla}\cdot\mathbf{D}+\mathbf{H}\underbrace{\boldsymbol{\nabla}\cdot\mathbf{B}}_{0}+(\boldsymbol{\nabla}\times\mathbf{H})\times\mathbf{B}+(\underbrace{\boldsymbol{\nabla}\times\mathbf{E}}_{-\partial\mathbf{B}/\partial t})\times\mathbf{D}\right]dV.\end{aligned} \tag{7.79}$$

It can now be shown (we leave the verification to Example 7.10) that the terms can be combined in the following way so that the component $j$ of the equation is given by:

$$\left\{\frac{d\,(\text{mechanical momentum})}{dt}\right\}_j = -\frac{\partial}{\partial t}\int(\mathbf{D}\times\mathbf{B})_j dV + \int\frac{\partial}{\partial x_i}T_{ij}dV, \tag{7.80}$$

where

$$T_{ij} = (D_iE_j + B_iH_j) - \delta_{ij}\frac{1}{2}(\mathbf{D}\cdot\mathbf{E}+\mathbf{B}\cdot\mathbf{H}) \quad \text{joules/meter}^3. \tag{7.81}$$

This expression is equal to $T_{ji}$ for $D = \epsilon E, B = \mu H$. The expression $T_{ij}$ is called *Maxwell stress tensor* or *stress-energy-momentum tensor*.[6] In a self-evident way we have (on application of Gauss' divergence theorem)

$$\frac{d}{dt}(\mathbf{p}^{\text{mech}} + \mathbf{P}^{\text{field}})_j - \int_F T_{ij}(d\mathbf{F})_i = 0. \tag{7.82}$$

The term on the right represents the *flow* of momentum through the surface $F$. If this surface is shifted to infinity, so that the term on the right is zero, we have

$$\mathbf{p}^{\text{mech}} + \mathbf{P}^{\text{field}} = \text{constant}, \tag{7.83}$$

where $\mathbf{P}^{\text{field}} = \int dV\mathbf{p}^{\text{field}}, \mathbf{p}^{\text{field}} = \mathbf{D}\times\mathbf{B}$ = volume density of the field momentum $\mathbf{P}^{\text{field}}$. This is the *law of conservation of momentum in electrodynamics.* Note that the field momentum density is not given by the Poynting vector $\mathbf{E}\times\mathbf{H}$! The expression

$$K_j = T_{ij}(d\mathbf{F})_i \tag{7.84}$$

is the $j$-component of a force acting on the area $d\mathbf{F}$. Since pressure = force/area, we see that

$$-\frac{K_jn_j}{dF} = -T_{ij}n_in_j \tag{7.85}$$

($n_i$: $i$-component of the unit vector) is the pressure of the radiation, i.e. the *radiation pressure.* The tails of comets are observed to point away from the

[6] Cook [33] remarks on pp. 183–184 on this terminology: "...the terminology hanging over from the days when the electromagnetic field was viewed as a distortion or stress in an elastic, mechanical ether."

sun; this is a consequence of the pressure of the radiation from the sun. We have here really been considering the microscopic field momentum. Considerations of the macroscopic bulk field momentum lead to difficulties which are pointed out in Jackson [73] and discussed in more detail in literature cited there.

**Example 7.2 (Separating a charge-anticharge pair).** A charge $q$, originally together with a charge $-q$ at rest at some point, is moved away from the charge $-q$. Show that the mechanical momentum $\mathbf{p}^{\text{mech}}$ of the charge $q$ is exactly cancelled by the field momentum, i.e. their sum is zero. (Compare with Examples 17.5 and 21.2).

**Solution:** The Lorentz force acting on a charge $q$ in a flux density of $\mathbf{B}$ teslas (due to some current or moving charge) is $\mathbf{F}_q = q\mathbf{v}\times\mathbf{B}$ newtons. We give the charge $q$ a mechanical momentum $\mathbf{p}^{\text{mech}}$ by pulling it away from $-q$ with a force equal and opposite to that of the Lorentz force. The charge $-q$, moving relatively in the opposite direction, provides the flux density $\mathbf{B}$. In fact from the Biot–Savart law we see that the fields $\mathbf{B}$ generated by the motion of the two opposite charges moving in opposite directions are opposite, i.e. $\mathbf{B}_q = -\mathbf{B}_{-q}$. In moving the charge it generates field momentum which has to be such that the total momentum is conserved, i.e. equal to that initially, i.e. zero. Thus from $\mathbf{F}_q$ and Newton's equation (with $\mathbf{v} = d\mathbf{x}/dt$)

$$\frac{d}{dt}\mathbf{p}^{\text{mech}} = -q\frac{d\mathbf{x}}{dt}\times\mathbf{B}_{-q}, \quad \mathbf{p}^{\text{mech}} = -q\int d\mathbf{x}\times\mathbf{B}_{-q} \quad \text{meters/second.} \tag{7.86}$$

From the Gauss law we obtain, now with only the one charge $q$ to be considered, $q = \int \mathbf{D}\cdot d\mathbf{a} = \int D_l da_l$, so that with $dx_j da_l = \delta_{jl} dV$ for element of area $d\mathbf{a}$, we have

$$\begin{aligned} p_i^{\text{mech}} &= -q\int \epsilon_{ijk} dx_j B_k = -\int \epsilon_{ijk} dx_j B_k D_l da_l = -\int \epsilon_{ijk} B_k D_l \delta_{jl} dV \\ &= -\int \epsilon_{ijk} B_k D_j dV = -\int dV(\mathbf{D}\times\mathbf{B})_i \quad \text{meters/second.} \end{aligned} \tag{7.87}$$

Hence $\mathbf{p}^{\text{mech}} + \mathbf{P}^{\text{field}} = 0$. The components of the field momentum perpendicular to the line joining the charges cancel and leave only the component along that line.

## 7.8 Further Examples

**Example 7.3 (The variable conductor loop).** A rectangular current loop with constant width $y = b$ consists of three fixed conducting rods and one movable conducting rod which allows to change the length $x$ of the loop. This movable rod is moved with constant velocity $\mathbf{v}$ in the direction of length of the loop (i.e. in the direction of positive $x$). The time-dependent field of induction $\mathbf{B} = \mathbf{e}_z B_0 \cos\omega t$ acts in the overall space perpendicular to the plane of the loop.

(a) Calculate the total induced electromotive force $V$ in the loop.
(b) What is obtained for $V$, if the movable conducting rod is held fixed at a distance $x$ from the opposite side of the rectangle, but the entire current loop moves with velocity $\mathbf{v} = v\mathbf{e}_x$?

**Solution:**
(a) We have here a case, in which the area of the loop, as well as the field $\mathbf{B}$ vary with time. The variations of both contribute to the induced potential $V$. We use Faraday's law (7.50),

$$V = \oint_C \mathbf{E} \cdot d\mathbf{l} = -\frac{d}{dt} \int_{F(C)} \mathbf{B} \cdot d\mathbf{F} \quad \text{volts.} \tag{7.88}$$

If the area $F$ to be integrated over varies with time, we have to differentiate with respect to time here. We assume first that $\mathbf{B}$ is constant with respect to time. (Here only so that we do not have to drag its time derivative all the way along!). We then have

$$\begin{aligned}
\frac{d}{dt} \int_{\mathbf{F}(t)} \mathbf{B} \cdot d\mathbf{F} &= \lim_{\delta t \to 0} \left[ \frac{\int_{\mathbf{F}(t+\delta t)} \mathbf{B} \cdot d\mathbf{F} - \int_{\mathbf{F}(t)} \mathbf{B} \cdot d\mathbf{F}}{\delta t} \right] \\
&= \lim_{\delta t \to 0} \frac{1}{\delta t} \left[ \int_{\mathbf{F}(t)+\delta t d\mathbf{F}/dt} \mathbf{B} \cdot d\mathbf{F} - \int_{\mathbf{F}(t)} \mathbf{B} \cdot d\mathbf{F} \right] \\
&= \lim_{\delta t \to 0} \left[ \frac{1}{\delta t} \delta t \frac{d\mathbf{F}}{dt} \cdot \frac{d}{d\mathbf{F}} \int \mathbf{B} \cdot d\mathbf{F} \Big|_{\mathbf{F}(t)} \right] \\
&= (\mathbf{B})_{\mathbf{F}} \cdot \frac{d\mathbf{F}}{dt}.
\end{aligned} \tag{7.89}$$

Including now also the time dependence of $\mathbf{B}$, we have

$$\frac{d}{dt} \int_{\mathbf{F}(t)} \mathbf{B} \cdot d\mathbf{F} = \int_F \frac{\partial \mathbf{B}(t)}{\partial t} \cdot d\mathbf{F} + (\mathbf{B})_{\mathbf{F}(t)} \cdot \frac{d\mathbf{F}}{dt}. \tag{7.90}$$

In the case of our example we have

$$V = \oint_{C(F)} \mathbf{E} \cdot d\mathbf{l} = -\frac{d}{dt} \int_{F(C)} \mathbf{B} \cdot d\mathbf{F} = -\mathbf{B} \cdot \frac{d\mathbf{F}(C)}{dt} - \int \frac{\partial \mathbf{B}}{\partial t} \cdot d\mathbf{F}. \quad (7.91)$$

We choose the area $\mathbf{F}$ to point in the direction of $z$, as indicated in Fig. 7.5 ($\mathbf{F} \parallel \mathbf{e}_z$). The voltage $V$ is given by:

$$V = -\mathbf{B}bv \cdot (\mathbf{e}_y \times \mathbf{e}_x) + \int_0^x \omega B_0 \sin \omega t \underbrace{b dx}_{dF} \quad \text{volts} \quad (7.92)$$

and hence by

$$\begin{aligned} V &= -B_0 vb \cos \omega t + \omega B_0 bx \sin \omega t \\ &= B_0 b\{x\omega \sin \omega t - v \cos \omega t\} \\ &= B_0 b\sqrt{v^2 + x^2\omega^2} \sin(\omega t - \delta) \quad \text{volts}, \end{aligned} \quad (7.93)$$

where $\delta = \tan^{-1}(v/x\omega)$, i.e. $\cos \delta = x\omega/\sqrt{v^2 + x^2\omega^2}$ and correspondingly for $\sin \delta$.

(b) We now assume, that the movable part of the rectangular loop is fixed to the other part of the loop (at position $x$) so that the area of the loop remains constant in time but the entire current loop moves with velocity $\mathbf{v}$ in the direction of $\mathbf{e}_x$. Then

$$V = -\frac{d}{dt} \int_{F(C)} \mathbf{B} \cdot d\mathbf{F} = -\int_{F(C)} \frac{d\mathbf{B}}{dt} \cdot d\mathbf{F} = -\int_{F(C)} \left[\frac{\partial \mathbf{B}}{\partial t} + (\mathbf{v} \cdot \mathbf{\nabla})\mathbf{B}\right] \cdot d\mathbf{F}, \quad (7.94)$$

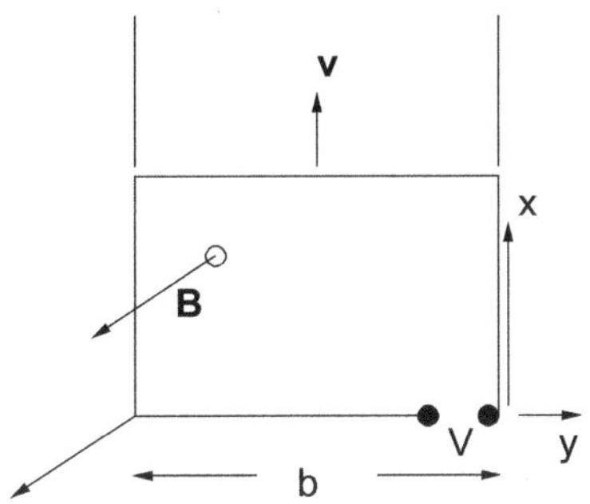


**Fig. 7.5.** The conducting loop with movable rod.

where $\mathbf{v}$ represents the velocity of the charges in the field $\mathbf{B}$. But now (for $\mathbf{v}$ constant with $\boldsymbol{\nabla}\cdot\mathbf{B}=0$), see Appendix C,

$$\boldsymbol{\nabla}\times(\mathbf{B}\times\mathbf{v}) = (\mathbf{v}\cdot\boldsymbol{\nabla})\mathbf{B} - (\mathbf{B}\cdot\boldsymbol{\nabla})\mathbf{v} + \mathbf{B}(\boldsymbol{\nabla}\cdot\mathbf{v}) - \mathbf{v}(\boldsymbol{\nabla}\cdot\mathbf{B}) = (\mathbf{v}\cdot\boldsymbol{\nabla})\mathbf{B}. \tag{7.95}$$

For given $\mathbf{B} = B\mathbf{e}_z$, we have $\boldsymbol{\nabla}B_z = 0$. Although we know that this contribution vanishes, we drag it along, so that (using Stokes's theorem in the second step):

$$\begin{aligned} V &= -\int_{F(C)} \left[\frac{\partial\mathbf{B}}{\partial t} + \boldsymbol{\nabla}\times(\mathbf{B}\times\mathbf{v})\right]\cdot d\mathbf{F} \\ &= -\int_{F(C)} \frac{\partial\mathbf{B}}{\partial t}\cdot d\mathbf{F} - \oint_{C(F)} (\mathbf{B}\times\mathbf{v})\cdot d\mathbf{l} \\ &= -\int_{F(C)} \frac{\partial\mathbf{B}}{\partial t}\cdot d\mathbf{F} + \oint_{C(F)} (\mathbf{v}\times\mathbf{B})\cdot d\mathbf{l}. \end{aligned} \tag{7.96}$$

In our case

$$\mathbf{v}\times\mathbf{B} = vB\mathbf{e}_x\times\mathbf{e}_z = vB\mathbf{e}_y,$$

$$+\oint_{C(F)} (\mathbf{v}\times\mathbf{B})\cdot d\mathbf{l} = vB\oint \mathbf{e}_y\cdot d\mathbf{l} = vB(b-b) = 0 \tag{7.97}$$

(since the field $\mathbf{B}$ is constant, it has the same value along $\mathbf{b}$ as along $-\mathbf{b}$; for this reason these contributions cancel). As expected, also the contribution of $\mathbf{v}\times\mathbf{B}$ is zero, and

$$V = -\int \frac{\partial\mathbf{B}}{\partial t}\cdot d\mathbf{F} = \omega B_0 bx\sin\omega t = B_0 b(\omega x)\sin\omega t \quad \text{volts}. \tag{7.98}$$

This means, the induced potential is independent of the velocity. This is, what one expects, because in the rest frame of the conductor one observes — in the case of the constant field $\mathbf{B}$ — at every instant of time the same situation.

**Example 7.4 (The wire pendulum).** A wire pendulum (of length $R$) oscillates with velocity $\dot{x} = \omega d\cos\omega t$, where $d$ is the largest horizontal deflection of the pendulum as shown in Fig. 7.6. The constant magnetic field of B webers/meter$^2$ points out of the plane of the pendulum. Determine the induced voltage $V$.

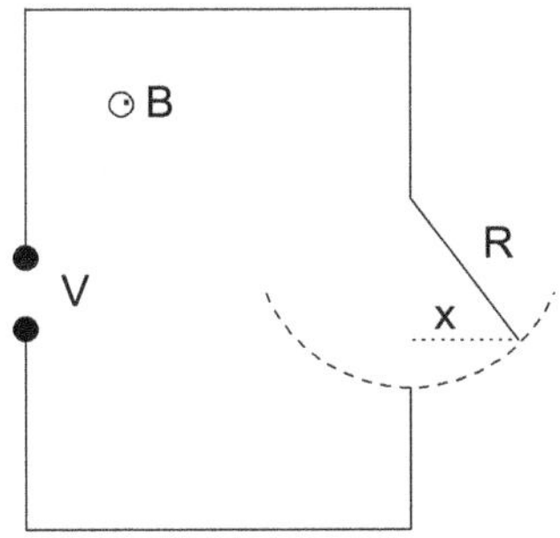


**Fig. 7.6.** The wire pendulum.

**Solution:** We start with Faraday's law (7.20),

$$V = -\frac{d}{dt}\int \mathbf{B}\cdot d\mathbf{F} = \frac{d}{dt}[B \times \text{area}(t)]$$
$$= \frac{d}{dt}[B(\text{constant area} + \text{varying area})]. \tag{7.99}$$

Set $\sin\theta = x/R, \sin\theta_2 = d/R$, so that

$$\frac{\sin\theta}{\sin\theta_2} = \frac{x}{d}, \quad x(t) = \frac{d\sin\theta}{\sin\theta_2} = \frac{d\sin\omega t}{\sin\theta_2}. \tag{7.100}$$

The variable part of the area is then $\simeq \frac{1}{2}Rx(t)$. It follows that the requested voltage $V$ is

$$V = \frac{d}{dt}\left[\frac{BR}{2}\frac{d\sin\omega t}{\sin\theta_2}\right] = \frac{1}{2}B\omega R^2\cos\omega t \quad \text{volts.} \tag{7.101}$$

**Example 7.5 (The revolving coil[7]).** In Example 6.3 the field in the gap of a circularly shaped electromagnet was calculated. A small circularly shaped coil of radius $b$ and $n$ turns revolves fully within the gap of the electromagnet with angular velocity $\omega$ about a diameter perpendicular to the magnetic field which is assumed to be homogeneous. $R$ is the resistance and $L$ the self-inductance of the coil. Formulate an equation from which the current induced in the coil can be calculated.

[7]This example assumes familiarity with the concept of self-inductance which will be introduced in Chapter 8.

**Solution:** We begin with Faraday's law (7.20),

$$\nabla \times \mathbf{E} = -\frac{\partial \mathbf{B}}{\partial t}. \tag{7.102}$$

The induced voltage $V_i$ is

$$V_i = \int \mathbf{E} \cdot d\mathbf{l} = \int \nabla \times \mathbf{E} \cdot d\mathbf{F} = -\frac{d}{dt}\int \mathbf{B} \cdot d\mathbf{F}. \tag{7.103}$$

The induced flux is $\phi = \int \mathbf{B} \cdot d\mathbf{F}$, i.e. $V_i = -d\phi/dt$. The inductances $L_{kj}$ of the circuit are defined by $\phi_k = \sum_j L_{kj} I_j$ (cf. Chapter 8). Here we have two circuits — the one of the coil, as well as another one which is responsible for the varying flux through the area of the coil, i.e.

$$\phi = \underbrace{LI}_{\text{self-induction part}} + \underbrace{\tilde{\phi}}_{\text{the varying flux}} . \tag{7.104}$$

Here $\tilde{\phi} = nB \times$ projection of area $F$ of one turn of the coil perpendicular to the magnetic field. Since (cf. Eq. (8.1))

$$IR = \underbrace{V_{\text{generator}}}_{0} - \frac{d\phi}{dt}, \tag{7.105}$$

we have

$$IR = -L\dot{I} - \frac{d\tilde{\phi}}{dt}. \tag{7.106}$$

In other words,

$$IR + L\dot{I} = -\frac{d\tilde{\phi}}{dt}, \tag{7.107}$$

where $\tilde{\phi} = nBF\sin\omega t$, with $B$ the field through area $F$ and $F = \pi b^2$, as indicated in Fig. 7.7. The study of such equations is the subject of Chapter 8.

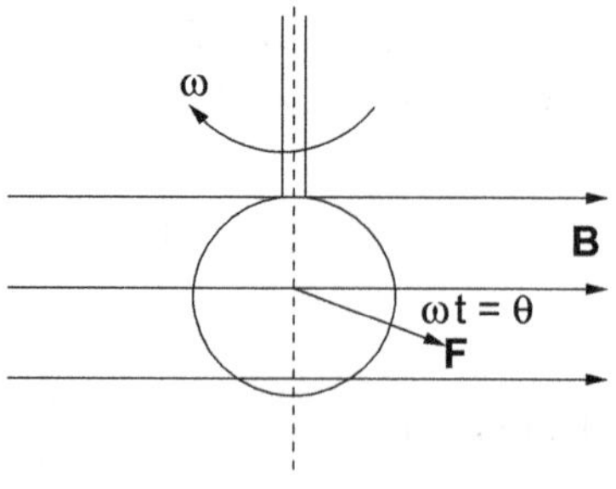


**Fig. 7.7.** The revolving coil.

**Example 7.6 (The rectangular conductor loop).** A magnetic field varying with time is given by the following expression: $\mathbf{B} = \mathbf{e}_y B_0 \cos\omega t$, with $B_0$ constant. Determine the voltage induced in a rectangular conductor loop in the $(x, z)$-plane with side lengths $a$ in the direction of $x$, and $b$ in the direction of $z$.

**Solution:** Let the unit vector $\mathbf{e}_y$ point into the paper. Then the magnetic flux into the paper is

$$\Phi = \int \mathbf{B} \cdot d\mathbf{F} = \int_{x=0}^{a} \int_{z=0}^{b} B_0 \cos\omega t \mathbf{e}_y \cdot dx dz \mathbf{e}_y, \tag{7.108}$$

i.e. $\Phi = B_0 \cos\omega tab$. The induced voltage is therefore

$$V = \oint_C \mathbf{E} \cdot d\mathbf{l} = -\frac{d}{dt}\Phi = -\frac{d}{dt}[B_0 ab \cos\omega t] = abB_o\omega \sin\omega t \quad \text{volts.} \tag{7.109}$$

**Example 7.7 (The metal trolley[8]).** A metal trolley (mass $m$) is placed on two parallel, circular, conducting rails (a distance of L meters apart), as indicated in Fig. 7.8. A uniform magnetic field $\mathbf{B_0}$ points vertically out of the plane. With a rocket engine the trolley is given a constant thrust (i.e. force) of $F_0$ newtons. The resistance between the rails is $R$. Determine the voltage

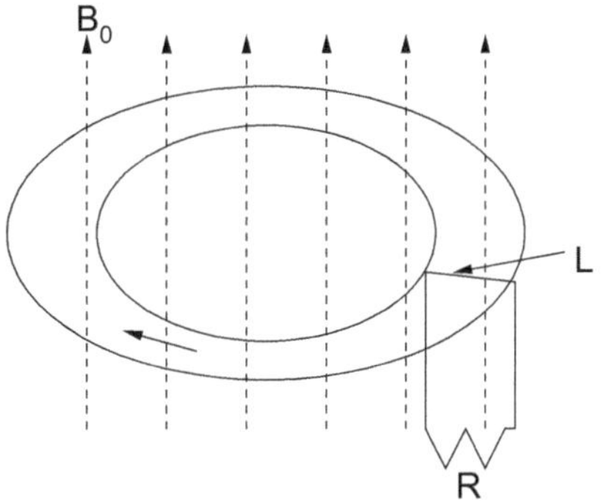


**Fig. 7.8.** The rails of the trolley with induced circuit.

[8]See also *The Electromagnetic Problem Solver* [47], Example 11–36, p. 621.

$V$ induced in the circuit linking the trolley with the resistance, and discuss its time dependence.
(Hint: It is advisable to start from Ampère's law of force).

**Solution:** Ampère's law of force is given by Eq. (5.23), i.e.

$$\mathbf{F} = I \int d\mathbf{s} \times \mathbf{B}(\mathbf{r}) \quad \text{newtons.} \tag{7.110}$$

In the present case we need to know the forces on the right side of Newton's equation which are responsible for the motion of the trolley, i.e. in

$$m\frac{dv}{dt} = F_0 - F_{\text{applied}}. \tag{7.111}$$

We obtain the force $F_{\text{applied}}$ from Ampère's law (see discussion after Eq. (5.26)) as $B_0IL$ newtons, by considering the current element with resistance $R$ between the rails and identifying $ds$ with $L$. We obtain the current $I$ from Faraday's law or correspondingly from the formula for the induced potential difference, see e.g. Eq. (7.5),

$$V = \oint \left(d\mathbf{s} \times \frac{d\mathbf{l}}{dt}\right) \cdot \mathbf{B} = B_0 L v, \quad \text{i.e.} \quad I = \frac{V}{R} = \frac{B_0 v L}{R}, \tag{7.112}$$

so that

$$m\frac{dv}{dt} = F_0 - \frac{B_0^2 L^2 v}{R} \quad \text{or} \quad m\frac{dv}{dt} + \frac{B_0^2 L^2 v}{R} = F_0. \tag{7.113}$$

This is an inhomogeneous differential equation of the first order which is readily solved (cf. Eqs. (8.87) and (8.88)) with the initial condition $v = 0$ at time $t = 0$. One obtains after simple integrations the physically plausible result

$$v = \frac{RF_0}{L^2 B_0^2}\left[1 - e^{-t/(mR/L^2 B_0^2)}\right] \quad \text{m/s.} \tag{7.114}$$

We see that in the course of time ($t \to \infty$) the velocity becomes constant.

## 7.9 Exercises

**Example 7.8 (Coulomb's law).** Starting from the Lorentz force and Maxwell's equations in differential form derive the Coulomb law.

**Example 7.9 (Energy of a spherical condenser).** The energy of a spherical condenser of radii $a$ and $b$, $a < b$, with charges $q, -q$, was derived in Example 2.19 (see Eq. (2.209)). This energy resides in the electric field

$E$ between the spheres. Rederive the result from the formula for the energy density.

**Example 7.10 (The Maxwell stress tensor).** Show that

$$[\mathbf{E}(\boldsymbol{\nabla}\cdot\mathbf{D}) + (\boldsymbol{\nabla}\times\mathbf{H})\times\mathbf{B} + \mathbf{H}(\boldsymbol{\nabla}\cdot\mathbf{B}) + (\boldsymbol{\nabla}\times\mathbf{E})\times\mathbf{D}]_i = \sum_j \frac{\partial}{\partial x_j}T_{ij}, \tag{7.115}$$

where

$$T_{ij} = (D_iE_j + B_iH_j) - \delta_{ij}\frac{1}{2}(\mathbf{D}\cdot\mathbf{E} + \mathbf{B}\cdot\mathbf{H}). \tag{7.116}$$

[Hint: Consider first $[\mathbf{E}(\boldsymbol{\nabla}\cdot\mathbf{D}) + (\boldsymbol{\nabla}\times\mathbf{E})\times\mathbf{D}]_i$, and prove the relation for this part, or equivalently that

$$[\mathbf{Q}(\boldsymbol{\nabla}\cdot\mathbf{Q}) + (\boldsymbol{\nabla}\times\mathbf{Q})\times\mathbf{Q}]_i = \sum_j \frac{\partial}{\partial x_j}R_{ij}, \tag{7.117}$$

where $R_{ij} = Q_iQ_j - Q^2\delta_{ij}/2$].

**Example 7.11 (Radiation pressure).** Show that a diagonal element of $T_{ij}$ can be written:

$$T_{ii} = \frac{1}{2}\epsilon_0(E_i^2 - E_j^2 - E_k^2) + \frac{1}{2\mu_0}(B_i^2 - B_j^2 - B_k^2), \tag{7.118}$$

where $i, j, k = x, y, z$ and all different. Hence show that for a closed surface containing homogeneous and isotropic radiation the magnitude of the pressure exerted on the surface is equal to one-third of the energy density of the radiation.
[Hint: Set $E_i^2 = E^2/3$ etc.].

**Example 7.12 (A charged particle in an electromagnetic field).** The Lagrangian $L$ of a particle of charge $e$ and mass $m$ in an electromagnetic vector potential $\mathbf{A}(q_i, t)$ and scalar potential $\phi(q_i), i = 1, 2, 3$, is defined by

$$L(q_i, \dot{q}_i) = \frac{1}{2}m\sum_i \dot{q}_i^2 + e\sum_i \dot{q}_iA_i(q_i, t) - \phi(q_i) \tag{7.119}$$

with electric field $\mathbf{E}$ and the magnetic induction $\mathbf{B}$ given by

$$\mathbf{E} = -\frac{\partial\mathbf{A}}{\partial t} - \boldsymbol{\nabla}\phi \quad \text{newtons/coulomb}, \quad \mathbf{B} = \boldsymbol{\nabla}\times\mathbf{A} \quad \text{teslas}. \tag{7.120}$$

Derive the equation of motion of the particle, i.e.[9]

$$\frac{d}{dt}(m\dot{\mathbf{r}}) = e[\mathbf{E} + \dot{\mathbf{r}} \times \mathbf{B}] \quad \text{newtons.} \tag{7.121}$$

**Example 7.13 (The Meissner effect).** In a superconducting state the electrons of a metal form pairs called *Cooper pairs* of charge $q = 2e$, and their current, called *Cooper current*, is such that $\mathbf{j} = \rho_0[\boldsymbol{\nabla}\varphi - q\mathbf{A}/\hbar]$ amperes/m$^2$, where $\rho_0$ is a constant, $\varphi$ a scalar potential and $\mathbf{A}$ the electromagnetic vector potential. Considering a constant magnetic field $\mathbf{B} = B_0(0, 0, 1)$ teslas and a superconductor occupying the half-plane $z > 0$, show that

$$B_z = B_0 \exp(-z/\lambda_L) \quad \text{teslas}, \quad \lambda_{\rm L} = \sqrt{\mu_0\rho_0 \mathrm{q}} \quad \text{meters}, \tag{7.122}$$

where $\lambda_L$ is called *London length.* What do you conclude from the result? [Hint: Use $\boldsymbol{\nabla} \times \mathbf{B} = \mu_0\mathbf{j}, \boldsymbol{\nabla} \cdot \mathbf{B} = 0$].

**Example 7.14 (Time average of Poynting vector).** Consider the Poynting vector for real fields

$$\mathbf{E}(\mathbf{r}, t) = \mathbf{E}_0(\mathbf{r}) \cos \omega t, \quad \mathbf{H}(\mathbf{r}, t) = \mathbf{H}_0(\mathbf{r}) \cos \omega t. \tag{7.123}$$

Show that the time average of the Poynting vector is $\mathbf{E}_0(\mathbf{r}) \times \mathbf{H}_0(\mathbf{r})/2$.

**Example 7.15 (Field of steady currents within a solid sphere).** A sphere of radius $a$ has a magnetic dipole moment $\mathbf{m}$ due to steady currents of $\mathbf{j}$ amperes/m$^2$ within. Use Eq. (5.39), i.e.

$$\mathbf{m} = \frac{1}{2}\int d\mathbf{r}[\mathbf{r} \times \mathbf{j}(\mathbf{r})] \quad \text{ampere} \cdot \text{meters}^2, \tag{7.124}$$

together with Eq. (6.24) and

$$\langle \mathbf{B} \rangle = \frac{1}{4\pi a^3/3}\int \mathbf{B} dV \quad \text{teslas}, \tag{7.125}$$

to show that the average of the field $\mathbf{B}$ over the sphere is given by

$$\langle \mathbf{B} \rangle = \frac{\mu_0}{4\pi}\frac{2\mathbf{m}}{a^3} \quad \text{teslas.} \tag{7.126}$$

**Example 7.16 (Dipole moment of spinning sphere).** A uniformly charged solid sphere of radius $a$ and total charge $q$ is set spinning with angular velocity $\omega$ about the $z$-axis. What is the dipole moment of the sphere? What is the average magnetic field within the sphere?

[9]See e.g. Müller–Kirsten [100], p. 76.

**Example 7.17 (Path of a particle in E, B fields).** A particle of mass $m$ and charge $q$ moves under the influence of fields $\mathbf{E} = E_0\mathbf{e}_x$ and $\mathbf{B} = B_0 \cos\theta\, \mathbf{e}_x + B_0 \sin\theta\, \mathbf{e}_y$, with $E_0, B_0$ constant. Determine the path $\mathbf{r}(t)$ of the particle for initial condition $\mathbf{r}(0) = 0$.

**Example 7.18 (The Hertz vector $\mathbf{\Pi}$).** The Hertz vector $\mathbf{\Pi}$ is defined by the relation

$$\phi = -\nabla \cdot \mathbf{\Pi}, \tag{7.127}$$

where $\phi$ is the scalar potential in Eq. (7.33), i.e.

$$-\nabla\phi = \mathbf{E} + \frac{\partial \mathbf{A}}{\partial t}. \tag{7.128}$$

Assume that the vector potential $\mathbf{A}$ is chosen to be related to $\mathbf{\Pi}$ by the equation

$$\mathbf{A} = \mu\sigma\mathbf{\Pi} + \epsilon\mu\frac{\partial \mathbf{\Pi}}{\partial t}, \tag{7.129}$$

with $\mathbf{j} = \sigma\mathbf{E}$. Show with the help of the Maxwell equation $\nabla \times \mathbf{H} = \mathbf{j} + \partial\mathbf{D}/\partial t$, that the Hertz vector satisfies the wave equation

$$\triangle\mathbf{\Pi} - \epsilon\mu\frac{\partial^2 \mathbf{\Pi}}{\partial t^2} = \mu\sigma\frac{\partial \mathbf{\Pi}}{\partial t}, \tag{7.130}$$

and that $\mathbf{E}$ and $\mathbf{H}$ are given by

$$\mathbf{E} = \nabla \times (\nabla \times \mathbf{\Pi}), \quad \mathbf{H} = \nabla \times \left\{\epsilon\frac{\partial \mathbf{\Pi}}{\partial t} + \sigma\mathbf{\Pi}\right\}. \tag{7.131}$$

**Example 7.19 (Unit of H).** Check, by using the energy density contained in Eq. (7.65), that the magnetic field strength $\mathbf{H}$ is given in amperes/meter. Check that a magnetic charge would be given in coulombs (cf. Eq. (21.28)).

**Example 7.20 (Poynting vector in a coaxial cable).** A coaxial cable has an inner conductor of radius $r_1$ and the outer conductor of (inner) radius $r_2 > r_1$, both conductors being nonmagnetic (i.e. $\mu = \mu_0$) and lossless (no resistance). The inner conductor is held at voltage $V$ providing a current $I$ in it along the axial direction $z$. What are the electric and magnetic fields $E_r$ and $H_\theta$ and their directions, and hence the Poynting vector $S$?

(Answers: $E_r = V/r\ln(r_2/r_1)$, $H_\theta = I/2\pi r$, $S = VI/2\pi r^2 \ln(r_2/r_1)$.

Chapter 8

# Application to Coils, Circuits and Transmission Lines

## 8.1 Introductory Remarks

In this chapter we consider circuits of currents, we define mutual and self-inductances $L$, and consider coils and solenoids, and return once again to gauge transformations. Again examples are given for purposes of illustration. An interesting result will be that the product of self-inductance $L$ and capacity $C$ of a transmission line per unit length is equal to the product of dielectric constant and magnetic permeability of the surrounding medium. This result will be encountered once again in Chapter 14 in our treatment of wave guides. In Chapter 9 it will be shown that the latter product is, in fact, $1/c^2$, where $c$ is the velocity of light. Thus the quantity $1/\sqrt{LC}$ represents the velocity of the electromagnetic wave travelling along the transmission line. Extending the concept of resistance to that of *impedance*, we also consider more complicated circuits and networks, such as those used in microphones and speakers and known generally as *transducers* and *wave filters*, and we consider the extension of these to *transmission lines*.[1] The networks are generally evaluated with the help of *Kirchhoff's laws* which will be elucidated here. The consideration of the mechanical analogy to an electromagnetic circuit is left to exercises. We treat transmission lines here in terms of currents and voltages. For a treatment of reflection and absorption in transmission lines equivalently in terms of electric and magnetic fields we refer to Jackson [73], Section 8.12. In quantum theory periodic electric fields act essentially like wave filters.

[1]Like e.g. Schelkunoff [129], p. 148, we use the term "transmission line" in a restricted sense for wave guides with small transverse dimensions.

We add a note on currents. In practice on distinguishes between direct currents (DC) and alternating currents (AC). In DC circuits the current generated e.g. by a battery is constant and flows in one direction. The currents in most electronic devices like computers are of this type. In AC circuits the currents vary periodically (i.e. sinusoidally) with time, thereby also reversing their direction and passing through zero voltage. These AC source voltages are those commonly supplied to homes and businesses. One of their advantages is that they can be transmitted more easily over large distances at high voltage and with less loss of power, also a disruption at some point is not as dangerous as in the case of a DC transmission line. For use in a home a high AC voltage has to be reduced (e.g. to 120 V in many places) with the help of a transformer (see Example 8.15).

## 8.2 Inductances $L_{ij}$

We assume that several current-carrying circuits are given. For each, say the $k$th circuit, we have

$$I_k R_k = (V_e)_k - \frac{\partial \Phi_k}{\partial t} \quad \text{volts.} \tag{8.1}$$

Here $\Phi_k = \int_{F_k} \mathbf{B} \cdot d\mathbf{F}$ is the magnetic *flux* or *flux linkage* through the $k$th circuit (originating from currents in other circuits) with enclosed area $F_k$, and external or generator voltage or e.m.f. $V_e$. Equation (8.1) represents a combination of the usual Ohm's law with Faraday's law. If now the current $I_j$ in circuit $j$ is varied with time, then the magnetic flux through circuit $k$ varies and hence according to the relation $(V_{\text{induced}})_k = -\partial\Phi_k/\partial t$ also the induced voltage $(V_{\text{induced}})_k$, and we can write (using the superposition principle, i.e. the summability of contributions of all circuits)

$$(V_{\text{induced}})_k = -\sum_j L_{kj} \frac{\partial I_j}{\partial t} \quad \text{volts,} \tag{8.2}$$

where the coefficients of inductance, $L_{kj}$, are constants. Hence we can write

$$-\frac{\partial \Phi_k}{\partial t} = -\sum_j L_{kj} \frac{\partial I_j}{\partial t},$$

or

$$\Phi_k(t) = \sum_j L_{kj} I_j(t) \quad \text{webers.} \tag{8.3}$$

Thus we have, for instance, $\Phi_2(I_1) = L_{21}I_1$. If the current $I = I(t)$ varies with time $t$, also $\mathbf{j}$ and $\rho$ must vary with $t$, i.e.

$$I(t) = \int \mathbf{j}(\mathbf{r},t)\cdot d\mathbf{F}(\mathbf{r}) = \int \rho(\mathbf{r},t)\frac{d\mathbf{s}\cdot d\mathbf{F}(\mathbf{r})}{dt} \quad \text{amperes.} \tag{8.4}$$

The equation of continuity must be preserved (in view of charge conservation), i.e.

$$\boldsymbol{\nabla}\cdot\mathbf{j} + \frac{\partial\rho}{\partial t} = 0. \tag{8.5}$$

We can obtain the relation (8.3) also as follows, thereby determining the coefficients $L_{kj}$. We have

$$\Phi_k(t) = \int_{F_k}\mathbf{B}(\mathbf{r},t)\cdot d\mathbf{F} = \int_{F_k}\boldsymbol{\nabla}\times\mathbf{A}(\mathbf{r},t)\cdot d\mathbf{F}(\mathbf{r}) \overset{\text{Stokes}}{=} \oint_k \mathbf{A}(\mathbf{r},t)\cdot d\mathbf{s}_k. \tag{8.6}$$

But, as we saw previously (cf. Eq. (6.43)), the vector potential at a point $\mathbf{r}$ resulting from current densities $\mathbf{j}(\mathbf{r},t)$ is given by the following expression:

$$\mathbf{A}(\mathbf{r},t) = \frac{\mu}{4\pi}\int\frac{\mathbf{j}(\mathbf{r}',t)d\mathbf{r}'}{|\mathbf{r}-\mathbf{r}'|} = \frac{\mu}{4\pi}\sum_i I_i(t)\int\frac{d\mathbf{s}_i}{|\mathbf{r}-\mathbf{r}_i|} \quad \text{tesla}\cdot\text{meters} \tag{8.7}$$

with $\mathbf{j}(\mathbf{r}',t)d\mathbf{r}' = I(t)d\mathbf{s}(\mathbf{r}')$, and $ds_i$ is the length of the current element at $\mathbf{r}_i$. Inserting this into Eq. (8.6), we obtain

$$\Phi_k(t) = \sum_i L_{ki}I_i(t) \quad \text{webers,} \quad L_{ki} = \frac{\mu}{4\pi}\int_i\int_k\frac{d\mathbf{s}_i\cdot d\mathbf{s}_k}{|\boldsymbol{r}_k-\boldsymbol{r}_i|} \quad \text{henries.} \tag{8.8}$$

Obviously $L_{ki} = L_{ik}$. The coefficient $L_{kk}$ is called the *self-inductance* of conductor $k$ (measured in henries (H) or Wb/A). In its case, one has to take into account its cross section, since otherwise the expression $1/|\mathbf{r}_k - \mathbf{r}_i|$ becomes problematic. Inserting Eq. (8.8) into Eq. (8.1), we obtain a relation of great importance for circuits:

$$I_kR_k + \sum_i L_{ki}\dot{I}_i = (V_e)_k \quad \text{volts.} \tag{8.9}$$

The coefficients $L_{ik}$ for $i \neq k$ are called *mutual inductances*. In the following examples we calculate inductances of both types, self-inductances and mutual inductances. The relation (8.9) expresses what is called *Kirchhoff's first law*, namely that the applied voltage or electromotive force, e.m.f., in a closed circuit is equal to the sum of the potentials across the various resistances, inductances and condensers (as we shall see below) in the circuit. We now consider examples. *Kirchhoff's second law* expresses the conservation of the current (at junctions of the circuit).

**Example 8.1 (The long solenoid).** Determine the self-inductance of a long solenoid.

**Solution:** We let the length of the solenoid be $l$ and assume it has $n$ turns per unit length, and cross-sectional area $S$. The current in the solenoid is taken to be $I$. The magnetic flux through $nl$ turns is $nl$ times the flux through one turn and is $\Phi = \int \mathbf{B} \cdot d\mathbf{F} = \mu_0 HSnl$. The magnetic field strength $H$ (in italics!) is according to Example 5.6 $H = nI$. Thus the self-inductance $L_{11}$ is given by

$$L_{11} = \frac{\Phi}{I} = n^2 \mu_0 Sl \quad \mathrm{H}, \tag{8.10}$$

where H (upright type!) stands for *henry* or *henries*, the unit in the MKSA-system of units, in which magnetic flux $\Phi$ is given in *webers* (Wb). Evidently the condition for the existence of a nonvanishing inductance $L$ is the existence of a nonvanishing flux $\Phi$. If the current carrying conductors are "thin", i.e. of practically vanishing cross-sectional area, only the flux $\Phi$ resulting from the turns of the conductor is of practical significance. On the other hand, if the conductor is not "thin", i.e. "thick", the flux through its cross-sectional area also has to be taken into account. This, of course, complicates the considerations. We therefore consider an example of this case. As a matter of interest we mention that the calculation of the inductance of a rectangle of $n$ turns closely wrapped can be found in the literature.[2]

**Example 8.2 (Self-inductance of a coaxial cable).** Figure 8.1 shows the cross section through the cable with total current zero whose self-inductance is to be calculated (see also Example 8.5).

**Solution:** (Important aspects of coaxial cables will be treated under the topic of Wave Guides and Resonators). We consider an element of the cable of unit length. The inner conductor carrying current $I$ is taken to have radius $r_1$ and magnetic permeability $\mu_1$. The outer conductor carrying current $I$ in the opposite direction is taken to be shell-like, i.e. thin, and of radius $r_2$. The space in-between is filled with a substance of magnetic permeability $\mu_2$. We have

$$I = \int_{r=0}^{r=r_1} \mathbf{j} \cdot d\mathbf{F}, \quad I = \pi r_1^2 j. \tag{8.11}$$

[2] Attwood [5], p. 282.

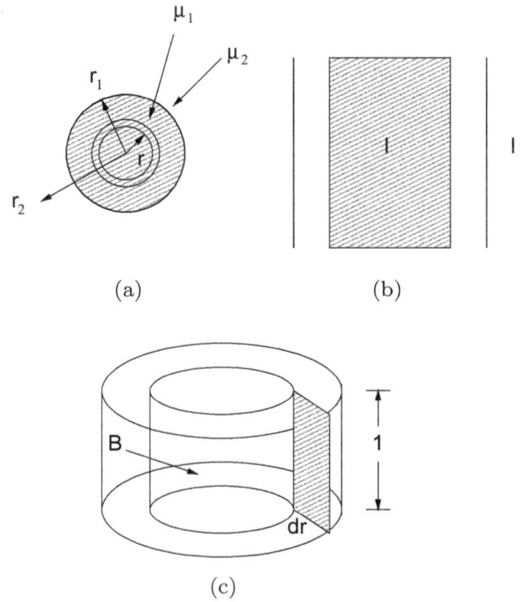


**Fig. 8.1.** The coaxial cable: (a) transverse cross section, (b) longitudinal cross section, (c) view to direction of flux.

We consider a circular path of radius $r, r \leq r_1$, and apply the integral form of Ampère's law. Then

$$\oint_{\text{rad } r} \mathbf{H} \cdot d\mathbf{l} = \int_{F=\pi r^2} \mathbf{j} \cdot d\mathbf{F}, \quad \text{i.e.} \quad 2\pi r H_r = \pi r^2 j = \frac{\pi r^2}{\pi r_1^2} I, \tag{8.12}$$

i.e. the field is determined by this inner fraction of the current $I$ and is given by

$$H_r = \frac{Ir}{2\pi r_1^2} \quad \text{amperes/meter.} \tag{8.13}$$

The lines of force of $\mathbf{B}$, are circles around the axis of the cable, as indicated in Fig. 8.1 (recall that according to the Biot–Savart law $d\mathbf{B} \propto I d\mathbf{s} \times \mathbf{r}$). The magnetic flux through the area of breadth $dr$ and of unit length is

$$\delta\Phi = \mathbf{B} \cdot d\mathbf{F} = \mu_1 H_r dr \times 1 = \frac{\mu_1 I r dr}{2\pi r_1^2} \quad \text{webers.} \tag{8.14}$$

We recall that in the case of a solenoid with (say) $n$ turns per unit length, we calculate the flux through one turn carrying current $I$ and multiply by $n$ to obtain the flux through $n$ turns. In the present case we do not have multiples of $I$, but rather a fraction of $I$. Thus instead of multiplying $\delta\Phi$ by a number $n$ representing a number of turns, we have to multiply $\delta\Phi$ by the fraction $\pi r^2/\pi r_1^2$, since the flux results only from this fraction of current. In order to understand this point better it may help to visualize the inner part of the cable as a section of a torus-like structure and to compare this with our treatment of the solenoid. The self-inductance of the inner part of the cable is therefore given by $L_1 = \Phi/I$, where

$$\Phi = \int_0^{r_1} I \frac{r^2}{r_1^2} \frac{\mu_1 r dr}{2\pi r_1^2} = I \int_0^{r_1} \frac{\mu_1 r^3 dr}{2\pi r_1^4} = I \left[ \frac{\mu_1 r^4}{2\pi 4 r_1^4} \right]_0^{r_1} = I \frac{\mu_1}{8\pi}, \text{ i.e. } L_1 = \frac{\mu_1}{8\pi}. \tag{8.15}$$

We now consider a circle of radius $r > r_1$:

$$2\pi r H_r = \oint \mathbf{H} \cdot d\mathbf{l} = \int \mathbf{j} \cdot d\mathbf{F} = I, \tag{8.16}$$

i.e.

$$H_r = \frac{I}{2\pi r} \quad \text{amperes/meter.} \tag{8.17}$$

The magnetic flux through an area of width $dr$ and length 1 meter in the intermediate space (with no current) is therefore

$$\Phi = \int \mathbf{B} \cdot d\mathbf{F} = \int_{r_1}^{r_2} \mu_2 H_r dr \times 1 = \frac{\mu_2 I}{2\pi} \int_{r_1}^{r_2} \frac{dr}{r} = \frac{\mu_2 I}{2\pi} \ln \frac{r_2}{r_1}. \tag{8.18}$$

Hence *per unit length*

$$L_2 = \frac{\mu_2}{2\pi} \ln \frac{r_2}{r_1} \quad \text{henries,} \tag{8.19}$$

and the total self-inductance *per unit length* is

$$L = L_1 + L_2 = \frac{1}{4\pi} \left( \frac{1}{2}\mu_1 + 2\mu_2 \ln \frac{r_2}{r_1} \right) \quad \text{henries.} \tag{8.20}$$

This result is derived by a different method in Example 8.5. cf. also Example 7.20.

**Example 8.3 (Self-inductance of a parallel wire cable of radius a).** Calculate the self-inductance of a pair of parallel wires of radius $a$ and a distance $d$ apart. Figure 8.2 shows transverse and longitudinal cross sections through such a cable.

**Solution:** The space between the two parallel wires is taken to have magnetic permeability $\mu$. The self-inductance is to be calculated. We imagine the two

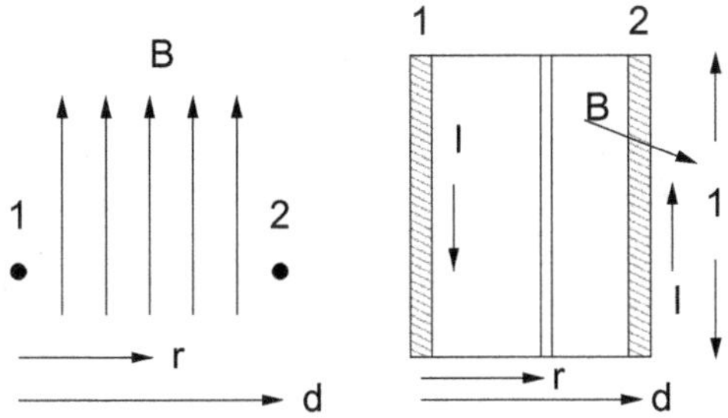


**Fig. 8.2.** The parallel wire cable.

wires connected at infinity to form a closed circuit. We first calculate the field $H_r$ in a radial domain $dr$ as indicated in Fig. 8.2. This field receives contributions from both wires. We saw above (Eq. (8.17)) that the field $H$ at a distance $r$ from the center of a wire is given by $I/2\pi r$. Thus the field in the region between $r$ and $r + dr$ is

$$H_r = \frac{I}{2\pi r} + \frac{I}{2\pi(d-r)} \quad \text{amperes/meter.} \tag{8.21}$$

Hence we have

$$\begin{aligned}\Phi &= \int_{d\times 1} \mathbf{B}\cdot d\mathbf{F} = \mu \int_a^{d-a} dr \times 1\left(\frac{I}{2\pi r} + \frac{I}{2\pi(d-r)}\right) \\ &= \frac{\mu I}{2\pi}\left[\ln r - \ln(d-r)\right]_a^{d-a} = 2\frac{\mu I}{2\pi}\ln\frac{d-a}{a}.\end{aligned} \tag{8.22}$$

The self-inductance $L = \Phi/I$ is therefore given by

$$L = \frac{\mu}{\pi}\ln\left[\frac{d-a}{a}\right] \quad \text{henries} \tag{8.23}$$

*per unit length.* This result neglects the flux through the cross sections of the wires (assumed to be small or complete skin effect). In Example 8.11 the result (8.23) is obtained by a different method.

**Example 8.4 (Mutual inductance of one coil wound around another).** A coil having $n_2$ turns per unit length and length $l_2$ is wound around a long solenoid with $n_1$ turns per unit length, cross-sectional area $S_1$ and current $I_1$. What is the mutual inductance $L_{12}$?

**Solution:** According to Eq. (8.4) the flux $\Phi_2(I_1)$ through circuit 2 as a consequence of the current $I_1$ in circuit 1 is given by

$$\begin{aligned}\Phi_2(I_1) = L_{21}I_1 = n_2l_2\int_{S_1}\mathbf{B}\cdot d\mathbf{F} &= \mu_0(I_1n_1)S_1n_2l_2\\ &= n_1n_2\mu_0S_1l_2I_1.\end{aligned} \tag{8.24}$$

It follows that

$$L_{21} = n_1n_2\mu_0S_1l_2 \quad \text{henries.} \tag{8.25}$$

This mutual inductance can be calculated from its definition and as such is independent of whether an induced current actually flows in the other circuit. See also discussion below.

Consider, as in the preceding example, a coil of finite length wound around an infinitely long solenoid. In this case of the infinitely long inner solenoid, the field **B** is zero at all points of the coil wound around it (see Example 5.6). Can an observable current be induced in the outer coil? Since the field **B** is everywhere zero along the outer coil, there is no Lorentz force to produce the current. However, the vector potential **A** resulting from the current in the long solenoid does not vanish everywhere outside this solenoid (as will be shown below). The problem therefore requires a more detailed treatment and leads eventually to the so-called Aharonov–Bohm effect (see Chapter 19). We shall see that the effect of the vector potential around the long solenoid is a quantum mechanical phase effect which is observable only in appropriate interference experiments. We therefore investigate first the vector potential **A** of a long solenoid.

## 8.3 The Vector Potential of a Long Solenoid

We consider a coil with coordinates as indicated in Fig. 8.3. Inside the solenoid the field is $\mathbf{B} = (0, 0, B_0)$, i.e. homogeneous along the $z$-axis. We obtain this expression with $\mathbf{B} = \nabla \times \mathbf{A}$ for the vector potential

$$\mathbf{A} = \frac{B_0}{2}(-y, x, 0) \equiv \frac{1}{2}\mathbf{B}\times\mathbf{r} \quad \text{tesla}\cdot\text{meters.} \tag{8.26}$$

The vector potential **A** is therefore (a) proportional to $r$, and (b) perpendicular to the $z$-axis and to **r**. In the case of the infinitely long solenoid the field **B** outside is zero (cf. Example 5.6). Does this mean that **A** is also zero? No!

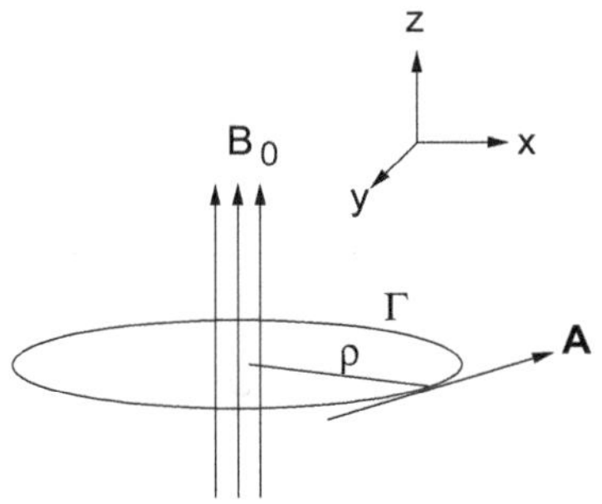


**Fig. 8.3.** The long solenoid with fields $B_0$ and $\mathbf{A}$.

We consider the path $\Gamma$ of radius $\rho > a$ as in Fig. 8.3. The flux through the area with the circle $\Gamma$ as boundary is

$$\Phi = \int_{F(\Gamma)} \mathbf{B} \cdot d\mathbf{F} = B_0 \pi a^2 + B_{\text{outside}} \pi (\rho^2 - a^2) = B_0 \pi a^2, \tag{8.27}$$

so that

$$\Phi = \oint_\Gamma \mathbf{A} \cdot d\mathbf{l} = B_0 \pi a^2 \quad \text{webers.} \tag{8.28}$$

Thus the field $\mathbf{A}$ along the path $\Gamma$ can be nonzero! We know the direction of $\mathbf{A}$ is that of the tangent to the path $\Gamma$. This suggests taking for $\mathbf{A}_{\text{outside}}$ the following ansatz:

$$\mathbf{A}_{\text{outside}} = k \left( -\frac{y}{x^2 + y^2}, \frac{x}{x^2 + y^2}, 0 \right), \quad k = \text{const.}, \tag{8.29}$$

because we know

$$\mathbf{A}_{\text{inside}} = \frac{1}{2} B_0 (-y, x, 0). \tag{8.30}$$

First we have to verify that $\mathbf{B}_{\text{outside}} = \boldsymbol{\nabla} \times \mathbf{A}_{\text{outside}} = 0$. Thus we consider

$$\boldsymbol{\nabla} \times \mathbf{A}_{\text{outside}} = k \begin{vmatrix} \mathbf{e}_x & \mathbf{e}_y & \mathbf{e}_z \\ \partial/\partial x & \partial/\partial y & \partial/\partial z \\ -y/(x^2+y^2) & x/(x^2+y^2) & 0 \end{vmatrix}. \tag{8.31}$$

For instance for the $x$-component:

$$\frac{\partial}{\partial y}(0) - \frac{\partial}{\partial z}\left(\frac{x}{x^2 + y^2}\right) = 0.$$

Similarly we obtain zero for the $y$-component. In the case of the $z$-component we require

$$\begin{aligned}
&\frac{\partial}{\partial x}\left(\frac{x}{x^2+y^2}\right)+\frac{\partial}{\partial y}\left(\frac{y}{x^2+y^2}\right)\\
&=\frac{1}{x^2+y^2}-\frac{2x^2}{(x^2+y^2)^2}+\frac{1}{x^2+y^2}-\frac{2y^2}{(x^2+y^2)^2}\\
&=\frac{2(x^2+y^2)-2x^2-2y^2}{(x^2+y^2)^2}=0,
\end{aligned} \tag{8.32}$$

hence $\mathbf{B}_{\text{outside}}=0$. We determine the constant $k$ such that $\mathbf{A}_{\text{outside}}$ yields the correct magnetic flux, i.e.

$$\int_\Gamma \mathbf{A}_{\text{outside}}\cdot d\mathbf{l}=B_0\pi a^2. \tag{8.33}$$

But now

$$\rho=(x^2+y^2)^{1/2} \quad \text{and} \quad |\mathbf{A}|=\frac{k}{\rho}, \tag{8.34}$$

i.e.

$$\left(-\frac{y}{\sqrt{x^2+y^2}},\frac{x}{\sqrt{x^2+y^2}},0\right)$$

is a *unit vector parallel to* $\mathbf{A}$. Hence

$$\int_\Gamma \mathbf{A}_{\text{outside}}\cdot d\mathbf{l}=\frac{k}{\rho}2\pi\rho=2\pi k=B_0\pi a^2, \tag{8.35}$$

so that

$$k=\frac{1}{2}B_0a^2 \tag{8.36}$$

and

$$\mathbf{A}_{\text{outside}}=\frac{1}{2}B_0\frac{a^2}{x^2+y^2}(-y,x,0) \quad \text{tesla}\cdot\text{meters}. \tag{8.37}$$

It should be noted that at $r=a$ the field $\mathbf{A}_{\text{inside}}$ passes continuously over into $\mathbf{A}_{\text{outside}}$. We thus have a potential $\mathbf{A}_{\text{outside}}\neq 0$, where $\mathbf{B}=0$. Is this strange?

Why is it that $\mathbf{A}_{\text{outside}}$ does not vanish? Can we gauge it away, i.e. can we perform a gauge transformation so that the resulting vector potential vanishes? A suitable gauge transformation

$$\mathbf{A} \to \mathbf{A}' = \mathbf{A} + \boldsymbol{\nabla}\chi \tag{8.38}$$

seems to be one with the following function $\chi$:

$$\chi(t, x, y, z) = -\frac{1}{2}B_0 a^2 \phi(x, y) \tag{8.39}$$

with

$$\phi(x, y) = \tan^{-1}\left(\frac{y}{x}\right), \quad \frac{y}{x} = \tan\phi. \tag{8.40}$$

The quantity $\phi$ is the polar angle shown in Fig. 8.4. At $x < 0, y > 0$ we have $\tan\phi < 0, \phi \to \pi$. At $x < 0, y < 0$ we have $\tan\phi > 0, \phi \to -\pi$. The values of $\phi$ on both sides of $y = 0, x < 0$ therefore differ by $2\pi$. We put the cut as indicated in Fig. 8.4. The singularity at $x = 0$ with the cut along the negative semi-axis is described as a *Dirac string.*

We now demonstrate that $\mathbf{A}' = 0$, except along $y = 0, x < 0$. We have

$$\begin{aligned}
\frac{\partial\phi}{\partial x} &= \frac{\partial}{\partial x}\tan^{-1}\left(\frac{y}{x}\right) = \frac{-y/x^2}{1 + (y/x)^2} = -\frac{y}{x^2 + y^2}, \\
\frac{\partial\phi}{\partial y} &= \frac{\partial}{\partial y}\tan^{-1}\left(\frac{y}{x}\right) = \frac{1/x}{1 + (y/x)^2} = \frac{x}{x^2 + y^2}.
\end{aligned} \tag{8.41}$$

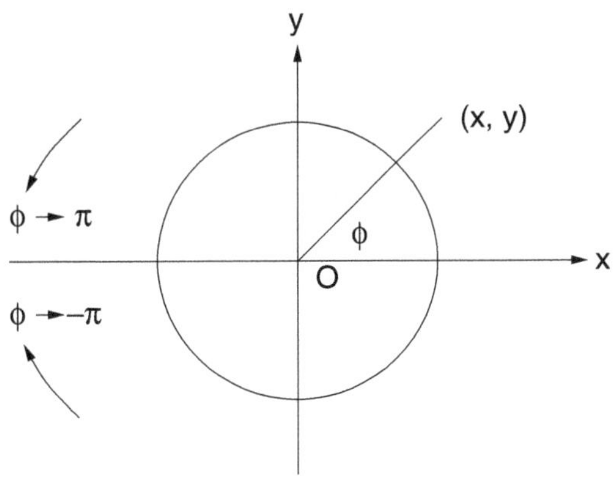


**Fig. 8.4.** The $(x, y)$-plane with discontinuity along the negative $x$-axis.

These expressions are very similar to $A_x, A_y$. The gauge transformation with $\chi(t,x,y,z)$ as in Eq. (8.39) therefore leads to the vector potential

$$\begin{aligned}\mathbf{A}' &= \mathbf{A} + \boldsymbol{\nabla}\chi \\ &= \frac{B_0 a^2}{2}\begin{pmatrix}-y/(x^2+y^2)\\ x/(x^2+y^2)\\ 0\end{pmatrix} - \frac{B_0 a^2}{2}\begin{pmatrix}-y/(x^2+y^2)\\ x/(x^2+y^2)\\ 0\end{pmatrix} \\ &= \begin{pmatrix}0\\0\\0\end{pmatrix}. \end{aligned} \tag{8.42}$$

Thus $\mathbf{A}$ has been "*gauged away*", except where $\chi$ is singular: In the half-plane $y = 0, x < 0$. This means that at $y = 0$ the function $\chi$ possesses a discontinuity as indicated in Fig. 8.5. We recall that in the case of the step function $\theta$: $d\theta(y)/dy = \delta(y)$, so that for $x < 0$:

$$\chi = f(x,y)\theta(-y) + g(x,y)\theta(y), \quad f(x,0) \neq g(x,0).$$

Hence

$$\begin{aligned}\boldsymbol{\nabla}\chi &= \left(\frac{\partial\chi}{\partial x}, \frac{\partial\chi}{\partial y}, \frac{\partial\chi}{\partial z}\right) \\ &= \left\{\frac{\partial f(x,y)}{\partial x}\theta(-y) + \frac{\partial g(x,y)}{\partial x}\theta(y),\right. \\ &\quad \left.\frac{\partial f(x,y)}{\partial y}\theta(-y) + \frac{\partial g(x.y)}{\partial x}\theta(y) + \delta(y)(g(x,y) - f(x,y)), 0\right\}.\end{aligned}$$

Thus the new gauge potential $\mathbf{A}' = \mathbf{A} + \boldsymbol{\nabla}\chi$ has a singularity of the type of a delta function in the $y$-component, and step function behavior in the

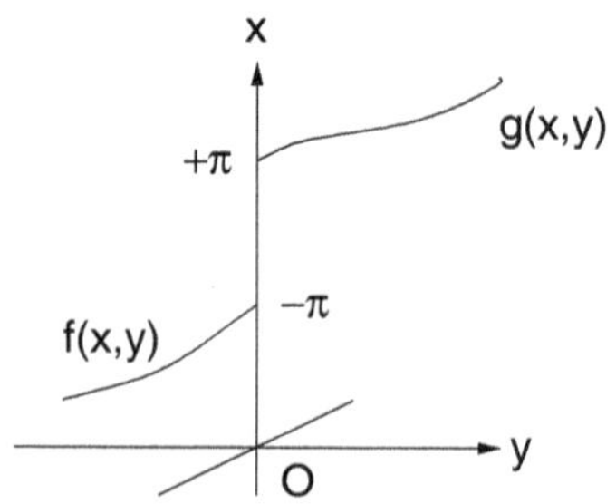


**Fig. 8.5.** The discontinuity of $\chi$ at $y = 0$.

other components. It should be noted, however, that there is no necessity to resort to a singular gauge transformation!

The observation of this effect, called the *Aharonov–Bohm effect*, i.e. that an electron moving in a part of space where the field $\mathbf{B}$ is zero but the vector potential is not, was achieved in quantum mechanical experiments. Older books do not cover this topic. For additional information we refer to Chapter 19 where the effect is treated in more detail and to literature cited there. The "physical meaning" of the vector potential is discussed in detail by Konopinski [79], pp. 158–160.

## 8.4 Energy of a Self-Inductance

In many cases the self-inductance $L$ can also be derived from the magnetic energy $W^{\rm mag}$, in fact even more easily than otherwise. We have

$$\begin{aligned} -\frac{\partial W^{\rm mag}}{\partial t} &= \text{loss of magnetic energy per unit time} \\ &= \text{energy in form of Joule heat per unit time} \\ &= IV_{\rm in} \;\; (V_{\rm in} = \text{induced voltage}) \\ &= -I\frac{\partial \Phi}{\partial t} = -I\frac{\partial}{\partial t}(LI) \\ &= -\frac{1}{2}L\frac{\partial I^2}{\partial t} \quad \text{joules/second} \end{aligned} \tag{8.43}$$

provided the conductor was not deformed. The following relation therefore follows:

$$W^{\rm mag} = \frac{1}{2}LI^2 \quad \text{joules.} \tag{8.44}$$

In the case of several circuits the magnetic energy is correspondingly

$$\begin{aligned} W^{\rm mag} &= \frac{1}{2}\sum_i L_i I_i^2 + \frac{1}{2}\sum_i \sum_j L_{ij} I_i I_j \\ &= \frac{1}{2}L_1 I_1^2 + \frac{1}{2}L_2 I_2^2 + L_{12} I_1 I_2 \quad \text{joules,} \end{aligned} \tag{8.45}$$

if $L_{12} = L_{21}, L_{11} \equiv L_1, L_{22} \equiv L_2$, etc.

The *magnetic force* $\mathbf{F}$, which gives rise to a change of magnetic energy, is given by

$$dW^{\rm mag} = -\mathbf{F} \cdot d\mathbf{l} = \boldsymbol{\nabla} W \cdot d\mathbf{l} \quad \text{joules.} \tag{8.46}$$

Here $d\mathbf{l}$ is the virtual displacement[3] of a circuit for a fixed current or fixed magnetic flux. Correspondingly large mechanical forces $\mathbf{F}^{\rm mech}$ are needed, in order to prevent an actual displacement (so that for the virtual displacement in statics $-\delta W = \mathbf{F}\cdot\delta\mathbf{l} + \mathbf{F}^{\rm mech}\cdot\delta\mathbf{l}^{\rm mech} = 0$). We now consider examples.

**Example 8.5 (Self-inductance of the coaxial cable).** Derive the self-inductance of the coaxial cable defined in Example 8.2 from its magnetic energy.

**Solution:** The expression for the magnetic energy is (cf. Eq. (7.28))

$$W = \frac{1}{2}\int d\mathbf{r}\,\mathbf{H}\cdot\mathbf{B} \quad \text{joules}, \tag{8.47}$$

where (in the present case) $d\mathbf{r} = r dr d\theta \times$ unit length. We calculate separately the magnetic energy of the inner and outer parts of the cable. Using the notation of Example 8.2 and the fields calculated there, we have for the inner part from $r = 0$ to $r = r_1$:

$$W_1 = \frac{1}{2}\int d\mathbf{r}\mu_1\left(\frac{Ir}{2\pi r_1^2}\right)^2 = \frac{2\pi\mu_1 I^2}{2(2\pi r_1^2)^2}\int_0^{r_1} r^3 dr = \frac{\mu_1 I^2}{2}\frac{1}{8\pi}. \tag{8.48}$$

Similarly we obtain for the part from $r_1$ to $r_2$:

$$W_2 = \frac{1}{2}\int d\mathbf{r}\mu_2\left(\frac{I}{2\pi r}\right)^2 = \frac{2\pi I^2\mu_2}{2(2\pi)^2}\int_{r_1}^{r_2}\frac{rdr}{r^2} = \frac{\mu_2 I^2}{2}\frac{\ln(r_2/r_1)}{2\pi}. \tag{8.49}$$

Comparing the sum of $W_1$ and $W_2$ with Eq. (8.45), i.e.

$$W = \frac{1}{2}LI^2 \quad \text{joules}, \tag{8.50}$$

we obtain the result of Example 8.2, i.e. *per unit length*

$$L = \frac{1}{4\pi}\left(\frac{1}{2}\mu_1 + 2\mu_2\ln\frac{r_2}{r_1}\right) \quad \text{henries}. \tag{8.51}$$

From Eq. (8.50) we can deduce the following interpretation of the role played by the inductance $L$. The equation reminds us of the formula for the kinetic energy of a particle of mass $m$ and velocity $v$. Here $W$ is the magnetic energy $W$ of Eq. (8.47). Thus the inductance $L$ acts on the current $I$ like the inertia of a mass on the velocity $v$ of the particle — i.e. the larger $L$

[3]Discussions of this can be found for example in Cheng [29], p. 252, and particularly in Attwood [5], pp. 313–322. Note also the analogous situation in electrostatics involved in Example 2.23.

the larger the inertia, which here means the electromagnetic wave is slowed down. Compare formula (8.50) also with Eq. (2.222) for the energy in a capacitor.

**Example 8.6 (Coaxial coils).** Two coaxial coils with respectively $N_1$ and $N_2$ turns and radii $a_1$ and $a_2$ carry currents $I_1$ and $I_2$. The distance separating their centers is $z \gg a_1, a_2$. Determine the magnetic force between the coils. Re-express the result in terms of the magnetic moments of the coils.

**Solution:** We calculate first the vector potential $\mathbf{A}(\mathbf{r})$. We can calculate this in several ways. The quickest way is to use Eq. (6.4), i.e.

$$\mathbf{A}(\mathbf{r}) = k' \frac{\mathbf{m} \times \mathbf{r}}{r^3}, \quad k' = \frac{\mu_0}{4\pi}, \tag{8.52}$$

(where the moment $\mathbf{m}$ is located at the origin, which has to be modified in our application here!) together with the expression for the magnetic moment as the product of current times area of the conducting loop.

It is instructive to derive the expression from first principles, i.e. from the Biot–Savart law or correspondingly from the expression for the vector potential derived therefrom, i.e.

$$\mathbf{A}(\mathbf{r}) = \frac{\mu_0}{4\pi} \int \frac{I d\mathbf{s}(\mathbf{r}')}{|\mathbf{r} - \mathbf{r}'|} \quad \text{tesla} \cdot \text{meters}. \tag{8.53}$$

We distinguish between the planes of the two coils or rather first between the plane of the coil with current $I_1$ and a point P with position vector $\mathbf{r}$ on the circle with radius $a_2 = R \sin\theta$ in the plane of conductor “2”, where $R$ is the distance between the point $P$ and the center of the circular conductor “1” with radius $a_1$, and $\theta$ is the angle between $\mathbf{R}$ and the axis of the two circular conductors. The point $P'(\mathbf{r}')$ at element $d\mathbf{s}$ of the circular conductor “1” has coordinates in the latter's plane given by

$$\mathbf{r}'' = \mathbf{e}_{x''} a_1 \cos\phi' + \mathbf{e}_{y''} a_1 \sin\phi'. \tag{8.54}$$

Then

$$d\mathbf{s} = \mathbf{r}''(\phi' + d\phi') - \mathbf{r}''(\phi') = d\phi' \frac{d\mathbf{r}''}{d\phi'} = d\phi' \left[ -\mathbf{e}_{x''} a_1 \sin\phi' + \mathbf{e}_{y''} a_1 \cos\phi' \right]. \tag{8.55}$$

Using the cosine-theorem we have

$$|\mathbf{r} - \mathbf{r}'|^2 = (PP')^2 = R^2 + a_1^2 - 2a_1 R \cos\psi \tag{8.56}$$

and (as may be deduced from the geometry)

$$R\cos\psi = R\sin\theta\cos(\phi - \phi'). \tag{8.57}$$

Observe that when $\phi = \phi'$ the point $P'$ has the same angular position on the upper circle as the point $P$ on the lower, and angle $\psi$ is in that case the complement of $\theta$, i.e. $\theta = \pi/2 - \psi$. When $\phi \neq \phi'$, the factor $\cos(\phi - \phi')$ provides the appropriate projection. Hence with Eq. (8.53)

$$\mathbf{A}(\mathbf{r}) = \frac{\mu_0 I}{4\pi}\int \frac{d\phi'(-\mathbf{e}_{x''}a_1\sin\phi' + \mathbf{e}_{y''}a_1\cos\phi')}{\sqrt{R^2 + a_1^2 - 2a_1R\sin\theta\cos(\phi - \phi')}} \quad \text{tesla}\cdot\text{meters}. \tag{8.58}$$

With polar coordinates $\phi, \rho$ in the plane of conductor "1" and

$$\mathbf{e}_{x''} = -\mathbf{e}_\phi\sin\phi + \mathbf{e}_\rho\cos\phi, \quad \mathbf{e}_{y''} = \mathbf{e}_\phi\cos\phi + \mathbf{e}_\rho\sin\phi, \tag{8.59}$$

and choosing the angle $\phi = 0$, $\mathbf{A}(\mathbf{r})$ becomes

$$\mathbf{A}(\mathbf{r}) = \frac{\mu_0 I}{4\pi}\int_0^{2\pi} \frac{d\phi'(-a_1\sin\phi'\mathbf{e}_\rho + a_1\cos\phi'\mathbf{e}_\phi)}{\sqrt{R^2 + a_1^2 - 2a_1R\sin\theta\cos\phi'}}. \tag{8.60}$$

Replacing $\phi'$ by $\phi' + \pi$ we obtain

$$\begin{aligned}\mathbf{A}(\mathbf{r}) &= \frac{\mu_0 I}{4\pi}\int_{-\pi}^{\pi} \frac{d\phi'(a_1\sin\phi'\mathbf{e}_\rho - a_1\cos\phi'\mathbf{e}_\phi)}{\sqrt{R^2 + a_1^2 + 2a_1R\sin\theta\cos\phi'}} \\ &= -\frac{\mu_0 I}{4\pi}\int_0^{2\pi} \frac{d\phi' a_1\cos\phi'\mathbf{e}_\phi}{\sqrt{R^2 + a_1^2 + 2a_1R\sin\theta\cos\phi'}}.\end{aligned} \tag{8.61}$$

For $R^2 \gg a_1^2$ this is

$$\begin{aligned}\mathbf{A}(\mathbf{r}) &\simeq -\frac{\mu_0 I}{4\pi}\int_0^{2\pi} \frac{a_1\cos\phi'\mathbf{e}_\phi d\phi'}{R}\left[1 - \frac{a_1^2 + 2a_1R\sin\theta\cos\phi'}{2R^2}\right] \\ &\simeq \frac{\mu_0 I}{4\pi}\frac{a_1^2}{R^2}\mathbf{e}_\phi\int_0^{2\pi} d\phi'\sin\theta\cos^2\phi'.\end{aligned} \tag{8.62}$$

Since $\int_0^{2\pi}\cos^2\theta d\theta = \pi$, we have

$$\mathbf{A}(\mathbf{r}) \simeq \frac{\mu_0 I}{4\pi}\frac{\pi a_1^2}{R^2}\sin\theta\mathbf{e}_\phi + O(1/R^4) \quad \text{tesla}\cdot\text{meters}. \tag{8.63}$$

With a different expansion of the denominator in Eq. (8.61) (for $R^2 + a_1^2 \gg a_1 R$) we can write

$$\begin{aligned}\mathbf{A}(\mathbf{r}) &\simeq -\frac{\mu_0 I}{4\pi}\int_0^{2\pi}\frac{a_1\cos\phi'\mathbf{e}_\phi d\phi'}{(R^2+a_1^2)^{1/2}}\left[1-\frac{a_1 R\sin\theta\cos\phi'}{(R^2+a_1^2)}\right]\\ &\simeq \frac{\mu_0 I}{4\pi}\frac{\pi a_1^2 R\sin\theta\mathbf{e}_\phi}{(R^2+a_1^2)^{3/2}}\\ &= \frac{\mu_0 I}{4\pi}\frac{\pi a_1^2\rho\mathbf{e}_\phi}{(R^2+a_1^2)^{3/2}} + O[1/(R^2+a_1^2)^{5/2}] \quad \text{tesla}\cdot\text{meters} \qquad (8.64)\end{aligned}$$

with $\rho$ now in the plane of conductor "2" as in Fig. 8.7.

We now return to our original problem. We assume the two coils to be arranged as depicted in Fig. 8.6. The electromagnetic vector potential $\mathbf{A}(\mathbf{r})$ at point $P(\mathbf{r})$ due to the current $I_1$ in the $N_1$ turns of the first coil is given by $N_1$ superpositions of the field of one turn, i.e. from Eq. (8.63)

$$\mathbf{A}(\mathbf{r}) = N_1\frac{\mu_0 I_1}{4\pi}\frac{\pi a_1^2}{R^2}\sin\theta\mathbf{e}_\phi. \qquad (8.65)$$

The magnetic flux $\Phi_2$ through one turn of the second coil is given by

$$\begin{aligned}\Phi_2 &= \int_{F_2}\mathbf{B}\cdot d\mathbf{F} = \int_{F_2}\boldsymbol{\nabla}\times\mathbf{A}\cdot d\mathbf{F} = \oint_{C_2}\mathbf{A}(\mathbf{r})\cdot d\mathbf{l}_2\\ &= \oint_{C_2}\frac{N_1\mu_0 I_1\pi a_1^2 a_2}{4\pi(z^2+a_2^2)^{3/2}}a_2 d\phi_2 = \frac{2\mu_0 I_1 N_1\pi a_1^2\pi a_2^2}{4\pi(z^2+a_2^2)^{3/2}} \quad \text{webers}, \qquad (8.66)\end{aligned}$$

where $dl_2 = a_2 d\phi_2$ and we used $\sin\theta = a_2/R$ and $R^2 = z^2 + a_2^2$.

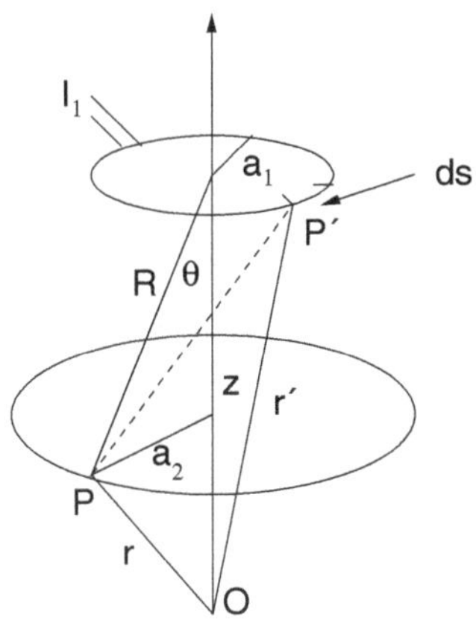


**Fig. 8.6.** Two coaxial coils.

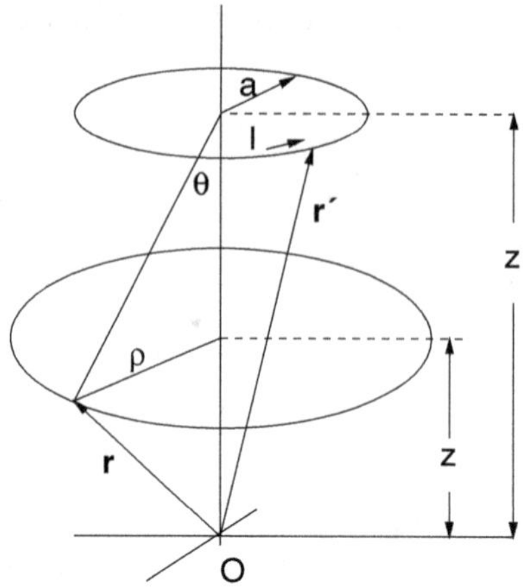


**Fig. 8.7.** Obtaining **A** at **r**.

The mutual inductance $L_{12} = L_{21}$ is given by

$$N_2\Phi_2 = L_{21}I_1, \tag{8.67}$$

i.e.

$$L_{21} = \frac{N_1N_2 2\mu_0\pi a_1^2\pi a_2^2}{4\pi(z^2+a_2^2)^{3/2}} \simeq \frac{2\mu_0 N_1N_2\pi a_1^2\pi a_2^2}{4\pi z^3} \simeq L_{12} \quad \text{for} \quad z \gg a_1, a_2. \tag{8.68}$$

The interaction energy of the coils is given by

$$W = \frac{1}{2}[L_{12}I_1I_2 + L_{21}I_1I_2] \simeq L_{12}I_1I_2 \quad \text{joules}. \tag{8.69}$$

Hence the magnetic force in the direction $z$ is

$$\mathbf{F} = -\boldsymbol{\nabla}W = \mathbf{e}_z 6\mu_0 N_1N_2\frac{I_1\pi a_1^2 I_2\pi a_2^2}{4\pi z^4} = \mathbf{e}_z\frac{6\mu_0}{4\pi}\frac{m_1m_2}{z^4}, \tag{8.70}$$

where

$$m_1 = N_1I_1\pi a_1^2, \quad m_2 = N_2I_2\pi a_2^2 \quad \text{ampere}\cdot\text{meters}^2 \tag{8.71}$$

are the magnetic moments of the coils.[4]

[4]Similar considerations can be found in Clemmow [32], p. 154. The coefficient of mutual induction of two coaxial circular loops has also been calculated in Ferraro [54], p. 245.

**Example 8.7 (The solenoid of finite length).** Determine for a solenoid of finite length the radial component of the magnetic induction, i.e. magnetic flux density, near the center of the solenoid and near the axis.[5]

**Solution:** In solving this problem we use the result of Eq. (8.64) obtained above for the vector potential $\mathbf{A}(\mathbf{r})$ at a point with position vector $\mathbf{r}$ due to current $I$ in a circular conductor of radius $a$, as indicated in Fig. 8.7, i.e. (apart from a sign)

$$\mathbf{A}(\mathbf{r}) = \frac{\mu_0 I}{4\pi}\frac{\pi a^2 \rho \mathbf{e}_\phi}{[(z-z')^2+a^2+\rho^2]^{3/2}} \quad \text{tesla} \cdot \text{meters}. \tag{8.72}$$

This is an approximation for large $(z-z')^2+a^2$.

We consider a coil with $N$ turns per unit length and of total length $L$. We imagine the current $I$ as circling around the cylindrical surface of the coil. Per unit length of the coil the current is then $NI$. The current in one slice of the solenoid of thickness $dz'$ is $NIdz'$. This expression is to be identified with $I$ in Eq. (8.72), if we compare the circular conductor with an infinitesimal ring of the solenoid. We then sum (i.e. integrate) over all ring-like elements of the solenoid, in order to obtain the expression for the total vector potential at the point $\mathbf{r}$. Then

$$\begin{aligned}\mathbf{A}(\mathbf{r}) &= \int_{-L/2}^{L/2}\frac{\mu_0 NIdz'\pi a^2\rho\mathbf{e}_\phi}{4\pi[(z-z')^2+a^2+\rho^2]^{3/2}}\\ &= \frac{\mu_0 NI\pi a^2\rho}{4\pi}\int_{-L/2}^{L/2}\frac{dz'\mathbf{e}_\phi}{4\pi[(z-z')^2+a^2+\rho^2]^{3/2}}.\end{aligned} \tag{8.73}$$

From Tables of Integrals we have

$$\int\frac{dz'}{[z'^2+\rho^2]^{3/2}} = \frac{z'}{\rho^2\sqrt{z'^2+\rho^2}},$$

i.e.

$$\begin{aligned}\text{Int} :&= \int_{z'=-L/2}^{z'=L/2}\frac{d(z'-z)}{[(z-z')^2+a^2+\rho^2]^{3/2}}\\ &= \left[\frac{1}{(\rho^2+a^2)}\frac{(z'-z)}{\sqrt{(z-z')^2+a^2+\rho^2}}\right]_{z'=-L/2}^{z'=L/2}\end{aligned}$$

[5]See also Jackson [73], problem 5.1, p. 205.

$$= \frac{(L/2 - z)}{(\rho^2 + a^2)\sqrt{(z - L/2)^2 + a^2 + \rho^2}} + \frac{(L/2 + z)}{(\rho^2 + a^2)\sqrt{(z + L/2)^2 + a^2 + \rho^2}}$$

$$= \frac{(L/2 - z)}{(\rho^2 + a^2)\sqrt{z^2 + L^2/4 + a^2 + \rho^2}} \left[1 - \frac{Lz}{z^2 + L^2/4 + \rho^2 + a^2}\right]^{-1/2}$$

$$+ \frac{(L/2 + z)}{(\rho^2 + a^2)\sqrt{z^2 + L^2/4 + a^2 + \rho^2}} \left[1 + \frac{Lz}{z^2 + L^2/4 + \rho^2 + a^2}\right]^{-1/2}. \tag{8.74}$$

This can be expanded as

$$\frac{(L/2 - z)}{(\rho^2 + a^2)\sqrt{z^2 + L^2/4 + a^2 + \rho^2}} \left[1 + \frac{Lz}{2(z^2 + L^2/4 + \rho^2 + a^2)} + \frac{3}{8}\frac{(Lz)^2}{(z^2 + L^2/4 + \rho^2 + a^2)^2} + \cdots\right]$$

$$+ \frac{(L/2 + z)}{(\rho^2 + a^2)\sqrt{z^2 + L^2/4 + a^2 + \rho^2}} \left[1 - \frac{Lz}{2(z^2 + L^2/4 + \rho^2 + a^2)} + \frac{3}{8}\frac{(Lz)^2}{(z^2 + L^2/4 + \rho^2 + a^2)^2} + \cdots\right],$$

so that

$$\text{Int} = \frac{L}{(\rho^2 + a^2)\sqrt{z^2 + L^2/4 + a^2 + \rho^2}} - \frac{Lz^2}{(\rho^2 + a^2)(z^2 + L^2/4 + a^2 + \rho^2)^{3/2}} + \frac{3}{8}\frac{L(Lz)^2}{(\rho^2 + a^2)(z^2 + L^2/4 + a^2 + \rho^2)^{5/2}}, \tag{8.75}$$

or

$$\text{Int} = \frac{L(z^2 + L^2/4 + a^2 + \rho^2) - Lz^2}{(\rho^2 + a^2)(z^2 + L^2/4 + a^2 + \rho^2)^{3/2}} + \frac{3}{8}\frac{L(Lz)^2}{(\rho^2 + a^2)(z^2 + L^2/4 + a^2 + \rho^2)^{5/2}}. \tag{8.76}$$

Thus

$$\mathbf{A}(\mathbf{r}) = \frac{\mu_0 N I \pi a^2 \rho}{4\pi(\rho^2 + a^2)} \left[\frac{L(L^2/4 + \rho^2 + a^2)}{(z^2 + L^2/4 + a^2 + \rho^2)^{3/2}} + \frac{3}{8}\frac{L(Lz)^2}{(z^2 + L^2/4 + a^2 + \rho^2)^{5/2}}\right] \mathbf{e}_\phi. \tag{8.77}$$

However,

$$\mathbf{B}(\mathbf{r}) = \boldsymbol{\nabla} \times \mathbf{A}(\mathbf{r}), \quad B_\rho = \frac{1}{\rho}\frac{\partial A_z}{\partial \phi} - \frac{\partial A_\phi}{\partial z}. \tag{8.78}$$

Hence in our case

$$\begin{aligned} B_\rho = -\frac{\partial A_\phi}{\partial z} &= -\frac{\partial}{\partial z}\frac{\mu_0 N I \pi a^2 \rho}{4\pi(\rho^2+a^2)}\left[\frac{L(L^2/4+\rho^2+a^2)}{(z^2+L^2/4+a^2+\rho^2)^{3/2}}\right. \\ &\quad \left. + \frac{3}{8}\frac{L(Lz)^2}{(z^2+L^2/4+a^2+\rho^2)^{5/2}}\right] \\ &= \frac{\mu_0}{4\pi}\frac{3}{2}\frac{2zNI\pi a^2 \rho L(L^2/4+\rho^2+a^2)}{(\rho^2+a^2)(z^2+L^2/4+a^2+\rho^2)^{5/2}} \\ &\quad - \frac{\mu_0}{4\pi}\frac{NI\pi a^2 \rho \frac{3}{8} 3zL^3}{(\rho^2+a^2)(z^2+L^2/4+a^2+\rho^2)^{5/2}} \\ &\quad + O[1/(z^2+L^2/4+a^2+\rho^2)^{7/2}]. \end{aligned} \tag{8.79}$$

For $\rho \ll a, |z| \ll L$ this becomes

$$\begin{aligned} B_\rho &\simeq \frac{\mu_0}{4\pi}\frac{96zNI\pi\rho(L^2/4+a^2)}{L^4} - \frac{\mu_0}{4\pi}\frac{24NI\pi\rho z}{L^2} \\ &= \frac{\mu_0}{4\pi}\left[\frac{24zNI\pi\rho}{L^2} + \frac{96zNI\pi\rho a^2}{L^4} - \frac{24NI\pi\rho z}{L^2}\right] \\ &= \frac{\mu_0}{4\pi}\left[\frac{96zNI\pi\rho a^2}{L^4}\right] \quad \text{teslas.} \end{aligned} \tag{8.80}$$

This result agrees with that stated by Jackson (see above).

## 8.5 Simple Current Circuits

### 8.5.1 *Current circuits with R and L*

In the case of a single (meaning one-mesh) circuit with resistance $R$ and self-inductance $L$ we obtain from the general equation (8.9) established earlier,

$$I_k R_k + \sum_j L_{kj}\dot{I}_j = V_k, \tag{8.81}$$

the relation

$$RI + L\frac{dI}{dt} = V. \tag{8.82}$$

We consider two possible cases of the applied *alternating potential* or e.m.f. (an e.m.f. $E = E_0 \cos(\omega t + \delta)$ or corresponding voltage of this type is supplied by an alternator, in which a coil is rotated in a magnetic field):

1. $V = V_0 = \text{const}$. In this case the solution of Eq. (8.16) is the transient current

$$I = \frac{V_0}{R}(1 - e^{-Rt/L}) \quad \text{amperes,} \tag{8.83}$$

so that after a longer period of time the current $I$ assumes the value of Ohm's law.
2. $V = V_0 \cos \omega t$. Here $\omega = 2\pi\nu$ radians per second is the alternating voltage circular frequency, $\nu$ is the frequency expressed in hertz. The current is the real part of the solution $\tilde{I}$ of the equation[6]

$$R\tilde{I} + L\frac{d\tilde{I}}{dt} = V_0 e^{i\omega t}. \tag{8.84}$$

We set

$$\tilde{I} = F(t)e^{i\omega t}, \tag{8.85}$$

so that

$$RF + L(\dot{F} + i\omega F) = V_0,$$

i.e.

$$L\dot{F} + (R + Li\omega)F = V_0. \tag{8.86}$$

The general solution of the first order differential equation

$$\frac{dy}{dx} + Py = Q \tag{8.87}$$

is

$$y = e^{-\int^x P dx} \int dx Q e^{\int^x P dx}. \tag{8.88}$$

[6]Note that in this chapter we choose for convenience the time dependence $\exp(+i\omega t)$, elsewhere $\exp(-i\omega t)$. In the first form with plane waves $\exp[i(\omega t - kx)]$ decaying waves occur when $\Im k < 0$.

This can be verified by simple differentiation. Hence the solution of Eq. (8.86) is

$$\begin{aligned} F &= e^{-\int(i\omega+R/L)dt}\int \frac{V_0}{L}e^{\int(i\omega+R/L)dt}dt \\ &= e^{-(i\omega+R/L)t}\frac{V_0}{L}\int e^{(i\omega+R/L)t}dt \\ &= \frac{V_0}{L}\frac{1}{i\omega+R/L} = \frac{V_0(-i\omega+R/L)}{L(\omega^2+R^2/L^2)}. \end{aligned} \tag{8.89}$$

Consequently

$$\tilde{I} = \frac{V_0(R-i\omega L)}{(\omega^2L^2+R^2)}e^{i\omega t}, \tag{8.90}$$

and the real part is

$$\begin{aligned} I = \Re(\tilde{I}) &= \frac{V_0}{\omega^2L^2+R^2}[R\cos\omega t + \omega L\sin\omega t] \\ &= \frac{V_0}{\sqrt{R^2+\omega^2L^2}}\cos(\omega t-\delta), \end{aligned} \tag{8.91}$$

with

$$\cos\delta = \frac{R}{\sqrt{R^2+\omega^2L^2}}, \quad \sin\delta = \frac{\omega L}{\sqrt{R^2+\omega^2L^2}}. \tag{8.92}$$

The effective "resistance " $\sqrt{R^2+\omega^2L^2}$ is called *impedance*, and $\omega L$ is called *inductive resistance.*

### 8.5.2 *Circuits with L, C and R*

In the case of a one-mesh circuit with $L, C$ and $R$, the Eq. (8.9) or simply Kirchhoff's first law is

$$IR + L\dot{I} = V_e + (V_1 - V_2), \tag{8.93}$$

where $(V_1 - V_2)$ is the potential difference of the condenser in the circuit. (Note that the "current" from one condenser plate to the other is the *displacement current*). Thus the applied e.m.f. $V_e$ in a closed circuit is equal to the sum of the potentials across the $R$'s, $L$'s and $C$'s in the circuit. We saw earlier that, $q$ being the charge of a condenser,

$$V_2 - V_1 = \frac{q}{C} = \frac{1}{C}\int Idt, \quad I = \frac{dq}{dt}. \tag{8.94}$$

Hence Eq. (8.93) becomes

$$L\ddot{I} + R\dot{I} = \dot{V}_e - \frac{I}{C}, \quad L\ddot{I} + R\dot{I} + \frac{1}{C}I = \dot{V}_e,$$

or integrated

$$L\frac{dI}{dt} + RI + \frac{1}{C}\int Idt = V_e. \tag{8.95}$$

This equation should be compared with its mechanical equivalent, i.e. the equation of motion of a mass $m$ attached to a spring with spring constant $1/n$ and subjected to an external force $F(t)$ and a frictional force $rdx/dt = rv$ with frictional coefficient $r$, i.e.

$$m\frac{dv}{dt} + rv + \frac{1}{n}\int vdt = m\frac{d^2x}{dt^2} + r\frac{dx}{dt} + \frac{x}{n} = F(t). \tag{8.96}$$

Comparison of this equation with Eq. (8.95) shows that the inductance $L$ corresponds to the mass $m$, the current $I$ to the velocity $v = dx/dt$, the capacity $C$ to $n$ (the inverse of the spring constant), the Ohm resistance $R$ to the frictional coefficient $r$, and the applied voltage $V_e(t)$ to the applied force $F(t)$. We set $V_e = V_0 e^{i\omega t}$ (real part $V_0 \cos\omega t$). Then

$$L\ddot{I} + R\dot{I} + \frac{1}{C}I = i\omega V_0 e^{i\omega t}. \tag{8.97}$$

We also set $I = I_0 e^{i\omega t}$. Then

$$I_0\left(-\omega^2 L + i\omega R + \frac{1}{C}\right) = i\omega V_0, \tag{8.98}$$

i.e.

$$I_0 = \frac{V_0}{R + i(\omega L - 1/\omega C)}, \quad I = \frac{V_0 e^{i\omega t}}{R + i(\omega L - 1/\omega C)}. \tag{8.99}$$

For $V = V_0 \cos\omega t$ the real solution is similar to that in the previous case,

$$I = \frac{V_0 \cos(\omega t - \delta)}{\sqrt{R^2 + (\omega L - 1/\omega C)^2}} \quad \text{amperes}, \tag{8.100}$$

where now

$$\tan\delta = \frac{1}{R}\left(\omega L - \frac{1}{\omega C}\right). \tag{8.101}$$

The angle $\delta$ is the phase by which the current lags behind $V$. The expression $1/\omega C$ is called *capacitive resistance*, the expression

$$Z = R + i\left(\omega L - \frac{1}{\omega C}\right) \quad \text{ohms} \tag{8.102}$$

is called the *complex impedance* (here with $R, L, C$ arranged one after the other, i.e. in series), and the expression

$$Y = \omega L - \frac{1}{\omega C} \quad \text{ohms} \tag{8.103}$$

is called the *reactance* of the circuit. A circuit with $R = 0$ is said to be *purely reactive.* The frequency $\tilde{\omega}$ at which the impedance $Z$ is a minimum, i.e. where (for dimensions see Eq. (14.92))

$$\omega L = \frac{1}{\omega C}, \quad \tilde{\omega} = \frac{1}{\sqrt{LC}} \quad \text{second}^{-1},$$

is called the *resonant frequency* of the circuit (the quantity $\sqrt{L/C}$ is called the *characteristic impedance* of the circuit which will be discussed in a wider context in Section 8.7). We note that with the impedance defined by Eq. (8.102) we can rewrite Eq. (8.98) as (observe that the impedance $Z$ is given in ohms)

$$I = \frac{V}{Z} \quad \text{or} \quad V = ZI. \tag{8.104}$$

The mechanical equivalent to $V/I$ is the ratio $F$/velocity, and permits the definition of a *mechanical impedance.* From Eq. (8.102) we deduce that in special cases $Z$ can be $R$ or $i\omega L$ or $1/i\omega C$ or $R + i\omega L$ or ....

Consider as an example the network shown in Fig. 8.8 and apply Kirchhoff's laws to it. Current conservation is evident from the figure. Considering the potentials of the three meshes *ABDFA*, *ADFA* and *BDFB* with their respective mesh currents $I_1, I_2$ and $I_3$, we obtain the somewhat rearranged equations:

$$I_1\left(\frac{1}{i\omega C} + R + 2i\omega L\right) + I_2(2i\omega L) + I_3(R + i\omega L) = 0, \tag{8.105a}$$

$$I_1(2i\omega L) + I_2(2i\omega L) + I_3(i\omega L) = V_e, \tag{8.105b}$$

$$I_1(R + i\omega L) + I_2(i\omega L) + I_3(R + 2i\omega L) = 0. \tag{8.105c}$$

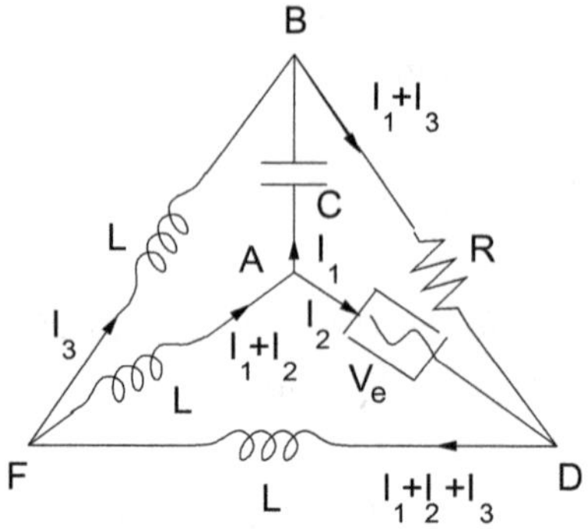


**Fig. 8.8.** A 4-node circuit.

Rewriting these equations respectively in the form

$$Z_{11}I_1 + Z_{12}I_2 + Z_{13}I_3 = 0, \tag{8.106a}$$

$$Z_{21}I_1 + Z_{22}I_2 + Z_{23}I_3 = V_e, \tag{8.106b}$$

$$Z_{31}I_1 + Z_{32}I_2 + Z_{33}I_3 = 0, \tag{8.106c}$$

we see that $Z_{12} = Z_{21}, Z_{13} = Z_{31}, Z_{23} = Z_{32}$. Thus the impedance matrix $(Z_{ij}), i, j = 1, 2, 3$, of the equation

$$\sum_j Z_{ij}I_j = V_i, \tag{8.107}$$

is symmetric. Provided the matrix $(Z_{ij})$ is nonsingular, we can reverse the equation and obtain

$$I_i = \sum_j Y_{ij}V_j. \tag{8.108}$$

The coefficients $Y_{ij}$ (which are also symmetric, $Y_{ij} = Y_{ji}$) are called *admittance coefficients* of the network. One can now prove that the e.m.f. $V_i'$ in mesh $i$, due to unit current $I_j'$ in mesh $j$ is equal to the e.m.f. $V_j''$ in mesh $j$ due to unit current $I_i''$ in mesh $i$. A corresponding relation holds for the currents. These results follow from the following result known as the *reciprocity theorem*, i.e.

$$\sum_i V_i'I_i'' = \sum_j V_j''I_j'. \tag{8.109}$$

We prove the theorem as follows. We have:

$$\sum_i V_i' I_i'' = \sum_i \left( \sum_j Z_{ij} I_j' \right) I_i'' = \sum_{j,i} Z_{ji} I_i' I_j''$$
$$= \sum_i \left( \sum_j Z_{ij} I_j'' \right) I_i' = \sum_j V_j'' I_j'. \tag{8.110}$$

We illustrate with the following example the evaluation of networks[7] and the calculation of their impedances $Z_{ij}$.[8]

**Example 8.8 (A network of impedances).** Consider the network of Fig. 8.9 and determine the impedances $Z_{ij}, i, j = 1, 2, 3$, in terms of the impedances shown.

**Solution:** We choose the currents $I_i, i = 1, 2, 3$, as shown and we write out the potential equations (like Eq. (8.93)) for the following three meshes:

1. *FGABCF* (mesh current $I_1$):

$$I_1(Z_{GF} + Z_{AB} + Z_{BC} + Z_{CF}) + I_2(Z_{BC} + Z_{CF}) - I_3(Z_{CF}) = V_1, \tag{8.111}$$

2. *FBCF* (mesh current $I_2$):

$$I_2(Z_{BF} + Z_{BC} + Z_{CF}) + I_1(Z_{BC} + Z_{CF}) - I_3(Z_{CF}) = 0, \tag{8.112}$$

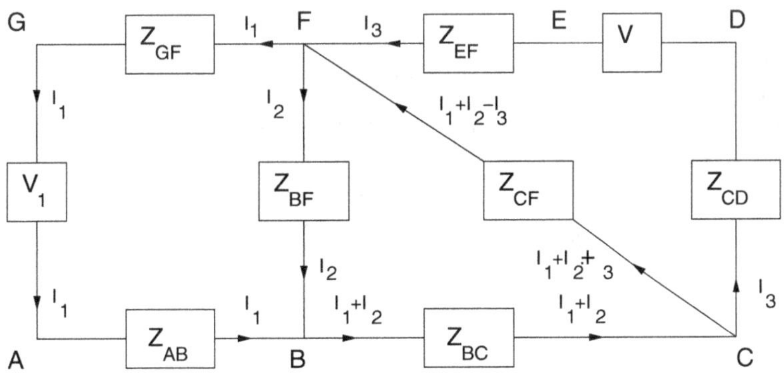


**Fig. 8.9.** A network of impedances.

[7]For a very readable introductory treatment of this topic see also Lorrain and Corson [89], Chapters 5 and 16–18.
[8]This is the network shown in Schelkunoff [129], p. 102, but not evaluated there.

3. *FCDF* (mesh current $I_3$):

$$-I_1(Z_{CF}) - I_2(Z_{CF}) + I_3(Z_{CF} + Z_{CD} + Z_{EF}) = V. \tag{8.113}$$

We note that if we insert the last equation into the first, we obtain:

$$I_2(Z_{BF}) + V_1 - I_1(Z_{GF} + Z_{AB}) = 0. \tag{8.114}$$

This is the equation one obtains if one considers the mesh *FGABF*. Rewriting the equations in matrix form we obtain:

$$\begin{pmatrix} Z_{GF} + Z_{AB} + Z_{BC} + Z_{CF} & Z_{BC} + Z_{CF} & -Z_{CF} \\ Z_{BC} + Z_{CF} & Z_{BF} + Z_{BC} + Z_{CF} & -Z_{CF} \\ -Z_{CF} & -Z_{CF} & Z_{CF} + Z_{CD} + Z_{EF} \end{pmatrix}$$
$$\times \begin{pmatrix} I_1 \\ I_2 \\ I_3 \end{pmatrix} = \begin{pmatrix} V_1 \\ 0 \\ V \end{pmatrix}. \tag{8.115}$$

Comparison of this matrix equation with the equation

$$\sum_j Z_{ij} I_j = V_i \tag{8.116}$$

permits us to read off the results, i.e. the impedances $Z_{ij}$. Note again that $Z_{ij} = Z_{ji}$.

## 8.6 Self-Inductances: Conjugate Function Method

In Section 3.7 we saw how the capacities of essentially 2-dimensional systems can be calculated by the method of conjugate functions $U$ and $V$ combined to the complex function $W = U + iV$. Here we consider the corresponding steps to obtain self-inductances. We start from Eq. (5.53),

$$\mathbf{A}(\mathbf{r}) = k' \int \frac{\mathbf{j}(\mathbf{r}')d\mathbf{r}'}{|\mathbf{r} - \mathbf{r}'|} \quad \text{tesla} \cdot \text{meters}, \tag{8.117}$$

and choose $\mathbf{A}$ along the $z$-direction with $A_x = 0,\ A_y = 0,\ A_z = V$. Then

$$\mathbf{B} = \boldsymbol{\nabla} \times \mathbf{A} = \begin{vmatrix} \hat{x} & \hat{y} & \hat{z} \\ \partial/\partial x & \partial/\partial y & \partial/\partial z \\ 0 & 0 & V \end{vmatrix} = \left(\frac{\partial V}{\partial y}, -\frac{\partial V}{\partial x}, 0\right) \quad \text{teslas}, \tag{8.118}$$

and we obtain

$$B^2 = \left(\frac{\partial V}{\partial x}\right)^2 + \left(\frac{\partial V}{\partial y}\right)^2 = |\nabla V|^2, \quad B = |\nabla V|. \tag{8.119}$$

Then the magnetic energy in vacuum can be written

$$W = \frac{1}{2}\int d\mathbf{r}\,\mathbf{H}\cdot\mathbf{B} = \frac{1}{2\mu_0}\int d\mathbf{r}|\nabla V|^2 \quad \text{joules}. \tag{8.120}$$

However,

$$\nabla\cdot(V\nabla V) = (\nabla V)^2 + V\nabla^2 V, \tag{8.121}$$

and so with Gauss' divergence theorem and neglecting the surface integral at infinity we obtain

$$W = -\frac{1}{2\mu_0}\int d\mathbf{r}V\nabla^2 V. \tag{8.122}$$

Using Eq. (5.65), i.e. (since $A_z = V$)

$$\triangle\mathbf{A}(\mathbf{r}) = -\mu_0\mathbf{j}(\mathbf{r}), \quad \text{and so} \quad \triangle V = -\mu_0 j(\mathbf{r}), \tag{8.123}$$

we obtain

$$W = \frac{1}{2}\int d\mathbf{r}\, j(\mathbf{r})V(\mathbf{r}) \quad \text{joules}. \tag{8.124}$$

Assuming the wires are thin, so that the flux through them is negligible and hence $V$ constant across them, we can write

$$W = \frac{1}{2}V\int d\mathbf{r}\, j(\mathbf{r}) = \frac{1}{2}VI \tag{8.125}$$

per wire and per unit length in the $z$-direction. In the case of two wires with antiparallel currents $I_1, I_2$, we have

$$W = \frac{1}{2}I[V_1 - V_2] \equiv \frac{1}{2}I[V]. \tag{8.126}$$

But, by definition of the inductance $L$,

$$W = \frac{1}{2}LI^2 \quad \text{joules}. \tag{8.127}$$

Hence

$$L = \frac{[V]}{I} \quad \text{henries}. \tag{8.128}$$

Our next aim is to express $I$ in terms of the conjugate function $U$ defined as in Section 3.7. We recall Eq. (6.38) and Eq. (3.88):

$$\begin{aligned} I &= \oint_C \mathbf{H} \cdot d\mathbf{s} = \frac{1}{\mu_0} \oint \mathbf{B} \cdot d\mathbf{s} = \frac{1}{\mu_0} \oint |\boldsymbol{\nabla} V| ds \\ &= \frac{1}{\mu_0} \oint |\boldsymbol{\nabla} U| ds = \frac{[U]}{\mu_0} \quad \text{amperes.} \end{aligned} \tag{8.129}$$

The direction of $\boldsymbol{\nabla} U$ is that of constant $V$. Since for both wires $|I_1| = |I_2| \equiv I$, $[U]$ is also the same for both wires. Hence we obtain

$$L = \frac{[V]}{I} = \mu_0 \frac{[V]}{[U]} \quad \text{henries} \tag{8.130}$$

per unit length in the $z$-direction. We now apply this result in some Examples. An example we do not discuss here is that with $W = -(2\pi/I)$ $[A_z/\mu_0 + iV_m]$, where $V_m$ is the magnetic scalar potential of Chapter 5. For details see Bradley [22], p. 66.

**Example 8.9 (The parallel plate condenser).** Calculate the self-inductance by the method of conjugate functions.

**Solution:** We let the plates be a distance $d$ apart, i.e. $d = [y]$. We choose (cf. Section 3.7)

$$W := U + iV = x + iy, \quad \text{i.e.} \quad U = x, \quad V = y. \tag{8.131}$$

Thus we obtain for the self-inductance

$$L = \mu_0 \frac{[V]}{[U]} = \frac{\mu_0 d}{[x]} = \mu_0 d \quad \text{henries} \tag{8.132}$$

per unit area perpendicular to this paper. Using the result of Example 3.9 for the capacity $C$, where $k = 1/4\pi\epsilon_0$, we observe that

$$LC = \frac{\mu_0 d}{4\pi k d} = \epsilon_0 \mu_0 \quad (\text{meters/second})^{-2}. \tag{8.133}$$

**Example 8.10 (Coaxial cylinders).** Calculate the self-inductance of coaxial cylinders with radii $a, b, a < b$.

**Solution:** As in Example 3.11 we choose

$$W := V + iU = \ln(re^{i\theta}), \quad U = \theta, \quad V = \ln r. \tag{8.134}$$

Then

$$L = \frac{\mu_0[\ln r]}{[\theta]} = \frac{\mu_0 \ln(b/a)}{2\pi} \quad \text{henries} \tag{8.135}$$

per unit length. The result should be compared with that of Example 8.2. Multiplying $L$ by $C$ of Example 3.11 we again obtain $LC = \epsilon_0\mu_0$.

**Example 8.11 (Parallel wires, a distance D apart).** Calculate the self-inductance of two parallel wires of radius $a$ and a distance $D$ apart.

**Solution:** We can use the result of Example 3.12,

$$C = \frac{1}{4k\cosh^{-1}(D/2a)} \quad \text{farads} \tag{8.136}$$

together with

$$L = \mu_0 \frac{[V]}{[U]}, \quad C = \frac{1}{4\pi k}\frac{[U]}{[V]} = \epsilon_0 \frac{[U]}{[V]}. \tag{8.137}$$

Thus

$$L = \frac{\mu_0}{4\pi k C} = \frac{\mu_0}{\pi}\cosh^{-1}\left(\frac{D}{2a}\right) \quad \text{henries} \tag{8.138}$$

with $LC = \mu_0\epsilon_0$. Using $\cosh^{-1} x = \pm[\ln(2x) + O(1/x^2)]$ for $x > 1$, the self-inductance may be approximated as

$$L \simeq \frac{\mu_0}{\pi}\ln\left(\frac{D}{a}\right), \quad D \gg a. \tag{8.139}$$

This result should be compared with the result of Example 8.3.

How circuits function can be illustrated by considering some simple cases. The most elementary direct current (DC) circuits are *series circuits* and *parallel circuits*. In series circuits the current $I$ emanating from an e.m.f. or battery passes through each and every component of the circuit — whether resistance $R$, capacitor $C$ or inductor $L$, or a light bulb — as the word series implies the components are all in the same line and that of the battery of voltage $V$. If there are two or more light bulbs, each will have its fraction $V_i$ of $V$, i.e. $V = V_1 + V_2 + V_3 + \cdots$ but with the same current $I = I_1 = I_2 = I_3 = \cdots$. This follows from Kirchhoff's laws Eq. (8.9). In the case of parallel circuits the components are arranged in the circuit like the rungs of a ladder. In such cases the voltage $V_i$ across each component is the same, i.e. $V = V_1 = V_2 = V_3 = \cdots$, but the sum of the currents adds up to that in the battery, i.e. $I = I_1 + I_2 + I_3 + \cdots$. Thus in this case a light bulb enjoys the full voltage of the battery and is more likely to glow if this is strong enough.

To prevent a possible breakdown of a bulb one can connect a resistor with it in parallel which will eat up part of the current.

## 8.7 Transmission Lines and 4-Terminal Networks

### 8.7.1 *Preliminary remarks*

The propagation of electromagnetic radiation in a controlled manner from one place to another and over large distances is an important application of electrodynamics. It implies that electromagnetic waves have to be conducted in a contained non-diffusing way and at many different frequencies in cables called transmission lines, or, when power in transmission lines become too great, in wave guides or similar structures. This is the topic we are concerned with in this chapter as well as in the later chapter on wave guides. We are all familiar with cables connecting lamps to power sockets, and with radio transmitters and receivers, and remote control units to switch on TV. But how does the transmission take place and what goes on in the cable? How can one ensure that the radiation does not get lost on the way?

One distinguishes between electromagnetic waves having long wavelengths $\lambda$ (even of several thousand km as in submarine communication) and hence short or low frequencies $\nu = c/\lambda$ of a few Hz, and electromagnetic waves having short wavelengths (of around 10 m and less as in TV, WLAN and microwave ovens) and hence high frequencies of several GHz ($10^9$ Hz) and more. The former are normally treated under the topic of transmission lines and the latter under that of wave guides. Roughly speaking, the intermediate frequency domain, i.e. that between frequencies termed low and high, is that of medium frequencies of the order of 0.3–3 MHz or of wavelengths 1000–100 m, the visible light of lamps and light wave guides having wavelengths of the order of 500 nm ($10^{-9}$ m) or frequencies of around 500 THz ($10^{12}$ Hz).

In communication work[9] the energy associated with sound waves must be converted into electrical energy, in which form it is transferred from one point to another and then reconverted into acoustical energy. The devices for accomplishing these interconversions, such as microphones and speakers or radio transmitters and receivers, are known generally as *transmission lines* or *transducers*. There are several types of transducers, such as electrostatic,

[9]The communication part of the electromagnetic spectrum is that with wavelengths $\lambda$ expressed in meters in the domain $10^5 > \lambda > 10^{-3}$ corresponding to frequencies $\nu$ in hertz in the region $10^3 < \nu < 10^{11}$.

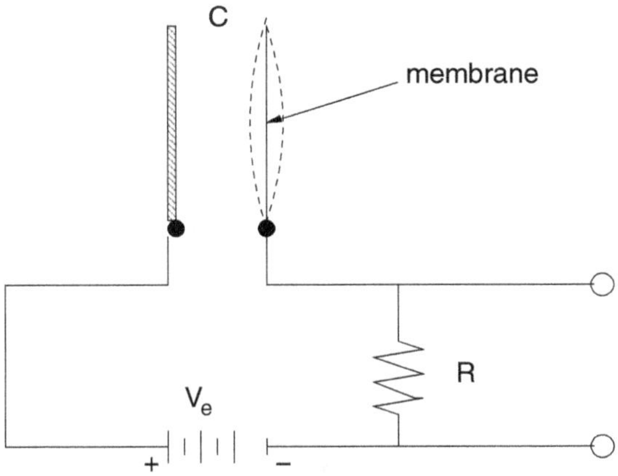


**Fig. 8.10.** Conversion of acoustic energy into electric energy.

electromagnetic, magnetostrictive and Piezoelectric, and other types. The principle of an electrostatic sound transducer is illustrated in Fig. 8.10.[10] In essence the conversion of sound into an electric current is effected by a condenser consisting of a very thin membrane electrode and an oppositely placed solid electrode (the current acting between the two being Maxwell's displacement current as in Example 7.1). When the membrane, made of electrically conducting material is exposed to sound waves, it vibrates and thereby changes the capacity $C$ of the condenser which depends on the distance between the two electrodes. This latter separation is normally of the order of 20 μm (micrometer) and is contained in the formula for $C$ (cf. the capacity of the parallel plate condenser, Eq. (2.101)). The capacity of a condenser microphone is of the order of 10–200 pF. Depending on the use of the device the frequencies of the current transmitted must be selected or amplified by appropriate additions to the network which we will be concerned with here.

Before we consider 4-terminal networks, we define first a 2-terminal electromagnetic system. Such a system is shown schematically in Fig. 8.11. There with circles as in Fig. 8.10 only the two input terminals or ports of this system protrude with currents flowing to the right and to the left.[11] At these terminals the harmonic input current $I_i$ and the input voltage $V_i$ define the *input impedance* $Z$ by the relation

$$V_i = ZI_i. \tag{8.140}$$

[10]We consulted here: Veit [154], p. 84.
[11]See also Jackson [73], p. 243.

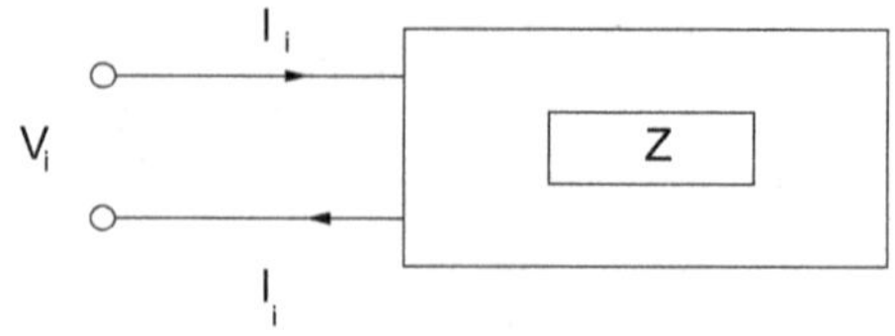


**Fig. 8.11.** A 2-terminal network with input impedance $Z$.

A simple example of this relation is Eq. (8.104). The diagram assumes that radiation losses are negligible (no radiation resistance). The concept of impedance has been considered in depth by Schelkunoff [129], who argues that the concept does not even depend on the existence of terminals.[12]

In the following we will be concerned with 4-terminal networks, also called two-port networks. Networks with more than 4 terminals (also called n-gate networks) are, of course, also considered in the literature.[13]

### 8.7.2 *Simple types of 4-terminal networks*

We consider a 4-terminal network with input voltage $V_1$ and input current $I_1$, and output voltage $V_0$ and output current $I_0$ as indicated in Fig. 8.12.

The pairs of terminals may be those of a transformer or of a telephone transmission line or of two antennas or of something similar. The case $V_0 = 0$ means "*short-circuiting*". We now suppose an impedance $Z$ introduced as in Fig. 8.13(a).

For this network we have from Kirchhoff's laws, that $I_1 = I_0$ and $V_0 = V_1 + ZI_0$. We write these equations in the following matrix form:

$$\begin{pmatrix} V_1 \\ I_1 \end{pmatrix} = \begin{pmatrix} 1 & -Z \\ 0 & 1 \end{pmatrix} \begin{pmatrix} V_0 \\ I_0 \end{pmatrix}. \tag{8.141}$$

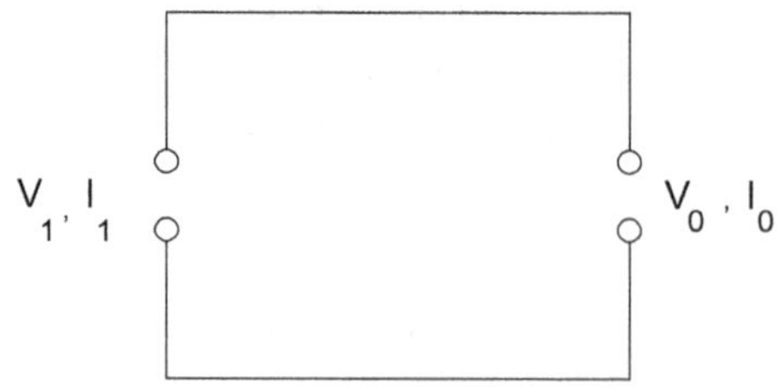


**Fig. 8.12.** A 4-terminal network.

[12]Schelkunoff [129], pp. 480–496. The word impedance was coined by Oliver Heavyside.
[13]See e.g. Stock in Epprecht *et al.* [46].

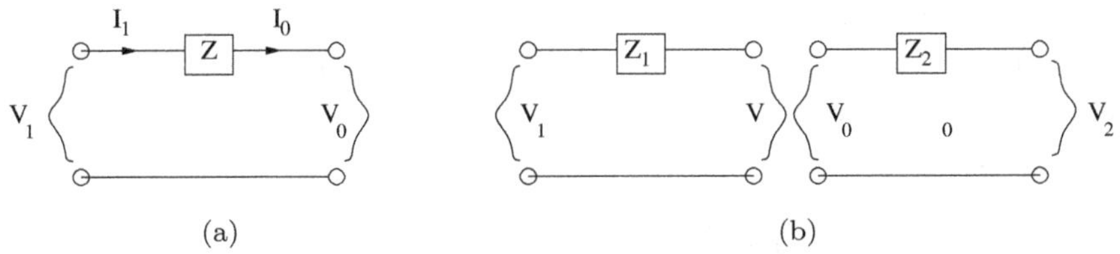


**Fig. 8.13.** (a) A network with one impedance, (b) one with two impedances in series.

Observe that the circuit section containing the impedance $Z$ is represented by the $2\times 2$ matrix here. Next we consider two such networks linked in series as in Fig. 8.13(b). We prove that the total impedance of this system is

$$Z = Z_1 + Z_2. \tag{8.142}$$

We prove this as follows. As before we have Eq. (8.141), but in addition

$$\begin{pmatrix} V_0 \\ I_0 \end{pmatrix} = \begin{pmatrix} 1 & -Z_2 \\ 0 & 1 \end{pmatrix} \begin{pmatrix} V_2 \\ I_2 \end{pmatrix}, \tag{8.143}$$

so that

$$\begin{aligned} \begin{pmatrix} V_1 \\ I_1 \end{pmatrix} &= \begin{pmatrix} 1 & -Z_1 \\ 0 & 1 \end{pmatrix} \begin{pmatrix} 1 & -Z_2 \\ 0 & 1 \end{pmatrix} \begin{pmatrix} V_2 \\ I_2 \end{pmatrix} \\ &= \begin{pmatrix} 1 & -(Z_1 + Z_2) \\ 0 & 1 \end{pmatrix} \begin{pmatrix} V_2 \\ I_2 \end{pmatrix}. \end{aligned} \tag{8.144}$$

Thus $Z = Z_1 + Z_2$ is now the total impedance.

Next we consider *admittances*. As we saw before with Eqs. (8.102) and (8.104), an *admittance* $Y$ is defined as the reciprocal of the corresponding impedance, i.e.

$$Y = \frac{1}{R}, \quad \frac{1}{i\omega L} \quad \text{or} \quad i\omega C, \tag{8.145}$$

according as

$$Z = R,\ i\omega L \quad \text{or} \quad \frac{1}{i\omega C}. \tag{8.146}$$

In general

$$Z = \sqrt{R^2 + (\omega L - 1/\omega C)^2}. \tag{8.147}$$

The real part of $Z = R + iX$ is called the *resistance*, and its imaginary part the *reactance*. Correspondingly the real part of $Y = G + iB$ is called the *conductance* and its imaginary part the *susceptance*.[14] Consider the case of one admittance as in the 4-terminal network of Figs. 8.14(a) and 8.14(b). Again applying Kirchhoff's laws we have the two equations $V_1 = V_0$ and $I_0 = I_1 + YV_0$, which we rewrite as

$$\begin{pmatrix} V_1 \\ I_1 \end{pmatrix} = \begin{pmatrix} 1 & 0 \\ -Y & 1 \end{pmatrix} \begin{pmatrix} V_0 \\ I_0 \end{pmatrix}. \tag{8.148}$$

Observe that the section in Fig. 8.14(a) containing the admittance $Y$ is represented by the $2 \times 2$ matrix here. Next we consider the case of two admittances $Y_1, Y_2$ in parallel, as in Fig. 8.14(b). Then

$$\begin{aligned} \begin{pmatrix} V_1 \\ I_1 \end{pmatrix} &= \begin{pmatrix} 1 & 0 \\ -Y_1 & 1 \end{pmatrix} \begin{pmatrix} 1 & 0 \\ -Y_2 & 1 \end{pmatrix} \begin{pmatrix} V_2 \\ I_2 \end{pmatrix} \\ &= \begin{pmatrix} 1 & 0 \\ -(Y_1 + Y_2) & 1 \end{pmatrix} \begin{pmatrix} V_2 \\ I_2 \end{pmatrix}, \end{aligned} \tag{8.149}$$

so that the total admittance is

$$Y = Y_1 + Y_2. \tag{8.150}$$

We now consider combinations of various types which are used in circuits called *filters* (meaning frequency filters, as we shall see below). We consider two types, known as (1) the *symmetrical* T-*junction*, and (2) the *symmetrical* $\pi$-*junction*.[15]

1. *The symmetrical T-junction*

The network of this type has the form shown in Fig. 8.15(a). In actual fact, in so-called filters there are generally more than one element as in Figs. 8.15(b)

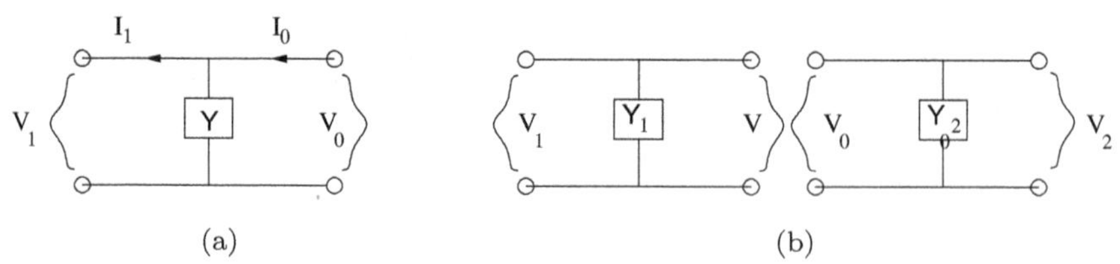


**Fig. 8.14.** (a) A single admittance, (b) two admittances in parallel.

[14]Schelkunoff [129], p. 27.
[15]Schelkunoff [129], pp. 110–111.

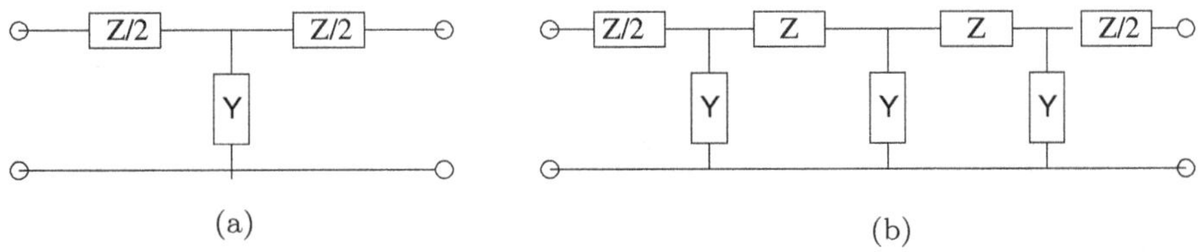


**Fig. 8.15.** (a) A single $T$-junction, (b) a more general case.

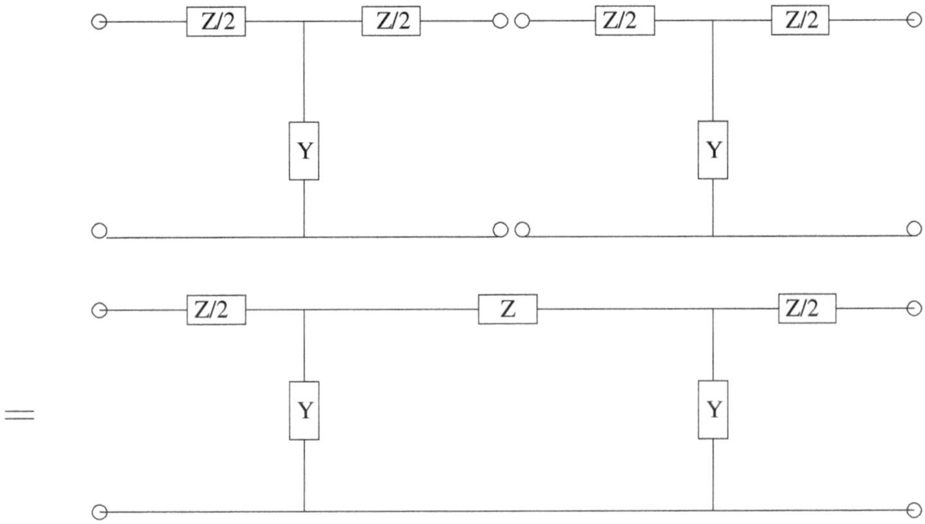


**Fig. 8.16.** Two $T$-junctions joined together.

and 8.16. The matrix representing the circuit shown in Fig. 8.15(a) is

$$\begin{aligned}
&\begin{pmatrix} 1 & -Z/2 \\ 0 & 1 \end{pmatrix}\begin{pmatrix} 1 & 0 \\ -Y & 1 \end{pmatrix}\begin{pmatrix} 1 & -Z/2 \\ 0 & 1 \end{pmatrix} \\
&= \begin{pmatrix} 1 & -Z/2 \\ 0 & 1 \end{pmatrix}\begin{pmatrix} 1 & -Z/2 \\ -Y & 1+YZ/2 \end{pmatrix} \\
&= \begin{pmatrix} 1+YZ/2 & -Z[1+YZ/4] \\ -Y & 1+YZ/2 \end{pmatrix} \equiv \begin{pmatrix} a & b \\ c & d \end{pmatrix}.
\end{aligned} \tag{8.151}$$

Below we require $a, b, c$ from this relation. Observe that in all these examples (also in those below) the skew diagonal elements of the matrices representing the junction are normally negative.

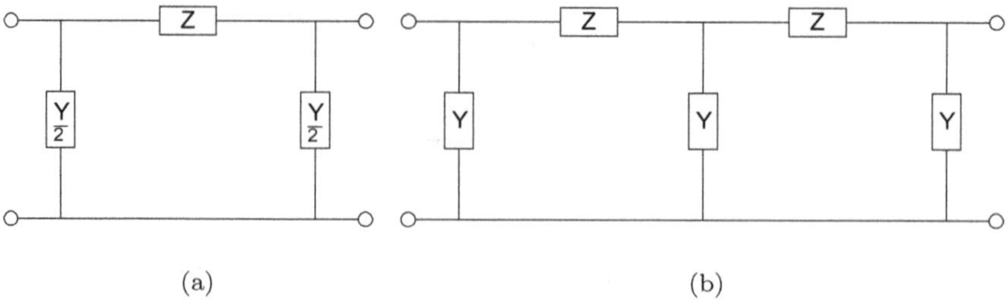


**Fig. 8.17.** (a) A single $\pi$-junction, (b) a more general case.

2. *The symmetrical $\pi$-junction*

The junction or network of this type is shown in Fig. 8.17(a). The matrix representing the circuit shown in Fig. 8.17(a) is

$$\begin{pmatrix} 1 & 0 \\ -Y/2 & 1 \end{pmatrix} \begin{pmatrix} 1 & -Z \\ 0 & 1 \end{pmatrix} \begin{pmatrix} 1 & 0 \\ -Y/2 & 1 \end{pmatrix}$$
$$= \begin{pmatrix} 1+ZY/2 & -Z \\ -Y[1+YZ/4] & 1+YZ/2 \end{pmatrix} \equiv \begin{pmatrix} a & b \\ c & d \end{pmatrix}. \tag{8.152}$$

Below we require $a, b, c$ from this relation. We observe that here as in Eq. (8.151) we have $a = 1 + ZY/2$. We make the following observations.

1. The determinants of the matrices above are all equal to 1 (also in unsymmetrical cases), i.e. $ad - bc = 1$. This follows from the reciprocity theorem.
2. For symmetrical transducers

$$\begin{pmatrix} a & b \\ c & d \end{pmatrix}, \quad a = d. \tag{8.153}$$

For proofs and more details we refer to Schelkunoff [129].

### 8.7.3 *Chains of transducers*

We consider a semi-infinite chain[16] of identical transducers as indicated in Fig. 8.18. Each transducer or wave filter is represented by a box which is supposed to contain circuits of the type discussed above.

[16]Schelkunoff [129], p. 108.

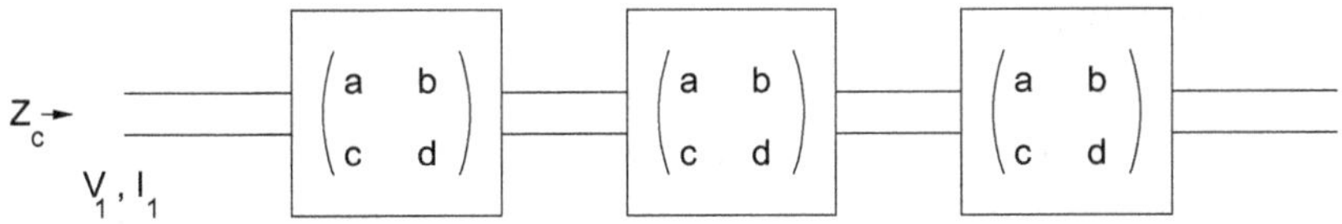


(a)

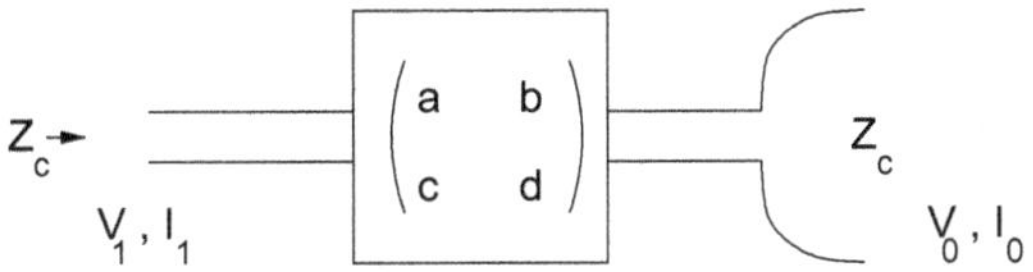


(b)

**Fig. 8.18.** (a) The semi-infinite chain of transducers, (b) one transducer cut off.

One defines as *characteristic impedance* the (input) impedance "*looking into the chain*" given by the relation $V_1 = Z_C I_1$. If the first transducer is cut off, then the impedance looking into the rest is still $Z_C$. Hence as far as this transducer is concerned, the rest of the chain can be replaced by a single impedance $Z_C$ and this transducer is said to be "*matched*" (by this impedance). Thus since both input and output impedances are then $Z_C$ the situation is that depicted in Fig. 8.18(b), and we have

$$\begin{pmatrix} V_1 \\ I_1 \end{pmatrix} = \begin{pmatrix} a & b \\ c & d \end{pmatrix} \begin{pmatrix} V_0 \\ I_0 \end{pmatrix}, \tag{8.154}$$

and so

$$\begin{pmatrix} Z_C I_1 \\ I_1 \end{pmatrix} = \begin{pmatrix} a & b \\ c & d \end{pmatrix} \begin{pmatrix} Z_C I_0 \\ I_0 \end{pmatrix}. \tag{8.155}$$

Recall, as noted before, that the elements $b$ and $c$ are normally negative. Thus

$$\begin{aligned} Z_C I_1 &= a Z_C I_0 + b I_0, \\ I_1 &= c Z_C I_0 + d I_0, \end{aligned} \tag{8.156}$$

and hence, if we take the ratio of these equations,

$$Z_C = \frac{a Z_C + b}{c Z_C + d}. \tag{8.157}$$

Solving this quadratic in $Z_C$ we obtain

$$Z_C^{\mp} := \frac{1}{2c}\left[(a-d) \pm \sqrt{(a-d)^2 + 4bc}\right]. \tag{8.158}$$

Note that since $c$ is negative, $Z_C^+$ is the positive root and $Z_C^-$ is the negative root. Note also that for symmetrical transducers $a = d$.

The *current transfer ratio* is defined as the quantity

$$e^{-\Gamma^{\mp}} := \frac{I_1}{I_0} \overset{(8.156)}{=} cZ_C^{\mp} + d = \frac{1}{2}\left[(a+d) \pm \sqrt{(a-d)^2 + 4bc}\right]. \tag{8.159}$$

The reciprocal current transfer ratio is

$$e^{\Gamma^{\pm}} = \frac{2}{(a+d) \mp \sqrt{(a-d)^2 + 4bc}}. \tag{8.160}$$

Rationalizing this expression and remembering that $ad - bc = 1$, one finds that

$$e^{\Gamma^{\pm}} = e^{-\Gamma^{\mp}}, \quad -\Gamma^{\mp} = \Gamma^{\pm}. \tag{8.161}$$

Hence we write, defining now the *propagation constant* $\Gamma$,

$$\Gamma^+ = \Gamma, \quad \Gamma^- = -\Gamma. \tag{8.162}$$

An expression for $\Gamma$ not involving roots is obtained by averaging. Thus

$$\cosh(\Gamma) = \frac{1}{2}\left[e^{\Gamma} + e^{-\Gamma}\right] = \frac{1}{2}(a+d), \tag{8.163}$$

so that

$$\Gamma = \cosh^{-1}\left(\frac{a+d}{2}\right) = i\cos^{-1}\left(\frac{a+d}{2}\right). \tag{8.164}$$

We shall in particular be concerned with the latter case when

$$-1 < \frac{a+d}{2} < 1, \quad \Gamma \text{ imaginary}.$$

We distinguish between *forward* and *backward* waves. In the case of forward waves with $\Gamma^+ = \Gamma$ and $Z_C^+$ both positive we have with the generator on the right of the chain of transducers

$$\frac{I_1}{I_0} = \frac{I_2}{I_1} = \cdots = \frac{I_n}{I_{n-1}} = e^{-\Gamma} \tag{8.165}$$

with

$$\frac{V_0}{I_0} = \frac{V_1}{I_1} = \cdots = \frac{V_n}{I_n} = Z_C^+. \tag{8.166}$$

Thus the forward current at the $n$th transducer is $I_n = Ae^{-\Gamma n}$ with forward voltage $V_n = Z_C^+ I_n = AZ_C^+ e^{-\Gamma n}$. Correspondingly one has in the case of backward waves (or generator on the left) with $\Gamma^-(= -\Gamma)$ and $Z_C^-$ negative: Backward current at the $n$th transducer $= I_n = Be^{\Gamma n}$ (in each case the current decays away from the generator, as one would expect) and forward voltage $V_n = Z_C^- I_n = BZ_C^- e^{\Gamma n}$. Thus after n symmetrical transducers ($Z_C^+ = -Z_C^-$) altogether:

$$\begin{aligned} I_n &= Ae^{-\Gamma n} + Be^{\Gamma n}, \\ V_n &= AZ_C^+ e^{-\Gamma n} + Z_C^- Be^{\Gamma n}. \end{aligned} \tag{8.167}$$

These equations may also be arrived at by considering a general member of the chain, i.e. that of Fig. 8.19, and solving the recurrence relations by the method of difference equations as in Example 8.24. Note also that a forward time-dependence factor $\exp(-i\omega t)$ is suppressed as also e.g. in Eq. (8.175).

In the case of symmetrical transducers $a = d$, so that

$$\Gamma = \cosh^{-1}(a) = i\cos^{-1}(a), \tag{8.168}$$

and therefore

$$Z_C^{\mp} = \frac{1}{2c}\left[\pm\sqrt{4bc}\right] = \frac{1}{2(-c)}\left(\mp\sqrt{4(-b)(-c)}\right) = \mp\sqrt{\frac{b}{c}}. \tag{8.169}$$

In the case of the symmetrical $T$-junction (with $b$ and $c$ obtained from Eq. (8.151)) this is (remember $Z_C^+$ was defined as positive)

$$Z_C^{\pm} = \pm\left[Z\left(\frac{1}{4}Z + Y^{-1}\right)\right]^{1/2} \tag{8.170}$$

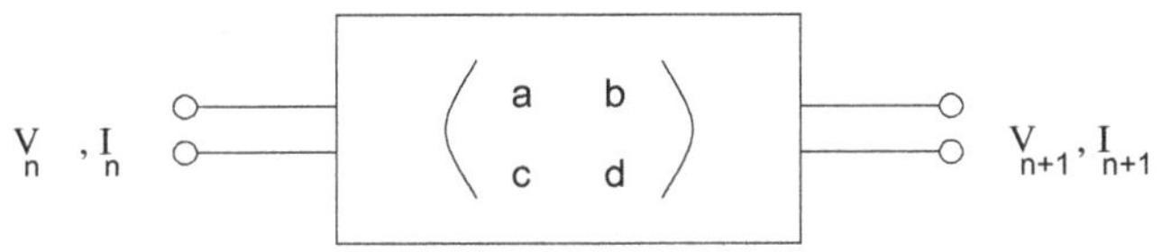


**Fig. 8.19.** A single segment of the chain.

with ($a$ obtained from Eq. (8.151))

$$\Gamma = \cosh^{-1}\left(1 + \frac{1}{2}YZ\right). \tag{8.171}$$

In the case of the symmetrical $\pi$-junction we obtain correspondingly from Eq. (8.152)

$$Z_C^{\pm} = \pm\left[Y\left(\frac{1}{4}Y + Z^{-1}\right)\right]^{-1/2}, \tag{8.172}$$

and

$$\Gamma = \cosh^{-1}\left(1 + \frac{1}{2}YZ\right). \tag{8.173}$$

We note here that for $\mu$ small one has the expansion

$$\cosh^{-1}(1+\mu) = \sqrt{2\mu}\left(1 + \frac{1}{4}\mu + O(\mu^2)\right), \tag{8.174}$$

so that for $YZ < 2$, one has $\Gamma \simeq \sqrt{YZ}$.

### 8.7.4 *Finite chains of transducers and filters*

We consider a finite chain as depicted in Fig. 8.20.

The decaying wave has

$$I_n = I_0 e^{-n\Gamma}. \tag{8.175}$$

In general one has for the current and potential in the $n$th transducer:

$$I_n = Ae^{-\Gamma n} + Be^{\Gamma n}, \quad V_n = Z_C(Ae^{-n\Gamma} - Be^{n\Gamma}), \tag{8.176}$$

where $A$ and $B$ are determined from boundary conditions. Then for a *semi-infinite* chain of transducers with generator at one end only there is no backward (i.e. reflected) wave, and hence

$$I_n, V_n \propto e^{-\Gamma n}, \ \Gamma > 0.$$

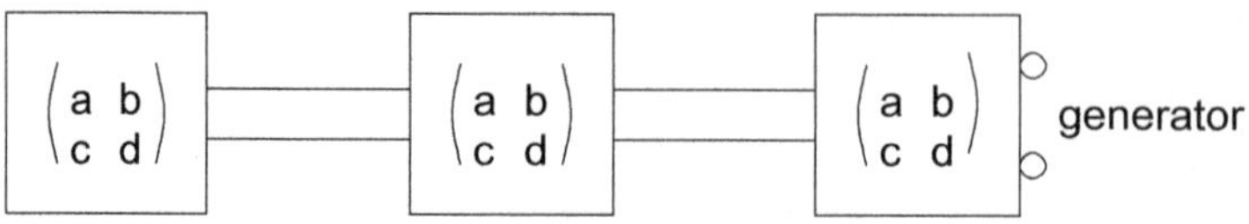


**Fig. 8.20.** A finite chain of transducers with generator on the right.

Similarly, for a *finite* chain with impedance $Z_c$ at one end — in this case the chain is said to be "*matched*" by the impedance — and the generator at the other, there will also be no backward wave. If $\Gamma$ is real and positive as defined, $I_n$ decreases rapidly with $n$. If $\Gamma$ is pure imaginary, there is only a phase shift and the wave passes down the chain of transducers unattenuated. The range of frequencies permitted to pass through is called the *pass band*. The pass band is such that $\Gamma$ is pure imaginary (sometimes $\Gamma$ is complex but this case is not of much importance). One distinguishes between 1. low-pass filters, 2. high-pass filters, and 3. band-pass filters, which we now consider in this order.[17]

1. *Low-pass filter*

For a symmetrical $T$- or $\pi$-junction we have with the parameter $a$ from Eqs. (8.151) and (8.152)

$$\Gamma = i\cos^{-1}(a) = i\cos^{-1}\left(1+\frac{1}{2}YZ\right), \tag{8.177}$$

i.e. $\Gamma$ is imaginary provided that

$$-1 < 1+\frac{1}{2}YZ < 1. \tag{8.178}$$

This is the condition which has to be satisfied for the pass-band. Consider the system of Fig. 8.21(a), the transducer of a low-pass filter. In this case we have

$$Z = i\omega L, \quad Y = i\omega C.$$

It follows that

$$-1 < 1-\frac{1}{2}\omega^2 LC < 1, \quad 4 > \omega^2 LC, \quad \text{i.e.} \quad 0 < \omega < \frac{2}{\sqrt{LC}}. \tag{8.179}$$

2. *High-pass filter*

As the name implies, this type of filter allows only high frequencies to pass through. Consider the system of Fig. 8.21(b). In this case

$$Z = \frac{1}{i\omega C}, \quad Y = \frac{1}{i\omega L}.$$

[17] Such networks are also treated in e.g. Lorrain and Corson [89], pp. 377–382, and, of course, in Schelkunoff [129]. Filters and transmission lines are also treated in Chapter 9 of Bleaney and Bleaney [12].

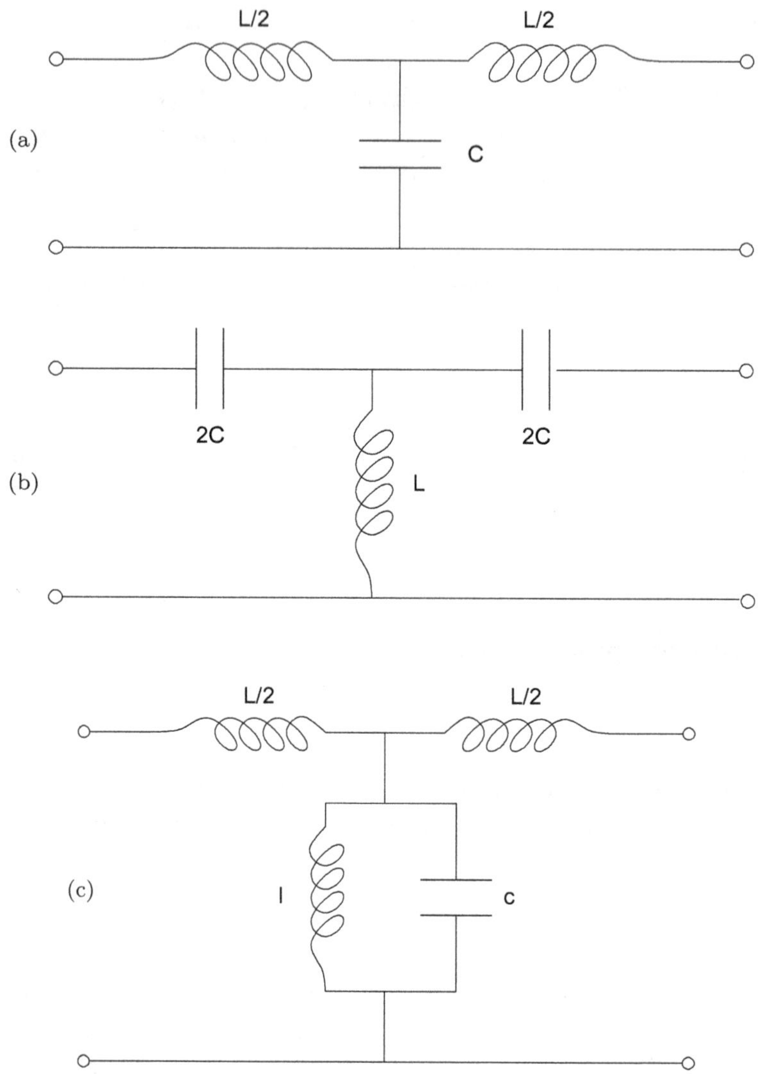


**Fig. 8.21.** Elements of (a) low-pass, (b) high-pass, (c) band-pass filters.

The frequency range of the pass-band of this system is restricted to

$$-1 < 1 + \frac{1}{2}YZ < 1, \quad -1 < 1 - \frac{1}{2\omega^2 LC} < 1,$$

i.e. $\omega$ is restricted to the high frequency region,

$$\frac{1}{2\sqrt{LC}} < \omega < \infty. \tag{8.180}$$

3. *Band-pass filter*

This type of filter allows frequencies in a certain range (as for example in tuning a radio). Consider the system of Fig. 8.21(c). In this case[18]

$$Z = i\omega L, \quad Y = i\omega c + \frac{1}{i\omega l}.$$

The pass-band allows frequencies for which

$$-1 < 1 + \frac{1}{2}YZ < 1, \quad -1 < 1 + \frac{1}{2}i\omega L\left(i\omega c + \frac{1}{i\omega l}\right) < 1,$$

which implies

$$\frac{1}{\sqrt{cl}} < \omega < \frac{2}{\sqrt{Lc}}\sqrt{1 + \frac{L}{4l}}. \tag{8.181}$$

For $l \to \infty$ this case gives the low-pass filter. For $l$ very small the difference between the frequencies is very small, i.e.

$$\frac{1}{\sqrt{cl}} < \omega < \frac{2}{\sqrt{Lc}}\sqrt{\frac{L}{4l}}\left[1 + \frac{4l}{L}\right]^{1/2} \simeq \frac{1}{\sqrt{cl}}\left[1 + \frac{2l}{L}\right]. \tag{8.182}$$

### 8.7.5 *Transmission lines*

In this and the following subsections we apply the foregoing considerations to real transmission lines and investigate such properties as conditions for attenuation losses, short-circuiting and open-circuiting, and finally we consider reflection and transmission of waves along such lines.

We consider now filters forming transmission lines. Real transmission lines are e.g. (a) a parallel wire cable insulated from one another, and (b) a cylindrical, i.e. coaxial wire cable. The connection to our previous considerations is effected by considering an infinite number of transducers per unit length. Let's select the frequency domain of the symmetrical $T$-junction. We consider in the first place n *such sections per unit length.* Let $z$ be the series impedance per unit length, $z = nZ$, and $y$ the parallel admittance per unit length, $y = nY$, and $\gamma$ the propagation constant per unit length, $\gamma = n\Gamma$ corresponding to the logarithm of the ratio current on one side/current on

[18]We hope there is no confusion between $c$ here as a capacitance, earlier as an element of the matrix (8.153), and elsewhere as the speed of light.

the other side. Hence

$$\gamma = n\Gamma \overset{(8.171)}{=} n\cosh^{-1}\left(1+\frac{1}{2}YZ\right) = n\cosh^{-1}\left(1+\frac{yz}{n^2}\right) \quad \text{meters}^{-1}. \tag{8.183}$$

Using the small-$\mu$ expansion of $\cosh^{-1}(1+\mu)$ of Eq. (8.174) since we are interested in large values of $n$, we obtain

$$\gamma \simeq n\sqrt{\frac{2yz}{2n^2}} = \sqrt{yz} \quad \text{meters}^{-1} \quad (\text{for } n \to \infty). \tag{8.184}$$

The characteristic impedance for a symmetrical $T$-junction is (Eq. (8.170))

$$Z_C = \sqrt{Z\left(\frac{1}{4}Z+\frac{1}{Y}\right)} = \sqrt{\frac{z}{n}\left(\frac{1}{4}\frac{z}{n}+\frac{n}{y}\right)} \simeq \sqrt{\frac{z}{y}} \quad \text{for} \quad \text{n} \to \infty. \tag{8.185}$$

For a direct calculation of the current $I$ and the voltage $V$ at length $x$ we consider an infinitesimal element of length $dx$ of the transmission line as indicated in Fig. 8.23. We have (with decrease of $V$ and $I$)

$$dV = -zdxI, \quad dI = -ydxV, \tag{8.186}$$

(cf. Eq. (8.104)) i.e.

$$\frac{\partial V}{\partial x} = -zI, \quad \frac{\partial I}{\partial x} = -yV. \tag{8.187}$$

Assuming $z$ and $y$ are constant over $x$, we obtain the fundamental equations of transmission line theory, i.e.

$$\frac{\partial^2 I}{\partial x^2} = -y\frac{\partial V}{\partial x} = yzI, \quad \frac{\partial^2 I}{\partial x^2} - yzI = 0, \tag{8.188}$$

and similarly

$$\frac{\partial^2 V}{\partial x^2} - yzV = 0.$$

These single variable equations are known as the *telegrapher's equations* in the frequency domain (in the time domain the independent variables are distance and time, as in the equations for wave guides — see Eq. (14.83)). Then in particular

$$I \propto e^{\pm x\sqrt{yz}}, \quad \cosh(x\sqrt{yz}), \quad \sinh(x\sqrt{yz}).$$

We choose

$$I = Ae^{-x\sqrt{yz}} + Be^{x\sqrt{yz}}. \tag{8.189}$$

Then

$$V = -\frac{1}{y}\frac{\partial I}{\partial x} = \sqrt{\frac{z}{y}}\left\{Ae^{-x\sqrt{yz}} - Be^{x\sqrt{yz}}\right\}. \tag{8.190}$$

Since

$$Z_C = \sqrt{\frac{z}{y}} \quad \text{and} \quad \gamma = \sqrt{yz},$$

$V$ is the same as we had the equation we had earlier in Eq. (8.176).

As typical examples of real transmission lines we consider Fig. 8.22(a) the parallel wire cable (the two wires being insulated from one another) and the cylindrical, i.e. coaxial wire cable, as illustrated in Figs. 8.22 and 8.23. Let $R$ be the resistance per unit length of both wires. The flux between

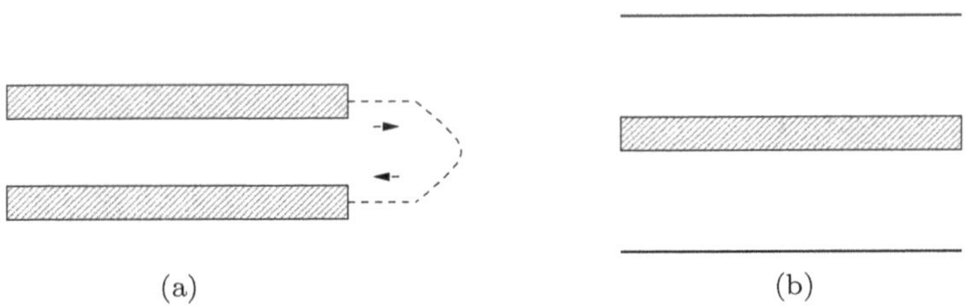


**Fig. 8.22.** Two real transmission lines, (a) the parallel wire cable and (b) the coaxial cylinder cable.

**Fig. 8.23.** The infinitesimal section of a transmission line.

the wires gives rise to a self-inductance $L$ per unit length. Then the series impedance per unit length of the cable is

$$z = i\omega L + R \quad \text{ohms}. \tag{8.191}$$

The leakage through the insulation between the wires gives rise to a parallel conductance per unit length $G$ which with the capacity per unit length $C$ adds up to the parallel admittance per unit length,

$$y = G + i\omega C \quad \text{ohms}^{-1}. \tag{8.192}$$

These relations imply a characteristic impedance per unit length $Z_C$ given by

$$Z_C = \sqrt{\frac{z}{y}} = \sqrt{\frac{i\omega L + R}{G + i\omega C}} = \sqrt{\frac{L}{C}}\left(\frac{1 - iR/\omega L}{1 - iG/\omega C}\right)^{1/2}. \tag{8.193}$$

The square-root factor in brackets is the effect of losses and changes slightly the phase between $V$ and $I$. The complex propagation constant $\gamma$ per unit length is given by

$$\begin{aligned}
\gamma = \sqrt{yz} = y\sqrt{\frac{z}{y}} &= (G + i\omega C)\sqrt{\frac{L}{C}}\left(\frac{1 - iR/\omega L}{1 - iG/\omega C}\right)^{1/2} \\
&= i\omega C\left(1 - \frac{iG}{\omega C}\right)\sqrt{\frac{L}{C}}\left(\frac{1 - iR/\omega L}{1 - iG/\omega C}\right)^{1/2} \\
&= i\omega\sqrt{LC}\sqrt{(1 - iR/\omega L)(1 - iG/\omega C)} \\
&= \sqrt{(R + i\omega L)(G + i\omega C)} \\
&\simeq i\omega\sqrt{LC}\left\{1 + \frac{1}{8\omega^2}\left(\frac{R}{L} - \frac{G}{C}\right)^2\right\} + \frac{1}{2}\sqrt{LC}\left(\frac{R}{L} + \frac{G}{C}\right).
\end{aligned} \tag{8.194}$$

Thus, setting

$$\gamma = i\beta + \delta \quad \text{meters}^{-1}, \tag{8.195}$$

the *phase constant* $\beta$ and *attenuation constant* $\delta$ are given by

$$\beta \simeq \omega\sqrt{LC}\left\{1 + \frac{1}{8\omega^2}\left(\frac{R}{L} - \frac{G}{C}\right)^2\right\}, \tag{8.196}$$

and

$$\delta \simeq \frac{1}{2}\sqrt{LC}\left(\frac{R}{L} + \frac{G}{C}\right). \tag{8.197}$$

Hence for the forward wave (with insertion of time dependence)

$$I \propto e^{i\omega t}e^{-x(i\beta+\delta)} = e^{i(\omega t-\beta x)}e^{-\delta x}. \tag{8.198}$$

Here the *wave number* $\beta = 2\pi/\lambda$, where $\lambda$ is the wavelength, and $\omega = 2\pi\nu$, where $\nu$ is the frequency. For the backward wave the sign of $x$ has to be changed. In general we have a combination of both, i.e. forward wave and reflected wave from the end.

In considering the velocity of waves one distinguishes between the phase velocity and the group velocity. The *phase velocity* $v_p$ is defined as

$$v_p = \text{frequency } \times \text{ wavelength } = \nu\lambda = \frac{\omega}{\beta}. \tag{8.199}$$

In the present case of electromagnetic waves in the cable we have

$$v_p \simeq \frac{1}{\sqrt{LC}}\left\{1 - \frac{1}{8\omega^2}\left(\frac{R}{L} - \frac{G}{C}\right)^2\right\} \quad \text{meters/second}, \tag{8.200}$$

where $R$ and $G$ are small. We note that

$$v_p \simeq \frac{1}{\sqrt{LC}} \simeq \frac{1}{\sqrt{\mu_0\epsilon_0}} \quad \text{meters/second}, \tag{8.201}$$

which is the velocity of the wave in the medium between the wires. We also observe that the phase velocity is a function of the angular frequency $\omega$ (giving distortions) except for one particular case when

$$\frac{R}{L} = \frac{G}{C}. \tag{8.202}$$

This relation is known as the *Heaviside condition* for no distortion.

The other velocity called *group velocity*, $v_g$, is defined by the relation

$$v_g = \frac{d\omega}{d\beta} = \frac{1}{d\beta/d\omega}, \tag{8.203}$$

and is here (recall $L$ and $C$ are here per unit length)

$$v_g \simeq \frac{1}{\sqrt{LC}}\left\{1 + \frac{1}{8\omega^2}\left(\frac{R}{L} - \frac{G}{C}\right)^2\right\} \quad \text{meters/second}. \tag{8.204}$$

In systems with losses the velocity of energy propagation is not the same as the group velocity so that the result does not imply that energy travels faster than the velocity of light.

One distingusihes between the following three cases.

1. *The lossless line*

In this case $R = G = 0$. Then

$$Z_C = \sqrt{\frac{z}{y}} = \sqrt{\frac{L}{C}} \quad \text{and} \quad \gamma = \sqrt{yz} = i\omega\sqrt{LC}, \tag{8.205}$$

and

$$v_p = \frac{\omega}{\beta} = \frac{1}{\sqrt{LC}}. \tag{8.206}$$

There is no attenuation or dispersion (frequency dependence). The electric field is directed perpendicular to the direction of motion (i.e. the cable) and the magnetic field is perpendicular to both, see Example 7.20.

2. *The lossy line*

This is the case discussed above with

$$\frac{R}{L} - \frac{G}{C} \neq 0.$$

In the case of the lossy line, there are different frequency waves at different times since the wave or phase velocity depends upon $\omega$, and there is distortion due to dispersion. There is very little distortion due to attenuation, since $\delta$ does not depend upon $\omega$ in the first approximation.

3. *The distortionless line*

However, the distortion vanishes if $R/L = G/C$, the Heaviside condition. In fact it vanishes exactly, since then (see first line of Eq. (8.194))

$$\gamma = i\omega\sqrt{LC}\left(1 - \frac{iR}{\omega L}\right) = i\omega\sqrt{LC} + R\sqrt{\frac{C}{L}}, \tag{8.207}$$

so that both $v_p = \omega/\beta$ and $\delta$ are then independent of the angular frequency $\omega$.[19]

### 8.7.6 *Input and output impedances of a transmission line*

Instead of writing the current in terms of exponentials it is convenient to write it now as

$$I \propto \cosh\gamma(X - x) + \alpha\sinh\gamma(X - x), \tag{8.208}$$

[19]For further discussion see e.g. Attwood [5], p. 425.

where $\gamma = \sqrt{yz}$ and $\alpha =$ constant. Then with $Z_c = \gamma/y$,

$$V = -\frac{1}{y}\frac{\partial I}{\partial x} \propto Z_C\{\sinh\gamma(X - x) + \alpha\cosh\gamma(X - x)\}. \tag{8.209}$$

Thus at any point $x$ as in Fig. 8.24

$$Z_x = \frac{V}{I} = Z_C\frac{\sinh\gamma(X - x) + \alpha\cosh\gamma(X - x)}{\cosh\gamma(X - x) + \alpha\sinh\gamma(X - x)}, \tag{8.210}$$

and therefore in particular at $x = X$, $Z_x \to Z_X$, this $Z_X$ being the load,

$$Z_X = \alpha Z_C, \quad \alpha = \frac{Z_X}{Z_C}. \tag{8.211}$$

It follows that

$$Z_x = Z_C\frac{Z_C\sinh\gamma(X - x) + Z_X\cosh\gamma(X - x)}{Z_C\cosh\gamma(X - x) + Z_X\sinh\gamma(X - x)}. \tag{8.212}$$

The input impedance $Z_0$ is therefore given by (setting $x = 0$)

$$Z_0 = Z_C\frac{Z_C\sinh\gamma(X) + Z_X\cosh\gamma(X)}{Z_C\cosh\gamma(X) + Z_X\sinh\gamma(X)}. \tag{8.213}$$

For a *lossless line* $R = 0$ and $G = 0, \delta = 0, \gamma = i\beta, \beta = \omega\sqrt{LC}$ and hence[20]

$$Z_0 = Z_C\frac{iZ_C\sin\beta X + Z_X\cos\beta X}{Z_C\cos\beta X + iZ_X\sin\beta X}. \tag{8.214}$$

For a *short-circuited lossless line*, i.e. one with $Z_X = 0, V_X = 0$ (no voltage at $X$), one obtains

$$Z_0 = iZ_C\tan\beta X. \tag{8.215}$$

If $V = 0$ at $x = 0$, i.e. if $Z_0 = 0$, one has

$$\beta X = n\pi. \tag{8.216}$$

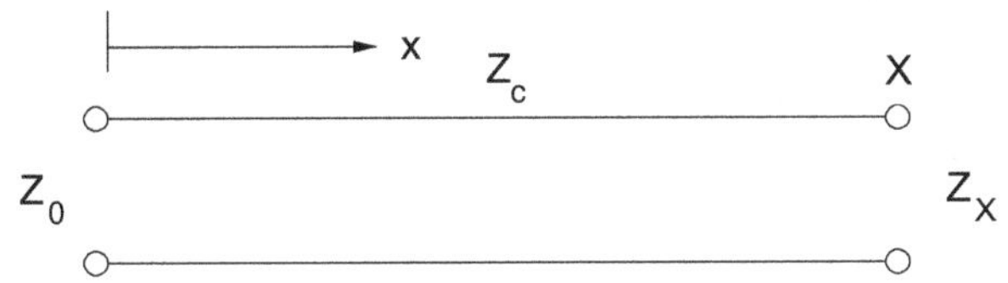


**Fig. 8.24.** Input and output impedances.

[20]Another reference where this formula is derived and discussed is Robinson [125], p. 85.

But $\beta = 2\pi/\lambda$, so that in this case

$$\frac{2\pi X}{\lambda} = n\pi, \quad \text{or} \quad \frac{X}{\lambda/2} = n, \tag{8.217}$$

where $n$ is an integer. Thus in this case the length $X$ of the line is an integral number of half-wavelengths.

In the case of an *open-circuited lossless line* $Z_X = \infty$ (no current at $X$) and

$$Z_0 = -iZ_C \cot\beta X. \tag{8.218}$$

If $V = 0$ at $x = 0$, i.e. if $Z_0 = 0$, one now has

$$\beta X = \frac{n+1}{2}\pi, \quad \text{or} \quad \frac{2\pi}{\lambda}X = \frac{n+1}{2}\pi, \quad \text{or} \quad \frac{4X}{\lambda} = n+1. \tag{8.219}$$

If $Z_0 = \infty$, we have $\beta X = n\pi = (2\pi/\lambda)X$, i.e.

$$\frac{X}{\lambda/2} = n. \tag{8.220}$$

In this case $X$ is an integral number of half-wavelengths.

Finally we consider the "*quarter wave lossless matching line*". This is a transmission line of length $X = \lambda/4$, so that we have $\beta X = (2\pi/\lambda)X = (2\pi/\lambda)(\lambda/4) = \pi/2$, where $\lambda$ is the wavelength of the waves. From Eq. (8.214) we obtain for this case

$$Z_0 = \frac{Z_C^2}{Z_X}. \tag{8.221}$$

This result will be used below.

### 8.7.7 *Reflection and transmission*

We consider two transmission lines with input impedances $Z_1$ and $Z_2$ at their junction as indicated in Fig. 8.25. Let $t$ be the amount of current (i.e. fraction) which passes through the junction in the direction indicated. However, not all of the current will pass through. There will also be some which is reflected, say the fraction $r$ of the current of strength unity. Then $Z_1$ is the impedance seen on looking towards the left from the junction, and $Z_2$ is the impedance seen on looking towards the right from the junction. There are two boundary conditions which the currents have to satisfy.

(a) The currents are continuous. Hence we must have

$$1 - r = t. \tag{8.222}$$

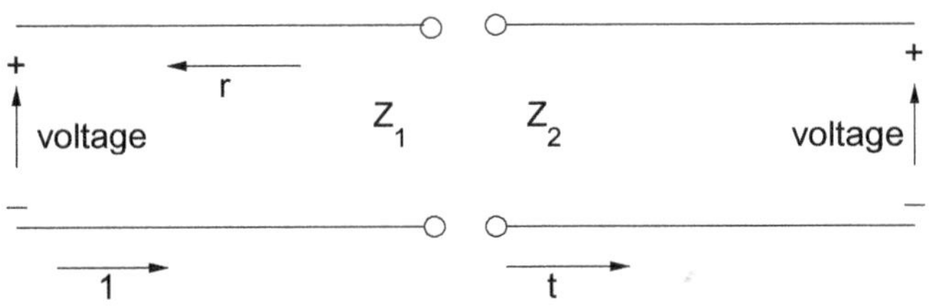


**Fig. 8.25.** Two transmission lines with a junction in between.

(b) The voltage across the terminals at the junction is continuous, i.e. is the same whether looking to the right or to the left. Hence (since voltage is current times impedance)

$$(1 + r)Z_1 = tZ_2. \tag{8.223}$$

Note that voltages are scalar quantities and add. From the two conditions we obtain the *reflection coefficient* $r$ or amount of current reflected back and *transmission coefficient* $t$, the amount of current transmitted:

$$r = \frac{Z_2 - Z_1}{Z_2 + Z_1}, \quad t = \frac{2Z_1}{Z_2 + Z_1} \tag{8.224}$$

(i.e. $r$ is $r/1$ and $t$ is $t/1$).

One can also define *reflection* and *transmission coefficients* $R$ and $T$ in terms of voltages. These represent voltages. Clearly $R$ is the ratio of reflected voltage to incident voltage, i.e.

$$R = \frac{-Z_1.r}{Z_1.1} = -r, \tag{8.225}$$

and the transmitted voltage divided by the incident voltage is

$$T = \frac{Z_2.t}{Z_1.1} = \frac{Z_2}{Z_1}t. \tag{8.226}$$

We now consider a number of examples.

(a) Two semi-infinite transmission lines joined together as in Fig. 8.26(a). Here $Z_1 = Z_{C1}$ on looking into $Z_{C1}$ and $Z_2 = Z_{C2}$ on looking into $Z_{C2}$. The coefficients $r, t, R, T$ follow from the equations above.
(b) As in Fig. 8.26(b), here $Z_1 = Z_C, Z_2 = Z$, and the coefficients follow again as above.
(c) In this case of Fig. 8.26(c) the impedance seen to the right is $Z_{C2}$. Here therefore, in series

$$Z_1 = Z_{C1}, \quad Z_2 = Z + Z_{C2}.$$

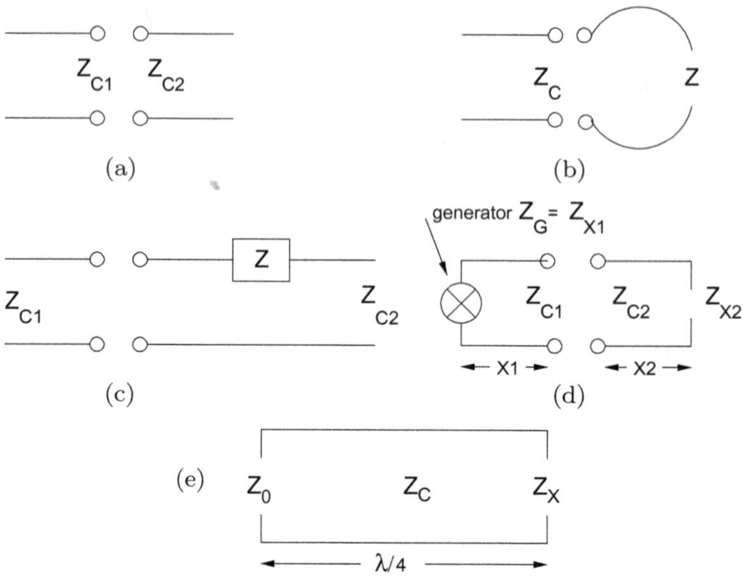


**Fig. 8.26.** Junctions (a), (b), (c), (d), (e).

(d) In the case of Fig. 8.26(d) the generator impedance is the impedance $Z_{X1}$. What are the reflection coefficients at the junction? Here $Z_1$, the input impedance of a transmission line of characteristic impedance $Z_{C1}$ and terminated by a load $Z_{X1}$ at distance $X_1$ is (from the general result above for a lossless line, i.e. Eq. (8.214)):

$$Z_1 = Z_{C1}\frac{\{iZ_{C1}\sin(\beta X_1) + Z_{X1}\cos(\beta X_1)\}}{\{Z_{C1}\cos(\beta X_1) + iZ_{X1}\sin(\beta X_1)\}}. \tag{8.227}$$

The impedance $Z_2$ is similarly obtained by looking to the right.

(e) Fig. 8.26(e) is a sketch of the so-called "*quarter-wave matching line*". In this case the line has length $\lambda/4$, and the input impedance $Z_0$ is given by (cf. Eq. (8.221))

$$Z_0 = \frac{Z_C^2}{Z_X}. \tag{8.228}$$

This quarter-wave matching line is of use in connecting two transmission lines. Consider the case of two transmission lines with impedances $Z_0$ and $Z_X$. In order to connect the two transmission lines, insert a quarter wave matching line, i.e. insert into the junction of Fig. 8.27(a) the element of Fig. 8.27(b). The relation which has to be satisfied is therefore, as we know from Eq. (8.221), $Z_0 = Z_C^2/Z_X, Z_C^2 = Z_0Z_X$. Thus $Z_C$ is the *geometric mean*

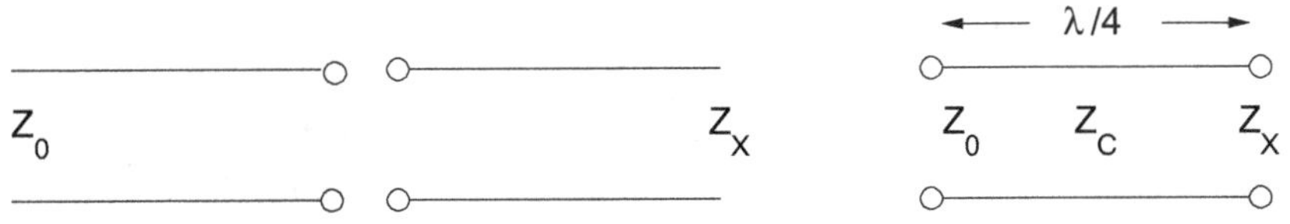


(a) (b)

**Fig. 8.27.** Insert into the junction (a) the element (b).

between $Z_0$ and $Z_X$; it matches the two impedances (not very good at high frequencies due to the skin effect). There is no reflection at the junction if the impedances looking each way are the same. This is easily arranged, since by inserting the quarter-wave line of characteristic impedance $Z_C$ before (say) $Z'$, one can change the input impedance from this $Z'$ to (say) $Z'' = Z_C^2/Z'$. For a more extensive treatment of this topic we refer to Schelkunoff [129].

## 8.8 Exercises

**Example 8.12 (Square circuits on a cube).** Two square circuits form the boundaries of opposite faces of a cube of side lengths $a$. Show that the mutual inductance $L_{12}$ is given by[21]

$$L_{12} = \frac{2\mu_0 a}{\pi}\left[\sinh^{-1} 1 - \sinh^{-1}\left(\frac{1}{\sqrt{2}}\right) + 1 + \sqrt{3} - 2\sqrt{2}\right] \quad \text{henries.} \tag{8.229}$$

[Hint: Start from Eq. (8.4)].

**Example 8.13 (The toroidal solenoid).** A toroidal solenoid of external radius $b + a, b \gg a$, and internal radius $a$ (i.e. of its circular cross section), has $N$ turns, $N$ being assumed large. Show that its self-inductance $L_{11}$ is given by[22]

$$L_{11} = \mu_0 N^2[b - \sqrt{b^2 - a^2}] \quad \text{henries.} \tag{8.230}$$

How does the result compare with that of Example 8.1?
[Hint: Use cylindrical polar coordinates $(r, \phi, z)$ given by $z = a\sin\theta, r = n + a\cos\theta, \phi = \text{const.}$].[23]

[21] See also Bradley [22], p. 96.
[22] See e.g. Starling and Woodall [144], p. 1104.
[23] See also Bradley [22], p. 101.

**Example 8.14 (Series and parallel circuits).** Consider 2-terminal direct current networks, and more specifically those connected either in the so-called series form or parallel form. In the case of circuits connected in series the components resistor, capacitor, inductor, as well as battery and switches are all connected one after the other in the form of a closed circuit. In the case of the parallel connections the components are connected to the two terminals like the rungs of a ladder. Recalling the definition of capacitance $C$ ($V = q/C$) and Kirchhoff's laws show that the total capacitance $C$ of $n$ capacitors with capacitances $C_i$ connected in series is

$$C = \left(\sum_{i=1}^{n} \frac{1}{C_i}\right)^{-1}.$$

Show that in the case of resistances $R_i$ connected in series the total resistance is

$$R = \sum_{i=1}^{n} R_i.$$

Show that the total conductance $G$ follows the formula for the total capacitance $C$, whereas the total inductance $L$ follows the formula for the total resistance $R$. Considering similarly the circuits connected in parallel form, show that the total capacitance $C$ is given by

$$C = \sum_{i=1}^{n} C_i$$

and the total resistance $R$ is given by

$$R = \left(\sum_{i=1}^{n} \frac{1}{R_i}\right)^{-1}.$$

Show also that then the total conductance $G$ follows the form of $C$ and the total inductance the form of $R$.

**Example 8.15 (The voltage transformer).**

(a) A transformer designed to convert an alternating current from a high (low) to a low (high) voltage consists essentially of two independent coils of several turns wound around an iron core. An e.m.f. or $V_e$ is applied to the primary circuit (index 1). Write down Kirchhoff's first law for both circuits and solve the equations by setting $I_1 = Ae^{i\omega t}, I_2 = Be^{i\omega t}$.

(b) Using the results of (a) show that the current in the primary circuit is always increased by the existence of the secondary current.
[Hint: The equation for the primary circuit is $L_1\dot{I}_1 + L_{12}\dot{I}_2 + R_1 I_1 = V_e$, that for the secondary circuit: $L_2\dot{I}_2 + L_{21}\dot{I}_1 + R_2 I_2 = 0$].

$$\text{(Answers: } I_1 = C\cos(\omega t - \phi')/\sqrt{L_1^2\omega^2 + R_1^2},$$
$$I_2 = -L_{21}\omega A e^{i(\omega t+\theta)}/\sqrt{L_2^2\omega^2 + R_2^2},$$
where $C$ is the maximum value of $V_e$).[24]

**Example 8.16 (Application of Kirchhoff's two laws).** Consider the circuit with junctions $A, B, D, F$ (also called *nodes* or sometimes *terminals*) shown in Fig. 8.28. The currents in lines or branches $AB, AD$ and $FB$ are respectively taken to be $I_1, I_2$ and $I_3$. Verify Kirchhoff's second law (that the currents flowing in sum up to those flowing out) at each node of the 4-node network. Then consider the three circuits (or *meshes*) *ABDFA*, *ADFA* and *BDFB* with mesh currents $I_1, I_2$ and $I_3$, and apply Kirchhoff's first law, assuming an e.m.f. $E_0 \cos\omega t$ applied between $A$ and $D$. Finally calculate the current $I_3$ in $FB$.[25]

**Example 8.17 (Discharge of a condenser).** Derive a formula for the frequency and damping of the oscillations produced when a charged condenser of capacity $C$ is allowed to discharge through a long solenoid of length $l$, wound with $n$ turns and possessing a small resistance $R$.

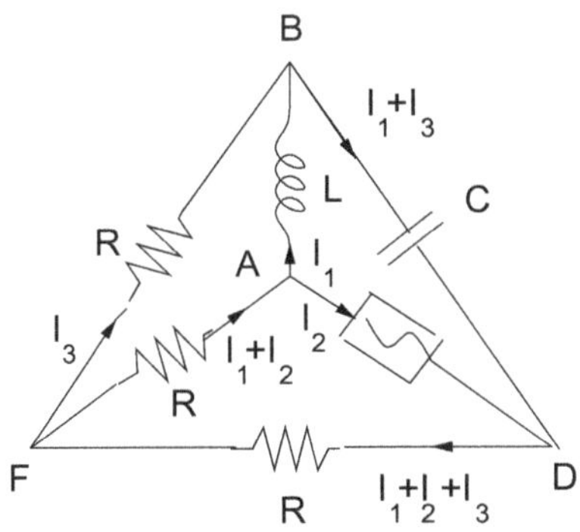


**Fig. 8.28.** A 4-node circuit.

[24]See e.g. Starling and Woodall [144], p. 1104.
[25]For a different formulation see Bradley [22], p. 110.

**Example 8.18 (The charge of a condenser).** Consider a circuit consisting of the generator of the e.m.f. $V$ with negligible resistance, a capacitance $C$, an inductance $L$, and a resistance $R$, all in series. Show that

$$L\ddot{q} + R\dot{q} + \frac{q}{C} = V(t), \tag{8.231}$$

where $q$ is the charge on the condenser. If the e.m.f. is $V(t) = ae^{i\omega t}, a =$ const., and at time $t = 0$ both $q(0) = 0, \dot{q}(0) = 0$, show how to determine $q(t)$. Indicate how the results could be used to find the current produced by $V(t) = a\sin(\omega t + \delta)$, switched on at $t = 0$.

**Example 8.19 (Parallel and series arrangements in mechanics).** Consider the mechanical analogy to an electromagnetic circuit and assume an impressed force $F(t) \propto \exp(i\omega t)$ acting on a system consisting of two masses $m_1, m_2$ or two springs with inverse spring constants $n_1, n_2$, or two frictional forces with parameters $r_1, r_2$, in each case the two being arranged *in parallel*. Then as in Eq. (8.96) $\dot{x} = n\dot{F}(t), m\dot{x} = \int F(t)dt$ and $\dot{x} = F(t)/r$, and hence $F = \dot{x}/i\omega n$ and $F = i\omega m\dot{x}$. Show that the resulting mechanical impedance $Z_{mp} := F(t)/\dot{x}$ ($m$ for mechanical and $p$ for parallel) implies

$$m = m_1 + m_2, \quad \frac{1}{n} = \frac{1}{n_1} + \frac{1}{n_2}, \quad r = r_1 + r_2. \tag{8.232}$$

Sketch appropriate diagrams indicating that $F = F_1 + F_2, \dot{x} = \dot{x}_1 + \dot{x}_2$. How do these relations change when the arrangements are *in series*?

(Answer: $1/m = 1/m_1 + 1/m_2, n = n_1 + n_2, 1/r = 1/r_1 + 1/r_2$).

**Example 8.20 (Vector potential of a solenoid).** Show that the vector potential $A$ inside an infinitely long solenoid of radius $a$ and $n$ turns per meter and carrying a current $I$ is at a distance $r$ from its axis:

$$A = \mu_0 n\frac{rI}{2} \quad \text{tesla} \cdot \text{meters}. \tag{8.233}$$

What is the vector potential at a distance $r$ from its axis outside the solenoid?

(Answer: $A = \mu_0 na^2 I/2r$).

**Example 8.21 (The toroidal coil).** Show that the self-inductance of a toroidal coil of square cross-sectional area $a^2$ and mean radius $R$ and a total of $N$ turns is given by

$$L = \frac{\mu_0 N^2 a}{2\pi}\ln\left[\frac{R + a/2}{R - a/2}\right] \quad \text{henries}. \tag{8.234}$$

**Example 8.22 (Mutual inductance).** What is the mutual inductance between the toroidal coil of Example 8.21 and a long straight wire along the axis of the toroidal coil?
[Hint: Assume current $I$ either in the coil or in the wire].

(Answer: $(\mu_0 aN/2\pi)\ln[(R+a/2)/(R-a/2)]$).

**Example 8.23 (Pass-bands of a wave filter).** Calculate the value of the propagation constant $\Gamma$ at angular frequency $\omega$ when each section of a chain of similar four-terminal sections is constructed as in Fig. 8.29, and show how the result could be used to examine the pass-bands of a filter composed of such sections.

**Example 8.24 (Propagation constants derived from difference equation).** Starting from the equations relating the current $I_n$ and voltage $V_n$ at the $n$th transducer to $I_{n+1}$ and $V_{n+1}$ at the neighboring transducer, i.e. starting from the equations

$$\begin{pmatrix} V_{n+1} \\ I_{n+1} \end{pmatrix} = (M)\begin{pmatrix} V_n \\ I_n \end{pmatrix}, \quad M = \begin{pmatrix} a & b \\ c & d \end{pmatrix}, \quad ad - bc = 1, \tag{8.235}$$

and observing that

$$V_n = \frac{1}{c}(I_{n+1} - dI_n), \quad \text{and hence} \quad V_{n+1} = \frac{1}{c}(I_{n+2} - dI_{n+1}), \tag{8.236}$$

obtain the following equation:

$$I_{n+2} - (a+d)I_{n+1} + (ad - bc)I_n = 0. \tag{8.237}$$

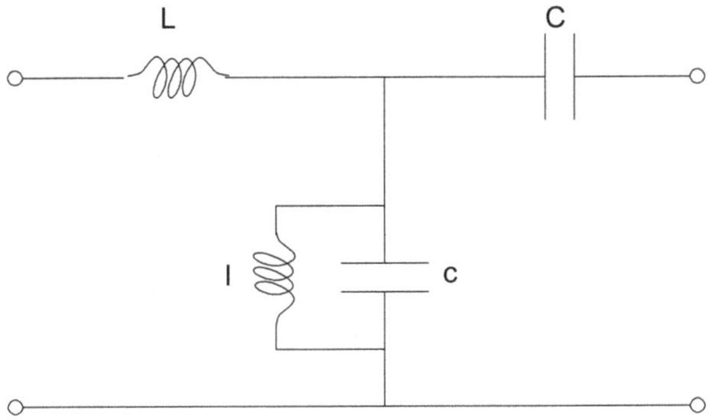


**Fig. 8.29.** Section of a filter chain.

Solve this equation as a difference equation[26] with a trial solution $I_n \propto \exp(-\Gamma n)$ and thus obtain the solution

$$I_n = Ae^{-\Gamma^+ n} + Be^{-\Gamma^- n}, \tag{8.238}$$

where

$$\lambda^{\pm} := e^{-\Gamma^{\pm}} \tag{8.239}$$

are the eigenvalues of the matrix $M$, i.e. solutions $\lambda$ of the *characteristic equation* $\det(M - \lambda 1\!\!1_{2\times 2}) = 0$. Finally obtain $V_n$ in terms of the propagation exponentials.

**Example 8.25 (Characteristic impedance of a chain of $\pi$-networks).** By applying Kirchhoff's laws to the symmetric $\pi$-network of Fig. 8.30 show that the characteristic impedance is[27]

$$Z_C = 2Z_2\sqrt{\frac{Z_1}{Z_1 + 4Z_2}} \quad \text{ohms.} \tag{8.240}$$

What is the propagation constant?

**Example 8.26 (A band-pass filter).** Determine the pass-band of the filter sketched in Fig. 8.31.

(Answer: $1/\sqrt{L_1C_1} \leq \omega \leq \sqrt{(1/L_1)\{(1/C_1) + (4/C_2)\}}$).

**Example 8.27 (Electric charge on a condenser).** A circuit consisting of a resistance $R$ and a capacitance $C$ has a steady e.m.f. impressed in the time interval $0 \leq t \leq T$. Obtain the current $I(t)$ and from this the electric charge $q(t)$ on the condenser.

**Example 8.28 (Transmission lines terminated by resonant circuits).** Determine the input impedance of a loss-less line one-eighth of a wavelength

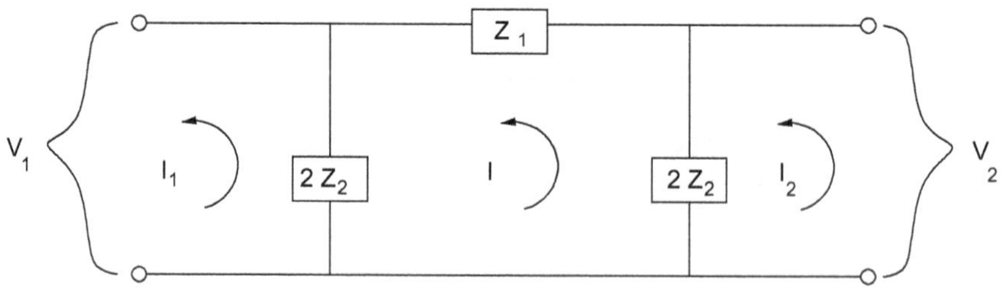


**Fig. 8.30.** The $\pi$-network.

[26] See e.g. Durell and Robson [41], p. 229.
[27] See also Schelkunoff [129], p. 111.

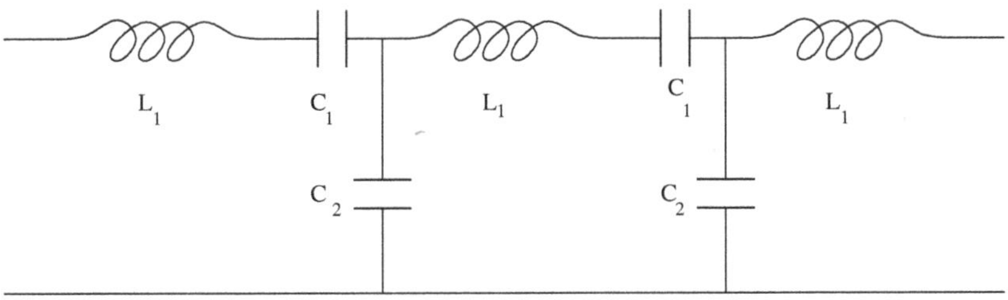


**Fig. 8.31.** A band-pass filter.

long (a) when terminated by a series resonant circuit consisting of a coil with inductance $L$ and a condenser with capacitance $C$, and (b) when terminated with the circuit in a parallel arrangement.

**Example 8.29 (Solving the circuit equation with Laplace transform[28]).** A constant voltage $V_0$ is applied from time $t = 0$ to a circuit comprising a resistance $R$, and inductance $L$ and a capacity $C$, all in series. Find the current $I$ at time $t > 0$ (i) if $C < 4L/R^2$, (ii) if $C = 4L/R^2$, and (iii) if $C > 4L/R^2$.
[Hint: Use the method of Laplace transforms, i.e.

$$A(t) = \frac{1}{2\pi i}\int_{p=c-i\infty}^{c+i\infty} e^{pt}S(p)dp, \quad S(p) = \int_{t'=0}^{\infty} e^{-pt'}A(t')dt', \tag{8.241}$$

where the constant $c$ must be greater than the real part of the value of $p$ at which the singularity occurs in $S(p)$. In the case of the following differential equation with constant coefficients $\alpha, \beta, \gamma$ (as in the equation $LI'' + RI' + I/C = V_0'$),

$$\alpha A'' + \beta A' + \gamma A = f(t), \tag{8.242}$$

multiply throughout by $\exp(-pt)$ and integrate with respect to $t$ from 0 to $\infty$. Then

$$(\alpha p^2 + \beta p + \gamma)S_A(p) = \alpha[A'(0) + pA(0) + \beta A(0)] + S_f(p). \tag{8.243}$$

The boundary conditions come in right from the start. Whence $S_A(p)$ and $A(t)$. Thus the problem here is to obtain $S_I(p)$ and hence $I(t)$. Alternatively one can obtain $I(t)$ from Tables of Laplace Transforms[29]].

[28]Dingle [39].
[29]Magnus and Oberhettinger [90], p. 125.

(Answers:

$$(i)\quad V_0\sqrt{\frac{C/L}{1-R^2C/4L}}e^{-tR/2L}\sin\left\{\frac{t}{\sqrt{LC}}\sqrt{1-\frac{R^2C}{4L}}\right\},$$

$$(ii)\quad \frac{V_0t}{L}e^{-Rt/2L},$$

$$(iii)\quad \frac{2V_0}{R\sqrt{1-4L/R^2C}}e^{-tR/2L}\sinh\left\{\frac{tR}{2L}\sqrt{1-\frac{4L}{R^2C}}\right\} \tag{8.244}$$

$S(p)$ is called Laplace transform and $A(t)$ its inversion).

**Example 8.30 (Circuit with alternating voltage[30]).** For the same circuit as in Example 8.29, find the charge on the condenser with capacity $C$ if an alternating voltage $V_0\exp(i\omega t)$ is applied from $t=0$ on. What happens to the charge when (a) $t=0$, and (b) $t\to\infty$?

(Answers:

$$\begin{aligned}(i)\quad &\frac{V_0C}{1-\omega^2LC+i\omega RC}\left[e^{i\omega t}-e^{-tR/2L}\left\{\cos\left(\frac{t}{\sqrt{LC}}\sqrt{1-\frac{R^2C}{4L}}\right)\right.\right.\\ &\left.\left.+\frac{1}{2}\sqrt{\frac{C}{L}}\frac{R+2i\omega L}{\sqrt{1-R^2C/4L}}\sin\left(\frac{t}{\sqrt{LC}}\sqrt{1-\frac{R^2C}{4L}}\right)\right\}\right],\\ (ii)\quad &\frac{4V_0L}{(R+2i\omega L)^2}\left[e^{i\omega t}-e^{tR/2L}\left\{1+\frac{t}{2L}(R+2i\omega L)\right\}\right],\\ (iii)\quad &\frac{V_0C}{1-\omega^2LC+i\omega RC}\left[e^{i\omega t}-e^{-tR/2L}\left\{\cosh\left(\frac{tR}{2L}\sqrt{1+\frac{4L}{R^2C}}\right)\right.\right.\\ &\left.\left.+\frac{1+2i\omega L/R}{\sqrt{1-4L/R^2C}}\sinh\left(\frac{tR}{2L}\sqrt{1-\frac{4L}{R^2C}}\right)\right\}\right],\end{aligned} \tag{8.245}$$

$$(a)\ \text{charge}\to 0,\quad (b)\ \text{charge}=\frac{1}{i\omega}\frac{V_0}{R+i\omega L+1/i\omega C}e^{i\omega t} \tag{8.246}$$

Thus the transient terms die away leaving a charge).

[30]Dingle [39].

**Example 8.31 (Equivalent representations of a reactive network).** Show by drawing appropriate simple finite reactive circuits that such a circuit may be represented either as a series combination of a series resonant circuit and a succession of parallel resonant circuits, or as a parallel combination of a parallel resonant circuit and a succession of series resonant circuits.[31]

**Example 8.32 (Force law on inductance for constant current).** Consider an inductance $L$ composed of separate self-inductances $L_1$ and $L_2$ and mutual inductance $M$ so connected that current $I$ flows in the same relative direction and thus develops an attractive force. Consider an infinitesimal move of the system with condition $I =$ constant. Show that the force $F$ on $L$ is given by $\mathbf{F} = I^2\boldsymbol{\nabla}L/2$.

**Example 8.33 (Magnetic field of coaxial coils).** Find the magnetic field **B** associated with the system of coaxial coils considered in Example 8.6.

**Example 8.34 (Mutual inductance of two circuits).** A long, straight lightning conductor in the flat wall of a building is $d$ meters away from a rectangular window having a metal frame of area $a\cdot$ meters times $b$ meters, $a < b$, the sides of lengths $a$ being parallel to the lightning conductor. What is the mutual inductance of the circuits?
[Hint: Look at Eq. 8.51].

(Answer: $(\mu_0 a/2\pi)\ln((b+d)/d)$ henries).

**Example 8.35 (The transducer as a wave filter).** Consider an electron of mass $m$ and energy $E$ moving through a periodic electric field. Consider this quantum mechanically with the use of the one-dimensional Schrödinger equation for the periodic *Kronig–Penney potential* of height $V_0$ over a distance $\delta$ as shown in Fig. 8.32.

The solution $\psi$ of the Schrödinger equation has to satisfy two boundary conditions at the junctions: Corresponding to the voltage $V$ in Eq. (8.154)

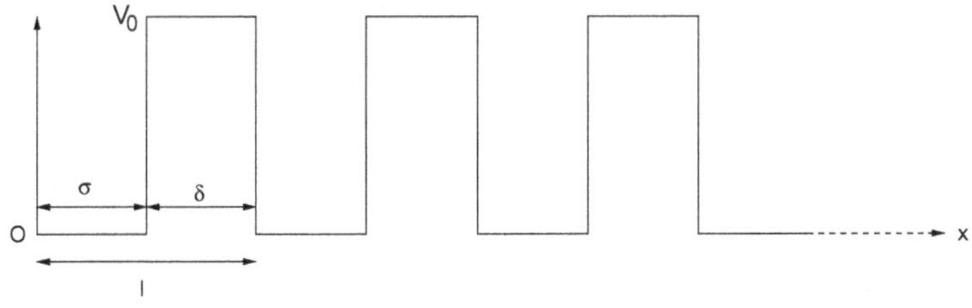


**Fig. 8.32.** The Kronig–Penney potential.

[31]See also Schelkunoff [129], pp. 124–125.

the wave function $\psi$ has to be continuous, and corresponding to the current $I$ in Eq. (8.154) the derivative $\psi' = d\psi/dx$ has to be continuous, i.e. in complete analogy we have at a junction

$$\begin{pmatrix}\psi\\ \psi'\end{pmatrix} = \begin{pmatrix}a & b\\ c & d\end{pmatrix}\begin{pmatrix}\psi_0\\ \psi'_0\end{pmatrix}. \tag{8.247}$$

Writing

$$\frac{d^2\psi}{dx^2} + \beta^2\psi = 0, \quad \frac{d^2\psi}{dx^2} - \gamma^2\psi = 0, \tag{8.248}$$

where

$$\beta^2 = \frac{2mE}{\hbar^2}, \quad \gamma^2 = \frac{2m}{\hbar^2}(V_0 - E), \tag{8.249}$$

and

$$\Gamma := i\cos^{-1}\left(\frac{a+d}{2}\right) \equiv i\beta l, \tag{8.250}$$

show that

$$\cos\kappa l = \cos\gamma\delta.\cos\beta\sigma + \frac{\gamma^2-\beta^2}{2\gamma\beta}\sinh\gamma\delta.\sin\beta\sigma, \tag{8.251}$$

where $l$ and $\sigma$ are the distances indicated in Fig. 8.32.
[Hint: Write out the solutions and their derivatives of the two Schrödinger equations].

**Example 8.36 ($\mathbf{L_{12}^2 \le L_1L_2}$).** Show that the square of the mutual inductance $L_{12}$ of two circuits cannot exceed the product of the self-inductances $L_1, L_2$ of the two circuits.

**Example 8.37 (Change of power supply).** An electrical appliance is stated to be used on a supply of 110 volts and to allow a current of 5 amperes. If the supply is 240 volts, what additional resistance has to be put in series with the appliance so that the current is 5 amperes?

(Answer: 26 ohms).

**Example 8.38 (Lamps in parallel).** How many 40 watt light bulbs can be used in parallel with a 10 ampere fuse without blowing this up if the voltage of the circuit is 240 volts?

(Answer: 60).

# Chapter 9

# Electromagnetic Waves

## 9.1 Introductory Remarks

In the following we consider electromagnetic waves in vacuum and in media. This means, we first obtain the wave equations from Maxwell's equations and then obtain the very important vacuum relation already referred to in Chapter 1 (with $\epsilon_0 = 8.854 \times 10^{-12}$ farads/m, $\mu_0 = 4\pi \times 10^{-7}$ henries/m or newtons/ampere$^2$),

$$c^2 = \frac{1}{\epsilon_0 \mu_0} \quad \mathrm{m}^2/\mathrm{s}^2, \tag{9.1}$$

which determines the velocity of the electromagnetic wave in the vacuum (or more generally with $\epsilon_0 \to \epsilon, \mu_0 \to \mu$ in a medium) as the reciprocal of the square root of the product of dielectric constant and magnetic permeability. One should note the difference to the case of Chapter 8, where the velocity was that of the wave in a transmission cable. We then investigate solutions in vacuum and in conducting media, and in the latter case we encounter the important *skin effect*. Various more specific aspects of electromagnetic waves are dealt with in Chapter 12 (on metals), in Chapter 13 (on charged particles in the stratosphere — the case of no lattice), and in Chapter 15.

## 9.2 Electromagnetic Waves in Vacuum

In this vacuum case we have $\mathbf{D} = \epsilon_0 \mathbf{E}, \mathbf{B} = \mu_0 \mathbf{H}, \mathbf{j} = 0, \rho = 0$, since we consider first the case of electromagnetic waves without charges and currents (simply as a matter of simplicity we leave out the source terms here, considered to be far away). With these provisions we write down Maxwell's equations:

$$\nabla \times \mathbf{B} = \mu_0\epsilon_0\frac{\partial \mathbf{E}}{\partial t}, \quad \nabla \times \mathbf{E} = -\frac{\partial \mathbf{B}}{\partial t}, \quad \nabla\cdot\mathbf{E} = 0, \quad \nabla\cdot\mathbf{B} = 0. \tag{9.2}$$

It is a general procedure to apply to the first two equations, which are "*curl*" or "rotational" equations, the curl operator once again in order to reduce them to Laplace form. This is achieved with the help of the formula

$$\nabla \times (\nabla \times \mathbf{B}) = \nabla(\nabla\cdot\mathbf{B}) - \triangle\mathbf{B}. \tag{9.3}$$

Applying this operation to the first equation and using the second, we obtain

$$\underbrace{\nabla \times (\nabla \times \mathbf{B})}_{\mu_0\epsilon_0\partial(\nabla\times\mathbf{E})/\partial t} = \nabla(\nabla\cdot\mathbf{B}) - \triangle\mathbf{B} = -\mu_0\epsilon_0\frac{\partial^2\mathbf{B}}{\partial t^2}, \tag{9.4}$$

i.e.

$$-\triangle\mathbf{B} = -\mu_0\epsilon_0\frac{\partial^2\mathbf{B}}{\partial t^2} \tag{9.5}$$

or in compact form

$$\Box\mathbf{B} = 0, \quad \Box = \triangle - \frac{1}{c^2}\frac{\partial^2}{\partial t^2}, \quad c^2 = \frac{1}{\mu_0\epsilon_0}. \tag{9.6}$$

In the wave equation the parameter $c$ obviously has the significance and dimension of the velocity of a wave of the field $\mathbf{B}$. In an analogous way we obtain from Eqs. (9.2) also $\Box\mathbf{E} = 0$. Thus the fields $\mathbf{E}, \mathbf{B}$ appear as waves propagating with the velocity $c$ in vacuum. This velocity is independent of the source and depends only on $\epsilon_0, \mu_0$, whose product connects electric and magnetic properties to the *velocity of light* (here in vacuum). We write[1]

$$\Box = \sum_{\mu=0}^{3}\partial^\mu\partial_\mu = \sum_{\mu,\nu=0,1,2,3}\partial^\mu g_{\mu\nu}\partial^\nu \tag{9.7}$$

with

$$\partial^\mu \equiv \frac{\partial}{\partial x_\mu}, \quad x_0 = ct, \quad g_{\mu\nu} = \begin{pmatrix} -1 & 0 & 0 & 0 \\ 0 & 1 & 0 & 0 \\ 0 & 0 & 1 & 0 \\ 0 & 0 & 0 & 1 \end{pmatrix}. \tag{9.8}$$

[1]The justification will be seen later in the context of the Special Theory of Relativity.

One can also write

$$\Box = \sum_{\mu=1}^{4} \partial^\mu \partial^\mu, \quad x_4 = ict. \tag{9.9}$$

The differential operator $\Box$ is called *D'Alembert operator*; this operator is the spacetime generalization of the Laplace operator $\triangle$.

Solutions $\Psi$ of equations $\Box\Psi = 0$ are of the form $e^{\pm i(\mathbf{k}\cdot\mathbf{r}\mp\omega t)}$ with $k^2 = \omega^2/c^2$, since

$$\begin{aligned}\Box e^{\pm i(\mathbf{k}\cdot\mathbf{r}\mp\omega t)} &= \left[-k^2 - \frac{1}{c^2}(-i\omega)^2\right] e^{\pm i(\mathbf{k}\cdot\mathbf{r}\mp\omega t)} \\ &= \left[-k^2 + \frac{\omega^2}{c^2}\right] e^{\pm i(\mathbf{k}\cdot\mathbf{r}\mp\omega t)}.\end{aligned} \tag{9.10}$$

A wave for which

$$f(\mathbf{r}, t) \equiv \mathbf{k}\cdot\mathbf{r} - \omega t = \text{const.} \tag{9.11}$$

is called a *plane wave*. In its case

$$\boldsymbol{\nabla} f(\mathbf{r}, t) = \mathbf{k}, \tag{9.12}$$

i.e. *the wave vector* $\mathbf{k}$ is orthogonal to the plane defined by $f(\mathbf{r}, t) = \text{const.}$ From $k^2 = \omega^2/c^2$ we deduce that $k = \omega/c = 2\pi/\lambda$, $\lambda$ the wavelength in meters. However, for an arbitrary curve $\mathbf{r} = \mathbf{r}(t)$ in the plane $f(\mathbf{r}, t) = \text{const.}$ one has $\mathbf{k}\cdot\dot{\mathbf{r}} = \omega$, i.e. $\mathbf{k}\cdot\mathbf{v} = \omega$, $v = \omega/k$ for $\dot{\mathbf{r}} \parallel \mathbf{k}$, as we can choose. The vector $\mathbf{k}$ points in the direction of propagation of the wave. For electromagnetic waves in vacuum $v = c$.

We now consider a *general solution* of the equation $\Box\mathbf{E} = 0$. We can write this

$$\mathbf{E} = \mathbf{E}_{01} e^{i(\mathbf{k}\cdot\mathbf{r}-\omega t)} + \mathbf{E}_{02} e^{-i(\mathbf{k}\cdot\mathbf{r}+\omega t)} \tag{9.13}$$

("one time, several space directions"). We can restrict ourselves first to

$$\mathbf{E} = \mathbf{E}_0 e^{i(\mathbf{k}\cdot\mathbf{r}-\omega t)} \quad \text{newtons/coulomb} \tag{9.14}$$

(or to $\mathbf{E} = \mathbf{E}_0 \cos(\mathbf{k}\cdot\mathbf{r} - \omega t)$, since only the real part has physical significance). We let the vector $\mathbf{E}_0$ be constant in space and time. Then

$$\boldsymbol{\nabla}\cdot\mathbf{E} = \mathbf{E}_0\cdot(i\mathbf{k}) e^{i(\mathbf{k}\cdot\mathbf{r}-\omega t)}, \quad \frac{\partial\mathbf{E}}{\partial t} = \mathbf{E}_0(-i\omega) e^{i(\mathbf{k}\cdot\mathbf{r}-\omega t)}, \tag{9.15}$$

i.e. comparing left and right sides of each equation we have the replacements

$$\boldsymbol{\nabla} \to i\mathbf{k}, \quad \frac{\partial}{\partial t} \to -i\omega. \tag{9.16}$$

We consider $\mathbf{B}$ in a similar way. Maxwell's equations (9.2) then become

$$\left.\begin{aligned} \mathbf{k}\times\mathbf{B} &= -(\omega/c^2)\mathbf{E}, \\ \mathbf{k}\times\mathbf{E} &= \omega\mathbf{B}, \end{aligned}\right\} \therefore \mathbf{E}\perp\mathbf{B},\mathbf{k} \quad \text{and} \quad \mathbf{B}\perp\mathbf{E},\mathbf{k},$$

$$\left.\begin{aligned} \mathbf{k}\cdot\mathbf{E} &= 0, \\ \mathbf{k}\cdot\mathbf{B} &= 0, \end{aligned}\right\} \therefore \mathbf{E},\mathbf{B}\perp\mathbf{k}. \tag{9.17}$$

It follows that the electromagnetic waves in vacuum are *transverse waves*, i.e. that the fields $\mathbf{E}$ and $\mathbf{B}$ are perpendicular to the propagation vector $\mathbf{k}$, and in addition the fields are orthogonal to each other, as illustrated in Fig. 9.1. We infer from the first two of the four relations (9.17) by scalar multiplication by $\mathbf{E}$ and $\mathbf{B}$ the important (but not universally valid) consequence that

$$\mathbf{k}\cdot(\mathbf{B}\times\mathbf{E}) = -\omega\mathbf{B}^2 = -\frac{\omega}{c^2}\mathbf{E}^2, \quad \therefore \quad \frac{1}{c^2}|\mathbf{E}|^2 = |\mathbf{B}|^2. \tag{9.18}$$

The vector $\mathbf{E}_0$ in the expression $\mathbf{E} = \mathbf{E}_0 e^{i(\mathbf{k}\cdot\mathbf{r}-\omega t)}$ defines the direction of oscillation. If $\mathbf{E}_0$ is constant in time one says, the wave is *linearly polarized.* The transport of energy per unit area (cf. the Poynting vector) takes place in the direction of the propagation vector $\mathbf{k} \propto \mathbf{E}\times\mathbf{B}$.

We let $\boldsymbol{\epsilon}_1$ and $\boldsymbol{\epsilon}_2$ be unit vectors, called *polarization vectors*, which span the plane orthogonal to $\mathbf{k}$. Then, in a general case,

$$\mathbf{E} = \mathbf{E}_1 + \mathbf{E}_2, \quad |\mathbf{E}| = |\mathbf{E}_0|, \quad \mathbf{B} = \mathbf{B}_1 + \mathbf{B}_2 \tag{9.19}$$

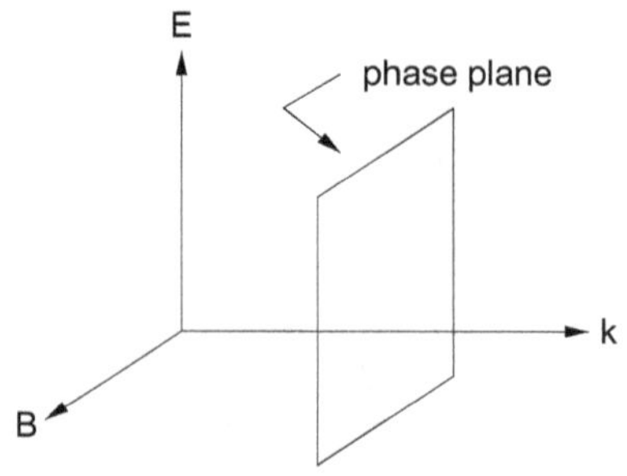


**Fig. 9.1.** The phase plane $\perp$ vector of propagation.

with $(i = 1, 2)$

$$\mathbf{E}_i(\mathbf{r}, t) = \epsilon_i E_i e^{i(\mathbf{k}\cdot\mathbf{r}-\omega t)}, \quad \mathbf{B}_i(\mathbf{r}, t) = \epsilon_i B_i e^{i(\mathbf{k}\cdot\mathbf{r}-\omega t)}. \tag{9.20}$$

Since $\mathbf{E} \perp \mathbf{B}$, it suffices and is convenient, to consider the vector $\mathbf{E}$ as a *representative of the entire wave.* $\mathbf{E}_1$, $\mathbf{E}_2$ have different phases, if we permit $E_1, E_2$ to be complex, i.e.

$$E_1 = |E_1|e^{i\theta_1}, \quad E_2 = |E_2|e^{i\theta_2}. \tag{9.21}$$

The wave sum is also a wave, and each wave has its own amplitude, phase and polarization, meaning direction of oscillation. When two light waves are added as electromagnetic vector fields, the amplitude of the sum depends on the amplitudes, phases and polarizations of the individual waves. In the case where two waves have the same polarization and equal amplitudes and are in phase, the resultant sum has twice the individual amplitude, whereas in their case with opposite phases the sum has zero amplitude due to cancellation. Since only the relative phase is of interest, we can choose $\theta_1 = 0$. If $\theta_2 = 0$, the wave is said to be *linearly polarized.* In this case we have

$$\mathbf{E} = (|E_1|\epsilon_1 + |E_2|\epsilon_2)e^{i(\mathbf{k}\cdot\mathbf{r}-\omega t)}, \tag{9.22a}$$

or

$$\Re\mathbf{E} = (|E_1|\epsilon_1 + |E_2|\epsilon_2)\cos(\mathbf{k}\cdot\mathbf{r} - \omega t). \tag{9.22b}$$

If $\theta_2 = \varphi \neq 0$, one says the wave is *elliptically polarized,*

$$\mathbf{E}(\mathbf{r}, t) = \epsilon_1 E_1 e^{i(\mathbf{k}\cdot\mathbf{r}-\omega t)} + \epsilon_2 E_2 e^{i(\mathbf{k}\cdot\mathbf{r}-\omega t+\varphi)}. \tag{9.23}$$

A *circularly polarized* wave is obtained, if $|E_1| = |E_2|$ *and* $\varphi = \pm\pi/2$, i.e.

$$\mathbf{E}(\mathbf{r}, t) = |E_1|(\epsilon_1 \pm i\epsilon_2)e^{i(\mathbf{k}\cdot\mathbf{r}-\omega t)} \tag{9.24}$$

(left and right polarized) or

$$\Re\mathbf{E}(\mathbf{r}, t) = |E_1|\epsilon_1\cos(\mathbf{k}\cdot\mathbf{r} - \omega t) \pm |E_1|\epsilon_2\sin(\mathbf{k}\cdot\mathbf{r} - \omega t). \tag{9.25}$$

For example, when $\mathbf{k}$ is parallel to the $z$-axis, $\epsilon_1$ parallel to $\mathbf{x}$, and $\epsilon_2$ parallel to $\mathbf{y}$, we have:

$$\Re E_x(\mathbf{r}, t) = |E_1|\cos(kz - \omega t), \quad \Re E_y(\mathbf{r}, t) = \pm|E_1|\sin(\omega t - kz). \tag{9.26}$$

In the case of the elliptically polarized wave $|E_1| \neq |E_2|$. These cases are illustrated in Fig. 9.2.

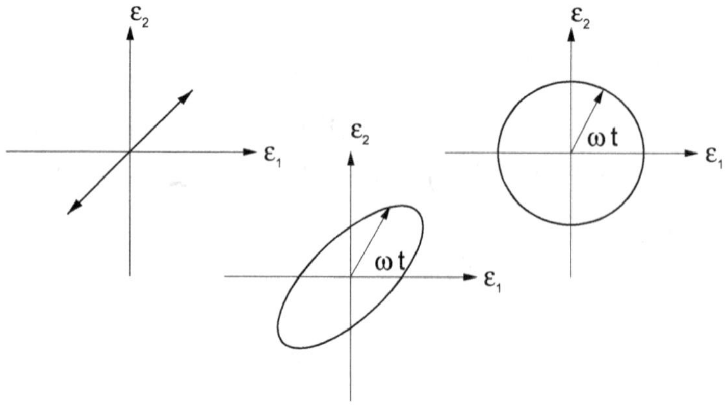


**Fig. 9.2.** Linear, circular and elliptic polarization.

## 9.3 Electromagnetic Waves in Media

We now consider the case of a *homogeneous, infinitely extended, uncharged but conducting medium* with medium parameters $\epsilon, \mu$ and a current density $\mathbf{j}$ as source. In this case we have

$$\mathbf{D} = \epsilon \mathbf{E}, \quad \mathbf{B} = \mu \mathbf{H}, \quad \mathbf{j} = \sigma \mathbf{E}, \quad \rho = 0. \tag{9.27}$$

The Maxwell equations (7.40)–(7.43) in this case are

$$\nabla \cdot \mathbf{E} = 0, \quad \nabla \cdot \mathbf{H} = 0, \quad \nabla \times \mathbf{E} = -\mu \frac{\partial \mathbf{H}}{\partial t}, \quad \nabla \times \mathbf{H} = \epsilon \frac{\partial \mathbf{E}}{\partial t} + \sigma \mathbf{E}. \tag{9.28}$$

It follows that:

$$\underbrace{\nabla \times (\nabla \times \mathbf{E})}_{\nabla \underbrace{(\nabla \cdot \mathbf{E})}_{0} - \nabla^2 \mathbf{E}} = -\mu \frac{\partial}{\partial t} \underbrace{\nabla \times \mathbf{H}}_{\epsilon(\partial \mathbf{E}/\partial t) + \sigma \mathbf{E}} . \tag{9.29}$$

We thus obtain the following equation for $\mathbf{E}$ and analogously the corresponding equation for $\mathbf{H}$, i.e.

$$\triangle \mathbf{E} = \epsilon \mu \frac{\partial^2 \mathbf{E}}{\partial t^2} + \mu \sigma \frac{\partial \mathbf{E}}{\partial t}, \quad \triangle \mathbf{H} = \epsilon \mu \frac{\partial^2 \mathbf{H}}{\partial t^2} + \mu \sigma \frac{\partial \mathbf{H}}{\partial t}. \tag{9.30}$$

These equations are also known as *telegraph equations.* In the case of *dielectrics*, also called *insulators* (with vanishing conductivity, $\sigma = 0$) they become:

$$\left(\triangle - \frac{1}{c'^2}\frac{\partial^2}{\partial t^2}\right)\mathbf{E} = 0, \quad \left(\triangle - \frac{1}{c'^2}\frac{\partial^2}{\partial t^2}\right)\mathbf{H} = 0, \tag{9.31}$$

where

$$c' = \frac{1}{\sqrt{\epsilon\mu}} = \sqrt{\frac{\epsilon_0\mu_0}{\epsilon\mu}}c \quad \text{meters/second.} \tag{9.32}$$

We can see that $c'$ is the *phase velocity*, i.e. $\omega/\sqrt{\mathbf{k}^2}$, by setting $\mathbf{E}, \mathbf{H} \propto e^{i(\mathbf{k}\cdot\mathbf{r}-\omega t)}$, so that $(\triangle + \mathbf{k}^2)\mathbf{E}, \mathbf{H} = 0$ with $\sqrt{\mathbf{k}^2} = \omega/c' = \omega\sqrt{\epsilon\mu}$, and the phase velocity is $\omega/k = 1/\sqrt{\epsilon\mu}$.

The *refractive index* $n$ known from optics, is defined by the ratio

$$n = \frac{c}{c'} = \frac{\sqrt{\epsilon\mu}}{\sqrt{\epsilon_0\mu_0}} = c\sqrt{\epsilon\mu}. \tag{9.33}$$

Applications of this relation will be considered in later chapters.

## 9.4 Frequency Dependence of $\epsilon$ and $\sigma$

At high frequencies of electromagnetic radiation the parameters $\epsilon, \mu$ and $\sigma$ become frequency dependent.[2] When we introduced these parameters we were concerned with static fields. We therefore face the problem of determining their frequency dependence. We consider here only the two more important cases, i.e. those of $\epsilon$ and $\sigma$, and take in most cases simply $\mu \sim$ constant. Again applications will be considered in more detail in later chapters.

### 9.4.1 *The generalized dielectric constant*

We set

$$\mathbf{E} = \mathbf{E}_0 e^{i(\mathbf{k}\cdot\mathbf{r}-\omega t)} \quad \text{newtons/coulomb}, \quad \mathbf{H} = \mathbf{H}_0 \mathrm{e}^{\mathrm{i}(\mathbf{k}\cdot\mathbf{r}-\omega t)} \quad \text{amperes/meter.} \tag{9.34}$$

Then from Maxwell's equations (9.28):

$$\boldsymbol{\nabla} \times \mathbf{H} = \epsilon\frac{\partial \mathbf{E}}{\partial t} + \sigma\mathbf{E} = -i\epsilon\omega\mathbf{E} + \sigma\mathbf{E} \equiv -i\omega\eta\mathbf{E}, \tag{9.35}$$

---

[2]Variation of $\sigma$ with temperature is treated in detail in Bleaney and Bleaney [12], pp. 378–382.

where the quantity

$$\eta = \epsilon + i\frac{\sigma}{\omega} \quad \text{farads/meter} \tag{9.36}$$

is called the *generalized (complex) dielectric constant.* Substituting now the Eqs. (9.34) and (9.35) into the four Maxwell equations, we obtain

$$\begin{aligned}
\nabla \times \mathbf{H} = \epsilon\frac{\partial \mathbf{E}}{\partial t} + \sigma\mathbf{E} &\to \mathbf{k} \times \mathbf{H} = -\omega\eta\mathbf{E}, \\
\nabla \times \mathbf{E} = -\mu\frac{\partial \mathbf{H}}{\partial t} &\to \mathbf{k} \times \mathbf{E} = \omega\mu\mathbf{H}, \\
\nabla \cdot \mathbf{H} = 0 &\to \mathbf{k} \cdot \mathbf{H} = 0, \\
\nabla \cdot \mathbf{E} = 0 &\to \mathbf{k} \cdot \mathbf{E} = 0.
\end{aligned} \tag{9.37}$$

Thus again we have $\mathbf{H}, \mathbf{E} \perp \mathbf{k}$, but *no longer* $|\mathbf{E}| = |\mathbf{H}|$. Instead:

$$\underbrace{\mathbf{k} \times (\mathbf{k} \times \mathbf{H})}_{\underbrace{(\mathbf{k}\cdot\mathbf{H})}_{0}\mathbf{k} - \mathbf{k}^2\mathbf{H}} = -\omega\eta\underbrace{\mathbf{k} \times \mathbf{E}}_{\omega\mu\mathbf{H}}. \tag{9.38}$$

It follows that

$$k^2\mathbf{H} = \omega^2\eta\mu\mathbf{H} = \mu\epsilon\omega^2\left(1 + i\frac{\sigma}{\epsilon\omega}\right)\mathbf{H} \quad \text{amperes/meter}^3. \tag{9.39}$$

Provided $\mathbf{H} \neq 0$ everywhere (hence the reference to infinitely extended media at the beginning) we have therefore

$$k^2 = \mu\epsilon\omega^2\left(1 + i\frac{\sigma}{\epsilon\omega}\right) \quad \text{meters}^{-2}. \tag{9.40}$$

In view of its frequency dependence this relation is called the *dispersion relation of the conducting medium.* We can solve the relation for $k = \alpha + i\beta$, where for conductivity $\sigma = 0$:

$$k = \omega\sqrt{\epsilon\mu} = \frac{\omega}{c}n \quad \text{meters}^{-1} \quad (\text{here } \sigma = 0), \tag{9.41}$$

where $n$ is the *refractive index* introduced by Eq. (9.33). As a *generalized refractive index* $p$ one defines (instead of $n = \sqrt{\epsilon\mu}$) the quantity

$$p = \sqrt{\eta\mu}c = p_1 + ip_2. \tag{9.42}$$

Setting $k = \alpha + i\beta$, a wave moving in the direction of $x$ contains the factor

$$e^{i(kx - \omega t)} = e^{i(\alpha x - \omega t)}e^{-\beta x}. \tag{9.43}$$

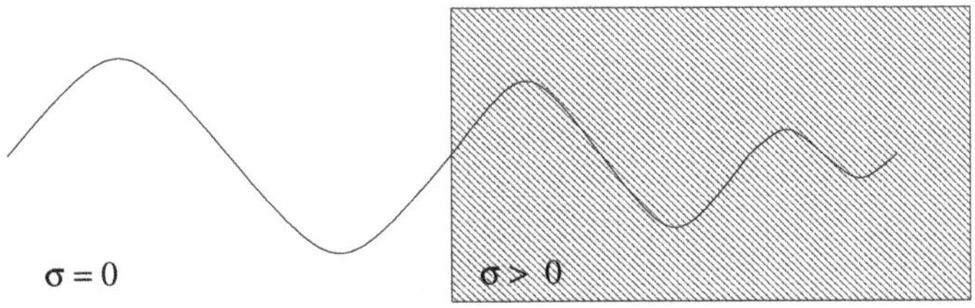


**Fig. 9.3.** Damping of a wave in a conducting medium.

Here $\beta, (\beta > 0)$, is of the order of $\sigma$ for small $\sigma$. We see therefore that the wave in the conductor is damped, as indicated in Fig. 9.3. The distance $|1/\beta|$ is termed the *penetration depth* or *skin depth.*

In the next subsection we determine the frequency dependence of the conductivity $\sigma$. The result will be inserted into Eq. (9.36) for the generalized dielectric constant.

### 9.4.2 *Frequency dependence of σ*

The conductivity of metals has its explanation in the existence of free electrons. We assume we have $n$ such electrons in a volume $V$ and $N = n/V$ per unit volume. Then Newton's equation of motion for one such electron, say electron "$k$" with velocity $\mathbf{v}_k$, is, multiplied by the charge $e$,

$$m\frac{d}{dt}(e\mathbf{v}_k) = e^2\mathbf{E}_k - \xi e\mathbf{v}_k, \tag{9.44}$$

where $\mathbf{E}_k$ is the field of the other electrons and $\xi$ is the frictional resistance of the lattice atoms. Summing over all such forces we have

$$\frac{1}{V}\sum_k m\frac{d}{dt}(e\mathbf{v}_k) = \frac{1}{V}\sum_k e^2\mathbf{E}_k - \frac{1}{V}\xi\sum_k e\mathbf{v}_k. \tag{9.45}$$

But the current (density) (current $I$ per unit area) is

$$\mathbf{j}(\mathbf{r}, t) = \sum_k e\dot{\mathbf{r}}_k(t)\delta(\mathbf{r} - \mathbf{r}_k(t)), \tag{9.46}$$

so that the volume integral is

$$\int_V \mathbf{j}(\mathbf{r}, t)d\mathbf{r} = \sum_k e\dot{\mathbf{r}}_k(t). \tag{9.47}$$

We now define a *mean current density* $\langle \mathbf{j}(t)\rangle$ by the expression (note the averaging over space, and the remaining time dependence of the mean value!)

$$\langle \mathbf{j}(t)\rangle V := \int_V \mathbf{j}(\mathbf{r},t)dV \equiv \int_V \mathbf{j}(\mathbf{r},t)d\mathbf{r}. \tag{9.48}$$

We also define a *mean field strength* $\mathbf{E}$ by the relations

$$\mathbf{E} = \frac{1}{n}\sum_k \mathbf{E}_k, \quad N\mathbf{E} := \frac{1}{V}\sum_k \mathbf{E}_k, \tag{9.49}$$

and obtain with Eq. (9.44),

$$m\frac{d}{dt}\langle \mathbf{j}(t)\rangle = \frac{m}{V}\frac{d}{dt}\sum_k e\mathbf{v}_k = \frac{1}{V}\sum_k \{e^2\mathbf{E}_k - \xi e\mathbf{v}_k\} = N\mathbf{E}e^2 - \xi\langle \mathbf{j}(t)\rangle. \tag{9.50}$$

In the following we write for convenience

$$\langle \mathbf{j}(t)\rangle \to \mathbf{j}(t). \tag{9.51}$$

Then Eq. (9.50) is

$$m\frac{d\mathbf{j}}{dt} = N\mathbf{E}e^2 - \xi\mathbf{j}. \tag{9.52}$$

For $\mathbf{j} = \text{const.}$, a stationary current (density), we set $\mathbf{j} = \sigma_0\mathbf{E}$ and obtain:

$$\mathbf{j} = \sigma_0\mathbf{E}, \quad \text{with} \quad \sigma_0 = \frac{Ne^2}{\xi}. \tag{9.53}$$

The quantity $\sigma_0$ is called *direct current conductivity.* It follows that $\xi = Ne^2/\sigma_0$. Setting

$$\frac{1}{\tau} := \frac{\xi}{m}, \tag{9.54}$$

we see that $\tau$ has the dimension of time and is interpreted as a mean time interval between successive collisions of the electron with the lattice, called the *relaxation time.* We thus obtain from Eq. (9.52) the equation

$$\frac{d\mathbf{j}}{dt} + \frac{1}{\tau}\mathbf{j} = \frac{Ne^2}{m}\mathbf{E}. \tag{9.55}$$

Assuming $\mathbf{E} \propto e^{-i\omega t}$, $\mathbf{j} \propto e^{-i\omega t}$, we have

$$-i\omega\mathbf{j} + \frac{1}{\tau}\mathbf{j} = \frac{Ne^2}{m}\mathbf{E}, \tag{9.56}$$

i.e.

$$\mathbf{j} = \sigma(\omega)\mathbf{E} \quad \text{amperes/meter}^2 \tag{9.57}$$

with

$$\sigma(\omega) = \frac{Ne^2/m}{-i\omega + 1/\tau} = \frac{\sigma(0)}{1 - i\omega\tau} \quad (\text{ohm} \cdot \text{meter})^{-1}, \tag{9.58}$$

where $\sigma(0) = \sigma_0$. For small frequencies $\omega$, we have $\sigma(0) =$ constant. The conductivity $\sigma(\omega)$ is real also for $\omega\tau \ll 1$, and almost purely imaginary for $\omega\tau \gg 1$ (this implies a phase difference of $\pi/2$ between $\mathbf{j}$ and $d\mathbf{j}/dt$, or between $\mathbf{j}$ and $\mathbf{E}$).

The linear relation (9.57) is known as *Ohm's law*. In situations in which $\mathbf{j}$ and $\mathbf{E}$ are not necessarily parallel, the ratios of their components in different directions depend on the detailed solution of the basic equations. Circumstances in which Ohm's law is or is not obeyed are discussed in specialized literature — e.g. in connection with the physics of a plasma as in Spitzer [143], pp. 31–41. We can now insert the expression (9.58) for $\sigma(\omega)$ into Eq. (9.36) in order to obtain the complete frequency dependence of the generalized dielectric constant, i.e.

$$\eta = \epsilon + i\frac{\sigma(\omega)}{\omega} \quad \text{farads/meter}. \tag{9.59}$$

Fields which are called "high frequency fields" in practice are those for which

$$\omega\tau \ll 1, \quad \omega \ll 1/\tau,$$

where $1/\tau$ is large in the case of strong damping. Thus the term "high frequency" used in practice does not correspond to the limiting case of $\omega \to \infty$. In the case $\omega\tau \ll 1$ we have $\sigma$ real (see Eq. (9.58)), so that damping occurs. This is the reason why these so-called "high frequency fields" hardly penetrate into a conductor. In general, however, $\epsilon$ and $\sigma$ are frequency dependent.

We consider now the ratio of *displacement current* $j_d$ to *conduction current* $j$, i.e. that of their densities

$$\frac{j_d}{j} = \frac{\partial D/\partial t}{\sigma E} = \frac{\epsilon \partial E/\partial t}{\sigma E}. \tag{9.60}$$

With $E = E_0 e^{-i\omega t}$ and thus $|j_d/j| = \epsilon\omega/\sigma$ and $\sigma \sim 5.8 \times 10^7\ \Omega^{-1}$ per meter (or older: mhos per meter) as for copper, and (say) frequency $\nu = \omega/2\pi = 5 \times 10^{11}$ hertz, one obtains with $\epsilon = \epsilon_0$ the value $|j_d/j| \simeq 4.8 \times 10^{-7}$. This is such a small value that for all frequencies in ordinary circuits $|j_d/j| < 1\%$.

We look at this relationship more closely. We define the *plasma frequency* $\omega_P$ by

$$\omega_P^2 = \frac{Ne^2}{\epsilon m}. \tag{9.61}$$

Then from Eqs. (9.36) and (9.58):

$$\begin{aligned}\eta(\omega) &= \epsilon\left[1+\frac{i\sigma}{\epsilon\omega}\right] = \epsilon\left[1+\frac{i}{\epsilon\omega}\frac{Ne^2/m}{-i\omega+1/\tau}\right]\\ &= \epsilon\left[1+\frac{i\omega_P^2}{\omega(-i\omega+1/\tau)}\right] = \epsilon\left[1-\frac{\omega_P^2}{\omega^2(1-1/i\omega\tau)}\right].\end{aligned} \tag{9.62}$$

*We consider the case* $\omega\tau \gg 1$.

In this case

$$\eta(\omega) = \epsilon\left[1-\frac{\omega_P^2}{\omega^2}\right]. \tag{9.63}$$

We distinguish between two subsidiary cases:

(a) $\omega \ll \omega_P$:

$$\eta(\omega) \simeq -\epsilon\frac{\omega_P^2}{\omega^2}. \tag{9.64}$$

The index of refraction

$$p = c\sqrt{\eta\mu} \simeq ic\sqrt{\epsilon\mu}\frac{\omega_P}{\omega} \tag{9.65}$$

is in this case pure imaginary. In order to obtain the penetration depth we have to calculate

$$k^2 = \omega^2\eta\mu \simeq \omega^2\left(-\epsilon\frac{\omega_P^2}{\omega^2}\right)\mu = -\epsilon\mu\omega_P^2, \tag{9.66}$$

so that

$$e^{ikz} = e^{i\sqrt{-\epsilon\mu\omega_P^2}z} = e^{\pm\sqrt{\epsilon\mu}\omega_P z} \equiv e^{\pm z/z_0}, \tag{9.67}$$

i.e. the penetration depth

$$z_0 = \frac{1}{\sqrt{\epsilon\mu}\omega_P} = \sqrt{\frac{m}{Ne^2\mu}} \quad \text{meters} \tag{9.68}$$

is *independent of the frequency*.

The conductivity is given by Eq. (9.58). In the case under consideration ($\omega\tau \gg 1$), we have

$$\sigma(\omega) = i\frac{Ne^2}{m\omega}. \tag{9.69}$$

The factor "$i$" implies, that $\mathbf{j}$ and $\mathbf{E}$ of Ohm's law $\mathbf{j} = \sigma(\omega)\mathbf{E}$ have a phase difference of $\pi/2$. Since $\tau$ does not appear here, we would have obtained this case also without the friction term in the equation for $\mathbf{j}$, i.e. from (cf. Eq. (9.52))

$$\frac{d\mathbf{j}}{dt} = \frac{Ne^2}{m}\mathbf{E}. \tag{9.70}$$

Thus this case is normally not applicable to rigid bodies or metals since these possess an atomic lattice which gives rise to the friction term (but see the more detailed treatment in Chapter 12). The case finds application, however, for instance in the case of ions (charge $e$) in the stratosphere (which is the subject of Chapter 13).

(b) $\omega \gg \omega_P$:

In this case we have from Eq. (9.63):

$$\eta(\omega) = \epsilon\left(1 - \frac{\omega_P^2}{\omega^2}\right), \tag{9.71}$$

and the refractive index

$$p = c\sqrt{\eta\mu} \simeq c\sqrt{\epsilon\mu}\sqrt{1 - \frac{\omega_P^2}{\omega^2}} \tag{9.72}$$

is real. Hence

$$k^2 = \omega^2\eta\mu = \omega^2\epsilon\mu\left(1 - \frac{\omega_P^2}{\omega^2}\right) \tag{9.73}$$

is positive, so that

$$k = \omega\sqrt{\epsilon\mu}\sqrt{1 - \frac{\omega_P^2}{\omega^2}} \quad \text{meter}^{-1} \tag{9.74}$$

is real. With $\sqrt{\epsilon\mu} = 1/c'^2$ this relation can be written

$$k^2c'^2 = \omega^2 - {\omega_P}^2. \tag{9.75}$$

This means, there is no damping. On the other hand, again

$$\sigma(\omega) = i\frac{Ne^2}{m\omega} = i\frac{\epsilon\omega_P^2}{\omega} \propto \frac{1}{\omega}, \tag{9.76}$$

so that

$$\mathbf{j} = \sigma(\omega)\mathbf{E} \sim 0 \quad \text{for} \quad \omega \gg \omega_P. \tag{9.77}$$

Then there is practically no conduction current. Since the resistance of the lattice is practically zero (no damping) ($1/\tau \ll \omega$, hence $1/\tau$ negligible), the explanation has to be found in the conduction electrons — one can say, at high frequencies $\omega$ they acquire "*inertia*", so that $\mathbf{j} \to 0$.[3] Thus fields called "high frequency fields" in practice are not those of our case (b) with $\omega \gg \omega_P$ and $\omega\tau \gg 1$, but those for which $\omega\tau \ll 1$. In the latter case $\sigma \sim$ is real as observed after Eq. (9.58), and damping occurs. We see therefore: For $\omega\tau \ll 1$ there is little dissipation of energy in the metal (the wave is almost completely reflected), and only for $\omega > \omega_P$ one has transmission and little reflection.

## 9.5 The (Normal) Skin Effect

Let electromagnetic radiation with frequency $\nu = \omega/2\pi$ fall in the direction of $x$ on an uncharged metal of conductivity $\sigma$ and magnetic permeability $\mu$. We assume the time dependence of the fields to be given by $\mathbf{E}, \mathbf{H} \propto e^{-i\omega t}$. The relevant Maxwell equations together with Ohm's law $\mathbf{j} = \sigma\mathbf{E}$ are then (neglecting the displacement current),

$$\begin{aligned} &(a) \quad \nabla \times \mathbf{H} = \mathbf{j}, \quad (b) \quad \nabla \times \mathbf{E} = -\mu\frac{\partial \mathbf{H}}{\partial t}, \\ &(c) \quad \nabla \cdot \mathbf{H} = 0, \quad (d) \quad \nabla \cdot \mathbf{E} = 0. \end{aligned} \tag{9.78}$$

Taking the curl of (b) of Eqs. (9.78) we obtain

$$\nabla \times (\nabla \times \mathbf{E}) = \nabla(\nabla \cdot \mathbf{E}) - \triangle\mathbf{E} = -\mu\frac{\partial}{\partial t}\underbrace{(\nabla \times \mathbf{H})}_{\mathbf{j}}, \tag{9.79}$$

i.e. with $\mathbf{E} = \mathbf{j}/\sigma$,

$$\triangle\mathbf{j} = \mu\sigma\frac{\partial \mathbf{j}}{\partial t}. \tag{9.80}$$

Since $\mathbf{j} \propto e^{-i\omega t}$ like $\mathbf{E}$, it follows that

$$\triangle\mathbf{j} = \gamma^2\mathbf{j}, \quad \gamma^2 = -i\omega\mu\sigma \tag{9.81}$$

[3]See the discussion in Greiner [61], p. 300: For $\omega \to \infty$ metals become "transparent", i.e. the metal behaves like a dielectric permitting the transmission of radiation.

with $\sqrt{-i} = (1-i)/\sqrt{2}$:

$$\gamma = (1-i)\sqrt{\frac{\mu\sigma\omega}{2}} \equiv (1-i)\lambda, \quad \lambda = \sqrt{\frac{\mu\sigma\omega}{2}} > 0. \tag{9.82}$$

For the case of one spatial dimension,

$$\mathbf{j}(x,t) = e^{-i\omega t}\mathbf{j}(x) \tag{9.83}$$

we therefore have from Eq. (9.81) (with $j_0 = j(0)$):

$$\frac{d^2}{dx^2}j(x) = \gamma^2 j(x), \quad j(x) = j_0 e^{-\gamma x} = j_0 e^{-\lambda x}e^{i\lambda x}. \tag{9.84}$$

For obvious physical reasons we have to choose the damped solution, i.e. that with the attenuation factor. The expression $1/\lambda$ is called *penetration* or *skin depth.*[4] We observe that at the skin depth both the magnitude and the phase of the oscillating electric field change exponentially. This implies that the velocity of an electron in this field results from an integration, and hence the current density is no longer directly related to the electric field, and thus the simple Ohm's law is not valid. The integrated effect of the electric field over a volume introduces a nonlocality which when taken into account leads to the "*anomalous skin effect*".[5]

## 9.6 Exercises

**Example 9.1 (The skin effect in a finite plate).** In the case of a conducting film of uniform thickness $d$, conductivity $\sigma$, and magnetic permeability $\mu$, the solution of the equation

$$\frac{d^2 j}{dx^2} = \gamma^2 j, \quad \gamma = \sqrt{\frac{\sigma\mu\omega}{2}}(1-i) \quad \text{meters}^{-1}, \tag{9.85}$$

for the current density $j$ as a function of $x$ (the distance from the central plane of the film) and of an alternating current of frequency $\nu = \omega/2\pi$ flowing along it, is given by

$$j = j_0\frac{\cosh(\gamma x)}{\cosh(\gamma d/2)} \quad \text{amperes/meter}^2, \tag{9.86}$$

---

[4]For detailed discussion with illustrations see also Ferrari [54], pp. 143–149.

[5]See Portis [121], p. 313. The relevant original investigations are those of Reuter and Sondheimer [123], Chambers [28] and Dingle [38].

so that $j$ is the same for negative $x$, as for positive $x$, $j_0$ being the current density at the surfaces $x = \pm d/2$. Show that the total current $I$ per unit breadth of the film is numerically

$$I = |j_0|d\sqrt{\frac{\tanh^2\xi + \tan^2\xi}{2\xi^2(1+\tanh^2\xi\tan^2\xi)}} \quad \text{amperes/meter}, \tag{9.87}$$

where $\xi = d\sqrt{\sigma\mu\omega/8}$. Discuss the limit $\xi \to 0$.

**Example 9.2 (Total current).** By integrating $j(x,t)$ of Eqs. (9.83) and (9.84) from $x = 0$ to infinity calculate the total current $I$ for unit breadth.

(Answer: $I = j_0 \exp[-i(\omega t + \pi/4)]/\sqrt{\sigma\mu\omega}$ amperes).

**Example 9.3 (Skin effect and heat produced in a medium).** A harmonically varying electric field in the $x$-direction is maintained at the plane surface $z = 0$ of an infinitely deep medium with electrical conductivity $\sigma$ and magnetic permeability $\mu$. Neglecting the displacement current, calculate the electric field $E(z)$ as a function of the depth $z$ from the surface in terms of $\sigma, \mu$, the angular frequency $\omega$, and $E(0)$. By finding the average over time of the quantity $\sigma E^2(z)$ and then integrating over $z$, determine the total heat produced per unit time in the medium. Correlate your result with the value obtained for the Poynting vector at the surface.

**Example 9.4 (Skin effect in a cylindrical conductor).** Show that in the case of a cylindrical conductor of radius $a$ with its axis along the $z$-axis of a Cartesian system of coordinates the current density is given by

$$j(\rho) = j(a)\frac{J_0(\alpha\rho)}{J_0(\alpha a)} \quad \text{amperes/meter}^2, \tag{9.88}$$

where $\rho$ is the radial coordinate and $J_0$ a Bessel function. What is the constant $\alpha$?

[Hint: Express Eq. (9.81), $\triangle\mathbf{j} = \gamma^2\mathbf{j}$, in cylindrical coordinates].

**Example 9.5 (Skin effect in a cylindrical wire).** Consider a wire of circular cross section of radius $a$. Since electromagnetic fields penetrate only into a film-like part of the surface ($z = 0$) of small thickness $\delta$, the actual shape of the cross section is immaterial. Show that an infinite half-plane coordinate system may be constructed on the wire so that planar results can be applied. Show that the total current obtained by integrating the current density from depth $z = 0$ to infinity is given by

$$I = \frac{2\pi a\delta}{1-i}j_0 \quad \text{amperes}. \tag{9.89}$$

What is the voltage drop along the cylinder (obtained by integrating $E$ along the wire)? Hence show that the cylindrical wire of length $l$ has a complex impedance $Z$ given by

$$Z = \frac{(1-i)l}{2\pi a\delta\sigma} \quad \text{ohms.} \tag{9.90}$$

# Chapter 10

# Moving Charges in Vacuum

## 10.1 Introductory Remarks

In the following we study moving charges in a vacuum, i.e. we consider the Maxwell equations with nonvanishing charge density (hence also with nonvanishing current density). We investigate the solutions of the wave equations and thus obtain potentials and fields of a moving charge. Finally we consider the radiation of an oscillating dipole (consisting of two charges). In particular we shall see that accelerations of charges produce the dominant field contributions at large distances (called "*radiation fields*") and that these are mutually orthogonal. These considerations are of eminent importance for an understanding of electrodynamical phenomena, also because the dipole is often referred to as a kind of idealized classical model of an atom and thus offers the explanation why this classical picture of an atom is wrong — the dipole radiates off energy and exhausts itself therewith.

## 10.2 Maxwell's Equations for Moving Charges

Here, for motion in the vacuum, $\mathbf{D} = \epsilon_0\mathbf{E}, \mathbf{B} = \mu_0\mathbf{H}$, so that the four Maxwell equations (7.40)–(7.43) become

$$\boldsymbol{\nabla}\cdot\mathbf{E} = \frac{1}{\epsilon_0}\rho(\mathbf{r},t), \quad \boldsymbol{\nabla}\cdot\mathbf{H} = 0 \tag{10.1}$$

and

$$\boldsymbol{\nabla}\times\mathbf{E} = -\mu_0\frac{\partial\mathbf{H}}{\partial t} = -\frac{\partial}{\partial t}(\mu_0\mathbf{H}) \tag{10.2}$$

and

$$\boldsymbol{\nabla} \times \mathbf{H} = \mathbf{j}(\mathbf{r}, t) + \epsilon_0 \frac{\partial \mathbf{E}}{\partial t}. \tag{10.3}$$

Expressing the magnetic flux density $\mathbf{B}$ in terms of the vector potential $\mathbf{A}$, we have

$$\mu_0 \mathbf{H} = \boldsymbol{\nabla} \times \mathbf{A}, \tag{10.4}$$

and from Eq. (10.2) we have

$$\boldsymbol{\nabla} \times \mathbf{E} = -\boldsymbol{\nabla} \times \frac{\partial \mathbf{A}}{\partial t}, \tag{10.5}$$

i.e.

$$\mathbf{E} = -\frac{\partial \mathbf{A}}{\partial t} - \boldsymbol{\nabla}\phi, \tag{10.6}$$

as already familiar. With $\boldsymbol{\nabla} \cdot \mathbf{E} = \rho/\epsilon_0$ we obtain

$$\frac{1}{\epsilon_0}\rho = -\frac{\partial}{\partial t}\boldsymbol{\nabla} \cdot \mathbf{A} - \boldsymbol{\nabla} \cdot \nabla\phi, \tag{10.7}$$

and with Eq. (10.3) we obtain

$$\boldsymbol{\nabla} \times \mathbf{H} = \frac{1}{\mu_0} \underbrace{\boldsymbol{\nabla} \times (\boldsymbol{\nabla} \times \mathbf{A})}_{\boldsymbol{\nabla}(\boldsymbol{\nabla}\cdot\mathbf{A}) - \triangle\mathbf{A}} = \mathbf{j} + \epsilon_0 \frac{\partial}{\partial t}\left(-\frac{\partial \mathbf{A}}{\partial t} - \boldsymbol{\nabla}\phi\right). \tag{10.8}$$

For use below we rewrite Eq. (10.7) as

$$-\triangle\phi - \frac{\partial}{\partial t}\boldsymbol{\nabla} \cdot \mathbf{A} = \frac{1}{\epsilon_0}\rho, \tag{10.9}$$

and Eq. (10.8) as

$$-\triangle\mathbf{A} + \mu_0\epsilon_0 \frac{\partial^2 \mathbf{A}}{\partial t^2} + \boldsymbol{\nabla}\left(\boldsymbol{\nabla} \cdot \mathbf{A} + \mu_0\epsilon_0 \frac{\partial \phi}{\partial t}\right) = \mu_0 \mathbf{j}, \tag{10.10}$$

where $\mu_0\epsilon_0 = 1/c^2$.

We have seen earlier that the observable fields $\mathbf{E}$ and $\mathbf{H}$ are independent of the particular choice or gauge of the vector potential. Previously, however, we encountered the transformation from one gauge potential to another only in the context of electrostatics and magnetostatics (see, however,

Section 7.5). If the fields depend also on time, the gauge transformations are given by

$$\mathbf{A} = \mathbf{A}' + \boldsymbol{\nabla}\chi, \quad \phi = \phi' - \frac{\partial \chi}{\partial t} \tag{10.11}$$

for arbitrary functions $\chi$. We apply these transformations to Eqs. (10.9) and (10.10) and obtain

$$-\triangle\left(\phi' - \frac{\partial \chi}{\partial t}\right) - \frac{\partial}{\partial t}\boldsymbol{\nabla}\cdot(\mathbf{A}' + \boldsymbol{\nabla}\chi) = \frac{1}{\epsilon_0}\rho, \tag{10.12}$$

$$\begin{aligned} &-\triangle(\mathbf{A}' + \boldsymbol{\nabla}\chi) + \frac{1}{c^2}\frac{\partial^2}{\partial t^2}(\mathbf{A}' + \nabla\chi) \\ &+\boldsymbol{\nabla}\left(\boldsymbol{\nabla}\cdot\mathbf{A}' + \boldsymbol{\nabla}\cdot\boldsymbol{\nabla}\chi + \frac{1}{c^2}\frac{\partial \phi'}{\partial t} - \frac{1}{c^2}\frac{\partial^2 \chi}{\partial t^2}\right) = \mu_0\mathbf{j}, \end{aligned} \tag{10.13}$$

i.e.

$$-\triangle\phi' - \frac{\partial}{\partial t}\boldsymbol{\nabla}\cdot\mathbf{A}' = \frac{1}{\epsilon_0}\rho \tag{10.14}$$

and

$$-\triangle\mathbf{A}' + \frac{1}{c^2}\frac{\partial^2}{\partial t^2}\mathbf{A}' + \boldsymbol{\nabla}\left(\boldsymbol{\nabla}\cdot\mathbf{A}' + \frac{1}{c^2}\frac{\partial \phi'}{\partial t}\right) = \mu_0\mathbf{j}. \tag{10.15}$$

We observe: The terms in $\chi$ cancel out, i.e. the equations are *invariant under the gauge transformation.* We also observe: For a special gauge, which means if we consider potentials $\mathbf{A}'$ and $\phi'$ which satisfy the *additionally chosen condition*

$$\boldsymbol{\nabla}\cdot\mathbf{A}' + \frac{1}{c^2}\frac{\partial \phi'}{\partial t} = 0, \quad \frac{1}{c}\frac{\partial}{\partial t}\left(\frac{1}{c}\phi'\right) + \boldsymbol{\nabla}\cdot\mathbf{A}' = 0, \tag{10.16}$$

(the second form for later reference) called *Lorenz gauge*,[1] Eqs. (10.14) and (10.15) simplify to

$$\Box\phi' = -\frac{1}{\epsilon_0}\rho, \quad \Box\mathbf{A}' = -\mu_0\mathbf{j}, \tag{10.17}$$

where the *D'Alembertian* $\Box$ is given by

$$\Box = \triangle - \frac{1}{c^2}\frac{\partial^2}{\partial t^2}. \tag{10.18}$$

[1]The Danish physicist L.V. Lorenz (1829–1891) implied here is often confused with the Dutch physicist H.A. Lorentz. For clarification see e.g. Baylis [8], p. 97, or Felsager [51], p. 10.

In fact, that it is even necessary to demand a condition like the Lorenz gauge[2] can be seen by a more detailed investigation of the solutions of Eqs. (10.14), (10.15) and (10.17). In order to see that the gauge condition implies the *transversality* of the electromagnetic field, and instead of the four components of the potentials $\mathbf{A}'$ and $\phi'$ the independence of only two mutually orthogonal field components $\mathbf{E}$ and $\mathbf{H}$, also requires further detailed consideration.[3] In the following we write $(\phi'/c, \mathbf{A}')$ again as $(\phi/c, \mathbf{A})$ (the factor $c$ will be explained in Chapter 17). Thus we have instead of Eq. (10.16) the relation

$$\boldsymbol{\nabla}\cdot\mathbf{A}+\frac{1}{c^2}\frac{\partial\phi}{\partial t}=0, \tag{10.19}$$

called *Lorenz gauge condition*, and the equations

$$\Box\phi=-\frac{1}{\epsilon_0}\rho, \quad \Box\mathbf{A}=-\mu_0\mathbf{j}. \tag{10.20}$$

**Example 10.1 (Coulomb gauge).** Starting from Maxwell's equations for moving charges in vacuum, derive the wave equations of the potentials $\phi, \mathbf{A}$ in Coulomb gauge ($\boldsymbol{\nabla}\cdot\mathbf{A}=0$) and show that a transverse current is the source of the potential $\mathbf{A}$.

**Solution:** From the above we have the Eqs. (10.14) and (10.15), i.e.

$$\triangle\phi+\frac{\partial}{\partial t}\boldsymbol{\nabla}\cdot\mathbf{A}=-\frac{1}{\epsilon_0}\rho, \tag{10.21}$$

and

$$\triangle\mathbf{A}-\frac{1}{c^2}\frac{\partial^2}{\partial t^2}\mathbf{A}-\boldsymbol{\nabla}\left(\boldsymbol{\nabla}\cdot\mathbf{A}+\frac{1}{c^2}\frac{\partial\phi}{\partial t}\right)=-\mu_0\mathbf{j}. \tag{10.22}$$

With Coulomb gauge $\boldsymbol{\nabla}\cdot\mathbf{A}=0$ and Eq. (10.18) it follows that (note the difference to Eq. (10.20))

$$\triangle\phi=-\frac{1}{\epsilon_0}\rho, \quad \Box\mathbf{A}=-\mu_0\mathbf{j}_t, \tag{10.23}$$

---

[2]A different gauge, called Coulomb gauge, $\boldsymbol{\nabla}\cdot\mathbf{A}=0$, is treated, for example, in Schwabl [131], p. 127.
[3]See Chapter 18.

where (with $\mathbf{D} = \epsilon_0 \mathbf{E}$) the current $\mathbf{j}_t$ is given by (using Eq. (10.3))

$$\begin{aligned}\mathbf{j}_t &= \mathbf{j} - \frac{1}{\mu_0 c^2}\boldsymbol{\nabla}\frac{\partial \phi}{\partial t} = \mathbf{j} - \frac{1}{\mu_0 c^2}\frac{\partial}{\partial t}(\boldsymbol{\nabla}\phi) \\ &= \mathbf{j} + \frac{\partial \mathbf{D}}{\partial t} + \epsilon_0 \frac{\partial^2 \mathbf{A}}{\partial t^2} \quad \left(\text{with} \quad \mathbf{E} = -\frac{\partial \mathbf{A}}{\partial t} - \boldsymbol{\nabla}\phi\right) \\ &= \boldsymbol{\nabla} \times \mathbf{H} + \epsilon_0 \frac{\partial^2 \mathbf{A}}{\partial t^2}. \end{aligned} \tag{10.24}$$

Together with Ampère's law expressed as in Eq. (5.69), $\boldsymbol{\nabla} \times \mathbf{B} + \triangle \mathbf{A} = 0$, we also have

$$\triangle \mathbf{A} = -\mu_0 \boldsymbol{\nabla} \times \mathbf{H} \equiv -\mu_0 \mathbf{j}_{tt} \tag{10.25}$$

(this defines $\mathbf{j}_{tt}$). Hence

$$\mathbf{j}_t = \mathbf{j}_{tt} + \epsilon_0 \frac{\partial^2 \mathbf{A}}{\partial t^2}. \tag{10.26}$$

We construct the divergence of this with the Coulomb gauge. The operator $-i\boldsymbol{\nabla}$ applied to $\mathbf{A} \propto e^{i\mathbf{k}\cdot\mathbf{r}}$ yields $\mathbf{k}$, the propagation vector. Since (see Appendix C)

$$\boldsymbol{\nabla} \cdot \mathbf{j}_{tt} = \boldsymbol{\nabla} \cdot (\boldsymbol{\nabla} \times \mathbf{H}) = 0, \tag{10.27}$$

we have $\mathbf{j}_{tt} \perp \mathbf{k}$, i.e. $\boldsymbol{\nabla} \cdot \mathbf{j}_t = 0$, which means $\mathbf{j}_t$ is transverse, i.e. $\mathbf{k} \cdot \mathbf{j}_t = 0$. Note that this requires the gauge condition $\boldsymbol{\nabla} \cdot \mathbf{A} = 0$.

In order to be able to solve Eq. (10.20) we need the appropriate *Green's function* $G(\mathbf{r}, t; \mathbf{r}', t')$, which is an inhomogeneous solution of the equation

$$\square G(\mathbf{r}, t; \mathbf{r}', t') = \delta(\mathbf{r} - \mathbf{r}')\delta(t - t'). \tag{10.28}$$

In the following Example 10.2 we calculate that particular Green's function, which represents the propagation of a light signal or disturbance of some kind as a spherical wave spreading out with velocity $c$ from $\mathbf{r}'$ at time $t'$ to $\mathbf{r}$ at time $t > t'$. This Green's function which takes into account *causality* is called the *retarded Green's function*, and the corresponding potentials are called the *retarded potentials*. According to the following Example 10.2 this *retarded Green's function* is given by

$$G(\mathbf{r}, t; \mathbf{r}', t') = -\frac{\delta(t - t' - |\mathbf{r} - \mathbf{r}'|/c)}{4\pi|\mathbf{r} - \mathbf{r}'|}. \tag{10.29}$$

Hence

$$\phi(\mathbf{r},t) = -\frac{1}{\epsilon_0}\int G(\mathbf{r},t;\mathbf{r}',t')\rho(\mathbf{r}',t')d\mathbf{r}'dt', \tag{10.30}$$

implies

$$\phi(\mathbf{r},t) = \frac{1}{4\pi\epsilon_0}\int d\mathbf{r}'\frac{\rho(\mathbf{r}',t-|\mathbf{r}-\mathbf{r}'|/c)}{|\mathbf{r}-\mathbf{r}'|}. \tag{10.31}$$

Similarly one obtains from the vector equation in Eq. (10.20)

$$\mathbf{A}(\mathbf{r},t) = \frac{\mu_0}{4\pi}\int d\mathbf{r}'\frac{\mathbf{j}(\mathbf{r}',t-|\mathbf{r}-\mathbf{r}'|/c)}{|\mathbf{r}-\mathbf{r}'|}. \tag{10.32}$$

We now derive the retarded Green's function.

**Example 10.2 (The retarded Green's function).** Calculate the Green's function $G$ which takes into account *causality*, i.e. that a cause must precede its effect.

**Solution:** We have

$$\Box = \triangle - \frac{1}{c^2}\frac{\partial^2}{\partial t^2} \tag{10.33}$$

and the Fourier representation of the 4-dimensional delta function, i.e.

$$\delta(\mathbf{r}-\mathbf{r}')\delta(t-t') = \frac{1}{(2\pi)^4}\int_{-\infty}^{\infty} d\mathbf{k}\int_{-\infty}^{\infty} d\omega e^{i\mathbf{k}\cdot(\mathbf{r}-\mathbf{r}')}e^{-i\omega(t-t')}. \tag{10.34}$$

The Fourier representation of the time-dependent Green's function can be written

$$G(\mathbf{r},t;\mathbf{r}',t') = \int_{-\infty}^{\infty} d\mathbf{k}\int_{-\infty}^{\infty} d\omega g(\mathbf{k},\omega)e^{i\mathbf{k}\cdot(\mathbf{r}-\mathbf{r}')}e^{-i\omega(t-t')}, \tag{10.35}$$

where the *spectral function* $g(\mathbf{k},\omega)$ is still to be determined. First we have

$$\Box G(\mathbf{r},t;\mathbf{r}',t') = \int_{-\infty}^{\infty} d\mathbf{k}\int_{-\infty}^{\infty} d\omega g(\mathbf{k},\omega)\left(\frac{\omega^2}{c^2}-k^2\right)e^{i\mathbf{k}\cdot(\mathbf{r}-\mathbf{r}')}e^{-i\omega(t-t')}. \tag{10.36}$$

Inserting Eqs. (10.34) and (10.36) into the equation for $G$, Eq. (10.28), and comparing coefficients, we obtain

$$g(\mathbf{k},\omega) = -\frac{1}{(2\pi)^4}\frac{1}{k^2-\omega^2/c^2} \tag{10.37}$$

and therefore by insertion into Eq. (10.35):

$$G(\mathbf{r}, t; \mathbf{r}', t') = -\frac{1}{(2\pi)^4}\int_{-\infty}^{\infty} d\mathbf{k}\int_{-\infty}^{\infty} d\omega \frac{c^2}{c^2k^2 - \omega^2} e^{i[\mathbf{k}\cdot(\mathbf{r}-\mathbf{r}')-\omega(t-t')]}. \quad (10.38)$$

The integrand of the integral on the right has simple poles at $\omega = \pm ck$. The integration is performed with the help of Cauchy's residue theorem, i.e. the contour integral relation

$$\oint f(z)dz = 2\pi i \sum_{k=1}^{n} \text{Res}_k f(z). \quad (10.39)$$

The Green's function represents an instantaneous and local disturbance or event or something similar which occurs at radius $\mathbf{r}'$ at time $t = t'$ and propagates as a spherical wave with velocity $c$. A causality-preserving wave propagating forward in time must be such that

$$G(\mathbf{r}, t; \mathbf{r}', t') = 0 \quad \text{for} \quad t < t'. \quad (10.40)$$

Thus for $t > t'$ the function $G$ represents a wave which spreads into the future. The singularities of $g(\mathbf{k}, \omega)$ are at $\omega = \pm ck$. We now consider the plane of complex $\omega$ with $\omega = \omega_R + i\omega_I$, and try to define the causal waves by a suitable choice of the integration contour in this plane. To ensure that $G = 0$ for $t < t'$, we have to circumvent the poles at $\omega = \pm ck$. This can be achieved by displacing them by an infinitesimal amount $(-i\epsilon)$, which means replacing in the above equation $\omega$ by $\omega + i\epsilon$, so that we write

$$G(\mathbf{r}, t; \mathbf{r}', t') = -\frac{1}{(2\pi)^4}\int_{-\infty}^{\infty} d\mathbf{k}\int_{-\infty}^{\infty} d\omega \frac{e^{i[\mathbf{k}\cdot(\mathbf{r}-\mathbf{r}')-\omega(t-t')]}}{k^2 - (\omega + i\epsilon)^2/c^2}. \quad (10.41)$$

We consider first the integration with respect to $\omega$, i.e. the integral

$$I(k) = \int_{-\infty}^{\infty} d\omega \frac{e^{-i\omega(t-t')}}{k^2 - (\omega + i\epsilon)^2/c^2}\theta(t - t'). \quad (10.42)$$

The idea is, to add to the integral from $-\infty$ to $+\infty$ the integral along the infinite semi-circle either in the upper or in the lower half-plane so that as a consequence of the displacement of the poles by $-i\epsilon, \epsilon > 0$, the integral along this semi-circle falls off exponentially towards infinity and hence does not contribute. This means, for $t - t' > 0$ we have to choose the semi-circle in the lower half-plane as indicated in Fig. 10.1, and for $t' - t > 0$ in the upper half-plane. Since we have two poles of the first order, we obtain two residues. We let the poles be $\omega_1 = ck - i\epsilon$ and $\omega_2 = -ck - i\epsilon$, both in the lower

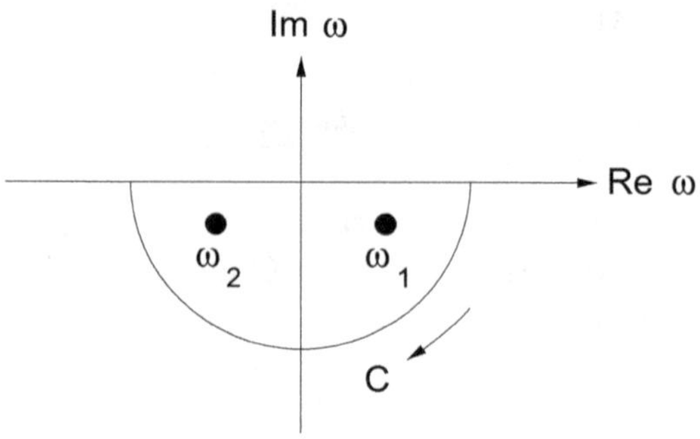


**Fig. 10.1.** The contour $C$ in the complex $\omega$-plane.

half-plane. Their residues are the coefficients of the factors $1/(\omega - \omega_1)$ and $1/(\omega - \omega_2)$ in the integrand and we have (by the method of partial fractions)

$$\frac{1}{k^2 - (\omega + i\epsilon)^2/c^2} = \frac{c}{2k}\left[-\frac{1}{\omega - \omega_1} + \frac{1}{\omega - \omega_2}\right]. \tag{10.43}$$

The residues are obtained as

$$\text{residue at } \omega_1 = -\frac{c}{2k}e^{-ick(t-t')}, \quad \text{residue at } \omega_2 = \frac{c}{2k}e^{ick(t-t')}. \tag{10.44}$$

We thus obtain with Cauchy's theorem for the integral around the contour $C$ shown in Fig. 10.1

$$\begin{aligned} I(k) &= \int_C d\omega \frac{1}{k^2 - (\omega + i\epsilon)^2/c^2} e^{-i\omega(t-t')}\theta(t - t') \\ &= -2\pi i\left(-\frac{c}{2k}e^{-ick(t-t')} + \frac{c}{2k}e^{ick(t-t')}\right) = \frac{2\pi c}{k}\sin(ck(t - t')) \end{aligned} \tag{10.45}$$

($-2\pi i$ since the contour is traversed in the clockwise direction). Hence we have thus far

$$G(\mathbf{r}, t; \mathbf{r}', t') = \begin{cases} 0 & : \ t < t', \\ -\frac{c}{(2\pi)^3}\int_{-\infty}^{\infty} d\mathbf{k} e^{i\mathbf{k}\cdot(\mathbf{r}-\mathbf{r}')}k^{-1}\sin(ck(t - t')) & : \ t > t'. \end{cases} \tag{10.46}$$

We still have to perform the integration with respect to $\mathbf{k}$. To this end we introduce spherical coordinates. Let $\theta$ be the polar angle to the vector $\mathbf{r} - \mathbf{r}'$ as $z$-axis. We then have

$$d\mathbf{k} = k^2 \sin\theta dk d\theta d\phi, \quad \mathbf{k}\cdot(\mathbf{r} - \mathbf{r}') = k|\mathbf{r} - \mathbf{r}'|\cos\theta. \tag{10.47}$$

Inserting this into the expression for $G$ we obtain

$$G(\mathbf{r},t;\mathbf{r}',t') = -\frac{c}{(2\pi)^3}\int d\phi d\theta dk k \sin\theta \sin(ck(t-t'))e^{ik\cos\theta|\mathbf{r}-\mathbf{r}'|}. \quad (10.48)$$

Integration with respect to $\phi$ yields the factor $2\pi$, and we have

$$G(\mathbf{r},t;\mathbf{r}',t') = \frac{c}{(2\pi)^2}\int_{k=0}^{\infty} k\sin(ck(t-t'))dk \int_{\theta=0}^{\pi} e^{ik\cos\theta|\mathbf{r}-\mathbf{r}'|}d\cos\theta. \quad (10.49)$$

The integral over $\theta$ yields $-2\sin(k|\mathbf{r}-\mathbf{r}'|)/k|\mathbf{r}-\mathbf{r}'|$, so that

$$G(\mathbf{r},t;\mathbf{r}',t') = \frac{-c}{2\pi^2|\mathbf{r}-\mathbf{r}'|}\int_0^{\infty}\sin\left(\frac{kc|\mathbf{r}-\mathbf{r}'|}{c}\right)\sin(ck(t-t'))dk. \quad (10.50)$$

Now we use the relation

$$\begin{aligned} c\int_0^{\infty} dk \sin Akc \sin Bkc &= \frac{1}{(2i)^2}\int_{-\infty}^{\infty} dk\, c[e^{i(A+B)kc} - e^{-i(A-B)kc}] \\ &= -\frac{\pi}{2}[\delta(A+B) - \delta(A-B)], \end{aligned} \quad (10.51)$$

which is thus verified by replacing the sines by exponentials and using the integral representation of the one-dimensional delta function. Thus $G$ can be expressed as the sum of two delta function contributions, i.e. we obtain

$$G(\mathbf{r},t;\mathbf{r}',t') = \frac{1}{4\pi|\mathbf{r}-\mathbf{r}'|}[\delta(t-t'+|\mathbf{r}-\mathbf{r}'|/c) - \delta(t-t'-|\mathbf{r}-\mathbf{r}'|/c)]. \quad (10.52)$$

Since the argument of the first delta function is nowhere zero for $t > t'$ this cannot contribute to $G$ and therefore drops out. We thus obtain

$$G(\mathbf{r},t;\mathbf{r}',t') = -\frac{\delta(t-t'-|\mathbf{r}-\mathbf{r}'|/c)}{4\pi|\mathbf{r}-\mathbf{r}'|}. \quad (10.53)$$

This Green's function describes the effect at time $t > t'$ and at radius $r > r'$ of a delta function-like cause at time $t = t'$ and radius $r = r'$ propagating from the source with the velocity of light.

## 10.3 The Liénard–Wiechert Potentials of a Moving Point Charge

Our aim is now to calculate the potentials $\phi(\mathbf{r},t), \mathbf{A}(\mathbf{r},t)$ for the case of a point charge moving with velocity $\mathbf{v}(\mathbf{t})$. The very important results we obtain are the so-called *Liénard–Wiechert potentials*. We expect, of course, to regain in the limit of vanishing velocity the well known static expressions we had before.

We let $\mathbf{r}_0(t)$ be the vector representing the path or trajectory of the point charge $e$ and set

$$\mathbf{v}(t) = \dot{\mathbf{r}}_0(t) \tag{10.54}$$

for its velocity. The charge density can then be written

$$\rho(\mathbf{r}, t) = e\delta(\mathbf{r} - \mathbf{r}_0(t)) \quad \text{coulombs/meter}^3, \tag{10.55}$$

and the current density

$$\mathbf{j}(\mathbf{r}, t) = \rho(\mathbf{r}, t)\mathbf{v}(t) = e\mathbf{v}(t)\delta(\mathbf{r} - \mathbf{r}_0(t)) \quad \text{amperes/meter}^2. \tag{10.56}$$

We consider first the scalar potential $\phi$. Inserting into Eq. (10.31) the expression for $\rho$, we obtain

$$\begin{aligned}\phi(\mathbf{r}, t) &= \frac{e}{4\pi\epsilon_0}\int d\mathbf{r}' \frac{\delta(\mathbf{r}' - \mathbf{r}_0(t - |\mathbf{r} - \mathbf{r}'|/c))}{|\mathbf{r} - \mathbf{r}'|} \\ &= \frac{e}{4\pi\epsilon_0}\int dt' \int d\mathbf{r}' \frac{\delta(\mathbf{r}' - \mathbf{r}_0(t'))\delta(t' - t + |\mathbf{r} - \mathbf{r}'|/c)}{|\mathbf{r} - \mathbf{r}'|} \\ &= \frac{e}{4\pi\epsilon_0}\int dt' \frac{\delta(t' - t + |\mathbf{r} - \mathbf{r}_0(t')|/c)}{|\mathbf{r} - \mathbf{r}_0(t')|} \quad \text{volts.}\end{aligned} \tag{10.57}$$

We define a time difference $u$ by

$$u = t' - t + \frac{|\mathbf{r} - \mathbf{r}_0(t')|}{c} = t' - t + \frac{\sqrt{(\mathbf{r} - \mathbf{r}_0(t'))^2}}{c}, \tag{10.58}$$

so that

$$\frac{du}{dt'} = 1 - \frac{\mathbf{n}(t') \cdot \mathbf{v}(t')}{c}, \quad \mathbf{n} = \frac{\mathbf{r} - \mathbf{r}_0(t')}{|\mathbf{r} - \mathbf{r}_0(t')|} \tag{10.59}$$

and

$$dt' = \frac{du}{1 - \mathbf{n}(t') \cdot \mathbf{v}(t')/c}. \tag{10.60}$$

It follows that

$$\phi(\mathbf{r}, t) = \frac{e}{4\pi\epsilon_0}\int \frac{du}{1 - \mathbf{n}(t') \cdot \mathbf{v}(t')/c} \frac{\delta(u)}{|\mathbf{r} - \mathbf{r}_0(t')|}, \tag{10.61}$$

i.e.

$$\phi(\mathbf{r},t) = \frac{e/4\pi\epsilon_0}{|\mathbf{r}-\mathbf{r}_0(t')|(1-\mathbf{n}(t')\cdot\mathbf{v}(t')/c)} \quad \text{volts}, \tag{10.62}$$

where (from $u = 0$ in the expression (10.58) for $t'$)

$$t' = t - \frac{1}{c}|\mathbf{r}-\mathbf{r}_0(t')|, \quad \mathbf{v}(t') \equiv \dot{\mathbf{r}}_0(t'). \tag{10.63}$$

In a similar way we obtain from Eqs. (10.32) and (10.56) the vector potential

$$\mathbf{A}(\mathbf{r},t) = \frac{e\mathbf{v}(t')\mu_0/4\pi}{|\mathbf{r}-\mathbf{r}_0(t')|(1-\mathbf{n}(t')\cdot\mathbf{v}(t')/c)} \quad \text{tesla}\cdot\text{meters}. \tag{10.64}$$

These expressions, called *Liénard–Wiechert potentials*, are *exact solutions* of the above wave equations *for a pointlike, charged particle* moving along the trajectory $\mathbf{r}_0(t')$. For

$$\mathbf{r}-\mathbf{r}_0(t') \equiv \mathbf{r}(t') \tag{10.65}$$

the potential $\phi$ is

$$\phi(\mathbf{r},t) = \left(\frac{e/4\pi\epsilon_0}{r(t')-\mathbf{r}(t')\cdot\mathbf{v}(t')/c}\right)_{t'=t-r(t')/c} \quad \text{volts}, \tag{10.66}$$

i.e. $|\mathbf{r}(t')| = c(t-t')$. One should note, that the vector $\mathbf{r}$ to the field point at which the results give the potentials is here to be considered as fixed and $\mathbf{r} = \mathbf{r}_0(t') + \mathbf{r}(t')$. The expression

$$t' = t - \frac{r(t')}{c} \tag{10.67}$$

is called *retarded time* and is defined implicitly by this equation (e.g. when $r(t') = vt' + a$ we have $t' = t - (vt'+a)/c, t - t' = (a/c + vt/c)/(1+v/c)$). If $c$ were infinite (which is *not* possible according to the theory of relativity), we would have $t' = t$ for $r(t') =$ finite. This means the radiation emitted by the charge would reach any other point in space instantaneously (which is nonsense). However, for $c =$ finite, the radiation emitted by the charge reaches other points only with delay, or, as one says, with *retarded time.* This means the radiation (i.e. the fields) observed at the point $\mathbf{r}$ at time $t$ was emitted by the charge at its earlier position $t - t'$ seconds before.

Analogously to Eq. (10.66) we obtain

$$\mathbf{A}(\mathbf{r},t) = \frac{e\mu_0}{4\pi}\left(\frac{\mathbf{v}(t')}{r(t') - \mathbf{r}(t')\cdot\mathbf{v}(t')/c}\right)_{t'=t-r(t')/c} \quad \text{tesla}\cdot\text{meters}. \qquad (10.68)$$

These are the potentials at position $\mathbf{r}$ (or $\mathbf{r}(t')$ away from $\mathbf{r}_0(t')$) and at time $t$ in the case of a point charge $e$. One should observe our notation, in order not to confuse $\mathbf{r}$ and $\mathbf{r}(t')$. In Section 10.5 we apply these potentials to the oscillating electric dipole, called *Hertz dipole*, with dipole moment $\mathbf{p}(t)$.

**Example 10.3 (Lorentz 4-form of the Liénard–Wiechert potentials).** (This example assumes familiarity with Chapter 17). Let the components of $\mathbf{r}_0(t')$ be the coordinates of the charge $e$ and the components of $\mathbf{r}$ those of an observation point with light vector (i.e. one with $c^2 = (\text{distance})^2/(\text{time})^2$) $U^\mu = (c(t-t'), \mathbf{r}-\mathbf{r}_0(t'))$ given by $U^\mu U_\mu = 0$. Verify that the Liénard–Wiechert potentials can be combined into the 4-form

$$A^\mu = \frac{e}{4\pi\epsilon_0 c}\frac{u^\mu}{U_\nu u^\nu}, \qquad (10.69)$$

where $u^\mu$ is the 4-velocity defined (later) by Eq. (17.64), i.e. with components $u_0 = c/\sqrt{1-\beta^2}$, and $u_i = -v_i/\sqrt{1-\beta^2}$, and $u^i = v_i/\sqrt{1-\beta^2}$, $\beta^2 = v^2/c^2$.

**Solution:** The light vector is given by $c^2(t-t')^2 - (\mathbf{r}-\mathbf{r}_0(t'))^2 = 0$, i.e. $c(t-t')/\sqrt{(\mathbf{r}-\mathbf{r}_0(t'))^2} = 1$. Hence Eqs. (10.62) and (10.64) with Eq. (17.31) and $\epsilon_0\mu_0 c^2 = 1$ yield the expression

$$\begin{aligned} A^\mu = \left(\frac{1}{c}\phi, \mathbf{A}\right) &= \frac{e}{4\pi\epsilon_0 c^2}\left(\frac{c}{|\mathbf{r}-\mathbf{r}_0|(1-\mathbf{n}\cdot\mathbf{v}/c)}, \frac{\mathbf{v}}{|\mathbf{r}-\mathbf{r}_0|(1-\mathbf{n}\cdot\mathbf{v}/c)}\right) \\ &= \frac{e}{4\pi\epsilon_0 c^2}\frac{(c,\mathbf{v})}{|\mathbf{r}-\mathbf{r}_0|(c(t-t')/|\mathbf{r}-\mathbf{r}_0| - \mathbf{n}\cdot\mathbf{v}/c)} \\ &= \frac{e}{4\pi\epsilon_0 c}\frac{(c,\mathbf{v})/\sqrt{1-\beta^2}}{(c^2(t-t') - \mathbf{v}\cdot(\mathbf{r}-\mathbf{r}_0))/\sqrt{1-\beta^2}} = \frac{e}{4\pi\epsilon_0 c}\frac{u^\mu}{U_\nu u^\nu}. \end{aligned} \qquad (10.70)$$

## 10.4 The Fields E, B of a Moving Point Charge

Our next task is to derive the electric and magnetic fields of a single moving charge from the potentials $\phi$ and $\mathbf{A}$ derived in Section 10.3. We obtain the fields $\mathbf{E}$ and $\mathbf{B}$ from the equations

$$\mathbf{E} = -\nabla\phi - \frac{\partial\mathbf{A}}{\partial t} \quad \text{newtons/coulomb}, \quad \mathbf{B} = \nabla\times\mathbf{A} \quad \text{teslas}. \qquad (10.71)$$

Thus we have to perform complicated differentiations. We start with **E**:

$$\begin{aligned}\mathbf{E} = -\boldsymbol{\nabla}&\frac{e/4\pi\epsilon_0}{|\mathbf{r}-\mathbf{r}_0(t')|(1-\mathbf{n}(t')\cdot\mathbf{v}(t')/c)}\\ &-\frac{\partial}{\partial t}\left[\frac{e\mathbf{v}(t')\mu_0/4\pi}{|\mathbf{r}-\mathbf{r}_0(t')|(1-\mathbf{n}(t')\cdot\mathbf{v}(t')/c)}\right].\end{aligned}\tag{10.72}$$

Here $t' = t - |\mathbf{r}-\mathbf{r}_0(t')|/c$, so that

$$\begin{aligned}\frac{\partial t'}{\partial t} &= 1-\frac{1}{c}\left(\frac{\partial}{\partial t'}|\mathbf{r}-\mathbf{r}_0(t')|\right)\frac{\partial t'}{\partial t},\\ \frac{\partial}{\partial t'}|\mathbf{r}-\mathbf{r}_0(t')| &= \frac{\partial}{\partial t'}\sqrt{(\mathbf{r}-\mathbf{r}_0(t'))^2} = -\mathbf{v}(t')\cdot\mathbf{n}(t'),\end{aligned}\tag{10.73}$$

and hence

$$\frac{\partial t'}{\partial t} = 1+\frac{1}{c}\mathbf{v}(t')\cdot\mathbf{n}(t')\frac{\partial t'}{\partial t},\quad \frac{\partial t'}{\partial t} = \frac{1}{(1-\mathbf{n}(t')\cdot\mathbf{v}(t')/c)}.\tag{10.74}$$

Similarly (since $\boldsymbol{\nabla}|\mathbf{r}| = \mathbf{r}/r$)

$$\begin{aligned}\boldsymbol{\nabla}t' &= -\frac{1}{c}(\boldsymbol{\nabla})_{t'=\text{const.}}|\mathbf{r}-\mathbf{r}_0(t')| - \frac{1}{c}\left(\frac{\partial}{\partial t'}|\mathbf{r}-\mathbf{r}_0(t')|\right)\boldsymbol{\nabla}t'\\ &= -\frac{\mathbf{n}(t')}{c} - \frac{1}{c}\underbrace{\left(\frac{\partial}{\partial t'}|\mathbf{r}-\mathbf{r}_0(t')|\right)}_{-\mathbf{v}(t')\cdot\mathbf{n}(t')}\boldsymbol{\nabla}t'\end{aligned}\tag{10.75}$$

and hence

$$\boldsymbol{\nabla}t' = \frac{-\mathbf{n}(t')}{c(1-\mathbf{n}(t')\cdot\mathbf{v}(t')/c)}.\tag{10.76}$$

Moreover, with Eq. (10.73),

$$\begin{aligned}\frac{\partial\mathbf{n}}{\partial t'} = \frac{\partial}{\partial t'}\left(\frac{\mathbf{r}-\mathbf{r}_0(t')}{|\mathbf{r}-\mathbf{r}_0(t')|}\right) &= \frac{(\mathbf{v}(t')\cdot\mathbf{n}(t'))(\mathbf{r}-\mathbf{r}_0(t'))}{|\mathbf{r}-\mathbf{r}_0(t')|^2} - \frac{\mathbf{v}(t')}{|\mathbf{r}-\mathbf{r}_0(t')|}\\ &= \frac{\mathbf{n}(t')\times(\mathbf{n}(t')\times\mathbf{v}(t'))}{|\mathbf{r}-\mathbf{r}_0(t')|}.\end{aligned}\tag{10.77}$$

We now introduce the acceleration $\mathbf{a}$ by setting $\mathbf{a}(t') = d\mathbf{v}(t')/dt'$. Using Eqs. (10.73) and (10.77), we obtain

$$\begin{aligned}
&\frac{\partial}{\partial t'}\left[|\mathbf{r}-\mathbf{r}_0(t')|\left\{1-\frac{\mathbf{v}(t')\cdot\mathbf{n}(t')}{c}\right\}\right] \\
&\overset{(10.73),(10.77)}{=} -\mathbf{v}(t')\cdot\mathbf{n}(t')\left\{1-\frac{\mathbf{v}(t')\cdot\mathbf{n}(t')}{c}\right\} - |\mathbf{r}-\mathbf{r}_0(t')|\frac{\mathbf{a}(t')\cdot\mathbf{n}(t')}{c} \\
&\qquad - \frac{\mathbf{v}(t')}{c}\cdot\underbrace{[\mathbf{n}(t')\times(\mathbf{n}(t')\times\mathbf{v}(t'))]}_{(\mathbf{n}(t')\cdot\mathbf{v}(t'))\mathbf{n}(t')-\mathbf{n}^2(t')\mathbf{v}(t')} \\
&= -\mathbf{v}(t')\cdot\left\{\mathbf{n}(t')-\frac{\mathbf{v}(t')}{c}\right\} - \frac{|\mathbf{r}-\mathbf{r}_0(t')|}{c}\mathbf{a}(t')\cdot\mathbf{n}(t').
\end{aligned} \tag{10.78}$$

Now, $\phi = \phi(\mathbf{r}, t'(\mathbf{r}))$, so that

$$\boldsymbol{\nabla}\phi = (\boldsymbol{\nabla}\phi)_{t'=\text{const.}} + \frac{\partial\phi}{\partial t'}\boldsymbol{\nabla}t'. \tag{10.79}$$

Thus we obtain with Eqs. (10.62) and (10.76) (as well as $\boldsymbol{\nabla}(1/|\mathbf{r}|) = -\mathbf{n}/r^2$, see e.g. Eq. (2.6)):

$$\begin{aligned}
-\boldsymbol{\nabla}\phi = &-\boldsymbol{\nabla}\left[\frac{e/4\pi\epsilon_0}{|\mathbf{r}-\mathbf{r}_0(t')|(1-\mathbf{v}(t')\cdot\mathbf{n}(t')/c)}\right]_{t'=\text{const.}} \\
&-\frac{\partial}{\partial t'}\left[\frac{e/4\pi\epsilon_0}{|\mathbf{r}-\mathbf{r}_0(t')|(1-\mathbf{v}(t')\cdot\mathbf{n}(t')/c)}\right]\left[\frac{-\mathbf{n}(t')}{c(1-\mathbf{v}(t')\cdot\mathbf{n}(t')/c)}\right],
\end{aligned} \tag{10.80}$$

and with Eq. (10.78) (skipping here two steps with rearrangements of terms):

$$\begin{aligned}
-\boldsymbol{\nabla}\phi = &\frac{(e/4\pi\epsilon_0)[\mathbf{n}(1-v^2/c^2)-\mathbf{n}(\mathbf{n}\cdot\mathbf{v}/c)(1-\mathbf{n}\cdot\mathbf{v}/c)]}{|\mathbf{r}-\mathbf{r}_0|^2(1-\mathbf{n}\cdot\mathbf{v}/c)^3} \\
&+\frac{(e/4\pi\epsilon_0)\mathbf{n}(t')[\mathbf{a}(t')\cdot\mathbf{n}(t')]}{c^2(1-\mathbf{n}\cdot\mathbf{v}/c)^3|\mathbf{r}-\mathbf{r}_0|}.
\end{aligned} \tag{10.81}$$

Next we have to calculate $\partial\mathbf{A}/\partial t$. We have first with Eq. (10.64):

$$\begin{aligned}\frac{\partial\mathbf{A}}{\partial t'} &= \frac{\partial}{\partial t'}\left[\frac{(e\mu_0/4\pi)\mathbf{v}(t')}{|\mathbf{r}-\mathbf{r}_0(t')|(1-\mathbf{v}(t')\cdot\mathbf{n}(t')/c)}\right]\\ &= \frac{(e\mu_0/4\pi)\mathbf{a}(t')}{|\mathbf{r}-\mathbf{r}_0(t')|(1-\mathbf{v}(t')\cdot\mathbf{n}(t')/c)}\\ &\quad -\frac{e\mu_0\mathbf{v}(t')\frac{\partial}{\partial t'}\{|\mathbf{r}-\mathbf{r}_0(t')|(1-\mathbf{v}(t')\cdot\mathbf{n}(t')/c)\}}{4\pi|\mathbf{r}-\mathbf{r}_0(t')|^2(1-\mathbf{v}(t')\cdot\mathbf{n}(t')/c)^2},\end{aligned} \tag{10.82}$$

and then with Eq. (10.78):

$$\begin{aligned}\frac{\partial\mathbf{A}}{\partial t'} &= -\frac{e\mu_0\mathbf{v}(t')\{-\mathbf{v}(t')\cdot(\mathbf{n}(t')-\mathbf{v}(t')/c)-|\mathbf{r}-\mathbf{r}_0(t')|\mathbf{a}(t')\cdot\mathbf{n}(t')/c\}}{4\pi|\mathbf{r}-\mathbf{r}_0(t')|^2(1-\mathbf{v}(t')\cdot\mathbf{n}(t')/c)^2}\\ &\quad +\frac{(e\mu_0/4\pi)\mathbf{a}(t')}{|\mathbf{r}-\mathbf{r}_0(t')|(1-\mathbf{v}(t')\cdot\mathbf{n}(t')/c)}.\end{aligned} \tag{10.83}$$

Using Eq. (10.74), we have

$$\frac{\partial\mathbf{A}}{\partial t} = \frac{\partial\mathbf{A}}{\partial t'}\frac{\partial t'}{\partial t} = \frac{\partial\mathbf{A}}{\partial t'}\frac{1}{(1-\mathbf{n}(t')\cdot\mathbf{v}(t')/c)}, \tag{10.84}$$

and hence

$$\begin{aligned}\frac{\partial\mathbf{A}}{\partial t} &= \frac{(e\mu_0/4\pi)\mathbf{v}(t')\{\mathbf{v}(t')\cdot(\mathbf{n}(t')-\mathbf{v}(t')/c)\}}{|\mathbf{r}-\mathbf{r}_0(t')|^2(1-\mathbf{v}(t')\cdot\mathbf{n}(t')/c)^3}\\ &\quad +\frac{(e\mu_0/4\pi)\{\mathbf{a}(t')-\mathbf{n}(t')\times(\mathbf{a}(t')\times\mathbf{v}(t')/c)\}}{|\mathbf{r}-\mathbf{r}_0(t')|(1-\mathbf{v}(t')\cdot\mathbf{n}(t')/c)^3}.\end{aligned} \tag{10.85}$$

With this result we can calculate the electric field strength $\mathbf{E}$, using also the relation $c^2\epsilon_0\mu_0 = 1$:

$$\begin{aligned}\mathbf{E} &= -\boldsymbol{\nabla}\phi - \frac{\partial\mathbf{A}}{\partial t}\\ &= \frac{e}{4\pi\epsilon_0}\left[\frac{\mathbf{n}(1-v^2/c^2)-\mathbf{n}(\mathbf{n}\cdot\mathbf{v}/c)(1-\mathbf{n}\cdot\mathbf{v}/c)-(\mathbf{v}/c)\{\mathbf{v}/c\cdot(\mathbf{n}-\mathbf{v}/c)\}}{|\mathbf{r}-\mathbf{r}_0|^2(1-\mathbf{n}\cdot\mathbf{v}/c)^3}\right.\\ &\quad \left.+\frac{\mathbf{n}\{\mathbf{a}\cdot\mathbf{n}\}-\{\mathbf{a}-\mathbf{n}\times(\mathbf{a}\times\mathbf{v}/c)\}}{c^2(1-\mathbf{n}\cdot\mathbf{v}/c)^3|\mathbf{r}-\mathbf{r}_0|}\right] \quad \text{newtons/coulomb.}\end{aligned} \tag{10.86}$$

For large $|\mathbf{r}-\mathbf{r}_0|$ and $v/c \ll 1$ the dominant contributions come from the second term as

$$\mathbf{E} \to \mathbf{E}_0 = \frac{e}{4\pi\epsilon_0}\frac{\mathbf{n}(t')\{\mathbf{a}(t')\cdot\mathbf{n}(t')\}-\mathbf{a}(t')}{c^2|\mathbf{r}-\mathbf{r}_0|}, \quad \mathbf{n}\cdot\mathbf{E}_0 = 0. \tag{10.87}$$

Thus the radial component of $\mathbf{E}_0 \propto 1/|\mathbf{r}-\mathbf{r}_0(t')|$ vanishes which means the radial field is of order $1/r^2$. For later considerations we observe here, that $\mathbf{n}\cdot\mathbf{E} = O(1/|\mathbf{r}-\mathbf{r}_0|^2)$ and that because $\mathbf{n}\times\mathbf{n}=0$ we have

$$\mathbf{n}\times\mathbf{E} = -\frac{e}{4\pi\epsilon_0}\frac{\mathbf{n}(\mathbf{t}')\times\{\mathbf{a}(t')-\mathbf{n}(t')\times(\mathbf{a}(t')\times\mathbf{v}(t')/c)\}}{c^2(1-\mathbf{n}\cdot\mathbf{v}/c)^3|\mathbf{r}-\mathbf{r}_0|} + O\left(\frac{1}{|\mathbf{r}-\mathbf{r}_0|^2}\right). \tag{10.88}$$

Our next step is to evaluate $\boldsymbol{\nabla}\times\mathbf{A}$ for the calculation of $\mathbf{B}$. First we observe that

$$\begin{aligned}\boldsymbol{\nabla}\times\mathbf{A} &= \boldsymbol{\nabla}_{t'=\text{const.}}\times\mathbf{A} + (\boldsymbol{\nabla}t')\frac{\partial}{\partial t'}\times\mathbf{A}\\ &= (\boldsymbol{\nabla}\times\mathbf{A})_{t'=\text{const.}} - \frac{\partial\mathbf{A}}{\partial t'}\times\boldsymbol{\nabla}t'.\end{aligned} \tag{10.89}$$

Here with Eq. (10.64)

$$(\boldsymbol{\nabla}\times\mathbf{A})_{t'=\text{const.}} = \left[\boldsymbol{\nabla}\times\frac{e\mathbf{v}(t')\mu_0/4\pi}{|\mathbf{r}-\mathbf{r}_0(t')|(1-\mathbf{n}(t')\cdot\mathbf{v}(t')/c)}\right]_{t'=\text{const.}}. \tag{10.90}$$

In the following vector product the gradient acts on the expression in the denominator; writing out the components and performing the differentiation one obtains (writing $|\mathbf{r}|$ instead of $|\mathbf{r}-\mathbf{r}_0(t')|$)

$$\boldsymbol{\nabla}\times\frac{\mathbf{v}(t')}{|\mathbf{r}|} = -\frac{1}{r^3}(\mathbf{r}\times\mathbf{v}(t')) = -\frac{1}{r^2}(\mathbf{n}\times\mathbf{v}(t')). \tag{10.91}$$

Hence we obtain

$$(\boldsymbol{\nabla}\times\mathbf{A})_{t'=\text{const.}} = -\frac{(e\mu_0/4\pi)(\mathbf{n}(t')\times\mathbf{v}(t'))}{|\mathbf{r}-\mathbf{r}_0(t')|^2(1-\mathbf{n}(t')\cdot\mathbf{v}(t')/c)}. \tag{10.92}$$

We also require in Eq. (10.89) the differentiation of $\mathbf{A}$ with respect to $t'$, i.e.

$$\frac{\partial\mathbf{A}}{\partial t'} = \frac{\partial}{\partial t'}\left[\frac{(e\mu_0/4\pi)\mathbf{v}(t')}{|\mathbf{r}-\mathbf{r}_0(t')|(1-\mathbf{n}(t')\cdot\mathbf{v}(t')/c)}\right], \tag{10.93}$$

which we already have with Eq. (10.83), i.e.

$$\frac{\partial \mathbf{A}}{\partial t'} = \frac{(e\mu_0/4\pi)\mathbf{v}(t')\{\mathbf{v}(t')\cdot(\mathbf{n}(t')-\mathbf{v}(t')/c)+|\mathbf{r}-\mathbf{r}_0(t')|\mathbf{a}(t')\cdot\mathbf{n}(t')/c\}}{|\mathbf{r}-\mathbf{r}_0(t')|^2(1-\mathbf{n}(t')\cdot\mathbf{v}(t')/c)^2} + \frac{(e\mu_0/4\pi)\mathbf{a}(t')}{|\mathbf{r}-\mathbf{r}_0(t')|(1-\mathbf{n}(t')\cdot\mathbf{v}(t')/c)}, \quad (10.94)$$

and finally (again skipping one intermediate rearrangement step)

$$\frac{\partial \mathbf{A}}{\partial t'} = \frac{(e\mu_0/4\pi)\mathbf{v}(t')\{\mathbf{v}(t')\cdot(\mathbf{n}(t')-\mathbf{v}(t')/c)\}}{|\mathbf{r}-\mathbf{r}_0(t')|^2(1-\mathbf{n}(t')\cdot\mathbf{v}(t')/c)^2} + \frac{(e\mu_0/4\pi)\{\mathbf{a}(t')-\mathbf{n}(t')\times(\mathbf{a}(t')\times\mathbf{v}(t')/c)\}}{|\mathbf{r}-\mathbf{r}_0(t')|(1-\mathbf{n}(t')\cdot\mathbf{v}(t')/c)^2}. \quad (10.95)$$

Using Eq. (10.89) we can now obtain the field **B**. With a rearrangement of terms we obtain

$$\mathbf{B} = \frac{e\mu_0}{4\pi}\frac{(\mathbf{v}(t')\times\mathbf{n}(t'))[(1-\mathbf{n}(t')\cdot\mathbf{v}(t')/c)^2+(\mathbf{v}(t')/c)\cdot(\mathbf{n}(t')-\mathbf{v}(t')/c)]}{|\mathbf{r}-\mathbf{r}_0(t')|^2(1-\mathbf{n}(t')\cdot\mathbf{v}(t')/c)^3} + \frac{e\mu_0}{4\pi}\frac{\{\mathbf{a}(t')-\mathbf{n}(t')\times(\mathbf{a}(t')\times\mathbf{v}(t')/c)\}\times\mathbf{n}(t')}{c|\mathbf{r}-\mathbf{r}_0(t')|(1-\mathbf{n}(t')\cdot\mathbf{v}(t')/c)^3} \text{ teslas.} \quad (10.96)$$

For large $|\mathbf{r}-\mathbf{r}_0|$ and $v/c \ll 1$ the dominant contributions come from the second term as

$$\mathbf{B}\rightarrow\mathbf{B}_0 = \frac{e\mu_0}{4\pi}\frac{\mathbf{a}(t')\times\mathbf{n}(t')}{c|\mathbf{r}-\mathbf{r}_0(t')|}, \quad \mathbf{n}\cdot\mathbf{B}_0 = 0. \quad (10.97)$$

Thus this radial component of order $1/|\mathbf{r}-\mathbf{r}_0(t')|$ vanishes which means the radial field is of order $1/r^2$.

We now have at our disposal the desired expressions for the field strengths **E** and **B** at a point **r** for the case when the charge $e$ moves with velocity **v** along its trajectory $\mathbf{r}_0$. We see that for velocities considerably smaller than $c$ the above expressions yield the nonrelativistic approximations

$$\mathbf{E}\rightarrow\frac{e}{4\pi\epsilon_0}\frac{\mathbf{n}}{|\mathbf{r}-\mathbf{r}_0(t)|^2}, \quad \mathbf{B}\rightarrow\frac{e\mu_0}{4\pi}\frac{(\mathbf{v}\times\mathbf{n})}{|\mathbf{r}-\mathbf{r}_0(t)|^2}, \quad \mathbf{n}=\frac{\mathbf{r}-\mathbf{r}_0(t')}{|\mathbf{r}-\mathbf{r}_0(t')|}, \quad (10.98)$$

which we recognize as the expressions of the Coulomb and Biot–Savart laws.

We also see, that both field strengths consist of two contributions — one decreasing like $1/|\mathbf{r}|$ with $r\rightarrow\infty$ and another decreasing like $1/|\mathbf{r}|^2$ with $r\rightarrow\infty$. Thus the dominant contributions are those of the first type.

These contributions contain the acceleration $\mathbf{a}$ and are therefore referred to as *radiation fields*. Hence *accelerated charges emit electromagnetic radiation.* The dominant contributions of these radiated fields are now — put together for convenience (and for later reference) —

$$\begin{aligned}\mathbf{E} &\simeq \frac{e}{4\pi\epsilon_0}\frac{\mathbf{n}(t')\{\mathbf{a}(t')\cdot\mathbf{n}(t')\}-\mathbf{a}(t')}{c^2|\mathbf{r}-\mathbf{r}_0(t')|} \quad \text{newtons/coulomb},\\ \mathbf{B} &\simeq \frac{e\mu_0}{4\pi}\frac{\mathbf{a}(t')\times\mathbf{n}(t')}{c|\mathbf{r}-\mathbf{r}_0(t')|} \quad \text{teslas.}\end{aligned} \tag{10.99}$$

Furthermore, one can convince oneself, that $\mathbf{E}$ and $\mathbf{B}$ are mutually orthogonal and that their moduli are equal. In the case of motion with vanishing acceleration, the velocity is constant. This is the case of a charge in an *inertial frame*; in this case the fields transform in accordance with the transformation from one inertial frame to another, these transformations being those of Lorentz transformations (see Chapter 17).

Obviously the next quantity to evaluate is the Poynting vector $\mathbf{S}$. Comparing the dominant contributions of order $1/|\mathbf{r}-\mathbf{r}_0(t')|$ in Eqs. (10.88) and (10.96), we see that (with $\epsilon_0\mu_0 c^2 = 1$)

$$c(\mathbf{n}\times\mathbf{E})\epsilon_0\mu_0 = \mathbf{B} + O(1/r^2), \quad \frac{\mathbf{n}\times\mathbf{E}}{\mu_0 c} = \mathbf{H} + O(1/r^2). \tag{10.100}$$

Hence the Poynting vector is

$$\begin{aligned}\mathbf{S} &= \mathbf{E}\times\mathbf{H} \overset{(10.100)}{=} \mathbf{E}\times\left(\frac{\mathbf{n}\times\mathbf{E}}{\mu_0 c}\right)\\ &= \frac{E^2}{c\mu_0}\mathbf{n} - \frac{(\mathbf{E}\cdot\mathbf{n})\mathbf{E}}{c\mu_0}\\ &\overset{(10.87)}{=} \frac{E^2}{c\mu_0}\mathbf{n} + O(1/|\mathbf{r}-\mathbf{r}_0|^3) \quad \text{watts/meter}^2.\end{aligned} \tag{10.101}$$

We deduce from the equation for $\mathbf{E}$, i.e. Eq. (10.86), that for $\mathbf{r}$ large and velocities small compared to $c$, and hence $t' \simeq t$ in the following:

$$\mathbf{E} = \frac{e}{4\pi\epsilon_0}\frac{\mathbf{a}^*}{c^2|\mathbf{r}-\mathbf{r}_0(t)|}, \quad \mathbf{a}^* := \mathbf{n}(t')\{\mathbf{a}(t')\cdot\mathbf{n}(t')\} - \mathbf{a}(t') \equiv \mathbf{n}\times(\mathbf{n}\times\mathbf{a}). \tag{10.102}$$

Since $\mathbf{a}^*\cdot\mathbf{n} = 0$, the quantity $\mathbf{a}^*$ is the acceleration perpendicular to $\mathbf{n}$, i.e. to the vector of the "direction of observation", i.e. the direction to the point $P$, at which we want to know the value of the fields. These directions are

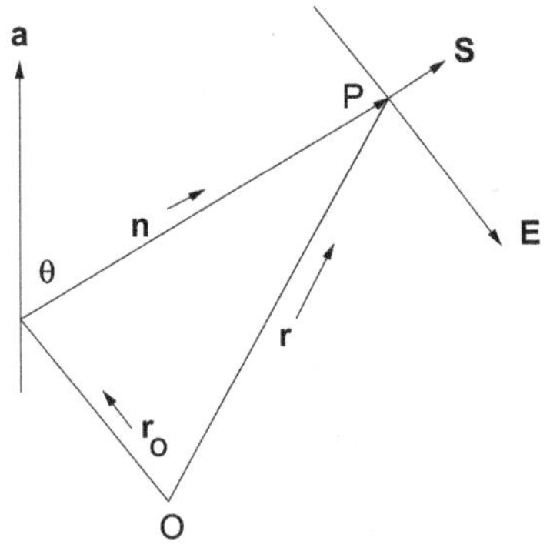


**Fig. 10.2.** The directions of **a** and **S**.

indicated in Fig. 10.2. The modulus of the electric field strength at the point $P$ is therefore given by the projection of **a** in the direction of observation, i.e., since $\mathbf{n} \times \mathbf{a}^* = 0 - \mathbf{n} \times \mathbf{a}$,

$$|\mathbf{E}| = \frac{e}{4\pi\epsilon_0} \frac{a \sin\theta}{c^2 |\mathbf{r} - \mathbf{r}_0(t)|} \quad \text{newtons/coulomb}, \tag{10.103}$$

which is zero along $\mathbf{n}$ (i.e. for $\theta = 0$), $\theta$ being the angle between $\mathbf{n}$ and $\mathbf{a}$, as in Fig. 10.2.

For the modulus of the Poynting vector we obtain (using $\epsilon_0\mu_0 c^2 = 1$)

$$|\mathbf{S}| = \frac{e^2 a^2 \sin^2\theta}{4\pi\epsilon_0 4\pi c^3 |\mathbf{r} - \mathbf{r}_0(t)|^2} \quad \text{watts/meter}^2. \tag{10.104}$$

The total energy $W$ radiated off per unit time at time $t'$, $\partial W/\partial t$, is obtained by integrating over all angles at radius $|\mathbf{r} - \mathbf{r}_0(t)|$, i.e. the *power* or rate is (cf. Eqs. (7.63) and (7.64))

$$\begin{aligned} P = \frac{dW}{dt} &= \left| \int \mathbf{S} \cdot d\mathbf{F} \right| = \int_0^\pi 2\pi \sin\theta d\theta |\mathbf{r} - \mathbf{r}_0(t)|^2 |\mathbf{S}| \quad \text{watts} \\ &= \frac{e^2 a^2}{8\pi\epsilon_0 c^3} \int_0^\pi \sin^3\theta d\theta = \frac{2}{3} \frac{e^2 a^2}{4\pi\epsilon_0 c^3} \quad \text{watts}, \end{aligned} \tag{10.105}$$

where we put $\cos\theta = z$ and evaluated the integral as

$$\int_0^\pi \sin^3\theta d\theta = \int_{-1}^1 dz(1 - z^2) = 4/3. \tag{10.106}$$

The final result, Eq. (10.105), is called *Larmor radiation formula* in the case of radiation of a single charged particle.[4] Since the main quantity appearing in this Larmor formula is the acceleration $a$, exercises or applications based on this formula generally require some acceleration. From Eq. (10.105) we deduce that the power loss per unit solid angle can be written

$$\frac{dP}{d\Omega} = \frac{e^2 a^2 \sin^2\theta}{4\pi\epsilon_0 4\pi c^3} \quad \text{watts per unit solid angle,} \tag{10.107}$$

where $\theta$ is the angle between the acceleration $\mathbf{a}$ and the direction $\mathbf{n}$ into the solid angle element $d\Omega = 2\pi \sin\theta d\theta$.

**Example 10.4 (Trick derivation of radiated fields).** The following trick calculation of radiated fields is due to Thompson.[5] We derive the central formula Eq. (10.109b) first with reference to Fig. 10.3. For a better understanding we then repeat the arguments with reference to Fig. 10.2.

**Solution:** Consider a point charge $e$, moving initially with velocity $\mathbf{u}$. During a short time interval $\delta t$ this charge is accelerated, and finally moves with velocity $\mathbf{u}'$. The acceleration is $\mathbf{a} = \delta\mathbf{u}/\delta t$. A signal, i.e. radiation, travels with the velocity of light $c$. Assume that a signal emitted by the charge $e$ at the *start* of acceleration reaches the sphere of radius $r + \delta r = r + c\delta t$ after time $t$ and thus the sphere of radius $r$ at an earlier time. Then a signal emitted

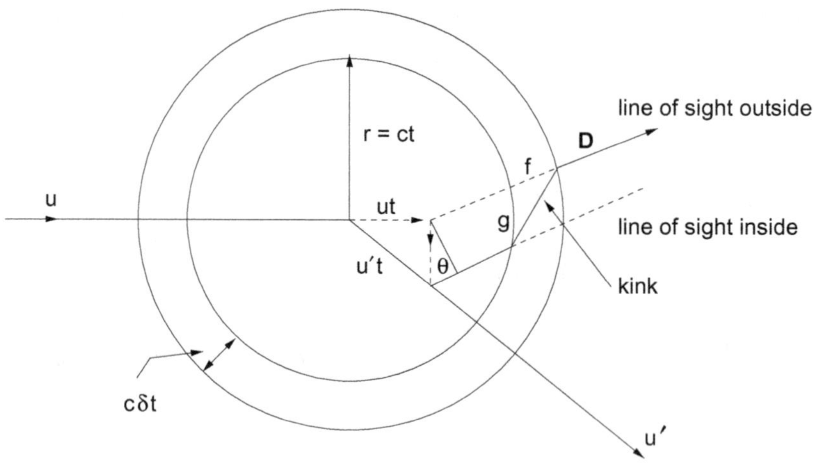


**Fig. 10.3.** The kink in **D**.

[4]Below another Larmor formula is obtained, but for dipole radiation.
[5]Dingle [37] and Thomson [148]. A somewhat different presentation can be found in Panofsky and Phillips [111], pp. 359–363.

at the *end* of acceleration reaches the sphere of radius $r + \delta r - c\delta t = r = ct$ after time $t$. Observers within the smaller sphere know that the charge is moving along a new path (since a signal of the change of path has already reached them), and hence know that the charge has moved the distance $u't$ along $\mathbf{u}'$. Observers outside the larger sphere still think that the charge is moving along the old path (since a signal of the change of path has not yet reached them) and hence think the charge has moved a distance $ut$ along $\mathbf{u}$. The direction of the radial component of the field must therefore shift sideways between the spheres, in other words, there is a *kink* in the line of the dielectric displacement field $\mathbf{D}$ there. This kink is indicated in Fig. 10.3. The ratio of the tangential and radial components of the field $\mathbf{D}$ along the kink is the same as the ratio of the lengths of the vector components along the tangential and radial directions. Thus (cf. Fig. 10.3, where $(\mathbf{u}' - \mathbf{u})t = (-\mathbf{a}\delta t)t$)

$$\left[\frac{D_{\text{tangential}}}{D_{\text{radial}}}\right]_{r=ct} := \frac{g}{f} = \left[\frac{(-a\delta t)t\sin\theta}{c\delta t}\right]_{r=ct} = -\frac{ar\sin\theta}{c^2}, \tag{10.108}$$

where $g$ and $f$ are the infinitesimal lengths indicated in Fig. 10.3. But (cf. Eqs. (2.3) and (4.115), and recall that $k = 1/4\pi\epsilon_0$)

$$D_{\text{radial}} = \epsilon_0 E_{\text{radial}} = \frac{\epsilon_0 ke}{r^2}, \quad E_{\text{radial}} = k\frac{e}{r^2} \tag{10.109a}$$

(in agreement with the comment after Eq. (10.87)), so that from Eq. (10.108)

$$E_{\text{tangential}} = -\frac{ar\sin\theta}{c^2}\frac{ke}{r^2} = -\frac{ea\sin\theta}{4\pi\epsilon_0 c^2 r} \quad \text{newtons/coulomb.} \tag{10.109b}$$

We observe that this result agrees with that of Eq. (10.103). In the text of Panofsky and Phillips [111] this equation corresponds to their Eq. (20.26), p.361.

We now reconsider the geometric arguments with reference to Fig. 10.2. We note that

$$\frac{d\mathbf{r}_0(t)}{dt} \equiv \mathbf{u}, \quad \frac{d\mathbf{r}'_0(t)}{dt} \equiv \mathbf{u}',$$

where $\mathbf{u}'$ is the velocity of the electron at the end of the acceleration $\mathbf{a}$ and hence $\mathbf{r}'_0(t)$ is in Fig. 10.2 a vector from $O$ to a point $Q$ on $\mathbf{a}$ further up (not shown). The line of the vector $\mathbf{n}$ to the point $P$ is the first line of sight, and a straight line drawn parallel to this from the acceleration endpoint $Q$ to a point $P'$ is the second line of sight, these lines being straight lines. The circles with origin $O$ and radial separation

$$f \equiv c\delta t = \delta r$$

and passing through $P$ (with $r(t) = ct$) and $P'$ (with $r = r(t + \delta t)$) are at these points infinitesimally straight line tangents there with the line $PP'$ being the kink in Fig. 10.3. The two lines of sight crossing the tangents form a triangle of which two side lengths are those of the kink and $f$. The third line of length $g$ is the intercept between the two lines of sight. This line with infinitesimal length is equal to the length of the line opposite angle $\theta$ of the triangle formed there by the vectors $\mathbf{a}$ and $\mathbf{n}$. Thus since $\mathbf{a}t$ is the velocity corresponding to acceleration $\mathbf{a}$, one obtains in the time interval $\delta t$ and at $r = ct$

$$|\sin\theta| = \frac{g}{(at)\delta t}$$

with $f = c\delta t$. The ratio $g/f$ of these lengths is also the ratio of the $D$-field components there which thus leads to Eqs. (10.109a) and (10.109b).

In order to obtain the radiated magnetic field we recall the Biot–Savart law (6.101) in which we insert the current density $\mathbf{j}$ of a single charge $e$ moving steadily with velocity $\mathbf{u}$ at position $\mathbf{r}'$, i.e. we consider (cf. Eq. (5.25))

$$\mathbf{B}(\mathbf{r}) = \frac{\mu}{4\pi}\int \frac{\mathbf{j}\times(\mathbf{r}-\mathbf{r}')}{|\mathbf{r}-\mathbf{r}'|^3}d\mathbf{r}' \quad \text{teslas} \quad \text{with} \quad \mathbf{j}(\mathbf{r}') = e\mathbf{u}\delta(\mathbf{r}'), \tag{10.110}$$

so that

$$\mathbf{B}(\mathbf{r}) = \frac{\mu}{4\pi}\frac{e\mathbf{u}\times\mathbf{r}}{r^3}, \quad \mathbf{H}(\mathbf{r}) = \frac{e}{4\pi}\frac{\mathbf{u}\times\mathbf{r}}{r^3}. \tag{10.111}$$

The change of $\mathbf{H}$ in the kink resulting from the acceleration $\mathbf{a} = \delta\mathbf{u}/\delta t$ is

$$\delta\mathbf{H} = \frac{e}{4\pi}\frac{\delta\mathbf{u}\times\mathbf{r}}{r^3} = \frac{e}{4\pi}\frac{\mathbf{a}\delta t\times\mathbf{r}}{r^3}. \tag{10.112}$$

If the observer in the kink projects back to the position of the charge (i.e. he will extrapolate to $t = 0$), he will think he sees a magnetic field ($t = r/c$)

$$\mathbf{H}_{\text{radiated}} = \frac{e}{4\pi}\frac{t\mathbf{a}\times\mathbf{r}}{r^3} = \frac{e}{4\pi}\frac{\mathbf{a}\times\mathbf{r}}{cr^2} \quad \text{amperes/meter.} \tag{10.113}$$

This means

$$\mathbf{H}_{\text{radiated}} = \frac{e}{4\pi}\frac{\mathbf{a}(t)\times\mathbf{n}}{cr} \quad \text{amperes/meter.} \tag{10.114}$$

The expression for $\mathbf{H}$ is seen to agree with that of $\mathbf{B}/\mu_0$ of Eq. (10.99).

**Example 10.5 (Radiation emitted by an accelerated proton).** A proton (charge $q$) is given a constant acceleration $a$ in a van de Graaff accelerator by a potential difference of 700 kilovolts. The acceleration region has a length of 3 meters. Calculate the ratio of the energy emitted by the proton, $U_{\text{radiation}}$, to its final kinetic energy, $U_k$, and derive with the help of the Larmor formula a relation between the emitted energy per unit time and the kinetic energy.

**Solution:** The energy emitted by a charge $q$ in time $t$ is according to Eq. (10.105) the energy

$$U_{\text{radiation}} = Pt = \frac{dU}{dt}t = \frac{2q^2a^2t}{4\pi\epsilon_0 3c^3} \quad \text{joules.} \tag{10.115}$$

Since velocity $v = at$, the length of the proton's runway to achieve its constant acceleration $a$ is $s = at^2/2 = vt/2$, so that $t = 2s/v, a = v/t = v^2/2s$, and $a^2 = (v^2/2s)(v/t), a^2t = v^3/2s$, and hence

$$U_{\text{radiation}} = \frac{q^2v^3}{4\pi\epsilon_0 3c^3 s} \quad \text{joules.} \tag{10.116}$$

The kinetic energy at the end is (with $M=$ mass of the proton, $V$ the potential):

$$U_k = \frac{1}{2}Mv^2 = qV, \quad v = \left(\frac{2qV}{M}\right)^{1/2}. \tag{10.117}$$

Hence

$$\frac{U_{\text{radiation}}}{U_k} = \frac{2q^2v}{4\pi\epsilon_0 3Mc^3 s} = \frac{2q^2}{4\pi\epsilon_0 3Mc^3 s}\left(\frac{2qV}{M}\right)^{1/2}. \tag{10.118}$$

In MKSA-units:

$$q = 1.6 \times 10^{-19} \text{ coulombs}, \quad V = 7 \times 10^5 \text{ volts},$$
$$c = 3 \times 10^8 \text{ meters per second}, \quad M = 1.67 \times 10^{-27}\text{C}/(\text{V} \cdot \text{m}). \tag{10.119}$$

It follows that

$$\frac{U_{\text{radiation}}}{U_k} = 1.31 \times 10^{-20}. \tag{10.120}$$

Thus this ratio of energy radiated off to the kinetic energy of the proton is extremely small!

Second part: $U_k = Mv^2/2$. Hence

$$\frac{dU_k}{dx} = Mv\frac{dv}{dx} = Mv\frac{dv}{dt}\frac{dt}{dx} = Ma, \quad \therefore a = \frac{1}{M}\frac{dU_k}{dx} \quad \text{meters/second}^2. \tag{10.121}$$

We insert this expression into the Larmor formula (10.105) and find:

$$\frac{dU_{\text{radiation}}}{dt} = \frac{2q^2}{4\pi\epsilon_0 3M^2c^3}\left(\frac{dU_k}{dx}\right)^2 \quad \text{watts.} \tag{10.122}$$

This is trivial for a constant acceleration; otherwise of general significance.

**Example 10.6**[6] **(Model of bremsstrahlung).** As a simplified model of bremsstrahlung[7] we consider the following scattering process. Electrons having velocity $\mathbf{v}$ are scattered off ions with charge $Q$. With the help of the Coulomb force law (which supplies the acceleration!) and the Larmor formula (10.105) derive an expression for the energy $W$ per unit time emitted as bremsstrahlung (let $N_e$ be the density of the electrons).

**Solution:** Let $m$ be the mass of an electron. The acceleration $a$ then follows from the Coulomb law as

$$a = \frac{1}{4\pi\epsilon_0}\frac{qQ}{mr^2} \quad \text{m/s}^2, \tag{10.123}$$

where $r$ is the distance between the electron and the ion.

The Larmor formula for the radiation power $P$ of a single charged particle is from Eq. (10.105):

$$P = \frac{2q^2a^2}{4\pi\epsilon_0 3c^3} \equiv \frac{dW}{dt} \quad \text{watts.} \tag{10.124}$$

With (see Fig. 10.4) $r^2 = x^2 + b^2, x = vt$, the formula yields the expression (by inserting a from Eq. (10.123))

$$P = \frac{1}{(4\pi\epsilon_0)^3}\frac{2q^4Q^2}{3m^2c^3(v^2t^2+b^2)^2} \quad \text{watts.} \tag{10.125}$$

Here $b$ is, as indicated in Fig. 10.4, the so-called *impact parameter* of the scattering process. In order to obtain the total energy emitted by one

[6] See also Bekefi and Barrett [9], pp. 278–283.

[7] This is the radiation from an electron which is decelerated (usually with no change in direction). See Panofsky and Phillips [111], p. 361.

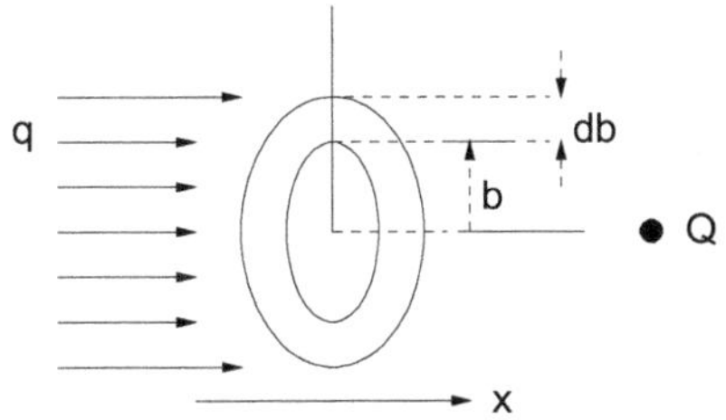


**Fig. 10.4.** The impact parameter $b$.

electron, we integrate over the time $t$ of this electron and obtain (using formula 120.2, p. 25 of Dwight [42])

$$W = \frac{1}{(4\pi\epsilon_0)^3}\frac{2q^4Q^2}{3m^2c^3}\int_{-\infty}^{\infty}\frac{dt}{(v^2t^2+b^2)^2} = \frac{1}{(4\pi\epsilon_0)^3}\frac{2q^4Q^2}{3m^2c^3}\frac{\pi}{2vb^3} \quad \text{joules.} \tag{10.126}$$

If $N_e$ is the number of electrons per unit volume as given, their current is $N_e v$. The entire energy emitted per unit time in the form of bremsstrahlung is therefore

$$P_{\text{brems}} = \int_{b_0}^{\infty} WN_e v \underbrace{2\pi b db}_{\text{cylindrical element of area}} \quad \text{watts,} \tag{10.127}$$

where $b_0$ is the smallest impact parameter. This result is incorrect only in the region of small impact parameters $b$, since for these the path of the electrons is not linear (in view of strong deflection close to the ion). With this approximation the result is

$$P_{\text{brems}} = \frac{1}{(4\pi\epsilon_0)^3}\frac{2N_eq^4Q^2\pi^2}{3m^2c^3}\int_{b_0}^{\infty}\frac{db}{b^2} = \frac{1}{(4\pi\epsilon_0)^3}\frac{2N_eq^4Q^2\pi^2}{3m^2c^3}\frac{1}{b_0} \quad \text{watts.} \tag{10.128}$$

**Example 10.7 (Cyclotron radiation).** Derive an expression for the energy per unit time emitted by an electron which is circling in a homogeneous magnetic field (cyclotron radiation).

**Solution:** An electron (charge $q = -e$) injected horizontally into a vertical magnetic field $B_0$ performs circular orbits around the lines of force of this field. Even with constant circular velocity the electron experiences a centripetal acceleration and hence emits electromagnetic radiation. In the case of low energy electrons, one refers to this as cyclotron radiation, in the case of relativistic electrons as synchrotron radiation. Let $v$ be the velocity of the electron on its circular orbit. The magnetic field subjects the electron to the Lorentz force $q(\mathbf{v} \times \mathbf{B}_0)$ resulting in an acceleration $v^2/R$, where $R$ is the radius of its circular orbit. Thus $qvB_0 = mv^2/R$ and hence $R = mv/qB_0$. The acceleration is therefore

$$a = \frac{v^2}{R} = \frac{v^2 qB_0}{mv} = v\frac{qB_0}{m} \quad \text{m/s}^2. \tag{10.129}$$

The *cyclotron frequency* $\omega_c$ is defined by $v = \omega_c R$, so that (note that the Larmor frequency, cf. Section 13.3, is frequently defined as $\omega_c/2$)

$$\omega_c = \frac{v}{R} = \frac{vqB_0}{mv} = \frac{qB_0}{m} \text{ rad/s}. \tag{10.130}$$

The acceleration is therefore $a = v\omega_c$. Inserting this expression into the single particle Larmor formula (10.105), we obtain for the power (i.e. energy per unit time) emitted by the electron the result

$$P_c = \frac{q^2\omega_c^2 v^2}{6\pi\epsilon_0 c^3} = \frac{1}{2}mv^2\frac{q^4}{3\pi\epsilon_0 c^3 m^3}B_0^2 \quad \text{watts/electron}. \tag{10.131}$$

If the emitting substance contains $N_e$ electrons per meter$^3$, the power per meter$^3$ is $P = N_e P_c$, i.e. $P = 6.21 \times 10^{-20} N_e B_0^2 u$ W/m$^3$, where $u$ is the energy of the electron expressed in electron volts (1 eV $= 1.6 \times 10^{-19}$ joules, see Appendix B).

## 10.5 The Hertz Dipole

We assume an arrangement of charges as depicted in Fig. 10.5 which is described as a *Hertz dipole*. Our notation corresponds to that used earlier in

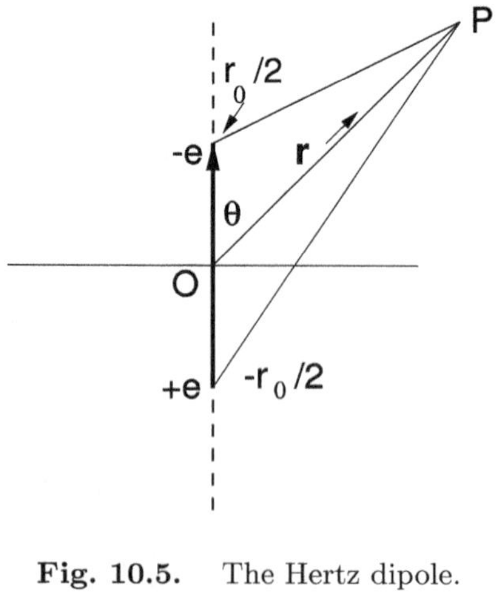


**Fig. 10.5.** The Hertz dipole.

this chapter. We assume that

$$\frac{v}{c} \ll 1, \quad \mathbf{r}(t') = \mathbf{r} - \mathbf{r}_0(t') \simeq \mathbf{r}. \tag{10.132}$$

for the distance to the field observation point $r \sim$ large, which is a plausible approximation. We consider again first the potentials $\mathbf{A}$ and $\phi$, and subsequently the field strengths $\mathbf{E}$ and $\mathbf{H}$.

### 10.5.1 *The potentials*

By superposition of the potentials of the two charges $+e, -e$ of the dipole at the points $\mathbf{r}_0/2, -\mathbf{r}_0/2$ we obtain from Eq. (10.64) for the vector potential

$$\begin{aligned}\mathbf{A}(\mathbf{r},t) = &-\frac{e\mu_0}{4\pi}\left[+\frac{\dot{\mathbf{r}}_0/2}{r - \mathbf{r}\cdot\dot{\mathbf{r}}_0/2c}\right]_{t'=t-r(t')/c\simeq t-r/c} \\ &+\frac{e\mu_0}{4\pi}\left[-\frac{\dot{\mathbf{r}}_0/2}{r + \mathbf{r}\cdot\dot{\mathbf{r}}_0/2c}\right]_{t'=t-r(t')/c\simeq t-r/c}.\end{aligned} \tag{10.133}$$

For large values of $r$ one would naturally approximate this by, setting $\mathbf{p} = -e\mathbf{r}_0$,

$$\mathbf{A}(\mathbf{r},t) \simeq -\frac{\mu_0}{4\pi}\left(\frac{e\dot{\mathbf{r}}_0(t')}{r}\right)_{t'=t-r/c} = \frac{\mu_0}{4\pi}\frac{\dot{\mathbf{p}}(t-r/c)}{r} + O(|\dot{\mathbf{r}}_0/c|). \tag{10.134}$$

Here we content ourselves first with *only this contribution*, because — as we shall see — we then obtain $\mathbf{H}$ of the same order as $\mathbf{E}$. In a similar manner we calculate with the help of Eq. (10.66) the scalar potential $\phi$ of the dipole[8]:

$$\begin{aligned}\phi(\mathbf{r},t) &= -\frac{e}{4\pi\epsilon_0}\left(\frac{1}{r - \mathbf{r}\cdot\dot{\mathbf{r}}_0/2c}\right)_{t'=t-r/c} + \frac{e}{4\pi\epsilon_0}\left(\frac{1}{r + \mathbf{r}\cdot\dot{\mathbf{r}}_0/2c}\right)_{t'=t-r/c} \\ &\overset{r\text{ large}}{\simeq} -\frac{e}{4\pi\epsilon_0 r}\left(1 + \frac{\mathbf{r}\cdot\dot{\mathbf{r}}_0}{2rc} + \cdots\right)_{t'=t-r/c} \\ &\quad + \frac{e}{4\pi\epsilon_0 r}\left(1 - \frac{\mathbf{r}\cdot\dot{\mathbf{r}}_0}{2rc} + \cdots\right)_{t'=t-r/c} \\ &\simeq \frac{1}{4\pi\epsilon_0}\left(\frac{\dot{\mathbf{p}}\cdot\mathbf{r}}{cr^2}\right)_{t'=t-r/c} = O(1/r) \quad \text{volts}.\end{aligned} \tag{10.135}$$

Thus in this case the dominant contributions of order $1/r$ cancel out — which is different from the case of $\mathbf{A}(\mathbf{r},t)$. It is therefore necessary to examine the approximations made in the denominator in more detail.

[8]In the second step we also neglect contributions of $O(|\dot{\mathbf{r}}_0/c|^n)$ for $n \geq 2$.

On the other hand, *for smaller values of* $r$ (i.e. of $O(1/r^2)$) *we expect the familiar expression for the potential of the static dipole*, i.e. (cf. Eq. (3.173))

$$\phi = k\frac{\mathbf{d}\cdot\mathbf{r}}{r^3}, \quad k = \frac{1}{4\pi\epsilon_0}. \tag{10.136}$$

We observe, however, that $\phi$ does not supply this expression! This is due to the fact, that in deriving this expression, we replaced the factor $1/|\mathbf{r}-\mathbf{r}_0(t')|$ by $1/|\mathbf{r}| = 1/r$, i.e. $|\mathbf{r}-\mathbf{r}_0(t')| \simeq r$. We therefore return to Eq. (10.66), i.e. to

$$\phi = \left(\frac{e/4\pi\epsilon_0}{r(t') - \mathbf{r}(t')\cdot\mathbf{v}(t')/c}\right)_{t'=t-r(t')/c}, \tag{10.137}$$

or more explicitly to Eq. (10.62):

$$\phi(\mathbf{r},t) = \frac{e}{4\pi\epsilon_0}\frac{1}{[|\mathbf{r}-\mathbf{r}_0(t')| - (\mathbf{r}-\mathbf{r}_0(t'))\cdot\mathbf{v}(t')/c]}\Bigg|_{t'=t-|\mathbf{r}-\mathbf{r}_0(t')|/c}. \tag{10.138}$$

Now,

$$|\mathbf{r}-\mathbf{r}_0(t')| = \sqrt{\mathbf{r}^2 - 2\mathbf{r}\cdot\mathbf{r}_0(t') + \mathbf{r}_0(t')^2},$$

so that for $r^2 \gg r_0^2$ we can expand:

$$|\mathbf{r}-\mathbf{r}_0(t')| \simeq \sqrt{\mathbf{r}^2 - 2\mathbf{r}\cdot\mathbf{r}_0(t')} \simeq r\left(1 - \frac{\mathbf{r}\cdot\mathbf{r}_0(t')}{r^2}\right) = r - \mathbf{r}_0(t')\cdot\frac{\mathbf{r}}{r}. \tag{10.139}$$

But from the definition (10.59),

$$\mathbf{n} = \frac{\mathbf{r}-\mathbf{r}_0(t')}{|\mathbf{r}-\mathbf{r}_0(t')|}, \tag{10.140}$$

we obtain by multiplication

$$\mathbf{r} = \mathbf{n}|\mathbf{r}-\mathbf{r}_0(t')| + \mathbf{r}_0(t') \simeq \mathbf{n}\left(r - \mathbf{r}_0(t')\cdot\frac{\mathbf{r}}{r}\right) + \mathbf{r}_0(t'), \tag{10.141}$$

and hence

$$\mathbf{r}_0(t')\cdot\frac{\mathbf{r}}{r} \simeq \mathbf{r}_0(t')\cdot\mathbf{n}\left(1 - \mathbf{r}_0(t')\cdot\frac{\mathbf{r}}{r^2}\right) + \frac{\mathbf{r}_0(t')^2}{r} \simeq \mathbf{r}_0(t')\cdot\mathbf{n} + O(1/r).$$

Thus from Eq. (10.139):

$$|\mathbf{r}-\mathbf{r}_0(t')| \simeq r - \mathbf{r}_0(t')\cdot\mathbf{n} + O(1/r). \tag{10.142}$$

Here the second term on the right (with the scalar product) was missing in the first approximations used for the potentials of the two charges in

Eq. (10.135). We insert this expression now in $\phi$, keeping in mind, that $\mathbf{v}(t') = \dot{\mathbf{r}}_0(t')$. Then (for the case of a single charge $e$) Eq. (10.138) becomes

$$\begin{aligned}\phi(\mathbf{r},t) &\simeq \frac{e}{4\pi\epsilon_0}\frac{1}{[r - \mathbf{r}_0(t')\cdot\mathbf{n} - (\mathbf{r}-\mathbf{r}_0(t'))\cdot\dot{\mathbf{r}}_0(t')/c]}\Bigg|_{t'=t-r(t')/c\simeq t-r/c} \\ &\simeq \frac{e}{4\pi\epsilon_0}\frac{1}{[r - \underbrace{\mathbf{r}_0(t')\cdot\mathbf{n}}_{\text{not in first approx.}} - \mathbf{r}\cdot\dot{\mathbf{r}}_0(t')/c]}\Bigg|_{t'\simeq t-r/c}. \end{aligned} \tag{10.143}$$

Applying this general formula to the special charges of the dipole we have to take into account that the coordinates of the charges, i.e. $\mathbf{r}_0(t')$, have to be replaced by $\pm\mathbf{r}_0(t')/2$. With these changes we obtain instead of Eq. (10.135):

$$\begin{aligned}\phi(\mathbf{r},t) = \Bigg[&-\frac{e}{4\pi\epsilon_0}\left(\frac{1}{r - \mathbf{r}_0\cdot\mathbf{n}/2 - \mathbf{r}\cdot\dot{\mathbf{r}}_0/2c}\right) \\ &+ \frac{e}{4\pi\epsilon_0}\left(\frac{1}{r + \mathbf{r}_0\cdot\mathbf{n}/2 + \mathbf{r}\cdot\dot{\mathbf{r}}_0/2c}\right)\Bigg]_{t'=t-r/c},\end{aligned}$$

and thus for $r$ large:

$$\begin{aligned}\phi(\mathbf{r},t) &\overset{r\text{ large}}{\simeq} -\frac{e}{4\pi\epsilon_0 r}\left(1 + \frac{\mathbf{r}_0\cdot\mathbf{n}}{2r} + \frac{\mathbf{r}\cdot\dot{\mathbf{r}}_0}{2rc} + \cdots\right)_{t'=t-r/c} \\ &\quad + \frac{e}{4\pi\epsilon_0 r}\left(1 - \frac{\mathbf{r}_0\cdot\mathbf{n}}{2r} - \frac{\mathbf{r}\cdot\dot{\mathbf{r}}_0}{2rc} + \cdots\right)_{t'=t-r/c} \\ &\simeq \frac{1}{4\pi\epsilon_0}\left(\frac{\mathbf{p}\cdot\mathbf{n}}{r^2} + \frac{\dot{\mathbf{p}}\cdot\mathbf{r}}{cr^2}\right)_{t'=t-r/c} \\ &= \frac{1}{4\pi\epsilon_0}\left(\frac{\mathbf{p}\cdot\mathbf{r}}{r^3} + \frac{\dot{\mathbf{p}}\cdot\mathbf{r}}{cr^2}\right)_{t'=t-r/c} \quad \text{volts.}\end{aligned} \tag{10.144}$$

We therefore have the situation, that for $r\to\infty$ the second term ($\propto 1/r$) dominates, but for smaller $r$ the first term ($\propto 1/r^2$). One should note that on the whole the expansion corresponds to the multi-pole expansion, i.e. Coulomb contribution + dipole contribution $+\cdots$.

### 10.5.2 *The field strengths*

We obtain the field strengths $\mathbf{E}$ and $\mathbf{H}$ from the equations

$$\mathbf{E} = -\frac{\partial\mathbf{A}}{\partial t} - \boldsymbol{\nabla}\phi, \quad \mu_0\mathbf{H} = \boldsymbol{\nabla}\times\mathbf{A}. \tag{10.145}$$

We evaluate first $-\nabla\phi$. For this we need

$$\frac{\partial}{\partial \mathbf{r}}(r) = \left(\frac{\partial}{\partial x}, \frac{\partial}{\partial y}, \frac{\partial}{\partial z}\right)\sqrt{x^2+y^2+z^2} = \frac{\mathbf{r}}{r},$$
$$\frac{\partial}{\partial \mathbf{r}}\left(\frac{1}{r^2}\right) = -\frac{2}{r^3}\frac{\partial r}{\partial \mathbf{r}} = -\frac{2\mathbf{r}}{r^4}, \quad \frac{\partial}{\partial \mathbf{r}}\left(\frac{1}{r^3}\right) = -\frac{3}{r^4}\frac{\partial r}{\partial \mathbf{r}} = -\frac{3\mathbf{r}}{r^5}. \tag{10.146}$$

Then

$$\begin{aligned}
-\nabla\phi &= -\frac{1}{4\pi\epsilon_0}\frac{\partial}{\partial \mathbf{r}}\left\{\left[\frac{\mathbf{p}\cdot\mathbf{r}}{r^3} + \frac{\dot{\mathbf{p}}\cdot\mathbf{r}}{cr^2}\right]_{t'=t-r(t')/c}\right\} \\
&= -\frac{1}{4\pi\epsilon_0}\frac{\partial}{\partial \mathbf{r}}\left\{\left[\frac{\mathbf{p}(t-r(t')/c)\cdot\mathbf{r}}{r^3} + \frac{\dot{\mathbf{p}}(t-r(t')/c)\cdot\mathbf{r}}{cr^2}\right]_{t'=t-r(t')/c}\right\} \\
&= \frac{1}{4\pi\epsilon_0}\left[-\frac{\dot{\mathbf{p}}\cdot\mathbf{r}}{r^3}\frac{\partial}{\partial \mathbf{r}}\left(-\frac{r(t')}{c}\right) - \frac{\mathbf{p}}{r^3} - \mathbf{p}\cdot\mathbf{r}\frac{\partial}{\partial \mathbf{r}}\left(\frac{1}{r^3}\right)\right. \\
&\quad \left. -\frac{\ddot{\mathbf{p}}\cdot\mathbf{r}}{cr^2}\frac{\partial}{\partial \mathbf{r}}\left(-\frac{r(t')}{c}\right) - \frac{\dot{\mathbf{p}}}{cr^2} - \frac{\dot{\mathbf{p}}\cdot\mathbf{r}}{c}\frac{\partial}{\partial \mathbf{r}}\left(\frac{1}{r^2}\right)\right],
\end{aligned} \tag{10.147}$$

where

$$r(t') = |\mathbf{r} - \mathbf{r}_0(t')| = \sqrt{(\mathbf{r} - \mathbf{r}_0(t'))^2} \tag{10.148}$$

and

$$\frac{\partial r(t')}{\partial \mathbf{r}} = \frac{\mathbf{r} - \mathbf{r}_0(t')}{|\mathbf{r} - \mathbf{r}_0(t')|} \simeq \frac{\mathbf{r}}{r}. \tag{10.149}$$

Hence

$$\begin{aligned}
-\nabla\phi &= \frac{1}{4\pi\epsilon_0}\left[\frac{\dot{\mathbf{p}}\cdot\mathbf{r}}{cr^3}.\frac{\mathbf{r}}{r} - \frac{\mathbf{p}}{r^3} - \mathbf{p}\cdot\mathbf{r}\left(-\frac{3\mathbf{r}}{r^5}\right) + \frac{\ddot{\mathbf{p}}\cdot\mathbf{r}}{c^2r^2}.\frac{\mathbf{r}}{r} - \frac{\dot{\mathbf{p}}}{cr^2} - \frac{\dot{\mathbf{p}}\cdot\mathbf{r}}{c}\left(-\frac{2\mathbf{r}}{r^4}\right)\right] \\
&= \frac{1}{4\pi\epsilon_0}\left[3\frac{(\dot{\mathbf{p}}\cdot\mathbf{r})\mathbf{r}}{cr^4} - \frac{\mathbf{p}}{r^3} + \frac{3(\mathbf{p}\cdot\mathbf{r})\mathbf{r}}{r^5} + \frac{(\ddot{\mathbf{p}}\cdot\mathbf{r})\mathbf{r}}{c^2r^3} - \frac{\dot{\mathbf{p}}}{cr^2}\right].
\end{aligned} \tag{10.150}$$

This together with Eqs. (10.134) and (10.145) yields

$$\mathbf{E} = -\frac{\mu_0}{4\pi}\frac{\ddot{\mathbf{p}}}{r} + \frac{1}{4\pi\epsilon_0}\left[-\frac{\dot{\mathbf{p}}}{cr^2} + \frac{3(\dot{\mathbf{p}}\cdot\mathbf{r})\mathbf{r}}{cr^4} + \frac{(\ddot{\mathbf{p}}\cdot\mathbf{r})\mathbf{r}}{c^2r^3} + \frac{3(\mathbf{r}\cdot\mathbf{p})\mathbf{r}}{r^5} - \frac{\mathbf{p}}{r^3}\right] \tag{10.151}$$

and

$$\begin{aligned}\mu_0 \mathbf{H} &= \boldsymbol{\nabla} \times \mathbf{A} \\ &= \frac{\mu_0}{4\pi} \boldsymbol{\nabla} \times \frac{\dot{\mathbf{p}}(t - r/c)}{r} = \frac{\mu_0}{4\pi} \frac{\mathbf{r}}{r} \frac{\partial}{\partial r} \times \frac{\dot{\mathbf{p}}(t - r/c)}{r} \\ &= -\frac{\mu_0}{4\pi c} \frac{\mathbf{r}}{r} \times \frac{\ddot{\mathbf{p}}}{r} - \frac{\mu_0}{4\pi} \frac{\mathbf{r}}{r} \times \frac{\dot{\mathbf{p}}}{r^2} \\ &= \frac{\mu_0}{4\pi} \frac{\ddot{\mathbf{p}} \times \mathbf{r}}{cr^2} + \frac{\mu_0}{4\pi} \frac{\dot{\mathbf{p}} \times \mathbf{r}}{r^3},\end{aligned} \tag{10.152}$$

so that

$$\mathbf{H} = \frac{1}{4\pi} \left[ \frac{\ddot{\mathbf{p}} \times \mathbf{r}}{cr^2} + \frac{\dot{\mathbf{p}} \times \mathbf{r}}{r^3} \right] \quad \text{amperes/meter.} \tag{10.153}$$

In the so-called *near field region*, (i.e. for $r \sim$ small compared with very large), we have

$$\mathbf{E} = \frac{1}{4\pi\epsilon_0} \left[ \frac{3(\mathbf{p} \cdot \mathbf{r})\mathbf{r}}{r^5} - \frac{\mathbf{p}}{r^3} \right], \quad \mathbf{H} = \frac{1}{4\pi} \frac{\dot{\mathbf{p}} \times \mathbf{r}}{r^3}. \tag{10.154}$$

However, in the so-called *far field region*, (i.e. for $r \sim$ large), we obtain

$$\mathbf{E} \simeq -\frac{\mu_0}{4\pi} \frac{\ddot{\mathbf{p}}}{r} + \frac{1}{4\pi\epsilon_0} \frac{(\ddot{\mathbf{p}} \cdot \mathbf{r})\mathbf{r}}{c^2 r^3} = \frac{1}{4\pi\epsilon_0} \frac{(\ddot{\mathbf{p}} \times \mathbf{r}) \times \mathbf{r}}{c^2 r^3} \tag{10.155}$$

and

$$\mathbf{H} \simeq \frac{1}{4\pi} \frac{\ddot{\mathbf{p}} \times \mathbf{r}}{cr^2} \quad \text{amperes/meter,} \tag{10.156}$$

where

$$p = p(t')|_{t' = t - r(t')/c \simeq t - r/c}.$$

Since $\epsilon_0 = 1/\mu_0 c^2$, we deduce from Eqs. (10.155) and (10.156), that in the far field region

$$\mathbf{E} \perp \mathbf{r}, \mathbf{H}.$$

We see that in this region (considered as a scalar triple product)

$$\mathbf{r} \cdot \mathbf{E} \propto [(\ddot{\mathbf{p}} \times \mathbf{r}) \times \mathbf{r}] \cdot \mathbf{r} = (\ddot{\mathbf{p}} \times \mathbf{r}) \cdot (\mathbf{r} \times \mathbf{r}) = 0, \tag{10.157}$$

and

$$\mathbf{r} \cdot \mathbf{H} \propto (\ddot{\mathbf{p}} \times \mathbf{r}) \cdot \mathbf{r} = \ddot{\mathbf{p}} \cdot (\mathbf{r} \times \mathbf{r}) = 0. \tag{10.158}$$

It follows that $\mathbf{r} \perp \mathbf{E}, \mathbf{H}$. Since also $\mathbf{E} \perp \mathbf{r}, \mathbf{H}$, the vectors $\mathbf{r}, \mathbf{E}, \mathbf{H}$ define at *large distances* an orthogonal system. At large distances

$$\mathbf{E} = \frac{1}{c\epsilon_0}\mathbf{H} \times \frac{\mathbf{r}}{r} \quad \text{newtons/coulomb.} \tag{10.159}$$

Thus

$$\begin{aligned} -\mathbf{E} \times \frac{\mathbf{r}}{r} &= -\frac{1}{c\epsilon_0}\left(\mathbf{H} \times \frac{\mathbf{r}}{r}\right) \times \frac{\mathbf{r}}{r} = \frac{1}{c\epsilon_0}\frac{\mathbf{r}}{r} \times \left(\mathbf{H} \times \frac{\mathbf{r}}{r}\right) \\ &= -\frac{1}{c\epsilon_0}\left(\underbrace{\left\{\mathbf{H} \cdot \frac{\mathbf{r}}{r}\right\}}_{=0}\frac{\mathbf{r}}{r} - \frac{r^2}{r^2}\mathbf{H}\right) = \frac{1}{c\epsilon_0}\mathbf{H}. \end{aligned} \tag{10.160}$$

## 10.6 Current Element $Id\mathbf{s}$ and Dipole Radiation

We now consider the following. An element of current $Id\mathbf{s}$ of a conducting wire with alternating current $I$ can be looked at as an electric dipole of length $ds$ with charges $\pm q = \pm q_0 e^{i\omega t}$ at its endpoints. Since

$$\mathbf{p} = qd\mathbf{s}, \quad \mathbf{p}_0 = q_0 d\mathbf{s}, \quad \mathbf{p} = \mathbf{p}_0 e^{i\omega t}, \tag{10.161}$$

we can put formally (here $d\mathbf{p} = dqd\mathbf{s}$)

$$Id\mathbf{s} = \frac{dq}{dt}d\mathbf{s} \equiv \dot{\mathbf{p}}. \tag{10.162}$$

Then

$$\dot{\mathbf{p}} = \dot{q}d\mathbf{s} = i\omega\mathbf{p} \equiv i\omega\mathbf{p}_0 e^{i\omega t}, \quad \ddot{\mathbf{p}} = i\omega\dot{\mathbf{p}} = i\omega Id\mathbf{s}. \tag{10.163}$$

Only the real part is physically meaningful, so that

$$\dot{\mathbf{p}} \to -\omega\mathbf{p}_0 \sin\omega t. \tag{10.164}$$

Thus in the far-away region[9] (cf. Eq. (10.155))

$$\mathbf{E} = \frac{(\ddot{\mathbf{p}} \times \mathbf{r}) \times \mathbf{r}}{4\pi\epsilon_0 c^2 r^3} = \frac{i\omega(Id\mathbf{s} \times \mathbf{r}) \times \mathbf{r}}{4\pi\epsilon_0 c^2 r^3}. \tag{10.165}$$

Using the relations $c = \nu\lambda, \omega = 2\pi\nu$, $\omega = 2\pi c/\lambda$, we obtain

$$\mathbf{E} = i\frac{2\pi(Id\mathbf{s} \times \mathbf{r}) \times \mathbf{r}}{4\pi\epsilon_0\lambda c r^3} \quad \text{newtons/coulomb,} \tag{10.166}$$

[9]Like previously $\mathbf{p} = \mathbf{p}(t')$, where $t' = t - r/c$, but for reasons of simplicity we write simply $p(t)$, where $t \to t - r/c$.

and similarly (cf. Eq. (10.156))

$$\mathbf{H} = \frac{\ddot{\mathbf{p}} \times \mathbf{r}}{4\pi c r^2} = \frac{i\omega I d\mathbf{s} \times \mathbf{r}}{4\pi c r^2} = i\frac{2\pi I d\mathbf{s} \times \mathbf{r}}{\lambda 4\pi r^2} \quad \text{amperes/meter.} \tag{10.167}$$

### 10.6.1 *The power of an oscillating dipole*

The radiation emitted by $Id\mathbf{s}$ or $\mathbf{p}$ is therefore called *dipole radiation.* In propagating radially away from the dipole, the wave transports energy away. The *power*, i.e. the loss of energy per unit time, can be calculated with the help of the Poynting vector $\mathbf{S}$. With Eqs. (10.158) and (10.159) we obtain

$$\begin{aligned} \mathbf{S} &= \mathbf{E} \times \mathbf{H} \quad \text{watts/meter}^2 \\ &\overset{\text{far region}}{=} \frac{1}{c\epsilon_0}\left(\mathbf{H} \times \frac{\mathbf{r}}{r}\right) \times \mathbf{H} = -\frac{\mathbf{H}}{c\epsilon_0} \times \left(\mathbf{H} \times \frac{\mathbf{r}}{r}\right) \\ &= \frac{1}{c\epsilon_0}(\mathbf{H}\cdot\mathbf{H})\frac{\mathbf{r}}{r} - \frac{1}{c\epsilon_0}\underbrace{\left(\mathbf{H}\cdot\frac{\mathbf{r}}{r}\right)}_{0}\mathbf{H}, \end{aligned} \tag{10.168}$$

and hence with Eq. (10.167):

$$\mathbf{S} = \frac{1}{c\epsilon_0}\mathbf{H}^2\frac{\mathbf{r}}{r} = \frac{1}{c\epsilon_0}\left(\frac{\ddot{\mathbf{p}} \times \mathbf{r}}{4\pi c r^2}\right)^2\frac{\mathbf{r}}{r} = \frac{(\ddot{\mathbf{p}})^2 \sin^2\theta}{(4\pi)^2\epsilon_0 c^3 r^2}\frac{\mathbf{r}}{r} \quad \text{watts/meter}^2. \tag{10.169}$$

We see that $\mathbf{S}$ points in the direction of $\mathbf{r}$. *The oscillating dipole does not emit radiation in the direction of its axis, but perpendicular to this maximally, i.e. in the equatorial plane*, as indicated in Fig. 10.6. We now put, along the lines given above,

$$\dot{\mathbf{p}} = -\omega\mathbf{p}_0 \sin\omega t, \quad \ddot{\mathbf{p}} = -\omega^2\mathbf{p}_0 \cos\omega t \tag{10.170}$$

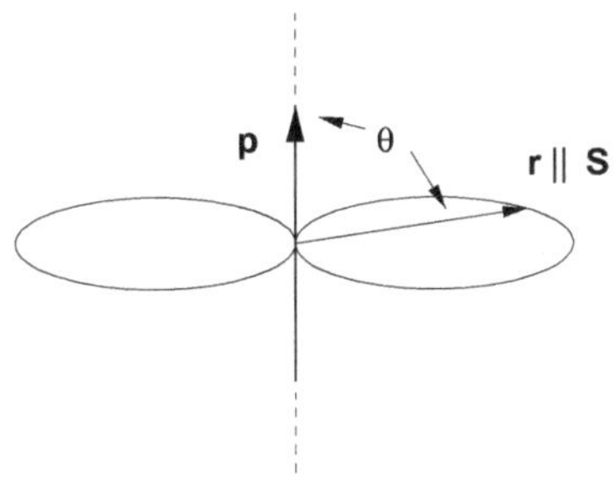


**Fig. 10.6.** The radiating dipole.

(actually here $t \to t - r/c$, called *retarded time*, but this is irrelevant here) and insert this in **S** of Eq. (10.169). Then

$$|\mathbf{S}| = \frac{p_0^2\omega^4}{\epsilon_0(4\pi)^2c^3r^2}\cos^2\omega t\sin^2\theta \quad \text{watts/meter}^2. \tag{10.171}$$

The *average energy emitted per unit time and unit area* is obtained by calculating the average of $\cos^2\omega t$ over one complete period of oscillation $T = 2\pi/\omega$, i.e. by considering on the one hand[10]

$$\frac{1}{T}\int_0^T \cos^2\omega t dt = \frac{1}{T}\int_0^T \cos^2\frac{2\pi t}{T}dt = \frac{1}{2\pi}\int_0^{2\pi}\cos^2 x dx = \frac{\pi}{2\pi} = \frac{1}{2}, \tag{10.172}$$

but also (required a little later)

$$\frac{1}{T}\int_0^T \{\cos^2\omega t + \sin^2\omega t\}\, dt = \frac{1}{T}T = 1 \quad \text{and} \quad \frac{1}{T}\int_0^T \sin^2\omega t dt = \frac{1}{2}. \tag{10.173}$$

On the other hand we have to integrate $|\mathbf{S}|$ over all directions of a sphere with radius $r$. Hence the average energy emitted per unit time is ($d\Omega = d\varphi\sin\theta d\theta$)

$$\begin{aligned} L &= \frac{1}{2}\int |\mathbf{S}|r^2 d\Omega = \frac{1}{2}\int_0^{2\pi} d\varphi\int_0^{\pi}|\mathbf{S}|r^2\sin\theta d\theta \\ &= \frac{1}{2}\frac{2\pi p_0^2\omega^4}{\epsilon_0(4\pi)^2c^3}\int_0^{\pi}\sin^3\theta d\theta = \frac{p_0^2\omega^4}{4c^3 4\pi\epsilon_0}\int_{-1}^{1}(1-z^2)dz. \end{aligned} \tag{10.174}$$

Here we put $z = \cos\theta$, so that $dz = -\sin\theta d\theta, (1-z^2)dz = -\sin^3\theta d\theta$. Moreover, $\int_{-1}^{1}(1-z^2)dz = [z - z^3/3]_{-1}^{1} = 4/3$, so that

$$L = \frac{p_0^2\omega^4}{4\pi\epsilon_0 3c^3} \quad \text{watts}. \tag{10.175}$$

This result is also known as a *Larmor formula*; we can call it the *Larmor dipole formula*,[11] in order to distinguish it from the Larmor formula (10.105) which we obtained for a *single* charged particle. The present formula implies that $L \propto p_0^2$, i.e. proportional to the square of the amplitude $p_0$, but also that

[10]See e.g. Dwight [42], formula 440.20, p. 95, $\int\cos^2 x dx = x/2 + (\sin x\cos x)/2$.

[11]A simple magnetic moment $\mathbf{m}$ due to a charge $q$ which is moving with simple harmonic motion does not radiate since (see Example 5.22) $\dot{\mathbf{m}} = 0$. For the detailed calculation see Podolsky and Kunz [118], p. 216.

$L \propto \omega^4$, the fourth power of the frequency. Expression (10.175) should be compared with the *Rayleigh–Jeans law* in statistical mechanics, which gives the classical limit of black body radiation.[12] The radiation emitted by atoms is of dipole-type; however, the classical picture of an atom as a dipole like the one discussed here, is — as we know — wrong (and in any case would not explain the stability of atoms).

### 10.6.2 *Radiation resistance*

We saw in connection with Eq. (10.170) that

$$Id\mathbf{s} = \dot{\mathbf{p}} = -\omega \mathbf{p}_0 \sin \omega t. \tag{10.176}$$

We now put

$$|\mathbf{p}_0\omega| \equiv |I_0 d\mathbf{s}|, \quad \text{or} \quad p_0^2\omega^2 = (I_0 d\mathbf{s})^2. \tag{10.177}$$

We can say therefore that on the average (this means here over one period) and using Eq. (10.173)

$$\begin{aligned} \frac{1}{T}\int_0^T (Id\mathbf{s})^2 dt &= \frac{1}{T}\int_0^T \omega^2 p_0^2 \sin^2 \omega t dt = \frac{1}{2}\omega^2 p_0^2 = \frac{1}{2} I_0^2 (d\mathbf{s})^2, \\ I_0^2 &= \frac{2}{T}\int_0^T I^2(t) dt \equiv 2\bar{I}^2, \end{aligned} \tag{10.178}$$

i.e. $I_0 \sim \sqrt{2}\times$ mean current $\bar{I}$. Then, inserting into Eq. (10.175),

$$L = \frac{(I_0 d\mathbf{s})^2\omega^2}{3c^3 4\pi\epsilon_0} = \frac{(\sqrt{2}\bar{I} d\mathbf{s})^2\omega^2}{3c^3 4\pi\epsilon_0} \quad \text{J/s}. \tag{10.179}$$

This result is to be compared with (see below) the *Ohmic power of a resistance* $R$, i.e. with

$$L = I^2 R \quad \text{watts}. \tag{10.180}$$

[12] This is a body with absorption coefficient $a(\omega) = 1$ in the case of thermal equilibrium with its surroundings so that because of the densely lying levels, at every frequency per unit area and per unit time and per direction of polarization the amount of energy absorbed is equal to the amount of energy emitted ($1 - a(\omega)$ is the fraction of reflected energy). Note: The number $n$ of photonic states in the wave number interval $(\kappa, \kappa + d\kappa)$ is (since there are 2 polarization directions) $n = 2V4\pi\kappa^2 d\kappa$. Since $\kappa \propto 1/\lambda$, it follows that $n \propto d\lambda/\lambda^4$. In the case of monochromatic radiation $\int \delta(\lambda - \lambda_0) d\lambda/\lambda^4 = 1/\lambda_0^4 \propto \omega_0^4$. For an elementary treatment see Müller–Kirsten [102], pp. 110–112.

This expression is obtained as follows. We obtain from Eq. (7.65) that:

$$\begin{aligned}
&\textit{Loss of energy per unit volume and per unit time}\\
&= -\frac{\partial}{\partial t}\left\{\frac{1}{2}(\mathbf{E}\cdot\mathbf{D}+\mathbf{H}\cdot\mathbf{B})\right\}\\
&= \boldsymbol{\nabla}\cdot(\mathbf{E}\times\mathbf{H}) + \frac{d^2\text{work}}{dV\,dt} = \boldsymbol{\nabla}\cdot(\mathbf{E}\times\mathbf{H}) + \mathbf{j}\cdot\mathbf{E}. \qquad (10.181)
\end{aligned}$$

Integrating this over the volume and using Gauss's divergence theorem, we obtain

Loss of energy per unit time

= Transport of energy through the surface + Ohmic power.

Hence the Ohmic power is (with $\mathbf{j} = \sigma\mathbf{E}$, $\mathbf{j}dV = Id\mathbf{s}$) the volume integral of $\mathbf{j}\cdot\mathbf{E}$, i.e. ($\sigma$ being the conductivity of the wire, $\rho = 1/\sigma$ its resistivity)

$$\begin{aligned}
\int \mathbf{j}\cdot\mathbf{E}dV &= \frac{1}{\sigma}\int \mathbf{j}^2 dV = \frac{1}{\sigma}\int \mathbf{j}\cdot Id\mathbf{s} = \frac{1}{\sigma}\int I^2\frac{d\mathbf{s}\cdot d\mathbf{s}}{dV}\\
&= \frac{I^2}{\sigma}\int\frac{dsds}{dV} = \frac{I^2}{\sigma}\int\frac{ds}{F} = I^2\rho\int\frac{ds}{F}\\
&= I^2R \quad \text{watts}, \qquad (10.182)
\end{aligned}$$

where $R = \rho s/F$ ohms as in Eq. (6.5). We thus have obtained the above expression for $L$. The quantity $L/\overline{I}^2$ in Eq. (10.179) is therefore (by comparison with $L = I^2R$) referred to as *radiation resistance.* The radiation resistance is a quantity which depends on the spatial dimensions of the radiator. The reader will have observed that in all cases considered here it is an accelerated charge which radiates. However, a charge travelling with a constant velocity in a dielectric medium radiates if its velocity exceeds the velocity of light therein. This type of radiation which we do not discuss here further is called *Cerenkov radiation.* For a lucid discussion we refer in particular to Hewson [70], pp. 82–84, and Portis [121], pp. 567–576.

## 10.7 Further Examples

**Example 10.8 (Comparison of powers of rod and loop antennas).** With a very approximate calculation compare the power of a rod (or electric dipole) antenna with that of a loop (or magnetic dipole) antenna.

**Solution:** We return to the far-field result for $\mathbf{E}$ given by Eq. (10.166), i.e.

$$\mathbf{E} = i\frac{2\pi(I_1 d\mathbf{s}_1 \times \mathbf{r}) \times \mathbf{r}}{4\pi\epsilon_0 \lambda c r^3} \quad \text{newtons/coulomb}. \tag{10.183}$$

This expression gives the electric field strength at a distance $\mathbf{r}$ far from the current element $I_1 d\mathbf{s}_1$. The electromotive force and hence *voltage* induced in a conductor element $d\mathbf{s}_2$ at the point $\mathbf{r}$ is

$$\mathbf{E} \cdot d\mathbf{s}_2 = V_2 \quad \text{volts}, \tag{10.184}$$

i.e.

$$V_2 = i\frac{2\pi(I_1 d\mathbf{s}_1 \times \mathbf{r}) \times \mathbf{r} \cdot d\mathbf{s}_2}{4\pi\epsilon_0 \lambda c r^3} = i\frac{2\pi I_1 (d\mathbf{s}_1 \times \mathbf{r}) \cdot (\mathbf{r} \times d\mathbf{s}_2)}{4\pi\epsilon_0 \lambda c r^3} \quad \text{volts}. \tag{10.185}$$

This expression is completely symmetric in $d\mathbf{s}_1$ and $d\mathbf{s}_2$; thus if $d\mathbf{s}_2$ were carrying current $I_1$, then $V_2$ would be the voltage induced in $d\mathbf{s}_1$.

The simplest way to receive radio waves is with the help of a loop antenna, i.e. a closed antenna (in general with several turns). The *induced electromotive force* or voltage is

$$V = \oint \underbrace{\mathbf{E}}_{\mathbf{E}\text{ of wave}} \cdot d\mathbf{s} = \int \boldsymbol{\nabla} \times \mathbf{E} \cdot d\mathbf{F} = \int \left(-\frac{\partial \mathbf{B}}{\partial t}\right) \cdot d\mathbf{F} \quad \text{volts}. \tag{10.186}$$

Let us assume the time dependence $\mathbf{B} \propto e^{i\omega t}$ with angular frequency $\omega$, i.e.

$$\frac{\partial \mathbf{B}}{\partial t} = i\omega \mathbf{B}. \tag{10.187}$$

Then, if the area $A$ of the loop is small, $A \sim l^2$, so that $\mathbf{B}$ is approximately uniform over $A$ (we assume: $l \ll \lambda \ll r$), and thus with Eq. (10.186),

$$V = -i\omega \mathbf{B} \cdot \mathbf{A}. \tag{10.188}$$

We attempt a very rough comparison of the power of the loop antenna with that of a rod antenna of length $l$. In the case of the loop

$$V_{\text{loop}} = -i\omega B A,$$

if $\mathbf{B} \parallel \mathbf{A}$ (then maximal — we assume an optimal orientation of the antenna). In the case of the rod-antenna (see Eq. (10.159)) we have

$$V_{\text{rod}} = \mathbf{E} \cdot \mathbf{l} = \left(\mathbf{H} \times \frac{\mathbf{r}}{r}\right) \cdot \frac{\mathbf{l}}{c\epsilon_0} = \frac{Hl}{c\epsilon_0} \quad \text{or} \quad \frac{Bl}{c\mu_0\epsilon_0}. \tag{10.189}$$

Thus the ratio is

$$\left|\frac{V_{\text{loop}}}{V_{\text{rod}}}\right| = \mu_0\epsilon_0 c\frac{\omega BA}{Bl}. \tag{10.190}$$

Since $c = \nu\lambda, \nu = \omega/2\pi$, and so $c = \omega\lambda/2\pi, \omega = 2\pi c/\lambda$, it follows that

$$\left|\frac{V_{\text{loop}}}{V_{\text{rod}}}\right| = \mu_0\epsilon_0 c\frac{BA2\pi c}{\lambda Bl} = \mu_0\epsilon_0 c^2\frac{2\pi A}{l\lambda} = \frac{2\pi A}{l\lambda} \tag{10.191}$$

for optimal reception. Now if $A \sim O(l^2)$, the ratio is $\sim l/\lambda$. Since we assumed that the dimensions of the loop are very small compared with $\lambda$, i.e. $l \ll \lambda$, the rod is the better receiver in these considerations. In view of our earlier treatment of ring-shaped conductors we can say, that the loop with oscillating current can also be looked at as a *magnetic dipole.*

**Example 10.9 (The Thomson scattering cross section[13]).** A plane, monochromatic wave with electric vector (the $k$ here must not be confused with our parameter $k = 1/4\pi\epsilon_0$)

$$\mathbf{E} = \mathbf{e}_x E_0 \cos(\omega t - kz) \tag{10.192}$$

(where $\mathbf{e}_x$ is a unit vector in the direction of $x$) falls on a *free* electron (charge $q$, mass $m$) at the origin of coordinates. Calculate:

(a) the radiation power $P$ of a single electron,
(b) the mean radiation power $\langle P\rangle$ of such an electron (averaged over one period of oscillation),
(c) the rate of incoming radiation $\mathbf{S}$,
(d) the mean rate of incoming radiation $\langle S_z\rangle$, and
(e) show that in vacuum $\langle P\rangle = \sigma\langle S_z\rangle$, where

$$\sigma = \frac{2}{3}\left(\frac{q^2}{mc^2}\right)^2\sqrt{\frac{\mu_0}{\epsilon}}\frac{c}{4\pi\epsilon_0} \quad \text{meters}^2, \tag{10.193}$$

is the so-called *Thomson scattering cross section*[14].

[13] A cross section $\sigma$ is a quantity with dimension of $(\text{length})^2$, i.e. $d\sigma =$ (energy radiated into $d\Omega$/unit time)/(incident energy/unit area and unit time). The obvious possibility to define this here can be seen from the definition of the Poynting vector as a quantity with dimension of $(\text{energy})(\text{time})^{-1}(\text{length})^{-2}$.

[14] Note that the formula (10.193) can be written, using $\mu\epsilon_0 c^2 = 1, \epsilon = \epsilon_0$, as $\sigma = (8\pi/3)R^2$ meters$^2$, where $R^2 = q^2/4\pi\epsilon_0 mc^2$ is the classical radius of the electron (cf. Example 4.7).

**Solution:** We have $E_x = E_0 \cos(\omega t - kz)$. An electron with charge $q$ at the origin $z = 0$ is therefore given an acceleration (in Newton's nonrelativistic treatment) given by the applied force $qE_x$ divided by $m$,

$$\ddot{x} \equiv a_x = \frac{q}{m} E_0 \cos \omega t \quad \text{meters} \cdot \text{second}^{-2} \tag{10.194}$$

with solution $x \propto \cos \omega t$, so that here in the case of the free (unbound) electron

$$a_x = -\omega^2 x, \quad q a_x = -q\omega^2 x \equiv -\omega^2 p, \quad p := qx. \tag{10.195}$$

The power radiated by a single electron is given by the single particle Larmor formula of Eq. (10.105) (see above, here still without the averaging)

$$P(t) = \frac{2q^2 a_x^2(t')}{4\pi\epsilon_0 3c^3}\bigg|_{t'=t-r/c} = \frac{2q^4}{3m^2c^3 4\pi\epsilon_0} E_0^2 \{\cos(\omega t - kz)\}^2 \quad \text{watts.} \tag{10.196}$$

*Averaging* over one period of oscillation $T = 2\pi/\omega$ we obtain the mean value

$$\langle P \rangle = \frac{2q^4 E_0^2}{4\pi\epsilon_0 3m^2 c^3} \langle \cos^2(\omega t - kz) \rangle \quad \text{watts,} \tag{10.197}$$

where (cf. Eq. (10.172)

$$\langle \cos^2(\omega t - kz) \rangle = \frac{1}{T} \int_0^T \cos^2(\omega t - kz) dt = \frac{1}{2},$$

i.e.

$$\langle P \rangle = \frac{q^4 E_0^2}{3m^2 c^3 4\pi\epsilon_0} \quad \text{watts.} \tag{10.198}$$

The rate at which radiation (this is the given electric field) falls on the electron is obtained with the help of the Poynting vector, $\mathbf{S} = \mathbf{E} \times \mathbf{H}$. Since $\mathbf{E} \to E_x \mathbf{e}_x = E_0 \cos(\omega t - kz)\mathbf{e}_x$, the Maxwell equation (7.41) implies (with $\mathbf{B} \propto \exp\{i(\omega t - kz)\}$)

$$-\frac{\partial \mathbf{B}}{\partial t} = -i\omega \mathbf{B} = \boldsymbol{\nabla} \times \mathbf{E} = \left(\mathbf{e}_z \frac{\partial}{\partial z}\right) \times \mathbf{E}, \quad \text{or}$$

$$-i\omega \mathbf{B} = -ik\mathbf{e}_z \times \mathbf{E}, \quad \mathbf{B} = \frac{k\mathbf{e}_z \times \mathbf{E}}{\omega}. \tag{10.199}$$

In Section 9.2 we saw that $k^2 = \omega^2 \mu_0 \epsilon_0$, so that

$$\mathbf{B} = \sqrt{\epsilon_0 \mu_0} \frac{\mathbf{k} \times \mathbf{E}}{k}, \quad \mathbf{k} = k\mathbf{e}_z. \tag{10.200}$$

Hence with $B = \mu_0 H$

$$\mathbf{B} \to B_y = \sqrt{\epsilon_0 \mu_0} E_0 \cos(\omega t - kz), \quad H_y = \sqrt{\frac{\epsilon_0}{\mu_0}} E_0 \cos(\omega t - kz). \tag{10.201}$$

It follows that

$$\mathbf{S} \to S_z = E_x H_y = \sqrt{\frac{\epsilon_0}{\mu_0}} E_0^2 \cos^2(\omega t - kz). \tag{10.202}$$

The time-averaged rate of incoming radiation is (with $T = 2\pi/\omega$)

$$\begin{aligned} \langle S_z \rangle &= \frac{1}{T} \int_0^T dt S_z = \sqrt{\frac{\epsilon_0}{\mu_0}} E_0^2 \frac{\omega}{2\pi} \int_0^{2\pi/\omega} dt \cos^2(\omega t - kz) \\ &= \frac{1}{2} \sqrt{\frac{\epsilon_0}{\mu_0}} E_0^2 \quad \mathrm{W/m^2}. \end{aligned} \tag{10.203}$$

Hence

$$E_0^2 = 2 \sqrt{\frac{\mu_0}{\epsilon_0}} \langle S_z \rangle,$$

so that with insertion of this in Eq. (10.198):

$$\langle P \rangle = \frac{2q^4}{3m^2c^3} \sqrt{\frac{\mu_0}{\epsilon_0}} \frac{\langle S_z \rangle}{4\pi\epsilon_0} \equiv \sigma \langle S_z \rangle, \tag{10.204}$$

where

$$\sigma = \frac{2}{3} \left( \frac{q^2}{mc^2} \right)^2 \sqrt{\frac{\mu_0}{\epsilon_0}} \frac{c}{4\pi\epsilon_0} \quad \text{meter}^2. \tag{10.205}$$

This cross section is seen to be independent of the frequency $\omega$ of the incident radiation. Generalizations are considered in Examples 10.10 and 10.11. For discussions of validity and applicability see Jackson [73], Section 14.7, pp. 679–683.

**Example 10.10 (The electron bound in an atom).** In Example 10.9 we considered the free electron. We now consider an *electron bound in an atom* subject to a binding force proportional to $x$ but neglecting damping effects. The problem is to derive the new expression for the Thomson cross section.

**Solution:** The equation of motion of the electron of Example 10.9 now has an additional term $\omega_0^2 x$ representing the binding of the electron to an atom (harmonic oscillator models are always the easiest to construct!), so that the equation of motion is

$$\frac{d^2x}{dt^2} + \omega_0^2 x = \frac{q}{m} E_0 \cos(\omega t), \quad \therefore x = \frac{q/m}{\omega_0^2 - \omega^2} E_0 \cos \omega t. \tag{10.206}$$

and the acceleration is

$$\therefore a_x = \ddot{x} = -\frac{(q/m)\omega^2}{\omega_0^2 - \omega^2} E_0 \cos \omega t = -\omega^2 x. \tag{10.207}$$

Thus we obtain now (replacing Eq. (10.204)) $\langle P \rangle = \sigma \langle S_z \rangle$ with

$$\sigma = \frac{2}{3} \left( \frac{q^2}{mc^2} \right)^2 \left( \frac{\omega^2}{\omega_0^2 - \omega^2} \right)^2 \sqrt{\frac{\mu_0}{\epsilon_0}} \frac{c}{4\pi\epsilon_0} \quad \text{meters}^2. \tag{10.208}$$

We see that $\sigma$ is now a function of $\omega$. In the case of scattering of light off atoms and molecules in air $\omega_0 \gg \omega$ (called *Rayleigh scattering*), and hence

$$\sigma \simeq \frac{2}{3} \left( \frac{q^2}{mc^2} \right)^2 \left( \frac{\omega}{\omega_0} \right)^4 \sqrt{\frac{\mu_0}{\epsilon_0}} \frac{c}{4\pi\epsilon_0} \quad \text{meters}^2. \tag{10.209}$$

Blue light with $\lambda_{\text{blue}} = 3900\text{Å} \ll \lambda_{\text{red}} = 7600\text{Å}$ is therefore scattered much more than red light, because:

$$c = \nu\lambda, \quad \omega = 2\pi\nu, \quad \therefore \omega \propto \frac{1}{\lambda}. \tag{10.210}$$

The scattering off *free* electrons is (see Example 10.9) independent of the frequency. As a consequence the light scattered by the corona of the sun is white.

**Example 10.11 (Line broadening).**

(a) Calculate $\langle W_0 \rangle$, the work done by the force $E_x$ of Example 10.9 applied to an electron averaged over one period of oscillation $T$. What is the meaning of the result?
(b) Add $\alpha \cos(\omega t)$ to the velocity of the electron, calculate the power $\langle W_0 \rangle / T$, which is the power of radiated energy. What is the corrected equation of motion of the electron?
(c) How does the loss of radiated energy become evident?

**Solution:**
(a) We have the equation of motion of an electron along the $x$-direction,

$$m\ddot{x} = qE_0 \cos \omega t = F_{\text{applied}}, \quad \dot{x} = \frac{qE_0}{m\omega} \sin(\omega t). \tag{10.211}$$

Hence the work done by the applied force is $W_0$:

$$W_0 = \int F_{\text{applied}} dx = \int F_{\text{applied}} \dot{x} dt, \tag{10.212}$$

and its average over one period $T = 2\pi/\omega$ is:

$$\begin{aligned} \frac{\langle W_0 \rangle}{T} &= \frac{1}{T} \int_0^T F_{\text{applied}} \frac{qE_0}{m\omega} \sin(\omega t) dt = \frac{q^2 E_0^2}{m\omega T} \int_0^T \cos(\omega t) \sin(\omega t) dt \\ &= \frac{q^2 E_0^2}{m\omega T} \int_0^T \frac{\sin(2\omega t)}{2} dt = \frac{q^2 E_0^2}{m\omega T} \frac{1}{2} \left[ -\frac{\cos(2\omega t)}{2\omega} \right]_0^{2\pi/\omega} = 0. \end{aligned} \tag{10.213}$$

Thus the work done by the applied force is unequal to the radiated energy. Where does the radiated energy come from? Equation (10.211) must contain a term, i.e. a force, whose work does not average out to zero.
(b) With addition of the extra term we have

$$\dot{x} = \frac{qE_0}{m\omega} \sin(\omega t) + \alpha \cos(\omega t), \tag{10.214}$$

so that (using Eq. (10.172))

$$\frac{\langle W_0 \rangle}{T} = \frac{1}{T} \int_0^T F_{\text{applied}} \alpha \cos(\omega t) dt = \frac{\alpha}{T} \int_0^T qE_0 \cos^2 \omega t \equiv \frac{1}{2} qE_0 \alpha. \tag{10.215}$$

In Example 10.9 we calculated the mean power of the radiated energy, see Eq. (10.198). We equate this to $\langle W_0 \rangle / T$ here and thus determine $\alpha$. Then

$$\langle P \rangle = \frac{q^4 E_0^2}{3m^2 c^3 4\pi\epsilon_0} \equiv \frac{1}{2} qE_0 \alpha = \frac{\langle W_0 \rangle}{T}, \quad \therefore \alpha = \frac{2q^3 E_0}{3m^2 c^3 4\pi\epsilon_0}. \tag{10.216}$$

Hence the velocity becomes

$$\dot{x} = \frac{qE_0}{m\omega} \sin(\omega t) + \frac{2q^3 E_0}{3m^2 c^3 4\pi\epsilon_0} \cos(\omega t), \tag{10.217}$$

so that

$$m\ddot{x} = \underbrace{qE_0 \cos(\omega t)}_{F_{\text{applied}}} \underbrace{- \frac{2q^3 E_0 \omega}{3mc^3 4\pi\epsilon_0} \sin(\omega t)}_{F_{\text{radiation}}}, \tag{10.218}$$

where $F_{\text{radiation}}$ is the radiative reaction. This equation is known as the *Abraham–Lorentz equation*[15] *of motion.* The ratio

$$\frac{F_{\text{radiation}}}{F_{\text{applied}}} = \frac{2q^2\omega}{3mc^3 4\pi\epsilon_0}\tan(\omega t) \tag{10.219}$$

is in general very small.
(c) The loss of radiated energy is due to a damping or "*line broadening*" (speaking spectroscopically). Let us write the radiation reaction as a "friction force" proportional to velocity, i.e. we put the friction force $m\beta\dot{x} = F_{\text{radiation}}$, i.e. from Eq. (10.218):

$$m\beta\frac{dx}{dt} \equiv \frac{2q^3 E_0\omega}{3mc^3 4\pi\epsilon_0}\sin(\omega t). \tag{10.220}$$

Here we replace $dx/dt$ by the leading term of Eq. (10.214) and obtain

$$\frac{dx}{dt} \sim \frac{qE_0}{m\omega}\sin(\omega t) + (\text{something very small}). \tag{10.221}$$

Inserting the latter into Eq. (10.220) we obtain

$$\beta = \frac{2q^2\omega^2}{3mc^3 4\pi\epsilon_0} \quad 1/\text{second}. \tag{10.222}$$

Next recall from Example 10.10 that $\omega_0$ represents (idealized) the frequency of oscillation of the electron bound to an atom. We now add to the equation of motion the "friction" term. We solve the equation of motion (the modified version of Eq. (10.206))

$$m\ddot{x}(t) + m\beta\dot{x}(t) + m\omega_0^2 x(t) = qE_0\cos(\omega t), \tag{10.223a}$$

by taking the real part of the equation

$$m\ddot{x} + m\beta\dot{x} + m\omega_0^2 x = qE_0 e^{i\omega t}. \tag{10.223b}$$

Setting $x = x_0\exp(i\omega t)$, we obtain the equation

$$\omega^2 - i\omega\beta - (\omega_0^2 - qE_0/mx_0) = 0. \tag{10.223c}$$

The roots of this equation are (observe the imaginary part is $i\beta/2$)

$$\omega_\pm = i\frac{\beta}{2} \pm \sqrt{(\omega_0^2 - qE_0/mx_0) - (\beta/2)^2}. \tag{10.224}$$

[15]For a different formulation of this equation see Cottingham and Greenwood [34], pp. 147–148. Their $\tau$ is our $\beta/\omega^2$ here.

The solution $x(t)$ is (this follows from Eq. (10.223b)

$$x(t) = x_0 e^{i\omega t} = \frac{qE_0}{m} \frac{1}{\omega_0^2 + i\beta\omega - \omega^2} e^{i\omega t}, \tag{10.225}$$

and the acceleration

$$\ddot{x}(t) = \frac{qE_0}{m} \frac{\omega^2}{\omega^2 - i\beta\omega - \omega_0^2} e^{i\omega t}, \tag{10.226}$$

with

$$\Re\, \ddot{x}(t) = \frac{qE_0}{m} \omega^2 \frac{(\omega^2 - \omega_0^2)\cos(\omega t) - \beta\omega \sin(\omega t)}{(\omega^2 - \omega_0^2)^2 + \beta^2\omega^2}. \tag{10.227}$$

For $\beta \to 0$ we regain the result (10.207) of Example 10.10. For $\omega \to \omega_0$ the denominator is minimized, and one expects the expression — which when squared enters the power $P$ — to have a maximum. For $\omega^2 \neq \omega_0^2$ the expression develops poles at $\omega_\pm$ in the $\omega$-plane with imaginary part $\omega_I = i\beta/2, \beta > 0$. Thus the power $P$ which involves the square of the acceleration when expressed in terms of $\omega$ and hence as a spectral function develops the damping factor $\exp(-\beta t)$. Since $\beta$ results from the radiation, this factor is interpreted as providing radiation damping. Thus we find that the radiation term $m\beta\dot{x}$ produces a damping factor $\exp(-\beta t)$ ("*radiation damping*") with so-called *line width* $\beta \equiv \Delta\omega$ at half maximum as indicated in Fig. 10.7. The quantity $\beta$ has the dimension of 1/time and represents the mean life of the energy of oscillations. Very narrow "lines" characterize long-lived states. (Note: Calculationally analogous considerations arise in Section 15.5).

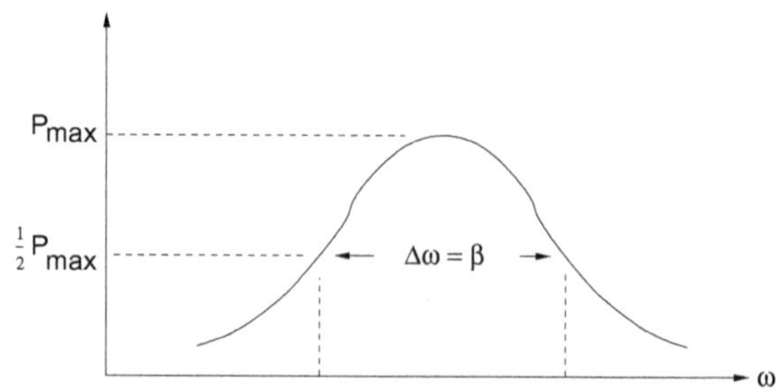


**Fig. 10.7.** Line broadening by radiation damping.

## 10.8 Exercises

**Example 10.12 (Vector potential in Coulomb gauge).** What is the vector potential corresponding to that of Eq. (10.32) in the Coulomb gauge?[16]

**Example 10.13 (Conversion to Gaussian units).** The Thomson scattering cross section (10.205) involves (with $q = e$) the factor $e^2/mc^2$. Convert this factor into Gaussian units and evaluate it in Gaussian units.

(Answer: $2.7 \times 10^{-13}$ (statcoulombs)$^2$/gramme(cm/sec)$^2$).

**Example 10.14 (Potentials of a charge moving with constant velocity).** Let $\mathbf{r} = (x, y, z)$ be the position of the point of observation and let $\mathbf{r}' = (x', y', z')$ be the position of the charge $e$ at the time of observation $t$. The charge $e$ travels with constant speed $v$ parallel to the $x$-axis. Show that the retarded denominator in the Liénard–Wiechert potentials is given by

$$[(x - x')^2 + \{(y - y')^2 + (z - z')^2\}(1 - v^2/c^2)]^{1/2}. \tag{10.228}$$

[Hint: Show first that the retarded denominator is $c\triangle t - \mathbf{c}\triangle t \cdot \mathbf{v}/c$, where $c\triangle t$ is the distance travelled by the light signal when the charge has reached its position of observation].

**Example 10.15 (Radiation from a charge moving along a circle[17]).** A point charge $e$ moves with constant angular velocity $\omega$ round a circle of radius $a$. What is the total rate of loss of energy of radiation?

**Example 10.16 (Relation between retarded and present volume elements).** The relation between retarded time $t'$ and present time $t$ is given by Eq. (10.63) for the case where $|\mathbf{r} - \mathbf{r}_0(t')|$ is the distance from the source point $\mathbf{r}_0(t')$ to the field point $\mathbf{r}$, the difference $t - t'$ being the time delay owing to the finite value $c$ of the speed of light. The relationship between times in terms of present position of the charge $e$, i.e. $\mathbf{r}_0(t)$, and its retarded position $\mathbf{r}_0(t')$ is correspondingly

$$t' = t - \frac{1}{c}|\mathbf{r}_0(t) - \mathbf{r}_0(t')|. \tag{10.229}$$

[16] See e.g. Robinson [124], p. 21.
[17] This is also an exercise in Oppenheimer [108], p. 66.

Show that the Jacobian $J$ relating the present volume element $dV$ to the retarded volume element $dV'$ so that $dV = JdV'$, is given by

$$J = 1 - \mathbf{n}(t') \cdot \mathbf{v}(t')/c. \tag{10.230}$$

as in the denominators of the potentials (10.62), (10.64).
[For a detailed derivation see e.g. Rossiter [126], pp. 109–115].

**Example 10.17 (Lorenz gauge satisfied).** Show that the Liénard–Wiechert potentials satisfy the Lorenz gauge condition (10.19).

**Example 10.18 (Comparison of radiation fields).** Compare or relate with one another the radiated fields $\mathbf{E}$ and $\mathbf{H}$ of an accelerated charge $e$, an oscillating dipole moment $p$ and a current $I$, taking the acceleration of the charge proportional to $\exp(i\omega t)$.

**Example 10.19 (Comparison of fields of charge e, oscillating pand I).** A point charge $e$ suffers an acceleration $a$ for a very short time $\delta t$. Draw and annotate a diagram showing the kink in the electric vector after a time $t$, and show from the angle of the kink and the known radial (Coulomb) field that the tangential electric field $E$ at $(r, \theta)$ is $E = \mu ea \sin\theta/4\pi r$. Assuming that the densities of transverse magnetic and electric energies are equal, deduce the transverse magnetic field and the total energy radiated per unit time. Taking $a \propto \exp(i\omega t)$, express the fields and the time-average of the energy emission in terms of (a) the dipole moment $p$, and (b) the current $I$.

**Example 10.20 (Evaluation of the Thomson formula).** Evaluate the Thomson total cross section formula (10.193) for the case of a free electron. (Answer: $6.6 \times 10^{-29}$ meters$^2$).

**Example 10.21 (Dipole radiation).** In the steps from Eq. (10.171) to Eq. (10.175) we performed the time averaging first and then the angular integration. Redo the calculations in the reverse order to show that the dipole radiation before averaging is proportional to $(\ddot{\mathbf{p}})^2$.

**Example 10.22 (Oscillating spherically symmetric charge distribution).** A spherically symmetric charge distribution oscillates radially. Show that there is no radiation. Comment: In nonrelativistic quantum mechanics this is nonzero. For this calculation see Oppenheimer [108], pp. 24–25.

# Chapter 11

# Optic Laws and Diffraction

## 11.1 Introductory Remarks

With the realization that light is nothing but electromagnetic radiation which reaches our eyes, it is clear that basic optical phenomena must also find their explanation in electrodynamics and hence in Maxwell's equations. The visible part of the electromagnetic spectrum is the small domain between the infrared region which has wavelength $\lambda$ expressed in meters in the region $10^{-3} > \lambda > 10^{-6}$, and the ultraviolet region which has $\lambda$ in the region $10^{-7} > \lambda > 10^{-8}$, i.e. more precisely between $3.9 \times 10^{-7}$ and $7.7 \times 10^{-7}$ meters. The subject of this chapter is therefore the derivation of the well known laws of reflection and refraction of optics from Maxwell's equations at an interface which in the present chapter is of dielectric–dielectric type (in Chapter 12 of dielectric–conductor type). In subsequent chapters these laws will then be applied to metals, radio waves and wave guides.[1] These applications will not be restricted to the visible section of the electromagnetic spectrum. We supplement this topic here with an introduction to the electromagnetic theory of diffraction.

## 11.2 Continuity Conditions

Our procedure will be to arrive at the laws of optics from a consideration of the behavior of an electromagnetic wave at the interface between two media. First, however, we recapitulate and summarize for use here the essential continuity conditions and those of their validity for normal and tangential

[1] Recommendable texts to supplement this chapter, particularly with regard to more practical aspects, are e.g. those of Guenther [64] and Born and Wolf [18].

components of the fields, with indices (1) and (2) referring to two different media.

(a) $\nabla \cdot \mathbf{D} = \rho$. Here $\rho$ is the density of the *"true" charges.* With this equation we have:

$$\int_V \nabla \cdot \mathbf{D} dV = \int_{F_V} \mathbf{D} \cdot d\mathbf{F} = \int_V \rho dV = Q. \tag{11.1}$$

In the case of the infinitesimal volume element of thickness $d \to 0$ as shown in Fig. 11.1 we obtain from $\int_{F_V} \mathbf{D} \cdot d\mathbf{F}$ (observe that in the limit $d \to 0$ the quantity $\triangle F$ is an element of area *in* the boundary surface)

$$(\mathbf{D}^{(1)} - \mathbf{D}^{(2)}) \cdot \triangle \mathbf{F} = (D_n^{(1)} - D_n^{(2)}) \triangle F = \triangle Q, \tag{11.2}$$

(the suffix $n$ referring to "normal" component) so that (see Fig. 11.1),

$$D_n^{(1)} - D_n^{(2)} = \sigma, \quad \sigma = \frac{\triangle Q}{\triangle F}. \tag{11.3}$$

In the case of the electric field strength $\mathbf{E}$ we now obtain ($Q_P$ being the polarization charge with density $\rho_P$) with Eqs. (4.112) and (4.113),

$$\mathbf{D} = \mathbf{E}\epsilon_0 + \mathbf{P}, \quad \nabla \cdot \mathbf{P} = -\rho_P, \tag{11.4}$$

the integrated expressions

$$\begin{aligned} \int_{F_V} \mathbf{E} \cdot d\mathbf{F} &= \frac{1}{\epsilon_0} \int_{F_V} (\mathbf{D} - \mathbf{P}) \cdot d\mathbf{F} = \frac{1}{\epsilon_0} \int_V (\nabla \cdot \mathbf{D} - \nabla \cdot \mathbf{P}) dV \\ &= \frac{1}{\epsilon_0} \int_V (\rho + \rho_P) dV = \frac{1}{\epsilon_0}(Q + Q_P). \end{aligned} \tag{11.5}$$

Hence at the interface

$$E_n^{(1)} - E_n^{(2)} = \frac{1}{\epsilon_0}(\sigma + \sigma_P). \tag{11.6}$$

Here $\sigma_P$ is the induced surface charge density.

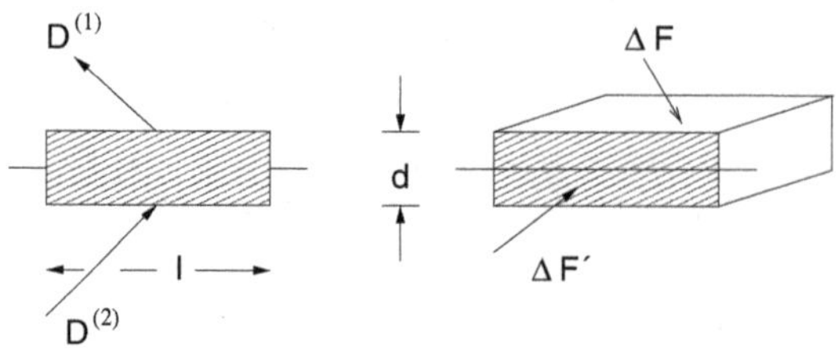


**Fig. 11.1.** Elements of area and volume.

(b) $\boldsymbol{\nabla}\times\mathbf{E} = -\partial\mathbf{B}/\partial t$ (Faraday). Here we have in the case of a cross-sectional area $F'$ with $\triangle F' = l \times d$ as in Fig. 11.1:

$$\int_{F'}(\boldsymbol{\nabla}\times\mathbf{E})\cdot d\mathbf{F} = -\frac{\partial}{\partial t}\int_{F'}\mathbf{B}\cdot d\mathbf{F} = -\frac{\partial}{\partial t}|\mathbf{B}_n|\triangle F' \to 0, \tag{11.7}$$

since $\triangle F' = ld \to 0$ with $d \to 0$. Using Stokes's theorem we obtain

$$\begin{aligned}\int_{F'}(\boldsymbol{\nabla}\times\mathbf{E})\cdot d\mathbf{F} &= \oint \mathbf{E}\cdot d\mathbf{s} = (\mathbf{E}\cdot\mathbf{s})^{(1)} + (\mathbf{E}\cdot\mathbf{s})^{(2)}\\ &= (\mathbf{E}^{(1)}\cdot\mathbf{l}) - (\mathbf{E}^{(2)}\cdot\mathbf{l}) = (E_t^{(1)} - E_t^{(2)})l,\end{aligned} \tag{11.8}$$

and with the preceding result

$$E_t^{(1)} - E_t^{(2)} = 0. \tag{11.9}$$

(c) $\boldsymbol{\nabla}\cdot\mathbf{B} = 0$. From this Maxwell equation we obtain immediately

$$\int_V \boldsymbol{\nabla}\cdot\mathbf{B}dV = \int_F \mathbf{B}\cdot d\mathbf{F} = 0, \tag{11.10}$$

so that

$$B_n^{(1)} - B_n^{(2)} = 0. \tag{11.11}$$

(d) $\boldsymbol{\nabla}\times\mathbf{H} = \partial\mathbf{D}/\partial t + \mathbf{j}$. If there are no *surface currents* $K$ (see below), i.e. if $\int \mathbf{j}\cdot d\mathbf{F} = 0$, then as under (b):

$$H_t^{(1)} - H_t^{(2)} = 0, \quad (K = 0). \tag{11.12}$$

The *definition of the above surface current* is given as follows. We consider the element of area $dF$ perpendicular to the interface between the two media as illustrated in Fig. 11.2. Then $d\mathbf{F}$ is a vector in the plane of the interface. We now consider equation ($d$) integrated over this area, i.e.

$$\int_{\triangle F}(\boldsymbol{\nabla}\times\mathbf{H})\cdot d\mathbf{F} = \int_{\triangle F} d\mathbf{F}\cdot\left[\frac{\partial\mathbf{D}}{\partial t} + \mathbf{j}\right]. \tag{11.13}$$

The first contribution on the right is infinitesimally

$$\frac{\partial}{\partial t}|\mathbf{D}|\triangle F = \frac{\partial}{\partial t}|\mathbf{D}|ld \to 0 \quad \text{with} \quad d \to 0. \tag{11.14}$$

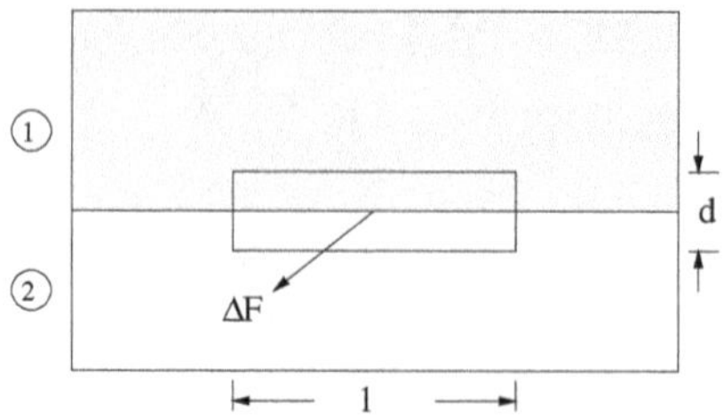


**Fig. 11.2.** The element of area ⊥ to the interface.

We rewrite the second contribution as

$$\int_{\triangle F} d\mathbf{F}\cdot\mathbf{j} = \mathbf{j}\cdot\triangle\mathbf{F} = jld. \tag{11.15}$$

For homogeneous conductors we have Ohm's law

$$\mathbf{j} = \sigma\mathbf{E}, \quad \text{i.e.} \quad j = \sigma E. \tag{11.16}$$

For perfect conductors $\sigma = \infty$ and $\mathbf{E} = -\boldsymbol{\nabla}\phi = 0$. Then it is possible that

$$jld = \sigma Eld \neq 0 \quad \text{and finite,} \tag{11.17}$$

and one defines

$$\lim_{\sigma\to\infty,\, d\to 0} \sigma Ed := K, \tag{11.18}$$

i.e. $jd \sim K$, where $K$ is a *surface current (density)* (see Jackson [73], p. 335); $K$ is also called "*linear current density*" (see Lim [88], p. 42).

If the conductivity $\sigma$ is finite, $K = 0$, i.e. in this case there is really no surface current (see Jackson [73], p. 336). The continuity condition for $H$ is therefore, treating $\int(\boldsymbol{\nabla}\times\mathbf{H})\cdot d\mathbf{F}$ as for $\mathbf{E}$, i.e. from Eq. (11.13) and Eq. (11.15),

$$\int_{\triangle F}(\boldsymbol{\nabla}\times\mathbf{H})\cdot d\mathbf{F} = \int_{\triangle F}\mathbf{H}\cdot d\mathbf{l} = ljd = Kl, \tag{11.19}$$

and it follows that

$$H_t^{(1)} - H_t^{(2)} = K \quad (= 0 \text{ for } \sigma \text{ finite}). \tag{11.20}$$

For later purposes we can write this also in vector form as

$$\mathbf{n}\times(\mathbf{H}^{(1)} - \mathbf{H}^{(2)}) = \mathbf{K}, \tag{11.21}$$

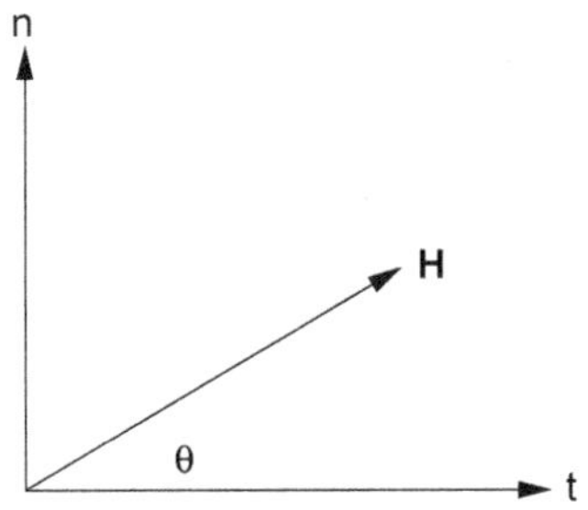


**Fig. 11.3.** Normal and tangential directions.

where $\mathbf{n}$ is a unit vector perpendicular to the interface, and $\mathbf{K}$ is a vector in the plane of the interface. One should note that as indicated in Fig. 11.3:

$$H_t = \mathbf{H} \cdot \mathbf{e}_t = |\mathbf{H}| \cos\theta = |\mathbf{H}| \sin\left(\frac{\pi}{2} - \theta\right) = |\mathbf{H} \times \mathbf{n}|. \tag{11.22}$$

The limits $d \to 0, \sigma \to \infty$ apply to an idealized case, and therefore the question arises whether a surface current does actually exist or not. We shall see later, in Chapter 14 on wave guides (see Eq. (14.243)), that a current density $\mathbf{j}$ which is restricted to a very thin region corresponds to an *effective surface current* $\mathbf{K}_{\text{eff}}$ (see also Jackson [73], p. 339).

**Example 11.1 (Change of direction of E at an interface).** The electric field $\mathbf{E}$ in medium (1) with dielectric constant $\epsilon_1 = 7$ farads/meter falls at an angle of 60° to the surface normal on the interface to a medium (2) with dielectric constant $\epsilon_2 = 2$ farads/meter. What is the angle between $\mathbf{E}$ and the surface normal in medium (2)?

**Solution:** We have

$$E_{t1} = E_{t2}, \quad D_{n1} - D_{n2} = 0 \text{ (charge)}, \quad \therefore \epsilon_1 E_{n1} = \epsilon_2 E_{n2}. \tag{11.23}$$

We are also given: $E_{t1} = \tan 60° E_{n1}$, $E_{t2} = \tan\theta_2 E_{n2}$. The quantity requested is $\theta_2$:

$$\tan\theta_2 = \frac{E_{t2}}{E_{n2}} = \frac{E_{t1}}{\epsilon_1 E_{n1}/\epsilon_2} = \frac{\tan 60° E_{n1}}{\epsilon_1 E_{n1}/\epsilon_2} = \frac{\epsilon_2}{\epsilon_1}\tan 60° = \frac{\tan 60°}{3.5} = 0.495, \tag{11.24}$$

so that $\theta_2 = 26.4°$.

## 11.3 Electromagnetic Waves in Media

In the following we consider a *plane, electromagnetic* wave, which falls at an angle from medium (1) on the interface to medium (2), as indicated in Fig. 11.4.

We recapitulate first some of our earlier considerations for the case of *electromagnetic waves in a medium* with dielectric constant $\epsilon$, magnetic permeability $\mu$ and conductivity $\sigma$. We had the equations

$$\mathbf{D} = \epsilon \mathbf{E}, \quad \mathbf{B} = \mu \mathbf{H}, \quad \mathbf{j} = \sigma \mathbf{E}. \tag{11.25}$$

In the case of charge density $\rho = 0$ the equations are

$$\begin{aligned} \boldsymbol{\nabla} \cdot \mathbf{H} = 0, &\quad \boldsymbol{\nabla} \times \mathbf{H} = \epsilon \frac{\partial \mathbf{E}}{\partial t} + \sigma \mathbf{E}, \\ \boldsymbol{\nabla} \cdot \mathbf{E} = 0, &\quad \boldsymbol{\nabla} \times \mathbf{E} = -\mu \frac{\partial \mathbf{H}}{\partial t}. \end{aligned} \tag{11.26}$$

From the first two equations we obtain after forming

$$\boldsymbol{\nabla} \times (\boldsymbol{\nabla} \times \mathbf{H}) = \boldsymbol{\nabla}(\boldsymbol{\nabla} \cdot \mathbf{H}) - (\boldsymbol{\nabla} \cdot \boldsymbol{\nabla})\mathbf{H} \tag{11.27}$$

with $\boldsymbol{\nabla} \cdot \mathbf{H} = 0$:

$$\triangle \mathbf{H} = \epsilon\mu \frac{\partial^2 \mathbf{H}}{\partial t^2} + \mu\sigma \frac{\partial \mathbf{H}}{\partial t}. \tag{11.28a}$$

With $\boldsymbol{\nabla} \times (\boldsymbol{\nabla} \times \mathbf{E})$, one obtains analogously

$$\triangle \mathbf{E} = \epsilon\mu \frac{\partial^2 \mathbf{E}}{\partial t^2} + \mu\sigma \frac{\partial \mathbf{E}}{\partial t}. \tag{11.28b}$$

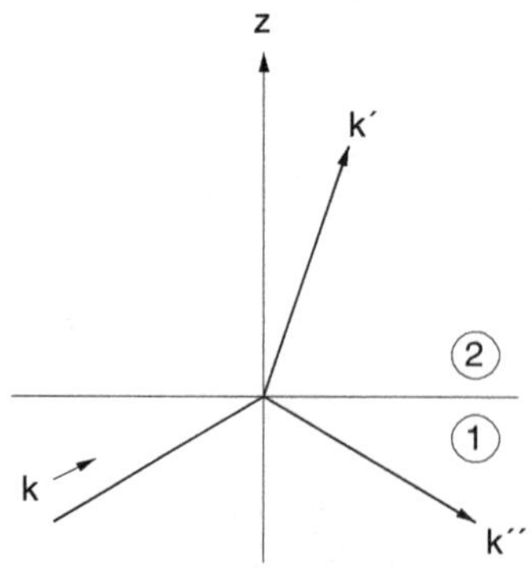


**Fig. 11.4.** An electromagnetic wave falling at an inclination on the interface between two media.

For $\sigma = 0$ and propagation velocity

$$c' = \frac{1}{\sqrt{\epsilon\mu}} = \sqrt{\frac{\epsilon_0\mu_0}{\epsilon\mu}}c \quad \text{meters/second} \tag{11.29}$$

this is

$$\triangle \mathbf{E} - \frac{1}{c'^2}\frac{\partial^2 \mathbf{E}}{\partial t^2} = 0. \tag{11.30}$$

In the case of *high frequency radiation*, the parameters $\epsilon, \mu, \sigma$ become frequency-dependent. Considering *plane waves*, we write for $\mathbf{E}$ and $\mathbf{H}$[2]:

$$\mathbf{E} = \mathbf{E}_0 e^{i(\mathbf{k}\cdot\mathbf{r}-\omega t)}, \quad \mathbf{H} = \mathbf{H}_0 e^{i(\mathbf{k}\cdot\mathbf{r}-\omega t)}. \tag{11.31}$$

Inserting these expressions into Eq. (11.26) we obtain

$$\boldsymbol{\nabla} \times \mathbf{H} = i\mathbf{k} \times \mathbf{H} = -\epsilon i\omega \mathbf{E} + \sigma \mathbf{E} \equiv -i\omega\eta \mathbf{E}, \tag{11.32}$$

where

$$\eta = \epsilon + i\frac{\sigma}{\omega} \quad \text{farads/meter}, \tag{11.33}$$

which is the *generalized dielectric constant*, in the case $\sigma \neq 0$ for a conducting medium. It follows that we can write Maxwell's equations — these are here Eqs. (11.26) —

$$\left.\begin{aligned} \mathbf{k} \times \mathbf{H} &= -\omega\eta \mathbf{E}, \\ \mathbf{k} \times \mathbf{E} &= \omega\mu \mathbf{H}, \\ \mathbf{k} \cdot \mathbf{E} &= 0, \\ \mathbf{k} \cdot \mathbf{H} &= 0. \end{aligned}\right\} \tag{11.34}$$

The last two equations show that here also we have *transversality* of the electromagnetic field (but not $|\mathbf{E}| = |\mathbf{H}|$). From these equations we determine $|\mathbf{k}|$:

$$\underbrace{\mathbf{k} \times (\mathbf{k} \times \mathbf{H})}_{(\mathbf{k}\cdot\mathbf{H})\mathbf{k} - k^2\mathbf{H}} = -\omega\eta \underbrace{\mathbf{k} \times \mathbf{E}}_{\omega\mu\mathbf{H}}, \tag{11.35}$$

where $\mathbf{k} \cdot \mathbf{H} = 0$. Hence for $\mathbf{H} \neq 0$:

$$k^2 = \omega^2\eta\mu = \omega^2\mu\left(\epsilon + i\frac{\sigma}{\omega}\right), \tag{11.36}$$

[2] Note we do not use throughout the same time factor with the minus sign. The present convention is that of Jackson [73].

i.e.

$$k^2 = \mu\epsilon\omega^2\left(1 + i\frac{\sigma}{\omega\epsilon}\right) \quad \text{meters}^{-2}. \tag{11.37}$$

This relation is called the *dispersion relation* of a conducting medium. Setting $k = \alpha + i\beta$, we obtain

$$k = \sqrt{\mu\epsilon\omega}\left[\left\{\frac{1}{2}\left(\sqrt{1+\left(\frac{\sigma}{\omega\epsilon}\right)^2}+1\right)\right\}^{1/2} + i\left\{\frac{1}{2}\left(\sqrt{1+\left(\frac{\sigma}{\omega\epsilon}\right)^2}-1\right)\right\}^{1/2}\right]. \tag{11.38}$$

For $\sigma = 0$ we have $k$ real; i.e. the wave is not damped, and the medium is said to be "*transparent*" (allows radiation to pass through).

We now put

$$k = |\mathbf{k}|e^{i\varphi}, \tag{11.39}$$

and find

$$|k| = \sqrt{\alpha^2+\beta^2} = \sqrt{\epsilon\mu}\omega\left[1+\left(\frac{\sigma}{\omega\epsilon}\right)^2\right]^{1/4},$$

$$\varphi = \tan^{-1}\frac{\beta}{\alpha} = \frac{1}{2}\tan^{-1}\left(\frac{\sigma}{\omega\epsilon}\right). \tag{11.40}$$

The *(generalized) refractive index* is defined as the quantity

$$p = c\sqrt{\mu\eta} = p_1 + ip_2. \tag{11.41}$$

Hence for the *vacuum* we have ($\sigma = 0$):

$$p = \sqrt{\mu_0\epsilon_0}c, \quad \text{i.e.} \quad p = 1 \quad \text{and} \quad k = \frac{\omega}{c}. \tag{11.42}$$

## 11.4 Kinematical Aspects of Reflection and Refraction: Snell's Law

### 11.4.1 *Preliminary remarks*

We now consider reflection and refraction of a wave at a boundary surface between two media (1) and (2) as indicated in Fig. 11.5 which are not metals. Thus we consider *dielectric–dielectric boundaries.*

For our kinematical considerations here we make the following four *assumptions*:

(a) There are no surface charges or surface currents,

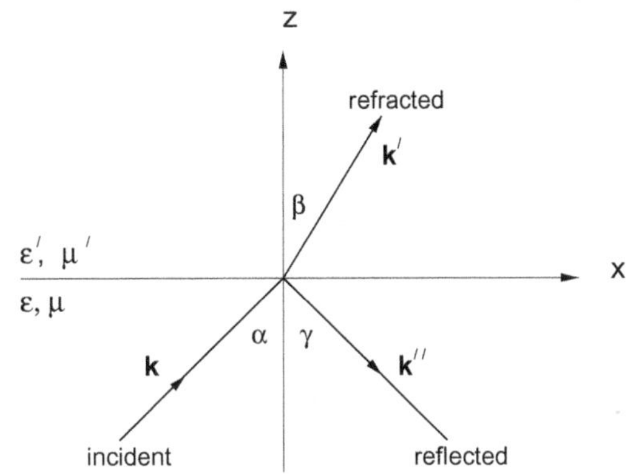


**Fig. 11.5.** Directions of incident, refracted and reflected waves with angular inclinations to the normal.

(b) both media have conductivity $\sigma = 0$,[3] (i.e. are transparent),
(c) in order that $B = \mu H$ be valid, the media are assumed not to be ferromagnetic ($B = \mu H$ is *not* valid for these!),
(d) the incident wave is a *plane wave* with

$$\mathbf{E} = \mathbf{E}_0 e^{i(\mathbf{k}\cdot\mathbf{r}-\omega t)} \tag{11.43}$$

(in the case of a plane wave $f(\mathbf{r}, t) := \mathbf{k}\cdot\mathbf{r} - \omega t = \text{const.}, \mathbf{k} = \boldsymbol{\nabla} f(\mathbf{r}, t)$).

The equations (cf. Eqs. (11.34))

$$\mathbf{k} \times \mathbf{E} = \omega\mu\mathbf{H} = \omega\mathbf{B}, \quad k = \omega\sqrt{\epsilon\mu} \tag{11.44}$$

then imply for the magnetic flux density and magnetic field strength

$$\mathbf{B} = \sqrt{\epsilon\mu}\frac{\mathbf{k}\times\mathbf{E}}{k}, \quad \mathbf{H} = \sqrt{\frac{\epsilon}{\mu}}\frac{\mathbf{k}\times\mathbf{E}}{k}. \tag{11.45}$$

We assume that there are waves in both media, *reflected waves* in one medium and *refracted waves* in the other. For these we set

(1) *in the case of reflection:*

$$\mathbf{E}'' = \mathbf{E}_0'' e^{i(\mathbf{k}''\cdot\mathbf{r}-\omega'' t)}, \quad \mathbf{B}'' = \sqrt{\epsilon\mu}\frac{\mathbf{k}''\times\mathbf{E}''}{k''}, \tag{11.46}$$

[3]Metals will be treated in Chapter 12. Most of the considerations here for $\sigma = 0$ are also valid for $\sigma \neq 0$ and finite. Then also the surface current density $K = 0$. For radiation from a dielectric into a metal one then has to consider a dielectric–metal or dielectric–conductor boundary.

(2) *in the case of refraction:*

$$\mathbf{E}' = \mathbf{E}'_0 e^{i(\mathbf{k}'\cdot\mathbf{r}-\omega' t)}, \quad \mathbf{B}' = \sqrt{\epsilon'\mu'}\frac{\mathbf{k}'\times\mathbf{E}'}{k'}. \tag{11.47}$$

In the *case of normal incidence* of the wave and in view of the transversality (see Eqs. (11.34)) we can write in self-evident notation using Eq. (11.45):

$$\sqrt{\frac{\mu}{\epsilon}}H_{\rm in} = E_{\rm in}, \quad \sqrt{\frac{\mu}{\epsilon}}H_{\rm refl} = E_{\rm refl}, \quad \sqrt{\frac{\mu'}{\epsilon'}}H_{\rm refr} = E_{\rm refr}. \tag{11.48}$$

### 11.4.2 *Kinematical aspects of reflection and refraction*

The kinematical aspects of reflection and refraction which we consider now are consequences of the *general boundary condition*, that the conditions of continuity summarized above (whatever they express in detail) must be valid *at every time $t$ and at every point of the interface.* Consider the conditions

$$H_t^{(1)} - H_t^{(2)} = 0, \quad E_t^{(1)} - E_t^{(2)} = 0 \quad \text{at} \quad z = 0. \tag{11.49}$$

If these conditions hold at every time $t$ and at all points of the interface plane $(x, y)$, then we must have for (e.g.) $\mathbf{E}$ (with vector addition on both sides of $z = 0$):

$$[(\mathbf{E}_0)_t e^{i(\mathbf{k}\cdot\mathbf{r}-\omega t)}]_{z=0} + [(\mathbf{E}''_0)_t e^{i(\mathbf{k}''\cdot\mathbf{r}-\omega'' t)}]_{z=0} = [(\mathbf{E}'_0)_t e^{i(\mathbf{k}'\cdot\mathbf{r}-\omega' t)}]_{z=0},$$

i.e.

$$E_{0t}e^{i(k_x x + k_y y - \omega t)} + E''_{0t}e^{i(k''_x x + k''_y y - \omega'' t)} = E'_{0t}e^{i(k'_x x + k'_y y - \omega' t)}. \tag{11.50}$$

If this is to hold at all times $t$, we must have (by multiplying both sides by $\exp(i\omega' t)$)

$$e^{i(\omega'-\omega)t} = 1 = e^{i(\omega'-\omega'')t},$$

i.e.

$$\omega = \omega' = \omega'' \tag{11.51}$$

(possibly for $\omega t = \omega'' t + 2m\pi, m = 0, \pm 1, \ldots$, but demanding validity for all $t$, hence also for $t = 0$, implies $m = 0$). Similarly we must have for projections onto the $(x, y)$ plane

$$\mathbf{k}\cdot\mathbf{r} = \mathbf{k}'\cdot\mathbf{r} = \mathbf{k}''\cdot\mathbf{r} \quad \text{at} \quad z = 0. \tag{11.52}$$

Thus the projections of $\mathbf{k}, \mathbf{k}', \mathbf{k}''$ on any arbitrarily chosen vector $\mathbf{r}$ in the interface plane must be equal. However, this is possible only if (see below) $\mathbf{k}, \mathbf{k}', \mathbf{k}''$ lie in one and the same plane, the so-called *plane of incidence.* This can be seen as follows. On the one hand, if we replace $\mathbf{r}$ by a unit vector along the $x$-axis, the $x$-components have to be equal, and analogously the $y$-components have to be equal. On the other hand, if two of the three vectors $\mathbf{k}, \mathbf{k}', \mathbf{k}''$ lie in the $(x, z)$-plane, and so have zero $y$-components, (also at $z = 0$), also the $y$-component of the third vector is zero, implying that this vector must also lie in the $(x, z)$-plane, i.e.

$$\{\mathbf{k} \cdot \mathbf{r} = \mathbf{k}' \cdot \mathbf{r} = \mathbf{k}'' \cdot \mathbf{r}\}_{z=0} = 0. \tag{11.53}$$

If $\mathbf{k}, \mathbf{k}''$ span the $(x, z)$-plane, i.e. if $k_y = 0 = k''_y$, then

$$k_x x = k'_x x + k'_y y = k''_x x; \tag{11.54}$$

but because $k_x = k'_x = k''_x$, it follows that $k'_y = 0$. Furthermore we deduce from Eq. (11.52) for the $x$-components (as we can see from Fig. 11.5)

$$k \sin\alpha = k' \sin\beta = k'' \sin\gamma. \tag{11.55}$$

Since $\omega = \omega''$ and in general for $\sigma = 0 : k = \sqrt{\epsilon\mu}\omega$, it follows that[4] $k = k''$ in the same medium, and hence

$$\alpha = \gamma, \tag{11.56}$$

which implies that *the incident angle is equal to the reflection angle.* In addition *Snell's law* must hold, i.e.[5]

$$\frac{\sin\alpha}{\sin\beta} = \frac{k'}{k} = \sqrt{\frac{\epsilon'\mu'}{\epsilon\mu}} = \frac{n'}{n} = \frac{c_1}{c_2} = \frac{\lambda_1}{\lambda_2}, \tag{11.57}$$

where $c_1, \lambda_1$ are velocity of light and wavelength in the medium with parameters $\epsilon, \mu$, and $c_2, \lambda_2$ these quantities of the medium with parameters $\epsilon', \mu'$, and the appropriate refractive indices are

$$n = \frac{c}{c_1} = \sqrt{\frac{\epsilon\mu}{\epsilon_0\mu_0}} \quad \text{and} \quad n' = \frac{c}{c_2} = \sqrt{\frac{\epsilon'\mu'}{\epsilon_0\mu_0}}. \tag{11.58}$$

---

[4]In the case of different media the $\omega$'s are equal, but the $k$'s, $\sqrt{\epsilon\mu}\omega$, are different.

[5]Thus if $\mu \simeq \mu'$ one has Snell's law as $\sin\alpha/\sin\beta = \sqrt{\epsilon'/\epsilon}$ for a dielectric–dielectric interface. In the case of a dielectric–conductor interface $\epsilon'$ becomes $\eta = \epsilon' + O(\sigma)$, where $\sigma$ is the conductivity.

## 11.5 Dynamical Aspects of Reflection and Refraction: The Fresnel Formulas

### 11.5.1 *The conditions*

The dynamical aspects of reflection and refraction follow from the *specific boundary conditions* for $\mathbf{D}, \mathbf{E}, \mathbf{B}, \mathbf{H}$. We assume that the media are isotropic (i.e. $\mathbf{D} = \epsilon\mathbf{E}$) and that there are neither surface charges nor surface currents. We use the relations found above (Eq. (11.37) with $\sigma = 0$)

$$\frac{\sqrt{\epsilon\mu}}{k} = \frac{1}{\omega} = \frac{1}{\omega'} = \frac{\sqrt{\epsilon'\mu'}}{k'}. \tag{11.59}$$

We begin with the relation $E_t^{(1)} - E_t^{(2)} = 0$. We let $\mathbf{n}$ be the unit vector parallel to $\mathbf{e}_z$ in the direction of the $z$-axis. Then for some vector $\mathbf{A}$

$$A_x = \mathbf{A}\cdot\mathbf{e}_x = |\mathbf{A}|\cos\theta = |\mathbf{A}|\sin\left(\frac{\pi}{2} - \theta\right) = |\mathbf{A}\times\mathbf{n}|. \tag{11.60}$$

With this way of writing the $x$-component of a vector quantity we can rewrite the tangential condition $E_t^{(1)} - E_t^{(2)} = 0$ as

$$(a) \qquad (\mathbf{E}_0 + \mathbf{E}_0'' - \mathbf{E}_0')\times\mathbf{n} = 0. \tag{11.61}$$

The next condition we consider is $B_n^{(1)} - B_n^{(2)} = 0$. With the expression (11.45) for $\mathbf{B}, \mathbf{B}'$ and $\mathbf{B}''$, as well as

$$\mathbf{E} = \mathbf{E}_0 e^{i(\mathbf{k}\cdot\mathbf{r}-\omega t)}, \quad \mathbf{E}' = \mathbf{E}_0' e^{i(\mathbf{k}'\cdot\mathbf{r}-\omega' t)}, \quad \mathbf{E}'' = \mathbf{E}_0'' e^{i(\mathbf{k}''\cdot\mathbf{r}-\omega'' t)}, \tag{11.62}$$

we have (re-expressing $\sqrt{\epsilon'\mu'}$ with Eq. (11.57))

$$(\mathbf{B}+\mathbf{B}''-\mathbf{B}')\cdot\mathbf{n} \propto \left[\sqrt{\epsilon\mu}\left(\frac{\mathbf{k}\times\mathbf{E}_0}{k} + \frac{\mathbf{k}''\times\mathbf{E}_0''}{k''}\right) - \underbrace{\sqrt{\epsilon'\mu'}}_{(k'/k)\sqrt{\epsilon\mu}}\frac{\mathbf{k}'\times\mathbf{E}_0'}{k'}\right]\cdot\mathbf{n} = 0. \tag{11.63}$$

With (11.59) and $k = k''$ (reflection) we obtain from this the relation

$$(b) \qquad (\mathbf{k}\times\mathbf{E}_0 + \mathbf{k}''\times\mathbf{E}_0'' - \mathbf{k}'\times\mathbf{E}_0')\cdot\mathbf{n} = 0. \tag{11.64}$$

The scalar product with $\mathbf{n}$ gives us the normal component of the vector in the bracket.

Next we consider the condition $H_t^{(1)} = H_t^{(2)}$. From this equality we obtain as in case (*b*) with $\mathbf{B} = \mu\mathbf{H}$ the relation

$$(c) \qquad \left[\frac{1}{\mu}(\mathbf{k}\times\mathbf{E}_0 + \mathbf{k}''\times\mathbf{E}_0'') - \frac{1}{\mu'}(\mathbf{k}'\times\mathbf{E}_0')\right]\times\mathbf{n} = 0. \tag{11.65}$$

The vector product with $\mathbf{n}$ gives us the tangential component of the vector in the bracket (recall Eq. (11.21)).

Finally we obtain from the condition $D_n^{(1)} = D_n^{(2)}$ the relation

$$(d) \qquad [\epsilon(\mathbf{E}_0 + \mathbf{E}_0'') - \epsilon'\mathbf{E}_0']\cdot\mathbf{n} = 0. \tag{11.66}$$

In the next section we consider two cases of linear polarization (linearly polarized implies: $\mathbf{E}, \mathbf{B}$ constant in direction and modulus).

### 11.5.2 *Two cases of linear polarization*

**Case 1: $\mathbf{E}_0$ perpendicular to the plane of incidence**

The plane of incidence is the plane spanned by the propagation vectors $\mathbf{k}, \mathbf{k}', \mathbf{k}''$. The field vectors are, as we saw, orthogonal to these (cf. transversality, Eq. (9.17)). According to the given condition the incident field vector $\mathbf{E}_0$ is perpendicular to this plane and therefore lies in the tangential plane. Since $E_t^{(1)} - E_t^{(2)} = 0$, there is no change of direction of the field $\mathbf{E}$ at the interface of the media for either reflection or refraction, i.e. with $E_{0x} = 0 = E_{0z}$ we also have $E_{0x}' = 0 = E_{0z}'$ and $E_{0x}'' = 0 = E_{0z}''$. The relation (*a*) implies

$$(\mathbf{E}_0 + \mathbf{E}_0'')_t = (\mathbf{E}')_t. \tag{11.67}$$

The tangential plane is the plane of the interface, i.e. the $(x, y)$-plane. A general vector in this plane would be $\mathbf{r}_t = x\mathbf{e}_x + y\mathbf{e}_y$. Since in the case of the $E$-fields here, only the $y$-components are not zero, we can rewrite the tangential relation also as

$$E_{0y} + E_{0y}'' = E_{0y}', \tag{11.68}$$

or

$$(A) \qquad E_0 + E_0'' = E_0'. \tag{11.69}$$

One should note that $E_0, E_0'', E_0'$ are *amplitudes*, which can be complex. We thus have the situation depicted in Fig. 11.6. The orientation of the $\mathbf{B}$-vectors has been chosen in Fig. 11.6 such that the transport of energy $\mathbf{E}\times\mathbf{H}$ is positive in the direction of the wave vectors.

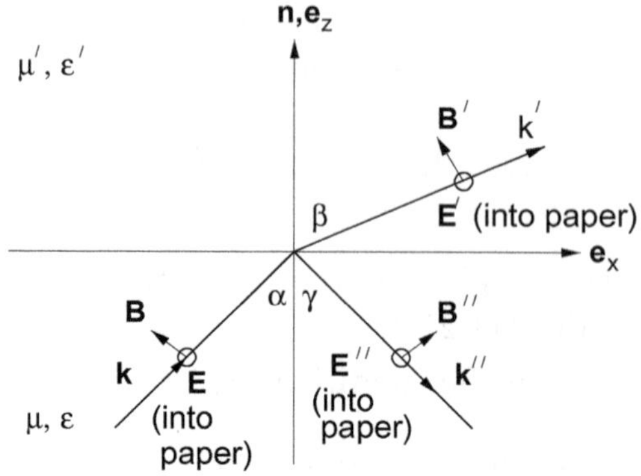


**Fig. 11.6.** Case 1: The directions of the field components at incidence, refraction and reflection.

We can rewrite the relation (*b*), i.e. Eq. (11.64), using the property of scalar triple products

$$(\mathbf{A}\times\mathbf{B})\cdot\mathbf{C}=\mathbf{C}\cdot(\mathbf{A}\times\mathbf{B})=(\mathbf{C}\times\mathbf{A})\cdot\mathbf{B}. \tag{11.70}$$

Thus Eq. (11.64) can be written as

$$(\mathbf{n}\times\mathbf{k})\cdot\mathbf{E}_0+(\mathbf{n}\times\mathbf{k}'')\cdot\mathbf{E}_0''-(\mathbf{n}\times\mathbf{k}')\cdot\mathbf{E}_0'=0, \tag{11.71}$$

i.e. with angles as in Fig. 11.6,

$$k\sin\alpha\mathbf{e}_y\cdot\mathbf{E}_0+k''\sin\gamma\mathbf{e}_y\cdot\mathbf{E}_0''-k'\sin\beta\mathbf{e}_y\cdot\mathbf{E}_0'=0. \tag{11.72}$$

However, according to Snell's law, Eqs. (11.55)–(11.57),

$$k\sin\alpha=k'\sin\beta,\quad \alpha=\gamma,\quad k=k'', \tag{11.73}$$

so that

$$k\sin\alpha\mathbf{e}_y\cdot(\mathbf{E}_0+\mathbf{E}_0''-\mathbf{E}_0')=0. \tag{11.74}$$

Since the sum of the vectors **E** (as above) lies in the tangential plane, we conclude as we concluded from (*a*) that $E_0+E_0''-E_0'=0$. Hence this case yields no new relation.

Consider now relation (*c*). We handle Eq. (11.65) as follows. We have

$$\frac{1}{\mu}(\mathbf{k}\times\mathbf{E}_0)\times\mathbf{n}=\frac{1}{\mu}[(\mathbf{n}\cdot\mathbf{k})\mathbf{E}_0-\mathbf{k}\underbrace{(\mathbf{n}\cdot\mathbf{E}_0)}_{0}]. \tag{11.75}$$

Hence for the incident wave

$$\frac{1}{\mu}(\mathbf{k}\times\mathbf{E}_0)\times\mathbf{n}=\frac{1}{\mu}\underbrace{k}_{\omega\sqrt{\epsilon\mu}}\mathbf{E}_0\cos\alpha=\omega\sqrt{\frac{\epsilon}{\mu}}\mathbf{E}_0\cos\alpha. \tag{11.76}$$

We obtain similar expressions for the reflected and refracted waves. First,

$$\frac{1}{\mu}(\mathbf{k}''\times\mathbf{E}_0'')\times\mathbf{n}=-\omega''\sqrt{\frac{\epsilon}{\mu}}\mathbf{E}_0''\cos\gamma. \tag{11.77}$$

The minus sign on the right follows from the fact that the projection of $\mathbf{k}''\times\mathbf{E}_0''$ is antiparallel to $\mathbf{n}$. Hence also

$$\frac{1}{\mu'}(\mathbf{k}'\times\mathbf{E}_0')\times\mathbf{n}=\omega'\sqrt{\frac{\epsilon'}{\mu'}}\mathbf{E}_0'\cos\beta. \tag{11.78}$$

But $\omega=\omega'=\omega''$ and $\alpha=\gamma$. Thus Eq. (11.65) yields the relation

$$\omega\sqrt{\frac{\epsilon}{\mu}}\mathbf{E}_0\cos\alpha-\omega\sqrt{\frac{\epsilon}{\mu}}\mathbf{E}_0''\cos\alpha-\omega\sqrt{\frac{\epsilon'}{\mu'}}\mathbf{E}_0'\cos\beta=0, \tag{11.79}$$

i.e. (since — see above — the direction of the field $\mathbf{E}$ does not change),

$$(B)\qquad (E_0-E_0'')\sqrt{\frac{\epsilon}{\mu}}\cos\alpha-E_0'\sqrt{\frac{\epsilon'}{\mu'}}\cos\beta=0. \tag{11.80}$$

This is a new relation.

Finally we consider relation $(d)$. The relation (11.66) yields nothing, since we have

$$(\mathbf{E}_0,\ \mathbf{E}_0'',\ \mathbf{E}_0')\cdot\mathbf{n}=0. \tag{11.81}$$

We have thus obtained the results $(A)$ and $(B)$.

Eliminating $E_0'$ from $(A)$ and $(B)$, we obtain

$$(E_0-E_0'')\sqrt{\frac{\epsilon}{\mu}}\cos\alpha-(E_0+E_0'')\sqrt{\frac{\epsilon'}{\mu'}}\cos\beta=0, \tag{11.82}$$

or

$$\begin{aligned}\frac{E_0''}{E_0}&=\frac{\sqrt{\epsilon/\mu}\cos\alpha-\sqrt{\epsilon'/\mu'}\cos\beta}{\sqrt{\epsilon/\mu}\cos\alpha+\sqrt{\epsilon'/\mu'}\cos\beta}\\&=\left[1-\sqrt{\frac{\epsilon'\mu}{\epsilon\mu'}}\frac{\cos\beta}{\cos\alpha}\right]\left[1+\sqrt{\frac{\epsilon'\mu}{\epsilon\mu'}}\frac{\cos\beta}{\cos\alpha}\right]^{-1}.\end{aligned} \tag{11.83}$$

Using the following relation obtained with Eq. (11.57),

$$\frac{\sin\alpha}{\sin\beta} = \sqrt{\frac{\epsilon'\mu'}{\epsilon\mu}}, \qquad \sqrt{\frac{\epsilon'}{\epsilon}} = \sqrt{\frac{\mu}{\mu'}\frac{\sin\alpha}{\sin\beta}}, \tag{11.84}$$

we can rewrite the ratio as

$$\begin{aligned}\frac{E_0''}{E_0} &= \left[1 - \frac{\mu}{\mu'}\frac{\sin\alpha\cos\beta}{\sin\beta\cos\alpha}\right]\left[1 + \frac{\mu}{\mu'}\frac{\sin\alpha\cos\beta}{\sin\beta\cos\alpha}\right]^{-1} \\ &= \left[1 - \frac{\mu}{\mu'}\frac{\tan\alpha}{\tan\beta}\right]\left[1 + \frac{\mu}{\mu'}\frac{\tan\alpha}{\tan\beta}\right]^{-1}.\end{aligned} \tag{11.85}$$

Eliminating $E_0''$ from $(A)$ and $(B)$, we obtain

$$2E_0\sqrt{\frac{\epsilon}{\mu}}\cos\alpha - E_0'\sqrt{\frac{\epsilon'}{\mu'}}\cos\beta = 0,$$

and hence

$$\begin{aligned}\frac{E_0'}{E_0} &= \frac{2\sqrt{\epsilon/\mu}\cos\alpha}{\sqrt{\epsilon/\mu}\cos\alpha + \sqrt{\epsilon'/\mu'}\cos\beta} = 2\left[1 + \sqrt{\frac{\epsilon'\mu}{\epsilon\mu'}}\frac{\cos\beta}{\cos\alpha}\right]^{-1} \\ &= 2\left[1 + \frac{\mu}{\mu'}\frac{\sin\alpha\cos\beta}{\cos\alpha\sin\beta}\right]^{-1} = 2\left[1 + \frac{\mu}{\mu'}\frac{\tan\alpha}{\tan\beta}\right]^{-1}.\end{aligned} \tag{11.86}$$

For the so-called *optical frequencies* (i.e. those in the visible range of frequencies, see Section 11.1) we have $\mu(\omega) \simeq \mu'(\omega)$. In this case we obtain the *Fresnel formulas* for light polarized vertically to the plane of incidence:

$$\frac{E_0''}{E_0} = \frac{\sin(\beta-\alpha)}{\sin(\alpha+\beta)}, \quad \frac{E_0'}{E_0} = \frac{2\sin\beta\cos\alpha}{\sin(\alpha+\beta)}. \tag{11.87}$$

### Case 2: $\mathbf{E}_0$ in the plane of incidence

In this case the directions of the fields are as indicated in Fig. 11.7. The relation $(a)$ implies at $z = 0$

$$E_0\cos\alpha - E_0''\cos\gamma - E_0'\cos\beta = 0, \tag{11.88}$$

or, since $\alpha = \gamma$,

$$(A') \qquad (E_0 - E_0'')\cos\alpha - E_0'\cos\beta = 0. \tag{11.89}$$

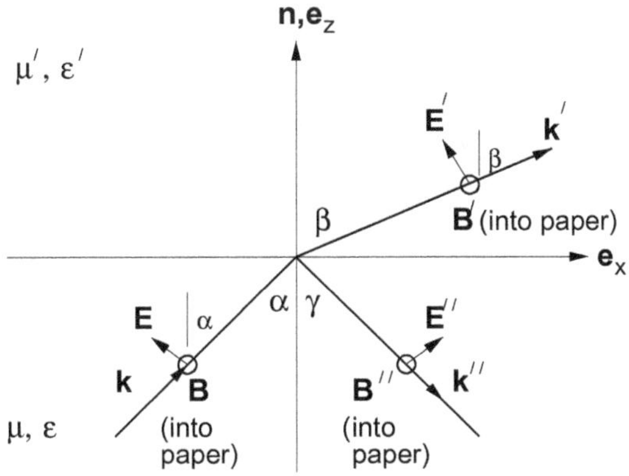


**Fig. 11.7.** Case 2: The directions of the field components at incidence, refraction and reflection.

The relation (*b*) gives nothing, since $\mathbf{k} \times \mathbf{E} \perp \mathbf{n}$. Relation (*d*) yields (with $\alpha = \gamma$)

$$\epsilon(E_0 + E_0'') \sin\alpha - \epsilon' E_0' \sin\beta = 0, \tag{11.90}$$

or since

$$\frac{\sin\alpha}{\sin\beta} = \sqrt{\frac{\epsilon'\mu'}{\epsilon\mu}} = \frac{n'}{n}, \tag{11.91}$$

we have

$$(E_0 + E_0'')\frac{\epsilon}{\epsilon'}\sqrt{\frac{\epsilon'\mu'}{\epsilon\mu}} - E_0' = 0, \tag{11.92}$$

so that

$$(B') \qquad (E_0 + E_0'')\sqrt{\frac{\epsilon}{\mu}} - E_0'\sqrt{\frac{\epsilon'}{\mu'}} = 0. \tag{11.93}$$

The relation (*c*) also yields $(B')$.

From both of the relations $(A')$ and $(B')$ we obtain the next pair of *Fresnel formulas*, using again the relation

$$\sqrt{\frac{\mu'}{\mu}} = \sqrt{\frac{\epsilon}{\epsilon'}\frac{\sin\alpha}{\sin\beta}}. \tag{11.94}$$

Eliminating $E'_0$ from the two relations $(A')$ and $(B')$, we obtain

$$\begin{aligned}\frac{E''_0}{E_0} &= \left[1-\sqrt{\frac{\epsilon\mu'\cos\beta}{\epsilon'\mu\cos\alpha}}\right]\left[1+\sqrt{\frac{\epsilon\mu'\cos\beta}{\epsilon'\mu\cos\alpha}}\right]^{-1} \\ &= \left[1-\frac{\epsilon}{\epsilon'}\frac{\tan\alpha}{\tan\beta}\right]\left[1+\frac{\epsilon}{\epsilon'}\frac{\tan\alpha}{\tan\beta}\right]^{-1}, \end{aligned} \tag{11.95}$$

and eliminating $E''_0$ we obtain

$$\frac{E'_0}{E_0} = 2\sqrt{\frac{\epsilon\mu'}{\epsilon'\mu}}\left[1+\sqrt{\frac{\epsilon\mu'\cos\beta}{\epsilon'\mu\cos\alpha}}\right]^{-1}. \tag{11.96}$$

For $\mu = \mu'$ we obtain (as can be verified by using Eq. (11.94) and replacing the tangent by sine and cosine and multiplying out)

$$\frac{E''_0}{E_0} = \frac{\tan(\alpha-\beta)}{\tan(\alpha+\beta)}, \quad \frac{E'_0}{E_0} = \frac{2\sin\beta\cos\alpha}{\sin(\alpha+\beta)\cos(\alpha-\beta)}. \tag{11.97}$$

For the discussion of "*total reflection*" below we observe here that if we set $\beta = \pi/2$ in Eqs. (11.87) and (11.97) (or alternatively in a preceding formula) we obtain (with $\mu = \mu'$)

$$\frac{E''_0}{E_0} = 1, \quad \frac{E'_0}{E_0} = 2\sqrt{\frac{\epsilon}{\epsilon'}} \neq 0. \tag{11.98}$$

### 11.5.3 *The Brewster angle and the case without reflection*

We observe that when $\alpha+\beta = \pi/2$ in the case of the vector $\mathbf{E}_0$ lying in the plane of incidence, we have (cf. (11.97)) $\tan(\alpha+\beta) = \infty$ and $E''_0 = 0$, but also $\sin\beta = \sin(\pi/2-\alpha) = \cos\alpha$. Snell's law (11.57) then implies

$$\frac{n'}{n} = \frac{\sin\alpha}{\sin\beta} = \tan\alpha_B. \tag{11.99}$$

The angle $\alpha_B$ is known as *Brewster's angle.* In particular when $\alpha = \pi/4$, also $\beta = \pi/4$, i.e. one has straight passage. On the other hand, if the refractive indices $n, n'$ are equal, we have $\alpha = \beta$ and hence (from Eq. (11.97))

$$E''_0 = 0, \quad E'_0 = E_0, \tag{11.100}$$

and there is no reflection — as expected. Thus at Brewster's angle with $\mathbf{E}_0$ in the plane of incidence there is no reflected ray whose electric vector oscillates in the plane of incidence. If arbitrary (i.e. unpolarized) light is incident at

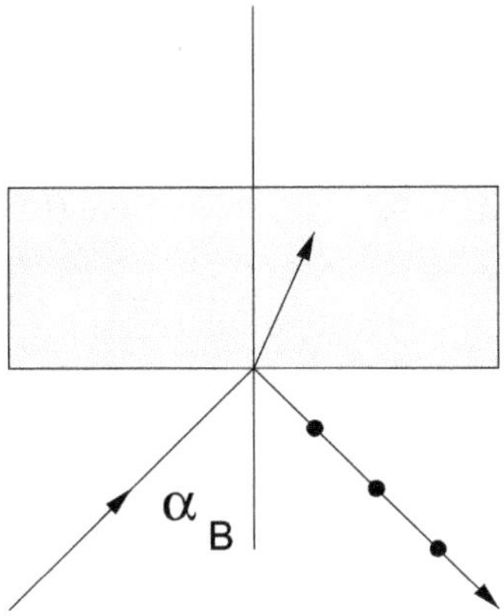


**Fig. 11.8.** Polarization at angle of incidence $=\alpha_B$.

Brewster's angle, the reflected ray is thus linearly polarized perpendicular to the plane of incidence, as indicated in Fig. 11.8.

### 11.5.4 *The general case*

The general case of an elliptically polarized wave can be dealt with by superposition of the cases 1 and 2, i.e. with

$$\mathbf{E}_1 = \boldsymbol{\epsilon}_1 E_1 e^{i(\mathbf{k}\cdot\mathbf{r}-\omega t)}, \quad \mathbf{E}_2 = \boldsymbol{\epsilon}_2 E_2 e^{i(\mathbf{k}\cdot\mathbf{r}-\omega t)}, \quad \boldsymbol{\epsilon}_1 \cdot \boldsymbol{\epsilon}_2 = 0, \tag{11.101}$$

where $\boldsymbol{\epsilon}_i, i = 1, 2$, are *polarization vectors* (i.e. unit vectors along the respective direction of polarization) and

$$\mathbf{B}_j = \sqrt{\mu\epsilon}\frac{\mathbf{k}\times\mathbf{E}_j}{k}, \quad j = 1, 2, \tag{11.102}$$

we have for the most general homogeneous wave in the direction of $\mathbf{k} = k\mathbf{n}_k$:

$$\mathbf{E}(\mathbf{r}, t) = (\boldsymbol{\epsilon}_1 E_1 + \boldsymbol{\epsilon}_2 E_2)e^{i(\mathbf{k}\cdot\mathbf{r}-\omega t)}. \tag{11.103}$$

Here $E_1, E_2$ are complex numbers, which also permit a phase difference between waves of different polarization. If $E_1$ and $E_2$ have the same phase, $E$ describes a linearly polarized wave with a polarization vector which makes with $\boldsymbol{\epsilon}_1$ an angle $\theta = \tan^{-1}(E_2/E_1)$ and has modulus $E = \sqrt{E_1^2 + E_2^2}$. If $E_1, E_2$ have different phases, $\mathbf{E}$ is elliptically polarized. If $|E_1| = |E_2|$ and if their phases differ by $\pi/2$, i.e. if $\mathbf{E}$ can be written

$$\mathbf{E} = E_0(\boldsymbol{\epsilon}_1 \pm i\boldsymbol{\epsilon}_2)e^{i(\mathbf{k}\cdot\mathbf{r}-\omega t)}, \tag{11.104}$$

then $\mathbf{E}$ is circularly polarized.

### 11.5.5 *Total reflection*

It remains to consider the case of total reflection. Let $n, n'$ be the refractive indices of the media on either side of an interface and $n' < n$. Then, see Fig. 11.9, there is a *critical angle* $\alpha_0$, at which $\beta \to \beta_0 = \pi/2$ (reflected and refracted fields in the same direction $x$).

Also (as we deduce from the above formulas) independent of the direction of polarization of the electric field (cf. Eq. (11.89), condition $(A')$)

$$E_0 = E_0'', \quad \text{but} \quad E_0' \neq 0. \tag{11.105}$$

This case is described as *total reflection.* Since

$$\frac{\sin\alpha'}{\sin\beta} = \frac{n'}{n}, \tag{11.106}$$

it follows that $\sin\alpha_0 = n'/n$.

We can consider this case also from another point of view, by demonstrating that there is no transport of energy into the other medium. For real angles $\beta$ the maximal value of $\sin\beta$ is, of course, 1 (for $\beta = \pi/2$). The maximal value of $\alpha$, which satisfies the condition

$$\frac{\sin\alpha}{\sin\beta} = \frac{n'}{n} < 1 \tag{11.107}$$

for real values of $\beta$, is therefore given by

$$\sin\alpha_0 = \frac{n'}{n} \tag{11.108}$$

($\sin\beta_0 = 1$). For $\alpha > \alpha_0$, hence for $n'/n = \sin\alpha_0$, we write the "angle of refraction" $\beta$ which is then complex as

$$\beta = \frac{\pi}{2} + i\delta. \tag{11.109}$$

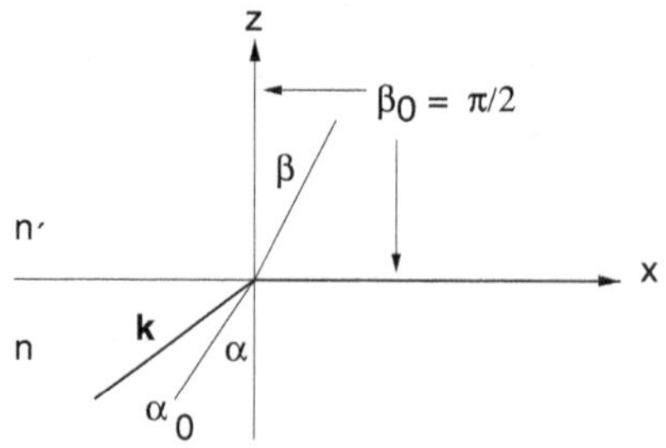


**Fig. 11.9.** The case of total reflection.

Then $\sin\beta = \cosh\delta$ and

$$\delta = \cosh^{-1}(\sin\beta) = \cosh^{-1}\left(\frac{n}{n'}\sin\alpha\right). \tag{11.110}$$

Moreover,

$$\cos\beta = \sqrt{1-\sin^2\beta} = +i\sqrt{\cosh^2\delta - 1} = i\sinh\delta, \tag{11.111}$$

and (cf. Fig. 11.9)

$$e^{i\mathbf{k}'\cdot\mathbf{r}} = e^{ik'z\cos\beta} = e^{-k'z\sinh\delta} \to 0 \quad \text{for} \quad z\to\infty. \tag{11.112}$$

It follows that the field of the refracted radiation, i.e. $\mathbf{E}'$, goes to 0 for $z\to\infty$, but for $z$ finite, as we can show, $\mathbf{n}\cdot(\mathbf{E}'\times\mathbf{H}') = 0$. This means, there is no transport of energy into the other medium. We see this as follows. Recalling the transversality condition and Eq. (11.102), we have, where $\mathbf{S}$ is the Poynting vector of the refracted radiation,

$$\mathbf{S}\cdot\mathbf{n} = \Re\left[\mathbf{n}\cdot\underbrace{(\mathbf{E}'\times\mathbf{H}')}_{\sqrt{\epsilon/\mu}\mathbf{E}'^2\mathbf{k}'/k'}\right] = \Re\left\{\sqrt{\frac{\epsilon}{\mu}}|E_0'|^2\frac{\mathbf{n}\cdot\mathbf{k}'}{k'}\right\} = 0, \tag{11.113}$$

the latter (i.e. taking the real part) because of "$i$" in

$$\mathbf{n}\cdot\mathbf{k}' = k'\cos\beta = ik'\sinh\delta. \tag{11.114}$$

We have total reflection. This total reflection is also evident from the fact that $E_0 = E_0''$, i.e. $\mathbf{E}_0^2 = \mathbf{E}_0''^2$.

We return once again to the general case:

$$\cos\beta = \pm(1-\sin^2\beta)^{1/2} = \pm\sqrt{1-\left(\frac{n}{n'}\right)^2\sin^2\alpha} = \pm\sqrt{1-\left(\frac{\sin\alpha}{\sin\alpha_0}\right)^2}. \tag{11.115}$$

For $\alpha < \alpha_0$ we have $\cos\beta$ real and hence $\beta$. For $\alpha > \alpha_0$, however, $\cos\beta$ is purely imaginary, i.e. $\beta$ complex:

$$\cos\beta = \pm i\left[\left(\frac{\sin\alpha}{\sin\alpha_0}\right)^2 - 1\right]^{1/2}. \tag{11.116}$$

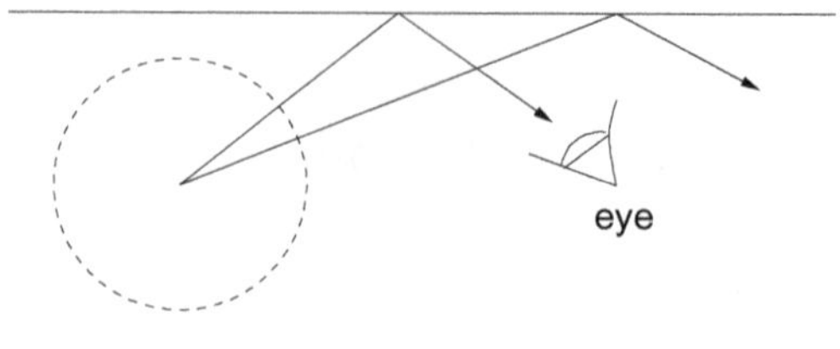


**Fig. 11.10.** Scattering of light at an interface.

The phase of the refracted wave is therefore ($\mathbf{k}'$ in the $(x, z)$-plane)

$$\begin{aligned} e^{i\mathbf{k}'\cdot\mathbf{r}} &= \exp\{i(k'_x x + k'_z z)\} = \exp\{i(k' x \sin\beta + k' z \cos\beta)\} \\ &= \exp\left\{ik' x \frac{n}{n'} \sin\alpha\right\} \exp\left\{\mp k' z \sqrt{\left(\frac{\sin\alpha}{\sin\alpha_0}\right)^2 - 1}\right\} \\ &\propto \exp(-z/z_0). \end{aligned} \tag{11.117}$$

Thus the wave is damped in the direction of $z$ if we make the physically plausible choice of selecting the exponentially decreasing function. In general

$$z_0 \equiv \frac{1}{k'\sqrt{(\sin\alpha/\sin\alpha_0)^2 - 1}} \quad \text{meters} \tag{11.118}$$

is of the order of several wavelengths, i.e. the refractive wave is damped so strongly, that in general there is no refractive wave, unless the thickness of the weaker medium is less than one wavelength. In such a case light can tunnel through the film, and one has to take a combination of both exponential functions. We see that when $\alpha = \alpha_0$, there is no damping. One has the scattering of light at the interface, which can also be observed, e.g. in an experiment with a sphere immersed in a medium, the sphere containing a lamp and permitting light to escape through some windows, as indicated in Fig. 11.10.

## 11.6 Useful Formulation of the Fresnel Formulas

We set

$$N = \sqrt{\frac{\epsilon'\mu}{\epsilon\mu'}}, \quad R = \frac{E''_0}{E_0}, \quad T = \frac{E'_0}{E_0}. \tag{11.119}$$

Then in the two cases of Section 11.5.2 we can rewrite the results as follows:

**Case (1): $\mathbf{E}_0 \perp$ plane of incidence**

$$(A)\quad 1+R-T=0, \quad (B)\quad (1-R)\cos\alpha - NT\cos\beta = 0, \tag{11.120}$$

so that $(1-R)\cos\alpha = N(1+R)\cos\beta$. Hence

$$R = \frac{\cos\alpha - N\cos\beta}{\cos\alpha + N\cos\beta}, \quad T = \frac{2\cos\alpha}{\cos\alpha + N\cos\beta}. \tag{11.121}$$

The expressions $R$ and $T$ are respectively called *reflection coefficient* and *transmission coefficient*.

**Case (2): $\mathbf{E}_0$ in the plane of incidence**

$$(A')\quad (1-R)\cos\alpha = T\cos\beta, \quad (B')\quad (1+R) = NT, \tag{11.122}$$

so that $(1-R)\cos\alpha = ((1+R)/N)\cos\beta$, and

$$R = \frac{N\cos\alpha - \cos\beta}{N\cos\alpha + \cos\beta}, \quad T = \frac{2\cos\alpha}{N\cos\alpha + \cos\beta}. \tag{11.123}$$

Here we have defined $R$ and $T$ as ratios of amplitudes; this is not the most common formulation. Some authors such as Lim [88] define $R$ and $T$ as ratios of the energy of the reflected ray, and that of the transmitted ray, to the energy of the incident ray. In the book of Jackson [73] (Chapter 7) the coefficients are not introduced. Since, as we emphasized, $\mathbf{E}_0, \mathbf{E}'_0, \mathbf{E}''_0$ are amplitudes, which can still be complex, also $R$ and $T$ can be complex. We introduce the coefficients $R$ and $T$ here in analogy to their counterparts in quantum mechanics or quantum mechanical scattering problems, which seems to suggest itself.

**Example 11.2 (Light falling on glass).**

(a) The refractive index of glass is $n' = 1.5$. Calculate what fraction of light intensity vertically incident on a surface of clean glass is reflected.
(b) What are the phases of the vectors $\mathbf{E}, \mathbf{H}$ at the interface between air $(n)$ and glass $(n', n' > n)$ in the case of vertical incidence?

**Solution:**

(a) We have (with $\mu = \mu', N = n'/n$) for vertical incidence $(\alpha = 0, \beta = 0)$ and irrespective of whether $\mathbf{E}_0$ lies in the plane of incidence or is perpendicular to this

$$\left|\frac{E_0''}{E_0}\right|^2 = |R|^2 = \left|\frac{n-n'}{n+n'}\right|^2 = \left(\frac{1/2}{5/2}\right)^2 = \left(\frac{1}{5}\right)^2 = \frac{1}{25} = 4\%. \quad (11.124)$$

We see that this is a very small fraction.

(b) For $\mathbf{E}_0 \perp$ plane of incidence: $\alpha = \beta = 0$. Hence

$$R = \frac{1-N}{1+N} = \frac{n-n'}{n+n'} < 0, \quad R = \frac{E_0''}{E_0}. \quad (11.125)$$

It follows that

$$E_0'' = e^{i\pi}|R|E_0, \quad (11.126)$$

i.e. $\mathbf{E}_0''$ points in a direction opposite to that of $\mathbf{E}_0$. For the direction of transport of energy (given by $(\mathbf{E} \times \mathbf{H})$) of the reflected wave to be opposite to that of the incident wave, the phase of $\mathbf{H}$ must remain unchanged.

**Example 11.3**[6] **(Calculation of the angle of refraction).** Polystyrene has a relative permittivity of 2.7. An electromagnetic wave in air is incident at an angle of 30° to the normal on a surface of polystyrene. Calculate the angle of refraction and repeat the calculation for the reverse case.

**Solution:** For the passage from air to polystyrene we set:

$$\epsilon_1 := \epsilon_0, \quad \epsilon_2 := 2.7\epsilon_0, \quad \mu_1 := \mu_0, \quad \mu_2 := \mu_0. \quad (11.127)$$

From Snell's law, Eq. (11.57), $\sin\theta_2 = (n_1/n_2)\sin\theta_1$, it follows that (since $\sin 30° = 1/2$)

$$\sin\theta_2 = \sqrt{\frac{1}{2.7}}(0.5) = 0.304, \quad \theta_2 = 17.7°. \quad (11.128)$$

From polystyrene to air: $\epsilon_1 := 2.7\epsilon_0$, $\epsilon_2 := \epsilon_0$, $\mu_1 = \mu_0$, $\mu_2 = \mu_0$, and

$$\sin\theta_2 = \sqrt{2.7}(0.5) = 0.822, \quad \theta_2 = 55.2°. \quad (11.129)$$

[6]See also *The Electromagnetic Problem Solver* [47], p. 649.

## 11.7 Diffraction of Light

### 11.7.1 *Introductory remarks*

By the *diffraction of light* at the edge of an opaque body, i.e. at an obstacle to the light there, or at a narrow slit, one means the breaking up of a beam of light into series of dark and light bands or coloured spectra. Similarly the diffraction of a beam of radiation or particles means its breaking up into series of alternately high and low intensities. We consider here the diffraction of light in the context of electromagnetic theory. We consider as depicted in Fig. 11.11 a monochromatic source point $S_0$ from which electromagnetic radiation is emitted, a diaphragm or screen with a slit or hole in it, also called aperture, a distance of order $R$ away from $S_0$, and a parallel screen with an observation point $P$ a large distance $r$ away from the diaphragm on the other side of the slit or aperture.[7]

### 11.7.2 *Basic equations and assumptions*

Let $u$ be any amplitude of an harmonic electromagnetic wave, e.g. $E, H, D$ or $B$ (i.e. any of the Cartesian components of one of the fields). Then, as we deduced e.g. in Section 9.2 from Maxwell's equations (or see e.g. Eq. (11.30)),

$$\nabla^2 u + \beta^2 u = 0, \tag{11.130}$$

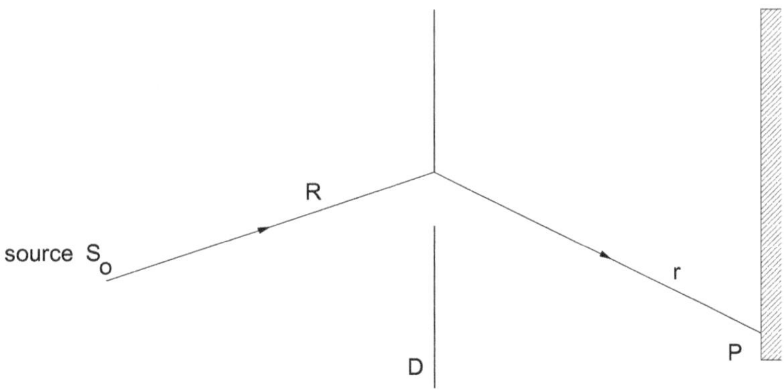


**Fig. 11.11.** Source $S_0$, diaphragm $D$ and screen with observation point $P$.

[7]An introductory treatment somewhat similar to that we give below may be found in the books of Hewson [70], pp. 86–101 and Slater and Frank [136], pp. 180–192. The treatment in the book of Jackson [73], Chapter 9, is on a more sophisticated level. See also Born and Wolf [18].

where ($\omega = 2\pi\nu, \nu\lambda = c = 1/\sqrt{\mu_0\epsilon_0}$, $\lambda$ being the wavelength)

$$\beta = \frac{2\pi}{\lambda} = \omega\sqrt{\mu_0\epsilon_0}. \tag{11.131}$$

In spherical polar coordinates, neglecting the angular variation in view of the spherical symmetry of the problem, we have (cf. Eq. (2.22b)) for $u = u(r)$,

$$\frac{1}{r^2}\frac{\partial}{\partial r}\left(r^2\frac{\partial u}{\partial r}\right) + \beta^2 u = 0, \tag{11.132}$$

or

$$\frac{\partial^2}{\partial r^2}(ru) + \beta^2(ru) = 0. \tag{11.133}$$

It follows that

$$u(r) = \frac{e^{-i\beta r}}{r} \tag{11.134}$$

is one possible solution. We select this solution with the minus sign in the argument of the exponential because this indicates that the wave is a forward moving or outgoing wave when combined with the time factor $\exp(i\omega t)$. We shall subsequently let

$$v = \frac{e^{-i\beta R}}{R} \tag{11.135}$$

be another solution which describes the spherically symmetrical waves emitted by the source $S_0$. We have first of all with $u = u(r), v = v(r)$,

$$\nabla^2 u + \beta^2 u = 0, \quad \nabla^2 v + \beta^2 v = 0. \tag{11.136}$$

Hence everywhere

$$v\nabla^2 u - u\nabla^2 v = 0 \quad \text{or} \quad \mathrm{u}\nabla^2\mathrm{v} - \mathrm{v}\nabla^2\mathrm{u} = 0. \tag{11.137}$$

Thus

$$\boldsymbol{\nabla}\cdot(u\boldsymbol{\nabla}v - v\boldsymbol{\nabla}u) = 0. \tag{11.138}$$

With this equation we also have, using Gauss' divergence theorem,

$$\begin{aligned} 0 &= \int_V dV(u\nabla^2 v - v\nabla^2 u) = \int_V dV\boldsymbol{\nabla}\cdot(u\boldsymbol{\nabla}v - v\boldsymbol{\nabla}u) \\ &= \int_{\text{enclosing surface S}} d\mathbf{S}\cdot(u\boldsymbol{\nabla}v - v\boldsymbol{\nabla}u). \end{aligned} \tag{11.139}$$

This is actually the second of Green's theorems, i.e. Eq. (3.35). We apply this relation to the volume between the spheres $S$ and $s$ as in Fig. 11.12. As indicated in Fig. 11.12 we exclude with the little sphere $s$ the point $P$ in order to prevent infinities from entering the solutions (i.e. those arising from $1/r$). Then[8]

$$\int_S (u\nabla v - v\nabla u)\cdot d\mathbf{S} = \int_s (u\nabla v - v\nabla u)\cdot d\boldsymbol{s}. \tag{11.140}$$

We insert here the solution (11.134) noting that (cf. Eq. (2.72))

$$\nabla u = \frac{\mathbf{r}}{r}\left\{-\left(\frac{i\beta}{r}+\frac{1}{r^2}\right)e^{-i\beta r}\right\}, \tag{11.141}$$

and

$$\int_{\text{sphere of radius r}} ds = 4\pi r^2. \tag{11.142}$$

Then the right hand side of Eq. (11.140) becomes:

$$\text{r.h.s.} = 4\pi\mathrm{r}^2\left\{\frac{\mathrm{e}^{-\mathrm{i}\beta\mathrm{r}}}{\mathrm{r}}\nabla\mathrm{v} + \frac{\mathbf{r}}{\mathrm{r}}\mathrm{v}\left(\frac{\mathrm{i}\beta}{\mathrm{r}}+\frac{1}{\mathrm{r}^2}\right)\mathrm{e}^{-\mathrm{i}\beta\mathrm{r}}\right\}. \tag{11.143}$$

In the limit $r \to 0$ this becomes

$$\lim_{r\to 0}(\text{r.h.s.}) \to 4\pi \mathrm{v}_{\mathrm{P}}\frac{\mathbf{r}}{\mathrm{r}}, \tag{11.144}$$

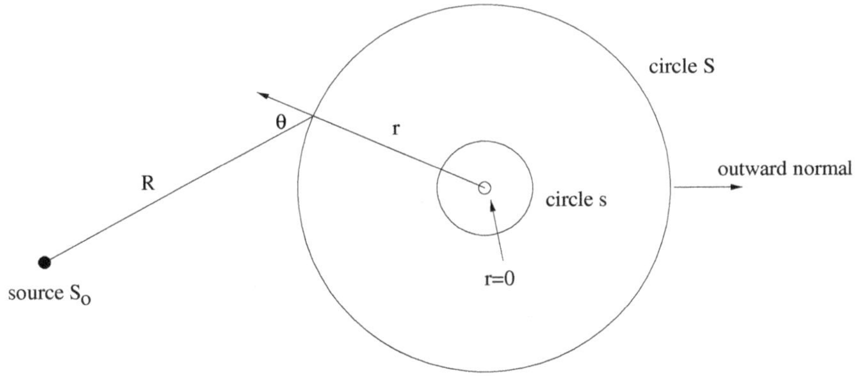


**Fig. 11.12.** The small and big circles around the observation point $P$.

[8]Note that frequently one writes $\nabla u(r)\cdot d\mathbf{S} = (\partial u(r)/\partial \mathbf{r})\cdot d\mathbf{S} = \mathbf{e}_r(\partial u(r)/\partial r)\cdot \mathbf{n}dS \equiv (\partial u(r)/\partial n)dS$.

where $v_P$ is the value of $v$ at $P$. Hence Eq. (11.140) becomes:

$$v_P = \frac{1}{4\pi}\int_S \left\{\frac{e^{-i\beta r}}{r}\boldsymbol{\nabla}v - v\boldsymbol{\nabla}\left(\frac{e^{-i\beta r}}{r}\right)\right\}\cdot d\mathbf{S}. \tag{11.145}$$

The actual amplitude of light in the screen depends on the material of which the screen is made. The relation (11.145) is known as *Kirchhoff's diffraction formula.*

In order to be able to proceed we make an assumption, called *Kirchhoff's assumption.* We assume that if the linear dimensions of the hole or aperture are much greater than the wavelength $\lambda$ of the radiation, we can take for a point source

$$v = \frac{1}{R}e^{-i\beta R}, \tag{11.146}$$

i.e. unmodified by the diaphragm or screen with the hole. That this assumption is, in general incorrect can be seen by taking as an example sound waves. Sound perceptible by humans has frequencies from about 20 Hz to 20 000 Hz (17 m to 17 mm). The sound wave on entering the hole produces secondary waves which produce their own effects.

Thus here with Kirchhoff's assumption we have

$$v_P = \frac{1}{4\pi}\int_{\text{holes}} \left\{\frac{e^{-i\beta r}}{r}\boldsymbol{\nabla}_{\mathbf{R}}\frac{e^{-i\beta R}}{R} - \frac{e^{-i\beta R}}{R}\left(\boldsymbol{\nabla}_{\mathbf{r}}\frac{e^{-i\beta r}}{r}\right)\right\}\cdot d\mathbf{S}. \tag{11.147}$$

It is now convenient to make the following approximations (as stated, only for convenience here).

1. We assume that the wave number $\beta$ is very large, i.e.

$$\beta = \frac{2\pi}{\lambda}, \quad \lambda \ \text{ small.}$$

   This assumption excludes sound waves.
2. We assume that in taking gradients we need only differentiate exponentials (i.e. other contributions are of nonleading order).[9]
3. We assume that the aperture in the diaphragm is small in comparison with the distance from the hole to the point of observation $P$. Thus the distances $r$ and $R$ do not vary very much and we may take these factors out of the integral sign.

[9]See Jackson [73], p. 441: "Only the phase factor $kR$ in $\exp(ikR)$ needs to be handled with some care".

With these assumptions Eq. (11.147) becomes (the circumflex indicating a unit vector)

$$v_P \simeq -\frac{1}{4\pi}\frac{i\beta}{rR}(\hat{\mathbf{R}}\cdot\hat{\mathbf{S}} - \hat{\mathbf{r}}\cdot\hat{\mathbf{S}})\int_{\text{holes}} e^{-i\beta(R+r)}dS, \tag{11.148}$$

and hence

$$v_P \propto \int_{\text{holes}} e^{-i\beta(R+r)}dS. \tag{11.149}$$

Here $R + r$ is the path difference of the light (i.e. from source to point of observation). The virtues of this analysis are the assumptions we got out of it. The assumptions are independent of the type of waves (i.e. spherical, cylindrical or planar) emitted by the source. This is true for the two classes of diffraction, *Fraunhofer* and *Fresnel diffraction* that we specialize to below. We now consider the geometry with Cartesian coordinates as shown in Fig. 11.13.

From the geometry of Fig. 11.13 we obtain:

$$\begin{aligned} r &= \sqrt{X^2 + (Y-y)^2 + (Z-z)^2} \\ &= [X^2 + Y^2 + Z^2 - 2Yy - 2Zz + y^2 + z^2]^{1/2} \\ &= [a^2 - 2(Yy + Zz) + y^2 + z^2]^{1/2} \\ &= a\left[1 - \left\{\frac{2(Yy+Zz)}{a^2} - \frac{y^2+z^2}{a^2}\right\}\right]^{1/2}, \end{aligned}$$

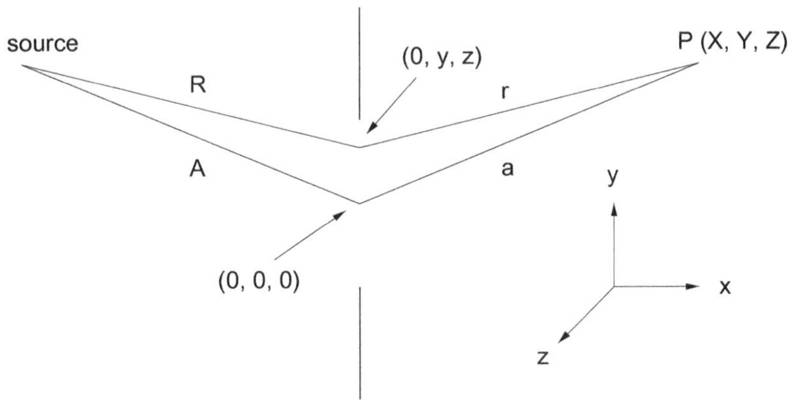


**Fig. 11.13.** The slit in the $(y, z)$-plane.

and on expanding for large $a$,

$$r \simeq a\left[1-\frac{Yy+Zz}{a^2}+\frac{y^2+z^2}{2a^2}\right]=a-\frac{Yy+Zz}{a}+\frac{y^2+z^2}{2a}. \tag{11.150}$$

In terms of the two *direction cosines* $m, n$ of the three of the point of observation $P$,[10]

$$m=\frac{Y}{a}, \quad n=\frac{Z}{a}, \quad l^2+m^2+n^2=1, \tag{11.151}$$

Eq. (11.150) can be written

$$r=a-(my+nz)+\frac{1}{2a}(y^2+z^2). \tag{11.152}$$

Similarly we obtain for $R$ the expression

$$R=A-(My+Nz)+\frac{1}{2A}(y^2+z^2), \tag{11.153}$$

where $M$ and $N$ are two of the three direction cosines of the line to the source,

$$M=\frac{Y'}{A}, \quad N=\frac{Z'}{A}, \quad L^2+M^2+N^2=1. \tag{11.154}$$

We also have for the area of the hole in the $(y, z)$ plane of Fig. 11.14,

$$dS=dydz. \tag{11.155}$$

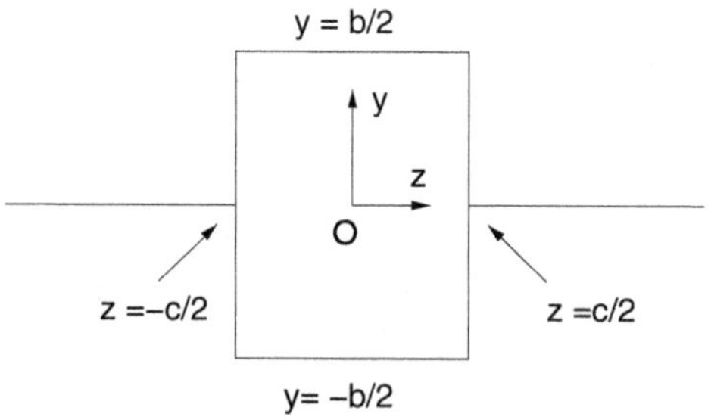


**Fig. 11.14.** The rectangular aperture.

[10]The direction cosines $l, m, n$ are the cosines of the angles $\delta_1, \delta_2, \delta_3$, which an oriented line makes with the positive $x$-, $y$-, $z$-axes at unit distance from the origin. Thus in a two-dimensional Cartesian system in polar representation (polar angle $\theta$) we have $l = \cos\delta_1 = \cos\theta, m = \cos\delta_2 = \cos(\pi/2-\theta) = \sin\theta$. These direction cosines are the elements of a row or column of a matrix describing the rotation around the third axis. In three dimensions with spherical polar angles $\theta, \phi$, one has $l = \cos\delta_1 = \cos\phi\sin\theta, m = \cos\delta_2 = \sin\phi\sin\theta, n = \cos\delta_3 = \cos\theta$ with $l^2+m^2+n^2 = 1$.

Hence, inserting these expressions into the expression (11.149) for $v_P$, we obtain:

$$v_P \propto \iint \exp\left[i\beta\left\{y(m+M)+z(n+N) - \frac{1}{2}(y^2+z^2)\left(\frac{1}{a}+\frac{1}{A}\right)\right\}\right] dydz. \tag{11.156}$$

We now become more specific and consider two classes of diffraction.

### 11.7.3 *Fraunhofer diffraction*

The Fraunhofer class of diffraction deals with plane wave fronts at the aperture — implying source and screen at infinite distances from the diffracting aperture. This reasoning is called the *far field approximation.* This aperture is here chosen of rectangular shape with lengths of sides $b$ and $c$ in the $y$ and $z$ directions respectively and centered at the origin, as indicated in Fig. 11.14. Thus we have here $A, a \to \infty$ with $1/a, 1/A \to 0$, or that quadratic terms in $y$ and $z$ can be neglected,[11] and hence $v_P$ reduces to

$$\begin{aligned} v_P &\propto \iint dydz \exp[i\beta\{y(m+M)+z(n+N)\}] \\ &= \int_{-b/2}^{b/2} e^{i\beta y(m+M)}dy \int_{-c/2}^{c/2} e^{i\beta z(n+N)}dz, \end{aligned} \tag{11.157}$$

since $y$ is independent of $z$, i.e. the limits of $y$ are independent of the limits of $z$. Thus

$$\begin{aligned} v_P &\propto \frac{1}{i\beta(m+M)}\left\{e^{i\beta(m+M)b/2} - e^{-i\beta(m+M)b/2}\right\} \\ &\quad \times \frac{1}{i\beta(n+N)}\left\{e^{i\beta(n+N)c/2} - e^{-i\beta(n+N)c/2}\right\} \\ &= bc\frac{\sin\{\beta(m+M)b/2\}\sin\{\beta(n+N)c/2\}}{\{\beta(m+M)b/2\}\{\beta(n+N)c/2\}}. \end{aligned} \tag{11.158}$$

The square of this quantity, $I$, is a positive periodic function with intensity maxima and minima typical of diffraction, i.e. $I \propto v_P^2$. Note that this quantity depends only on the direction cosines, so that the magnitude of

[11]Another possibility would be that light passing through the aperture is kept parallel by setting lenses on either side, one at the focal length from the source, the other at the focal length from $P$; this is equivalent to the limit $A \to \infty, a \to \infty$. When the quadratic terms have to be taken into account the diffraction is called Fresnel diffraction.

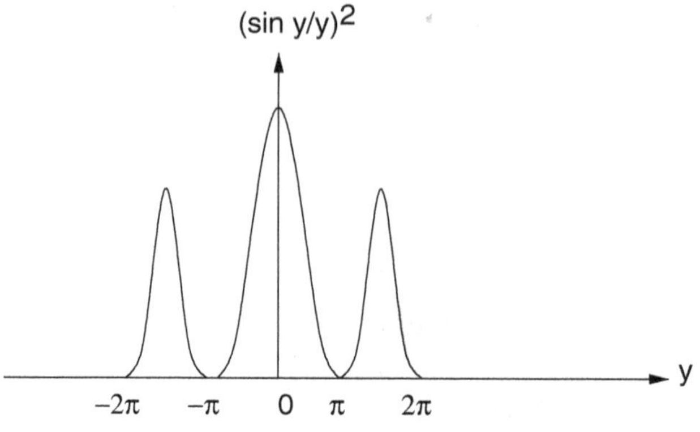


**Fig. 11.15.** The function $(\sin y/y)^2$.

the diffraction pattern on the observation screen depends on the distance between this screen and the aperture. The function

$$\left(\frac{\sin y}{y}\right)^2 \tag{11.159}$$

contained in the intensity $I$ is plotted schematically in Fig. 11.15. Thus this Fraunhofer diffraction pattern from a rectangular aperture has typical minima at points $y = \pm n\pi$. Since a corresponding behavior occurs also in the $z$-direction, the resulting pattern on the parallel observation screen consists of a rectangular arrangement of spots of varying intensities, with the brightest in the center.

### 11.7.4 *Fresnel diffraction*

The Fresnel class of diffraction deals with cylindrical and spherical wave fronts, i.e. a cylindrical source parallel to the $z$-axis. Consider cylindrical symmetry.[12] Then everything out of the paper is the same (for a cylindrical wave), and the diffraction ensues at the edge of a wedge (out of the paper). We consider diffraction by such a wedge. The wedge when placed in the position shown in Fig. 11.16 has the same effect when placed obliquely. Thus the wedge is approximately perpendicular to the straight line connecting the source with the observation point $P$. Since everything is independent of $z$

[12]Diffraction by a circular cylinder and a wedge and variants of these is considered at a higher mathematical level in Chapter 10 of Tyras [152].

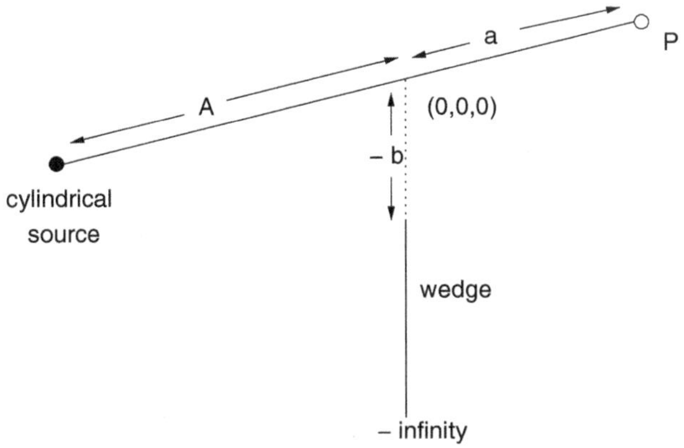


**Fig. 11.16.** Scattering of a cylindrical wave at a wedge.

we have no direction cosines and therefore from Eq. (11.156):

$$v_P \propto \int_{-b}^{\infty} dy \exp\left[-\frac{1}{2}i\beta y^2\left(\frac{1}{a}+\frac{1}{A}\right)\right]. \tag{11.160}$$

This case in which the quadratic terms in Eq. (11.156) cannot be neglected is called the *near field approximation* of Kirchhoff's equation. We set (this defines the quantity $u$)

$$\frac{\pi u^2}{2} = \frac{1}{2}\beta y^2\left(\frac{1}{a}+\frac{1}{A}\right), \tag{11.161}$$

and

$$b' = b\sqrt{\frac{\beta}{\pi}\left(\frac{1}{a}+\frac{1}{A}\right)}. \tag{11.162}$$

One now defines the following integrals, known as *Fresnel's integrals*:

$$C(u) := \int_0^u \cos\left(\frac{\pi}{2}u^2\right) du, \quad S(u) := \int_0^u \sin\left(\frac{\pi}{2}u^2\right) du. \tag{11.163}$$

In terms of these integrals the integral in the amplitude $v_P$ of Eq. (11.160) can be rewritten as

$$\int_0^u e^{-i\pi u^2/2} du = \int_0^u \left[\cos\left(\frac{\pi}{2}u^2\right) - i\sin\left(\frac{\pi}{2}u^2\right)\right] = C(u) - iS(u), \tag{11.164}$$

and the light intensity at $P$ is

$$|v_P|^2 \propto \left[\int_{-b'}^{\infty} du \cos\left(\frac{\pi}{2}u^2\right)\right]^2 + \left[\int_{-b'}^{\infty} du \sin\left(\frac{\pi}{2}u^2\right)\right]^2. \tag{11.165}$$

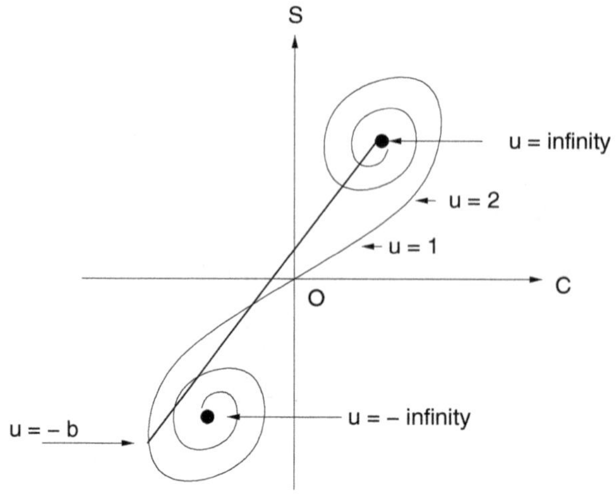


**Fig. 11.17.** The Cornu spiral.

The absolute amplitude of light is the positive square root

$$|v_P| = \sqrt{[C]^2 + [S]^2}. \tag{11.166}$$

This expression is given by the *chord* joining two points on the plotted curve in Fig. 11.17.

Since

$$(dC)^2 + (dS)^2 = \cos^2\left(\frac{\pi}{2}u^2\right)(du)^2 + \sin^2\left(\frac{\pi}{2}u^2\right)(du)^2 = (du)^2, \tag{11.167}$$

the straight line joining $u = -b'$ to $u = \infty$ gives the intensity in the plot of $S$ versus $C$ as in Fig. 11.17. The curve with parametric representation $(C(u), S(u))$ is known as *Cornu's spiral.*[13] Here $u$ is the length of the curve measured from the origin as indicated in Fig. 11.17 for cases $u = 1, u = 2, \ldots$. Thus the distance measured from the origin $O$ along the *arc* of the curve is equal to $\int du = u$ as a consequence of Eq. (11.167). Finally we add as a reminder that this case of Fresnel diffraction applies only to cylindrically symmetrical light.

## 11.8 Exercises

**Example 11.4 (Reflected, transmitted energies via Poynting vector).** Derive expressions for the amplitude reflection coefficient $R$ and transmission coefficient $T$ of plane waves falling obliquely upon a plane

[13]See e.g. Ohanian [106], pp. 514–515 and Jahnke and Emde [74], p. 37.

dielectric interface between two media, treating the two polarizations separately and expressing $R$ and $T$ in terms of the angle of incidence $\theta$ in medium (1), the angle of refraction in medium (2), and the ratio

$$N = \sqrt{\frac{\epsilon_2\mu_1}{\epsilon_1\mu_2}}, \tag{11.168}$$

where $\epsilon$ = dielectric constant and $\mu$ = permeability. Determine by means of the Poynting vector the incident, reflected and transmitted energy flows at the interface per unit area of the interface, and hence show that energy conservation requires that

$$(1 - R^2)\cos\theta = NT^2\cos\phi. \tag{11.169}$$

Verify for each polarization that your values of $R$ and $T$ conform to this. [Hint: Observe that Eq. (11.169) follows also from the two pairs of equations of Section 11.6].

**Example 11.5 (Verification of R and T).** Verify that $R$ and $T$ of Eq. (11.123) agree for $\mu = \mu'$ with the expressions of Eq. (11.97).

**Example 11.6 (Reflected and transmitted magnetic amplitudes).** Derive the amplitudes of the magnetic fields of reflected and transmitted waves which correspond to the amplitudes of the electric fields of Eq. (11.97).

**Example 11.7 (Fraunhofer diffraction at N rectangular slits).** Consider the case of $N$ equally spaced rectangular slits, each of the size of the single slit as in Fig. 11.14, and the centers of neighboring slits at a distance $d$ apart. Show that the diffraction pattern is given by that of a single slit multiplied by the factor

$$\frac{\sin^2(\pi mNd/\lambda)}{\sin^2(\pi md/\lambda)} \tag{11.170}$$

for a plane wave of wavelength $\lambda$ incident from the perpendicular direction. Show that if the slit separation is twice the slit width, the even-order spectra will be missing.[14]

**Example 11.8 (Fraunhofer diffraction at a circular aperture).** In the text a rectangular aperture in the $(y, z)$-plane was considered. Consider in its place a circular aperture of radius $a$ and obtain an expression for the intensity of light at the observation point $P$, and show — by using properties

[14]Hewson [70], Chapter 6, problem 2, p. 101. The solution may be found in Slater and Frank [136], p. 191.

of Bessel functions from Tables of Special Functions — that the diffraction pattern consists of alternate light and dark concentric circular fringes. [Hint: Use planar polar coordinates $y = \rho\cos\vartheta, z = \rho\sin\vartheta$, and the following integral representation $J_0(\rho) = (1/2\pi)\int_0^{2\pi}\exp(-i\rho\sin\vartheta)d\vartheta$, together with the relation $d(\rho J_1(\rho))/d\rho = \rho J_0(\rho)$, where $J_0, J_1$ are Bessel functions].

**Example 11.9 (Fraunhofer diffraction maxima).** It was shown that the function $(\sin y/y)^2$ of Eq. (11.159) determines the Fraunhofer diffraction pattern from a rectangular aperture. Show that the maxima of the function are determined by the transcendental equation $y = \tan y$, and determine the first few solutions of this equation.

**Example 11.10 (Laws of optics from Fermat's principle).** Derive (a) the law of reflection, Eq. (11.56), and (b) the law of refraction, Eq. (11.57) (Snell's law), from *Fermats principle* which may be expressed in the form

$$\int_{\text{along ray}} n(s)ds = \text{minimum}, \tag{11.171}$$

where $n$ is the refractive index of the medium and $s$ the variable which parametrizes the path of the ray. In case (a) choose coordinates as in Fig. 11.18 in which the ray starts from the origin $(0,0,0)$, hits a mirror at $(x,h,z)$ and is reflected to the fixed point $(X,0,0)$ in one and the same homogeneous medium with index of refraction $n(s) = \text{const}$. Thus

$$\int n(s)ds = n\{\sqrt{x^2+h^2+z^2} + \sqrt{(X-x)^2+h^2+z^2}\}, \tag{11.172}$$

has to be minimized. Show that $\partial/\partial z = 0$ of the result gives $z = 0$, implying that the ray is in the plane of the initial and final points. Then minimize with respect to $x$, i.e. take $\partial/\partial x = 0$, and show that $\sin i = \sin r$, where $i$ and $r$ are the angles shown in Fig. 11.18.

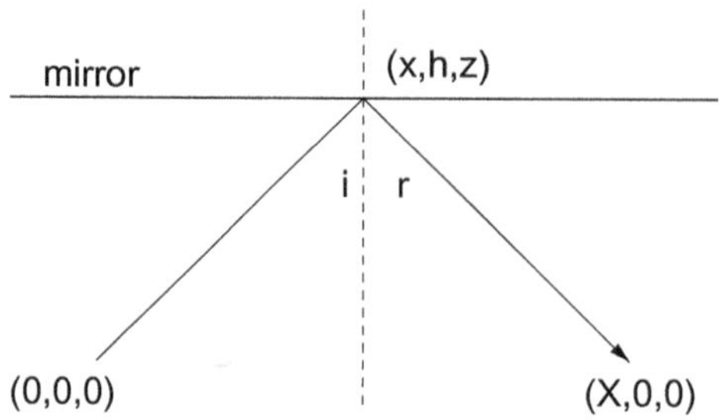


**Fig. 11.18.** Reflection.

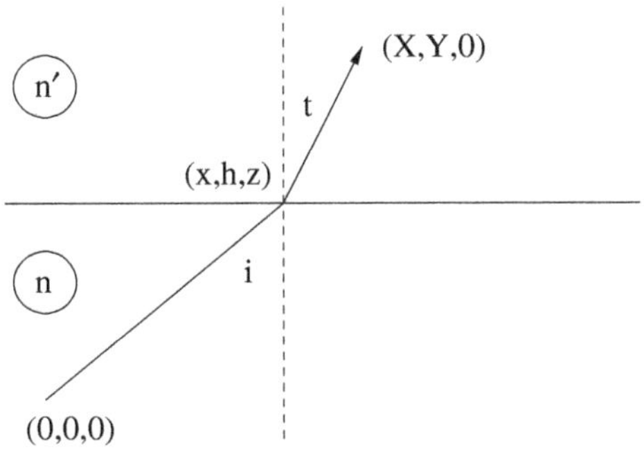


**Fig. 11.19.** Refraction.

In case (b) choose coordinates as in Fig. 11.19. Then Fermat's principle implies

$$\int n(s)ds = n\sqrt{x^2 + h^2 + z^2} + n'\sqrt{(X - x)^2 + (Y - h)^2 + z^2}. \quad (11.173)$$

Show that $\partial/\partial z = 0$ of this relation implies $z = 0$ and hence that the ray is in the plane of the initial and final points. Then show that $\partial/\partial x = 0$ yields $n \sin i = n' \sin r$, Snell's law.

# Chapter 12

# Metals

## 12.1 Introductory Remarks

Our considerations in the preceding chapter were restricted to media with conductivity $\sigma = 0$. We now proceed to corresponding considerations of reflection and absorption of electromagnetic waves by metals, i.e. the case $\sigma \neq 0$, and we shall see that we can resort to the formulas derived earlier.[1] The classical considerations developed below are very crude but have proved in practice to be reasonably good. It is a remarkable fact that the quantum theory — though providing correction, as we shall see — allows essentially a justification of the classical picture.[2] The correct explanation of the difference between conductors and insulators is given by the quantum mechanical band theory (cf. Bleaney and Bleaney [12], Chapter 12).

## 12.2 Reflection and Absorption of Electromagnetic Waves

In the case of the propagation of light in conducting media, primarily in metals now under consideration, we have $\boldsymbol{\nabla} \cdot \mathbf{E} = 0$ since all charges are neutralized by adjacent charges inside the metal implying charge density $\rho = 0$, and $\boldsymbol{\nabla} \cdot \mathbf{H} = 0$ as for any homogeneous medium since $\boldsymbol{\nabla} \cdot \mathbf{B} = 0$. There remain of Maxwell's equations Ampère's law

$$\boldsymbol{\nabla} \times \mathbf{H} = \frac{\partial \mathbf{D}}{\partial t} + \mathbf{j}, \tag{12.1}$$

[1]Although now old, the most important monograph on this subject was for a long time that of Wilson [163]. For literature directly related to the topic of this chapter see e.g. Cook [33], p. 258, 349–350 and Cheston [30], pp. 290–296.

[2]Slater and Frank [136], pp. 111–114.

and $\nabla \times \mathbf{E} = -\partial \mathbf{B}/\partial t$. With insertion of Ohm's law, $\mathbf{j} = \sigma \mathbf{E}$, where $\sigma$ is the direct or steady current electrical conductivity, we have in addition

$$\mathbf{D} = \epsilon \mathbf{E}, \quad \mathbf{j} = \sigma \mathbf{E}, \quad \mathbf{E} \propto e^{-i\omega t}, \tag{12.2}$$

and Eq. (12.1) becomes

$$\nabla \times \mathbf{H} = -i\omega\epsilon \mathbf{E} + \sigma \mathbf{E} = -i\omega \mathbf{E}\left(\epsilon + \frac{i\sigma}{\omega}\right), \tag{12.3}$$

i.e. in the case of conducting media we have the *generalized dielectric constant* (cf. Eq. (11.26))

$$\eta = \epsilon + \frac{i\sigma}{\omega}, \tag{12.4}$$

and we can pass from the expression for a dielectric to that of a metal by the replacement $\epsilon \to \eta$. Note that here the first contribution $\epsilon$ is that of the displacement current $\partial \mathbf{D}/\partial t$. Thus, although now $\mathbf{j} \neq 0$, $\int \mathbf{j} \cdot d\mathbf{F} \neq 0$, we can again proceed as before because we saw earlier, that for finite conductivity $\sigma$ the "linear current density" $K = 0$. Hence we have the case with no surface currents.

We consider now *reflection in the case of vertical incidence* of the electromagnetic wave from a dielectric into a metal, as indicated in Fig. 12.1.

In this case we have incident angle $\alpha = 0$ and (from Eq. (11.45), i.e. $\sin\alpha = (\epsilon'\mu'/\epsilon\mu)^{1/2}\sin\beta$) $\beta = 0$. Hence the *reflected intensity* is in both cases (1) and (2) considered in Section 11.5.2, and using formulas (11.65) and (11.66),

$$|R|^2 = \left|\frac{N-1}{N+1}\right|^2, \tag{12.5}$$

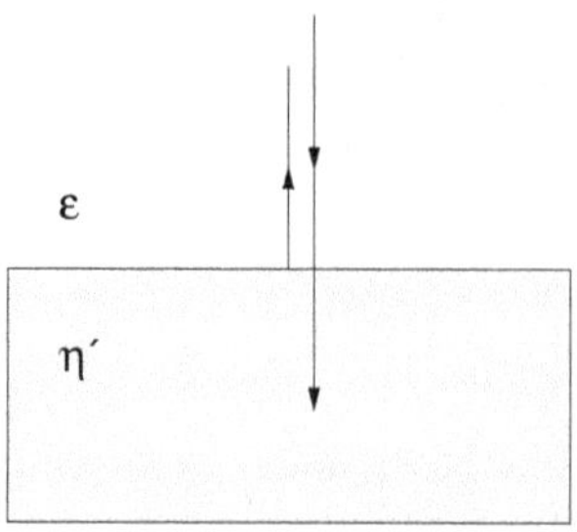


**Fig. 12.1.** Vertical incidence on a metal.

where with $\epsilon' \to \eta'$ in Eq. (11.64) for $N$[3]:

$$N = \sqrt{\frac{\eta' \mu}{\epsilon \mu'}}. \tag{12.6}$$

This quantity is the generalized refractive index. For *typical metals* $|N| \gg 1$, so that we can approximate $|R|^2$ (also called reflection coefficient[4]) as follows:

$$\begin{aligned} |R|^2 &= \left|\frac{1 - 1/N}{1 + 1/N}\right|^2 \simeq \left|1 - \frac{2}{N}\right|^2 \\ &= \left|1 - 2\left\{\Re\left(\frac{1}{N}\right) + i\Im\left(\frac{1}{N}\right)\right\}\right|^2 \\ &= \left\{1 - 2\Re\left(\frac{1}{N}\right)\right\}^2 + \left\{2\Im\left(\frac{1}{N}\right)\right\}^2, \end{aligned} \tag{12.7}$$

and hence

$$|R|^2 \simeq 1 - 4\Re\left(\frac{1}{N}\right) + O\left(\frac{1}{N^2}\right). \tag{12.8}$$

The *absorptivity* $A$ is a quantity defined by unity minus the reflected intensity (reflectivity),

$$A = \frac{\text{energy absorbed}}{\text{incident energy}} \equiv 1 - |R|^2, \tag{12.9}$$

so that for $|N| \gg 1$

$$A \simeq 4\Re\left(\frac{1}{N}\right). \tag{12.10}$$

The absorptivity can be measured by measuring the increase in heat energy of the metal. But now — note that the source of $\epsilon'$ is the displacement current

---

[3]Here we have metals in mind. In the case of isotropic diamagnetic substances (e.g. copper and silver) and paramagnetic substances (e.g. aluminium and air) $\mu$ differs in Gaussian units from unity by only a few parts in $10^5$, see Jackson [73], Section 5.8 and the table in Attwood [5], p. 205 where the relative permeabilities are given. In SIU units $\mu_0$ (vacuum) $= 4\pi \times 10^{-7}$ henries/meter. Then $|R|^2$ can be expressed in terms of velocities $v_1, v_2$ in the two media in view of the relation $v = 1/\epsilon\mu$.

[4]Slater and Frank [136], p. 120. There it is called $R$.

(which is negligible at lowest frequencies), whereas the source of $i\sigma'/\omega$ is the electron current (i.e. is due to movement of charges) —

$$N = \sqrt{\frac{\mu}{\epsilon\mu'}}\sqrt{\epsilon' + \frac{i\sigma'}{\omega}} = \sqrt{\frac{\mu}{\epsilon\mu'}}\sqrt{\frac{i\sigma'}{\omega}}\sqrt{1 + \frac{\omega\epsilon'}{i\sigma'}}. \tag{12.11}$$

Hence with

$$i^{-1/2} = e^{-i\pi/4} = \frac{1}{\sqrt{2}}(1 - i)$$

we have

$$4\frac{1}{N} = 4\sqrt{\frac{\omega\epsilon\mu'}{\sigma'\mu}}\frac{-i^{1/2}}{\sqrt{1 + \omega\epsilon'/i\sigma'}} = 4\sqrt{\frac{\omega}{2\sigma'}\frac{\epsilon\mu'}{\mu}}\frac{(1 - i)}{\sqrt{1 + \omega\epsilon'/i\sigma'}}. \tag{12.12}$$

For *low frequencies* ($\omega \to 0$) we have

$$\sigma = \text{const.} = \text{real}, \quad \text{i.e.} \quad \sigma(\omega) \simeq \sigma(0), \tag{12.13}$$

so that

$$A \simeq \sqrt{\frac{8\omega}{\sigma'}\frac{\epsilon\mu'}{\mu}} \propto \sqrt{\omega}\sqrt{\rho'}, \tag{12.14}$$

where $\rho' = 1/\sigma'$ is the *resistivity*. This is the so-called *Hagen–Rubens approximation* of the absorptivity of a metal at low frequencies (experimentally investigated by Hagen and Rubens).[5] The relation assumes the validity of Ohm's law $\mathbf{j} = \sigma\mathbf{E}$, where $\sigma$ is the direct current conductivity. This is normally true for metals if $\lambda > 30 \times 10^{-4}$cm, i.e. in the far infrared.

## 12.3 The Theory of Drude

At *low frequencies* ($\omega \to 0$) the equation of motion of a free electron (here a conduction electron, the valency electron of an atom) subjected to an applied electric field $E$ or electromagnetic plane wave polarized in the $x$-direction can be taken to be approximately given by the relevant Newton equation (in this case also the Lorentz force law for a particle of charge e acted upon by the electromagnetic field, here $E$)

$$m\ddot{x} + b\dot{x} = eE. \tag{12.15}$$

[5] Hagen and Rubens [66]. A discussion of the Hagen–Rubens formula can be found e.g. in Stratton [146], p. 508.

Here $m$ is the mass of the electron, and $b = m/\tau$, where $\tau$ has the dimension of time and is called *relaxation time* and $\dot{x}$ is the drift velocity. The parameter $\tau$ has the physical meaning of an average scattering or collision time, i.e. the average time interval between successive dissipative collisions of an electron with atoms of the metal lattice and $1/\tau$ may therefore be called *collision frequency.* Thus the viscous damping term $b\dot{x}$ takes into account the resistive force on the charge in the conducting material. The conduction electrons considered quantum mechanically as waves are scattered by the irregularities in the metal produced by thermal oscillations of the atoms. The theory is then known as the *free electron model.*[6] It is this scattering which produces the effect equivalent to friction in mechanics. Since the free charges are not bound to an atom there is no restoring force proportional to the displacement in Eq. (12.15). In the case of a bound electron, one would add a term $\omega_0^2 x$ to $b\dot{x}$ representing a binding of the electron to an atom through a force like that of an oscillator potential and permitting resonances (this is a purely classical, also crude, model at this stage but turns out to be quite successful in describing the significant aspects of the interaction of the electron with the field).

At *high frequencies* the field $\mathbf{E}$ alternates its direction so rapidly, that the electron has only a small chance to get sufficiently far away in order to be able to collide with an ion, as indicated in Fig. 12.2. This means, in this case the equation of motion can be approximated by[7]

$$m\ddot{x} = eE_0 e^{-i\omega t} = eE \quad \text{newtons.} \tag{12.16}$$

In the *intermediate domain* the equation of motion is therefore approximately

$$m\ddot{x} + \frac{m}{\tau}\dot{x} = eE_0 e^{-i\omega t} = eE \quad \text{newtons} \tag{12.17}$$

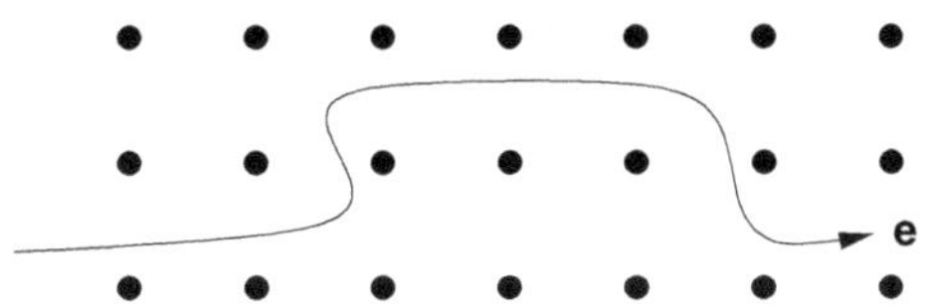


**Fig. 12.2.** A conduction electron in a metal lattice.

[6]Cf. Sommerfeld [140], Kittel [77], Sommerfeld and Bethe [141], Ashcroft and Mermin [4] and Ziman [168].

[7]Low and high frequency aspects of this theory of Drude are discussed in detail by Jackson [73], Section 7.5.

with solution[8]

$$\dot{x} = \frac{eE}{m(-i\omega + 1/\tau)} \quad \text{meters/second.} \tag{12.18}$$

But now[9] *current density* $\mathbf{j}$ = charge of one electron × number of electrons per unit volume × velocity = $ne\dot{\mathbf{x}}$, i.e. with $\sigma$ the electrical conductivity,

$$\mathbf{j} = \frac{ne^2\mathbf{E}}{m(-i\omega + 1/\tau)} = \sigma\mathbf{E} \quad \text{amperes/meter}^2, \tag{12.19}$$

so that ($n$ being the number of electrons per unit volume)[10]

$$\sigma = \frac{ne^2\tau}{m(1 - i\omega\tau)} \quad (\text{ohm} \cdot \text{meter})^{-1} \tag{12.20}$$

in the case of one dissipative (energy losing) mode. Thus at very low frequencies ($\omega\tau \ll 1$, microwave domain and lower) the conductivity is real and independent of the angular frequency. In the case of more than one dissipative mode the expression becomes

$$\sigma = \frac{e^2}{m}\sum_{i=1}^{N}\frac{n_i\tau_i}{1 - i\omega\tau_i} \quad (\text{ohm} \cdot \text{meter})^{-1}, \tag{12.21}$$

where $N$ is the number of dissipative (absorption) modes, and $n_i$ the number of electrons contributing to this mode $i$. We see therefore: The expression for $\sigma$ is in general complex and hence the Hagen–Rubens formula an unsatisfactory approximation. As mentioned before, the (today) correct and full theory is given by the quantum mechanical treatment. The first attempt in this direction was given by Sommerfeld [140] and Sommerfeld and Bethe [141]. It is found that the classical Drude result for the electrical conductivity $\sigma$, i.e. Eq. (12.20), is also obtained in the quantum mechanical treatment with the Fermi–Dirac distribution for electrons (for the derivation see Chapter 13 of Müller–Kirsten [102]). Various authors like Kittel [77] and Ashcroft and Mermin [4] therefore call it the Drude–Sommerfeld model.

[8] $\ddot{x} = -i\omega\dot{x}, \ \therefore m\ddot{x} + (m/\tau)\dot{x} = m(-i\omega + (1/\tau))\dot{x} = eE.$

[9] Initially the free electrons wander around arbitrarily and in any direction. After application of the field $\mathbf{E}$, however, they are forced to move antiparallel to the direction of $\mathbf{E}$.

[10] This formula is also discussed e.g. in Cheston [30], p. 293. The considerations leading to Eq. (12.19) can be looked at as an elementary derivation of Ohm's law.

We can write the conductivity $\sigma$ also in the form

$$\sigma = \frac{\sigma_{\rm dc}}{1 - i\omega\tau}, \quad \sigma_{\rm dc} \equiv \frac{ne^2\tau}{m}\,({\rm ohm}\cdot{\rm meter})^{-1}. \tag{12.22}$$

Here $\sigma_{\rm dc}$ is called the *direct current conductivity* and $\tau$ could be taken as $\bar{l}/\bar{v}$, the ratio of an average length of the mean free path, $\bar{l}$, divided by an average velocity, $\bar{v}$. We observe that — in the context of this model — the conductivity varies as $1/\omega$ at very high frequencies but is complex. Since this follows from Eq. (12.22) by neglecting 1 in the denominator, which means neglecting the term with $\dot{x}$ in Eq. (12.17), this means that at high frequencies the electron under consideration oscillates backwards and forwards between two atoms, but there are no collisions.

We return now to the calculation of the absorptivity and replace $\sigma'$ in the first of expressions (12.12) by

$$\sigma' = \frac{\sigma'_{dc}}{1 - i\omega\tau}, \tag{12.23}$$

so that from Eq. (12.12)

$$A = \left(\frac{16\omega\epsilon\mu'}{\sigma'_{dc}\mu}\right)^{1/2}\Re\left\{\sqrt{\frac{(1-i\omega\tau)}{i}}\frac{1}{\sqrt{1+\epsilon'\omega(1-i\omega\tau)/i\sigma'_{dc}}}\right\}. \tag{12.24}$$

But

$$\sqrt{-(i+i\omega\tau)} = \sqrt{\frac{\sqrt{1+\omega^2\tau^2}-\omega\tau}{2}} - i\sqrt{\frac{\sqrt{1+\omega^2\tau^2}+\omega\tau}{2}}, \tag{12.25}$$

so that for $\epsilon'\omega/\sigma'_{\rm dc} \ll 1$, we obtain the *Drude–Kronig relation*

$$\begin{aligned} A &\simeq \sqrt{\frac{8\omega\epsilon\mu'}{\sigma'_{\rm dc}\mu}}\sqrt{\sqrt{1+\omega^2\tau^2}-\omega\tau} \\ &= \text{Hagen} - \text{Rubens expression} \times \sqrt{\sqrt{1+\omega^2\tau^2}-\omega\tau}, \end{aligned} \tag{12.26}$$

and hence for $\omega\tau \gg 1$

$$\begin{aligned} A &= \sqrt{\frac{8\omega\epsilon\mu'}{\sigma'_{\rm dc}\mu}}\sqrt{\omega\tau\left(1+\frac{1}{2\omega^2\tau^2}+\cdots\right)-\omega\tau} \\ &\simeq \sqrt{\frac{8\omega\epsilon\mu'}{\sigma'_{\rm dc}\mu}\frac{\omega\tau}{2\omega^2\tau^2}} \\ &= \sqrt{\frac{4\epsilon\mu'}{\mu}\frac{1}{\tau\sigma'_{\rm dc}}} \equiv \overline{A}, \quad (\omega\tau \gg 1). \end{aligned} \tag{12.27}$$

The result (12.26) is known as *Drude–Kronig formula.* We infer from these results, that at low frequencies, i.e. $\omega\tau \ll 1$,

$$A \propto \sqrt{\omega} \quad \text{(this is the extreme infrared region)}, \tag{12.28}$$

whereas at high frequencies (infrared region instead of extreme infrared region) $A$ behaves like (12.27), i.e. is approximately constant. This behavior of the absorptivity is depicted in Fig. 12.3.

The *ultraviolet region* for metals is characterized by high frequencies. In this case the displacement current dominates over the electron current as we can see from Eq. (12.17). This means, in this case the metal behaves like a dielectric with reduced dielectric constant (see below). We saw that the effective dielectric constant of a metal is given by Eq. (12.4), and that at high angular frequencies $\omega$

$$\sigma = \frac{ne^2\tau}{m(1-i\omega\tau)} \simeq -\frac{ne^2}{im\omega} \quad (\Omega\cdot\mathrm{m})^{-1}, \tag{12.29}$$

so that

$$\eta = \underbrace{\epsilon - \frac{ne^2}{m\omega^2}}_{\text{reduced diel. const.}} \quad \mathrm{F}\cdot\mathrm{m}^{-1}. \tag{12.30}$$

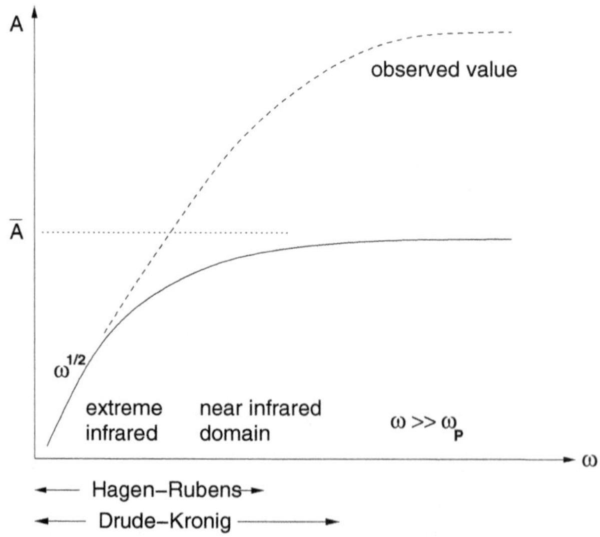


**Fig. 12.3.** Absorptivity $A$ as a function of angular frequency $\omega$. The dotted curve is the observed value.

Thus at very high frequencies (ultra-violet) a metal behaves like a dielectric with lowered dielectric constant. This consideration is also applicable to the case of a plasma (i.e. a highly ionized gas) whose electrical properties are dominated by those of free electrons.

In general the above *Drude–Kronig formula* cannot be correct. We know that due to the skin effect the electric field **E** varies considerably with penetration depth into the metal. This implies that the linear relation

$$\mathbf{j} = \sigma \mathbf{E} \quad \mathrm{A/m^2} \tag{12.31}$$

must be modified, if the length of the mean free path of the electrons is $\sim v/\omega$ and larger or of the order of the skin depth $z_0$, or if the relaxation time is very large. The above current density formula will then have to be replaced by a relation of the form (see comments after Eq. (9.84))

$$\mathbf{j}(\mathbf{r}) = \int d\mathbf{r}' g(\mathbf{r}, \mathbf{r}') \mathbf{E}(\mathbf{r}') \tag{12.32}$$

in order to take into account the rapid change of the field over the region of the mean path of an electron. In the course of motion of the electrons towards or away from the surface, electromagnetic energy is directly transferred to the conduction electrons. If one averages over all possible directions of these electrons, and if one takes into account, that the electrons are fermions — their maximum velocity $v_0$ at absolute temperature $T = 0$ being that of the Fermi boundary — one obtains a correction contribution. The Fermi velocity $v_0$ is determined as follows. Number of states = number of particles (according to the Pauli principle) $= N = 2V\kappa^3/(2\pi)^3, \mathbf{p} = \hbar\boldsymbol{\kappa}, \hbar = h/2\pi$ (the factor '2' takes into account the two possible spin directions of the electron as a fermion). This means[11]

$$N = 2\frac{V}{(2\pi)^3}\left(\frac{4}{3}\pi\frac{p^3 8\pi^3}{h^3}\right), \tag{12.33}$$

or, if $p = mv_0, n = N/V$,

$$v_0 = \frac{h}{m}\left(\frac{3n}{8\pi}\right)^{1/3} \quad \text{meters/second} \tag{12.34}$$

(as for the Fermi distribution at the absolute zero point of temperature). Taking these effects into account a further modification of the formula for the absorptivity of electromagnetic radiation in a metal results and is known

[11]For detailed derivation and discussion see Müller–Kirsten [102], Chapter 6.

as *Dingle–Holstein formula*.[12] This formula with the quantum statistical correction can be written

$$A_{\mathrm{DH}} = \text{Drude–Kronig expression} + \frac{3}{4}\frac{v_0}{c}, \tag{12.35}$$

where $c$ is the velocity of light in vacuum. The additional quantum theoretical term which is independent of temperature dominates at low temperatures. At higher temperatures and for good conductors like Cu and Ag both contributions of the formula are of comparable magnitude. Hence the observed value of the absorptivity $A$ is approximately twice the Drude–Kronig result (see curve in Fig. 12.3). Good *conductors* are characterized by a high conductivity $\sigma$ or correspondingly low *resistivity* $\rho$, e.g. copper has $\rho = 1.72 \times 10^{-8}$ ohm · meters, silver has $\rho = 1.59 \times 10^{-8}$ ohm · meters. Semi-conductors are materials of a resistivity of the order of $10^{-4}$ to $10^{2}$ ohm · meters, whereas *insulators* have resistivities greater than about $10^{8}$ ohm·meters, like glass ($10^{9}$–$10^{14}$ ohm·meters), and wood ($10^{8}$–$10^{11}$ ohm·meters).

**Example 12.1 (Absorptivity of silver).** Calculate the absorptivity of silver with direct current conductivity of $3 \times 10^{7}$ siemens/meter (siemens earlier mho) at a radiation frequency of $\nu = 10^{10}$ hertz (microwave region, $\lambda \simeq 3$ cm).

**Solution:** We use the Hagen–Rubens relation (12.14)

$$A \simeq \sqrt{\frac{8\omega\epsilon\mu'}{\sigma'\mu}}, \tag{12.36}$$

where we insert

$$\mu = \mu', \quad \omega = 2\pi \times 10^{10} \quad \text{radians/per second},$$
$$\epsilon = \epsilon_0 = 8.854 \times 10^{-12}, \quad \sigma' = 3 \times 10^{7}.$$

The result is $A = 3.9 \times 10^{-4}$.

**Example 12.2 (Skin effect and Drude theory).** Electromagnetic radiation with ($E_x$ is propagated along the $z$-axis)

$$\mathbf{E} = E(z)e^{-i\omega t}\mathbf{e}_x, \quad \mathbf{H} = H(z)e^{-i\omega t}\mathbf{e}_y \tag{12.37}$$

($\mathbf{e}_x, \mathbf{e}_y$ being unit vectors) falls vertically on a plate of metal. Show that the electric field in the plate at a distance $z$ from the surface and at low

[12] R.B. Dingle [37], R.B. Dingle [38], see in particular particular p. 342, and T. Holstein [72].

frequencies $\omega$ is given by the following expression:

$$E(z) = E(0)\exp\left\{-\frac{(1+i)z}{\delta\sqrt{1+i\omega\tau}}\right\} \quad \text{newtons/coulomb,} \tag{12.38}$$

where $\delta$, the penetration depth or attenuation length (also called inverse of attenuation) for small $\omega$, is given by

$$\delta = \sqrt{\frac{2}{\omega\mu\sigma_{\rm dc}}} \quad \text{meters.} \tag{12.39}$$

Here $\sigma_{\rm dc}$ is the direct current conductivity and $\tau$ the relaxation time. Finally evaluate $E(0)$ with the help of Fresnel's formulas.

**Solution:** We start from the macroscopic Maxwell equation or Ampère's law (observe that for $\mathbf{H} = 0$ the material current and the displacement current are oppositely directed)

$$\boldsymbol{\nabla} \times \mathbf{H} = \mathbf{j} + \frac{\partial \mathbf{D}}{\partial t}, \tag{12.40}$$

where in the present case for the left-hand side

$$\boldsymbol{\nabla} \times \mathbf{H} = \mathbf{e}_x\left(-\frac{\partial H_y}{\partial z}\right) + \mathbf{e}_z\left(\underbrace{\frac{\partial H_y}{\partial x}}_{0}\right), \tag{12.41}$$

since $H_y = H(z)$. Therefore,

$$-\frac{\partial H_y}{\partial z} = j_x + \epsilon\frac{\partial E_x}{\partial t}, \quad \text{or} \quad -\frac{dH(z)}{dz} = j_x - i\omega\epsilon E(z). \tag{12.42}$$

Similarly with

$$\boldsymbol{\nabla} \times \mathbf{E} = -\frac{\partial \mathbf{B}}{\partial t} = i\omega\mu\mathbf{H} = \mathbf{e}_y\frac{\partial E_x}{\partial z} + \mathbf{e}_z\left(-\underbrace{\frac{\partial E_x}{\partial y}}_{0}\right) \tag{12.43}$$

we obtain

$$\frac{dE(z)}{dz} = i\omega\mu H. \tag{12.44}$$

From Eqs. (12.44) and (12.42) we eliminate $H$ and obtain:

$$\frac{d^2E(z)}{dz^2} = i\omega\mu\frac{dH}{dz} = -i\omega\mu(j_x - i\omega\epsilon E), \ \text{i.e.} \ \frac{d^2E(z)}{dz^2} + \omega^2\epsilon\mu E(z) = -i\omega\mu j_x. \tag{12.45}$$

Thus we have *one* equation in $E(z)$ and $j_x$. We obtain a second relation from $\mathbf{j} = \sigma\mathbf{E}$, where $\mathbf{j} = ne\dot{\mathbf{r}}$ and with the simple Drude theory, i.e. Eq. (12.18),

$$m\ddot{\mathbf{r}} + \frac{m}{\tau}\dot{\mathbf{r}} = e\mathbf{E} = e\mathbf{E}_0 e^{-i\omega t}, \tag{12.46}$$

so that $\mathbf{r} \propto e^{-i\omega t}$ and

$$\dot{\mathbf{r}} = \frac{e\mathbf{E}\tau}{m(1 - i\omega\tau)}, \quad \mathbf{j} = \frac{ne^2\tau\mathbf{E}}{m(1 - i\omega\tau)} \tag{12.47}$$

and hence we can write (inserting above and below a factor $\bar{l}$)

$$\mathbf{j} = \frac{ne^2\bar{l}\tau\mathbf{E}}{m\bar{l}(1 - i\omega\tau)}, \tag{12.48}$$

where $\bar{l}$ is the length of a mean free path and $\bar{v} = \bar{l}/\tau$ a mean velocity. It follows that Ohm's law $\mathbf{j} = \sigma\mathbf{E}$ implies for the conductivity $\sigma$:

$$\sigma = \frac{ne^2\bar{l}}{m\bar{v}(1 - i\omega\tau)} \quad (\Omega \cdot \mathrm{m})^{-1}. \tag{12.49}$$

With $\sigma_{\mathrm{dc}} \equiv ne^2\bar{l}/m\bar{v}$ we can write $\mathbf{j} = \sigma_{\mathrm{dc}}\mathbf{E}/(1 - i\omega\tau)$. With this we obtain as our *second* equation

$$j_x = \frac{\sigma_{\mathrm{dc}} E(z)}{1 - i\omega\tau}. \tag{12.50}$$

Equations (12.45) and (12.50) now imply:

$$\frac{d^2E(z)}{dz^2} + \left(\epsilon\mu\omega^2 + \frac{i\mu\omega\sigma_{\mathrm{dc}}}{1 - i\omega\tau}\right) E(z) = 0. \tag{12.51}$$

We neglect the displacement current contribution $\epsilon\mu\omega^2$ and consider *low frequencies*. Moreover, $i = (\sqrt{i})^2 = [(1 + i)/\sqrt{2}]^2$. Hence the physically sensible, i.e. decreasing, solution is

$$\mathbf{E} = E(z)e^{-i\omega t}\mathbf{e}_x, \quad \text{with} \quad E(z) = E(0)\exp\left\{-\frac{(1 + i)z}{\delta\sqrt{1 - i\omega\tau}}\right\} \quad \mathrm{N/C}, \tag{12.52}$$

where[13]

$$\delta = \sqrt{\frac{2}{\mu\omega\sigma_{\mathrm{dc}}}} \quad \text{meters} \tag{12.53}$$

[13]See e.g. Cook [33], p. 360. A similar result applies in the case of plasmas, cf. Spitzer [143], pp. 47–52.

(this is the penetration depth of the refracted wave at low frequencies in agreement with Eq. (9.17)). We have here the case of the vector **E** lying in the plane of incidence. The appropriate Fresnel formulas of Section 11.6 are

$$(1-R)\cos\alpha = T\cos\beta, \quad (1+R) = NT,$$

$$\text{where} \quad R = \frac{E_0''}{E_0}, \quad T = \frac{E_0'}{E_0}, \quad N = \sqrt{\frac{\epsilon'\mu}{\epsilon\mu'}}, \tag{12.54}$$

where $E_0, E_0'', E_0'$ are the amplitudes of the incident, reflected and refracted waves. We are interested in $T$ with $E_0 = 1$, i.e. $T = E_0' = E(0)$. Vertical incidence ($\alpha = 0 \to \beta = 0$) implies $T = 1 - R$. $R$ follows from the above relations as $R = (N-1)/(N+1)$, so that $T = 1 - R = 2/(N+1) \simeq 2/N$ for $|N| \gg 1$. Neglecting in Eq. (12.11) the contribution of the displacement current, we have

$$N \simeq \sqrt{\frac{\mu}{\epsilon\mu'}}\sqrt{\frac{i\sigma'}{\omega}}, \tag{12.55}$$

or (with $\epsilon = \epsilon_0, \mu' = \mu, \sigma' \simeq \sigma_{\text{dc}}$)

$$\frac{1}{N} = \sqrt{\frac{\omega\epsilon_0}{\sigma'}}i^{-1/2}, \tag{12.56}$$

so that

$$E(0) = T \simeq \frac{2}{N} \simeq 2\sqrt{\frac{\omega\epsilon_0}{\sigma_{\text{dc}}}}i^{-1/2} \quad \text{N/C}. \tag{12.57}$$

## 12.4 Exercises

**Example 12.3 (Refractive index of a conducting gas).** Assuming $N$ absorption modes derive the real and imaginary parts of the index of refraction of a conducting gas.[14]

**Example 12.4 (The case of oblique incidence).** Discuss the problems arising in the calculation of the absorptivity of a metal for the case of nonvertical incidence of the electromagnetic wave.[15]

---

[14] See also Cheston [30], p. 293.

[15] This case is not easy. See discussion in Slater and Frank [136], p. 127, or the detailed treatment in Slater [137], Section 13.

**Example 12.5 (Penetration depth at high frequencies).** Derive the frequency dependence of the penetration depth $\delta$ of a good conductor at high angular frequencies $\omega$.

(Answer: $\exp(-\alpha z), \alpha \propto (1/\omega^2)$).[16]

**Example 12.6 (Absorption cross section).** Derive from the following definition of the absorption cross section $\sigma(\omega)$ an expression for this quantity in the context of the theory of Drude. The cross section $\sigma(\omega)$ is defined as the ratio

$$\sigma(\omega) := \frac{W}{nI}, \quad W = -\frac{1}{2}\Re\langle \mathbf{E}^* \cdot \mathbf{j}\rangle, \tag{12.58}$$

where $W$ is the average rate at which a medium absorbs energy per unit volume, and $I$ is equal to $c$ (velocity of light) times the electric energy density of the incident wave.

**Example 12.7 (Refractive index from wave velocity).** Obtain (e.g. from Eq. (12.51)) the wave number $k$ for a good conductor, from this the value of the velocity $v$ of the wave in the conductor and hence the index of refraction $p$ from the relation $p = c/v$.

**Example 12.8 (Reflectivity of copper[17]).** The resistivity of copper is about $1.7 \times 10^{-8}$ ohm · meters. Calculate the reflective power $|R|^2$ of copper for wavelengths of light $\lambda = 12 \times 10^{-8}$ m and $\lambda = 25.5 \times 10^{-8}$ m. The observed values of $1 - |R|^2$ are 1.6% and 1.17% at these wavelengths.

**Example 12.9 (Interpreting the dispersion relation).** Consider the context of Example 12.2. Show that for $\omega \gg 1/\tau$ the dispersion relation can be expressed as

$$v_p^2 = \frac{c^2}{1 - \omega_P^2/\omega^2}, \quad \omega_P^2 = \frac{ne^2}{m\epsilon}, \tag{12.59}$$

where $v_p$ is the phase velocity $\omega/k$ and $E(z) \propto \exp[i(kz - \omega t)]$. How do you interpret the case $\omega < \omega_P$? What is the penetration depth (when $v_p$ is imaginary)?[18] Can $v_p$ be larger than the velocity of light $c$?[19]

---

[16] Robinson [124], p. 146.
[17] Slater and Frank [136], p. 128.
[18] For related discussion but in the case of a plasma see Spitzer [143], pp. 50–52.
[19] For an elementary mechanical example illustrating that the phase velocity $v_p$ can exceed $c$ (different from the group velocity which transports energy) see Schwartz [132], pp. 253–256.

**Example 12.10 (Penetration depths of copper and earth).** Taking $\sigma$ of copper to be $5.7 \times 10^7$ ohm$^{-1}$ · meter$^{-1}$ and that of earth as $10^{-7}$ ohm$^{-1}$ · meter$^{-1}$, and $\omega/2\pi = 50$ hertz, calculate the respective penetration depths using $\mu = \mu_0 = 4\pi \times 10^{-7}$ henries · meter$^{-1}$.

(Answer: $9.4 \times 10^{-3}$ meters, $2.25 \times 10^5$ meters).

**Example 12.11 (Snell's law).** Derive Snell's law for a dielectric–conductor interface.
[Hint: See comments to Eq. (11.57)].

**Example 12.12 (Brewster's angle for a good conductor).** Show that for a good conductor Brewster's angle is approximately $\pi/2$ (so that unpolarized light incident on it becomes polarized on reflection only for grazing incidence).

# Chapter 13

# Propagation of Radio Waves in the Ionosphere

## 13.1 Introductory Remarks

The phenomenon we consider here,[1] takes place several hundred kilometers above the surface of the earth. The gaseous atoms in the upper atmosphere are ionized by extraterrestrial radiation (e.g. by cosmic rays, ultraviolet radiation from the sun etc.). The ionosphere consists of several layers of electrons and ions with first increasing and then decreasing densities.[2] Consider first only one layer. We let the dielectric constant near the surface of the earth be $\epsilon_0$ and in the ionized region ($\omega\tau \gg 1$, see Chapter 9)

$$\eta = \epsilon_0 - \frac{ne^2}{m\omega^2} \equiv \epsilon_0 - \frac{\omega_P^2}{\omega^2}\epsilon_0 \quad \text{farads/meter.} \tag{13.1}$$

Here

$n$ is the number of electrons per unit volume,

$m$ the mass of an electron,

$\omega_P$ the plasma frequency, $\omega_P^2 = ne^2/m\epsilon_0$.

The effect of the positively charged ions (i.e. their contribution to the conductivity and hence to the effective dielectric constant) can mostly be neglected, since the mass of the ions is approximately 2000 times larger than that of an electron. We also ignore at the beginning the effect of the

[1]A brief description — different from that here — can be found in Stratton [146]. A more complete account is given in Mimno [96].

[2]About 110 km above the earth the electron concentration suddenly becomes very large and extends vertically for several km. This layer of electrons is known as the *Kennelly–Heaviside layer* after its discoverers. Further up (some 150 km) is another such layer known as the *Appleton layer*. See e.g. Nelkon [104], pp. 342–343.

terrestrial magnetic field which in general cannot be neglected (see Jackson [73], pp. 292–294) and is considered in Section 13.3.

For the ray of a radio wave in the kHz to MHz frequency domain (i.e. ($10^3$–$10^6$ Hz) to return from the ionosphere back to earth, the density of electrons cannot increase indefinitely, since otherwise the ray would be bent more and more towards the vertical and hence would eventually escape. Thus beyond a certain maximum value, the refractive index must decrease again, and this means the density of electrons, enabling the ray to be bent away from the vertical. In any case the wave has a reflected part and a refracted part with coefficients $R$ and $T$ in the case of $\mathbf{E} \perp$ to the plane of incidence given by (cf. Eqs. (11.121))

$$R = \frac{\cos\alpha - N\cos\beta}{\cos\alpha + N\cos\beta}, \quad T = \frac{2\cos\alpha}{\cos\alpha + N\cos\beta}, \tag{13.2}$$

where $\alpha$ and $\beta$ are incident and refractive angles as in Figs. 13.1 and 13.2, and $N$ is (cf. Eq. (11.119)), now with generalized dielectric constants,

$$N = \sqrt{\frac{\eta'\mu}{\eta\mu'}}. \tag{13.3}$$

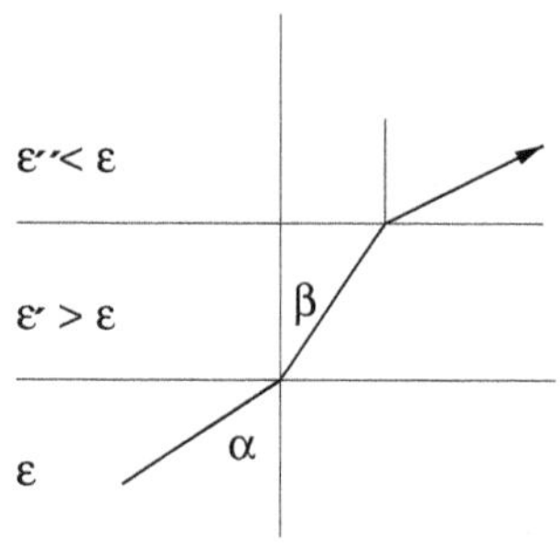


**Fig. 13.1.** Refraction of a ray.

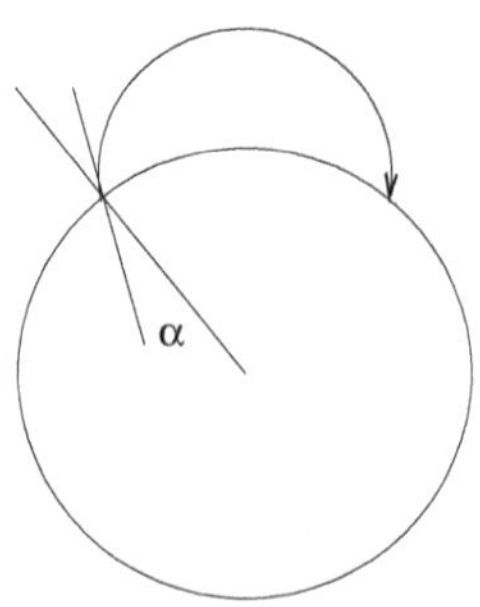


**Fig. 13.2.** Reflection after refraction.

## 13.2 Condition for Return of Waves

Along its ascent into the ionosphere the ray is refracted, as also on its way back. However, the highest point of the ray's path must be a point of reflection. Naturally this situation is somewhat idealized. For example some waves wander even directly from the emitter to the receiver. There are also waves which circle around the earth (with reflection from the ionosphere), and thus give rise to radio echos, which have been observed. Since the electron density increases with increasing height above the earth only very slowly, one has initially almost exclusively refraction. Since

$$N = \sqrt{\frac{\eta'\mu}{\eta\mu'}} \sim 1, \tag{13.4}$$

this means that for a ray starting vertically upwards $R = 0$ (since with Snell's law (11.57) also $\beta = 0$). After attaining its maximum value, the electron density decreases rapidly with further increase in height. In the plasma the dielectric constant is given by Eq. (13.1). We recall that we had defined the generalized refractive index by Eq. (11.41) as

$$p = c\sqrt{\eta\mu} \equiv p_1 + ip_2, \tag{13.5}$$

and that

$$k = \frac{\omega}{c}p, \quad e^{i(kx-\omega t)} = e^{i(\omega/c)(px-ct)}. \tag{13.6}$$

We observe that for $\eta < 0$, i.e. $p$ complex, the wave is strongly damped. For large masses $m$ (of the ions) $\eta \simeq \epsilon_0$. For $\eta < 0$ the wave transmitted beyond the reflection point is strongly damped (i.e. practically no transmitted wave), and one has effectively only reflection For this to occur the density must at least be large enough so that $\eta = 0$, i.e. (cf. Eq. (13.1))

$$\frac{n}{\epsilon_0} \gtrsim \frac{m\omega^2}{e^2}. \tag{13.7}$$

We now apply Snell's law to different layers of the atmosphere/ionosphere. For the case of three layers as illustrated in Fig. 13.3 we have (beginning with an angle of incidence whose angle of refraction is equal to the angle of incidence at the next layer and so on)

$$\sqrt{\epsilon\mu}\sin\alpha = \sqrt{\epsilon'\mu'}\sin\beta = \sqrt{\epsilon''\mu''}\sin\gamma = \cdots . \tag{13.8}$$

In the present case the electron density depends on the height $y$ above the ground, i.e. we have to write

$$n = n(y).$$

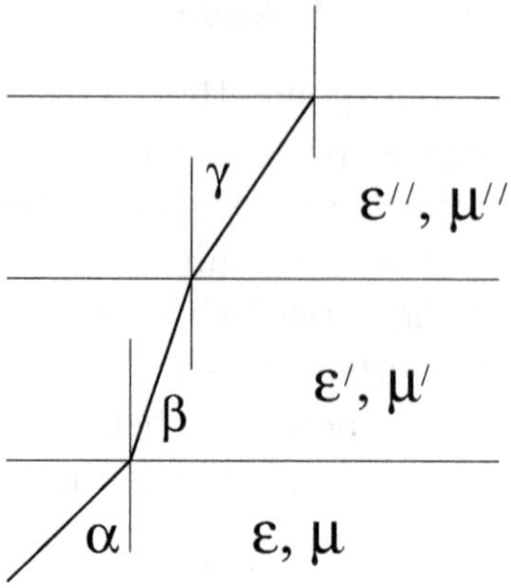


**Fig. 13.3.** Refraction of the ray on its way upwards.

Let $\theta_b$ be the angle of incidence at ground level and $\theta$ the angle of refraction at height $y$. Then we can write with the help of Eq. (13.1) and all $\mu's$ equal:

$$\left[\epsilon - \frac{n(y)e^2}{m\omega^2}\right]^{1/2} \sin\theta = \sqrt{\epsilon}\sin\theta_b, \tag{13.9}$$

and for $\epsilon = \epsilon_0$

$$\left[\epsilon_0 - \frac{n(y)e^2}{m\omega^2}\right]^{1/2} \sin\theta = \sqrt{\epsilon_0}\sin\theta_b, \tag{13.10}$$

so that

$$\left(1 - \frac{n(y)e^2}{\epsilon_0 m\omega^2}\right) \sin^2\theta = \sin^2\theta_b = 1 - \cos^2\theta_b, \tag{13.11a}$$

or

$$\cos^2\theta_b - \frac{n(y)e^2}{\epsilon_0 m\omega^2}(1 - \cos^2\theta) = 1 - \sin^2\theta = \cos^2\theta \tag{13.11b}$$

or

$$\cos^2\theta_b - \frac{n(y)e^2}{\epsilon_0 m\omega^2} = \cos^2\theta\left(1 - \frac{n(y)e^2}{\epsilon_0 m\omega^2}\right), \tag{13.11c}$$

or

$$\cos\theta\left[1 - \frac{n(y)e^2}{\epsilon_0 m\omega^2}\right]^{1/2} = \cos\theta_b\left[1 - \frac{n(y)e^2}{\epsilon_0 m\omega^2\cos^2\theta_b}\right]^{1/2}. \tag{13.12}$$

From Eqs. (13.9) and (13.12), we obtain

$$\tan\theta = \frac{\tan\theta_b}{\sqrt{1 - \frac{n(y)e^2}{\epsilon_0 m\omega^2 \cos^2\theta_b}}}. \tag{13.13}$$

We let $x$ and $y$ be horizontal and vertical coordinates as in Fig. 13.4. The highest point of the trajectory of the ray is a point of reflection; reflection of the ray therefore requires, as shown, a maximal height $y_{\max}$. Near the ground practically no reflection takes place, but only around $y_{\max}$. We have

$$\tan\theta = \frac{dx}{dy}, \tag{13.14}$$

so that with Eq. (13.13)

$$\int_0^x dx = x = \tan\theta_b \int_0^y \frac{dy}{\sqrt{1 - n(y)e^2/\epsilon_0 m\omega^2 \cos^2\theta_b}}. \tag{13.15}$$

With this we obtain for the range $R$ indicated in Fig. 13.4

$$R = 2\tan\theta_b \int_0^{y_{\max}} \frac{dy}{\sqrt{1 - n(y)e^2/\epsilon_0 m\omega^2 \cos^2\theta_b}}. \tag{13.16}$$

At $y_{\max}$ we have $\theta = \pi/2$, i.e. $\tan\theta = \infty$, i.e. at this point e.g. from Eq. (13.13)

$$\cos\theta_b = \sqrt{\frac{n(y_{\max})e^2}{\epsilon_0 m\omega^2}} = \frac{\omega_P}{\omega}. \tag{13.17a}$$

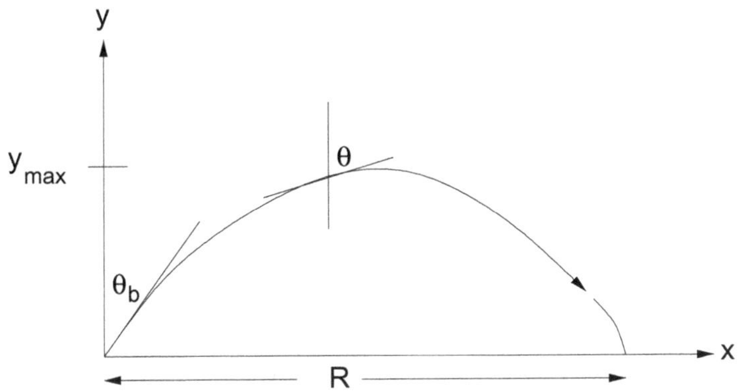


**Fig. 13.4.** The trajectory of the ray.

This equation may not always have solutions for any value of $\theta_b$. If for all $y$

$$1 - \frac{n(y)e^2}{\epsilon_0 m\omega^2 \cos^2\theta_b} > 0,$$

the angle $\theta$ of Eq. (13.13) is not $\pi/2$ and there is no $y_{\text{max}}$. In this case, however, $n(y)$ cannot grow indefinitely, and there must be a value $n_{\text{max}}$ with

$$\frac{n_{\text{max}}e^2}{\epsilon_0 m\omega^2} \equiv \cos^2\theta_0 \tag{13.17b}$$

(this expression defines $\theta_0$), at which

$$1 - \frac{\cos^2\theta_0}{\cos^2\theta_b} > 0, \tag{13.18}$$

i.e.

$$\cos^2\theta_b > \cos^2\theta_0, \quad \theta_b < \theta_0. \tag{13.19}$$

This means, with this condition the rays *cannot* return to earth. *For return* to be possible, we must have

$$\theta_b > \theta_0 \quad \left(\text{also} \quad 1 - \frac{n_{\text{max}}e^2}{\epsilon m\omega^2 \cos^2\theta_b} < 0\right). \tag{13.20}$$

This means, rays satisfying this condition, can be reflected back to earth. One can express this in another way by saying, that

$$\omega^2 < \frac{\omega_P^2}{\cos^2\theta_b} \tag{13.21}$$

for a fixed angle $\theta_b$. Thus for a fixed angle $\theta_b$ the frequency

$$\omega_{\text{max}} = \frac{\omega_P}{\cos\theta_b} \tag{13.22}$$

is the *maximum frequency* possible for a return to earth of the ray of radio waves. For normal incidence $\theta_b = 0, \cos\theta_b = 1$. The corresponding maximum or *critical frequency* $\nu_c$ at normal incidence is therefore

$$\nu_c = \frac{1}{2\pi}\sqrt{\omega_P} \quad \text{seconds}^{-1}. \tag{13.23}$$

The frequency range which allows reflection at the ionosphere is roughly that of short waves in the frequency range of 0.3–30 GHz or 100 m to 1 cm in wavelength.

## 13.3 Effect of Terrestrial Magnetic Field

We now want to take into account the effect of the magnetic field of the earth on the propagation of radio waves. The magnetic field strength of the earth is approximately $B_e \simeq 2 \times 10^{-5}$ tesla (0.2 gauss). In our treatment here we again consider only the electric component of the electromagnetic field of the waves. This suffices since the magnetic field component is much weaker, as we can see from the fact that $\mathbf{B} = O(E/c)$. We also simplify the terrestrial magnetic field to one along the $z$-direction parallel to the normal to the surface of the earth (the latter being assumed more or less flat for our purposes here). Hence we write for the terrestrial field

$$\mathbf{B}_e = (0, 0, B_e). \tag{13.24}$$

We can consider a more general case, i.e. that of the terrestrial field at some angle, by replacing in the following the $z$-component $\mathbf{B}_e$ by $B_e \cos\theta_e$. The equation of motion of a free particle with charge $q$ (i.e. an electron in the ionosphere) which is also subjected to the magnetic Lorentz force of Eq. (5.26) (and that was derived from Ampere's law) is then with $\mathbf{r}' = (x', y', z')$ given by

$$m\left(\ddot{\mathbf{r}}' + \frac{1}{\tau}\dot{\mathbf{r}}'\right) = q[\underbrace{\mathbf{E}}_{\text{ray}} + \dot{\mathbf{r}}' \times \underbrace{\mathbf{B}_e}_{\text{earth}}], \tag{13.25}$$

where $\tau$ is the relaxation time, i.e. the mean time interval between successive collisions of the particle with other particles. We put

$$\xi = x' + iy', \quad \text{so that} \quad (\dot{\mathbf{r}}' \times \mathbf{B}_e)_x = \dot{y}' B_e \tag{13.26}$$

and

$$E \equiv E_x + iE_y, \tag{13.27}$$

and define the *Larmor frequency* $\omega_L$ by

$$qB_e = m\omega_L \quad (\text{or} \quad m\omega_L \cos\theta_e) \quad \text{coulomb} \cdot \text{teslas}. \tag{13.28}$$

The frequency $\omega_L$ is the frequency of precession of a charged particle in a magnetic field (see Jackson [73], p. 293). Here we defined this frequency in agreement with the cyclotron frequency $\omega_c$ of Eq. (10.130). The Larmor precessional frequency is in some texts defined as $\omega_c/2$ (see e.g. Cheston [30],

p. 194). The components of the equation of motion (13.25) then combine to the complex equation

$$m\left(\ddot{\xi}+\frac{1}{\tau}\dot{\xi}\right)=qE-im\omega_L\dot{\xi}, \tag{13.29}$$

or

$$\ddot{\xi}+\left(\frac{1}{\tau}+i\omega_L\right)\dot{\xi}=\frac{q}{m}E. \tag{13.30}$$

Setting

$$E=\frac{1}{2}E_0e^{i(\omega t-kz)}=E_x+iE_y,$$

$$E_x=\frac{1}{2}E_0\cos(\omega t-kz),\quad E_y=\pm\frac{1}{2}E_0\sin(\omega t-kz), \tag{13.31}$$

where the latter are the $x$-, $y$-components of the real part of ($\mathbf{e}_x, \mathbf{e}_y$ again unit vectors)

$$\mathbf{E}_{\substack{R\\L}}(z,t)=\frac{1}{2}E_0(\mathbf{e}_x\mp i\mathbf{e}_y)e^{i(\omega t-kz)}, \tag{13.32}$$

i.e. the expressions ($R$, $L$ meaning right, left)

$$\begin{aligned}\Re\mathbf{E}_{\substack{R\\L}}(z,t)&=\frac{1}{2}E_0\cos(\omega t-kz)\mathbf{e}_x\pm\frac{1}{2}E_0\sin(\omega t-kz)\mathbf{e}_y\\&=E_x\mathbf{e}_x\pm E_y\mathbf{e}_y.\end{aligned} \tag{13.33}$$

These waves are described as right polarized and left polarized respectively. The sum

$$\mathbf{E}_R+\mathbf{E}_L=E_0e^{i(\omega t-kz)}\mathbf{e}_x, \tag{13.34}$$

or its real part

$$\Re(\mathbf{E}_R+\mathbf{E}_L)=E_0\cos(\omega t-kz)\mathbf{e}_x \tag{13.35}$$

represents a linearly polarized wave in the direction of $x$. Now with

$$E=\frac{1}{2}E_0e^{\pm i(\omega t-kz)} \tag{13.36}$$

we make the following ansatz for the solution of Eq. (13.30):

$$\xi=\xi_0e^{\pm i(\omega t-kz)} \tag{13.37}$$

and obtain

$$\left[-\omega^2 \pm i\omega\left(\frac{1}{\tau} + i\omega_L\right)\right]\xi = \frac{q}{m}E,$$

$$-\omega\left[\omega \pm \left(\omega_L - \frac{i}{\tau}\right)\right]\xi = \frac{q}{m}E,$$

and hence

$$\xi = -\frac{q}{m\omega}\frac{E(z,t)}{(\omega \pm \omega_L \mp i/\tau)}. \tag{13.38}$$

Recalling the current density

$$\mathbf{j} = (j_x, j_y, j_z),$$

as the charge of one electron multiplied by the number of electrons per unit volume $n$ times the velocity $= nq\dot{\mathbf{x}}$, we now define the following complex quantity (with $\omega_P^2 = nq^2/m\epsilon_0$):

$$j = j_x + ij_y = nq\dot{\xi} = -\frac{nq^2}{m\omega}\frac{\dot{E}}{(\omega \pm \omega_L \mp i/\tau)},$$

$$\dot{j} \equiv \frac{dj}{dt} = -\frac{nq^2}{m\omega}\frac{\ddot{E}}{(\omega \pm \omega_L \mp i/\tau)} = -\omega_P^2\frac{\epsilon_0\ddot{E}}{\omega(\omega \pm \omega_L \mp i/\tau)}. \tag{13.39}$$

Now, we know from Maxwell's equations (cf. e.g. Eq. (9.30)) that every component of the electric vector satisfies the equation (with $j = \sigma E$, Eq. (9.27))

$$\nabla^2 E_i - \epsilon\mu\frac{\partial^2 E_i}{\partial t^2} - \mu\sigma\frac{\partial E_i}{\partial t} = 0, \quad \nabla^2 E_i - \epsilon\mu\ddot{E}_i - \mu\dot{j}_i = 0, \tag{13.40}$$

so that for the complex quantities

$$E = E_x + iE_y, \quad \dot{j} = \dot{j}_x + i\dot{j}_y \tag{13.41}$$

the following equation results:

$$\nabla^2 E - \epsilon_0\mu_0\ddot{E} - \mu_0\dot{j} = 0, \tag{13.42}$$

or, with $\dot{j}$ from Eq. (13.39),

$$\nabla^2 E - \frac{1}{c'^2}\ddot{E} = 0, \quad \nabla^2 E - \epsilon_0\mu_0\left[1 - \frac{\omega_P^2}{\omega(\omega \pm \omega_L \mp i/\tau)}\right]\ddot{E} = 0, \tag{13.43}$$

where

$$\omega_P = \sqrt{\frac{nq^2}{m\epsilon_0}}, \quad p = \frac{c}{c'} = c\sqrt{\eta\mu} = \sqrt{\frac{\eta\mu}{\epsilon_0\mu_0}}, \quad \eta\mu = \frac{1}{c'^2}. \tag{13.44}$$

Here $\omega_P$ is the *plasma frequency* of the ionosphere. The definition of the generalized refractive index implies therefore for the refractive index of the ionosphere

$$\begin{aligned} p^2 &= \left[1 - \frac{\omega_P^2}{\omega(\omega \pm \omega_L \mp i/\tau)}\right] \\ &= 1 - \frac{\omega_P^2(\omega \pm \omega_L)}{\omega[(\omega \pm \omega_L)^2 + 1/\tau^2]} \mp i\frac{\omega_P^2}{\omega\tau[(\omega \pm \omega_L)^2 + 1/\tau^2]}. \end{aligned} \tag{13.45}$$

For the separation into real and imaginary parts,

$$p = p_1 + ip_2, \quad \text{and} \quad \omega \gg \frac{1}{\tau}, \tag{13.46}$$

it follows that

$$p_1 \simeq \left[1 - \frac{\omega_P^2}{\omega(\omega \pm \omega_L)}\right]^{1/2}, \quad p_2 \simeq \mp\frac{1}{2\omega\tau}\frac{\omega_P^2}{(\omega \pm \omega_L)^2}. \tag{13.47}$$

For right circularly polarized electromagnetic waves and left circularly polarized electromagnetic waves, i.e.

$$E = \frac{1}{2}E_0 e^{\pm i(\omega t - kz)}, \quad k = \frac{\omega}{c}p = \frac{\omega}{c}(p_1 + ip_2), \tag{13.48}$$

one has

$$E = \frac{1}{2}E_0 e^{\pm i\omega(t - p_1 z/c)} e^{\pm\omega p_2 z/c}. \tag{13.49}$$

The refractive index $p$ therefore differs for the two types of polarization. A linearly polarized wave, however, with

$$\mathbf{E} = (E_x, 0, 0), \quad E_x = E_0\cos(\omega t - kz) \tag{13.50}$$

can be considered as a superposition of two oppositely circularly polarized waves. We see therefore, that in the case of an originally linearly polarized wave, one of its circularly polarized components is absorbed more strongly (i.e. damped) than the other, so that the original wave becomes partially circularly polarized. A similar effect appears in the case of metals (cf. Example 12.12).

However here, the origin of $p_2$ is different from that in the case of metals with $\sigma \neq 0$. In the present case $p_2 \propto 1/\tau$, and $p_2 \to 0$ for $\tau \to \infty$. The parameter $\tau$ is the *relaxation time* and thus originates from a term (like friction) which implies a loss of energy in the form of heat. Observe that the magnetic field is contained in $\omega_L$.

**Example 13.1 (The pulsar).** A pulsar emits a *pulse* with frequencies $\omega_1, \omega_2$ (i.e. emitted simultaneously), which are considerably larger than the plasma frequency $\omega_P$ of the interstellar medium. The times of arrival of the pulses with frequencies $\omega_1, \omega_2$ are measured. Show that this permits the determination of the electron density of the medium integrated over a distance $L$, i.e. $\int_0^L n(l)dl$.

**Solution:** Since $\omega \gg \omega_P$, we have (cf. Eq. (9.73)) the dispersion relation

$$\frac{k^2}{\sqrt{\epsilon\mu}} = \omega^2 - \omega_P^2, \quad \text{or} \quad k^2c^2 = \omega^2 - \omega_P^2. \tag{13.51}$$

(with $\epsilon, \mu \sim \epsilon_0, \mu_0$). Moreover

$$\omega_P^2 = \frac{ne^2}{m\epsilon_0}, \quad n = n(l). \tag{13.52}$$

The *group velocity* $v_g$ is defined by (cf. Eq. (8.203))

$$v_g = \frac{d\omega}{dk}, \tag{13.53}$$

so that

$$v_g = \frac{d}{dk}\sqrt{k^2c^2 + \omega_P^2} = \frac{kc^2}{\omega} = \frac{c}{\omega}\sqrt{\omega^2 - \omega_P^2} = c\sqrt{1 - \left(\frac{\omega_P}{\omega}\right)^2}. \tag{13.54}$$

The time of flight $T$ of a pulse (or wave packet which travels with the group velocity, see Chapter 15) from its source to the observer is

$$T = \int_0^T \frac{dl}{v_g} = \frac{1}{c}\int_0^T \frac{dl}{\sqrt{1 - (\frac{\omega_P}{\omega})^2}} \simeq \frac{1}{c}\int_0^T \left(1 + \frac{\omega_P^2}{2\omega^2}\right) dl. \tag{13.55}$$

Let $T \to T_i$ for $\omega \to \omega_i$, $i = 1, 2$. Then

$$\Delta T = T_2 - T_1 = \frac{1}{2c}\int_0^T \left(\frac{\omega_P^2}{\omega_2^2} - \frac{\omega_P^2}{\omega_1^2}\right) dl = \frac{e^2}{2mc\epsilon_0}\left(\frac{1}{\omega_2^2} - \frac{1}{\omega_1^2}\right)\int_0^L n(l)dl. \tag{13.56}$$

This means

$$\int_0^L n(l)dl = \frac{2mc\epsilon_0}{e^2}\left(\frac{\omega_1^2\omega_2^2}{\omega_1^2 - \omega_2^2}\right)\Delta T. \tag{13.57}$$

Thus if $\triangle T$ is measured, one can obtain a measure of the electron density over a length $L$ of space. For $n(l)$ independent of $l$ one can thus obtain $L$. Note, however, that this requires two different frequencies.

**Example 13.2 (Scattering of radio waves).** Radio waves with frequency $\omega = 3 \times 10^7$ per second are emitted from a point $P$ on the surface of the earth (radius $6.37 \times 10^6$ m) and are reflected back from the ionosphere approximately $3 \times 10^5$ m above the surface of the earth. Calculate the largest angle of reflection $\theta$, at which the wave is reflected back to the surface of the earth. Also calculate the maximum density of the electrons determined by this angle.

**Solution:** Since the angle of reflection is equal to the angle of incidence, it is also the largest angle of incidence at layers in the stratosphere which is asked for, i.e. the extreme to vertical incidence.

Hence we draw a tangent $PP'$ to the surface of the earth, as shown in Fig. 13.5, and extend it to the point of reflection $Q$. We therefore have the geometry shown in Fig. 13.5. Hence

$$\sin\theta = \frac{6.37 \times 10^6}{6.37 \times 10^6 + 3 \times 10^5} = \frac{6.37}{6.37 + 0.3} = 0.9550, \quad \theta \simeq 73^\circ. \tag{13.58}$$

We use the following relation obtained above in the text (cf. Eq. (13.17a)) for $\theta = \theta_0$:

$$\cos\theta = \sqrt{\frac{ne^2}{m\epsilon_0\omega^2}}. \tag{13.59}$$

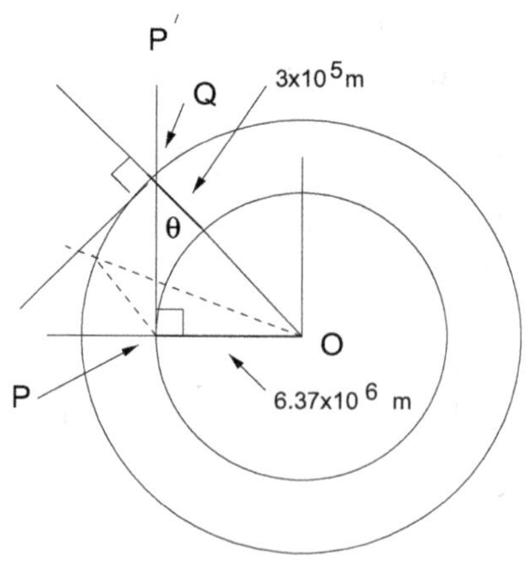


**Fig. 13.5.** The requested angle $\theta$.

With $\epsilon_0 = 8.854 \times 10^{-12}$ A $\cdot$ s/(V $\cdot$ m), this yields for $\omega = 3 \times 10^7$ per second the density $n \simeq 8.5 \times 10^{11}$ of free electrons per meter$^3$.

## 13.4 Exercises

**Example 13.3 (Maximum frequency for normal incidence).** What is the maximum frequency in the case of radio waves at normal incidence reflected from the ionosphere, assuming the number of electrons is $10^{12}$ per cubic meter.

(Answer: $8.98 \times 10^6$ Hz).

**Example 13.4 (Long waves of a dipole).** Very long waves of electromagnetic radiation can propagate almost undamped over long distances and even around the earth. Consider the following approximate situation. An oscillating dipole as in Section 10.6 of angular frequency $\omega$ is placed vertically upright at a point on the surface of the earth of radius $a$. The reflecting ionosphere begins at a height $h$ above the surface of the earth. Show that the electric field at a distance $r$ from the dipole is approximately given by

$$\mathbf{E} \simeq \frac{1}{i}\frac{Idsh}{C\omega r} \quad \text{newtons/coulomb,} \tag{13.60}$$

where $C$ is the capacity of the spherical condenser formed by the space between the earth and the ionosphere.
[Hint: Use Eq. (10.165)].

**Example 13.5 (Electron density of the ionosphere).** Given that a radio wave sent upwards at an angle of 30° to the normal and of frequency 10 GHz is to be reflected from the ionosphere, what is the number of electrons there per unit volume?

(Answer: $9.307 \times 10^{17}$ m$^{-3}$).

## Chapter 14

# Wave Guides and Resonators

### 14.1 What Are Wave Guides?

Wave guides are cylindrically shaped objects of metal (almost ideal conductors) which are open at both ends. Resonators, also called resonant cavities, differ from these in being closed at both ends. The interior of such objects is filled with some homogeneous material with electromagnetic constants $\epsilon, \mu$; alternatively the interior can be composed of layers of different materials. We have seen before (cf. Eqs. (9.84), (12.39) and discussions there), that the larger the conductivity $\sigma$ of a medium or the frequency of the incident electromagnetic wave, the smaller the skin depth (in Eq. (9.84) called $1/\lambda$, in Eq. (12.53) $\delta$, in Eq. (14.238) $\delta_0$), i.e. the depth of penetration of the radiation into the conducting medium. This follows from the dispersion relation that we encountered several times e.g. with Eq. (9.40), or Eq. (11.37), i.e.

$$\mathbf{k}^2 = \mu\epsilon\omega^2\left(1 + i\frac{\sigma}{\omega\epsilon}\right) \quad \text{meters}^{-2}, \tag{14.1}$$

together with the plane wave ansatz

$$\mathbf{E} = \mathbf{E}_0 e^{i(\mathbf{k}\cdot\mathbf{r}-\omega t)} \quad \text{newtons/coulomb}. \tag{14.2}$$

The square root of $\mathbf{k}^2$ is

$$k = \pm\sqrt{\mu\epsilon}\omega\sqrt{1 + i\frac{\sigma}{\omega\epsilon}} \quad \text{meters}^{-1}. \tag{14.3}$$

Here we have to choose the physically relevant sign. For instance for the wave $\exp(ikz)$ and a medium extended indefinitely in the direction of $z$, the sign has to be chosen such that the wave is damped, so that the current density $\mathbf{j} = \sigma\mathbf{E}$ cannot grow arbitrarily (which would be nonsensical). It is because of this skin effect that it is possible to confine radiation in hollow, pipe-like bodies or to use the latter for the propagation of waves (perfect conductors with $\sigma = \infty$ do not permit any penetration of radiation into the walls). Wave guides are used instead of transmission lines when power in the latter become too great as e.g. at microwave and higher frequencies, i.e. in the range of 0.3–3 GHz or wavelengths of the order of 100 cm to 10 cm. The radiation is fed into the wave guide or resonator with a sender or emitter and this induces in the walls surface charges and surface currents. We assume here that sender and receiver of electromagnetic waves or signals are effectively at infinity, so that there are no charge or current sources in any finite part of space which would have to be taken into account in our considerations. A signal is a wave packet propagating with group velocity, and is discussed in more detail in Section 15.3. Such a signal given at (say) time $t = 0$ and at a particular point of the conductor characterizes the external charges and currents which generate the corresponding fields $\mathbf{E}$ and $\mathbf{B}$ (charges for $\mathbf{E}$ and currents for $\mathbf{B}$). Correspondingly one can investigate the properties of wave guides and resonators in terms of current $I$ along the wave guide and the potential difference or voltage $V$ across it instead of the fields $E$ and $B$. We shall derive later the corresponding equations for the important case of the coaxial transmission cable, for which we calculated earlier the capacity and the self-inductance per unit length (see Examples 2.12, 4.4 and 8.2, 8.5). However the most notable difference between hollow wave guides and coaxial cables is that the former deliver energy by a guided electromagnetic field (hence the name) and not by a return current along an inner tube. The consideration of wave guides is therefore very instructive because this involves the full interplay between radiation and boundaries which will also reappear later in the consideration of the Casimir-effect.

We now consider electromagnetic waves in the hollow space enclosed by a wave guide (we consider resonators separately at the end). For $\mathbf{E}, \mathbf{H} \propto e^{-i\omega t}$, and (as explained above) $\mathbf{j} = 0$, $\rho = 0, \mathbf{D} = \epsilon\mathbf{E}$ we have the equations

$$\begin{aligned} \nabla \times \mathbf{E} &= i\omega\mathbf{B}, \\ \nabla \times \mathbf{B} &= -i\mu\epsilon\omega\mathbf{E}, \\ \nabla \cdot \mathbf{E} &= 0, \\ \nabla \cdot \mathbf{B} &= 0. \end{aligned} \tag{14.4}$$

We observe here the symmetry of the equations in exchanges $\mathbf{E} \leftrightarrow \mathbf{B}$ with $\omega \leftrightarrow -\omega\mu\epsilon$, which we will exploit later. We have seen in earlier chapters (Sections 9.2, 9.3 and 4.3) that by performing the "curl–curl" operation, i.e.

$\nabla \times \nabla$, on Maxwell's equations, the following equations result

$$\nabla^2 \mathbf{E} + \mu\epsilon\omega^2 \mathbf{E} = 0, \quad \nabla^2 \mathbf{B} + \mu\epsilon\omega^2 \mathbf{B} = 0. \tag{14.5}$$

In view of the cylindrical symmetry of the problem (cylinder axis along the $z$-axis) we expect waves, which travel along the $z$-direction, i.e. we set

$$\begin{aligned} \mathbf{E}(x, y, z, t) &= \mathbf{E}(x, y) e^{\pm ikz - i\omega t}, \\ \mathbf{B}(x, y, z, t) &= \mathbf{B}(x, y) e^{\pm ikz - i\omega t}. \end{aligned} \tag{14.6}$$

One should note that the directions of $\mathbf{E}$ and $\mathbf{B}$ are still arbitrary. In general it is also possible that $\mathbf{E}(x, y)$ and $\mathbf{B}(x, y)$ can depend on $z$. We now set for convenience

$$\nabla^2 = \nabla_\perp^2 + \nabla_z^2, \quad \text{where} \quad \nabla_\perp^2 = \frac{\partial^2}{\partial x^2} + \frac{\partial^2}{\partial y^2}, \tag{14.7}$$

and $\perp$ means "transverse", so that

$$(\nabla^2 + \mu\epsilon\omega^2)\mathbf{E} = (\nabla_\perp^2 + \mu\epsilon\omega^2 - k^2)\mathbf{E}(x, y) e^{\pm i(kz - \omega t)} = 0, \tag{14.8}$$

and hence we obtain

$$(\nabla_\perp^2 + \mu\epsilon\omega^2 - k^2)\mathbf{E}(x, y) = 0, \tag{14.9}$$

$$(\nabla_\perp^2 + \mu\epsilon\omega^2 - k^2)\mathbf{B}(x, y) = 0, \tag{14.10}$$

with the second equation for $\mathbf{B}$ resulting from the symmetry pointed out above (see Eq. (14.4)). We set in addition

$$\mathbf{E} = \mathbf{E}_z + \mathbf{E}_\perp, \quad \mathbf{B} = \mathbf{B}_z + \mathbf{B}_\perp. \tag{14.11}$$

## 14.2 Transverse Fields Derived from Longitudinal Fields

We now show that it suffices to know the $z$-components $\mathbf{E}_z, \mathbf{B}_z$; the transverse fields $\mathbf{E}_\perp, \mathbf{B}_\perp$ can then be obtained from these. We consider the two curl-equations of Eq. (14.4):

*Case* (1). We have

$$\nabla \times \mathbf{E} = i\omega \mathbf{B} \tag{14.12}$$

and with

$$\nabla_\perp = \left(\frac{\partial}{\partial x}, \frac{\partial}{\partial y}\right), \quad \nabla_z = \mathbf{e}_z \frac{\partial}{\partial z}, \tag{14.13}$$

the following equation:

$$(\boldsymbol{\nabla}_\perp + \boldsymbol{\nabla}_z) \times (\mathbf{E}_\perp + \mathbf{E}_z) = i\omega(\mathbf{B}_\perp + \mathbf{B}_z). \tag{14.14}$$

In the direction of $z$

$$\boldsymbol{\nabla}_\perp \times \mathbf{E}_\perp = i\omega\mathbf{B}_z \tag{14.15}$$

and in the $(x, y)$-plane

$$\boldsymbol{\nabla}_z \times \mathbf{E}_\perp + \boldsymbol{\nabla}_\perp \times \mathbf{E}_z = i\omega\mathbf{B}_\perp. \tag{14.16}$$

We multiply Eq. (14.16) from the left by $\boldsymbol{\nabla}_z \times$ and obtain (see comments at the end)

$$i\omega\boldsymbol{\nabla}_z \times \mathbf{B}_\perp = \boldsymbol{\nabla}_z \times (\boldsymbol{\nabla}_z \times \mathbf{E}_\perp) + \boldsymbol{\nabla}_z \times (\boldsymbol{\nabla}_\perp \times \mathbf{E}_z) \tag{14.17}$$

and thus

$$\begin{aligned} i\omega\boldsymbol{\nabla}_z \times \mathbf{B}_\perp &= \boldsymbol{\nabla}_z(\boldsymbol{\nabla}_z \cdot \mathbf{E}_\perp) - (\boldsymbol{\nabla}_z \cdot \boldsymbol{\nabla}_z)\mathbf{E}_\perp \\ &\quad + \boldsymbol{\nabla}_\perp(\boldsymbol{\nabla}_z \cdot \mathbf{E}_z) - (\boldsymbol{\nabla}_z \cdot \boldsymbol{\nabla}_\perp)\mathbf{E}_z. \end{aligned} \tag{14.18}$$

In the case of the first two expressions on the right of Eq. (14.18) we used the relation "*curl curl* = *grad div minus div grad*", in the case of the remaining terms the relation for a triple vector product

$$\mathbf{A} \times (\mathbf{B} \times \mathbf{C}) = (\mathbf{A} \cdot \mathbf{C})\mathbf{B} - (\mathbf{A} \cdot \mathbf{B})\mathbf{C}, \tag{14.19}$$

taking care of the ordering. The first and the last contributions on the right of Eq. (14.18) are scalar products of orthogonal vectors and therefore vanish. Hence we are left with

$$-\boldsymbol{\nabla}_z^2\mathbf{E}_\perp + \boldsymbol{\nabla}_\perp\left(\frac{\partial E_z}{\partial z}\right) = i\omega\boldsymbol{\nabla}_z \times \mathbf{B}_\perp. \tag{14.20}$$

But

$$-\boldsymbol{\nabla}_z^2\mathbf{E}_\perp = -\frac{\partial^2}{\partial z^2}(\mathbf{E}_\perp(x, y)e^{\pm ikz - i\omega t}) = k^2\mathbf{E}_\perp, \tag{14.21}$$

so that

$$k^2\mathbf{E}_\perp + \boldsymbol{\nabla}_\perp\left(\frac{\partial E_z}{\partial z}\right) = i\omega\boldsymbol{\nabla}_z \times \mathbf{B}_\perp. \tag{14.22}$$

*Case* (2). In a corresponding way we deal with the second of Eqs. (14.4),

$$\boldsymbol{\nabla} \times \mathbf{B} = -i\omega\mu\epsilon\mathbf{E}, \tag{14.23}$$

i.e.

$$(\nabla_\perp + \nabla_z) \times (\mathbf{B}_\perp + \mathbf{B}_z) = -i\omega\mu\epsilon(\mathbf{E}_\perp + \mathbf{E}_z), \tag{14.24}$$

where the left hand side is the following sum

$$(\nabla_\perp \times \mathbf{B}_\perp) + (\nabla_\perp \times \mathbf{B}_z) + (\nabla_z \times \mathbf{B}_\perp) + \underbrace{(\nabla_z \times \mathbf{B}_z)}_{0},$$

so that in the direction of $z$:

$$\nabla_\perp \times \mathbf{B}_\perp = -i\omega\mu\epsilon\mathbf{E}_z, \tag{14.25}$$

and transversely

$$\nabla_z \times \mathbf{B}_\perp = -\nabla_\perp \times \mathbf{B}_z - i\omega\mu\epsilon\mathbf{E}_\perp. \tag{14.26}$$

We insert this relation in Eq. (14.22) and obtain

$$k^2\mathbf{E}_\perp + \nabla_\perp\left(\frac{\partial E_z}{\partial z}\right) = i\omega[-\nabla_\perp \times \mathbf{B}_z - i\omega\mu\epsilon\mathbf{E}_\perp], \tag{14.27}$$

i.e. arranging the contributions in a different way,

$$\nabla_\perp\left(\frac{\partial E_z}{\partial z}\right) + i\omega(\nabla_\perp \times \mathbf{B}_z) = (\omega^2\mu\epsilon - k^2)\mathbf{E}_\perp. \tag{14.28}$$

Hence

$$(\mu\epsilon\omega^2 - k^2)\mathbf{E}_\perp = \left(\nabla_\perp\frac{\partial E_z}{\partial z} - i\omega(\mathbf{e}_z \times \nabla_\perp)B_z\right). \tag{14.29}$$

Owing to the symmetry $\omega \leftrightarrow -\omega\mu\epsilon$ of the Maxwell equations (14.4) in $\mathbf{E}$ und $\mathbf{B}$ we have also

$$(\mu\epsilon\omega^2 - k^2)\mathbf{B}_\perp = \left(\nabla_\perp\frac{\partial B_z}{\partial z} + i\omega\mu\epsilon(\mathbf{e}_z \times \nabla_\perp)E_z\right). \tag{14.30}$$

We see therefore: The transverse components of the fields, i.e. $\mathbf{E}_\perp, \mathbf{B}_\perp$, can be obtained from the longitudinal components $E_z, B_z$, in fact in each case from both longitudinal components. The relations (14.29) and (14.30) can also be derived in a different way, i.e. by immediate substitution from

$$\nabla \cdot \mathbf{E} = 0, \quad \nabla \cdot \mathbf{B} = 0 \tag{14.31a}$$

and

$$\nabla \times \mathbf{E} = -\frac{\partial \mathbf{B}}{\partial t}, \quad \nabla \times \mathbf{H} = \frac{\partial \mathbf{D}}{\partial t}. \tag{14.31b}$$

in the interior of the wave guide (i.e. not in the wall).[1]

## 14.3 Boundary Conditions

We now come to the important aspect of boundary conditions. We recall for this reason the continuity conditions at a boundary surface. We had in particular (cf. Eqs. (6.104) and (4.117))

$$B_n^{(1)} = B_n^{(2)}, \quad E_t^{(1)} = E_t^{(2)}, \quad \text{with} \quad \mathbf{k} \times \mathbf{E} = \omega\mu\mathbf{H} = \omega\mathbf{B}, \tag{14.32}$$

where "$n$" stands for "normal component" and "$t$" for "tangential component". The two conditions in Eq. (14.32) are exact. The other two conditions which we obtained and used previously are not needed in the present case since surface charge and surface current are here zero (for $\sigma = \infty$). The two conditions above for a perfectly conducting cylinder[2] can also be written ($\mathbf{e}_n \cdot \mathbf{e}_z = 0$)

$$\mathbf{n} \cdot \mathbf{B} = 0, \quad \mathbf{n} \times \mathbf{E} = 0 \tag{14.33}$$

with $B^{(1)} = B, E^{(1)} = E$ in the interior of the wave guide, and $B^{(2)} = 0 = E^{(2)}$ outside (again for $\sigma = \infty$, i.e. ideal conductors). We assume here that the wave guides are infinitely long in order to avoid finite end effects.

We can write the two boundary conditions:

$$\mathbf{n} \cdot (\mathbf{B}_\perp + \mathbf{B}_z) = 0 \quad \text{and} \quad \mathbf{n} \times (\mathbf{E}_\perp + \mathbf{E}_z) = 0, \tag{14.34}$$

i.e. (since $\mathbf{n} \perp \mathbf{B}_z$):

$$\mathbf{n} \cdot \mathbf{B}_\perp = 0 \tag{14.35}$$

and

$$\underbrace{\mathbf{n} \times \mathbf{E}_\perp}_{\text{vector}||\mathbf{e}_z} = 0, \tag{14.36}$$

$$\underbrace{\mathbf{n} \times \mathbf{E}_z}_{\text{vector}\perp\mathbf{e}_z} = 0, \quad \text{i.e.} \quad E_z = 0. \tag{14.37}$$

---

[1]In these equations the derivatives with respect to $t$ and $z$ are replaced by $-i\omega$ and $ik$ respectively, and the equations are then solved for the transverse components of the fields.
[2]Cf. Jackson [73], p. 342.

*These conditions hold at the boundary surface* (i.e. not anywhere else). Since we saw that the fields $\mathbf{E}_\perp, \mathbf{B}_\perp$ follow from $E_z, B_z$, we are interested in the boundary conditions and the differential equations of $E_z, B_z$. Thus we have to find the boundary condition of $B_z$. This is our next step.

Vector multiplication of the Maxwell equation (14.26) by $\mathbf{e}_z$ gives

$$\mathbf{e}_z \times (\boldsymbol{\nabla}_z \times \mathbf{B}_\perp) = -\mathbf{e}_z \times (\boldsymbol{\nabla}_\perp \times \mathbf{B}_z) - i\omega\mu\epsilon(\mathbf{e}_z \times \mathbf{E}_\perp). \tag{14.38}$$

Using Eq. (14.19) it follows that

$$0 - (\mathbf{e}_z \cdot \boldsymbol{\nabla}_z)\mathbf{B}_\perp = -\boldsymbol{\nabla}_\perp B_z + 0 - i\omega\mu\epsilon(\mathbf{e}_z \times \mathbf{E}_\perp), \tag{14.39}$$

i.e.

$$\frac{\partial \mathbf{B}_\perp}{\partial z} = \boldsymbol{\nabla}_\perp B_z + i\omega\mu\epsilon(\mathbf{e}_z \times \mathbf{E}_\perp). \tag{14.40}$$

Scalar multiplication of this equation by $\mathbf{n}$ yields

$$\frac{\partial}{\partial z}\mathbf{n} \cdot \mathbf{B}_\perp = \mathbf{n} \cdot \boldsymbol{\nabla}_\perp B_z + i\omega\mu\epsilon\mathbf{n} \cdot (\mathbf{e}_z \times \mathbf{E}_\perp). \tag{14.41}$$

Since

$$\mathbf{B} = \mathbf{B}(x,y)e^{\pm ikz - i\omega t}, \quad i.e. \quad \frac{\partial \mathbf{B}_\perp}{\partial z} = \pm ik\mathbf{B}_\perp, \tag{14.42}$$

we obtain ($\mathbf{n} \cdot \boldsymbol{\nabla}_\perp \equiv \partial/\partial n$)

$$\pm ik(\mathbf{n} \cdot \mathbf{B}_\perp) = \frac{\partial B_z}{\partial n} - i\omega\mu\epsilon(\mathbf{n} \times \mathbf{E}_\perp) \cdot \mathbf{e}_z. \tag{14.43}$$

Thus at the surface of the wave guide we have with Eqs. (14.35) and (14.36)

$$\left.\frac{\partial B_z}{\partial n}\right|_{\text{surface}} = 0, \quad \left(\frac{\partial}{\partial n} \equiv \mathbf{n} \cdot \boldsymbol{\nabla}_\perp\right). \tag{14.44}$$

Above we reduced the Maxwell equations to Eqs. (14.9) and (14.10), i.e. to the following two equations which apply to each of the three components of the fields, i.e.

$$(\boldsymbol{\nabla}_\perp^2 + \mu\epsilon\omega^2 - k^2)\mathbf{E}(x,y) = 0, \quad (\boldsymbol{\nabla}_\perp^2 + \mu\epsilon\omega^2 - k^2)\mathbf{B}(x,y) = 0. \tag{14.45}$$

Moreover, we saw that the transverse components $\mathbf{E}_\perp, \mathbf{B}_\perp$ can be derived from a knowledge of the longitudinal components $E_z, B_z$. Thus there remains the investigation of the equations

$$(\nabla_\perp^2 + \mu\epsilon\omega^2 - k^2)E_z(x,y) = 0, \quad (\nabla_\perp^2 + \mu\epsilon\omega^2 - k^2)B_z(x,y) = 0 \quad (14.46)$$

with the boundary conditions (14.37) and (14.44),

$$E_z\Big|_{\text{surface}} = 0, \quad \frac{\partial B_z}{\partial n}\Big|_{\text{surface}} = 0, \quad (14.47)$$

where the surfaces here are two-dimensional. Equations (14.46) with these conditions define an *eigenvalue problem.* In general the boundary conditions cannot be satisfied simultaneously.

## 14.4 Wave Guides and Their TEM Fields

We consider wave guides. The frequency $\omega$ is given; the eigenvalue problem then determines the permissible (axial) wave numbers $k$. Depending on the boundary conditions one distinguishes between fields of different types:

**TM**, *transverse magnetic*: $B_z = 0$ everywhere (hence the terminology) and $E_z|_{\text{surface}} = 0$,
**TE**, *transverse electric*: $E_z = 0$ everywhere (hence the terminology) and $\partial B_z/\partial n|_{\text{surface}} = 0$,
**TEM**, *transverse electric-magnetic*: $B_z = 0 = E_z$ everywhere.

We consider first the case of TEM fields.

### 14.4.1 *TEM fields*

In these cases we obtain from Eqs. (14.29) and (14.30) the equations

$$(\mu\epsilon\omega^2 - k^2)\mathbf{E}_\perp = 0, \quad (\mu\epsilon\omega^2 - k^2)\mathbf{B}_\perp = 0. \quad (14.48)$$

These equations have the trivial solution

$$\mathbf{E}_\perp = 0 = \mathbf{B}_\perp \quad \text{for} \quad \mu\epsilon\omega^2 - k^2 \neq 0. \quad (14.49)$$

Hence for a nontrivial solution we have in general

$$\mu\epsilon\omega^2 - k^2 = 0, \quad k = \sqrt{\mu\epsilon}\omega. \quad (14.50)$$

Equations (14.29) and (14.30) imply $0 \times \mathbf{E}_\perp, \mathbf{B}_\perp = 0$, i.e. $\mathbf{E}_\perp, \mathbf{B}_\perp$ remain undetermined. For this reason we now write the TEM fields

$$\mathbf{E}_{\text{TEM}} = (\mathbf{E}_\perp, E_z = 0), \quad \mathbf{B}_{\text{TEM}} = (\mathbf{B}_\perp, B_z = 0). \quad (14.51)$$

According to Eqs. (14.46) we now have (with $\nabla_\perp^2 \equiv \triangle_\perp$):

$$\triangle_\perp \mathbf{E}_\perp = 0, \quad \triangle_\perp \mathbf{B}_\perp = 0, \tag{14.52}$$

i.e. $\mathbf{E}_{\rm TEM}, \mathbf{B}_{\rm TEM}$ are solutions of the two-dimensional Laplace equation. Before we investigate these equations we show that

$$\mathbf{E}_{\rm TEM} \perp \mathbf{B}_{\rm TEM}. \tag{14.53}$$

This is *not* trivial, because so far we established the transversality of electromagnetic radiation only for the case of an unlimited medium (see e.g. the discussion after Eq. (9.17) and Section 9.3).

From the equation

$$\nabla \times \mathbf{E} = -\frac{\partial \mathbf{B}}{\partial \mathbf{t}} \quad \text{with} \quad \mathbf{B} \propto e^{-i\omega t} \tag{14.54}$$

we get

$$i\omega\mathbf{B} = \nabla \times \mathbf{E} = (\nabla_\perp + \nabla_z) \times \mathbf{E} = \nabla_\perp \times \mathbf{E} + \frac{\partial}{\partial z}(\mathbf{e}_z \times \mathbf{E}). \tag{14.55}$$

For $\mathbf{E}, \mathbf{B} \to \mathbf{E}_{\rm TEM}, \mathbf{B}_{\rm TEM}$, i.e. $B_z = 0 = E_z$, the fields $\mathbf{E}, \mathbf{B}$ are vectors in the $(x, y)$-plane so that $\nabla_\perp \times \mathbf{E}_{\rm TEM} \perp (x, y)$-plane, i.e. parallel to the $z$-axis, i.e. (since "curl grad $V = 0$")

$$\nabla_\perp \times \mathbf{E}_{\rm TEM} = 0 \quad (\to \mathbf{E}_{\rm TEM} = -\nabla_\perp \phi), \tag{14.56}$$

and

$$\mathbf{B}_{\rm TEM} = \frac{1}{i\omega}\frac{\partial}{\partial z}(\mathbf{e}_z \times \mathbf{E}_{\rm TEM}). \tag{14.57}$$

With

$$\mathbf{E}_{\rm TEM} = \mathbf{E}_{0\rm TEM}(x, y) e^{i(kz - \omega t)}, \tag{14.58}$$

we obtain

$$\mathbf{B}_{\rm TEM} = \frac{k}{\omega}(\mathbf{e}_z \times \mathbf{E}_{\rm TEM}), \quad \text{i.e.} \quad \mathbf{B}_{\rm TEM} \overset{(14.50)}{=} \sqrt{\mu\epsilon}(\mathbf{e}_z \times \mathbf{E}_{\rm TEM}). \tag{14.59}$$

Thus $\mathbf{E}_{\rm TEM}$ and $\mathbf{B}_{\rm TEM}$ obey the same transversality connection as in the case of the unlimited medium.

However, *TEM waves do not propagate in hollow wave guides, which are made of perfect conductor material* (i.e. have $\sigma = \infty$ and hence represent equipotential surfaces). This can be seen as follows. Inside the wave guide

$$\nabla \cdot \mathbf{E} = 0, \quad \nabla \cdot \mathbf{B} = 0, \tag{14.60}$$

i.e. for $E_z = 0 = B_z$:

$$\frac{\partial E_x}{\partial x} + \frac{\partial E_y}{\partial y} = 0, \quad \frac{\partial B_x}{\partial x} + \frac{\partial B_y}{\partial y} = 0. \tag{14.61}$$

On the other hand from Eq. (14.56) or from the $z$-components of the equations

$$\nabla \times \mathbf{E} = i\omega\mathbf{B}, \quad \nabla \times \mathbf{B} = -i\mu\epsilon\omega\mathbf{E} \quad \text{for} \quad E_z = 0 = B_z,$$

we obtain

$$\frac{\partial E_y}{\partial x} - \frac{\partial E_x}{\partial y} = i\omega B_z = 0, \quad \frac{\partial B_y}{\partial x} - \frac{\partial B_x}{\partial y} = -i\mu\epsilon\omega E_z = 0. \tag{14.62}$$

We write the solutions of these equations with potentials $\phi = \phi(x, y), \psi = \psi(x, y)$

$$\left.\begin{array}{ll} E_x = -\partial\phi/\partial x, & E_y = -\partial\phi/\partial y \\ B_x = -\partial\psi/\partial x, & B_y = -\partial\psi/\partial y \end{array}\right\}, \quad \mathbf{E}_\perp = -\nabla_\perp\phi, \quad \mathbf{B}_\perp = -\nabla_\perp\psi. \tag{14.63}$$

Then Eqs. (14.61) and (14.62) become

$$\left.\begin{array}{ll} \frac{\partial^2\phi}{\partial x^2} + \frac{\partial^2\phi}{\partial y^2} = 0, & \frac{\partial^2\psi}{\partial x^2} + \frac{\partial^2\psi}{\partial y^2} = 0, \\ \frac{\partial^2\phi}{\partial x\partial y} = \frac{\partial^2\phi}{\partial y\partial x}, & \frac{\partial^2\psi}{\partial x\partial y} = \frac{\partial^2\psi}{\partial y\partial x}, \end{array}\right\} \tag{14.64}$$

i.e. $\phi$ and $\psi$ are solutions of the 2-dimensional Laplace equations

$$\triangle_\perp\phi = 0, \quad \triangle_\perp\psi = 0. \tag{14.65}$$

Since, however, the conducting surface of the wave guide is an equipotential surface, on which the potential of $\mathbf{E}$ is constant, we deduce from Gauss' theorem and the fact that the interior of the wave guide does not contain any charges, that inside the wave guide

$$\mathbf{E}_\perp = -\nabla_\perp\phi = 0 \tag{14.66}$$

(from $\int_{F(V)} \mathbf{E} \cdot d\mathbf{F} = \int_V \nabla \cdot \mathbf{E} dV = \int_V \rho dV/\epsilon = 0$). The argument here is analogous to that which we used in considerations of spherical condensers and cylindrical condensers in electrostatics (see Section 2.9). Since the components of $\mathbf{B}$ are linked with components of $\mathbf{E}$ via the transversality condition (14.59), it follows that also $B_x = 0 = B_y$. Thus *TEM waves do not propagate in hollow and empty pipes.*

However, this is different if the wave guide contains inside another symmetric surface, like a concentric metal cylinder of circular cross section; a construction of this type which is no longer simply connected is known as

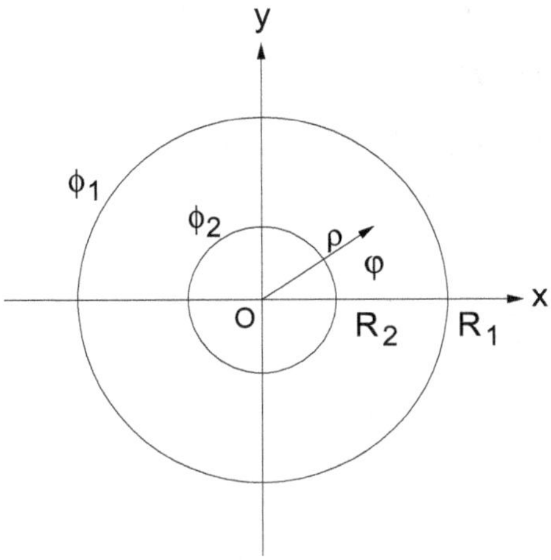


**Fig. 14.1.** Cross section of a coaxial cable.

a *coaxial cable.* In this case according to Fig. 14.1 $\phi_2 \neq \phi_1$. In this case the Laplace equation has a solution (for the fields $\mathbf{E}, \mathbf{B}$) which does not vanish. We have considered coaxial cylinder cables earlier in Chapter 8.

### 14.4.2 *The coaxial transmission cable*

We have for the TEM wave of the coaxial cable (see Section 14.4.1)

$$k = \sqrt{\mu\epsilon}\omega, \quad \triangle_\perp \mathbf{E}_{\text{TEM}} = 0, \quad \mathbf{B}_{\text{TEM}} = \sqrt{\mu\epsilon}(\mathbf{e}_z \times \mathbf{E}_{\text{TEM}}). \tag{14.67}$$

Also from Eq. (14.56), but now with exponential factor,

$$\mathbf{E}_{\text{TEM}} = -(\boldsymbol{\nabla}_\perp \phi) e^{-i(\omega t - kz)}, \tag{14.68}$$

and we have to solve the equation

$$\triangle_\perp \phi = 0. \tag{14.69}$$

We choose a cable with circular cross section and radii $R_1, R_2$ (with $R_2 < R_1$) as shown in Fig. 14.1. The cylindrical symmetry of the cable suggests separation of the equation in cylindrical coordinates, here in polar coordinates, so that we have from Eq. (2.21b)

$$\frac{1}{\rho}\frac{\partial}{\partial \rho}\left(\rho \frac{\partial \phi}{\partial \rho}\right) + \frac{1}{\rho^2}\frac{\partial^2 \phi}{\partial \varphi^2} = 0. \tag{14.70}$$

This equation has to be solved with the boundary conditions

$$\phi(\rho = R_1, \varphi) = \phi_1 = \text{const.},$$
$$\phi(\rho = R_2, \varphi) = \phi_2 = \text{const.} \quad (R_2 < R_1). \tag{14.71}$$

Because of the symmetry of the cylinder

$$\phi(\rho, \varphi_1) = \phi(\rho, \varphi_2) = \cdots = \phi(\rho), \tag{14.72}$$

i.e. $\partial\phi/\partial\varphi = 0$, so that

$$\frac{1}{\rho}\frac{\partial}{\partial\rho}\left(\rho\frac{\partial\phi}{\partial\rho}\right) = 0, \tag{14.73}$$

and hence

$$\phi(\rho) = A\ln\rho + B, \tag{14.74}$$

and so

$$\phi_1 = \phi(R_1) = A\ln R_1 + B, \quad \phi_2 = \phi(R_2) = A\ln R_2 + B,$$

i.e.

$$A = \frac{\phi_1 - \phi_2}{\ln(R_1/R_2)}, \quad B = \frac{\phi_2\ln R_1 - \phi_1\ln R_2}{\ln R_1 - \ln R_2}. \tag{14.75}$$

Thus we have

$$\boldsymbol{\nabla}_\perp\phi = \frac{\partial\phi}{\partial\rho}\mathbf{e}_\rho,$$

i.e. (since $\mathbf{e}_z \times \mathbf{e}_\rho = -\mathbf{e}_\varphi$)

$$\mathbf{E}_{\text{TEM}}(\rho, \varphi, z, t) = -\frac{A}{\rho}\mathbf{e}_\rho e^{-i(\omega t - kz)},$$
$$\mathbf{B}_{\text{TEM}}(\rho, \varphi, z, t) = -\sqrt{\mu\epsilon}\frac{A}{\rho}\mathbf{e}_\varphi e^{-i(\omega t - kz)}. \tag{14.76}$$

Expressed in terms of Cartesian coordinates the relations are

$$\mathbf{E}_{\text{TEM}}(x, y, z, t) = -\frac{A}{x^2 + y^2}(x\mathbf{e}_x + y\mathbf{e}_y)e^{-i(\omega t - kz)},$$
$$\mathbf{B}_{\text{TEM}}(x, y, z, t) = -\frac{\sqrt{\mu\epsilon}A}{x^2 + y^2}(-y\mathbf{e}_x + x\mathbf{e}_y)e^{-i(\omega t - kz)}. \tag{14.77}$$

One should note that, for $R_2 \to 0$ (removal of the inner cylinder) $A \to 0$ and $B \to \phi_1$, i.e. in this limiting case $\phi = \text{const.}$, and there is no TEM wave in the wave guide, as we observed in Section 14.4.1.

## 14.5 Fundamental Equations for the Coaxial Cable

We mentioned already in Section 14.1 that the investigation of properties of wave guides can also be carried out by focusing on the induced charges and currents, i.e. we infer from the fields **E** and **H** the presence of charge $q$ and current $I$, described as induced, which are the same as those which give rise to the fields **E** and **H**. We now derive the corresponding equations. These describe the propagation of the potential difference or voltage $V = V(t, z)$ and the current $I = I(t, z)$ along the cable. The equations that we obtain, are more general than our derivation here, and retain their validity for many other types of wave guides. Thus we consider a coaxial cable with radii $a, b$ (with $a < b$). We let $V(t, z)$ be the voltage between the outer and the inner cylinders at $z$-coordinate $z$, and we let $I(t, z)$ be the current induced in the inner cylinder (or the current corresponding to an appropriate surface current). Then we know that at distance $r$ with $r > a$, from the axis of this cylinder, as indicated in Fig. 14.2, the magnetic field strength **H** is given by

$$\mathbf{H} = \frac{I(t, z)}{2\pi r}\mathbf{e}_\varphi \quad \text{amperes/meter}, \tag{14.78}$$

as may be verified, for instance, by referring back to Eq. (8.17) of Example 8.2. We now apply Faraday's law

$$\oint_C \mathbf{E} \cdot d\mathbf{l} = -\frac{\partial}{\partial t}\int_{F(C)} \mathbf{B} \cdot d\mathbf{F} \tag{14.79}$$

to the area (1234) in the interior of the cable as shown in Fig. 14.2.

We obtain

$$\int_1^2 \mathbf{E} \cdot d\mathbf{l} + \int_2^3 \mathbf{E} \cdot d\mathbf{l} + \int_3^4 \mathbf{E} \cdot d\mathbf{l} + \int_4^1 \mathbf{E} \cdot d\mathbf{l} = -\frac{\partial}{\partial t}\int_a^b \mu \frac{I(t, z)}{2\pi r} dr dz. \tag{14.80}$$

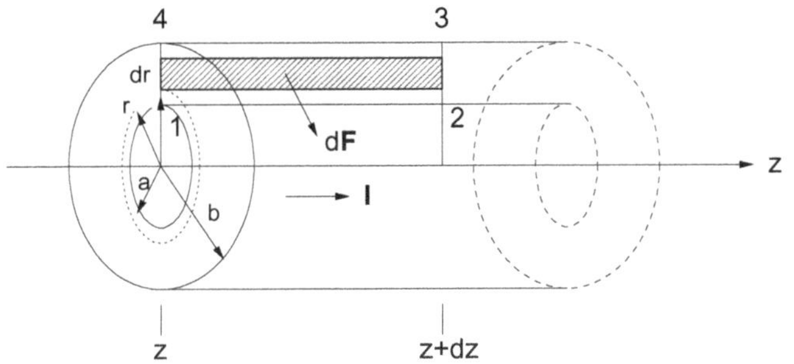


**Fig. 14.2.** The coaxial transmission cable.

We set

$$\int_4^1 \mathbf{E}\cdot d\mathbf{l} \equiv -V(t,z), \quad \int_2^3 \mathbf{E}\cdot d\mathbf{l} \equiv V(t,z+dz). \tag{14.81}$$

Then, since the field **E** is radial and inside the conductor material zero,

$$V(t,z+dz) - V(t,z) = -\frac{\partial}{\partial t}\int_a^b \mu\frac{I(t,z)}{2\pi r}drdz \tag{14.82}$$

implying

$$\frac{\partial V(t,z)}{\partial z} = -\frac{\mu}{2\pi}\ln(b/a)\frac{\partial I(t,z)}{\partial t},$$

or

$$\frac{\partial V(t,z)}{\partial z} = -L_0\frac{\partial I(t,z)}{\partial t} \tag{14.83}$$

with *self-inductance* (see Eq. (8.19) of Example 8.2):

$$L_0 = \frac{\mu}{2\pi}\ln(b/a) \quad \text{henries per unit length.} \tag{14.84}$$

This result is independent of our choice of directions in Fig. 14.3.

A second equation is obtained from Ampère's law to which Maxwell added his displacement current, i.e. from the equation

$$\oint \mathbf{B}\cdot d\mathbf{l} = \mu\int\left(\mathbf{j} + \frac{\partial \mathbf{D}}{\partial t}\right)\cdot d\mathbf{F}. \tag{14.85}$$

We apply this relation to a section of the cylindrical surface of the cable at radius $r, a < r < b$, as indicated in Fig. 14.3. The magnetic field **B** is

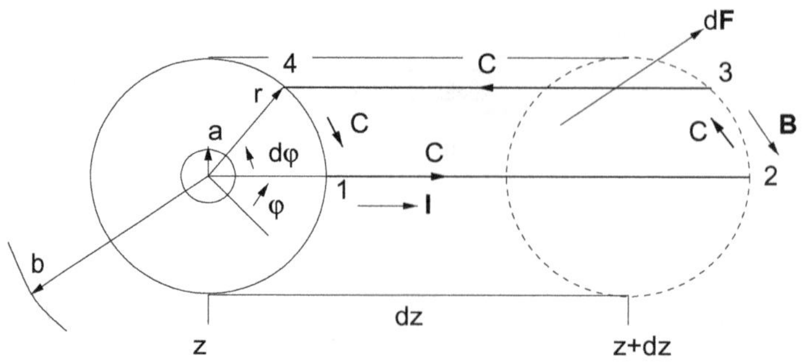


**Fig. 14.3.** Surface piece of the coaxial transmission cable.

entirely due to the current on the internal cylinder (as we observed earlier), and is given by $\mu\mathbf{H}$, i.e. with Eq. (14.78), by

$$\mathbf{B}(t,z) = \mu\frac{I(t,z)}{2\pi r}\mathbf{e}_\varphi \quad \text{teslas.} \tag{14.86}$$

We thus have along the path $C$ indicated in Fig. 14.3, where $d\mathbf{F} = -rd\varphi\mathbf{e}_\varphi \times dz\mathbf{e}_z$,

$$\begin{aligned}\oint_C \mathbf{B}\cdot d\mathbf{l} &= \left(\int_1^2 + \int_2^3 + \int_3^4 + \int_4^1\right)\mathbf{B}\cdot d\mathbf{l} \\ &= 0 + rd\varphi\frac{\mu}{2\pi r}I(t,z+dz) + 0 - \frac{\mu}{2\pi r}I(t,z)(rd\varphi) \\ &= \frac{\mu}{2\pi r}dzrd\varphi\frac{\partial I(t,z)}{\partial z}.\end{aligned} \tag{14.87}$$

Evaluating the other side of the Ampère–Maxwell equation (14.85) we obtain, since $\mathbf{j} = 0$ in the intermediate space and in the case of the cylinder $E = q/2\pi\epsilon r, q$ the induced charge per unit length (see Eq. (2.116) of Example 2.12),

$$\begin{aligned}\mu\int\frac{\partial\mathbf{D}}{\partial t}\cdot d\mathbf{F} &= \epsilon\mu\frac{\partial}{\partial t}\int_{\triangle F}\mathbf{E}\cdot d\mathbf{F} = -\epsilon\mu\frac{\partial}{\partial t}\left(\frac{q}{2\pi\epsilon r}dzrd\varphi\right) \\ &= -\epsilon\mu dzd\varphi\frac{\partial}{\partial t}\left(\frac{q}{2\pi\epsilon}\right) \\ &= -\epsilon\mu dzd\varphi\frac{\partial}{\partial t}\left(\frac{V(t,z)}{\ln(b/a)}\right),\end{aligned} \tag{14.88}$$

since $V = q\ln(b/a)/2\pi\epsilon$ per unit length, as may be checked by consulting Eq. (2.117) of Example 2.12. From the results (14.87) and (14.88) we obtain

$$\frac{\mu}{2\pi r}dzrd\varphi\frac{\partial I(t,z)}{\partial z} = -\epsilon\mu dzd\varphi\frac{\partial}{\partial t}\left(\frac{V(t,z)}{\ln(b/a)}\right), \tag{14.89}$$

and so

$$\frac{\partial I(t,z)}{\partial z} = -\frac{2\pi\epsilon}{\ln(b/a)}\frac{\partial V(t,z)}{\partial t},$$

or

$$\frac{\partial I(t,z)}{\partial z} = -C_0\frac{\partial V(t,z)}{\partial t}, \tag{14.90}$$

where

$$C_0 = \frac{2\pi\epsilon}{\ln(b/a)} \tag{14.91}$$

is the capacity of the cable per unit length obtained in Example 2.12. One should note again the relation[3]

$$L_0C_0 = \frac{\mu}{2\pi}\ln(b/a)\frac{2\pi\epsilon}{\ln(b/a)} = \mu\epsilon = \left(\frac{1}{c'}\right)^2 \tag{14.92}$$

(in vacuum $c' = c$). Differentiation of the *fundamental equations* (14.83) and (14.90) with respect to $t$ and $z$ yields

$$\begin{aligned} \frac{\partial^2 V(t,z)}{\partial t\partial z} &= -L_0\frac{\partial^2 I(t,z)}{\partial t^2}, \quad & \frac{\partial^2 V(t,z)}{\partial z^2} &= -L_0\frac{\partial^2 I(t,z)}{\partial t\partial z}, \\ \frac{\partial^2 I(t,z)}{\partial z\partial t} &= -C_0\frac{\partial^2 V(t,z)}{\partial t^2}, \quad & \frac{\partial^2 I(t,z)}{\partial z^2} &= -C_0\frac{\partial^2 V(t,z)}{\partial z\partial t}, \end{aligned} \tag{14.93}$$

and hence the equations

$$\begin{aligned} \frac{\partial^2 V(t,z)}{\partial z^2} - L_0C_0\frac{\partial^2 V(t,z)}{\partial t^2} &= 0, \\ \frac{\partial^2 I(t,z)}{\partial z^2} - L_0C_0\frac{\partial^2 I(t,z)}{\partial t^2} &= 0. \end{aligned} \tag{14.94}$$

These equations show that current and voltage propagate wavelike along the cable with the phase velocity $c' = 1/\sqrt{L_0C_0}$ because

$$I(t,z), V(t,z) \propto e^{i(kz-\omega t)}, \quad k^2 = L_0C_0\omega^2 = \frac{\omega^2}{c'^2}. \tag{14.95}$$

Equations (14.94) are the *telegrapher's equations* we encountered earlier in Section 8.7.5. There, see Eqs. (8.186)–(8.190), the time-dependence has been avoided by passing to the frequency domain. The relation (14.92), i.e. $L_0C_0 = \mu\epsilon$, is not accidental but results from similar geometrical factors appearing in the calculations of $L_0$ and $1/C_0$, so that they cancel in the product $L_0C_0$.[4]

## 14.6 TM and TE Waves in Wave Guides

### 14.6.1 *General considerations*

We first demonstrate that TM waves ($B_z = 0$) and TE waves ($E_z = 0$) are not transversal, i.e. they *do* possess longitudinal components. In the cases

[3]Observe that $L_0, C_0$ are inductance, capacitance *per unit length*. Correspondingly the vacuum values $\mu_0, \epsilon_0$ are given in henries (H), farads (F) *per unit length*, cf. Appendix B, Table B.2 and Section 9.1. Thus the dimensions of $\epsilon_0\mu_0$ are those of $\mathrm{F\cdot H/m^2}$ and hence those of $\mathrm{(C/V)\cdot(Wb/A)/m^2}$ and of $\mathrm{s^2/m^2}$.
[4]See Bleaney and Bleaney [12], pp. 139–140.

of TE and TM waves the expressions for $\mathbf{E}_\perp$ and $\mathbf{B}_\perp$ in Eqs. (14.29) and (14.30) also simplify. Setting

$$\gamma^2 = \mu\epsilon\omega^2 - k^2, \quad \mathbf{E},\, \mathbf{B} \propto e^{ikz}, \quad k^2 > 0, \tag{14.96}$$

we obtain from Eqs. (14.29) and (14.30) for $\gamma^2 \neq 0$:

**TM**: $B_z = 0$ everywhere,

$$\mathbf{E}_\perp = \frac{1}{\gamma^2} ik \boldsymbol{\nabla}_\perp E_z, \quad \mathbf{B}_\perp = \frac{1}{\gamma^2} i\mu\epsilon\omega(\mathbf{e}_z \times \boldsymbol{\nabla}_\perp) E_z; \tag{14.97}$$

**TE**: $E_z = 0$ everywhere,

$$\mathbf{E}_\perp = -\frac{i\omega}{\gamma^2}(\mathbf{e}_z \times \boldsymbol{\nabla}_\perp) B_z, \quad \mathbf{B}_\perp = \frac{ik}{\gamma^2} \boldsymbol{\nabla}_\perp B_z. \tag{14.98}$$

**TM**: Inserting in the second of Eqs. (14.97) the expression for $\boldsymbol{\nabla}_\perp E_z$ from the first of Eqs. (14.97), we obtain

$$\mathbf{B}_\perp = \frac{\mu\epsilon\omega}{k} \mathbf{e}_z \times \mathbf{E}_\perp. \tag{14.99}$$

**TE**: Inserting similarly in Eqs. (14.98) the expression of the equation for $\boldsymbol{\nabla}_\perp B_z$ into the other, we obtain

$$\mathbf{E}_\perp = -\frac{\omega}{k} \mathbf{e}_z \times \mathbf{B}_\perp. \tag{14.100}$$

We conclude: $E_z, B_z \neq 0$, because from Eqs. (14.97) for TM:

$$\boldsymbol{\nabla}_\perp E_z = \frac{\gamma^2}{ik} \mathbf{E}_\perp \tag{14.101}$$

and from Eqs. (14.98) for TE:

$$\boldsymbol{\nabla}_\perp B_z = \frac{\gamma^2}{ik} \mathbf{B}_\perp. \tag{14.102}$$

Thus the components $E_z$ (TM), $B_z$ (TE) are not both zero; i.e. *the TM and TE waves are not transversal waves*, they possess longitudinal components $E_z$ and $B_z$.

We saw previously (see Eqs. (14.45)), that $\mathbf{E}(x,y), \mathbf{B}(x,y)$ satisfy the equations

$$(\boldsymbol{\nabla}_\perp^2 + \gamma^2)\mathbf{E}(x,y) = 0, \quad (\boldsymbol{\nabla}_\perp^2 + \gamma^2)\mathbf{B}(x,y) = 0 \tag{14.103}$$

(i.e. for each of the the three components of $\mathbf{E}$ and $\mathbf{B}$). We have therefore in particular for the $z$-components:

**TM**:

$$(\boldsymbol{\nabla}_\perp^2+\gamma^2)\mathbf{E}_z(x,y)=0, \quad E_z|_{\text{surface}}=0, \quad (B_z=0 \text{ everywhere}), \quad (14.104)$$

**TE**:

$$(\boldsymbol{\nabla}_\perp^2+\gamma^2)\mathbf{B}_z(x,y)=0, \quad \left.\frac{\partial B_z}{\partial n}\right|_{\text{surface}}=0, \quad (E_z=0 \text{ everywhere}). \quad (14.105)$$

Equations (14.104) and (14.105) together with their boundary conditions define an eigenvalue problem for the determination of $E_z$ and $B_z$; we observed earlier that once both of these are known, one can derive from them the transverse field components. For the solution of Eqs. (14.104) and (14.105) one requires precise information about the geometry of the wave guide.

### 14.6.2 *Wave guides with rectangular cross section*

As an example we consider a wave guide with rectangular cross section as illustrated in Fig. 14.4. We show in particular, how the waves propagate through the wave guide by *continuous reflection from the walls.* We begin with **TM modes.**

The hollow inside of the wave guide is bounded by the planes

$$x=0, \quad x=a, \quad y=0, \quad y=b. \quad (14.106)$$

These are the surfaces on which the *boundary* condition $E_z|_{\text{surface}}=0$ has to be imposed. In Eq. (14.104) for $E_z$ the variables $x$ and $y$ can be separated. We therefore make for $E_z(x,y)$ the ansatz

$$E_z(x,y)=f(x)g(y), \quad (14.107)$$

so that

$$\left(\frac{\partial^2}{\partial x^2}+\frac{\partial^2}{\partial y^2}+\gamma^2\right)f(x)g(y)=0, \quad (14.108)$$

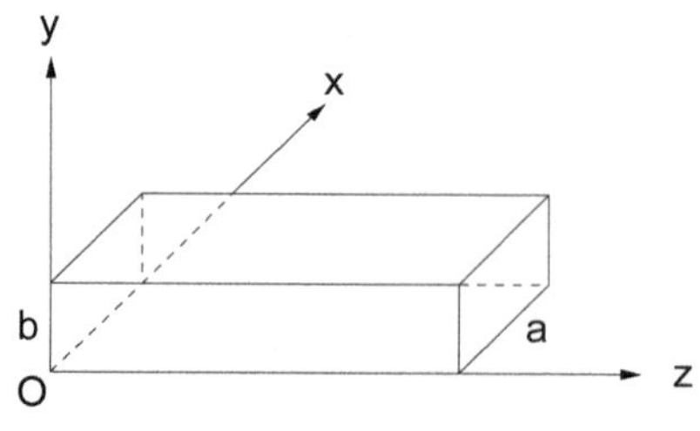


**Fig. 14.4.** Wave guide with rectangular cross section.

or for $x \neq 0, a$ and $y \neq 0, b$:

$$\frac{1}{f(x)} \frac{\partial^2 f(x)}{\partial x^2} + \frac{1}{g(y)} \frac{\partial^2 g(y)}{\partial y^2} + \gamma^2 = 0, \tag{14.109}$$

or

$$\frac{\partial^2 f(x)}{\partial x^2} + \rho^2 f(x) = 0 \tag{14.110}$$

and

$$\frac{\partial^2 g(y)}{\partial y^2} + (\gamma^2 - \rho^2) g(y) = 0, \tag{14.111}$$

where $\rho^2 =$ const. is a separation constant.

In order to solve the *first equation*, Eq. (14.110), we write

$$f(x) = A \cos \rho(x + x_0), \quad A \neq 0. \tag{14.112}$$

$A$ and $x_0$ are the two integration constants. These are determined with the help of the boundary conditions. Thus $E_z|_{\text{surface}} = 0$ implies

$$f(0) = 0, \quad f(a) = 0, \quad \text{so that} \quad \cos \rho \mathrm{x}_0 = 0, \quad \cos \rho(\mathrm{a} + \mathrm{x}_0) = 0, \tag{14.113}$$

and hence

$$\rho x_0 = \pm\frac{\pi}{2}, \pm\frac{3\pi}{2}, \ldots, \quad \rho(a + x_0) = \pm\frac{\pi}{2}, \pm\frac{3\pi}{2}, \ldots, \tag{14.114}$$

and

$$\rho = \frac{n\pi}{a}, \quad n = 0, \pm 1, \pm 2, \ldots. \tag{14.115}$$

(If we had instead of $\rho^2 \rightarrow -\rho^2$ and hence $f \propto \cosh \rho(x + x_0)$ or $\sinh \rho(x + x_0)$, the boundary conditions could not be satisfied in a nontrivial way). We conclude that

$$\begin{aligned} f(x) &= A \cos \rho(x + x_0) = A(\cos \rho x \cos \rho x_0 - \sin \rho x \sin \rho x_0) \\ &= A' \sin\left(\frac{n\pi x}{a}\right), \quad n = 0, \pm 1, \pm 2, \pm 3, \ldots. \end{aligned} \tag{14.116}$$

We now consider the *second equation*, Eq. (14.111). Proceeding as in the above case we obtain

$$g(y) = B \cos \sqrt{\gamma^2 - \rho^2}(y + y_0), \quad B \neq 0, \tag{14.117}$$

and $E_z|_{\text{surface}} = 0$ implies $g(0) = 0, \;\; g(b) = 0$, i.e.

$$\begin{aligned}\cos\sqrt{\gamma^2-\rho^2}y_0 &= 0 = \cos\sqrt{\gamma^2-\rho^2}(b+y_0)\\ &= -\sin\sqrt{\gamma^2-\rho^2}b\sin\sqrt{\gamma^2-\rho^2}y_0,\end{aligned} \tag{14.118}$$

so that

$$\sqrt{\gamma^2-\rho^2}y_0 = \frac{\pi}{2}, \frac{3\pi}{2}, \ldots, \qquad \sqrt{\gamma^2-\rho^2}b = m\pi, \tag{14.119}$$

and, inserting into $g(y) = -B\sin\sqrt{\gamma^2-\rho^2}y\sin\sqrt{\gamma^2-\rho^2}y_0$,

$$g(y) = B'\sin\left(\frac{m\pi y}{b}\right), \qquad m = 0, \pm 1, \ldots. \tag{14.120}$$

Summarizing we find: *TM modes have the eigensolutions* of Eq. (14.104) given by

$$E_z(x,y) = C'\sin\left(\frac{n\pi x}{a}\right)\sin\left(\frac{m\pi y}{b}\right) \tag{14.121a}$$

with

$$\gamma^2 - \rho^2 = \frac{m^2\pi^2}{b^2}, \qquad \rho = \frac{n\pi}{a}. \tag{14.121b}$$

From the last two relations we obtain the *eigenvalue relation*

$$\gamma^2 = \gamma_\lambda^2, \quad \gamma_\lambda^2 \equiv \frac{m^2\pi^2}{b^2} + \frac{n^2\pi^2}{a^2}, \qquad \lambda \equiv \{m,n\}. \tag{14.122}$$

The $x$- and $y$-components of $\mathbf{E}$ can now be obtained from $E_z$ with the help of Eq. (14.29) (recall that we are considering here TM waves which have $B_z = 0$ everywhere), or one can obtain these in some other way. Similar considerations apply in the case of the components of $\mathbf{B}$.

We saw in Section 14.3 that the following conditions hold at the boundary:

$$B_n^{(1)} = B_n^{(2)}, \quad E_t^{(1)} = E_t^{(2)}, \tag{14.123}$$

i.e. for the fields from inside the wave guide

$$(\mathbf{n}\cdot\mathbf{B})_{\text{surface}} = 0, \quad (\mathbf{n}\times\mathbf{E})_{\text{surface}} = 0. \tag{14.124}$$

We thus obtain

(a) for planes $x = 0, a$, $\mathbf{n}$ parallel and antiparallel to $\mathbf{e}_x$:

$$B_x = 0, \quad E_y = 0 = E_z, \quad \text{i.e.} \quad E_x \neq 0; \tag{14.125a}$$

(b) for planes $y = 0, b$, $\mathbf{n}$ parallel and antiparallel to $\mathbf{e}_y$:

$$B_y = 0, \quad E_x = 0 = E_z, \quad \text{i.e.} \quad E_y \neq 0. \tag{14.125b}$$

We assume

$$\mathbf{E}(x, y, z, t) = \mathbf{E}(x, y)e^{i(kz-\omega t)}, \quad \mathbf{B}(x, y, z, t) = \mathbf{B}(x, y)e^{i(kz-\omega t)}. \tag{14.126}$$

The equations to be solved are, cf. Eqs. (14.103), (14.96), and Eqs. (14.104), (14.105), (for $\mu = \mu_0, \epsilon = \epsilon_0, \epsilon_0\mu_0 = 1/c^2$)

$$\left(\boldsymbol{\nabla}_\perp^2 + \frac{\omega^2}{c^2} - k^2\right)\begin{cases}\mathbf{E}(x, y) = 0, \\ \mathbf{B}(x, y) = 0,\end{cases} \quad \text{with} \quad E_z|_{\text{surface}} = 0, \quad \left.\frac{\partial B_z}{\partial n}\right|_{\text{surface}} = 0. \tag{14.127}$$

As an example we consider $E_x$ (other components can be dealt with similarly). We therefore select the equation

$$\left(\boldsymbol{\nabla}_\perp^2 + \frac{\omega^2}{c^2} - k^2\right)E_x(x, y) = 0. \tag{14.128}$$

The variables $x, y$ can be separated, i.e. we put

$$E_x \propto (\sin\rho x \text{ or } \cos\rho x)(y\text{-part}),$$

so that

$$\left(\frac{\partial^2}{\partial y^2} - \rho^2 + \frac{\omega^2}{c^2} - k^2\right)(y\text{-part}) = 0. \tag{14.129}$$

Since $E_x$ is nonzero for $x = 0, a$ (see boundary conditions (14.125a)), we choose $E_x \propto \cos\rho x, \rho = m\pi/a$. Since, however, $E_x$ has to vanish for $y = 0, b$ (see boundary conditions (14.125b)), we choose for the $y$-part

$$\sin\left(\frac{y\pi n}{b}\right), \quad n = 0, 1, 2, \ldots,$$

so that

$$E_x(x, y) = \alpha\cos\left(\frac{m\pi x}{a}\right)\sin\left(\frac{n\pi y}{b}\right). \tag{14.130}$$

We obtain the same expression with Eq. (14.29) from $E_z$. For $E_x$ to satisfy Eq. (14.128), we must have[5]

$$-\left(\frac{m^2\pi^2}{a^2}+\frac{n^2\pi^2}{b^2}\right)+\frac{\omega^2}{c^2}-k^2=0. \tag{14.131}$$

For the wave $\mathbf{E}(x,y,z,t)\propto e^{ikz}$ to propagate in the direction of $z$, we must have $k^2>0$, i.e. $\mathbf{E}$ must be periodic in $z$. This means

$$k^2>0:\frac{\omega^2}{c^2}\geq\frac{\omega_0^2}{c^2},\quad \omega_0=c\sqrt{\frac{m^2\pi^2}{a^2}+\frac{n^2\pi^2}{b^2}}. \tag{14.132}$$

We see: $\omega$ must be larger than some *limiting frequency* $\omega_0$. Proceeding similarly with all components $E_x, E_y, E_z, B_x, B_y, B_z$, we obtain the following expressions

$$\left.\begin{aligned}E_x&=\alpha\cos(m\pi x/a)\sin(n\pi y/b)\\E_y&=\beta\sin(m\pi x/a)\cos(n\pi y/b)\\E_z&=\gamma\sin(m\pi x/a)\sin(n\pi y/b)\\B_x&=\alpha'\sin(m\pi x/a)\cos(n\pi y/b)\\B_y&=\beta'\cos(m\pi x/a)\sin(n\pi y/b)\\B_z&=\gamma'\cos(m\pi x/a)\cos(n\pi y/b)\end{aligned}\right\}\times e^{i(kz-\omega t)}. \tag{14.133}$$

We refer to these expressions again later.[6]

**TM waves** have everywhere $B_z=0$, i.e. $\gamma'=0$. We deduce from Eq. (14.133) that these waves are different from zero if (to be shown) $m\neq 0, n\neq 0$. In order to see this, we recall: The expressions for $\mathbf{E},\mathbf{B}$ must satisfy Maxwell's equations. Hence we consider these:

(a) $\boldsymbol{\nabla}\cdot\mathbf{E}=0$ yields (apart from the factor $\sin(n\pi x/a)\sin(n\pi y/b)$)

$$-\alpha\frac{m\pi}{a}-\beta\frac{n\pi}{b}+ik\gamma=0; \tag{14.134}$$

---

[5]It is sometimes helpful to set $k^2\equiv\tilde{\beta}^2$ and $\omega^2/c^2\equiv\beta^2$. Then $\tilde{\beta}\equiv 2\pi/\tilde{\lambda}$ is the wave number *inside* the guide, and $\beta\equiv\omega\sqrt{\mu\epsilon}$ is the wave number $2\pi/\lambda$ *outside* the guide. Then Eq. (14.131) is $\tilde{\beta}^2=\beta^2-\pi^2(m^2/a^2+n^2/b^2)$. The wave number $\tilde{\beta}$ gives the forward wave travelling down the wave guide. See Example 14.9.

[6]Later, in Section 14.7, we write

$$\mathbf{B}=\left(\frac{\partial\Psi}{\partial y},-\frac{\partial\Psi}{\partial x},0\right),\quad \mathbf{E}=\left(\frac{\partial\Pi}{\partial y},-\frac{\partial\Pi}{\partial x},0\right),$$

so that according to Eq. (14.133) $\Psi\propto\sin\ldots\sin\ldots$ and $\Pi\propto\cos\ldots\cos\ldots$.

(b) $\boldsymbol{\nabla}\cdot\mathbf{B}=0$ yields

$$\alpha'\frac{m\pi}{a}+\beta'\frac{n\pi}{b}+ik\gamma'=0; \tag{14.135}$$

(c) $\boldsymbol{\nabla}\times\mathbf{E}=i\omega\mathbf{B}$ yields[7]

$$\begin{aligned}\alpha' i\omega &= \gamma\frac{n\pi}{b}-ik\beta,\\ \beta' i\omega &= ik\alpha-\gamma\frac{m\pi}{a},\\ \gamma' i\omega &= \beta\frac{m\pi}{a}-\alpha\frac{n\pi}{b};\end{aligned} \tag{14.136}$$

(d) $\mu\boldsymbol{\nabla}\times\mathbf{H}=-i\omega\epsilon\mu\mathbf{E}$ yields

$$\begin{aligned}0 &= i\omega\epsilon\mu\alpha-\gamma'\frac{n\pi}{b}-\beta' ik,\\ 0 &= i\omega\epsilon\mu\beta+ik\alpha'+\gamma'\frac{m\pi}{a},\\ 0 &= i\omega\epsilon\mu\gamma-\beta'\frac{m\pi}{a}+\alpha'\frac{n\pi}{b}.\end{aligned} \tag{14.137}$$

For TM waves $\gamma'=0$. We show: For $m=0, n\neq 0$ the entire field is zero. From (b) we have

$$\alpha'\frac{m\pi}{a}+\beta'\frac{n\pi}{b}=0. \tag{14.138}$$

For $m=0, n\neq 0\rightarrow\beta'=0$. It then follows from the first of Eqs. (14.137), since $\gamma'=0, \beta'=0\rightarrow\alpha=0$. Then $E_x=0$ (since $\alpha=0$), $E_y, E_z, B_x=0$ (since $m=0$), $B_y=0$ (since $\beta'=0$), $B_z=0$ (since $\gamma'=0$). Hence the entire field $=0$ (can also be obtained faster from Eq. (14.99)). Thus (since our arguments assumed $n\neq 0$) there is only one limiting frequency $\omega_{11}$ which is such, that TM waves are possible only if

$$\omega>\omega_{11},\quad \omega_{11}=c\sqrt{\frac{\pi^2}{a^2}+\frac{\pi^2}{b^2}}. \tag{14.139}$$

**TE waves** have everywhere $E_z=0$, i.e. $\gamma=0$. In this case a solution exists also in the case when either $m$ is zero *or* $n$ is zero. The lowest limiting frequency for $a>b, m=1, n=0$ is (for $m=0, n=0$, according to (b) $\gamma'=0$, implies $E_x=0, E_y=0$, hence no field):

$$\omega_{10}=c\frac{\pi}{a}<\omega_{11}. \tag{14.140}$$

[7]Observe that for $m=1, n=0$ one obtains $\alpha'=-k\beta/\omega, \gamma'=\beta\pi/ia\omega$. These expressions will be used in Example 14.2.

With this $(m = 1, n = 0)$ the *fundamental wave* is given by (cf. Eq. (14.133))

$$\begin{aligned} E_x &= 0, \quad E_z = 0, \\ E_y &= \beta \sin(\pi x/a) e^{i(kz-\omega t)} \\ &= \frac{\beta}{2i}\left[e^{i(kz+\pi x/a-\omega t)} - e^{i(kz-\pi x/a-\omega t)}\right], \end{aligned} \tag{14.141}$$

which is a superposition of two waves with $k_x = \pi/a$.

Previously (cf. Eq. (9.11)) we had defined a plane wave by the expression

$$e^{i(\mathbf{k}\cdot\mathbf{r}-\omega t)} \quad \text{for} \quad \mathbf{k}\cdot\mathbf{r} - \omega t = \text{const.} \tag{14.142}$$

The wave vector $\mathbf{k}$ is perpendicular to the plane with constant phase and points in the direction of propagation of the wave. The equations

$$y = 0, \quad kz \pm \frac{\pi}{a}x - \omega t = \text{const.} = \alpha \tag{14.143}$$

describe *wave-fronts* of plane waves. The gradients of the wave-fronts are therefore, see Fig. 14.5, according to the waves of these wave-fronts and at $t = 0$, i.e.

$$z = \frac{\alpha}{k} \mp \frac{\pi}{a}\frac{x}{k},$$

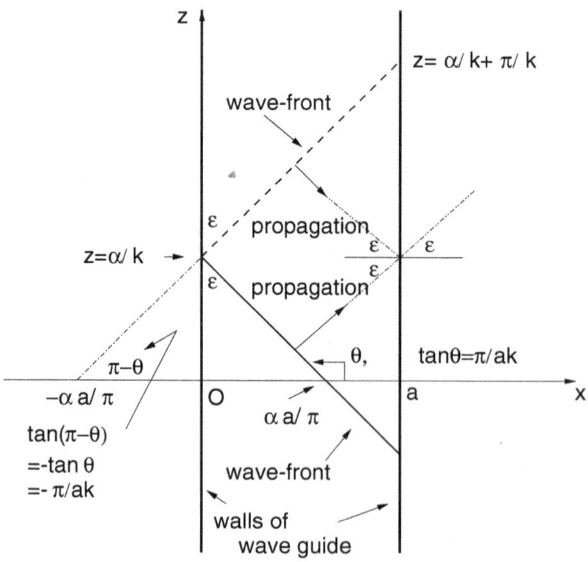


**Fig. 14.5.** Reflection at the walls of the wave guide.

given by the derivative

$$\frac{dz}{dx} = \mp\frac{\pi}{ak} \equiv \mp\tan\theta \tag{14.144}$$

as indicated in Fig. 14.5. One should note here the angles $\epsilon$ of both wave-fronts with respect to the $z$-axis, and the angles $|\pi/2-\epsilon|$ with respect to the $x$-axis. One wave-front can be looked at as the reflection of the other at the wall of the wave guide. The direction of propagation (i.e. of the normal to the wave-front) is in both cases given by

$$\cos\epsilon = \pm\frac{\alpha/k}{\sqrt{(\alpha/k)^2+(a\alpha/\pi)^2}} = \pm\frac{\pi/a}{\sqrt{(\pi/a)^2+k^2}}. \tag{14.145}$$

However, cf. Eq. (14.131),

$$\frac{\omega^2}{c^2} = k^2 + \frac{m^2\pi^2}{a^2} + \frac{n^2\pi^2}{b^2}. \tag{14.146}$$

In the case of the *fundamental* TE *wave* with $m=1, n=0$, we have

$$\frac{\omega^2}{c^2} = k^2 + \frac{\pi^2}{a^2} = k^2 + \frac{{\omega_{10}}^2}{c^2}, \quad \omega_{10}^2 = \frac{\pi^2}{a^2}c^2, \quad \frac{\pi}{a} = \frac{\omega_{10}}{c},$$
$$kc = \sqrt{\omega^2 - {\omega_{10}}^2}. \tag{14.147}$$

It follows that

$$\cos\epsilon = \mp\frac{\omega_{10}/c}{\omega/c} = \mp\frac{\omega_{10}}{\omega}. \tag{14.148}$$

(Later we shall need $1-\cos^2\epsilon = 1-(\omega_{10}/\omega)^2$). This means

$$\text{for} \quad e^{i(kz\pm\pi x/a-\omega t)} \quad \text{we have} \quad \cos\epsilon = \mp\frac{\omega_{10}}{\omega}. \tag{14.149}$$

We see therefore: We can imagine the field $\mathbf{E}$ as arising from a continuous reflection of two superposed plane waves at the walls $x=0,a$ which are incident on the walls at the angle $\epsilon$, one wave permitting interpretation as the reflection of the other (with the correct phase difference). We note that at the critical frequency $\omega_{10}$ we have $k=0$. In this case $\cos\epsilon = \pm1$, i.e. $\epsilon = 0, \pm\pi, \pm2\pi, \ldots$.

In our further exploration of wave guides with rectangular cross section we restrict ourselves to some general remarks. We let the eigenvalues of the eigenvalue problem of $E_{z\lambda}$ or $B_{z\lambda}$ be

$$\gamma^2 = \gamma_\lambda^2 > 0, \quad \lambda = 1, 2, \ldots. \tag{14.150}$$

Since (cf. Eq. (14.96)) $\gamma^2 = \mu\epsilon\omega^2 - k^2$, it follows that $\gamma_\lambda^2 = \mu\epsilon\omega^2 - k^2$, or

$$k^2 = \mu\epsilon\left(\omega^2 - \frac{\gamma_\lambda^2}{\mu\epsilon}\right). \tag{14.151}$$

Thus

$$k = \sqrt{\mu\epsilon}\sqrt{\omega^2 - \omega_\lambda^2}, \qquad \omega_\lambda^2 \equiv \frac{\gamma_\lambda^2}{\mu\epsilon}. \tag{14.152}$$

In order for the waves to propagate in the wave guide proportional to $e^{i(kz-\omega t)}$ $k$ must be real, i.e. $\omega^2 > \omega_\lambda^2$. Thus $\omega_\lambda$ is a *critical* or *limiting frequency.* This is indicated in Fig. 14.6. This means that if $\omega_2 < \omega$, then $\lambda_2 > \lambda$, i.e. the wavelength must be less than the appropriate $\lambda_2$, if it is to propagate in the wave guide.

### 14.6.3 *Wave guides with circular cross section*

As another example we consider a wave guide with a circular cross section of radius $\rho = a$, as illustrated in Fig. 14.7, and made of ideal conductor material.

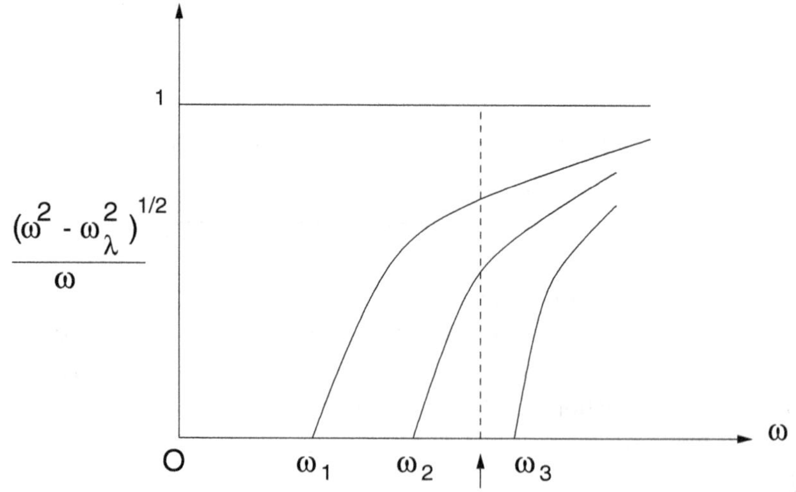


**Fig. 14.6.** At a given frequency $\omega$ only modes with eigenfrequency $\omega_\lambda < \omega$ are transmitted.

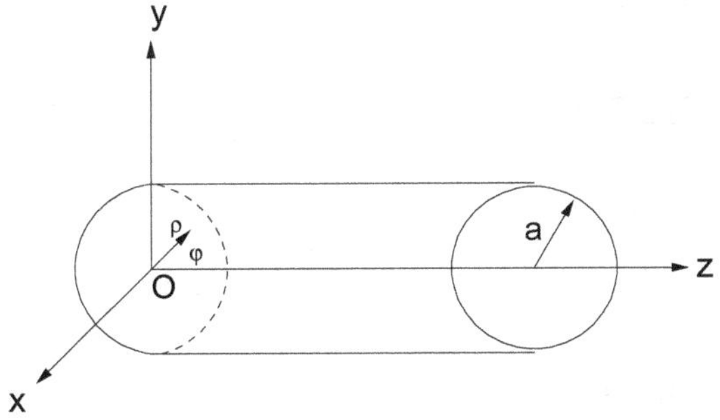


**Fig. 14.7.** Wave guide with circular cross section.

The boundary condition on $E_z$ is $E_z|_{\text{surface}} = 0$ for $\rho = a$, the angle $\varphi$ being arbitrary. In planar polar coordinates Eq. (14.104) is (see Eq. (2.86))

$$\left(\frac{\partial^2}{\partial\rho^2} + \frac{1}{\rho}\frac{\partial}{\partial\rho} + \frac{1}{\rho^2}\frac{\partial^2}{\partial\varphi^2} + \gamma^2\right) E_z(\rho,\varphi) = 0. \tag{14.153}$$

We write

$$E_z(\rho,\varphi) = g(\rho)h(\varphi), \tag{14.154}$$

so that

$$\frac{1}{g(\varphi)}\left(\frac{\partial^2 g(\rho)}{\partial\rho^2} + \frac{1}{\rho}\frac{\partial g(\rho)}{\partial\rho}\right) + \frac{1}{\rho^2}\frac{1}{h(\varphi)}\frac{\partial^2 h(\varphi)}{\partial\varphi^2} + \gamma^2 = 0. \tag{14.155}$$

The variables $\rho$ and $\varphi$ can be separated by setting

$$\frac{1}{h(\varphi)}\frac{\partial^2 h(\varphi)}{\partial\varphi^2} = -\mu^2 \tag{14.156}$$

and

$$\frac{\partial^2 g(\rho)}{\partial\rho^2} + \frac{1}{\rho}\frac{\partial g(\rho)}{\partial\rho} + \left(\gamma^2 - \frac{\mu^2}{\rho^2}\right) g(\rho) = 0. \tag{14.157}$$

The solution of the first equation is

$$h(\varphi) = e^{\pm i\mu\varphi}, \quad \mu = \text{const.} \tag{14.158}$$

Clearly we must demand that $E_z(\varphi) = E_z(\varphi + 2\pi)$, i.e. $h(\varphi) = h(\varphi + 2\pi)$, and therefore we must have

$$e^{\pm i\mu 2\pi} = 1, \quad \text{i.e.} \quad \mu = m = 0, \pm 1, \pm 2, \ldots. \tag{14.159}$$

The second equation is then

$$\frac{\partial^2 g(\rho)}{\partial \rho^2} + \frac{1}{\rho}\frac{\partial g(\rho)}{\partial \rho} + \left(\gamma^2 - \frac{m^2}{\rho^2}\right) g(\rho) = 0. \tag{14.160}$$

The equation

$$\frac{\partial^2 Z(x)}{\partial x^2} + \frac{1}{x}\frac{\partial Z(x)}{\partial x} + \left(1 - \frac{m^2}{x^2}\right) Z(z) = 0 \tag{14.161}$$

is the differential equation of the *cylinder functions* $J_m(x), Y_m(x)$ of integral order given in Tables of Special Functions, so that the general solution is

$$g(\rho) = AJ_m(\gamma\rho) + BY_m(\gamma\rho). \tag{14.162}$$

The solution $Y_m(x)$, known as *Neumann function*, is given by[8]

$$\begin{aligned} Y_m(x) &= \frac{2}{\pi} J_m(x)\ln(e^C x/2) - \frac{1}{\pi}\left(\frac{2}{x}\right)^m \sum_{l=0}^{m-1} \frac{(m-l-1)!}{l!}\left(\frac{x}{2}\right)^{2l} \\ &\quad + \frac{1}{\pi}\left(\frac{x}{2}\right)^m \sum_{l=0}^{\infty} \frac{(-1)^l}{l!(m+l)!}\left(\frac{x}{2}\right)^{2l}\left(\sum_{n=1}^{l}\frac{1}{n} + \sum_{n=1}^{l+m}\frac{1}{n}\right), \\ C &= \text{Euler constant} = 0.5772157\ldots, \end{aligned} \tag{14.163}$$

the *Bessel function* $J_m(x)$ by

$$J_m(x) = \left(\frac{x}{2}\right)^m \sum_{l=0}^{\infty} \frac{(ix/2)^{2l}}{l!(m+l)!}. \tag{14.164}$$

The field $E_z(x, y)$ or $E_z(\rho, \varphi)$ must be finite for $\rho \to 0$. Since $Y_m(\gamma\rho)$ becomes infinite for $\rho \to 0$, we are compelled to impose the finiteness condition

$$B = 0, \quad m \geq 0. \tag{14.165}$$

We thus obtain for the TM modes

$$E_z(\rho, \varphi) = AJ_m(\gamma\rho)e^{im\varphi}. \tag{14.166}$$

The boundary condition $E_z = 0$ for $\rho = a$ implies the eigenvalue equation

$$J_m(\gamma a) = 0. \tag{14.167}$$

This is a transcendental equation with infinitely many roots. For the $s$th positive root $\gamma a$ with $s \gg m$, an expansion is again given in books on Special

[8]See e.g. Magnus and Oberhettinger [90], p. 16, 17.

Functions, i.e.[9]

$$\gamma a \sim \beta - \frac{\mu - 1}{2^3 \beta} - \frac{(\mu - 1)(7\mu - 31)}{3.2^7 \beta^3} - \frac{(\mu - 1)(83\mu^2 - 982\mu + 3779)}{15.2^{10} \beta^5}$$
$$- \frac{(\mu - 1)(6949\mu^3 - 153855\mu^2 + 1585743\mu - 6277237)}{105.2^{15} \beta^7}$$
$$- O(1/\beta^9), \tag{14.168}$$

where

$$\beta = \left( s + \frac{1}{2}m - \frac{1}{4} \right)\pi, \quad \mu = 4m^2. \tag{14.169}$$

In order to understand the reflection of the wave at the walls, we can argue in a way analogous to our treatment of the previous case (cf. after Eq. (14.140)), where we expressed the trigonometric function in terms of exponentials. This means in the present case, we have to re-express the Bessel function $J_m$ in terms of Hankel functions $H^{(1)}, H^{(2)}$, i.e. we have to use the following formulas from books on Special Functions

$$J_m(R) = \frac{1}{2}\left[ H_m^{(1)}(R) + H_m^{(2)}(R) \right] \tag{14.170}$$

with ($|\arg R| < \pi$)

$$H_m^{(1)}(R) = \sqrt{\frac{2}{\pi R}} e^{i(R - m\pi/2 - \pi/4)}[1 + O(1/R)],$$
$$H_m^{(2)}(R) = \sqrt{\frac{2}{\pi R}} e^{-i(R - m\pi/2 - \pi/4)}[1 + O(1/R)], \tag{14.171}$$

for $R \neq 0$, i.e. away from the axis of the wave guide. We do not enter into further details.

## 14.7 Alternative Treatment Using Scalar and Vector Potentials

Our first step is to search for equations for $\mathbf{E}, \mathbf{B}$ in terms of the scalar potential $\phi$ and the vector potential $\mathbf{A}$. Again we assume that the interior of the wave guide is filled with some homogeneous material with electromagnetic

[9] The fullest account of properties of Bessel and related functions is that of Watson [158]. See p. 506. Summaries of all known properties of these and other Special Functions can be found in A. Erdélyi, W. Magnus, F. Oberhettinger and F.G. Tricomi [48].

constants $\epsilon, \mu$ (e.g. those of the vacuum). Then inside the wave guide the following equations apply

$$\nabla \times \mathbf{E} = -\frac{\partial \mathbf{B}}{\partial t}, \quad \nabla \times \mathbf{H} = \frac{\partial \mathbf{D}}{\partial t} = \epsilon \dot{\mathbf{E}}, \tag{14.172}$$

$$\mathbf{B} = \mu \mathbf{H}, \quad \mathbf{D} = \epsilon \mathbf{E}, \quad \nabla \cdot \mathbf{E} = 0, \quad \nabla \cdot \mathbf{B} = 0. \tag{14.173}$$

But, we also have (cf. Eq. (7.33))

$$\mathbf{E} = -\dot{\mathbf{A}} - \nabla \phi, \tag{14.174}$$

and

$$\nabla \times \mathbf{B} = \nabla \times (\nabla \times \mathbf{A}) = \nabla(\nabla \cdot \mathbf{A}) - \nabla^2 \mathbf{A} \tag{14.175}$$

and

$$\nabla \times \mathbf{B} = \mu \nabla \times \mathbf{H} = \mu \epsilon \dot{\mathbf{E}}. \tag{14.176}$$

Hence we can write

$$\mu \epsilon \dot{\mathbf{E}} = \nabla(\nabla \cdot \mathbf{A}) - \nabla^2 \mathbf{A} = \mu \epsilon [-\ddot{\mathbf{A}} - \nabla \dot{\phi}]. \tag{14.177}$$

Now that we are using the vector potential $\mathbf{A}$, we still have the freedom to choose some *gauge fixing condition.* We choose again (as in our treatment of the Liénard–Wiechert potentials) the *Lorenz gauge*, i.e. we set (see Eq. (10.19))

$$\nabla \cdot \mathbf{A} = -\mu \epsilon \dot{\phi}. \tag{14.178}$$

Then, by insertion in Eq. (14.177),

$$\left( \nabla^2 - \mu \epsilon \frac{\partial^2}{\partial t^2} \right) \mathbf{A} = 0. \tag{14.179}$$

With the ansatz $\mathbf{A} \propto e^{-i\omega t}$ (as above for $\mathbf{E}, \mathbf{H}$) the following equation results

$$(\nabla^2 + \mu \epsilon \omega^2) \mathbf{A} = 0. \tag{14.180}$$

Hence from Eq. (14.177):

$$-i \mu \epsilon \omega \mathbf{E} = \nabla(\nabla \cdot \mathbf{A}) + \mu \epsilon \omega^2 \mathbf{A} \tag{14.181}$$

with

$$\mathbf{B} = \mu \mathbf{H} = \nabla \times \mathbf{A}. \tag{14.182}$$

These Eqs. (14.181) and (14.182) show: $\mathbf{E}, \mathbf{B}$ can be derived from $\mathbf{A}$, which is a solution of Eq. (14.180).

We consider now the **TM** and **TE** cases separately.

(a) **TM**: $B_z = 0$ everywhere.
Setting $\mathbf{A} = (0, 0, \Psi)$ such that (observe the zero on the right means $B_z = 0$)

$$\mathbf{B} = \boldsymbol{\nabla} \times \mathbf{A} = \begin{vmatrix} \mathbf{e}_x & \mathbf{e}_y & \mathbf{e}_z \\ \partial/\partial x & \partial/\partial y & \partial/\partial z \\ 0 & 0 & \Psi \end{vmatrix} = \left(\frac{\partial \Psi}{\partial y}, -\frac{\partial \Psi}{\partial x}, 0\right) \tag{14.183}$$

(with $\boldsymbol{\nabla} \cdot \mathbf{B} = 0$), we obtain from Eq. (14.181),

$$-i\mu\epsilon\omega\mathbf{E} = \boldsymbol{\nabla}\left(\frac{\partial \Psi}{\partial z}\right) + \mu\epsilon\omega^2(0, 0, \Psi),$$

i.e.

$$-i\mu\epsilon\omega\mathbf{E} = \left(\frac{\partial^2 \Psi}{\partial x \partial z}, \frac{\partial^2 \Psi}{\partial y \partial z}, \frac{\partial^2 \Psi}{\partial z^2} + \mu\epsilon\omega^2\Psi\right). \tag{14.184}$$

However, we still have the equation $\boldsymbol{\nabla} \cdot \mathbf{E} = 0$, i.e. from this last Eq. (14.184) (interchanging $\triangle$ and $\partial/\partial z$)

$$-i\mu\epsilon\omega\boldsymbol{\nabla} \cdot \mathbf{E} \equiv \frac{\partial}{\partial z}\left(\frac{\partial^2 \Psi}{\partial x^2} + \frac{\partial^2 \Psi}{\partial y^2} + \frac{\partial^2 \Psi}{\partial z^2} + \mu\epsilon\omega^2\Psi\right) = 0, \tag{14.185}$$

and this is satisfied in view of Eq. (14.180), for $\mathbf{A} = (0, 0, \Psi)$, i.e.

$$\nabla^2\Psi + \mu\epsilon\omega^2\Psi = 0. \tag{14.186}$$

All fields can now be derived from the one scalar function $\Psi$, which is a solution of this equation (**H** from Eq. (14.183), **E** from Eq. (14.184)). The function $\Psi$ is frequently called "*stream function*". The associated boundary condition is, as we saw,

$$\mathbf{E}_z|_{\text{surface}} = 0, \quad \left\{\text{also } \left.\frac{\partial B_z}{\partial n}\right|_{\text{surface}} = 0, \text{ since } B_z = 0\right\}, \tag{14.187}$$

i.e. on the surface $E_{\text{tang}} = 0$. For the rectangular wave guide we have (see comment after Eq. (14.133))

$$\Psi \propto \sin\left(\frac{n\pi x}{a}\right)\sin\left(\frac{m\pi y}{b}\right)e^{ikz}e^{-i\omega t}. \tag{14.188}$$

(b) **TE**: $E_z = 0$ everywhere.

From the time derivative of Eq. (14.178) together with Eq. (14.174) we obtain the equation for the scalar potential $\phi$, i.e.

$$(\nabla^2 + \mu\epsilon\omega^2)\phi = 0. \tag{14.189}$$

Setting this time (observe the zero on the right means $E_z = 0$)

$$\mathbf{E} = \left(\frac{\partial \Pi}{\partial y}, -\frac{\partial \Pi}{\partial x}, 0\right), \tag{14.190}$$

we obtain from the first of Eqs. (14.172)

$$\begin{aligned} i\omega\mu\mathbf{H} = \nabla \times \mathbf{E} &= \nabla \times \left(\frac{\partial \Pi}{\partial y}, -\frac{\partial \Pi}{\partial x}, 0\right) \\ &= \left(\frac{\partial^2 \Pi}{\partial z \partial x}, \frac{\partial^2 \Pi}{\partial z \partial y}, -\frac{\partial^2 \Pi}{\partial x^2} - \frac{\partial^2 \Pi}{\partial y^2}\right) \\ &= \left(\frac{\partial^2 \Pi}{\partial z \partial x}, \frac{\partial^2 \Pi}{\partial z \partial y}, \frac{\partial^2 \Pi}{\partial z^2} + \mu\epsilon\omega^2\Pi\right), \end{aligned} \tag{14.191}$$

if we demand that $\Pi$ satisfies the same equation as $\Psi$, i.e.

$$(\nabla^2 + \mu\epsilon\omega^2)\Pi = 0. \tag{14.192}$$

We observe that $\nabla \cdot \mathbf{E} = 0$ and $\nabla \cdot \mathbf{B} = 0$ are automatically satisfied. Again all fields can be derived from the one scalar function $\Pi$ which is a solution of the above equation with the boundary condition

$$\left.\frac{\partial B_z}{\partial n}\right|_{\text{surface}} = 0 \quad \text{and} \quad E_z = 0 \quad \text{everywhere.} \tag{14.193}$$

For the rectangular wave guide this means (check with Eq. (14.133))

$$\Pi \propto \cos\left(\frac{m\pi x}{a}\right)\cos\left(\frac{n\pi y}{b}\right)e^{ikz}e^{-i\omega t}. \tag{14.194}$$

## 14.8 Wave Velocities

We return to our considerations of rectangular wave guides. The wave or phase velocity of the TE fundamental mode with $m = 1, n = 0$ is (cf. Eq. (14.147) and thereafter)

$$\begin{aligned} v_{\text{phase}} = \frac{\omega}{k} &\overset{(14.147)}{=} \frac{\omega c}{\sqrt{\omega^2 - \omega_{10}^2}} = \frac{c}{\sqrt{1 - (\omega_{10}/\omega)^2}} \\ &\overset{(14.148)}{=} \frac{c}{\sqrt{1 - \cos^2\epsilon}} = \frac{c}{\sin\epsilon} > c. \end{aligned} \tag{14.195}$$

The wavelength is $\lambda = 2\pi c/\omega$.

Next we consider the *phase velocity*[10] in the general case, this means for

$$\omega_{10} \to \omega_0 = c\sqrt{\frac{m^2\pi^2}{a^2} + \frac{n^2\pi^2}{b^2}}, \quad \text{i.e.} \quad \frac{\omega^2}{c^2} = k^2 + \frac{\omega_0^2}{c^2}. \tag{14.196}$$

In this general case the phase velocity is

$$v_{\text{phase}} = \frac{\omega}{k} = \frac{\omega}{\sqrt{\omega^2 - \omega_0^2}} c > c. \tag{14.197}$$

Thus the phase velocity is always larger than the velocity of light! In fact it becomes infinite at the critical frequency $\omega_0$.

The velocity of energy transport is here in the case without attenuation, i.e. without loss of energy, the socalled *group velocity*. This group velocity $v_g$ is defined by

$$v_g = \frac{d\omega}{dk} = \frac{d}{dk}\sqrt{k^2c^2 + \omega_0^2} = \frac{kc^2}{\sqrt{k^2c^2 + \omega_0^2}} = \frac{c\sqrt{\omega^2 - \omega_0^2}}{\omega} < c. \tag{14.198}$$

We see that the group velocity of the wave in the wave guide is always less than the velocity of light and vanishes at the critical frequency. Finally we observe that

$$v_{\text{phase}} v_g = c^2. \tag{14.199}$$

This is again a relation which is of wider generality than our derivation here.

## 14.9 Energy Transport in Wave Guides

### 14.9.1 *The complex Poynting vector*

In our earlier treatment of the Poynting vector in Section 7.7 we always assumed real quantities. We now wish to allow for complex fields, and complex $\epsilon, \mu$.

For the calculation of the transport of energy in wave guides we consider as a first step time averaged products of real vectors, which result from complex vectors and lead to the definition of the *complex Poynting vector*. This complex Poynting vector takes into account loss of energy due to

[10]The phase velocity was introduced with Eq. (9.32) and is discussed alongside the group velocity in Section 15.2.

complex $\epsilon$ and $\mu$. Consider the following real product,[11] in which we again separate the time dependence with the factor $e^{-i\omega t}$:

$$\begin{aligned}
&\mathbf{j}(x,t)\cdot\mathbf{E}(x,t)\\
&= \Re[\mathbf{j}(x)e^{-i\omega t}]\cdot\Re[\mathbf{E}(x)e^{-i\omega t}]\\
&= \frac{1}{2}[\mathbf{j}(x)e^{-i\omega t}+\mathbf{j}^*(x)e^{i\omega t}]\cdot\frac{1}{2}[\mathbf{E}(x)e^{-i\omega t}+\mathbf{E}^*(x)e^{i\omega t}]\\
&= \frac{1}{4}[\mathbf{j}(x)\cdot\mathbf{E}(x)e^{-2i\omega t}+\mathbf{j}^*(x)\cdot\mathbf{E}^\star(x)e^{2i\omega t}+\mathbf{j}(x)\cdot\mathbf{E}^*(x)+\mathbf{j}^*(x)\cdot\mathbf{E}(x)]\\
&= \frac{1}{2}\Re[\mathbf{j}(x)\cdot\mathbf{E}(x)e^{-2i\omega t}+\mathbf{j}^*(x)\cdot\mathbf{E}(x)]. \qquad (14.200)
\end{aligned}$$

The time average of this quantity is (averaged over one oscillation period $T = 2\pi/\omega$)

$$\begin{aligned}
\frac{1}{T}\int_0^T \mathbf{j}(x,t)\cdot\mathbf{E}(x,t)dt &= \frac{1}{T}\int_0^T dt\frac{1}{2}\Re[\mathbf{j}(x)\cdot\mathbf{E}(x)e^{-2i\omega t}+\mathbf{j}^*(x)\cdot\mathbf{E}(x)]\\
&= \frac{1}{2}\Re[\mathbf{j}^*(x)\cdot\mathbf{E}(x)]+\frac{1}{2}\Re\left\{\mathbf{j}(x)\cdot\mathbf{E}(x)\frac{1}{T}\int_0^T e^{-2i\omega t}dt\right\}. \qquad (14.201)
\end{aligned}$$

However,

$$\frac{1}{T}\int_0^T e^{-2i\omega t}dt = \frac{1}{-2i\omega T}[e^{-2i\omega t}]_0^{t=2\pi/\omega} = 0. \qquad (14.202)$$

Therefore

$$\frac{1}{T}\int_0^T \mathbf{j}(x,t)\cdot\mathbf{E}(x,t)dt = \frac{1}{2}\Re[\mathbf{j}^*(x)\cdot\mathbf{E}(x)]. \qquad (14.203)$$

Next we consider this expression integrated over a spatial volume $V$, i.e. the time-averaged Ohmic power (see below),

$$\begin{aligned}
\int_V dV\frac{1}{T}\int_0^T \mathbf{j}(x,t)\cdot\mathbf{E}(x,t)dt &= \int_V dV\frac{1}{2}\Re[\mathbf{j}^*(x)\cdot\mathbf{E}(x)]\\
&= \int_V dV\frac{1}{2}\Re[\{\nabla\times\mathbf{H}^*(x)-i\omega\mathbf{D}^*(x)\}\cdot\mathbf{E}(x)].
\end{aligned} \qquad (14.204)$$

In this foregoing step we replaced $\mathbf{j}$ by the Maxwell equation

$$\nabla\times\mathbf{H} = \mathbf{j}+\frac{\partial\mathbf{D}}{\partial t} = \mathbf{j}-i\omega\mathbf{D}(x) \quad \text{with} \quad \mathbf{D}(x,t) = e^{-i\omega t}\mathbf{D}(x). \qquad (14.205)$$

[11]The Ohmic power is $\int dV(1/T)\int_0^T \mathbf{j}\cdot\mathbf{E}dt$ where $\mathbf{j}dV = Id\mathbf{s} = (dq/dt)d\mathbf{s}$.

To go to the next line we use (see Appendix C):

$$\begin{aligned}\boldsymbol{\nabla}\cdot(\mathbf{u}\times\mathbf{v}) &= \mathbf{v}\cdot\mathrm{curl}\,\mathbf{u}-\mathbf{u}\cdot\mathbf{curl}\,\mathbf{v},\\ \boldsymbol{\nabla}\cdot(\mathbf{E}\times\mathbf{H}^*) &= \mathbf{H}^*\cdot(\boldsymbol{\nabla}\times\mathbf{E})-\mathbf{E}\cdot(\boldsymbol{\nabla}\times\mathbf{H}^*),\end{aligned} \tag{14.206}$$

so that we obtain

$$\begin{aligned}\int_V dV\frac{1}{T}\int_0^T \mathbf{j}(x,t)\cdot\mathbf{E}(x,t)dt = \int_V dV\frac{1}{2}\Re[\{-\boldsymbol{\nabla}\cdot(\mathbf{E}(x)\times\mathbf{H}^*(x))\\ +\mathbf{H}^*(x)\cdot(\boldsymbol{\nabla}\times\mathbf{E}(x))\}-i\omega\mathbf{D}^*(x)\cdot\mathbf{E}(x)].\end{aligned} \tag{14.207}$$

In the next step we use the following Maxwell equation divided by $\exp(-i\omega t)$:

$$\boldsymbol{\nabla}\times\mathbf{E}(x,t) = -\frac{\partial\mathbf{B}(x,t)}{\partial t} = i\omega\mathbf{B}(x,t), \tag{14.208}$$

and obtain the expression

$$\begin{aligned}\int_V dV\frac{1}{T}\int_0^T \mathbf{j}(x,t)\cdot\mathbf{E}(x,t)dt = \int_V dV\frac{1}{2}\Re[-\boldsymbol{\nabla}\cdot(\mathbf{E}(x)\times\mathbf{H}^*(x))\\ -i\omega(\mathbf{D}^*(x)\cdot\mathbf{E}(x)-\mathbf{H}^*(x)\cdot\mathbf{B}(x))].\end{aligned} \tag{14.209}$$

One now defines the *complex Poynting vector*[12]

$$\mathbf{S}^c := \frac{1}{2}(\mathbf{E}(x)\times\mathbf{H}^*(x)) \quad \text{watt}\cdot\text{meter}^{-2}, \tag{14.210}$$

and the *complex energy densities*

$$\omega_E = \frac{1}{2}\mathbf{E}(x)\cdot\mathbf{D}^*(x), \quad \omega_M = \frac{1}{2}\mathbf{B}(x)\cdot\mathbf{H}^*(x). \tag{14.211}$$

---

[12]The reader may be irritated by the factor of 1/2 here which we did not have in the definition of the real Poynting vector $S$ in Eq. (7.64). As Jackson [73], p. 242, remarks, for time-averages of products the convention is (cf. Eq. (14.203)) to take one-half of the real part of the product of one complex quantity with the complex conjugate of the other. Cf. Jackson's equations (6.109), p. 237 and (8.47), p. 347. For better clarity we here attach the superscript $c$ to $S$. If complex fields are used to represent the real fields, then the real part of $S^c$ is the time average of $S$; cf. Eq. (14.203). Some authors do not adhere to this convention, e.g. Felsen and Marcuvitz [52], p. 79.

Hence we obtain

$$\int_V dV \frac{1}{T}\int_0^T \mathbf{j}(x,t)\cdot\mathbf{E}(x,t)dt = -\Re\left[\int_V dV\boldsymbol{\nabla}\cdot\mathbf{S}^c + i\omega\int_V dV(\omega_E-\omega_M)\right], \tag{14.212}$$

or

$$\begin{aligned}\int_V dV \frac{1}{T}\int_0^T \mathbf{j}(x,t)\cdot\mathbf{E}(x,t)dt &= \int_V dV\frac{1}{2}\Re[\mathbf{j}^*(x)\cdot\mathbf{E}(x)]\\ &= -\Re\left[\int_{F(V)}\mathbf{S}^c\cdot d\mathbf{F} + i\omega\int_V dV(\omega_E-\omega_M)\right].\end{aligned} \tag{14.213}$$

This result is to be compared with the expression we had in Eq. (7.66). The minus sign in the second contribution originates from the appearance of $\mathbf{j}^*$ (instead of $\mathbf{j}$) in the preceding line. In the case of *lossless conductors* (i.e. with no damping) or dielectric media we have

$$\mathbf{B}(x) = \mu\mathbf{H}(x), \quad \mathbf{D}(x) = \epsilon\mathbf{E}(x), \quad \epsilon,\mu \text{ real}, \tag{14.214}$$

and the energy densities $\omega_E, \omega_M$ are real. In these cases we obtain

$$\int_V dV \frac{1}{T}\underbrace{\int_0^T \mathbf{j}(x,t)\cdot\mathbf{E}(x,t)dt}_{\text{Ohmic power}} = -\int_{F(V)}\Re\mathbf{S}^c\cdot\mathbf{n}dF \quad \text{watts}. \tag{14.215}$$

The left side of the equation represents the work (averaged over one period) done by the field $\mathbf{E}$ on the source current density $\mathbf{j}$ in the volume $V$. The expression $+\int\Re\mathbf{S}^c\cdot\mathbf{n}dF$ represents the corresponding transport of energy through the boundary surface to outside, so that $-\int\Re\mathbf{S}^c\cdot\mathbf{n}dF$ is the corresponding flow into the volume $V$. In other words, the energy flow $-\int\Re\mathbf{S}^c\cdot\mathbf{n}dF$ into the volume $V$, corresponds to the work done by the field $\mathbf{E}$ on a corresponding current density in $V$. If $\epsilon$ and $\mu$ are complex, there are losses of energy which appear in the energy equivalence relation through $\Re[i\omega\int dV(\omega_E-\omega_M)]$.

### 14.9.2 *Application of the complex Poynting vector*

Our first considerations of energy transport assume that the wave guide is made of perfect or ideal conductor material. The primary step is to evaluate the complex Poynting vector for TM and TE waves. In the case of **TM waves** we had in Eq. (14.97):

$$B_z = 0 \text{ everywhere}, \quad \mathbf{B}_\perp = \frac{i\mu\epsilon\omega}{\gamma^2}(\mathbf{e}_z \times \boldsymbol{\nabla}_\perp)E_z, \quad \mathbf{E}_\perp = \frac{ik}{\gamma^2}\boldsymbol{\nabla}_\perp E_z, \tag{14.216}$$

with $k^2 = \mu\epsilon\omega^2 - \gamma^2, \gamma^2 \to \gamma_\lambda^2 = \mu\epsilon\omega_\lambda^2$ as in Eq. (14.122). With these relations and Eq. (14.210), we obtain $(\mathbf{E}(x) := \mathbf{E}(x,y))$

$$\begin{aligned} 2\mathbf{S}^c &= \mathbf{E}(x) \times \mathbf{H}^*(x) = (\mathbf{E}_\perp(x), E_z(x)\mathbf{e}_z) \times (\mathbf{H}_\perp^*(x), 0) \\ &= (E_z(x)\mathbf{e}_z \times \mathbf{H}_\perp^*(x), \mathbf{E}_\perp(x) \times \mathbf{H}_\perp^*(x)) \\ &\equiv (2\mathbf{S}_\perp, 2S_z\mathbf{e}_z), \end{aligned} \tag{14.217}$$

where

$$\begin{aligned} 2\mathbf{S}_\perp^c &= E_z(x)\mathbf{e}_z \times \mathbf{H}_\perp^*(x) \overset{(14.216)}{=} -(E_z(x)\mathbf{e}_z) \times \frac{i\epsilon\omega}{\gamma^2}(\mathbf{e}_z \times \boldsymbol{\nabla}_\perp)E_z \\ &= E_z(x)\frac{i\epsilon\omega}{\gamma^2}\boldsymbol{\nabla}_\perp E_z, \end{aligned} \tag{14.218}$$

where in the last step the vector triple product is multiplied out. This contribution is, in general, purely imaginary, i.e. when $\epsilon, \gamma^2, E_z$ are real. The real part of $\mathbf{S}$ yields — as we have seen with Eq. (14.215) — the flow of energy averaged over one period of oscillation; in general $\mathbf{S}_\perp^c$ therefore does not contribute to this.

We restrict ourselves in the first place to wave guides made of perfect conductor material and *calculate the flow of energy in the longitudinal direction*, i.e. along the axis of the wave guide. In this case (cf. Eq. (14.217))

$$\begin{aligned} 2S_z^c\mathbf{e}_z &= \mathbf{E}_\perp(x) \times \mathbf{H}_\perp^*(x) \overset{(14.216)}{=} \frac{\epsilon\omega k}{\gamma^4}(\boldsymbol{\nabla}_\perp E_z) \times (\mathbf{e}_z \times \boldsymbol{\nabla}_\perp)E_z^* \\ &= \frac{\epsilon\omega k}{\gamma^4}(\boldsymbol{\nabla}_\perp E_z) \cdot (\boldsymbol{\nabla}_\perp E_z)^*\mathbf{e}_z. \end{aligned} \tag{14.219}$$

The flow of energy through the cross section $A$ of the wave guide is therefore

$$\Re\int_A S_z^c\mathbf{e}_z \cdot d\mathbf{F} = \Re\frac{\epsilon\omega k}{2\gamma^4}\int_A (\boldsymbol{\nabla}_\perp E_z) \cdot (\boldsymbol{\nabla}_\perp E_z)^* dF. \tag{14.220}$$

In Chapter 3 we used two Green's theorems of which the first was that of Eq. (3.34),

$$\int_V [\phi\nabla^2\psi + (\boldsymbol{\nabla}\phi) \cdot (\boldsymbol{\nabla}\psi)]dV = \oint_F \phi(\boldsymbol{\nabla}\psi) \cdot d\mathbf{F}. \tag{14.221}$$

Reducing the space dimension by one and making the replacement $\phi \to \phi^*$ we obtain as in Example 3.35:

$$\int_A [\phi^* \nabla_\perp^2 \psi + (\boldsymbol{\nabla}_\perp \phi)^* \cdot (\boldsymbol{\nabla}_\perp \psi)] dF = \oint_C \phi^* (\boldsymbol{\nabla}_\perp \psi) \cdot d\mathbf{l}. \tag{14.222}$$

Hence

$$\int_A S_z^c \mathbf{e}_z \cdot d\mathbf{F} = \frac{\epsilon \omega k}{2\gamma^4} \left[ \underbrace{\oint_C E_z^* (\boldsymbol{\nabla}_\perp E_z) \cdot d\mathbf{l}}_{0} - \int_A E_z^* \nabla_\perp^2 E_z dF \right], \tag{14.223}$$

where the path is to be taken along the surface. In view of the boundary condition $E_z|_{\text{surface}} = 0$ the first integral vanishes. The second integral can be rewritten with the help of Eq. (14.46), i.e.

$$(\boldsymbol{\nabla}_\perp^2 + \gamma^2) E_z(x, y) = 0, \tag{14.224}$$

so that the energy transmitted by the wave guide per unit time is the *power*

$$\langle L \rangle = \Re \int_A S_z^c \mathbf{e}_z \cdot d\mathbf{F} = \Re \frac{\epsilon \omega k}{2\gamma^2} \int_A E_z^*(x, y) E_z(x, y) dF \quad \text{watts}. \tag{14.225}$$

The integral on the right is a normalization integral of $E_z$. Since, as we saw, cf. Eq. (14.152),

$$k = \sqrt{\mu\epsilon}\sqrt{\omega^2 - \omega_\lambda^2}, \quad \gamma^2 \to \gamma_\lambda^2 = \mu\epsilon\omega_\lambda^2, \tag{14.226}$$

it follows that the power of the flow of energy along the axis of the wave guide is

$$\int_A S_z^c \mathbf{e}_z \cdot d\mathbf{F} = \frac{1}{2}\sqrt{\frac{\epsilon}{\mu}} \left(\frac{\omega}{\omega_\lambda}\right)^2 \sqrt{1 - \frac{\omega_\lambda^2}{\omega^2}} \int_A E_z^*(x, y) E_z(x, y) dF \quad \text{watts}. \tag{14.227}$$

For **TE waves** one obtains an analogous expression; in this case, however, the boundary condition which enters is $\partial B_z / \partial n|_{\text{surface}} = 0$.

### 14.9.3 *Attenuation of wave guides ($\sigma \neq 0$)*

Our next step is to consider the *attenuation* of wave guides, i.e. the loss of energy along the wave guide when the walls are made of a metal with *finite conductivity* (implying complex $k$). In the case of finite conductivity we observed earlier, the surface current (density) $K$ does not exist (see discussion after Eq. (11.18)). However, we then have the current density $\mathbf{j}$

in the non-perfect conductor, and we can define an *effective surface current* (density) $\mathbf{K}_{\text{eff}}$ by the relation

$$\mathbf{K}_{\text{eff}} = \int \mathbf{j}(\rho)d\rho, \qquad (14.228)$$

where $\rho$ is the depth of the current $\mathbf{j}(\rho)$ below the surface. We derived earlier (cf. Eq. (7.61)) the following expression for the Ohmic power of a conductor with conductivity $\sigma$:

$$L = \int \mathbf{j} \cdot \mathbf{E} dV \quad \text{watts.} \qquad (14.229)$$

Averaging this over one period of oscillation for real $\epsilon$ and $\mu$ of the interior of the wave guide, and independent of the frequency, we obtain with the complex Poynting vector (14.210)

$$\langle L \rangle = -\int_{F(V)} \Re \mathbf{S}^c \cdot \mathbf{n} dF = -\frac{1}{2}\int \Re \mathbf{n} \cdot \mathbf{E} \times \mathbf{H}^* dF \quad \text{watts,} \qquad (14.230)$$

so that

$$\frac{d\langle L \rangle}{dF} = -\frac{1}{2}\Re[\mathbf{n} \cdot \mathbf{E} \times \mathbf{H}^*] \quad \text{watts/meter}^2. \qquad (14.231)$$

For $\mathbf{E}$ we have to substitute here the field beyond the (hollow) interior of the wave guide. We calculate this as follows. Neglecting the displacement current we obtain from Maxwell's two "curl" equations (for inside the metal)

$$\boldsymbol{\nabla} \times \mathbf{H} = \mathbf{j}, \quad \boldsymbol{\nabla} \times \mathbf{E} = -\frac{\partial \mathbf{B}}{\partial t} \qquad (14.232)$$

with

$$\mathbf{E}, \mathbf{H} \propto e^{-i\omega t}, \quad \mathbf{B} = \mu \mathbf{H}, \quad \mathbf{j} = \sigma \mathbf{E},$$

the relations

$$\mathbf{E} = \frac{1}{\sigma}\boldsymbol{\nabla} \times \mathbf{H}, \quad \mathbf{H} = \frac{1}{i\mu\omega}\boldsymbol{\nabla} \times \mathbf{E}. \qquad (14.233)$$

Inserting one equation into the other we obtain

$$\mathbf{H} = \frac{1}{i\omega\mu\sigma}\boldsymbol{\nabla} \times (\boldsymbol{\nabla} \times \mathbf{H}) = \frac{1}{i\omega\mu\sigma}[\boldsymbol{\nabla} \underbrace{(\boldsymbol{\nabla} \cdot \mathbf{H})}_{0} - \nabla^2 \mathbf{H}],$$

so that

$$\boldsymbol{\nabla}^2 \mathbf{H} + i\omega\mu\sigma \mathbf{H} = 0. \qquad (14.234)$$

We consider a *wave guide with circular cross section* of radius $a$. We let $-\rho\mathbf{n}$ be the coordinate pointing *into* the conductor or wall, and we make the assumption that the variation of $\mathbf{H}$ in the plane perpendicular to $\mathbf{n}$ is so slow that we have in good approximation

$$\boldsymbol{\nabla}^2 \simeq \frac{\partial^2}{\partial \rho^2} \quad \text{(assumption).} \tag{14.235}$$

We then have for the field *inside* the conductor material

$$\frac{\partial^2}{\partial \rho^2}\mathbf{H} + i\omega\mu\sigma\mathbf{H} = 0 \tag{14.236}$$

and (with $\alpha = \text{const.} = a$ and $\sqrt{i} = (1+i)/\sqrt{2}$)

$$\begin{aligned}\mathbf{H}(x,y,z,t) &= \mathbf{H}_0 e^{-i\omega t} e^{ikz} e^{\pm i\sqrt{i\omega\mu\sigma}(\rho - a)} \\ &= \mathbf{H}_0 e^{-i\omega t} e^{ikz} e^{-(\rho-a)/\rho_0} e^{i(\rho-a)/\rho_0} \\ &\propto \exp[(i-1)(\rho-a)/\rho_0],\end{aligned} \tag{14.237}$$

where $\mathbf{H}_0 = \text{const.}$ and

$$\rho_0 = \frac{\sqrt{2}}{\sqrt{\omega\mu\sigma}} \quad \text{meters} \tag{14.238}$$

is the *skin depth*. For $\mathbf{E}$ we have to the same approximation

$$\mathbf{E} = \frac{1}{\sigma}\boldsymbol{\nabla} \times \mathbf{H} \simeq \frac{1}{\sigma}\boldsymbol{\nabla}_\rho \times \mathbf{H} \parallel \mathbf{e}_z \quad \text{for TM.} \tag{14.239}$$

We note that $\mathbf{H}_t = \mathbf{H}_0 e^{-i\omega t} e^{ikz}$ is the tangential field at (just outside) the surface of the conductor with $\rho = a$ as indicated in Fig. 14.8.

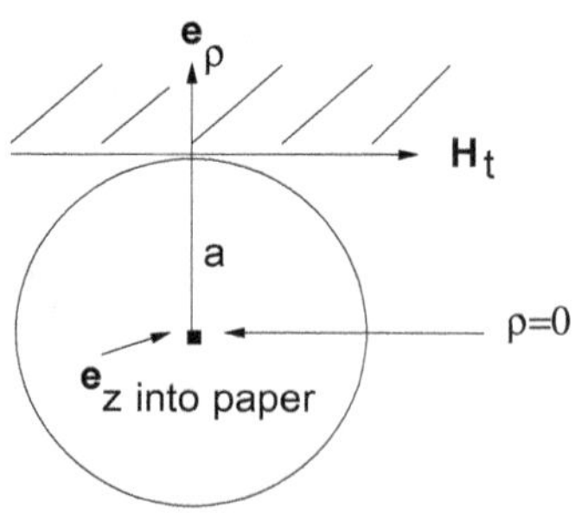


**Fig. 14.8.** The directions of vectors.

Now,

$$\nabla_\rho = -\mathbf{n}\frac{\partial}{\partial\rho} \equiv \mathbf{e}_\rho\frac{\partial}{\partial\rho}. \tag{14.240}$$

Here **n** is the unit vector pointing vertically out of the conductor. Then in the conductor (and just outside for $\rho = a$) with Eq. (14.237)

$$\mathbf{E}_t = \underbrace{\mathbf{E}}_{\parallel\, \mathbf{e}_z \text{ for TM}} = -\frac{1}{\sigma}(\mathbf{n}\times\mathbf{H}_t)\frac{(i-1)}{\rho_0} \propto \frac{1}{\sqrt{\sigma}}e^{[(i-1)(\rho-a)/\rho_0]} \tag{14.241}$$

(which is a very small electric field since $\sigma$ is large but finite). The effective surface current density is

$$\begin{aligned}\mathbf{K}_{\text{eff}} &= \int_a^\infty \mathbf{j}d\rho = \int_a^\infty \sigma\mathbf{E}d\rho = -\int_a^\infty \frac{\sigma}{\sigma\rho_0}(\mathbf{n}\times\mathbf{H})(i-1)d\rho \\ &= -\left[\frac{1}{\rho_0}(\mathbf{n}\times\mathbf{H})\frac{(i-1)}{(i-1)}\rho_0\right]_a^\infty,\end{aligned} \tag{14.242}$$

where in the last step one integrates over the two exponentials in Eq. (14.237) with respect to $\rho$ leaving exponentials and hence again **H**, so that with $\rho_0 = a$,

$$\mathbf{K}_{\text{eff}} = (\mathbf{n}\times\mathbf{H}_t), \quad \mathbf{H}_t = \mathbf{H}_0e^{-i\omega t}e^{ikz}, \quad \mathbf{H}_t = \mathbf{H}(\rho = a, z, t). \tag{14.243}$$

One may observe that this relation has the same form as that for the surface current density $K$ in the case of conductivity $\sigma = \infty$ (cf. Eq. (11.21)).[13] We now have

$$\begin{aligned}|\mathbf{K}_{\text{eff}}|^2 &= (\mathbf{n}\times\mathbf{H}_t)\cdot(\mathbf{n}\times\mathbf{H}_t^*) \equiv \mathbf{n}\cdot[\mathbf{H}_t\times(\mathbf{n}\times\mathbf{H}_t^*)] \\ &= \mathbf{n}\cdot[(\mathbf{H}_t\cdot\mathbf{H}_t^*)\mathbf{n} - (\mathbf{H}_t\cdot\mathbf{n})\mathbf{H}_t^*] \\ &\simeq |\mathbf{H}_t|^2.\end{aligned} \tag{14.244}$$

Since from (the reverse of Eq. (14.241)) the second of Eqs. (14.233),

$$\mathbf{H}_t = \frac{1}{i\omega\mu}\nabla\times\mathbf{E}_t \tag{14.245}$$

in the conductor, it follows that in this leading approximation

$$\mathbf{H}_t\cdot\mathbf{n} \simeq \frac{1}{i\omega\mu}\left\{\left(-\mathbf{n}\frac{\partial}{\partial\rho}\right)\times\mathbf{E}_t\right\}\cdot\mathbf{n} = 0, \tag{14.246}$$

whereas the vector product with **n** is $K_{\text{eff}}$ above so that $\mathbf{H}_t = \mathbf{H}$.

---

[13]Thus here at $\rho = a$ (surface) $\mathbf{E}_t$ (from Eq. (14.241) $= Z_s\mathbf{K}_{\text{eff}}$ (from Eq. (14.243)), where $Z_s = (1-i)\sqrt{\mu\omega/2\sigma} = (1-i)/\sigma\rho_0$. This quantity $Z_s$ is known as *surface impedance.*

We now calculate $\langle L\rangle$, the average of the Ohmic power of the wave guide over one period of oscillation, by inserting in Eq. (14.231) the expression (14.241) for $\mathbf{E}$ in the conductor. We have

$$\frac{d\langle L\rangle}{dF} = -\frac{1}{2}\Re[\mathbf{n}\cdot(\mathbf{E}\times\mathbf{H}^*)] \quad \text{watts/meter}^2, \tag{14.247}$$

and so (since $\mathbf{H} = \mathbf{H}_t$)

$$\begin{aligned}\frac{d\langle L\rangle}{dF} &\overset{(14.241)}{=} \frac{1}{2\sigma\rho_0}\Re[\,\underbrace{\underbrace{\mathbf{n}\cdot\{(\mathbf{n}\times\mathbf{H})\times\mathbf{H}^*\}}_{-\mathbf{n}\cdot\{\mathbf{H}^*\times(\mathbf{n}\times\mathbf{H})\}}}_{-\mathbf{n}\cdot\{(\mathbf{H}^*\cdot\mathbf{H})\mathbf{n}-(\mathbf{H}^*\cdot\mathbf{n})\mathbf{H}\}}(i-1)] \\ &= \frac{1}{2\sigma\rho_0}\Re[-(i-1)|\mathbf{H}|^2] = \frac{1}{2\sigma\rho_0}|\mathbf{H}|^2,\end{aligned} \tag{14.248}$$

and hence

$$\frac{d\langle L\rangle}{dF} \simeq \frac{1}{2\sigma\rho_0}|\mathbf{H}|^2 \quad \text{watts/meter}^2. \tag{14.249}$$

Here, as indicated in Fig. 14.9, $dF = dzdl$, where $dl$ is an element of the path around the surface of the wave guide, so that (with the minus sign indicating that the expression represents a rate of loss)

$$\begin{aligned}\frac{d\langle L\rangle}{dz} &= -\frac{1}{2\sigma\rho_0}\oint_C |\mathbf{H}|^2 dl = -\frac{1}{2\sigma\rho_0}\oint_C |\mathbf{H}_t|^2 dl \\ &= -\oint_C \frac{1}{2\sigma\rho_0}|\mathbf{K}_{\text{eff}}|^2 dl = -\oint_C \frac{1}{2\sigma\rho_0}|\mathbf{n}\times\mathbf{H}_t|^2 \underbrace{dl}_{ad\varphi}.\end{aligned} \tag{14.250}$$

This is the loss of energy per unit time and per unit length of the wave guide as a consequence of the finite conductivity $\sigma$ of its walls (remember $k$ is complex). For $\sigma \to \infty$ the right hand side of the attenuation rate (14.250) vanishes, as expected.

The integral is a line integral in the surface around the wave guide as indicated in Fig. 14.9. The expression for $d\langle L\rangle/dz$ involves only the value of the field $\mathbf{H}$ at the surface of the conductor.

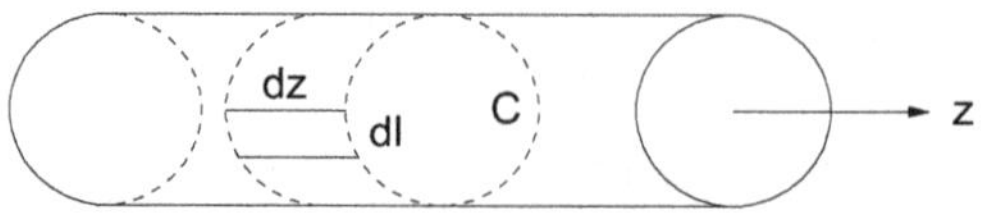


**Fig. 14.9.** The surface element $dF = dzdl$.

We have seen in Section 9.4 that the dispersion relation for conductors is (cf. Eq. (9.40))

$$k^2 = \mu\epsilon\omega^2 \left(1 + i\frac{\sigma}{\omega\epsilon}\right). \tag{14.251}$$

We now have to distinguish between the material of the walls and that of the dielectric inside the wave guide. The walls have a high conductivity so that the skin depth is as small as possible and the wave guide is comparable to one made of ideal conductor material. On the other hand one wants the dielectric of the interior to have a conductivity as small as possible. Our earlier investigation of the behavior of the conductivity $\sigma$ in Section 9.4.2 was based on a simple model consideration for the calculation of the frequency dependence of $\sigma$. These considerations do not take into account for instance a possible dependence on boundary conditions or on the geometry of the conductor. In the following we shall argue that if $\sigma$ is finite and large, and (as we assume) independent of the frequency, the propagation vector $\mathbf{k}$ of the electromagnetic wave develops an imaginary part due to the fact that — in this case of finite conductivity $\sigma$ — the resistivity is finite so that damping is present and hence attenuation takes place. Thus we write

$$k = \sqrt{k^2} = k_R + ik_I \tag{14.252}$$

in the case of penetration into the walls of the wave guide. The propagation of the fields $\mathbf{E}, \mathbf{H} \propto e^{ikz}$ is therefore damped. The field $\mathbf{H}$ enters the power $\langle L\rangle$ quadratically. Hence we can write

$$\langle L\rangle = e^{-2k_I z}\langle L\rangle_0, \tag{14.253}$$

or

$$\frac{d\langle L\rangle}{dz} = -2k_I\langle L\rangle, \quad \textit{i.e.} \quad k_I = -\frac{1}{2\langle L\rangle}\frac{d\langle L\rangle}{dz} \quad \text{meter}^{-1}. \tag{14.254}$$

The quantity $k_I$ is the socalled *attenuation constant*. The intention is, of course, to achieve $k_I$ as small as possible for an optimal use of the wave guide. If the interior of the wave guide is filled with a dielectric which has a small conductivity $\sigma_0$, this naturally also contributes to some damping, which we can take into account in the above dispersion relation. In this case we have *inside the wave guide* ("combining" Eq. (14.251) and $k^2 = \mu\epsilon\omega^2(1-\omega_\lambda^2/\omega^2)$)

$$k^2 = \mu\epsilon\omega^2 \left[1 - \frac{\omega_\lambda^2}{\omega^2}\right](1 + O(\sigma_0)). \tag{14.255}$$

In the next section we investigate an evaluation of the attenuation constant, and how this can be minimized for an optimal use of the wave guide.

### 14.9.4 *Optimal use of a wave guide*

Our next objective is a model evaluation of the attenuation constant $k_I$ given by Eq. (14.254) for the case of a wave guide with circular cross section and filled with an ideal dielectric ($\sigma_0 = 0$ in Eq. (14.255)). This requires the calculation of $d\langle L\rangle/dz$. Assuming the conductivity of the walls of the wave guide is close to that of a perfect conductor, we can take (14.225) for the power $\langle L\rangle$. As an example we consider the case of **TM waves.** In this case we have (see Eq. (14.97))

$$\mu\mathbf{H}_\perp = \mathbf{B}_\perp = \frac{i\mu\epsilon\omega}{\gamma^2}(\mathbf{e}_z \times \boldsymbol{\nabla}_\perp)E_z, \quad B_z = 0. \tag{14.256}$$

Here $E_z$ is to be obtained as solution of the equation (cf. Eq. (14.104))

$$(\boldsymbol{\nabla}_\perp^2 + \gamma^2)E_z(x,y) = 0 \tag{14.257}$$

with boundary condition $E_z|_{\text{surface}} \simeq O(1/\sqrt{\sigma})$ (see Eq. (14.47)).[14] Thus our calculations will be concerned with $E_z$ and its derivatives. In the case of the wave guide with circular cross section of radius $a$ we had (cf. Eq. (14.166))

$$E_z(\rho,\varphi) = AJ_m(\gamma\rho)e^{im\varphi}, \quad E_z|_{\text{surface}} \propto J_m(\gamma a) \simeq 0. \tag{14.258}$$

Hence we have in this case (TM: with $\mathbf{H}_z = 0$)

$$-\mathbf{n}\times\mathbf{H}_\perp = -\mathbf{n}\times\frac{i\epsilon\omega}{\gamma^2}(\mathbf{e}_z\times\boldsymbol{\nabla}_\perp)E_z = -\frac{i\epsilon\omega}{\gamma^2}\mathbf{e}_z(\mathbf{n}\cdot\boldsymbol{\nabla}_\perp E_z). \tag{14.259}$$

Here the transversal gradient is given by

$$\boldsymbol{\nabla}_\perp = \mathbf{e}_\rho\frac{\partial}{\partial\rho} + \mathbf{e}_\varphi\frac{1}{\rho}\frac{\partial}{\partial\varphi}, \tag{14.260}$$

and hence (cf. Eq. (14.240))

$$-\mathbf{n}\times\mathbf{H}_\perp = \frac{i\epsilon\omega}{\gamma^2}\mathbf{e}_z\frac{\partial E_z}{\partial\rho}, \tag{14.261}$$

---

[14]The condition $E_t^{(1)} - E_t^{(2)} = 0$ implies when $\sigma \neq \infty$, $E_t^{(1)} = O(1/\sqrt{\sigma})$ as we observed in Eq. (14.241). For very good conductors we can replace the right hand side by zero (Jackson [73], p. 351). For inclusion of such corrections see Jackson [73], Section 8.6, Perturbation of Boundary Conditions.

which is a vector along $\mathbf{e}_z$ derived from $E_z$ (recall we have $H_z = 0$ here). We can call this expression $\mathbf{H}^{\text{eff}}$ and insert it into Eq. (14.250) to obtain the loss of power which this gives rise to along the wave guide. Thus

$$\begin{aligned}\frac{d\langle L\rangle}{dz} &\overset{(14.250)}{=} -\oint_C \frac{1}{2\sigma\rho_0}\left(\frac{\epsilon\omega}{\gamma^2}\right)^2 \left|\frac{\partial E_z}{\partial\rho}\right|^2_{\rho=a} ad\varphi \\ &= -2\pi\frac{a(\epsilon\omega)^2}{2\sigma\rho_0\gamma^4}\left|\frac{\partial E_z}{\partial\rho}\right|^2_{\rho=a} \\ &\overset{(14.258)}{=} -\frac{\pi a(\epsilon\omega)^2}{\sigma\rho_0\gamma^2}|A|^2\left(\frac{\partial J_m(z)}{\partial z}\right)^2_{z=\gamma a} \\ &\equiv \tilde{A}. \end{aligned}\tag{14.262}$$

For the power $\langle L\rangle$ we derived the expression given by Eq. (14.225). We use this here with $dF = \rho d\varphi d\rho$ for the calculation of the main part of power arising from $E_z$ of Eq. (14.258):

$$\begin{aligned}L &= \Re\int_A S_z^c \mathbf{e}_z \cdot d\mathbf{F} = \Re\frac{\epsilon\omega k}{2\gamma^2}\int_A \rho d\varphi d\rho E_z{}^*(\rho,\varphi)E_z(\rho,\varphi) \\ &= \Re\frac{\epsilon\omega k}{2\gamma^2}2\pi\int_{\rho=0}^a \rho d\rho |A|^2[J_m(\gamma\rho)]^2 \equiv \langle L\rangle \equiv \tilde{B},\end{aligned}\tag{14.263}$$

where (see Eq. (14.255)) $\Re k \simeq \sqrt{\mu\epsilon}\sqrt{\omega^2 - \omega_\lambda^2}$, if we neglect a (generally small) boundary-dependent contribution to $\Re k$. We obtain therefore with Eq. (14.254):

$$\begin{aligned}k_I &= -\frac{1}{2\langle L\rangle}\frac{d\langle L\rangle}{dz} \equiv -\frac{1}{2}\frac{\tilde{A}}{\tilde{B}} = \frac{1}{2}\frac{\pi a(\epsilon\omega)^2}{\sigma\rho_0\gamma^2}\frac{\gamma^2}{\pi\epsilon\omega\Re k}\frac{(\partial J_m(z)/\partial z)^2_{z=\gamma a}}{\int_0^a \rho d\rho[J_m(\gamma\rho)]^2} \\ &= \frac{\epsilon\omega a}{2\sigma\rho_0\Re k}\frac{(\partial J_m(z)/\partial z)^2_{z=\gamma a}}{\int_0^a \rho d\rho[J_m(\gamma\rho)]^2} \quad \text{meter}^{-1}.\end{aligned}\tag{14.264}$$

In order to keep the attenuation, i.e. the loss of energy, as small as possible, one wants to minimize this expression or, put differently, one wants to determine those modes for which $k_I$ is minimal. We are therefore interested in the behavior of this expression as a function of the frequency $\omega$ ($\gamma^2 = \mu\epsilon\omega^2 - k^2$). Thus we obtain with the substitution of

$$\rho_0 = \rho_{0\lambda}\sqrt{\frac{\omega_\lambda}{\omega}}\tag{14.265}$$

(since the skin depth $\rho_0 \propto 1/\sqrt{\omega}$, as we saw in Eq. (14.238)) and

$$\gamma \to \gamma_\lambda, \quad \lambda = \{m, n\},$$

the expression

$$k_I = \frac{\epsilon a}{2\sigma\rho_{0\lambda}} \frac{\omega}{\sqrt{\omega_\lambda/\omega}} \frac{1}{\sqrt{\mu\epsilon}\sqrt{\omega^2 - \omega_\lambda^2}} f(\gamma_\lambda), \tag{14.266}$$

where

$$f(\gamma_\lambda) = \left[\frac{(\partial J_m(z)/\partial z)^2_{z=\gamma a}}{\int_0^a \rho d\rho [J_m(\gamma\rho)]^2}\right]_{\gamma_\lambda}. \tag{14.267}$$

Thus

$$k_I = \frac{a\sqrt{\epsilon/\mu}}{2\sigma\rho_{0\lambda}} \frac{\omega\sqrt{\omega/\omega_\lambda}}{\sqrt{\omega^2 - \omega_\lambda^2}} f(\gamma_\lambda) = \frac{a\sqrt{\epsilon/\mu}}{2\sigma\rho_{0\lambda}} \frac{\sqrt{\omega/\omega_\lambda}}{\sqrt{1 - \omega_\lambda^2/\omega^2}} f(\gamma_\lambda) \overset{\sigma\to\infty}{\to} 0. \tag{14.268}$$

The best use is made of the wave guide if it is used for those modes whose eigenfrequencies are closest to the frequency at the minimum of $k_I$. The frequency at the minimum of the curve in Fig. 14.10 is determined by equating the derivative of $k_I$ to zero, i.e.

$$\frac{d}{d\omega}\left[\frac{\sqrt{\omega}}{\sqrt{1 - \omega_\lambda^2/\omega^2}}\right] = 0, \tag{14.269}$$

i.e.

$$\left(1 - \frac{\omega_\lambda^2}{\omega^2}\right) = \omega\left(2\frac{\omega_\lambda^2}{\omega^3}\right), \quad \frac{\omega}{\omega_\lambda} = \sqrt{3}. \tag{14.270}$$

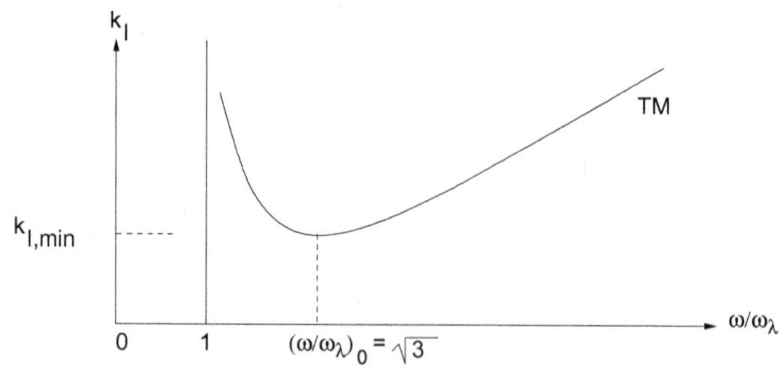


**Fig. 14.10.** The behavior of $k_I$ for TM modes.

The behavior of the attenuation constant $k_I$ as a function of the frequency $\omega$ near this minimum is shown in Fig. 14.10.

## 14.10 Resonators (Closed Wave Guides)

In the case of resonators or cavities we have additional boundary conditions at values of $z$, where the two ends of the resonator have their closures, and the boundary conditions there supply a third integer (recall that our considerations in Section 14.6 with the enforcement of boundary conditions, led to wave guide modes characterized by two integers). All these integers arise in analogy to quantum numbers in quantum mechanics, there each corresponding to quantization of one degree of freedom. Here, of course, we are dealing with macroscopic physics, and the analogy is restricted to that of the mathematical eigenvalue problem. We assume that the cylindrical resonator is closed at both ends by plates made of the same conductor material as the body of the resonator. The electromagnetic waves can now also be reflected at the two ends, hence the name resonator. Of course, a resonator can have any other shape, but an arbitrary shape is difficult to handle calculationally. A novel resonant cavity is, as also Jackson points out (Jackson [73], p. 334), the earth–ionosphere system (here treated in the preceding chapter).

We begin with **TM modes**, with time-dependence $e^{-i\omega t}$ (as before). In this case we have $\mathbf{B}_z = 0$ (everywhere), and we make the ansatz

$$\mathbf{E}_z = \psi(x,y)(A\sin kz + B\cos kz)e^{-i\omega t}\mathbf{e}_z. \tag{14.271}$$

As before we obtain the transverse components from the relations derived earlier, i.e. Eqs. (14.29) and (14.30), and hence here, with $\gamma^2 = \epsilon\mu\omega^2 - k^2$, from

$$\begin{aligned}\mathbf{E}_\perp &= \frac{k}{\gamma^2}\boldsymbol{\nabla}_\perp\psi(x,y)(A\cos kz - B\sin kz)e^{-i\omega t},\\ \mathbf{B}_\perp &= \frac{1}{\gamma^2}i\mu\epsilon\omega\mathbf{e}_z\times\boldsymbol{\nabla}_\perp\psi(x,y)(A\sin kz + B\cos kz)e^{-i\omega t}.\end{aligned} \tag{14.272}$$

In view of the general boundary condition $E_t^{(1)} = E_t^{(2)}$ we have at the ends: Inside $\mathbf{E}_\perp = 0$ for ideal conductors. Hence we have, with the geometry of the resonator as shown in Fig. 14.11, but not necessarily one of circular cross section,

$$\mathbf{E}_\perp(z=0) = 0, \quad \mathbf{E}_\perp(z=d) = 0, \tag{14.273}$$

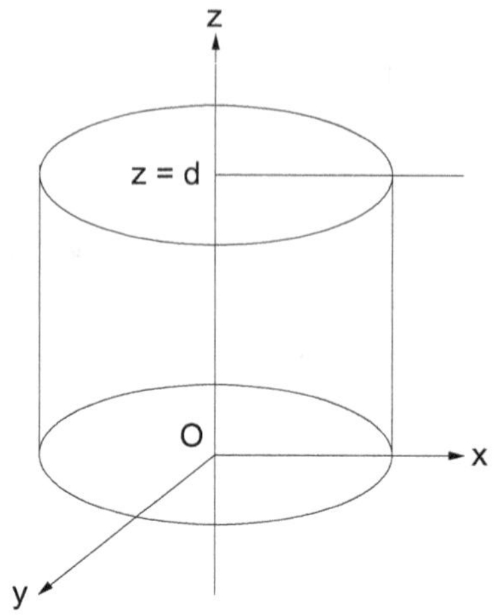


**Fig. 14.11.** The resonator.

and hence from Eq. (14.272):

$$A = 0 \quad \text{and} \quad \sin kd = 0, \quad \text{i.e.} \quad k = \frac{l\pi}{d},\ l = 0, \pm 1, \pm 2, \ldots. \tag{14.274}$$

With the parameters $A$ and $k$ thus determined we have for the electric field from Eqs. (14.271) and (14.272),

$$\begin{aligned} \mathbf{E}_z &= B \cos\left(\frac{l\pi z}{d}\right) \psi(x,y) e^{-i\omega t} \mathbf{e}_z, \\ \mathbf{E}_\perp &= -B \frac{l\pi}{d\gamma^2} \sin\left(\frac{l\pi z}{d}\right) \boldsymbol{\nabla}_\perp \psi(x,y) e^{-i\omega t}, \end{aligned} \tag{14.275}$$

and for the magnetic field

$$\begin{aligned} \mathbf{B}_z &= 0, \\ \mathbf{B}_\perp &= B \frac{i\mu\epsilon\omega}{\gamma^2} \cos\left(\frac{l\pi z}{d}\right) \mathbf{e}_z \times \boldsymbol{\nabla}_\perp \psi(x,y) e^{-i\omega t}. \end{aligned} \tag{14.276}$$

For **TE modes** we have $\mathbf{E}_z = 0$ (everywhere), and we make the ansatz

$$\mathbf{B}_z = \psi(x,y)(A \sin kz + B \cos kz) \mathbf{e}_z e^{-i\omega t}. \tag{14.277}$$

At the end faces of the cylindrical resonator $\mathbf{B}_n$ is continuous (this condition $B_n^{(1)} = B_n^{(2)}$ is again of general validity), but outside the field is zero (as a consequence of the skin effect the fields do not penetrate into the metal except at extremely high frequencies). Hence we have $\mathbf{B}_z(z=0)=0,\ \mathbf{B}_z(z=d)=0$ implying

$$B=0, \quad k=\frac{l\pi}{d}, \quad l=0,\pm 1,\pm 2,\ldots. \tag{14.278}$$

We thus obtain for the electric field with Eq. (14.29)

$$\begin{aligned}\mathbf{E}_z &= 0,\\ \mathbf{E}_\perp &= -A\frac{i\omega}{\gamma^2}\sin\left(\frac{l\pi z}{d}\right)\mathbf{e}_z\times\nabla_\perp\psi(x,y)e^{-i\omega t},\end{aligned} \tag{14.279}$$

and for the magnetic field with Eq. (14.30)

$$\begin{aligned}\mathbf{B}_z &= A\psi(x,y)\sin\left(\frac{l\pi z}{d}\right)e^{-i\omega t}\mathbf{e}_z,\\ \mathbf{B}_\perp &= A\frac{l\pi}{d\gamma^2}\cos\left(\frac{l\pi z}{d}\right)\nabla_\perp\psi(x,y)e^{-i\omega t}.\end{aligned} \tag{14.280}$$

Since $\mathbf{E}_z$ (TM) and $\mathbf{B}_z$ (TE) satisfy the equation (cf. Eqs. (14.9), (14.10))

$$(\triangle_\perp+\gamma^2)\left\{\begin{matrix}\mathbf{E}_z\\ \mathbf{B}_z\end{matrix}\right. = 0, \tag{14.281}$$

with boundary conditions $E_z|_{\text{surface}}=0$ and $(\partial B_z/\partial n)|_{\text{surface}}=0$ (cf. Eqs. (14.47)), it follows, that the scalar function $\psi(x,y)$ is in these cases solution of

$$(\triangle_\perp+\gamma^2)\psi(x,y)=0 \tag{14.282}$$

with the boundary condition

$$\psi|_{\text{surface}}=0 \quad \text{or} \quad \left.\frac{\partial\psi}{\partial n}\right|_{\text{surface}}=0, \tag{14.283}$$

where in both cases

$$\gamma^2=\epsilon\mu\omega^2-k^2=\epsilon\mu\omega^2-\left(\frac{l\pi}{d}\right)^2. \tag{14.284}$$

The equation for $\psi$ is explicitly

$$\left(\frac{d^2}{dx^2}+\frac{d^2}{dy^2}+\gamma^2\right)\psi(x,y)=0. \tag{14.285}$$

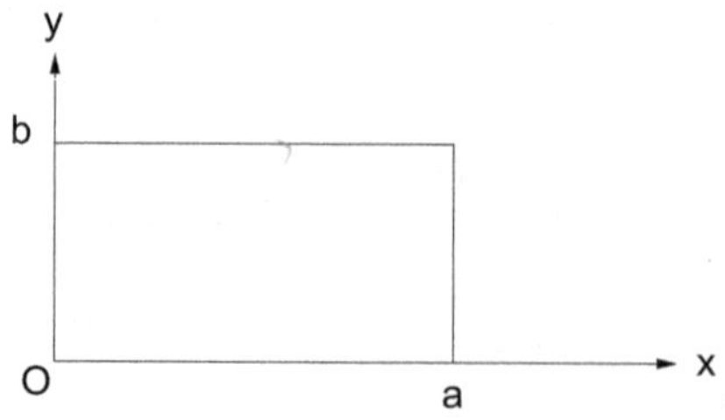


**Fig. 14.12.** Cross section of a rectangular resonator.

We now choose a *resonator of rectangular cross section* as illustrated in Fig. 14.12. The condition $\psi|_{\text{surface}} = 0$ implies that

- $\psi = 0$ on planes $x = 0, a, \quad y \in [0, b]$, and
- $\psi = 0$ on planes $y = 0, b, \quad x \in [0, a]$.

We choose the solution such that these conditions are satisfied, i.e.

$$\psi \propto \sin\left(\frac{m\pi x}{a}\right) \sin\left(\frac{n\pi y}{b}\right), \tag{14.286}$$

where $m, n$ are integers. Inserting this expression into Eq. (14.285) for $\psi$, we obtain

$$-\left(\frac{m\pi}{a}\right)^2 - \left(\frac{n\pi}{b}\right)^2 + \gamma^2 = 0,$$

or, with Eq. (14.284),

$$\epsilon\mu\omega^2 = \left(\frac{l\pi}{d}\right)^2 + \left(\frac{m\pi}{a}\right)^2 + \left(\frac{n\pi}{b}\right)^2. \tag{14.287}$$

We infer from this relation that the *eigenfrequencies* $\omega = \omega_{lmn}$ can be changed by changing the lengths $a, b$.

In the case of a *cylindrical resonator with circular cross section* of radius $R$ as illustrated in Fig. 14.13, the function $\psi = \psi(\rho, \varphi)$ is given by the two-dimensional Laplace equation

$$(\triangle_\perp + \gamma^2)\psi(\rho, \varphi) = 0, \quad \triangle_\perp = \frac{\partial^2}{\partial \rho^2} + \frac{1}{\rho}\frac{\partial}{\partial \rho} + \frac{1}{\rho^2}\frac{\partial^2}{\partial \varphi^2}. \tag{14.288}$$

Using again the method of separation of variables and *demanding*, that

$$\psi(\rho, \varphi) = \psi(\rho, \varphi + 2m\pi), \tag{14.289}$$

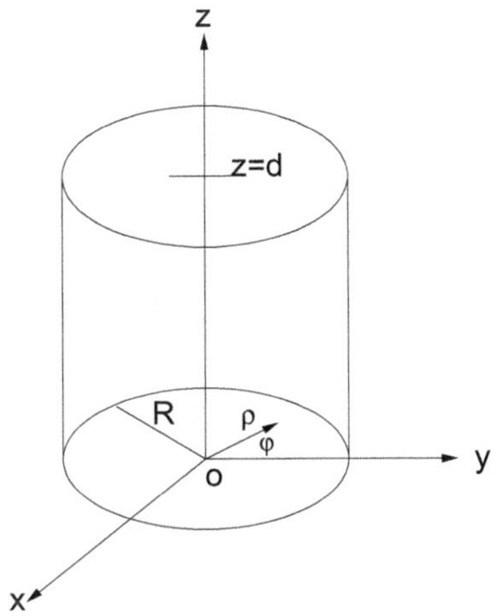


**Fig. 14.13.** The resonator with circular cross section.

we obtain

$$\psi(\rho,\varphi)=\psi(\rho)e^{im\varphi},\quad m=0,1,2,\ldots \tag{14.290}$$

with

$$\left(\frac{\partial^2}{\partial\rho^2}+\frac{1}{\rho}\frac{\partial}{\partial\rho}+\gamma^2-\frac{m^2}{\rho^2}\right)\psi(\rho)=0, \tag{14.291}$$

which is the *Bessel differential equation*. We thus have

$$\psi(\rho,\varphi)\propto J_m(\gamma\rho)e^{im\varphi}, \tag{14.292}$$

and with Eq. (14.275):

$$\mathbf{TM}:\quad \mathbf{E}_z=BJ_m(\gamma\rho)e^{im\varphi}\cos\left(\frac{l\pi z}{d}\right)e^{-i\omega t}\mathbf{e}_z. \tag{14.293}$$

The boundary condition

$$\psi(\gamma\rho)|_{\rho=R}=0\quad \text{implies}\quad J_m(\gamma R)=0. \tag{14.294}$$

We let $x_{mn}$ be the $n$th zero $\neq 0$ ($n=1,2,3,\ldots$) of $J_m(x)$. Then $\gamma R=x_{mn}$, i.e.

$$\gamma\equiv\gamma_{mn}=\frac{x_{mn}}{R}. \tag{14.295}$$

For $\omega$ we obtain, since $\gamma^2 = \epsilon\mu\omega^2 - (l\pi/d)^2$,

$$\omega^2 = \frac{1}{\epsilon\mu}\left[\gamma^2 + \left(\frac{l\pi}{d}\right)^2\right] = \frac{1}{\epsilon\mu}\left[\left(\frac{x_{mn}}{R}\right)^2 + \left(\frac{l\pi}{d}\right)^2\right]. \tag{14.296}$$

The *lowest frequency* is $\omega_0$:

$$\omega_0^2 = \frac{1}{\epsilon\mu}\left(\frac{x_{01}}{R}\right)^2, \quad (m = 0, n = 1, l = 0), \quad \text{where} \quad x_{01} \sim 2.4048. \tag{14.297}$$

The zeros of the lowest order Bessel functions are conveniently obtained from formulas given by Jahnke and Emde [74], p. 143 (actually the asymptotic expansion (14.168)):

$$\begin{aligned} J_0(x_n) = 0, \quad \frac{x_n}{\pi} &= n - \frac{1}{4} + \frac{0.050661}{4n-1} - \frac{0.053041}{(4n-1)^3} + \frac{0.262051}{(4n-1)^5} - \cdots, \\ J_1(x_n) = 0, \quad \frac{x_n}{\pi} &= n + \frac{1}{4} - \frac{0.151982}{4n+1} + \frac{0.015399}{(4n+1)^3} - \frac{0.245270}{(4n+1)^5} + \cdots. \end{aligned} \tag{14.298}$$

We note, that $\omega_0$ is independent of $d$. We obtain therefore from Eqs. (14.293) and (14.272) for the corresponding **TM mode**:

$$\mathbf{E}_z = BJ_0\left(\frac{2.4\rho}{R}\right)e^{-i\omega t}\mathbf{e}_z, \quad \mathbf{B}_\perp = \frac{i\mu\epsilon\omega_0}{\gamma_{01}^2}B\mathbf{e}_z \times \boldsymbol{\nabla}_\perp J_0\left(\frac{2.4\rho}{R}\right)e^{-i\omega t}. \tag{14.299}$$

In cylindrical coordinates

$$\boldsymbol{\nabla}_\perp = \mathbf{e}_\rho\frac{\partial}{\partial\rho} + \frac{1}{\rho}\mathbf{e}_\varphi\frac{\partial}{\partial\varphi}. \tag{14.300}$$

Hence

$$\boldsymbol{\nabla}_\perp J_0\left(\frac{2.4\rho}{R}\right) = \mathbf{e}_\rho\frac{\partial}{\partial\rho}J_0\left(\frac{2.4\rho}{R}\right), \tag{14.301}$$

so that (with $\mathbf{e}_z \times \mathbf{e}_\rho = \mathbf{e}_\varphi$)

$$\mathbf{B}_\perp \equiv \mathbf{B}_\varphi = \frac{i\mu\epsilon\omega_0}{\gamma_{01}^2}B\frac{\partial}{\partial\rho}J_0(\gamma_{01}\rho)e^{-i\omega t}\mathbf{e}_\varphi, \quad \left(\gamma_{01} \simeq \frac{2.4}{R}\right). \tag{14.302}$$

However (see Tables of Special Functions or Dwight [42], formula 801.4, p. 174),

$$\frac{d}{dx}J_m(x) = \frac{1}{2}(J_{m-1}(x) - J_{m+1}(x)) \tag{14.303}$$

and $J_{-m}(x) = (-1)^m J_m(x)$, so that

$$\frac{d}{dx}J_0(x) = \frac{1}{2}(J_{-1}(x) - J_1(x)) = -J_1(x). \tag{14.304}$$

Hence

$$\mathbf{B}_\varphi = -i\frac{\mu\epsilon\omega_0}{\gamma_{01}^2}B\gamma_{01}J_1(\gamma_{01}\rho)e^{-i\omega t}\mathbf{e}_\varphi, \tag{14.305}$$

where from Eq. (14.297):

$$\gamma_{01} = \sqrt{\epsilon\mu}\omega_0, \tag{14.306}$$

and therefore

$$\mathbf{B}_\varphi = -i\sqrt{\mu\epsilon}BJ_1(\gamma_{01}\rho)e^{-i\omega t}\mathbf{e}_\varphi. \tag{14.307}$$

For **TE modes** we had Eqs. (14.279) and (14.280). These yield in the present case

$$B_z = AJ_m(\gamma_{mn}\rho)e^{im\varphi}\sin\left(\frac{l\pi z}{d}\right)e^{-i\omega t}. \tag{14.308}$$

The boundary condition

$$\left.\frac{\partial B_z}{\partial n}\right|_{\rho=R} = \left.\frac{\partial B_z}{\partial \rho}\right|_{\rho=R} = 0$$

implies

$$\left.\frac{\partial}{\partial\rho}J_m(\gamma_{mn}\rho)\right|_{\rho=R} = 0. \tag{14.309}$$

We thus obtain the quantities $\gamma_{mn}$ from the zeros of the derivative of the Bessel function. For the lowest frequency[15] we need the lowest root; this is ($l \neq 0$ from Eq. (14.308), so that $B_z \neq 0$)

$$J_1'(x_{11}) = 0 \quad \text{with} \quad x_{11} = 1.841, \tag{14.310}$$

so that $m = 1, n = 1$ (the number 1.841 can immediately be inferred from formula (14.315) below by setting $p = 1$). A convenient formula to obtain

[15]The roots of $J_0'(x) = 0$ are $x = 3.8, 7.0, 10.1, \ldots$, i.e. higher than $x_{11}$.

this zero from is again given by Jahnke and Emde [74], p. 143, i.e.

$$\begin{aligned}
J_p'(x_n) &= 0, \quad (\textit{i.e.} \quad J_p(x_n) = \text{max.}), \\
x_n &= \gamma - \frac{q+3}{8\gamma} - \frac{1}{6(4\gamma)^3}(7q^2+82q-9) \\
&\quad - \frac{1}{15(4\gamma)^5}(83q^3+2075q^2-3039q+3537) - \cdots, \qquad (14.311)
\end{aligned}$$

where

$$q = 4p^2, \quad \gamma = \left(\frac{1}{2}p + \frac{1}{4} + n - 1\right)\pi, \quad n = 1, 2, 3, \ldots. \qquad (14.312)$$

For large values of $p$ one has for $x_1$ the expansion:

$$x_1 \simeq p + 0.808618p^{1/3} + \cdots. \qquad (14.313)$$

One can test these formulas for the case of $x_{11}$. Since $\omega^2$ is given by Eq. (14.296) with $\gamma_{mn} = x_{mn}/R$, it follows that in the case of this lowest frequency,

$$\omega_{111}^2 = \frac{1}{\epsilon\mu}\left[\left(\frac{1.8}{R}\right)^2 + \left(\frac{\pi}{d}\right)^2\right] \qquad (14.314)$$

($l = 1$; $B_z = 0$ for $l = 0$). We see that the fundamental frequency $\omega_{111}$ of the TE mode can be changed by changing $d$, the height of the cylindrical resonator (contrary to the fundamental frequency of the TM mode). Finally from Eq. (14.280) we have

$$B_z = AJ_1(\gamma_{11}\rho)\sin\left(\frac{\pi z}{d}\right)e^{-i\omega t}e^{i\varphi} \qquad (14.315)$$

(for the factor in $\varphi$ see Eq. (14.292)). The resonator modes we investigated here find application for instance in accelerator technology.[16]

## 14.11 Examples

**Example 14.1 (Vector potentials and rectangular wave guides).** Define for the charge-free interior of a wave guide the electric vector potential $\mathbf{G}$ by $\mathbf{E} = \boldsymbol{\nabla} \times \mathbf{G}$. Similarly we have $\mathbf{B} = \boldsymbol{\nabla} \times \mathbf{A}$. Find the wave equations of the vector potentials and sketch qualitatively the trajectories of constant field strengths.

---

[16]See e.g. Palumbo and Vaccaro [110]. See also Slater [138] and Maier and Slater [92].

**Solution:** In the charge-free space ($\rho = 0, \mathbf{j} = 0$) we can define an electric vector potential $\mathbf{G}$ by setting $\mathbf{E} = \boldsymbol{\nabla}\times\mathbf{G}$, so that $\boldsymbol{\nabla}\cdot\mathbf{E} = 0$ (see Appendix C). Maxwell's curl–equations are here

$$\boldsymbol{\nabla}\times\mathbf{E} = -\frac{\partial\mathbf{B}}{\partial t}, \quad \boldsymbol{\nabla}\times\mathbf{B} = \mu\frac{\partial\mathbf{D}}{\partial t}. \tag{14.316}$$

From these equations we deduce (so that $\mathbf{E} = \boldsymbol{\nabla}\times\mathbf{G}, \mathbf{B} = \boldsymbol{\nabla}\times\mathbf{A}$)

$$\mathbf{E} = -\frac{\partial\mathbf{A}}{\partial t}, \quad \mathbf{B} = \mu\epsilon\frac{\partial\mathbf{G}}{\partial t}. \tag{14.317}$$

Moreover,

$$\boldsymbol{\nabla}\times(\boldsymbol{\nabla}\times\mathbf{G}) = -\frac{\partial}{\partial t}(\boldsymbol{\nabla}\times\mathbf{A}), \quad \boldsymbol{\nabla}\times(\boldsymbol{\nabla}\times\mathbf{A}) = \mu\epsilon\frac{\partial}{\partial t}(\boldsymbol{\nabla}\times\mathbf{G}). \tag{14.318}$$

With the relation "*curl curl* = *grad div* − *div grad* " of Appendix C we obtain

$$\boldsymbol{\nabla}(\boldsymbol{\nabla}\cdot\mathbf{G}) - \triangle\mathbf{G} = -\frac{\partial\mathbf{B}}{\partial t} = -\mu\epsilon\frac{\partial^2\mathbf{G}}{\partial t^2}, \tag{14.319}$$

and

$$\boldsymbol{\nabla}(\boldsymbol{\nabla}\cdot\mathbf{A}) - \triangle\mathbf{A} = \mu\epsilon\frac{\partial\mathbf{E}}{\partial t} = -\mu\epsilon\frac{\partial^2\mathbf{A}}{\partial t^2}. \tag{14.320}$$

We choose the gauge fixing conditions

$$\boldsymbol{\nabla}\cdot\mathbf{A} = 0, \quad \boldsymbol{\nabla}\cdot\mathbf{G} = 0. \tag{14.321}$$

Then

$$-\triangle\mathbf{G} + \mu\epsilon\frac{\partial^2\mathbf{G}}{\partial t^2} = 0, \tag{14.322}$$

and a similar equation holds for $\mathbf{A}$. TE modes are defined by $E_z = 0$. With $\mathbf{G} = (0, 0, \Pi)$, it follows that

$$\mathbf{E} = \boldsymbol{\nabla}\times\mathbf{G} = \left(\frac{\partial\Pi}{\partial y}, -\frac{\partial\Pi}{\partial x}, 0\right). \tag{14.323}$$

It may be noted that the vector potential $\mathbf{G}$ has in this case only a $z$-component (compare with $\mathbf{B}$ in Section 8.4). As an example we take, with

$m, n = 0, 1, \ldots$, as in Eq. (14.133),

$$\Pi \propto \cos\left(\frac{m\pi x}{a}\right)\cos\left(\frac{n\pi y}{b}\right)e^{ikz}e^{-i\omega t},$$
$$\frac{\partial\Pi}{\partial y} \propto -\frac{n\pi}{b}\cos\left(\frac{m\pi x}{a}\right)\sin\left(\frac{n\pi y}{b}\right)e^{ikz}e^{-i\omega t},$$
$$\frac{\partial\Pi}{\partial x} \propto -\frac{m\pi}{a}\sin\left(\frac{m\pi x}{a}\right)\cos\left(\frac{n\pi y}{b}\right)e^{ikz}e^{-i\omega t}. \qquad (14.324)$$

Lines of constant electric field strength are given by $|\mathbf{E}|^2 = \text{const.}$, i.e.

$$\left(\frac{n\pi}{b}\right)^2\cos^2\left(\frac{m\pi x}{a}\right)\sin^2\left(\frac{n\pi y}{b}\right) + \left(\frac{m\pi}{a}\right)^2\sin^2\left(\frac{m\pi x}{a}\right)\cos^2\left(\frac{n\pi y}{b}\right) = \text{const.} \qquad (14.325)$$

The closed curves given by this equation are ellipse-like in the $(x, y)$-plane, as one can see by considering small values of $x$ and $y$, so that the sine can be approximated by its argument, and are rectangular for larger values, becoming tangential to the boundaries. Then the Cartesian equation of an ellipse is obtained, and since lines of different $|\mathbf{E}|^2$ cannot cut, such closed curves also result for other values of $x$ and $y$. Analogous considerations apply to TM modes. In their case $B_z = 0$ and $\mathbf{A} = (0, 0, \Psi)$ with

$$\mathbf{B} = \left(\frac{\partial\Psi}{\partial y}, -\frac{\partial\Psi}{\partial x}, 0\right). \qquad (14.326)$$

A side-view of the electric field only of the TE mode for $m = 1$ and $n = 0$, i.e.

$$E_y \propto -\frac{\pi}{a}\sin\left(\frac{\pi x}{a}\right)\exp[i(kz - \omega t)],$$

is sketched in Fig. 14.14.[17]

**Example 14.2 (Induced charges and currents).** Consider the fundamental TE mode of a rectangular wave guide and calculate all charges and currents induced in its surfaces (the wave guide is made of perfect conductor material).

[17]Numerically calculated field patterns of the TE(10) mode in a rectangular wave guide are shown in Ferrari [54], p. 163, and Schanda in Epprecht *et al.* [46], p. 48. Field lines on a computer graphic display have been given by Shintake [134]. See also Bleaney and Bleaney [12], pp. 287–289.

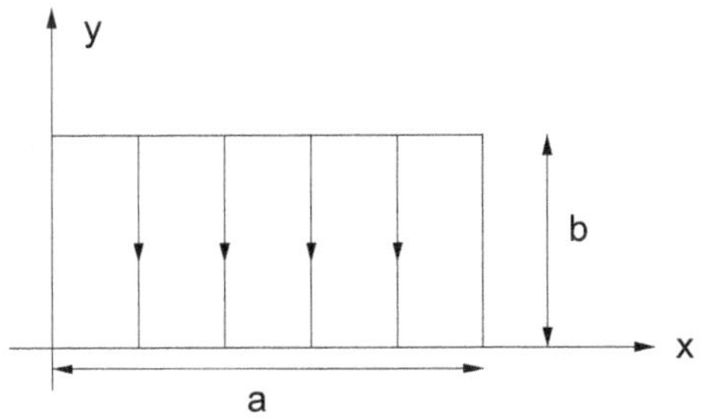


**Fig. 14.14.** The electric field of the TE ($m = 1, n = 0$) mode.

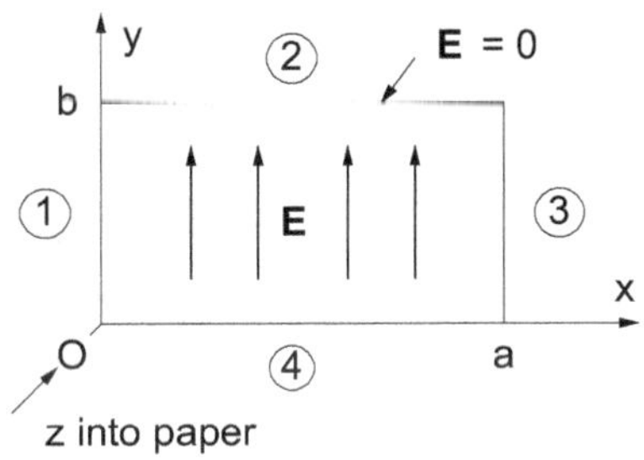


**Fig. 14.15.** Cross section through the rectangular wave guide.

**Solution:** We have the equations (see summaries of formulas in Sections 4.6, 6.8 and Eq. (4.40))

$$\begin{aligned} &1. \quad D_n^{(1)} - D_n^{(2)} = \sigma, \quad \text{physical} \Rightarrow \Re(\epsilon_1 E_n^{(1)} - \epsilon_2 E_n^{(2)}) = \sigma, \\ &2. \quad \mathbf{n} \times (\mathbf{H}^{(2)} - \mathbf{H}^{(1)}) = \mathbf{K}, \end{aligned} \tag{14.327}$$

from which the surface charge density $\sigma$ and $\mathbf{K}$ are to be calculated (definition of surface current density $\mathbf{K}$ in Section 11.2). In Section 14.6.2 we obtained the fundamental TE wave for $m = 1$ and $n = 0$. We have therefore (cf. Eq. (14.141)):

$$\mathbf{E} = \mathbf{e}_y \beta \sin\left(\frac{\pi x}{a}\right) e^{i(kz - \omega t)}. \tag{14.328}$$

Since $\mathbf{E} \parallel \mathbf{e}_y$, it follows that (see Fig. 14.15 for explanation of subscripts) $\sigma_{①} = 0 = \sigma_{③}, \sigma_{②} = -\sigma_{④}$. Calculation of $\sigma_{②}$: Analogous to the case of the parallel plate condenser we have $D_n^{(1)} = 0$. Hence the surface charge density on surface ② is (with $\epsilon$ the dielectric constant of the inside of the wave guide and the surface normal pointing in the direction of $-\mathbf{e}_y$)

$$\sigma_{②} = \Re(-D_n^{(2)}) = \Re(-)(-)\epsilon\beta \sin\left(\frac{\pi x}{a}\right) e^{i(kz-\omega t)} = \epsilon\beta \sin\left(\frac{\pi x}{a}\right) \cos(kz - \omega t), \tag{14.329}$$

For the field $\mathbf{B}$ we have (see Eqs. (14.133) and (14.136) for $m = 1, n = 0$)

$$\mathbf{B} = \beta\left[-\mathbf{e}_x \frac{k}{\omega} \sin\left(\frac{\pi x}{a}\right) + \mathbf{e}_z \frac{\pi}{ai\omega} \cos\left(\frac{\pi x}{a}\right)\right] e^{i(kz-\omega t)}, \tag{14.330}$$

and we have to evaluate $\mathbf{n} \times (\mathbf{H}^{(2)} - \mathbf{H}^{(1)})$. Since for the TE wave (cf. Eq. (14.100))

$$\mathbf{E}_\perp = -\frac{\omega}{k}(\mathbf{e}_z \times \mathbf{B}_\perp), \quad \text{with} \quad \mathbf{e}_z \nparallel \mathbf{B}_\perp, \tag{14.331}$$

we have $\mathbf{B}_\perp = 0$ for $\mathbf{E}_\perp = 0$, i.e. outside of the wave guide $\mathbf{B} = 0$, i.e. $\mathbf{H}^{(1)} = 0$. It remains to evaluate $\mathbf{n} \times \mathbf{H}$ for the respective surfaces as shown in Fig. 14.15 (therewith obtaining the surface current densities $\mathbf{K}$).
*Surface* ②: $\mathbf{n} = \mathbf{e}_y, y = b$. We obtain $\mathbf{H}$ from Eq. (14.330):

$$\begin{aligned}\mathbf{K} = \mathbf{n} \times \mathbf{H} &= \Re\left\{\frac{1}{\mu}\mathbf{e}_y \times \beta\left[-\mathbf{e}_x \frac{k}{\omega} \sin\left(\frac{\pi x}{a}\right) + \mathbf{e}_z \frac{\pi}{ai\omega} \cos\left(\frac{\pi x}{a}\right)\right] e^{i(kz-\omega t)}\right\} \\ &= -\frac{\beta}{\mu}\left[\frac{k}{\omega} \sin\left(\frac{\pi x}{a}\right) \cos(kz-\omega t)\mathbf{e}_z + \frac{\pi}{a\omega} \cos\left(\frac{\pi x}{a}\right) \sin(kz-\omega t)\mathbf{e}_x\right].\end{aligned} \tag{14.332}$$

*Surface* ①: $\mathbf{n} = \mathbf{e}_x, x = 0$.

$$\begin{aligned}\mathbf{K} &= \mathbf{n} \times \mathbf{H} \\ &= \Re\left\{\frac{1}{\mu}\mathbf{e}_x \times \beta\left[-\mathbf{e}_x \frac{k}{\omega} \sin\left(\frac{\pi x}{a}\right) + \mathbf{e}_z \frac{\pi}{ai\omega} \cos\left(\frac{\pi x}{a}\right)\right] e^{i(kz-\omega t)}\right\}\Bigg|_{x=0} \\ &= -\frac{\beta}{\mu}\frac{\pi}{a\omega} \sin(kz-\omega t)\mathbf{e}_y.\end{aligned} \tag{14.333}$$

**Example 14.3 (Determination of modes).** Consider two parallel, infinitely extended, perfectly conducting walls in the planes $y = 0, b$, as indicated in Fig. 14.16. The electric vector of the incident TE wave is polarized in the direction of $x$, i.e.

$$\mathbf{E}_0 = \mathbf{e}_x E_0^0 \exp[-i(\omega t - \mathbf{k}_0 \cdot \mathbf{r})] \tag{14.334}$$

with $\mathbf{r} = (0, y, z)$. Show that only waves with angular frequency

$$\omega_n = \frac{n\pi c}{b\cos\theta_0}, \qquad n = 0, 1, 2, \ldots \tag{14.335}$$

are propagated between the planes, where $\theta_0$ is the angle of incidence between $\mathbf{k}_0$ and $\mathbf{r}$.

**Solution:** According to the geometry depicted in Fig. 14.16 the vector $\mathbf{k}_0$ is perpendicular to the plane of equal phases $\mathbf{k}_0 \cdot \mathbf{r} = \text{const}$. This means with $\alpha_0 = \pi - \theta_0$ and $\beta_0 = \pi/2 - \theta_0$,

$$\begin{aligned} & k_0 y\cos(\mathbf{k}_0, \mathbf{y}) + k_0 z\cos(\mathbf{k}_0, \mathbf{z}) = \text{const.}, \\ & i.e. \quad -k_0 y\cos\theta_0 + k_0 z\sin\theta_0 = \text{const.}, \end{aligned} \tag{14.336}$$

so that

$$\mathbf{E}_0 = \mathbf{e}_x E_0^0 e^{-i\omega t} e^{ik_0(-y\cos\theta_0 + z\sin\theta_0)}. \tag{14.337}$$

Since $\mathbf{E}_0$ has to satisfy the wave equation, and there is no propagation of the wave in $x$, we see that its dispersion relation is

$$\frac{\omega^2}{c^2} = k_y^2 + k_z^2 = k_0^2\cos^2\theta_0 + k_0^2\sin^2\theta_0 = \mathbf{k}_0^2. \tag{14.338}$$

In vacuum $\omega = k_0 c$. (Here we have total reflection since the wall is a perfect conductor with $\sigma = \infty$, i.e. no radiation or energy penetrates into the wall). The $\mathbf{E}$ wave totally reflected from the wall, can also be written with the help of Fig. 14.16 as

$$\mathbf{E}_0'' = \mathbf{e}_x E_0'' e^{-i\omega t} e^{ik_0(y\cos\theta_0 + z\sin\theta_0)}. \tag{14.339}$$

The continuity condition of the field $\mathbf{E}$, i.e.

$$(\text{total } \mathbf{E}_{\text{tang}})|_{\text{wall}} = 0 \tag{14.340}$$

($E_{\text{tang}}^{(1)} = E_{\text{tang}}^{(2)} = 0$ in the perfect conductor, see skin effect in Section 9.5) implies, as shown in Section 11.4.2 (cf. Eqs. (11.68) and (11.69)),

$$E_0'' = -E_0^0, \quad (E_0' = 0). \tag{14.341}$$

Hence

$$\mathbf{E}_0'' = -\mathbf{e}_x E_0^0 e^{-i\omega t} e^{ik_0(y\cos\theta_0 + z\sin\theta_0)}. \tag{14.342}$$

The entire field $\mathbf{E}$ at $(x, y, z, t)$ is then:

$$\begin{aligned} \mathbf{E} = \mathbf{E}_0 + \mathbf{E}_0'' &= \mathbf{e}_x E_0^0 e^{-i\omega t} e^{ik_0 z\sin\theta_0}[e^{-ik_0 y\cos\theta_0} - e^{ik_0 y\cos\theta_0}] \\ &= -2i\mathbf{e}_x E_0^0 e^{-i\omega t} e^{ik_0 z\sin\theta_0}\sin(k_0 y\cos\theta_0). \end{aligned} \tag{14.343}$$

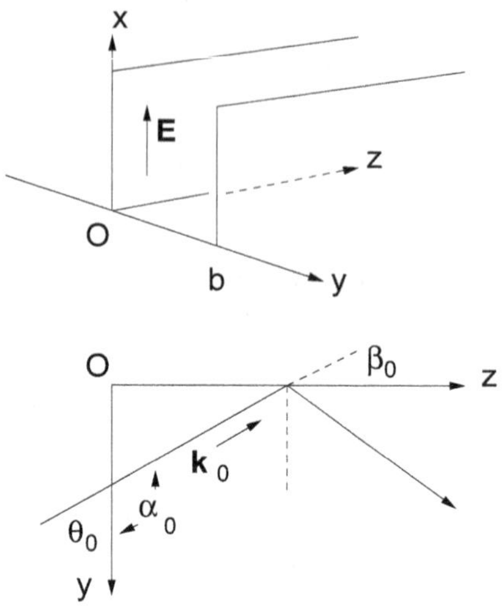


**Fig. 14.16.** Reflection between planes.

The boundary condition (14.340) at the wall implies at $y = 0, b$:

$$\sin(k_0 b\cos\theta_0) = 0, \quad \therefore \quad k_0 b\cos\theta_0 = n\pi, \quad n = 1, 2, \ldots.$$

Here $n = 1, 2, 3, \ldots$ are the mode numbers. The dispersion relation is now seen to be

$$\frac{\omega^2}{c^2} = k_0^2 = \left(\frac{n\pi}{b\cos\theta_0}\right)^2. \tag{14.344}$$

For wavelength $\lambda \to \lambda_0 = 2\pi/k_0$ we have

$$\lambda_0 = \frac{2\pi b\cos\theta_0}{n\pi} = \frac{2b\cos\theta_0}{n} \quad \text{meters}, \tag{14.345}$$

and frequency

$$\omega_{0n} = k_0 c = \frac{n\pi c}{b\cos\theta_0} \quad \text{seconds}^{-1}. \tag{14.346}$$

For a given angle of incidence $\theta_0$ only waves with these frequencies for $n = 1, 2, 3, \ldots,$ are propagated between the walls. In the text we considered cylindrical wave guides with modes characterized by two integers, and closed resonators (or cavities) with modes characterized by three integers. In the above case the waves are confined only in one dimension, and hence its modes are characterized by only one integer. The region between the surface of the earth and the upper ionosphere can be regarded — very roughly — as a resonator with reflection of radio waves from either limit. In the limit of an infinitely large radius of the earth, the wave propagation approaches the case discussed here.

## 14.12 Exercises

**Example 14.4 (TM waves in a rectangular wave guide).** For the rectangular wave guide bounded by $x = 0, a$ and $y = 0, b$, show from the relation (14.99), i.e.

$$\mathbf{B}_\perp = \frac{\mu\epsilon\omega}{k}\mathbf{e}_z \times \mathbf{E}_\perp, \tag{14.347}$$

that TM waves are possible only for frequencies $\omega$ such that

$$\omega > c\sqrt{\frac{\pi^2}{a^2} + \frac{\pi^2}{b^2}}. \tag{14.348}$$

**Example 14.5 (Elliptic wave guides).** Show that for elliptic wave guides (i.e. cylindrical but of elliptic cross section) the transverse modes are solutions of the Mathieu equation $\psi'' + (A + B\cos(2x))\psi = 0, A, B$ constants. (There are in general no other cases for which the problem of wave guides can be solved exactly apart from the cases treated here and excepting the case of cross-sections having the shape of equilateral or isosceles triangles).[18]

**Example 14.6 (Waves between coaxial circular cylinders).** In the case of waves between two coaxial circular cylinders of radii $a$ and $b$, $b > a$, one cannot exclude the second cylinder function $Y_m$ (as we could in Section 14.6.3). Thus a linear combination of both types of such functions must then vanish on the boundaries $\rho = a, b$. Assuming the radii are nearly equal (the curvature effect on the fields then being small) obtain the permissible modes using dominant approximations of the cylinder functions.[19]

**Example 14.7 (Tangential electric fields).** The surfaces of a wave guide are perfectly conducting. Explain why the tangential electric fields vanish, i.e. what unphysical result a nonvanishing of the tangential electric fields would imply.

**Example 14.8 (Maximum wavelength outside a rectangular wave guide).** Consider the wave guide of rectangular cross section treated in Section 14.6.2 for transverse electric (TE) waves. Show that for the

[18]Schelkunoff [129], pp. 393, 394.
[19]See Schelkunoff [129], p. 391.

wavelength $\tilde{\lambda}$ inside the guide to be real there is a maximum wavelength $\lambda$ outside the guide given by the upper limit of the inequality

$$\lambda < \frac{2}{\sqrt{m^2/a^2 + n^2/b^2}}. \tag{14.349}$$

Suppose $a > b$. What is the biggest wavelength one can "stick" into the wave guide, i.e. what is the maximum outside wavelength that could be propagated inside the wave guide?

**Example 14.9 (Phase and group velocities without and within).** Consider again the wave guide of rectangular cross section in Fig. 14.4 and TE waves. The phase velocities within and without the wave guide are respectively defined by

$$v^i_{\text{phase}} = \frac{\omega}{\tilde{\beta}}, \quad \tilde{\beta} = k \quad \text{and} \quad v^{ou}_{\text{phase}} = \frac{\omega}{\beta}, \quad \beta = \frac{\omega}{c}, \tag{14.350}$$

where

$$\tilde{\beta}^2 = \beta^2 - \pi^2\left(\frac{m^2}{a^2} + \frac{n^2}{b^2}\right). \tag{14.351}$$

The respective group velocities are correspondingly defined by

$$v^i_{\text{group}} = \frac{d\omega}{d\tilde{\beta}}, \quad v^{ou}_{\text{group}} = \frac{d\omega}{d\beta}. \tag{14.352}$$

Here "outside" the wave guide means in free space. Show that

$$v^i_{\text{group}} v^i_{\text{phase}} = (v^{ou}_{\text{group}})^2 = (v^{ou}_{\text{phase}})^2 \equiv (v^{ou})^2, \tag{14.353}$$

and that

$$v^i_{\text{group}} = v^{ou}\sqrt{1 - \frac{\lambda^2}{4}\left(\frac{m^2}{a^2} + \frac{n^2}{b^2}\right)}, \tag{14.354}$$

where $\lambda = 2\pi/\beta$ is the wavelength in free space.

**Example 14.10 (Number of modes in a perfectly conducting box).** Rewriting Eq. (14.287) as

$$1 = \left(\frac{l}{\beta d/\pi}\right)^2 + \left(\frac{m}{\beta a/\pi}\right)^2 + \left(\frac{n}{\beta b/\pi}\right)^2, \tag{14.355}$$

we recognize this equation as the equation of an ellipsoid, $\beta$ being the wave number. Since the spacing between the l's is unity, and similarly that for the m's and n's, the size of a unit cell is 1. Hence the number of positive (l,m,n) values (assumed large) is approximately equal to the volume of the octant $V_{1/8}$ with $V = 8V_{1/8}$. Calculate the number of normal modes $N$ in the volume $V$, and hence the number in an infinitesimal frequency range $d\nu$. This result is of considerable importance in statistical mechanics.[20]

(Answer: $N = V(4\pi/3)(\beta/2\pi)^3 = V(4\pi/3)(\nu/c)^3$ to be multiplied by 2 for 2 directions of polarization of the photon).

**Example 14.11 (Fields near the surface of a good conductor).** Consider the fields near the surface of a good, but not perfect, conductor. Show that the ratio of the normal component of the magnetic field to its tangential component is of the order of the skin depth to the wavelength of the oscillating fields.
[Hint: Use $\mathbf{E}_t$ of Eq. (14.241) and Faraday's law as in Eq. (14.233)].

**Example 14.12 (Fields inside and just outside a good conductor).** Verify Eq. (14.245). Explain why inside the conductor the magnetic field is much larger than the electric field. How is this just outside?

[20] See e.g. Müller–Kirsten [102], pp. 74, 111.

# Chapter 15

# Propagation of Waves in Homogeneous Media

## 15.1 Introductory Remarks

In this chapter we introduce the concepts of *signals* and *wave packets* and demonstrate the intricate connection between the analytic properties of the Fourier transform of the wave packet (also called its *spectral function*) and the limitation of its velocity by the velocity of light. Generally we assume the medium under discussion is homogeneous, meaning e.g. (for $i, j = 1, 2, 3$) $j_i = \sigma E_i, D_i = \epsilon E_i$, and not $j_i = \sum_j \sigma_{ij} E_j, D_i = \sum_j \epsilon_{ij} E_j$.

## 15.2 Dispersion Relation: Normal and Anomalous Dispersion

In preceding chapters we derived and became familiar with the dispersion relation of a conducting medium, i.e. in Section 11.3 we obtained with the sign convention there the relation (cf. Eq. (11.37) or Eq. (9.40))

$$k^2 = \omega^2 \eta \mu = \mu \epsilon \omega^2 \left(1 + i\frac{\sigma}{\omega\epsilon}\right) \quad \text{meters}^{-2}, \tag{15.1}$$

which also defines the generalized dielectric constant $\eta$ (one should note that in general the relation is more complicated, in particular, as we saw in Chapter 9, if the frequency dependence of the conductivity $\sigma$ is taken into account). We arrived at the expression (15.1) by assuming

$$\mathbf{j} = \sigma \mathbf{E}, \quad \mathbf{H} \propto e^{-i\omega t} e^{i\mathbf{k}\cdot\mathbf{r}}, \tag{15.2}$$

(note the sign convention) together with the two Maxwell curl equations, and we obtained

$$\mathbf{k}^2\mathbf{H} = \mu\epsilon\omega^2\left(1 + i\frac{\sigma}{\omega\epsilon}\right)\mathbf{H} \quad \text{amperes/meter}^3. \tag{15.3}$$

The relation (15.1) is a general consequence if we assume that $\mathbf{H}$ does not vanish anywhere in space. We now want to study the dispersion relations as functions of the frequency $\omega$.[1]

We had defined previously the generalized refractive index

$$n(k) \equiv p = \frac{\sqrt{\mu\eta}}{\sqrt{\mu_0\epsilon_0}}, \quad \sqrt{\mu_0\epsilon_0} = \frac{1}{c}, \quad c' = \frac{1}{\sqrt{\mu\eta}}, \quad p = \frac{c}{c'}, \tag{15.4}$$

so that

$$n(k) = \frac{ck}{\omega} \equiv \tilde{n}(\omega) \quad \text{or} \quad \omega(k) = \frac{ck}{n(k)} = \frac{ck}{\tilde{n}(\omega)}. \tag{15.5}$$

We note that from Eq. (15.4) we obtain the relation

$$n^2 = \frac{\mu\eta}{\mu_0\epsilon_0} \simeq \frac{\eta}{\epsilon_0}, \tag{15.6}$$

since for many substances $\mu \simeq \mu_0$. Thus if $\eta$ is written

$$\eta = \epsilon_0 + \epsilon_0\chi, \quad \text{we have} \quad n^2 = 1 + \chi, \tag{15.7}$$

where $\chi$ is the susceptibility defined earlier by Eqs. (4.23) and (4.25). The version (15.7) of the refractive index squared is frequently used in this context.

The *phase velocity* $v_p$ and the *group velocity* $v_g$ of a lossless medium[2] defined previously for real $n(k)$ or $\tilde{n}(\omega)$ are given by

$$\begin{aligned} v_p &= \frac{\omega}{k} = \frac{c}{n(k)} = \frac{c}{\tilde{n}(\omega)} \quad \text{meters/second}, \\ v_g &= \frac{d\omega}{dk} \overset{(15.5)}{=} \frac{c}{n(k)} - \frac{ck(dn(k)/dk)}{[n(k)]^2} \\ &= \frac{c}{\tilde{n}(\omega)} - \frac{ck(d\tilde{n}(\omega)/d\omega)(d\omega/dk)}{[\tilde{n}(\omega)]^2} \quad \text{meters/second}. \end{aligned} \tag{15.8}$$

---

[1] See also Jackson [73], Chapter 7.

[2] In the case of a dissipative medium the group velocity is restricted to a tiny frequency range in view of distortion of the wave packet.

Solving the latter for $d\omega/dk$ we obtain

$$v_g = \frac{d\omega}{dk} = \frac{c}{\tilde{n}(\omega) + \omega(d\tilde{n}(\omega)/d\omega)} \quad \text{meters/second.} \tag{15.9}$$

One says, there is no dispersion if $d\tilde{n}(\omega)/d\omega = 0$. For plane waves in vacuum $v_p = v_g = c$, but in a material medium where $\epsilon$ and $\mu$ may depend on $\omega$, the two velocities need not be equal. In dissipative media the group velocity loses its significance.

We now distinguish between two cases:

1. The case of *normal dispersion* defined by

$$\frac{d\tilde{n}(\omega)}{d\omega} > 0, \quad \text{so that} \quad v_g < v_p < c, \ \tilde{n}(\omega) > 1, \tag{15.10}$$

(in general $\epsilon/\epsilon_0 \geq 1$), and
2. the case of *anomalous dispersion* defined by

$$\frac{d\tilde{n}(\omega)}{d\omega} < 0, \quad \text{for which} \quad v_g > v_p, \tag{15.11}$$

which would imply $v_g > c$. This really means that the relation (15.9) then breaks down (see below), i.e. one has to take the real part when $\tilde{n}$ is complex!

The amplitude of a wave group moves (as one deduces from a superposition of plane waves) with the group velocity. Since the energy of a wave is determined by its amplitude, this implies that the energy of the wave is transmitted with the group velocity. However, we shall see in the following that no energy can be transported with a velocity $> c$. In the following we consider frequency superpositions of waves with phases

$$\delta(\omega) = k(\omega)x - \omega t. \tag{15.12}$$

A *stationary phase* is given by $\delta'(\omega) = 0$, i.e.

$$\frac{dk(\omega)}{d\omega}x - t = 0, \quad \text{or} \quad \frac{d\omega(k)}{dk} = \frac{x}{t}. \tag{15.13}$$

The ratio $x/t$ is described as *signal velocity*. In the present case this is equal to the group velocity. In general the group velocity of the dominant frequency can be treated as the signal velocity and hence as the velocity of the *transport of energy*. The Theory of Relativity teaches, that no physical velocity can exceed the velocity of light $c$ (often cited as phenomenological input of the

Special Theory of Relativity, but could also be considered as a consequence of the principle of relativity). In cases where

$$\frac{d\omega}{dk} = \frac{c}{n(k) + \omega(dn(k)/d\omega)} > c, \tag{15.14}$$

the signal velocity can not be identified with $d\omega/dk$. The wave packet then has a more complicated structure, so that the expansion about a stationary phase is not possible. The aim of this expansion is to use the value $k_0$, for which $\delta'(\omega) = 0$ as a point around which $\omega(k)$ is to be expanded:

$$\omega(k) = \omega(k_0) + (k - k_0)\omega'(k_0) + \cdots . \tag{15.15}$$

If $\tilde{n}$ is complex, we equate

$$\begin{aligned} v_g := \Re\frac{d\omega}{dk} &= \Re\frac{c}{\tilde{n}(\omega) + \omega(d\tilde{n}(\omega)/d\omega)} \\ &= \Re\frac{c}{\{\Re\tilde{n}(\omega) + \omega(d\Re\tilde{n}(\omega)/d\omega)\} + i\{\Im\tilde{n}(\omega) + \omega(d\Im\tilde{n}(\omega)/d\omega)\}} \\ &= c\frac{\{\Re\tilde{n}(\omega) + \omega(d\Re\tilde{n}(\omega)/d\omega)\}}{\{\Re\tilde{n}(\omega) + \omega(d\Re\tilde{n}(\omega)/d\omega)\}^2 + \{\Im\tilde{n}(\omega) + \omega(d\Im\tilde{n}(\omega)/d\omega)\}^2} \\ &\leq c. \end{aligned} \tag{15.16}$$

Since absorption effects are related to imaginary parts, we see that the validity of the relation has its explanation in the damping effects resulting from these imaginary parts. In the following we investigate the behavior of $\tilde{n}(\omega)$ in more detail. At the end of the chapter we shall see that when $d\Re\tilde{n}(\omega)/d\omega < 0$, the imaginary part $\Im\tilde{n}(\omega)$ is particularly large.

## 15.3 Absorption, Causality and Analyticity

We return to the consideration of a plane electromagnetic wave which is vertically incident on a dispersive medium as indicated in Fig. 15.1.

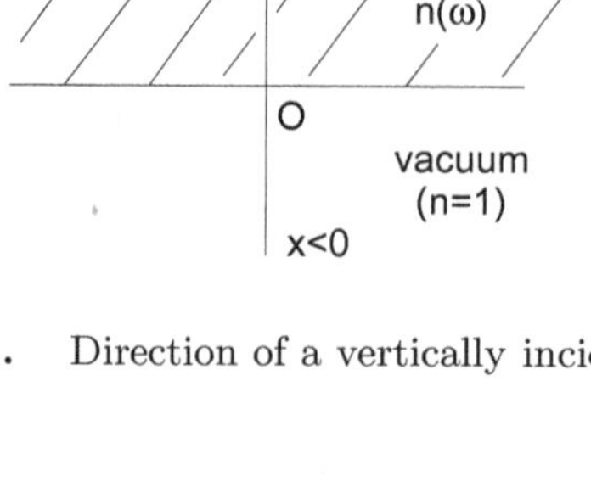


**Fig. 15.1.** Direction of a vertically incident wave.

A superposition of (or integral over) such waves represents a wave train or *wave packet*. We write this superposition in vacuum ($n = 1$)

$$u_V(x,t) = \int_{-\infty}^{\infty} A(\omega)e^{i(kx-\omega t)}d\omega \tag{15.17}$$

(we shall see later that the Fourier representation of a real quantity depends only on positive frequencies).

As before we deduce from Maxwell's curl equations:

- *In vacuum* ($x < 0$) we have $k = \omega/c$.
- *In the medium* ($x > 0$) we have $k(\omega) = \omega\tilde{n}(\omega)/c$.

At the boundary plane but on the side of the vacuum

$$u_V(0,t) = \int_{-\infty}^{\infty} A(\omega)e^{-i\omega t}d\omega, \quad \text{i.e.} \quad A(\omega) = \frac{1}{2\pi}\int_{-\infty}^{\infty} u_V(0,t)e^{i\omega t}dt. \tag{15.18}$$

Every single wave is subjected to absorption and reflection at the interface as discussed in Chapter 11. The transmission coefficient $T$ defined there by Eqs. (11.121) and (11.96) for $\mathbf{E}_0 \perp$ plane of incidence, i.e.

$$T = \frac{2\cos\alpha}{\cos\alpha + N\cos\beta}, \tag{15.19}$$

is in the present case of a wave vertically incident given by

$$T = \frac{2}{1+\tilde{n}(\omega)}, \quad (\mu = \mu_0). \tag{15.20}$$

Hence for the wave packet in the dispersive medium ($x > 0$) into which the waves are transmitted we have

$$u(x,t) = \int_{-\infty}^{\infty}\left(\frac{2}{1+\tilde{n}(\omega)}\right)A(\omega)e^{i(k(\omega)x-\omega t)}d\omega. \tag{15.21}$$

For a further consideration of this integral we require information about the analytical properties of $A(\omega)$ and $k(\omega)$.

### 15.3.1 *Properties of $A(\omega)$*

We consider a wave packet $u(x,t)$ having a bell-shaped form, and starting from the origin $x = 0$, as indicated in Fig. 15.2. Then $u(0,t)$ has

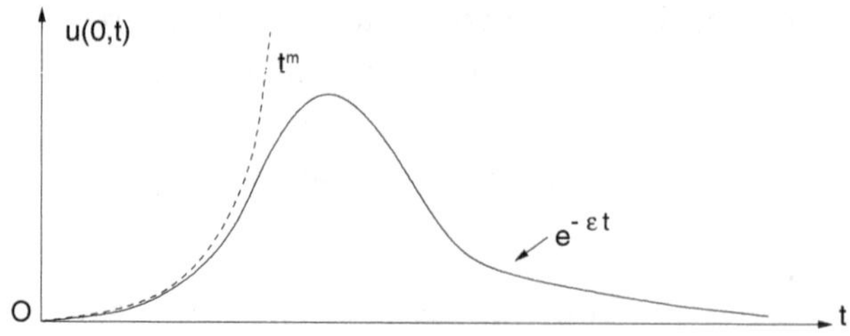


**Fig. 15.2.** The wave packet.

approximately the shape of

$$u(0,t) \simeq \frac{at^m}{m!}\theta(t)e^{-\epsilon t}, \quad \epsilon > 0 \tag{15.22}$$

($m$ integral and larger than zero). Thus the wave packet appears at the point $x = 0$ at time $t = 0$, grows to a maximum strength and ultimately disappears from $x = 0$. We have added the factor $\theta(t)$, in order to avoid the sector with $e^{-\epsilon(t \text{ negative})}$ in view of the integration we have to perform. Of course, one could achieve this also for example with

$$|t|^m e^{-\epsilon|t|}, \quad \epsilon > 0 \tag{15.23}$$

or other means. On the other hand, one can argue that expression (15.22) has a well defined boundary with $x = 0$ at $t = 0$.

The property $u(0,t) = 0$ for $t < 0$ is described as a *causality condition.* We then obtain, from Eq. (15.21) with $x = 0$ and Fourier transformation, the following relation

$$\begin{aligned}
\frac{2}{1+\tilde{n}(\omega)}A(\omega) &= \frac{1}{2\pi}\int_{-\infty}^{\infty} u(0,t)e^{i\omega t}dt \\
&= \frac{1}{2\pi}\frac{a}{m!}\int_0^{\infty} t^m e^{-(\epsilon - i\omega)t}dt \\
&= \frac{1}{2\pi}\frac{a}{m!}\frac{m!}{(\epsilon - i\omega)^{m+1}} \\
&= \frac{ai^{m+1}}{2\pi(\omega + i\epsilon)^{m+1}},
\end{aligned} \tag{15.24}$$

where we used the integral

$$(z-1)! \equiv \Gamma(z) = \int_0^{\infty} e^{-t}t^{z-1}dt. \tag{15.25}$$

Hence

$$A(\omega) \propto \frac{1}{2}\frac{(1+\tilde{n}(\omega))}{(\omega+i\epsilon)^{m+1}} \overset{\text{in vacuum}}{=} \frac{1}{(\omega+i\epsilon)^{m+1}}. \tag{15.26}$$

It is a matter of experience that the transmission of light through transparent media (i.e. those for which $\sigma = 0$) does not lead to resonance effects. We saw between Eqs. (11.38) and (11.39) and in Section 9.4.1 that in the case of such media $\tilde{n}(\omega)$ is real. Hence in these cases $\tilde{n}(\omega)$ is an analytic function without poles. However, if there are resonance effects, $\tilde{n}(\omega)$ will not be everywhere an analytic function of $\omega$, as we shall see. We infer from the above result that in that case $A(\omega)$ is, in general, an analytic function of $\omega$ with poles, i.e. a meromorphic function. The poles at

$$\omega + i\epsilon = 0$$

lie in the lower half of the $\omega$-plane. We infer this from the above result. But we can see this also more generally. Consider the vacuum wave packet

$$u_V(x,t) = \int_{-\infty}^{\infty} A(\omega)e^{i(kx-\omega t)}d\omega. \tag{15.27}$$

In vacuum $k = \omega/c$ (this is why we can write the integral also as $\int dk$). Thus in vacuum all waves (the partial waves as well as the group) propagate with the velocity of light. According to the Special Theory of Relativity (see later) a signal or wave packet is propagated only into the *future light cone* (see Fig. 15.3), which means that for $x > ct$ the integral must vanish, i.e. we must have

$$0 = [u_V(x,t)]_{x>ct} = \int_{-\infty}^{\infty} A(\omega)e^{i\omega(x-ct)/c}d\omega\theta(x-ct). \tag{15.28}$$

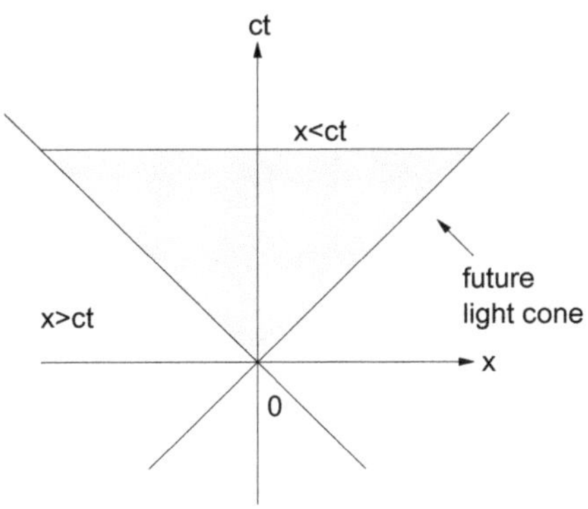


**Fig. 15.3.** The future light cone.

We investigate this equation with the help of a contour integral in the plane of complex $\omega$. If we choose the contour of integration $C$ as indicated in Fig. 15.4, the infinite semicircle with $\Im\omega > 0$, and thus the factor

$$e^{i\omega(x-ct)/c}\theta(x-ct)$$

with exponentially decreasing behavior, does not contribute to the integral and we have

$$0 = \int_{-\infty}^{\infty} \cdots + \int_{\curvearrowleft} \cdots ,$$

i.e.

$$0 = \int_C A(\omega)e^{i\omega(x-ct)/c}d\omega\theta(x-ct). \tag{15.29}$$

The Cauchy residue theorem now implies that $A(\omega)$ cannot possess poles in the plane $\Im\omega > 0$ (i.e. their residues would be zero). Thus $A(\omega)$ is analytic in that part of the complex plane. This means the poles of the amplitude $A(\omega)$ can only be in the lower half of the $\omega$-plane. We obtained this result from the condition that propagation takes place only into the future light cone or time-like domain, where $x/t \leq c$ (in the following we use this result that $A(\omega)$ is a meromorphic function with poles in the lower half of the complex plane). This property of no propagation into the space-like domain depends on the other property of poles located in the lower half-plane.

We saw at the beginning that in regions of anomalous dispersion it is possible that $v_G > c$. We recall that $v_G$ is quite generally (except when $v_G > c$) the velocity with which energy is transported, and hence a signal with velocity $x/t$. As such it cannot be larger than $c$ (according to the Special Theory of Relativity).

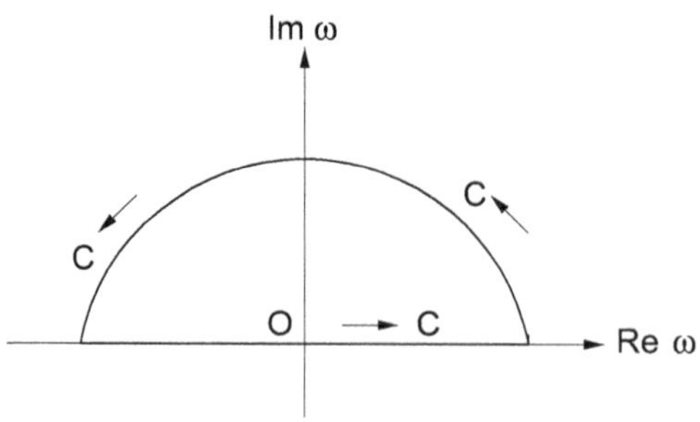


**Fig. 15.4.** The integration contour $C$.

We return to the expression of Eq. (15.21), i.e. to

$$u(x,t) = \int_{-\infty}^{\infty} \left(\frac{2}{1+\tilde{n}(\omega)}\right) A(\omega) e^{i(k(\omega)x-\omega t)} d\omega. \tag{15.30}$$

In order to be able to perform the integration we have to know the analytic properties of $\tilde{n}(\omega)$.

### 15.3.2 *Properties of $\tilde{n}(\omega)$*

In the discussion of the Drude theory in Section 12.3 we used the classical nonrelativistic equation of motion of a free electron interacting with a monochromatic electromagnetic plane wave polarized in the $x$-direction, i.e. (neglecting the magnetic component which is generally of order $1/c$ times the electric contribution)

$$m\ddot{x} + \frac{m}{\tau}\dot{x} = eE, \quad E = E_0 e^{-i\omega t}. \tag{15.31}$$

This would be the equation of one of the electrons of a nonconducting gas. If the electron is not completely free, but bound to some atom with eigenfrequency $\omega_k$ (and induced electric dipole moment of amplitude $ex$), then the equation is in the case of a simple one-dimensional but easily generalizable oscillator model

$$m\ddot{x} + \frac{m}{\tau}\dot{x} + m\omega_k^2 x = eE. \tag{15.32}$$

This equation takes into account that the (*almost* free conduction) electron does not follow the applied field without inertia. One should note that here we consider the electron as bound to the atom by an oscillator potential. From quantum mechanics we know that the Schrödinger equation with an oscillator potential $V_R = ax^2, a > 0$, allows only discrete bound states (i.e. states with infinite lifetime). The additional friction term we have introduced here has its counterpart in quantum mechanics as an imaginary part of the potential

$$V_I = i\Im V \quad \text{of} \quad V = V_R + i\Im V. \tag{15.33}$$

For $\Im V < 0$ the potential $V$ is absorptive, and violates unitarity (absorption of particles, energy into the target). A *resonance* is a state of *finite lifetime* $\tau = 1/\gamma$. It is not necessary to have $\Im V \neq 0$ in order to obtain resonances. Resonances are obtained, for instance, for screened Coulomb potentials or with potentials which permit the escape of particles to infinity through quantum tunneling. Such cases are, however, much more difficult to handle analytically, so that we restrict ourselves here for reasons of simplicity to a simple semi-classical model. Semi-classical means here, we start off classically, and then replace the frequency by discrete quantum mechanical frequencies.

Setting in Eq. (15.32):

$$x = x_0 e^{-i\omega t}, \tag{15.34}$$

we have (on the right side of the following equation we should strictly have the complete Lorentz force, but we assume that the magnetic contribution is negligible)[3]

$$\left(-m\omega^2 - \frac{m}{\tau} i\omega + m\omega_k^2\right) x = eE, \tag{15.35}$$

i.e. with $E(x,t) = E_0(x)e^{-i\omega t}$ the solution is

$$x = \frac{eE}{m(\omega_k^2 - \omega^2 - i\gamma\omega)}, \quad (\gamma = 1/\tau). \tag{15.36}$$

For $\gamma$ (called absorption coefficient or resonance width at half maximum) to represent a damping of $x$ or $E$, we must have $\Im\omega < 0$, i.e. at the resonance (at angular frequency $\omega_k$)

$$\omega = \frac{1}{2}\left[-i\gamma \pm \sqrt{4\omega_k^2 - \gamma^2}\right], \quad 4\omega_k^2 > \gamma^2,$$

and for $x \propto e^{-i\omega t}$ and $\gamma > 0$. We thus have

$$\mathbf{j} = ne\dot{\mathbf{x}} = -\frac{ne^2 i\omega \mathbf{E}}{m(\omega_k^2 - \omega^2 - i\gamma\omega)} \equiv \sigma \mathbf{E}, \tag{15.37}$$

so that the conductivity $\sigma$ is given by

$$\sigma = -\frac{ne^2 i\omega}{m(\omega_k^2 - \omega^2 - i\gamma\omega)}. \tag{15.38}$$

The generalized dielectric constant for a medium with conductivity $\sigma$ is as we know from Eq. (11.33).

$$\eta = \epsilon_0 + i\frac{\sigma}{\omega} = \epsilon_0 + \frac{ne^2}{m(\omega_k^2 - \omega^2 - i\gamma\omega)}, \tag{15.39}$$

[3]Observe that we can interpret the expression $p_m = ex, \mathbf{p}_m = \epsilon_0 \alpha \mathbf{E}$, where $\alpha = (e^2/m\epsilon_0)/(\omega_k^2 - \omega^2 - i\omega/\tau)$, as an electric molecular dipole moment induced by the field $\mathbf{E}$.

where $n$ is the number of particles (electrons) per unit volume. In general an electron of an electron gas may occupy one of many possible quantum mechanical states $k$ corresponding to resonant frequencies $\omega_k$ here, and there may be, say $n_k$ electrons per unit volume contributing to the $k$th resonant mode. It is then necessary to replace the last equation by

$$\eta = \epsilon_0 + \sum_k \frac{ne^2}{m} \frac{f_k}{\omega_k^2 - \omega^2 - i\gamma\omega}, \tag{15.40}$$

where the sum extends over the states of the electrons involved in the polarization of the medium. The factor[4]

$$\frac{e^2}{m(\omega_k^2 - \omega^2 - i\gamma\omega)}$$

is a measure of the polarizability of the medium since it can be related to the polarization $P$ in the relation (4.23), i.e. $D = \epsilon_0 E + P$ by writing $D = \eta E$, so that $\eta = \epsilon_0 + P/E$ and to the electric susceptibility $\chi_e$ defined by Eq. (4.21), i.e. $\mathbf{P} = \chi_e \epsilon_0 \mathbf{E}$. The sum in Eq. (15.40) divided by $\epsilon_0$ is therefore an expression for $\chi_e$.[5] Thus the considerations here can then also be considered to apply to a polarizable medium like a gas with conductivity zero. For strongly bound electrons $\omega_k \gg \omega$, and $\mathbf{j}$ is correspondingly small. We obtain therefore for the refractive index $p$ the result ($\sum_k f_k = 1$)

$$p = \sqrt{\frac{\eta}{\epsilon}} = \sqrt{1 + \frac{1}{\epsilon} \sum_k \frac{ne^2}{m} \frac{f_k}{\omega_k^2 - \omega^2 - i\gamma\omega}}. \tag{15.41}$$

The coefficients $f_k = n_k/n < 1$ are called *oscillator strengths*; they specify the probability of an electron with energy $\hbar\omega_k$ to be in state $k$, or, one can say, the proportionality to the number of electrons $n_k$ with energy $\hbar\omega_k$ per molecule. For atoms in vacuum $\epsilon$ is equal to $\epsilon_0$, and the refractive index is

$$p = \sqrt{\frac{\eta}{\epsilon_0}} \simeq 1 + \frac{ne^2}{2m\epsilon_0} \sum_k \frac{f_k}{\omega_k^2 - \omega^2 - i\gamma\omega} \equiv \tilde{n}(\omega). \tag{15.42}$$

For our purposes here it suffices to consider a dielectric, which permits only one resonance energy $\hbar\omega_k = \hbar\omega_0$, so that ($\omega_P^2$ with $f_0 = 1$ is the plasma

---

[4]This factor divided by $\epsilon_0$ is called the *molecular polarizability*, see e.g. Rossiter [126], p. 30, and is thus related to the electric susceptibility, Eq. (4.21).
[5]In some literature, Robinson [124], p. 132, the terms of this sum are referred to as Sellmeier terms.

frequency squared defined by Eq. (13.1))

$$p^2(\omega) = 1 + \frac{\omega_P^2}{\omega_0^2 - \omega^2 - i\gamma\omega}, \quad \omega_P^2 \equiv \frac{ne^2 f_0}{m\epsilon_0}. \tag{15.43}$$

We set

$$p^2(\omega) = \frac{\omega_0^2 - \omega^2 - i\gamma\omega + \omega_P^2}{\omega_0^2 - \omega^2 - i\gamma\omega} \equiv \frac{(\omega - \omega_a)(\omega - \omega_b)}{(\omega - \omega_c)(\omega - \omega_d)} \equiv \tilde{n}^2(\omega). \tag{15.44}$$

Then a calculation gives

$$\begin{aligned} \begin{matrix}\omega_a \\ \omega_b\end{matrix} &= -i\frac{\gamma}{2} \pm \sqrt{\omega_0^2 + \omega_P^2 - \frac{\gamma^2}{4}}, \\ \begin{matrix}\omega_c \\ \omega_d\end{matrix} &= -i\frac{\gamma}{2} \pm \sqrt{\omega_0^2 - \frac{\gamma^2}{4}}. \end{aligned} \tag{15.45}$$

We observe by looking at Eq. (15.45), that the poles of $p^2(\omega)$, like those of $A(\omega)$, are located in the lower half of the $\omega$-plane, as indicated in Fig. 15.5. We see that along the lines

$$\Re\omega_d \leq \Re\omega \leq \Re\omega_b, \quad \Re\omega_c \leq \Re\omega \leq \Re\omega_a \tag{15.46}$$

the refractive index $p(\omega)$ is pure imaginary. We also see that

$$\lim_{\omega\to\infty} \tilde{n}(\omega) = 1. \tag{15.47}$$

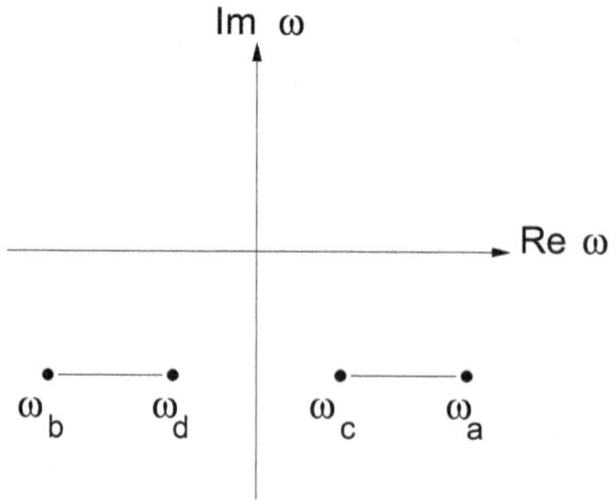


**Fig. 15.5.** The poles of $p(\omega)$ in the $\omega$-plane.

Next we consider the factor $1/(1+\tilde{n}(\omega))$ in the function $u(x,t)$ of Eq. (15.21), this being the wave packet in the dispersive medium. We have

$$\begin{aligned}\frac{1}{1+\tilde{n}(\omega)} &= \frac{1}{1+\sqrt{(\omega-\omega_a)(\omega-\omega_b)/(\omega-\omega_c)(\omega-\omega_d)}}\\ &= \frac{(\omega-\omega_c)(\omega-\omega_d)-\sqrt{(\omega-\omega_a)(\omega-\omega_b)(\omega-\omega_c)(\omega-\omega_d)}}{(\omega-\omega_c)(\omega-\omega_d)-(\omega-\omega_a)(\omega-\omega_b)}.\end{aligned} \tag{15.48}$$

The denominator is

$$\begin{aligned}(\omega-\omega_c)&(\omega-\omega_d)-(\omega-\omega_a)(\omega-\omega_b)\\ &= \omega^2-(\omega_c+\omega_d)\omega+\omega_c\omega_d-\omega^2+(\omega_a+\omega_b)\omega-\omega_a\omega_b\\ &= w(\omega_a+\omega_b-\omega_c-\omega_d)+(\omega_c\omega_d-\omega_a\omega_b)\\ &= -\frac{\gamma^2}{4}-\omega_0^2+\frac{2\gamma^2}{4}+\omega_0^2+\omega_P^2-\frac{\gamma^2}{4}=\omega_P^2.\end{aligned} \tag{15.49}$$

Hence

$$\frac{1}{1+\tilde{n}(\omega)} = \frac{(\cdots)(\cdots)-\sqrt{(\cdots)(\cdots)(\cdots)(\cdots)}}{\omega_P^2}, \tag{15.50}$$

i.e. $1/(1+\tilde{n}(\omega))$ has no pole in the $\omega$-plane. This example verifies that in the given case, and more generally (here without proof), the factor $2/(1+\tilde{n}(\omega))$ has no poles, i.e. is regular.

## 15.4 No Wave Packet with a Velocity $> c$

We are now in a position to prove that no wave packet in a dispersive medium with refractive index $\tilde{n}(\omega)$ can propagate with a velocity larger than the velocity of light. We consider the integral (see $u(x,t)$ of Eq. (15.21))

$$I = \int_{\overset{\curvearrowleft}{\rightarrow}} \frac{2}{1+\tilde{n}(\omega)} A(\omega) e^{i(k(\omega)x-\omega t)} d\omega \tag{15.51}$$

(physically the refractive index $\tilde{n}(\omega)$ can never be $-1$ for any value of $\omega$), in which the contour of integration $C = \overset{\curvearrowleft}{\rightarrow}$ is to be taken as indicated in Fig. 15.4. We have

$$I = \int_{-\infty}^{\infty} + \int_{\curvearrowleft} \cdots = \text{wave packet } u(x,t) + \int_{\curvearrowleft} \cdots. \tag{15.52}$$

We know: The integrand is a meromorphic function of $\omega$ with poles only in the lower half of the $\omega$-plane. Applying *Cauchy's residue theorem* we obtain

$$I = 2\pi i \sum \text{residues} = 0. \tag{15.53}$$

We have therefore

$$u(x,t) = -\int_{\curvearrowleft} \cdots . \tag{15.54}$$

We consider now the contour integral along the infinite semicircle, i.e. for $|\omega| \sim \infty$. Since

$$k(\omega) = \frac{\omega}{c}\tilde{n}(\omega) \to \frac{\omega}{c} \quad \text{for} \quad |\omega| \to \infty, \tag{15.55}$$

i.e.

$$\frac{d\omega}{dk} \to c, \quad \tilde{n}(\omega) \to 1,$$

we have

$$\int_{\curvearrowleft} \Rightarrow \int_{\curvearrowleft} \frac{2}{2} A(\omega) e^{i\omega(x/c-t)} d\omega. \tag{15.56}$$

This integral vanishes exponentially in the upper half of the $\omega$-plane, i.e. for $\Im\omega > 0$, *provided*

$$\frac{x}{c} - t > 0, \quad \text{i.e.} \quad v_{\text{signal}} = \frac{x}{t} > c. \tag{15.57}$$

This result implies that the propagation or spreading of waves into regions beyond the future light cone are excluded, or, phrased differently, in regions $x > ct$ the wave packet $u(x,t)$ vanishes.

In the next step we demonstrate, that for signal velocities

$$v_{\text{signal}} = \frac{x}{t} < c, \tag{15.58}$$

the wave packet has a definite value. This time we integrate in the lower half plane and consider the integral

$$I = \int_{\circlearrowright} \frac{2}{1+\tilde{n}(\omega)} A(\omega) e^{i(k(\omega)x - \omega t)} d\omega = \int_{-\infty}^{\infty} \cdots + \int_{\smile} \cdots . \tag{15.59}$$

We consider

$$\int_{\smile} \cdots \Rightarrow \int_{\smile} \frac{2}{2} A(\omega) e^{i\omega(x/c-t)} d\omega. \tag{15.60}$$

For $x < ct$, i.e. $v_{\text{signal}} \equiv x/t < c$, and $\Im\omega < 0$ we see that

$$e^{i\omega(x/c-t)}$$

is exponentially decreasing for $\omega \to \infty$. This means

$$\int_{\circlearrowright} \cdots = 0. \tag{15.61}$$

Hence

$$u(x,t) = \int_{-\infty}^{\infty} \cdots = I = 2\pi i \sum \text{residues}. \tag{15.62}$$

In order to be able to perform the integration explicitly, we require, of course, a knowledge of the function $A(\omega)$. Moreover, for the wave packet $u(x,t)$ to be nonvanishing, its spectral function $A(\omega)$ must possess poles in the *lower* half of the $\omega$-plane.

## 15.5 Explanation of the Anomalous Dispersion

We return to our approximate expression for the refractive index, i.e. to the square root of expression (15.43), i.e.

$$\tilde{n} = \sqrt{\frac{\eta}{\epsilon_0}} \simeq 1 + \frac{1}{2}\frac{\omega_P^2}{\omega_0^2 - \omega^2 - i\gamma\omega}, \tag{15.63}$$

taking into account only one pole at $\Re\omega = \omega_0$. With a real denominator we have

$$\tilde{n} = \sqrt{\frac{\eta}{\epsilon_0}} = 1 + \frac{\omega_P^2}{2}\frac{\omega_0^2 - \omega^2 + i\gamma\omega}{(\omega_0^2 - \omega^2)^2 + \gamma^2\omega^2} \tag{15.64}$$

and

$$\Re\tilde{n} = 1 + \frac{\omega_P^2}{2}\frac{(\omega_0^2 - \omega^2)}{(\omega_0^2 - \omega^2)^2 + \gamma^2\omega^2} \tag{15.65}$$

and

$$\Im\tilde{n} = \frac{\omega_P^2}{2}\frac{\gamma\omega}{(\omega_0^2 - \omega^2)^2 + \gamma^2\omega^2}. \tag{15.66}$$

Using a delta function representation of Example 2.5 (Eq. (2.65)), i.e.

$$\delta(x) = \lim_{\epsilon \to 0} \frac{1}{\pi}\frac{\epsilon}{\epsilon^2 + x^2}, \tag{15.67}$$

we see that in the limit of an infinitely sharp resonance:

$$\lim_{\gamma\omega\to 0}\Im\tilde{n} = \frac{\omega_P^2}{2}\pi\delta(\omega_0^2-\omega^2) \overset{(2.44)}{=} \frac{\omega_P^2}{2}\pi\frac{1}{2\omega_0}[\delta(\omega-\omega_0)+\delta(\omega+\omega_0)], \tag{15.68}$$

where the contribution of the second delta function vanishes since both $\omega$ and $\omega_0$ are positive. This is the case of a "*zero width resonance*" which we shall encounter again in Section 16.7. After 3 lines of calculations:

$$\frac{d\Re\tilde{n}}{d\omega} = \frac{\omega_P^2\omega[(\omega_0^2-\omega^2)^2-\gamma^2\omega_0^2]}{[(\omega_0^2-\omega^2)^2+\gamma^2\omega^2]^2}. \tag{15.69}$$

Now (for $\omega \neq 0$),

$$\frac{d\Re\tilde{n}}{d\omega} = 0, \quad \text{where} \quad (\omega_0^2-\omega^2)^2 = \gamma^2\omega_0^2, \tag{15.70}$$

i.e.

$$\omega^2 = \omega_0^2 \mp \gamma\omega_0, \tag{15.71}$$

and hence

$$\omega \simeq \omega_0 \mp \frac{1}{2}\gamma. \tag{15.72}$$

It follows that $\Re\tilde{n}$ has extrema at these points. Using Eq. (15.64) and setting $\omega \simeq \omega_0 \mp \gamma/2$, the refractive index $\tilde{n}$ at these points is:

$$\tilde{n} = 1 + \frac{\omega_P^2}{2}\frac{(\pm\gamma\omega_0)+i\gamma\sqrt{\omega_0^2\mp\gamma\omega_0}}{(\pm\gamma\omega_0)^2+\gamma^2(\omega_0^2\mp\gamma\omega_0)} \tag{15.73}$$

or

$$\Im\tilde{n} \overset{(15.66)}{\simeq} \frac{\omega_P^2}{2}\frac{\gamma\sqrt{\omega_0^2}}{2\gamma^2\omega_0^2} = \frac{\omega_P^2}{4\gamma\omega_0} \tag{15.74}$$

and

$$\Re\tilde{n} \overset{(15.73)}{\simeq} 1 + \frac{\omega_P^2}{2}\frac{(\pm\gamma\omega_0)}{2\gamma^2\omega_0^2} = 1 \pm \frac{\omega_P^2}{4\gamma\omega_0}. \tag{15.75}$$

Thus

$$|\Re\tilde{n}-1| = |\Im\tilde{n}| \quad \text{at} \quad \omega = \omega_0 \pm \frac{\gamma}{2}. \tag{15.76}$$

We see therefore that the case of the anomalous dispersion defined at the beginning of this chapter by Eq. (15.11) has

$$\frac{d\Re\tilde{n}}{d\omega} < 0, \quad \text{where} \quad \omega_0 - \frac{\gamma}{2} \leq \omega \leq \omega_0 + \frac{\gamma}{2}. \tag{15.77}$$

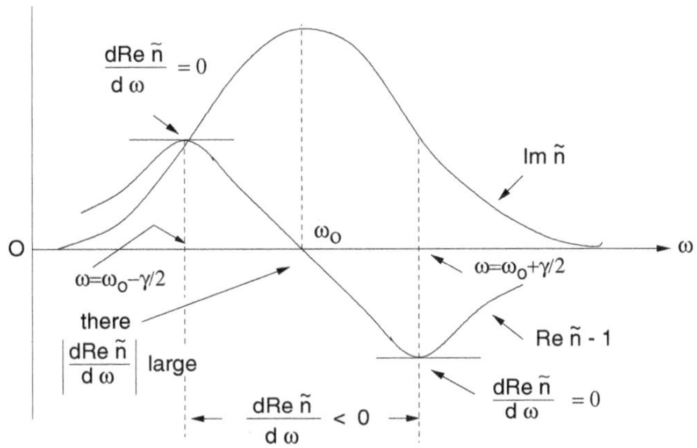


**Fig. 15.6.** $\Im\tilde{n}$ large where $d\Re\tilde{n}/d\omega$ is negative.

In this region, see Fig. 15.6, $\Im\tilde{n}$ is particularly large (this is also the region of the maximum).[6]

We shall see in Chapter 16, that a large imaginary part (of e.g. $\tilde{n}$) is connected with strong absorption effects. That physical velocities

$$v_g = \frac{c}{\tilde{n} + \omega(d\tilde{n}/d\omega)} > c \tag{15.78}$$

do not occur, i.e. when $d\tilde{n}/d\omega$ is negative, is now *qualitatively explained* by arguing that these mathematically arising velocities are removed by the absorptive effects, i.e. that in the presence of dissipation the group velocity loses its meaning. In Chapter 16 we return to these considerations from a different angle, that of dispersion relations.

**Example 15.1 (Method of stationary phase).** Consider the integral

$$I = \int F(\omega)e^{i\phi(\omega)}d\omega, \tag{15.79}$$

where $F(\omega)$ is a slowly varying function of $\omega$, whereas $\phi(\omega)$ is a rapidly varying phase with only one stationary point at $\omega_0$. Show that approximately

$$I \simeq \sqrt{\frac{2i\pi}{\phi''(\omega_0)}}F(\omega_0)e^{i\phi(\omega_0)}. \tag{15.80}$$

**Solution:** In the neighborhood of $\omega_0$ we can write

$$\phi(\omega) = \phi(\omega_0) + \frac{1}{2}\phi''(\omega_0) + \cdots . \tag{15.81}$$

[6]For analogous literature see e.g. Cheston [30], pp. 279–283, Jackson [73], Sections 7.5–7.8.

Since $F(\omega)$ varies in the neighborhood of $\omega_0$ only slowly, we have

$$I \simeq F(\omega_0)e^{i\phi(\omega_0)} \int_{-\infty}^{\infty} e^{\frac{i}{2}\phi''(\omega_0)(\omega-\omega_0)^2} d\omega. \tag{15.82}$$

The remaining integral is a Fresnel integral[7] (or can be evaluated with appropriate assumptions like an integral over a Gauss function) and leads to the above result.

## 15.6 Exercises

**Example 15.2 (Estimate of refractive index).** Verify that the quantity $ne^2/m\omega^2\epsilon_0$ is dimensionless. Obtain the order of magnitude of the refractive index of Eq. (15.40) by taking only $f_0$ with $f_0 = 1, \gamma = 0, \omega_0^2 - \omega^2 \simeq 10^{30}$ $(\text{rad/sec})^2$.

(Answer: Roughly of order unity).

**Example 15.3 (Limiting cases of generalized dielectric constant).** Consider the generalized dielectric constant $\eta$ of Eq. (15.40) for free electrons ($\omega_k^2 = 0$) and write out the formula for the cases of wave frequency much higher than the collision frequency, and for the collision frequency higher than the wave frequency.[8]

**Example 15.4 (Expansions of refractive index).** Show that in the case of absorption in the ultraviolet the square of the real part of the index of refraction can be expanded in an ascending power series of $\omega^2$. What does this expansion become if in addition there is absorption in the infrared?

**Example 15.5 (Displacement of a bound charge).** Show that the displacement $x$ of a bound charge as determined by Eq. (15.32) is given by

$$x = \frac{eE_0/m}{\sqrt{\omega^2\gamma^2 + (\omega^2 - \omega_k^2)^2}} \cos(\omega t + \beta) \quad \text{meters}, \quad \gamma = \frac{1}{\tau}, \tag{15.83}$$

where $\beta = \tan^{-1}(\omega\gamma/(\omega_k^2 - \omega^2))$.

---

[7]See e.g. Magnus and Oberhettinger [90], p. 96.
[8]Cf. Tyras [152], p. 53.

## Chapter 16

# Causality and Dispersion Relations

## 16.1 Introductory Remarks

In this chapter we continue the study of analytic properties of spectral functions, but concentrate on Green's functions and causality. In this connection important integral relations between real and imaginary parts of spectral functions are introduced, which are also known as *dispersion relations* or (in more mathematical contexts) as *Hilbert transforms*, and were first applied by Kramers to the dielectric susceptibility $\chi$.

## 16.2 Cause $U$ and Effect $E$

We let $U(t)$ represent a *cause*, e.g. a charge distribution which varies with time, and $E(t)$ an *effect*, e.g. an electromagnetic field. Since both quantities represent real physical quantities, we assume them to be real.

The *Fourier transforms* of the two quantities define other *representations* of $U(t)$ and $E(t)$:

$$\begin{aligned} U(t) &= \frac{1}{\sqrt{2\pi}}\int_{-\infty}^{\infty} d\omega e^{-i\omega t}u(\omega), \quad u(\omega) = \frac{1}{\sqrt{2\pi}}\int_{-\infty}^{\infty} dt e^{i\omega t}U(t), \\ E(t) &= \frac{1}{\sqrt{2\pi}}\int_{-\infty}^{\infty} d\omega e^{-i\omega t}e(\omega), \quad e(\omega) = \frac{1}{\sqrt{2\pi}}\int_{-\infty}^{\infty} dt e^{i\omega t}E(t). \quad (16.1) \end{aligned}$$

Since we demand that $U$ and $E$ are real, i.e.

$$U(t) = U^*(t), \quad E(t) = E^*(t), \quad (16.2)$$

it follows that

$$\int_{-\infty}^{\infty} d\omega e^{-i\omega t} u(\omega) = \int_{-\infty}^{\infty} d\omega e^{i\omega t} u^*(\omega) = \int_{-\infty}^{\infty} d\omega' e^{-i\omega' t} u^*(-\omega')$$
$$= \int_{-\infty}^{\infty} d\omega e^{-i\omega t} u^*(-\omega), \tag{16.3}$$

i.e.

$$u(\omega) = u^*(-\omega), \quad e(\omega) = e^*(-\omega). \tag{16.4}$$

These relations between the integrands of the Fourier transforms or spectral functions are referred to as "*crossing relations*". One should note that $\omega$ is assumed to be real. For $U(t)$, for instance, it follows that

$$U(t) = \frac{1}{\sqrt{2\pi}} \int_{-\infty}^{0} d\omega e^{-i\omega t} u(\omega) + \frac{1}{\sqrt{2\pi}} \int_{0}^{\infty} d\omega e^{-i\omega t} u(\omega)$$
$$= \frac{1}{\sqrt{2\pi}} \int_{0}^{\infty} d\omega' e^{i\omega' t} \underbrace{u(-\omega')}_{u^*(\omega')} + \frac{1}{\sqrt{2\pi}} \int_{0}^{\infty} d\omega e^{-i\omega t} u(\omega)$$
$$= \frac{1}{\sqrt{2\pi}} \int_{0}^{\infty} d\omega [u(\omega) e^{-i\omega t} + u^*(\omega) e^{i\omega t}]. \tag{16.5}$$

We set

$$u(\omega) = \frac{1}{2} r(\omega) e^{i\theta(\omega)}, \tag{16.6}$$

where $r(\omega), \theta(\omega)$ are real. Then

$$U(t) = \frac{1}{\sqrt{2\pi}} \int_{0}^{\infty} d\omega \left[\frac{1}{2} r(\omega) e^{i\theta(\omega)} e^{-i\omega t} + \frac{1}{2} r(\omega) e^{-i\theta(\omega)} e^{i\omega t}\right]$$
$$= \frac{1}{\sqrt{2\pi}} \int_{0}^{\infty} d\omega r(\omega) \cos[\omega t - \theta(\omega)], \tag{16.7}$$

i.e. the Fourier representation of $U(t)$ depends only on *positive* frequencies ($\omega > 0$). A corresponding result can be obtained for $E(t)$.

## 16.3 $(U, E)$-Linearity and Green's Functions

We let $U_i$ be the cause of the effect $E_i$. Then we assume that

$$\alpha U_i + \beta U_j$$

is cause of the effect

$$\alpha E_i + \beta E_j,$$

or more generally

$$E(t) = \int_{-\infty}^{\infty} dt' G(t,t') U(t') \tag{16.8}$$

(superposition principle). Here $G(t,t')$ is a *weight* or *Green's function.* We also assume that

$$G(t,t') = G(t-t'), \tag{16.9}$$

i.e. the process connecting $U$ and $E$ is independent of absolute times. Then $E$ is the inhomogeneous solution of a differential equation

$$DE(t) = U(t) \tag{16.10}$$

with

$$DG(t-t') = \delta(t-t') = \delta(t'-t). \tag{16.11}$$

We write $G(t,t')$ as $G(t-t')$ and not as $G(t'-t)$ for reasons of causality which lead us to the so-called *retarded Green's function.* Then

$$E(t) = \int_{-\infty}^{\infty} dt' G(t,t') U(t'). \tag{16.12}$$

This is precisely the relation given above. Since $U$ and $E$ are real, $G$ must also be real. Let $g(\omega)$ be the Fourier transform of $G$:

$$G(t) = \frac{1}{2\pi}\int_{-\infty}^{\infty} d\omega e^{-i\omega t} g(\omega), \quad g(\omega) = \int_{-\infty}^{\infty} G(t) e^{i\omega t} dt. \tag{16.13}$$

Then

$$\begin{aligned} e(\omega) &= \frac{1}{\sqrt{2\pi}}\int_{-\infty}^{\infty} dt e^{i\omega t} E(t) \\ &= \frac{1}{\sqrt{2\pi}}\int_{-\infty}^{\infty} dt e^{i\omega t}\int_{-\infty}^{\infty} dt' G(t-t') U(t'), \end{aligned} \tag{16.14}$$

and, inserting Eq. (16.13) for $G(t-t')$, this becomes

$$\begin{aligned} e(\omega) &= \frac{1}{(\sqrt{2\pi})^3}\int_{-\infty}^{\infty} dt e^{i\omega t}\int_{-\infty}^{\infty} dt' U(t')\int_{-\infty}^{\infty} d\omega' e^{-i\omega'(t-t')} g(\omega') \\ &= \frac{1}{\sqrt{2\pi}}\int_{-\infty}^{\infty} dt' U(t')\int d\omega' g(\omega') e^{i\omega' t'} \underbrace{\frac{1}{2\pi}\int_{-\infty}^{\infty} dt e^{i(\omega-\omega')t}}_{\delta(\omega-\omega')} \end{aligned} \tag{16.15}$$

and hence

$$e(\omega) = \frac{1}{\sqrt{2\pi}} \int_{-\infty}^{\infty} dt' U(t') g(\omega) e^{i\omega t'} \equiv g(\omega) u(\omega),$$

i.e.

$$e(\omega) = g(\omega)u(\omega), \quad e^*(-\omega) = g^*(-\omega)u^*(-\omega). \tag{16.16}$$

These relations are called *convolution theorems.*

Since

$$u(\omega) = u^*(-\omega), \quad e(\omega) = e^*(-\omega), \tag{16.17}$$

it follows that

$$e(\omega) = g^*(-\omega)u(\omega), \quad \text{i.e.} \quad g(\omega) = g^*(-\omega). \tag{16.18}$$

The latter of these relations implies again that $G(t)$ is real (see remarks at the beginning).

A *mono-frequent* cause $U_0 e^{-i\omega t}$ leads to a mono-frequent effect:

$$\begin{aligned} E(t) &= \int_{-\infty}^{\infty} dt' G(t-t') U(t') = \int_{-\infty}^{\infty} dt' G(t-t') U_0 e^{-i\omega t'} \\ &= U_0 e^{-i\omega t} \int_{-\infty}^{\infty} dt' G(t-t') e^{i\omega(t-t')} \\ &= U_0 e^{-i\omega t} \int_{-\infty}^{\infty} \underbrace{d(t'-t)}_{-d(t-t')} G(t-t') e^{i\omega(t-t')}, \end{aligned} \tag{16.19}$$

and hence

$$\begin{aligned} E(t) &= U_0 e^{-i\omega t} \int_{\infty}^{-\infty} (-) d(t-t') G(t-t') e^{i\omega(t-t')} \\ &= U_0 e^{-i\omega t} g(\omega) \equiv E_0 e^{-i\omega t}, \end{aligned} \tag{16.20}$$

where $E_0 = U_0 g(\omega)$.

## 16.4 Causality

The concept of causality expresses the fundamental idea that an effect $E(t)$ at time $t$ can only have and depend on causes $U(t)$ prior to time $t$. Since

$$E(t) = \int_{-\infty}^{\infty} dt' G(t-t') U(t'), \tag{16.21}$$

($G$ real for $U$ and $E$ real), this means $E(t)$ can only depend on such $U(t')$, for which $t' < t$. More concretely

$$G(t-t') = 0 \quad \text{for} \quad t-t' < 0, \tag{16.22}$$

i.e. $G(t-t') \propto \theta(t-t')$, i.e. only such values of the cause $U$ enter the effect $E(t)$ at time $t$, which appeared before the instant of time $t$, at which we are interested in the effect $E(t)$. This time ordering is described as *causality*. This causality is also described as *macrocausality* in contrast to *microcausality* or *locality*, which is a quantum theoretical concept concerning operators representing observables and implies that such observables commute if their separation is spacelike (because then they cannot exchange signals since no signal can travel with a velocity greater than that of light). The Green's function obeying macrocausality is the *retarded Green's function*. Consider once again a typical classical equation of motion

$$\ddot{x} + 2\gamma\dot{x} + \omega_0^2 x = f(t). \tag{16.23}$$

The friction term violates the time-reversal invariance of the simple Newton equation

$$\ddot{x} + \omega_0^2 x = 0 \tag{16.24}$$

(which is plausible, because a process involving friction, in which therefore energy is dissipated as heat, is irreversible). For $\gamma > 0$, the direction of the friction force $2\gamma\dot{x}$ is opposite to that of the motion since

$$\ddot{x} + \omega_0^2 x = -2\gamma\dot{x} + f(t). \tag{16.25}$$

The Green's function of this problem is, however, *uniquely defined by the friction term*, because with

$$\left(\frac{d^2}{dt^2} + 2\gamma\frac{d}{dt} + \omega_0^2\right) G(t,t') = \delta(t-t') \tag{16.26}$$

we obtain

$$G(t-t') = \frac{1}{2\pi}\int_{-\infty}^{\infty} d\omega \frac{e^{-i\omega(t-t')}}{\omega_0^2 - \omega^2 - 2i\gamma\omega}. \tag{16.27}$$

The poles of the integrand are located in the domain $\Im\omega < 0$, and are given by

$$\omega = -i\gamma \pm \sqrt{\omega_0^2 - \gamma^2}, \quad \gamma > 0. \tag{16.28}$$

In this way the integral with $\gamma \neq 0$ and the condition $G =$ finite is uniquely defined. Since there are no poles in the upper half-plane

$$\int_{\curvearrowright} \cdots = 2\pi i \sum \text{residues} = 0, \tag{16.29}$$

and it follows that

$$\int_{-\infty}^{\infty} \cdots = -\int_{\curvearrowleft} \cdots = 0 \quad \text{for} \quad t - t' < 0, \tag{16.30}$$

i.e.

$$G(t-t') = 0 \quad \text{for} \quad t - t' < 0 \tag{16.31}$$

and

$$G(t-t') \neq 0 \quad \text{for} \quad t - t' > 0. \tag{16.32}$$

This finite Green's function is the retarded Green's function. If

$$G(t-t') = 0 \quad \text{for} \quad t - t' < 0, \tag{16.33}$$

we have

$$x(t) = \int_{-\infty}^{\infty} G(t-t')f(t')dt' = \int_{-\infty}^{t} G(t-t')f(t')dt'. \tag{16.34}$$

The solution of the homogeneous equation which could be added to give the complete solution (with $x \to 0$ for $t \to \infty$), i.e. the solution of

$$\ddot{x} + 2\gamma\dot{x} + \omega_0^2 x = 0, \tag{16.35}$$

is

$$x_0(t) = \left(ae^{i\sqrt{\omega_0^2-\gamma^2}t} + be^{-i\sqrt{\omega_0^2-\gamma^2}t}\right)e^{-\gamma t}. \tag{16.36}$$

This diverges for $t \to -\infty$. To ensure that $x(t)$ is finite for all times $t$, we have to choose $a = 0 = b$, i.e. a contribution of the homogeneous equation to the solution is excluded by boundary conditions. In the case $\gamma \neq 0$ the integration contour is automatically fixed by the requirement of $G, x$ to be finite. In the case $\gamma = 0$, however, the contour must be specified separately, because then the poles $\omega = \pm\omega_0$ lie on the real axis (in the above simplified model). We consider this case by selecting a pole at $\omega = \omega_0 - i\epsilon, \epsilon > 0$, with $\epsilon \to 0$. It is possible to show rigorously that a dispersion without absorption contradicts causality.[1]

---

[1] Jost [75], p. 82.

## 16.5 Causality and Analyticity

We return to our original consideration and now permit *complex values of* $\omega$. Since

$$g(\omega) = \int_{-\infty}^{\infty} dt G(t) e^{i\omega t}, \quad G(t) \propto \theta(t), \tag{16.37}$$

it follows that

$$g(\omega) = \int_{0}^{\infty} dt G(t) e^{i\omega t}, \tag{16.38}$$

or (since $G(t)$ is real, see above)

$$g^*(\omega) = \int_{0}^{\infty} dt G(t) e^{-i\omega t} \neq g^*(\omega^*). \tag{16.39}$$

We now set

$$\omega = \omega_R + i\omega_I. \tag{16.40}$$

Then

$$g(\omega) \equiv g(\omega_R + i\omega_I) = \int_{0}^{\infty} dt G(t) e^{i\omega_R t} e^{-\omega_I t} \tag{16.41}$$

and

$$g^*(\omega) = g^*(\omega_R + i\omega_I) = \int_{0}^{\infty} dt G(t) e^{-i\omega_R t} e^{\omega_I t} = g(-\omega). \tag{16.42}$$

We assume that

$$G(t) \sim \text{ finite for } t \to \infty.$$

Then there exists a $g(\omega)$ for $\omega_I > 0$ and $g^*(\omega)$ for $\omega_I < 0$. The derivatives

$$\frac{d}{d\omega} g(\omega) \quad \text{and} \quad \frac{d}{d\omega} g^*(\omega)$$

also exist at every point with respectively $\omega_I > 0$ and $\omega_I < 0$. Hence $g(\omega)$ is analytic in the half-plane $\omega_I > 0$, and $g^*(\omega)$ is analytic in the half-plane $\omega_I < 0$. We see therefore that we can infer from the causality condition

$$G(t - t') = 0 \quad \text{for} \quad t - t' < 0 \tag{16.43}$$

the analyticity of the Fourier transforms $g(\omega)$ and $g^*(\omega)$ in the domains $\omega_I > 0$ and $\omega < 0$ respectively.

**Example 16.1 (The Cauchy–Riemann equations).** Starting from the integral

$$g(\omega) = \int_0^\infty dt G(t) e^{i\omega t}, \quad \omega = \omega_R + i\omega_I, \tag{16.44}$$

examine whether

$$g(\omega) = g_R(\omega_R, \omega_I) + i g_I(\omega_R, \omega_I) \equiv u + iv \tag{16.45}$$

satisfies the Cauchy–Riemann equations, i.e. with $\omega_R = x, \omega_I = y, \omega = x+iy$,

$$\frac{\partial}{\partial x} g_R = \frac{\partial}{\partial y} g_I, \quad \frac{\partial}{\partial x} g_I = -\frac{\partial}{\partial y} g_R, \tag{16.46}$$

or $u_x = v_y, v_x = -u_y$.

**Solution:** The solution is trivial. See also Section 3.7.

## 16.6 Principal Value Integrals and Dispersion Relations

From the fact that the spectral functions $g(\omega)$ and $g^*(\omega)$ possess simple poles in one half of the $\omega$-plane and are analytic in the other, one can derive relations between their real and imaginary parts called *dispersion relations.*

We consider the following integral taken along the infinite semicircle in the domain $\Im\omega' > 0$,

$$I = \int_{\curvearrowleft} d\omega' \frac{g(\omega')}{\omega' - \omega + i\epsilon}, \quad \epsilon > 0, \quad \omega \text{ real}. \tag{16.47}$$

With $\epsilon > 0$, which displaces the pole underneath the real axis, we could exclude it from within a contour along the real axis and then back along the semicircle at infinity. We can achieve the same effect, however, with a small semicircular distortion of the contour around the pole on the real axis as indicated in Fig. 16.1.

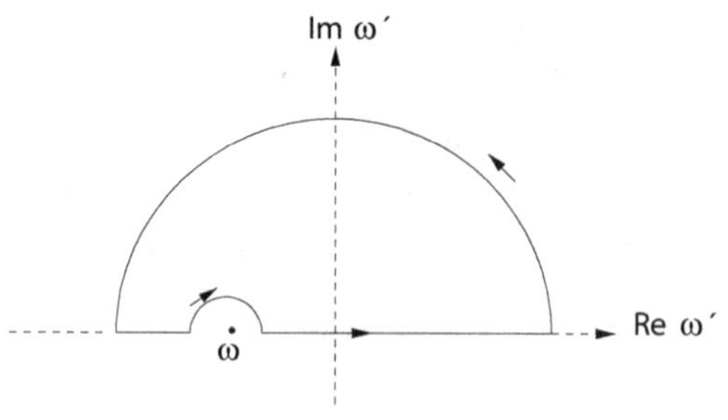


**Fig. 16.1.** The contour around the point $\omega$ in the complex $\omega'$-plane.

In the upper half-plane $\Im\omega' > 0$, as we saw, $g(\omega')$ is analytic. Thus *if* $g(\omega')$ decreases sufficiently fast towards zero for $|\omega'| \to \infty, \Im\omega' > 0$, so that $\int_{\curvearrowleft} \cdots = 0$, then

$$\int_{\circlearrowleft} d\omega' \cdots = 2\pi i \sum \text{residues} = 0 \tag{16.48}$$

and

$$\int_{-\infty}^{\infty} d\omega' \cdots = 0, \tag{16.49}$$

where the integration is meant to be taken along the contour shown in Fig. 16.1 from $-\infty$ to $+\infty$, i.e.

$$\begin{aligned} 0 &= \int_{-\infty}^{\infty} d\omega' \frac{g(\omega')}{\omega' - \omega} \\ &= P\int d\omega' \frac{g(\omega')}{\omega' - \omega} + \underbrace{\int_{\curvearrowright}}_{\substack{\text{infinitesimal} \\ \text{semicircle around } \omega'=\omega}} d\omega' \frac{g(\omega')}{\omega' - \omega}, \end{aligned} \tag{16.50}$$

where $P\int$ means "*principal value*" of the integral (which will be explained in detail below, see Eq. (16.83)). We note here that

$$P\int d\omega' \frac{g(\omega')}{\omega' - \omega} = \lim_{\epsilon \to 0}\left(\int_{-\infty}^{\omega-\epsilon} \cdots + \int_{\omega+\epsilon}^{\infty} \cdots\right), \tag{16.51}$$

and we shall return to this expression later. With

$$\omega' - \omega = \rho e^{i\varphi}, \quad d\omega' = i\rho e^{i\varphi} d\varphi, \tag{16.52}$$

we have

$$\begin{aligned} \underbrace{\int_{\curvearrowright}}_{\substack{\text{infinitesimal} \\ \text{semicircle around } \omega'=\omega}} d\omega' \frac{g(\omega')}{\omega' - \omega} &= \lim_{\rho \to 0}\left(ig(\omega)\int_{+\pi}^{0} \frac{e^{i\varphi}\rho d\varphi}{\rho e^{i\varphi}}\right) \\ &= ig(\omega)\int_{\pi}^{0} d\varphi = -ig(\omega)\pi. \end{aligned} \tag{16.53}$$

Hence

$$g(\omega) = \frac{1}{i\pi} P \int_{-\infty}^{\infty} d\omega' \frac{g(\omega')}{\omega' - \omega}, \tag{16.54}$$

or

$$g(-\omega) = \frac{1}{i\pi} P \int_{-\infty}^{\infty} d\omega' \frac{g(\omega')}{\omega' + \omega} = -\frac{1}{i\pi} P \int_{-\infty}^{\infty} d\omega' \frac{g(-\omega')}{\omega' - \omega}. \tag{16.55}$$

Comparison with

$$g^*(\omega) = -\frac{1}{i\pi} P \int_{-\infty}^{\infty} d\omega' \frac{g^*(\omega')}{\omega' - \omega} \tag{16.56}$$

verifies the relation obtained previously (see Eq. (16.4))

$$g(-\omega) = g^*(\omega). \tag{16.57}$$

The right side of the integral relations we thus obtained contains the factor "$i$". Taking real and imaginary parts of Eq. (16.54) we obtain:

$$\begin{aligned} \Re g(\omega) &= \frac{1}{\pi} P \int_{-\infty}^{\infty} d\omega' \frac{\Im g(\omega')}{\omega' - \omega}, \\ \Im g(\omega) &= -\frac{1}{\pi} P \int_{-\infty}^{\infty} d\omega' \frac{\Re g(\omega')}{\omega' - \omega}. \end{aligned} \tag{16.58}$$

We also have

$$\Re g^*(\omega) = -\frac{1}{\pi} P \int_{-\infty}^{\infty} d\omega' \frac{\Im g^*(\omega')}{\omega' - \omega}, \quad \Im g^*(\omega) = \frac{1}{\pi} P \int_{-\infty}^{\infty} d\omega' \frac{\Re g^*(\omega')}{\omega' - \omega}. \tag{16.59}$$

These relations are the promised *dispersion relations*. They are also called *Hilbert transforms*.

We can rewrite the dispersion relations in a form involving only positive frequencies. To achieve this, we write

$$\begin{aligned} \Re g(\omega) &= \frac{1}{\pi} P \int_{-\infty}^{0} d\omega' \frac{\Im g(\omega')}{\omega' - \omega} + \frac{1}{\pi} P \int_{0}^{\infty} d\omega' \frac{\Im g(\omega')}{\omega' - \omega} \\ &= \frac{1}{\pi} P \int_{0}^{\infty} d\omega' \frac{\Im g(-\omega')}{-\omega' - \omega} + \frac{1}{\pi} P \int_{0}^{\infty} d\omega' \frac{\Im g(\omega')}{\omega' - \omega}, \end{aligned} \tag{16.60}$$

(of course, the pole is contained only in one integral; in the other the principal value integral is an ordinary integral). Since

$$g(\omega) = g^*(-\omega), \quad g(-\omega) = g^*(\omega), \tag{16.61}$$

we have

$$\Re g(\omega) + i\Im g(\omega) = \Re g(-\omega) - i\Im g(-\omega), \tag{16.62}$$

i.e.

$$\Re g(\omega) = \Re g(-\omega), \quad \Im g(\omega) = -\Im g(-\omega). \tag{16.63}$$

Hence

$$\begin{aligned}
\Re g(\omega) &= -\frac{1}{\pi}P\int_0^\infty d\omega' \frac{\Im g(\omega')}{-\omega' - \omega} + \frac{1}{\pi}P\int_0^\infty d\omega' \frac{\Im g(\omega')}{\omega' - \omega} \\
&= \frac{1}{\pi}P\int_0^\infty d\omega' \Im g(\omega') \left[\frac{1}{\omega' + \omega} + \frac{1}{\omega' - \omega}\right] \\
&= \frac{1}{\pi}P\int_0^\infty \Im g(\omega') d\omega' \frac{2\omega'}{\omega'^2 - \omega^2} \\
&= \frac{2}{\pi}P\int_0^\infty d\omega' \omega' \frac{\Im g(\omega')}{\omega'^2 - \omega^2}.
\end{aligned} \tag{16.64}$$

Similarly

$$\begin{aligned}
\Im g(\omega) &= -\frac{1}{\pi}P\int_{-\infty}^\infty d\omega' \frac{\Re g(\omega')}{\omega' - \omega} \\
&= -\frac{1}{\pi}P\int_{-\infty}^0 d\omega' \frac{\Re g(\omega')}{\omega' - \omega} - \frac{1}{\pi}P\int_0^\infty d\omega' \frac{\Re g(\omega')}{\omega' - \omega} \\
&= -\frac{1}{\pi}P\int_0^\infty d\omega' \frac{\Re g(-\omega')}{-\omega' - \omega} - \frac{1}{\pi}P\int_0^\infty d\omega' \frac{\Re g(\omega')}{\omega' - \omega},
\end{aligned} \tag{16.65}$$

and using Eq. (16.63), this becomes

$$\begin{aligned}
\Im g(\omega) &= -\frac{1}{\pi}P\int_0^\infty d\omega' \frac{\Re g(\omega')}{-\omega' - \omega} - \frac{1}{\pi}P\int_0^\infty d\omega' \frac{\Re g(\omega')}{\omega' - \omega} \\
&= -\frac{1}{\pi}P\int_0^\infty d\omega' \Re g(\omega') \left[-\frac{1}{\omega' + \omega} + \frac{1}{\omega' - \omega}\right] \\
&= -\frac{1}{\pi}P\int_0^\infty d\omega' \Re g(\omega') \frac{(\omega' + \omega) - (\omega' - \omega)}{(\omega'^2 - \omega^2)} \\
&= -\frac{2\omega}{\pi}P\int_0^\infty d\omega' \frac{\Re g(\omega')}{\omega'^2 - \omega^2}.
\end{aligned} \tag{16.66}$$

Dispersion relations of this kind were first formulated by Kramers and Kronig (1926),[2] and, in fact, in application to the *dielectric susceptibility* $\chi_e(\omega)$, which we introduced in Chapter 4, i.e.

$$\chi_e(\omega) = \frac{\epsilon - \epsilon_0}{\epsilon_0} = \frac{\Re\epsilon(\omega) + i\Im\epsilon(\omega) - \epsilon_0}{\epsilon_0}. \tag{16.67}$$

We obtained the dispersion relations for $g(\omega)$ from the assumption that this function is analytic (i.e. free of poles) in the upper half of the complex $\omega$-plane. We saw earlier that the refractive index or rather its square or the dielectric constant have the same property. For this reason one can write down dispersion relations for this quantity. Thus, from the dispersion relation in terms of positive frequencies (by identifying $g(\omega)$ with $\epsilon(\omega)-\epsilon_0$), we obtain

$$\begin{aligned}\Re\epsilon(\omega) - \epsilon_0 &= \frac{2}{\pi}P\int_0^\infty dx \frac{x\Im\epsilon(x)}{x^2-\omega^2},\\ \Im\epsilon(\omega) &= -\frac{2\omega}{\pi}P\int_0^\infty dx \frac{\Re\epsilon(x)-\epsilon_0}{x^2-\omega^2}.\end{aligned} \tag{16.68}$$

We consider these relations in some applications in the next section.

## 16.7 Absorption: Special Cases

We consider three cases.

(a) The case $\Im\epsilon(\omega) = 0$ for all frequencies $\omega$ is the *case without absorption.* Recall: $\mathbf{E}, \mathbf{H} \propto e^{i(k(\omega)x-\omega t)}, k(\omega) = (\omega/c)\tilde{n}(\omega), \tilde{n}(\omega) = \sqrt{\epsilon\mu}/\sqrt{\epsilon_0\mu_0}$. If $\epsilon(\omega)$ contains an imaginary part, then also $\tilde{n}(\omega)$ and $k(\omega)$, and hence $\mathbf{E}, \mathbf{H}$ have exponentially decreasing amplitudes, i.e. there is damping. If $\Im\epsilon(\omega) = 0$, then $\Re\epsilon(\omega) = \epsilon_0$, i.e. $\Re(\omega)$ is independent of $\omega$, and

$$v_{\text{phase}} = \frac{\omega}{k} = \frac{c\sqrt{\mu_0\epsilon_0}}{\sqrt{\mu\epsilon}} = \frac{1}{\sqrt{\mu\epsilon}} = c'. \tag{16.69}$$

Thus there is no dispersion.[3] Effectively the dispersion relations say: If there is dispersion in some frequency region, then (in general) there is absorption in some other region.

(b) The case $\Im\epsilon(\omega) = \kappa\delta(\omega - \omega_n), \kappa > 0, 0 < \omega_n < \infty$. This case is called the case of the "*zero width resonance*" and corresponds to that of absorption

[2]Kramers [80], p. 333 and Kronig [81]. See also Hilgevoord [71].

[3]There is no dispersion when $d\tilde{n}(\omega)/d\omega = d(ck/\omega)/d\omega = 0$.

in a very narrow frequency region around $\omega = \omega_n$. Then for $\omega \neq \omega_n$:

$$\begin{aligned}\Re\epsilon(\omega) - \epsilon_0 &\stackrel{(16.68)}{=} \frac{2}{\pi}P\int_0^\infty xdx\frac{\Im\epsilon(x)}{x^2-\omega^2}\\ &= \frac{2}{\pi}P\int_0^\infty xdx\frac{\kappa\delta(x-\omega_n)}{x^2-\omega^2}\\ &\simeq \frac{2}{\pi}\frac{\kappa\omega_n}{\omega_n^2-\omega^2}.\end{aligned} \tag{16.70}$$

This behavior is illustrated in Fig. 16.2 and should be compared with the zero-width limit discussed in Section 15.5.
(c) The third case we consider is that with $\Im\epsilon(x) = \tau(x)\theta(x-\Omega)\theta(\Omega'-x)$ as in Fig. 16.3, i.e. absorption in the domain $0 < \Omega < x < \Omega'$ with $\tau(x)$

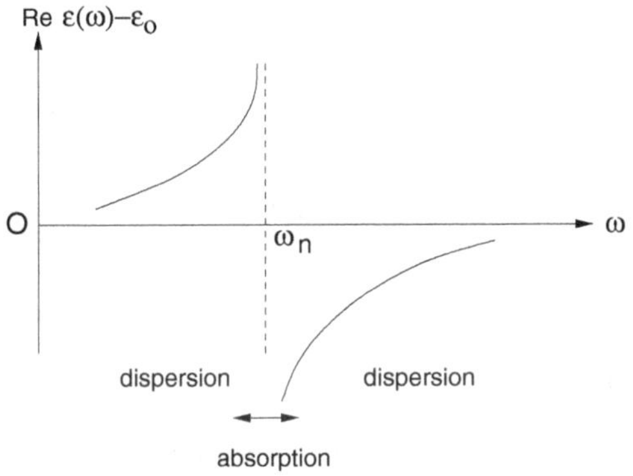


**Fig. 16.2.** Typical behavior of $\Re\epsilon$ for resonance absorption.

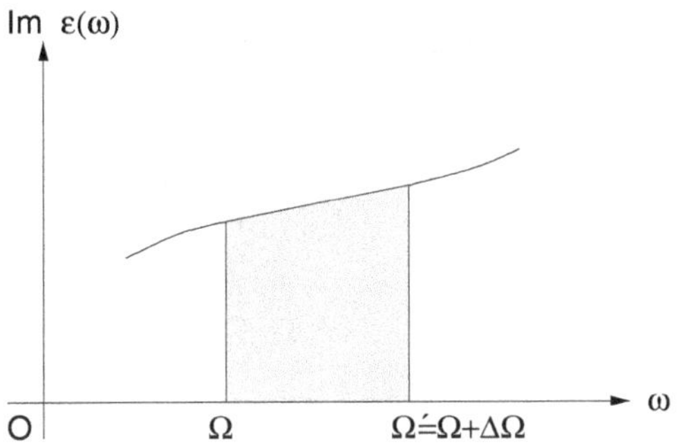


**Fig. 16.3.** Absorption in a finite frequency domain.

regular in this domain. Here

$$\Re\epsilon(\omega) - \epsilon_0 = \frac{2}{\pi}\int_{\Omega}^{\Omega'} dx x \frac{\tau(x)}{x^2 - \omega^2}. \tag{16.71}$$

We consider two cases here:

(i) If $\omega \gg \Omega', \Omega$, we have approximately

$$\Re\epsilon(\omega) - \epsilon_0 \simeq \frac{2}{\pi}\int_{\Omega}^{\Omega'} dx x \frac{\tau(x)}{-\omega^2}. \tag{16.72}$$

If $\Omega', \Omega$ are close together (a narrow band), we can write $\Omega' = \Omega + \triangle\Omega \ll \omega$ and

$$\Re\epsilon(\omega) - \epsilon_0 \simeq -\frac{2}{\pi}\frac{\Omega}{\omega^2}\int_{\Omega}^{\Omega+\triangle\Omega} dx \tau(x), \tag{16.73}$$

i.e.

$$\int_0^{\infty} dx \Im\epsilon(x) = -\int_{\Omega}^{\Omega+\triangle\Omega} dx \tau(x) \simeq \frac{\pi}{2\Omega}\left\{\omega^2(\Re\epsilon(\omega) - \epsilon_0)\right\}. \tag{16.74}$$

The left side represents the area under the curve $\Im\epsilon(x)$. The right side, which is valid for $\omega \gg \Omega', \Omega$, i.e. for

$$\lim_{\omega\to\infty}\left\{-\frac{\pi\omega^2}{2\Omega}(\Re\epsilon(\omega) - \epsilon_0)\right\},$$

represents the limiting value at high frequencies. Relations of this type are called *sum rules*. For instance, in our model consideration (see Eq. (15.40) for only one resonance):

$$\Re\epsilon(\omega) - \epsilon_0 = \frac{ne^2}{m}\frac{\omega_0^2 - \omega^2}{(\omega_0^2 - \omega^2)^2 + \gamma^2\omega^2} f_0 \overset{\omega\to\infty}{\simeq} -\frac{ne^2}{m}\frac{f_0}{\omega^2}. \tag{16.75}$$

Hence

$$\int_0^{\infty} dx \Im\epsilon(x) = -\int_{\Omega}^{\Omega+\triangle\Omega} dx \tau(x) \approx -\frac{\pi}{2\Omega}\frac{ne^2}{m} f_0. \tag{16.76}$$

(ii) If, on the other hand, we have $\omega \ll \Omega, \Omega' = \Omega + \triangle\Omega$, then

$$\Re\epsilon(\omega) - \epsilon_0 \approx \frac{2}{\pi}\int_{\Omega}^{\Omega'} dx x \frac{\tau(x)}{x^2} \approx \frac{2}{\pi}\frac{1}{\Omega}\int_{\Omega}^{\Omega'} dx \tau(x) \approx -\frac{2}{\pi}\frac{1}{\Omega}\int_0^{\infty} \Im\epsilon(x). \tag{16.77}$$

This relation relates the area under the curve of $\Im\epsilon(x)$ to $\Re\epsilon(\omega) - \epsilon_0$ for $\omega \ll \Omega, \Omega'$, i.e. for $\omega \to 0$. In the above example

$$\lim_{\omega \to 0} \{\Re\epsilon(\omega) - \epsilon_0\} \simeq \frac{ne^2}{m} \frac{1}{\omega_0^2} f_0, \tag{16.78}$$

so that

$$-\int_0^\infty dx \Im\epsilon(x) \approx \frac{\pi}{2} \frac{\Omega}{\omega_0^2} \frac{ne^2}{m} f_0. \tag{16.79}$$

Thus one can infer from the behavior of $\epsilon(x)$ the frequency $\omega_0$. Since $\omega_0 \simeq \Omega$ (see Fig. 15.6), the expressions (16.76) and (16.79) agree with each other.

## 16.8 Comments on Principal Values

The principal value of an integral can also be defined by

$$P\int_{-\infty}^{\infty} f(\omega)\frac{d\omega}{\omega} = \lim_{\epsilon \to 0} \frac{1}{2}\left[\int_{-\infty}^{\infty} f(\omega)\frac{d\omega}{\omega - i\epsilon} + \int_{-\infty}^{\infty} f(\omega)\frac{d\omega}{\omega + i\epsilon}\right]. \tag{16.80}$$

Here $f(\omega)$ must be a function which is analytic in the region of integration. The relation (16.80) is based on an identity, which is of considerable usefulness in many applications. This identity is *formally written*

$$\lim_{\epsilon \to 0}\left(\frac{1}{\omega \pm i\epsilon}\right) = P\left(\frac{1}{\omega}\right) \mp i\pi\delta(\omega). \tag{16.81}$$

The $\epsilon$-prescription on the left side is to be understood as saying that with $\epsilon > 0$ the pole is to be excluded from the region enclosed by the contour. It is with this $\epsilon$ prescription that the expression on the left is to be understood as equal to the expression on the right. The identity implies the relation

$$\lim_{\epsilon \to 0}\int_{-\infty}^{\infty} \frac{f(\omega)d\omega}{\omega \pm i\epsilon} = P\int_{-\infty}^{\infty} \frac{f(\omega)d\omega}{\omega} \mp i\pi f(0), \tag{16.82}$$

in which $f(\omega)$ is an analytic function in the domain of integration. In order to prove the identity, we recall the original definition of the principal value integral, i.e.

$$P\int_{-\infty}^{\infty} f(\omega)\frac{d\omega}{\omega} = \lim_{\epsilon \to 0}\left[\int_{-\infty}^{-\epsilon} f(\omega)\frac{d\omega}{\omega} + \int_{\epsilon}^{\infty} f(\omega)\frac{d\omega}{\omega}\right]. \tag{16.83}$$

Consider the term on the left side of the identity Eq. (16.81):

$$\frac{1}{\omega \pm i\epsilon} = \frac{\omega \mp i\epsilon}{\omega^2 + \epsilon^2} = \frac{\omega}{\omega^2 + \epsilon^2} \mp \frac{i\epsilon}{\omega^2 + \epsilon^2}. \tag{16.84}$$

In the limit $\epsilon \to 0$ we obtain with the help of a definition of the delta function (cf. Eq. (2.40)), i.e.

$$\delta(x) = \lim_{\epsilon \to 0} \frac{1}{\pi} \frac{\epsilon}{(\epsilon^2 + x^2)} \tag{16.85}$$

the relations

$$\lim_{\epsilon \to 0} \frac{1}{\omega \pm i\epsilon} = \lim_{\epsilon \to 0} \frac{\omega}{\omega^2 + \epsilon^2} \mp i\pi\delta(\omega). \tag{16.86}$$

The first term on the right of this identity is $1/\omega$ for $\omega \neq 0$. Multiplying this relation by $f(\omega)$ and integrating with respect to $\omega$, we obtain

$$\begin{aligned}
\lim_{\epsilon \to 0} \int_{-\infty}^{\infty} \frac{f(\omega)d\omega}{\omega \pm i\epsilon} &= \lim_{\epsilon \to 0} \int_{-\infty}^{\infty} \frac{\omega f(\omega)d\omega}{\omega^2 + \epsilon^2} \mp i\pi f(0) \\
&= \lim_{\epsilon \to 0} \left[ \int_{-\infty}^{-\epsilon} \frac{f(\omega)d\omega}{\omega} + \int_{\epsilon}^{\infty} \frac{f(\omega)d\omega}{\omega} \right] \\
&\quad + \lim_{\epsilon \to 0} \int_{-\epsilon}^{\epsilon} \frac{\omega f(\omega)d\omega}{\omega^2 + \epsilon^2} \mp i\pi f(0) \\
&= P \int_{-\infty}^{\infty} \frac{f(\omega)d\omega}{\omega} \mp i\pi f(0).
\end{aligned} \tag{16.87}$$

The integral from $-\epsilon$ to $\epsilon$ vanishes, since

$$\lim_{\epsilon \to 0} \int_{-\epsilon}^{\epsilon} \frac{\omega f(\omega)d\omega}{\omega^2 + \epsilon^2} = \lim_{\epsilon \to 0} f(0) \int_{-\epsilon}^{\epsilon} \frac{\omega d\omega}{\omega^2 + \epsilon^2} = 0, \tag{16.88}$$

since the integrand is an odd function of $\omega$. We have thus verified the formal relation (16.82). Equation (16.81) now follows from the addition of both cases, provided the contributions of the integrals around the infinite semicircles vanish.

The relation with the upper sign is more commonly used than the other because it is applicable when the function $f(\omega)$ is analytic in the upper half-plane, i.e. the relation

$$\lim_{\epsilon \to 0} \int_{-\infty}^{\infty} \frac{f(\omega)d\omega}{\omega + i\epsilon} = P \int_{-\infty}^{\infty} \frac{f(\omega)d\omega}{\omega} - i\pi f(0). \tag{16.89}$$

If

$$\int_{-\infty}^{\infty} \cdots = \int_{\curvearrowright} \cdots = 2\pi i \sum \text{residues} = 0, \tag{16.90}$$

it follows that

$$P\int_{-\infty}^{\infty} \frac{f(\omega)d\omega}{\omega} = i\pi f(0). \tag{16.91}$$

in agreement with $f(\omega) \equiv g(\omega)$ in Eq. (16.54).

**Example 16.2 (Evaluation of a principal value integral).** Evaluate with the help of the original definition of the principal value integral the following integral

$$P\int_{-\infty}^{\infty} \frac{\sin x' dx'}{x' - x}. \tag{16.92}$$

**Solution:** We separate the sine into its exponentials, so that

$$P\int_{-\infty}^{\infty} \frac{\sin x' dx'}{x' - x} = \frac{1}{2i} P\int_{-\infty}^{\infty} \frac{e^{ix'} dx'}{x' - x} - \frac{1}{2i} P\int_{-\infty}^{\infty} \frac{e^{-ix'} dx'}{x' - x}. \tag{16.93}$$

We consider the following integral along the contour $C_+$ in the complex $x'$-plane shown in Fig. 16.4. The contour $C_+$ excludes the pole at $x$. Hence $\int_{C_+} = 0$. Moreover $\int_{\curvearrowleft} = 0$ (along the infinite semicircle), so that

$$P\int_{-\infty}^{\infty} + \int_{\frown} = 0, \tag{16.94}$$

where the second contribution represents the integral along the small semicircle around the point $x$ of $x - \epsilon$ to $x + \epsilon$. It follows therefore

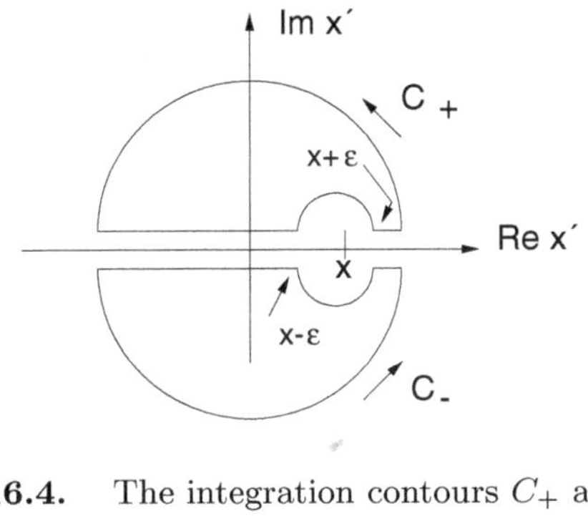


**Fig. 16.4.** The integration contours $C_+$ and $C_-$.

that

$$\frac{1}{2i}P\int_{-\infty}^{\infty}\frac{e^{ix'}dx'}{x'-x} = -\frac{1}{2i}\int_{x-\epsilon}^{x+\epsilon}\frac{e^{ix'}dx'}{x'-x}. \tag{16.95}$$

With

$$x'-x = re^{i\theta}, \qquad \frac{dx'}{x'-x} = id\theta \tag{16.96}$$

we obtain

$$\frac{1}{2i}P\int_{-\infty}^{\infty}\frac{e^{ix'}dx'}{x'-x} = -\frac{1}{2i}\int_{\pi}^{0} id\theta e^{ix} = \frac{\pi e^{ix}}{2}. \tag{16.97}$$

Similarly we obtain for the contour $C_-$ in Fig. 16.4 the result

$$\frac{1}{2i}\int_{C_-}\frac{e^{-ix'}dx'}{x'-x} = 0. \tag{16.98}$$

It follows that

$$\frac{1}{2i}P\int_{-\infty}^{\infty}\frac{e^{-ix'}dx'}{x'-x} = -\frac{1}{2i}\int_{x-\epsilon}^{x+\epsilon}\frac{e^{-ix'}dx'}{x'-x} = -\frac{1}{2i}\int_{-\pi}^{0} id\theta e^{-ix} = -\frac{\pi e^{-ix}}{2}. \tag{16.99}$$

Hence we obtain the result

$$P\int_{-\infty}^{\infty}\frac{\sin x' dx'}{x'-x} = \frac{\pi}{2}[e^{ix}+e^{-ix}] = \pi\cos x. \tag{16.100}$$

Above we obtained the dispersion relation (16.54). In the derivation we had assumed that

$$\int_{\curvearrowleft} d\omega' \frac{g(\omega')}{\omega'-\omega} = 0 \tag{16.101}$$

for $\omega' = Re^{i\theta}, R \to \infty$. It is possible, however, that this condition is not satisfied, e.g. it might happen that $g(\omega')$ approaches a nonzero constant $g_0$ for $|\omega'| \to \infty$. In the latter case, the integral implies

$$\int_{\curvearrowleft} d\omega' \frac{g(\omega')}{\omega'-\omega} = g_0\left(\lim_{R\to\infty}\int_{\curvearrowleft}\frac{d\omega'}{\omega'}\right) = g_0\int_0^{\pi} id\theta = i\pi g_0 \neq 0.$$

We then write

$$G(\omega) = g(\omega) - g_0, \quad \text{so that} \quad \lim_{|\omega|\to\infty} G(\omega) = 0, \tag{16.102}$$

and the dispersion relations derived previously now apply to $G(\omega)$. If

$$g(\omega) \overset{|\omega|\to\infty}{\longrightarrow} g_1\omega + g_0, \tag{16.103}$$

we set

$$G(\omega) = g(\omega) - (g_1\omega + g_0) \tag{16.104}$$

and so on. The constants $g_0, g_1, \ldots$ are called *subtraction constants* and the dispersion relations are called *subtracted dispersion relations.*

## 16.9 Exercises

**Example 16.3 (Principal value integral[4]).** Evaluate the principal value integral:

$$P\int_0^\infty \frac{\sin\sqrt{x'}dx'}{x'-x}. \tag{16.105}$$

What is the value of the integral if $x < 0$?

(Answers: $\pi\cos\sqrt{x}, 2\pi e^{-\sqrt{x}}$).

**Example 16.4 (Laplace's equation in two dimensions).** Show that if the partial derivatives of $u(x,y), v(x,y)$ exist and are connected by the Cauchy–Riemann equations, and $\phi_{xy} = \phi_{yx}$, it follows that both $u$ and $v$ satisfy Laplace's equation in two dimensions, i.e.

$$\nabla^2\phi = \frac{\partial^2\phi}{\partial x^2} + \frac{\partial^2\phi}{\partial y^2} = 0, \tag{16.106}$$

an equation which occurs frequently in mathematical physics.

**Example 16.5 (Evaluation of an integral with Cauchy's formula).** Use the method of contour integration to show that

$$\int_0^\infty \frac{dx}{x^4+a^4} = \frac{\pi}{2\sqrt{2}a^3}. \tag{16.107}$$

**Example 16.6 (Principal values).** Evaluate the following principal value integrals:

$$P\int_{-\infty}^\infty \frac{dx}{x(x-i)}, \tag{16.108a}$$

$$P\int_0^\infty \frac{\sin x}{x}dx, \tag{16.108b}$$

$$P\int_0^\infty \frac{\sin^2 x}{x^2}dx. \tag{16.108c}$$

(Answers: (a) $\pi$, (b) $\pi/2$, (c) $\pi/2$).

[4]A table of principal value integrals of a similar kind was distributed by Carprasse [25].

**Example 16.7 (Evaluation of a Kramers–Kronig relation).** Evaluate the Kramers–Kronig relation (16.66), i.e.

$$\Im g(\omega) = -\frac{2\omega}{\pi} P \int_0^\infty \frac{\Re g(\omega')d\omega'}{\omega'^2 - \omega^2} \tag{16.109}$$

for

$$\Re g(\omega') = \frac{R}{R^2 + \omega'^2 L^2}. \tag{16.110}$$

(Answer: $\Im g(\omega) = \omega L/(R^2 + \omega^2 L^2)$).

**Example 16.8 (A pulse propagating in a viscous medium).** Assume a wave amplitude of the following form propagating in the $x$-direction in a non-viscous medium:

$$\psi = e^{i\omega(t-x/c)}. \tag{16.111}$$

If the medium is viscous the velocity $c$ is related to the velocity $c_0$ when there is no absorption by the relation

$$c^2 = c_0^2 + \frac{4i\mu\omega}{3\rho}, \tag{16.112}$$

where $\mu$ is the (coefficient of) viscosity, $\rho$ the density of the medium, and $c_0^2 = \partial p/\partial \rho$, where $p$ is the pressure. Assuming the validity of the approximation

$$\frac{1}{c} \simeq \frac{1}{c_0} - \frac{2i\mu\omega}{3\rho c_0^3}, \tag{16.113}$$

and a delta function input pulse with amplitude $\delta(x)$ at time $t = 0$, show that the amplitude which for the delta function is

$$\frac{1}{2\pi} \int_{-\infty}^{\infty} e^{i\omega(t-x/c)} d\omega$$

gives the amplitude

$$\frac{1}{2}\sqrt{\frac{3\rho c_0^3}{2\pi\mu x}} \exp\left[-\frac{(t - x/c_0)^2}{8\mu x/3\rho c_0^3}\right]. \tag{16.114}$$

Sketch the shape of the pulse in $x$ or $t$.

Chapter 17

# Covariant Formulation of Electrodynamics

## 17.1 Introductory Remarks

In the following we consider first transformations from one reference frame of coordinates to another which moves with uniform velocity relative to the first; this is the topic of Einstein's Special Theory of Relativity. The significance and necessity of such considerations is immediately apparent if one recalls the field of an electric charge in its rest frame and then visualizes this from a moving frame: In the first case one has only the static electric field, but in the second — in view of the motion of the charge — one observes also a magnetic field. It therefore becomes necessary to formulate Maxwell's equations, and more generally all laws of physics, independent of the respective reference frame, and this means in *covariant formulation.*

## 17.2 The Special Theory of Relativity

### 17.2.1 *Introduction*

We recapitulate first some aspects of the Special Theory of Relativity,[1] which unifies Maxwell's electrodynamics with mechanics. The Special Theory of Relativity is based on the following two important principles or postulates of Einstein:

(1) *Einstein's principle of relativity* (1905). This principle says: The laws which describe the change of the state of a physical system are not affected

[1]The literature on this topic is too numerous to be reviewed here. Instead we refer only to a few sources which no doubt constitute a biased selection. Easily accessible are A. Einstein's paper of 1905 [44] and his own popular exposition [45]. All the main topics and applications can be found in Jackson [73]. For a modern perspective of various aspects we refer to Penrose [113].

by choosing one or another frame of reference which are related to each other by uniform translational motion (i.e. with constant relative velocity; systems which are accelerated to each other are treated in Einstein's General Theory of Relativity).

Reference frames in uniform translational motion to each other (uniform motion meaning moving with constant velocity) are also called *inertial frames*. These are local reference frames in which free motion (meaning potential or applied force equal to zero) is weightless (i.e. without gravitation, i.e. no planets or other massive bodies nearby which act on a mass $m$ through Newton's law of gravitation, hence the *Special* Theory of Relativity), and for which Newton's equation therefore implies[2]

$$m\ddot{x} = 0, \quad \therefore \quad m\dot{x} = \text{const.}, \quad \dot{x} = \frac{\text{const.}}{m}. \tag{17.1}$$

(2) *The velocity of light $c$ is independent of the source* and is in vacuum and in the presence of electromagnetic fields the constant $c$ in vacuum for all observers in inertial frames, as one would expect on the basis of the connection $c = 1/\sqrt{\epsilon_0\mu_0}$. (This means that light is a field with the characteristics of waves and thus does not conform with classical corpuscular ideas).

We recapitulate briefly the development which led to these postulates. The first postulate can also be formulated as saying: It is not possible to determine a so-called "absolute velocity" (neither with electromagnetic nor with optical methods). What initiated these considerations? We recall that Newton's laws are based on the idea of an inertial frame, i.e. an unaccelerated, not rotating frame referred to an *absolute space*. This absolute space was assumed to be immovable of its own nature and with no reference to anything "exterior" (meaning that in the case of an acceleration in space, the apparent acceleration of this space — viewed from the accelerated object — is not a consequence of Newton's laws but a consequence of our acceleration relative to the absolute space, the so-called absolute acceleration). The concept of an absolute space thus offers a criterion for an absolute acceleration ($\ddot{x} = \text{const.}$), but not for the concept of an absolute velocity or a state of absolute rest. In fact, an absolute velocity cannot be determined, as the first of the above postulates of Einstein implies (see also the Michelson–Morley experiment below).

---

[2]The rest frame of the particle is the frame with $\dot{x} = 0$; a particle with mass zero thus has no rest frame. "Uniform" motion means also independent of position.

In preceding chapters we had introduced $c$ as a constant which represents effectively the ratio between electric and magnetic units. Later we observed, that Maxwell's equations are wave equations for the fields $\mathbf{E}, \mathbf{H}$ with the phase velocity $c$ in vacuum. More than a century ago numerous observations had led to the conclusion that $c$ is the velocity of light. One then had to face the problem of understanding the propagation of electromagnetic waves in space, and it seemed suggestive to imagine their propagation similar to that of sound waves. Sound waves are vibrations in a medium like air. It was postulated therefore, that electromagnetic waves are propagated in a so-called "*ether*", which was assumed to be a space-filling medium introduced solely for the purpose of explaining the propagation of electromagnetic waves. The velocity $c$ would then be the velocity of light with respect to this "*ether*", which was assumed to be stationary. Referred to a frame which moves with velocity $u$ through the (stationary) *ether*, the velocity of light would — according to Galilean arguments — be $c' = c - u$ with respect to the moving frame. Then Maxwell's equations would have to be different in different reference frames in order to yield different velocities of light. One particular reference frame, that of the *ether*, would play a special role. This consideration is analogous to that of the velocity of sound waves relative to a train with velocity $u$, i.e.

$$v_{\text{relative train}} = v_{\text{in air}} - u_{\text{train}}.$$

It became necessary therefore to verify the possible existence of the *ether*, e.g. by measuring the absolute velocity $v$ of the earth in this *ether*. If the earth moves through this *ether*, which was assumed to be stationary, there would have to be an observable "*ether wind*" in the opposite direction. The experiment of Michelson and Morley[3] was expected to allow the observation of this *wind*, i.e. its (absolute) velocity. Instead of considering the setup of this optical experiment, we consider the analogous case of a swimmer (velocity $c$ in stationary water) in a river with current velocity $v \neq 0$. This consideration is simple and makes the issues solved by Einstein evident.[4] The simple geometry is sketched in Fig. 17.1. The distances $SR, ST$ in Fig. 17.1 are taken to be equal, i.e.

$$RS = ST = a. \tag{17.2}$$

[3]Michelson and Morley [93]. See also Panofsky and Phillips [111], Chapter 15, in which the experimental basis for the theory of special relativity is explored with great clarity.
[4]The swimmer analogy is also considered in Lieber [87].

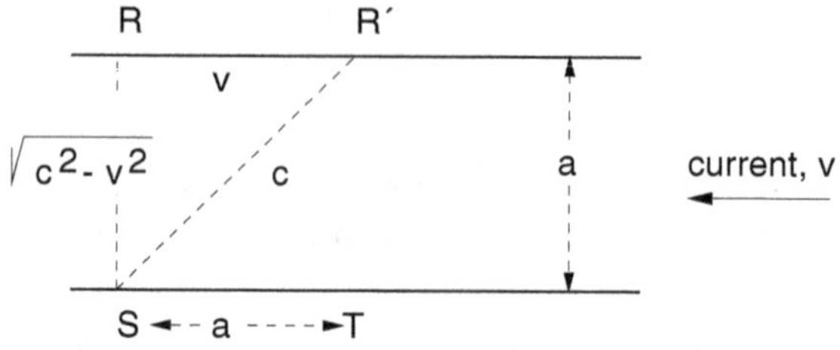


**Fig. 17.1.** The swimmer analogy to the experiment of Michelson and Morley.

Then the time the swimmer needs in order to swim from $S$ to $T$ and from $T$ back to $S$ (in the following we use the subscript "$l$" for "longitudinal") is

$$t_l = \frac{a}{c-v} + \frac{a}{c+v} = \frac{2ac}{c^2 - v^2}. \tag{17.3}$$

In order to reach $R$ from $S$, the swimmer must swim in the direction of $R'$. The time he needs in order to swim from $S$ to $R$ is the same as the time he needs in order to swim from $R$ to $S$, i.e.

$$t_\perp = 2\frac{a}{\sqrt{c^2 - v^2}}. \tag{17.4}$$

In the case of the swimmer (see below)

$$t_l > t_\perp \quad \text{for} \quad v < c, \tag{17.5}$$

since

$$\begin{aligned} t_l - t_\perp &= \frac{2ac}{c^2 - v^2} - 2\frac{a}{\sqrt{c^2 - v^2}} = 2\frac{a}{\sqrt{c^2 - v^2}}\left[\frac{c}{\sqrt{c^2 - v^2}} - 1\right] \\ &= 2\frac{a}{\sqrt{c^2 - v^2}}\left[\frac{1}{\sqrt{1 - v^2/c^2}} - 1\right] > 0 \quad \text{seconds}. \end{aligned} \tag{17.6}$$

In the Michelson–Morley experiment $c$ is the velocity of light referred to the *ether*, and $v$ the velocity of the earth relative to the *ether*. The idea was, to measure $t_l, t_\perp$, and to determine $v$ from $t_l - t_\perp$. The experiments of Michelson and Morley used light and mirrors and for the measurements interferometers of a high degree of precision. The results showed no difference between $t_l$ and $t_\perp$, i.e. one had to conclude from the experiments that $t_l = t_\perp (v \neq 0)$, i.e.

$$t_\perp = 2\frac{a}{\sqrt{c^2 - v^2}} \stackrel{?}{=} \frac{2ac}{c^2 - v^2} = t_l. \tag{17.7}$$

Since it was required that $v \neq 0$, the next problem was to explain this result. Lorentz observed that both expressions can be made to agree for $v \neq 0$, if one assumes that the quantity $a$ in $t_l$ had to be replaced by

$$\frac{a\sqrt{c^2-v^2}}{c} < a,$$

i.e. that parallel to $v$ matter shrinks (as a consequence of its electric properties). This was the so-called "Lorentz-contraction" hypothesis. Since all measuring rods would also shrink in the same way, the effect would not be observable. This kind of adjustment of explanations was, however, rejected in particular by Poincaré (who along with Lorentz was a significant forrunner of the originators of the theory of relativity). Lorentz (1904) therefore undertook a deeper investigation which led him to the transformation which today carries his name, however, without the appropriate explanation which was due to Einstein. The transformation including that of time is written (in the case of motion along parallel $x$-axes)

$$\begin{aligned} x' &= \gamma(x - vt), \quad y' = y, \quad z' = z, \\ t' &= \gamma\left(t - \frac{vx}{c^2}\right), \quad \gamma = \frac{1}{\sqrt{1 - v^2/c^2}}. \end{aligned} \tag{17.8}$$

Here $x$ is the position coordinate measured in a frame fixed in the ether, and $x'$ that in a frame fixed on earth at a time $t$, i.e. in Eq. (17.3) instead of

$$t = \frac{a}{c-v}, \quad (c-v)t = a = \underbrace{x'}_{\text{earth}}$$

one has

$$t = \frac{x'}{c-v}$$

(in the frame fixed on earth light travels in time $t$ the distance $a \equiv x'$), so that

$$x' = ct - vt = \underbrace{x}_{\text{ether}} - vt.$$

This expression assumes that the velocity of light $c$ is independent of the motion of the source. Consider again the case of the train T (with velocity $v$ with respect to the earth), from which a whistle sends out a sound wave with velocity $u$ as indicated in Fig. 17.2. Here

$$x' = (u - v)t = x - vt,$$

i.e. $u$ plays the role of $c$ and is independent of the motion of the source. One might think that by measuring $x, x', t$ the velocity $v$ could be determined as

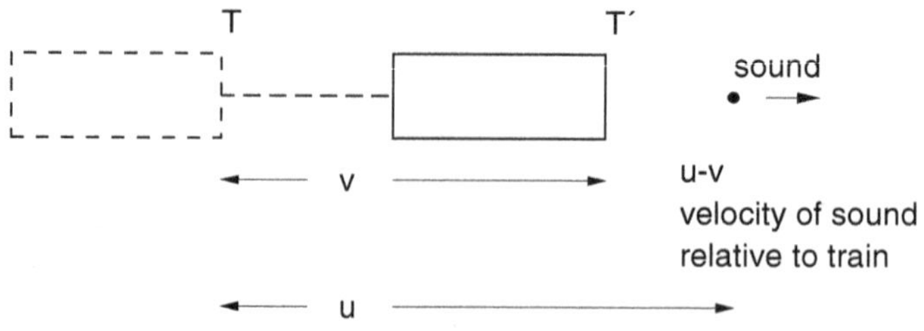


**Fig. 17.2.** Velocities of sound and the train.

an absolute velocity in contradiction to the result of the Michelson–Morley experiment. Hence a contradiction? This apparent contradiction was resolved by Einstein: Every observer has his own clock, and in order to compare times, light signals have to be exchanged. In the above $x = ct$ is the distance travelled by light in time $t$ (in the *ether*). The contraction hypothesis now says that $x'$ is to be replaced by

$$\sqrt{1 - \frac{v^2}{c^2}}x',$$

so that (as in Eq. (17.8))

$$x' = \frac{x - vt}{\sqrt{1 - v^2/c^2}}. \tag{17.9}$$

Lorentz reasoned that along with

$$x' = \gamma(x - vt), \tag{17.10}$$

also a time $t'$ given by

$$t' = \gamma\left(t - \frac{vx}{c^2}\right) \tag{17.11}$$

in his set of equations (17.8) had to be introduced. But what is $t'$? Lorentz looked at $t'$ as an artificial time, which had to be introduced for mathematical reasons. It became clear, therefore, that the fundamental concepts of length and time (also mass) had to be re-examined. This was done and achieved by Einstein in 1905. The statement of relativity, that for instance a material bullet fired with velocity $u$ from a train, which is moving with velocity $v$, has the velocity $u$ whether the train is moving or not, since

$$v_{\text{rel. train}} = (v + u) - v = u$$

was known before Einstein. Einstein's original intention was to extend this statement of classical mechanics to emitted light signals and hence to electrodynamics. In this Einstein realized that the laws of physics always have the same form, but are expressed in different reference frames (i.e. those of the respective observers). One consequence was a correction of Newtonian mechanics.

### 17.2.2 *Einstein's interpretation of Lorentz transformations*

The Lorentz transformation[5] from $(x, y, z, t)$ to $(x', y', z', t')$ thus describes the transformation from reference frame $K$ of one observer to the frame $K'$ of another observer, i.e. (in the case of relative motion parallel to the $x$-direction)

$$x' = \gamma(x - vt), \quad y' = y, \quad z' = z, \quad t' = \gamma\left(t - \frac{vx}{c^2}\right). \tag{17.12}$$

The Lorentz contraction therefore is not an actual contraction of the object (as Lorentz had suspected), but only an apparent shrinkage, which results from the *difference of measurements* made by observers in uniform relative motion to each other (i.e. with velocity $v$ constant). Moreover, $t'$ is simply the time measured in frame $K'$.

The reference to electrodynamics becomes particularly evident if we consider the propagation of light signals. If a bullet with velocity $u$ is fired from a train moving with velocity $v$ in the same direction, the velocity of the bullet with respect to a frame fixed on earth is $u + v$, because the bullet as a travelling object had the velocity $v$ before it was fired. If in a similar way a sound signal is sent out from a whistle on the train, i.e. with velocity $u$, then its velocity relative to the frame on earth is $u$, i.e. less than $u + v$ (the reason being the different nature of the bullet and sound). The velocity of the sound wave depends only on the medium (i.e. the air), but not on the velocity of the source. Applied to light, this is the contents of the second postulate of above, as follows also from $c = 1/\sqrt{\epsilon_0\mu_0}$. We deduce from these postulates now that for light observed in frame $K$:

$$x = ct \tag{17.13}$$

is the distance travelled by the signal in time $t$. But then we must have (if the velocity of light is independent of the motion of the source)

[5]The transformations today known as Lorentz transformations had already been formulated and published in 1887 by Voigt [156], however in relation to sound waves. In 1909 Lorentz wrote: "*which to my regret has escaped my notice all these years*"; see e.g. O'Rahilly [122], Vol. I, p. 325. In the chapter with title "*Voigt*" O'Rahilly attempts a critical analysis of these transformations.

that

$$x' = ct' \tag{17.14}$$

with the same $c$. We can infer from the Lorentz transformation that both equations, Eqs. (17.13) and (17.14), are compatible with each other, since with $x = ct$ we have from Eqs. (17.12):

$$x' = \gamma(c - v)t,$$

so that

$$t' = \gamma\left(t - \frac{vct}{c^2}\right) = \frac{\gamma(c-v)t}{c} = \frac{x'}{c}$$

and hence

$$x' = ct',$$

i.e. light has the same velocity in every frame of reference, i.e. the observed velocity of light is independent of the motion of the observer (contrary to the claim of the *ether* theory, in which $c$ would be the velocity of light with respect to the *ether* and hence the velocity of light observed by an observer travelling with velocity $v$ in *ether* $c - v$, in which case the observed velocity of light would depend on the velocity of the observer).

Finally we make the following observation. Solving the equations

$$x' = \gamma(x - vt), \quad t' = \gamma\left(t - \frac{vx}{c^2}\right) \tag{17.15}$$

for $x$ and $t$, we obtain

$$x = \gamma(x' + vt'), \quad t = \gamma\left(t' + \frac{vx'}{c^2}\right), \tag{17.16}$$

completely in agreement with the expectation that the frame with unprimed coordinates moves with velocity $-v$ away from the frame with primed coordinates if the frame with primed coordinates moves with velocity $v$ away from the frame with unprimed coordinates. Changing $v$ into $-v$ corresponds to the changes $x \longleftrightarrow x', t \longleftrightarrow t'$. Einstein's *principle of relativity* says that the laws of physics must transform "*covariantly*" with respect to Lorentz transformations, i.e. independent of the respective reference frame in uniform translational motion.[6]

[6] "Covariant" means "co-varying", i.e. "transforming in the same way". "Covariance" can also mean independent of a specific coordinate frame, i.e. applicable to one as for any other.

## 17.3 Minkowski Space

We now consider Lorentz transformations in general. If one adds to the Lorentz transformations which are elements of the Lorentz group $SO(1,3)$ which will be explained in more detail below, the simple translations in space and time, one obtains the *inhomogeneous Lorentz transformations*, which are also called *transformations of the Poincaré group* $\mathcal{P}$. A vector in *four-dimensional spacetime* is written in the so-called *covariant form*, which is characterized by *lower* indices, as

$$x_\mu, \quad \mu = 0, 1, 2, 3.$$

The path element $ds$ in this four-dimensional so-called *Minkowski space* $\mathbb{M}_4$ is with (observe the minus signs in $x_1 = -x$ etc.)

$$x_0 := ct, \quad x_1 := -x, \quad x_2 := -y, \quad x_3 := -z \tag{17.17}$$

given by

$$(ds)^2 \equiv ds^2 = (dx_0)^2 - \sum_{i=1}^{3}(dx_i)^2. \tag{17.18}$$

We are therefore using the *metric* (or so-called *metric tensor*) of this space $\mathbb{M}_4$ given by:

$$(\eta_{\mu\nu}) = \begin{pmatrix} 1 & 0 & 0 & 0 \\ 0 & -1 & 0 & 0 \\ 0 & 0 & -1 & 0 \\ 0 & 0 & 0 & -1 \end{pmatrix} \equiv (\eta^{\mu\nu}), \quad \det(\eta_{\mu\nu}) \neq 0. \tag{17.19}$$

On the other hand the unit matrix is defined by (note: no need for displacement of indices in the case of $\delta$)

$$(\mathbb{1}) = (\delta_\mu^\rho) \equiv (\eta_\mu{}^\rho), \quad \eta_\mu{}^\rho \eta_{\rho\nu} = \eta_{\mu\nu}, \tag{17.20}$$

so that

$$\delta_\mu^\rho \eta_{\rho\nu} \equiv \eta_{\mu\nu}, \quad \eta_{\mu\nu}\eta^{\nu\rho} = \delta_\mu^\rho. \tag{17.21}$$

The so-called *contravariant* vector $x^\mu$ with upper indices is then defined by

$$x^\mu = \eta^{\mu\nu} x_\nu, \tag{17.22}$$

so that in contravariant form

$$dx^\mu = \eta^{\mu\nu} dx_\nu, \tag{17.23}$$

or with $(dx_\nu) \equiv (dx_0, -dx_i), i = 1, 2, 3,$

$$(dx^\mu) = \begin{pmatrix} 1 & 0 & 0 & 0 \\ 0 & -1 & 0 & 0 \\ 0 & 0 & -1 & 0 \\ 0 & 0 & 0 & -1 \end{pmatrix} \begin{pmatrix} dx_0 \\ -dx_1 \\ -dx_2 \\ -dx_3 \end{pmatrix} = \begin{pmatrix} dx_0 \\ +dx_i \end{pmatrix}, \tag{17.24}$$

and therefore in terms of lower indices

$$dx_\mu dx^\mu = (dx_0, -dx_i) \begin{pmatrix} dx_0 \\ dx_i \end{pmatrix} = (dx_0)^2 - (dx_i)^2 = (ds)^2. \tag{17.25}$$

With $\eta^{\mu\nu}$ we can "raise" or "lower" indices, and indices of the same type (one upper, one lower) are summed over (if not explicitly stated, this is understood, and called Einstein's summation convention). We shall become familiar in the following with manipulations of this type. A quantity which is such that all indices are summed over is a *Lorentz invariant*, e.g.

$$dx^\mu dx_\mu = (ds)^2 \tag{17.26}$$

(the disappearance of the indices on the right side is described as their "contraction"). This invariance of $ds^2$ under Lorentz transformations is the extension to 4-dimensional spacetime of the invariance of the 3-dimensional $d\mathbf{s}^2$ under transformations of the rotational group $SO(3)$.

For later purposes we deviate a little and devote the remainder of this section to the *transformation from Cartesian to spherical polar coordinates* $(x_\mu) := (ct, r, \theta, \varphi)$. The square of the distance $(ds)^2$ between two neighboring spacetime points $P(ct, \mathbf{r})$ and $P(ct + cdt, \mathbf{r} + \mathbf{dr})$ with[7]

$$x_0 = ct, \quad x = r\sin\theta\cos\varphi, \quad y = r\sin\theta\sin\varphi, \quad z = r\cos\theta, \tag{17.27}$$

is, as we can also see geometrically,

$$\begin{aligned} ds^2 &= (cdt)^2 - (d\mathbf{r})^2 \\ &= (cdt)^2 - (dr)^2 - (rd\theta)^2 - r^2\sin^2\theta(d\varphi)^2 \\ &\equiv (cdt)^2 - dr^2 - r^2 d\Omega_2^2. \end{aligned} \tag{17.28}$$

[7]For $d\mathbf{s}^2 \equiv d\mathbf{r}^2$ see e.g. Magnus and Oberhettinger [90], p. 146.

This means the metric $\tilde{\eta}$ is given by

$$(\tilde{\eta}) = \begin{pmatrix} 1 & 0 & 0 & 0 \\ 0 & -1 & 0 & 0 \\ 0 & 0 & -r^2 & 0 \\ 0 & 0 & 0 & -r^2 \sin^2\theta \end{pmatrix}, \tag{17.29}$$

since then

$$ds^2 = \begin{pmatrix} cdt & dr & d\theta & d\varphi \end{pmatrix} \begin{pmatrix} 1 & 0 & 0 & 0 \\ 0 & -1 & 0 & 0 \\ 0 & 0 & -r^2 & 0 \\ 0 & 0 & 0 & -r^2 \sin^2\theta \end{pmatrix} \begin{pmatrix} cdt \\ dr \\ d\theta \\ d\varphi \end{pmatrix} \tag{17.30}$$

and so (the superscript $T$ in the second line meaning the expression is a column matrix)

$$\begin{aligned} ds^2 &= \begin{pmatrix} cdt & dr & d\theta & d\varphi \end{pmatrix} \\ &\quad \times (cdt \quad -dr \quad -r^2 d\theta \quad -r^2 \sin^2\theta d\varphi)^T \\ &= (cdt)^2 - (dr)^2 - (rd\theta)^2 - r^2 \sin^2\theta (d\varphi)^2. \end{aligned} \tag{17.31}$$

We can obtain the metric $(\tilde{\eta})$ also as follows. We start from the transformation

$$dx'_\mu = A_\mu{}^\kappa dx_\kappa \tag{17.32}$$

and set

$$\begin{aligned} (x'_\mu) &:= (x^0, x, y, z) \quad \text{(Cartesian spatial coordinates)}, \\ (x_\mu) &:= (x^0, r, \theta, \varphi) \quad \text{(Spherical spatial coordinates)}, \end{aligned} \tag{17.33}$$

so that

$$(A_\mu{}^\kappa) = \begin{pmatrix} \frac{\partial x^0}{\partial x^0} & \frac{\partial x^0}{\partial r} & \frac{\partial x^0}{\partial \theta} & \frac{\partial x^0}{\partial \varphi} \\ \frac{\partial x}{\partial x^0} & \frac{\partial x}{\partial r} & \frac{\partial x}{\partial \theta} & \frac{\partial x}{\partial \varphi} \\ \frac{\partial y}{\partial x^0} & \frac{\partial y}{\partial r} & \frac{\partial y}{\partial \theta} & \frac{\partial y}{\partial \varphi} \\ \frac{\partial z}{\partial x^0} & \frac{\partial z}{\partial r} & \frac{\partial z}{\partial \theta} & \frac{\partial z}{\partial \varphi} \end{pmatrix}. \tag{17.34}$$

Then ($T$ meaning transposition, and we have $(A_\nu{}^\rho)^T = A^\rho{}_\nu$, so that $A^{T\rho}_{\ \ \nu} = A_\nu{}^\rho$)

$$\begin{aligned} dx'^\mu dx'_\mu &= \eta^{\mu\nu} dx'_\nu dx'_\mu = \eta^{\mu\nu} A_\nu{}^\rho dx_\rho A_\mu{}^\kappa dx_\kappa \\ &= (A^{T\rho}{}_\nu \eta^{T\nu\mu} A_\mu{}^\kappa) dx_\rho dx_\kappa \\ &\equiv \tilde{\eta}^{\rho\kappa} dx_\rho dx_\kappa, \end{aligned} \tag{17.35}$$

where

$$\tilde{\eta}^{\rho\kappa} = A^{T\rho}{}_\nu \eta^{T\nu\mu} A_\mu{}^\kappa, \tag{17.36}$$

and explicitly

$$\begin{aligned} (\tilde{\eta}^{\rho\kappa}) = &\begin{pmatrix} \frac{\partial x^0}{\partial x^0} & \frac{\partial x}{\partial x^0} & \frac{\partial y}{\partial x^0} & \frac{\partial z}{\partial x^0} \\ \frac{\partial x^0}{\partial r} & \frac{\partial x}{\partial r} & \frac{\partial y}{\partial r} & \frac{\partial z}{\partial r} \\ \frac{\partial x^0}{\partial \theta} & \frac{\partial x}{\partial \theta} & \frac{\partial y}{\partial \theta} & \frac{\partial z}{\partial \theta} \\ \frac{\partial x^0}{\partial \varphi} & \frac{\partial x}{\partial \varphi} & \frac{\partial y}{\partial \varphi} & \frac{\partial z}{\partial \varphi} \end{pmatrix} \begin{pmatrix} 1 & 0 & 0 & 0 \\ 0 & -1 & 0 & 0 \\ 0 & 0 & -1 & 0 \\ 0 & 0 & 0 & -1 \end{pmatrix} \\ &\times \begin{pmatrix} \frac{\partial x^0}{\partial x^0} & \frac{\partial x^0}{\partial r} & \frac{\partial x^0}{\partial \theta} & \frac{\partial x^0}{\partial \varphi} \\ \frac{\partial x}{\partial x^0} & \frac{\partial x}{\partial r} & \frac{\partial x}{\partial \theta} & \frac{\partial x}{\partial \varphi} \\ \frac{\partial y}{\partial x^0} & \frac{\partial y}{\partial r} & \frac{\partial y}{\partial \theta} & \frac{\partial y}{\partial \varphi} \\ \frac{\partial z}{\partial x^0} & \frac{\partial z}{\partial r} & \frac{\partial z}{\partial \theta} & \frac{\partial z}{\partial \varphi} \end{pmatrix}. \end{aligned}$$

With the differentiated expressions obtained from the coordinates (17.27) this product implies

$$\begin{aligned} (\tilde{\eta}^{\rho\kappa}) = &\begin{pmatrix} 1 & 0 & 0 & 0 \\ 0 & x/r & y/r & z/r \\ 0 & xz/\sqrt{r^2-z^2} & yz/\sqrt{r^2-z^2} & -\sqrt{r^2-z^2} \\ 0 & -y & x & 0 \end{pmatrix} \\ &\times \begin{pmatrix} 1 & 0 & 0 & 0 \\ 0 & -1 & 0 & 0 \\ 0 & 0 & -1 & 0 \\ 0 & 0 & 0 & -1 \end{pmatrix} \begin{pmatrix} 1 & 0 & 0 & 0 \\ 0 & x/r & xz/\sqrt{r^2-z^2} & -y \\ 0 & y/r & yz/\sqrt{r^2-z^2} & x \\ 0 & z/r & -\sqrt{r^2-z^2} & 0 \end{pmatrix}, \end{aligned}$$

and thus yields

$$(\tilde{\eta}^{\rho\kappa}) = \begin{pmatrix} 1 & 0 & 0 & 0 \\ 0 & -1 & 0 & 0 \\ 0 & 0 & -r^2 & 0 \\ 0 & 0 & 0 & -r^2\sin^2\theta \end{pmatrix}. \tag{17.37}$$

Thus this result agrees with the matrix of Eq. (17.29). We can consider this as an exemplary verification of the relation ${A^T}^\rho{}_\nu = A_\nu{}^\rho$.

## 17.4 The 10-Parametric Poincaré Group

The 10-parametric Poincaré group consists of the following transformations:
1. *Translations* (4 independent parameters $a_\mu$):

$$T_4: \quad x'_\mu = x_\mu + a_\mu, \quad a_\mu \text{ real}. \tag{17.38}$$

The infinitesimal transformations are given by:

$$\delta x_\mu = x'_\mu - x_\mu = \alpha_\mu \quad \text{infinitesimal}. \tag{17.39}$$

2. *The proper homogeneous Lorentz transformations* (6 independent parameters):

$$L_6^+: \quad x'_\mu = l_\mu{}^\nu x_\nu, \quad x'^\mu = l^\mu_\rho x^\rho. \tag{17.40}$$

The matrices $(l)$ define a $4\times 4$ representation of the non-compact Lorentz group $SO(3,1)$; the number "3" refers to the three spatial dimensions with compact domains of validity (angles), and the number "1" to the additional dimension with the non-compact parameter of a Lorentz transformation. Strictly speaking the vectors $x^\mu, x_\mu$ belong to different spaces $T, T^*$. But since a metric is defined on the 4-dimensional manifold $\mathbb{R}^4$ (the Minkowski metric $\eta$, which together with $\mathbb{R}^4$ defines the Minkowski space $\mathbb{M}_4$), it suffices to focus on only one space. The covariant components of a contravariant vector $x^\nu \in T$ then follow from $x_\mu = \eta_{\mu\nu}x^\nu$. From the condition of invariance of $x^\mu x_\mu$ under Lorentz transformations, i.e. from

$$x'^\mu x'_\mu = l^\mu{}_\rho x^\rho l_\mu{}^\nu x_\nu = \delta^\nu_\rho x^\rho x_\nu = x^\nu x_\nu, \tag{17.41}$$

we obtain the orthogonality condition

$$l^\mu{}_\rho l_\mu{}^\nu = \delta^\nu_\rho, \quad \underbrace{\det(l_\mu{}^\nu) = +1,}_{\text{therefore } L^+} \quad l_\mu{}^\nu \text{ real}. \tag{17.42}$$

The real matrix $l$ consists of $4 \times 4 = 16$ real elements or parameters, which are restricted by the orthogonality condition in Minkowski space, $l^T l = \mathbb{1}_{4\times 4}$, i.e. by 4 (diagonal-) + 6 (off-diagonal-) = 10 conditions. Thus there remain $16-10 = 6$ independent parameters corresponding to 3 spatial rotations and 3 Lorentz transformations (and hence velocities) in these directions.

A special case is the case of motion of the reference frame system $K'(x'_\mu)$ with constant velocity $v$ along the $x$-axis of $K$ (as we considered earlier for simplicity in Section 17.2). We set

$$\beta = \frac{v}{c}, \quad \gamma = \frac{1}{\sqrt{1-\beta^2}}. \tag{17.43}$$

The Lorentz transformation (17.12) is now in contravariant and covariant forms respectively (with $x^0 = ct$, $x^1 = x$, $x^2 = y$, $x^3 = z$):

$$\begin{aligned} x'^0 &= \gamma(x^0 - \beta x^1), \\ x'^1 &= \gamma(x^1 - \beta x^0), \\ x'^2 &= x^2, \\ x'^3 &= x^3, \end{aligned} \tag{17.44a}$$

and (cf. Eq. (17.16))

$$\begin{aligned} x'_0 &= \gamma(x_0 + \beta x_1), \\ x'_1 &= \gamma(x_1 + \beta x_0), \\ x'_2 &= x_2, \\ x'_3 &= x_3. \end{aligned} \tag{17.44b}$$

In this special case of motion along parallel $x$-, $x'$-axes, $(l_\mu{}^\nu)$ is given by (since $x'_\mu = l_\mu{}^\nu x_\nu$)

$$(l_\mu{}^\nu) = \begin{pmatrix} \frac{\partial x'_0}{\partial x_0} & \frac{\partial x'_0}{\partial x_1} & \frac{\partial x'_0}{\partial x_2} & \frac{\partial x'_0}{\partial x_3} \\ \frac{\partial x'_1}{\partial x_0} & \frac{\partial x'_1}{\partial x_1} & \frac{\partial x'_1}{\partial x_2} & \frac{\partial x'_1}{\partial x_3} \\ \frac{\partial x'_2}{\partial x_0} & \frac{\partial x'_2}{\partial x_1} & \frac{\partial x'_2}{\partial x_2} & \frac{\partial x'_2}{\partial x_3} \\ \frac{\partial x'_3}{\partial x_0} & \frac{\partial x'_3}{\partial x_1} & \frac{\partial x'_3}{\partial x_2} & \frac{\partial x'_3}{\partial x_3} \end{pmatrix} \equiv \begin{pmatrix} \gamma & \gamma\beta & 0 & 0 \\ \gamma\beta & \gamma & 0 & 0 \\ 0 & 0 & 1 & 0 \\ 0 & 0 & 0 & 1 \end{pmatrix}. \tag{17.45}$$

Then, as expected and claimed in Eq. (17.42),

$$\det(l_\mu{}^\nu) = \gamma^2 - \gamma^2\beta^2 = \gamma^2(1-\beta^2) = +1. \tag{17.46}$$

Moreover from Eq. (17.40), i.e.

$$x'_\mu = l_\mu{}^\nu x_\nu, \tag{17.47}$$

i.e. the matrix relation we assumed above, i.e.

$$(x'_\mu) = \begin{pmatrix} ct' \\ -x' \\ -y' \\ -z' \end{pmatrix} = \begin{pmatrix} \gamma & \gamma\beta & 0 & 0 \\ \gamma\beta & \gamma & 0 & 0 \\ 0 & 0 & 1 & 0 \\ 0 & 0 & 0 & 1 \end{pmatrix} \begin{pmatrix} ct \\ -x \\ -y \\ -z \end{pmatrix}, \tag{17.48}$$

we obtain the equations

$$ct' = \gamma ct - \gamma\beta x, \quad t' = \frac{t - vx/c^2}{\sqrt{1-\beta^2}} \tag{17.49}$$

and

$$-x' = \gamma\beta ct - \gamma x, \quad x' = \frac{x - vt}{\sqrt{1-\beta^2}}. \tag{17.50}$$

We now compute, also for later reference purposes,

$$\begin{aligned} (l_{\mu\nu}) = (l_\mu{}^\rho \eta_{\rho\nu}) &= \begin{pmatrix} \gamma & \gamma\beta & 0 & 0 \\ \gamma\beta & \gamma & 0 & 0 \\ 0 & 0 & 1 & 0 \\ 0 & 0 & 0 & 1 \end{pmatrix} \begin{pmatrix} 1 & 0 & 0 & 0 \\ 0 & -1 & 0 & 0 \\ 0 & 0 & -1 & 0 \\ 0 & 0 & 0 & -1 \end{pmatrix} \\ &= \begin{pmatrix} \gamma & -\gamma\beta & 0 & 0 \\ \gamma\beta & -\gamma & 0 & 0 \\ 0 & 0 & -1 & 0 \\ 0 & 0 & 0 & -1 \end{pmatrix}. \end{aligned} \tag{17.51}$$

Moreover

$$(l^{\nu\rho}) = (\eta^{\nu\mu} l_\mu{}^\rho) = \begin{pmatrix} 1 & 0 & 0 & 0 \\ 0 & -1 & 0 & 0 \\ 0 & 0 & -1 & 0 \\ 0 & 0 & 0 & -1 \end{pmatrix} \begin{pmatrix} \gamma & \gamma\beta & 0 & 0 \\ \gamma\beta & \gamma & 0 & 0 \\ 0 & 0 & 1 & 0 \\ 0 & 0 & 0 & 1 \end{pmatrix},$$

which means

$$(l^{\nu\rho}) = \begin{pmatrix} \gamma & \gamma\beta & 0 & 0 \\ -\gamma\beta & -\gamma & 0 & 0 \\ 0 & 0 & -1 & 0 \\ 0 & 0 & 0 & -1 \end{pmatrix}, \tag{17.52}$$

and (again with $\det(l^{\nu}{}_{\mu}) = +1$)

$$(l^{\nu}{}_{\mu}) = (l^{\nu\rho}\eta_{\rho\mu}) = \begin{pmatrix} \gamma & -\gamma\beta & 0 & 0 \\ -\gamma\beta & \gamma & 0 & 0 \\ 0 & 0 & 1 & 0 \\ 0 & 0 & 0 & 1 \end{pmatrix}. \tag{17.53}$$

Also the product of Eq. (17.42) comes out as expected:

$$\begin{aligned}(l^{\nu}{}_{\mu})(l_{\nu}{}^{\rho}) &= \begin{pmatrix} \gamma & -\gamma\beta & 0 & 0 \\ -\gamma\beta & \gamma & 0 & 0 \\ 0 & 0 & 1 & 0 \\ 0 & 0 & 0 & 1 \end{pmatrix}\begin{pmatrix} \gamma & \gamma\beta & 0 & 0 \\ \gamma\beta & \gamma & 0 & 0 \\ 0 & 0 & 1 & 0 \\ 0 & 0 & 0 & 1 \end{pmatrix} \\ &= \begin{pmatrix} \gamma^2 - \gamma^2\beta^2 & 0 & 0 & 0 \\ 0 & -\gamma^2\beta^2 + \gamma^2 & 0 & 0 \\ 0 & 0 & 1 & 0 \\ 0 & 0 & 0 & 1 \end{pmatrix} = (\mathbb{1}). \end{aligned} \tag{17.54}$$

In applications it is sometimes useful to use yet another formulation of the Lorentz transformation. We set

$$\cosh\phi := \frac{1}{\sqrt{1-\beta^2}}, \quad \beta = \frac{v}{c}. \tag{17.55}$$

Since

$$\sinh^2\phi = \cosh^2\phi - 1 = -1 + \frac{1}{1-\beta^2} = \frac{\beta^2}{1-\beta^2}, \tag{17.56}$$

we have

$$\sinh\phi = \frac{\beta}{\sqrt{1-\beta^2}}. \tag{17.57}$$

Then the Eqs. (17.15),

$$x' = \frac{x - vt}{\sqrt{1-\beta^2}}, \quad t' = \frac{t - (vx/c^2)}{\sqrt{1-\beta^2}}, \tag{17.58}$$

can be written

$$x' = x\cosh\phi - ct\sinh\phi, \quad ct' = ct\cosh\phi - x\sinh\phi, \tag{17.59}$$

i.e.

$$(x'^{\mu}) = (A)(x^{\mu}) \tag{17.60}$$

and

$$\begin{pmatrix} x' \\ y' \\ z' \\ ct' \end{pmatrix} = \begin{pmatrix} \cosh\phi & 0 & 0 & -\sinh\phi \\ 0 & 1 & 0 & 0 \\ 0 & 0 & 1 & 0 \\ -\sinh\phi & 0 & 0 & \cosh\phi \end{pmatrix} \begin{pmatrix} x \\ y \\ z \\ ct \end{pmatrix} \tag{17.61}$$

with

$$\det(A) = \cosh^2\phi - \sinh^2\phi = 1. \tag{17.62}$$

This form of the transformation reminds one of rotations, however, with imaginary angles, which make $1/\sqrt{1-\beta^2}$ noncompact.

### 17.4.1 *Covariant and contravariant derivatives*

The (homogeneous) Lorentz transformations describe "rotations" in four-dimensional Minkowski space ("rotations" in the sense mentioned earlier). We now define contravariant and covariant derivatives as follows[8]:

$$(\partial^{\mu}) \equiv \left(\frac{\partial}{\partial x_{\mu}}\right) = \left(\frac{\partial}{\partial x_0}, \underbrace{-\nabla}_{\partial^i = \partial/\partial x_i}\right),$$
$$(\partial_{\mu}) \equiv \left(\frac{\partial}{\partial x^{\mu}}\right) = \left(\frac{\partial}{\partial x^0}, \underbrace{\nabla}_{\partial_i = \partial/\partial x^i}\right). \tag{17.63}$$

[8]Note the reason: $\partial^{\mu}(x^2) \equiv \partial^{\mu}(x^{\rho}x_{\rho}) = 2x^{\mu}$.

With these definitions we have (cf. Eq. (10.18) for $\Box$)

$$\partial^\mu \partial_\mu = \frac{\partial^2}{\partial {x_0}^2} - \nabla^2 \equiv -\Box. \tag{17.64}$$

The *chain rule* applies:

$$(\partial^\mu) \equiv \left(\frac{\partial}{\partial x_\mu}\right) = \underbrace{\frac{\partial x'^\rho}{\partial x_\mu}\frac{\partial}{\partial x'^\rho} = \eta^{\rho\kappa} l_\kappa{}^\mu \frac{\partial}{\partial x'^\rho}}_{x'^\rho = \eta^{\rho\kappa} x'_\kappa,\; x'_\kappa = l_\kappa{}^\nu x_\nu} = l^{\rho\mu}\partial'_\rho. \tag{17.65}$$

Hence

$$\partial^\mu = l^{\rho\mu}\partial'_\rho, \quad \text{and} \quad \therefore \;\; \partial_\nu = \eta_{\nu\mu}\partial^\mu = \eta_{\nu\mu} l^{\rho\mu}\partial'_\rho, \quad \text{so that } \partial_\nu = \partial'_\rho l^\rho_\nu. \tag{17.66}$$

We can show that this relation is identically satisfied, if $\partial'_\rho$ transforms like the vector $x'_\rho$, because then

$$\partial'_\rho l^\rho{}_\nu = l^\rho{}_\nu \partial'_\rho = l^\rho{}_\nu l_\rho{}^\kappa \partial_\kappa \overset{(17.42)}{=} \delta^\kappa_\nu \partial_\kappa = \partial_\nu. \tag{17.67}$$

This means the derivatives transform like the corresponding vectors.

## 17.5 Construction of the Field Tensor $F_{\mu\nu}$

One can show that for instance the four-divergence $\partial_\mu A^\mu$ and the d'Alembert operator $\Box \equiv \partial^\mu \partial_\mu$ are *Lorentz scalars, Lorentz invariants.* Since $\det(l_\mu{}^\nu) = 1$, it follows that the transformation of the *volume element in Minkowski space* is (cf. Eq. (17.45) and thereafter)

$$\begin{aligned} dx'_0 dx'_1 dx'_2 dx'_3 &= \begin{vmatrix} \frac{\partial x'_0}{\partial x_0} & \frac{\partial x'_0}{\partial x_1} & \frac{\partial x'_0}{\partial x_2} & \frac{\partial x'_0}{\partial x_3} \\ \frac{\partial x'_1}{\partial x_0} & \frac{\partial x'_1}{\partial x_1} & \frac{\partial x'_1}{\partial x_2} & \frac{\partial x'_1}{\partial x_3} \\ \frac{\partial x'_2}{\partial x_0} & \frac{\partial x'_2}{\partial x_1} & \frac{\partial x'_2}{\partial x_2} & \frac{\partial x'_2}{\partial x_3} \\ \frac{\partial x'_3}{\partial x_0} & \frac{\partial x'_3}{\partial x_1} & \frac{\partial x'_3}{\partial x_2} & \frac{\partial x'_3}{\partial x_3} \end{vmatrix} dx_0 dx_1 dx_2 dx_3 \\ &= dx_0 dx_1 dx_2 dx_3, \end{aligned} \tag{17.68}$$

i.e. *the volume element is invariant.*

In order to achieve the covariant formulation of Maxwell's equations, we consider first the *continuity equation* (5.12), i.e.

$$\boldsymbol{\nabla}\cdot\mathbf{j}+\frac{\partial\rho}{\partial t}=0. \tag{17.69}$$

We can write this now (with $(\partial_\mu)=(\partial_0,\partial_i)$)

$$\partial_\mu J^\mu\equiv\frac{\partial J^\mu}{\partial x^\mu}=0. \tag{17.70}$$

Here

$$J^\mu=(c\rho,\mathbf{j}) \tag{17.71}$$

is a four-vector (whether here $+\mathbf{j}$ or $-\mathbf{j}$, is a question of the direction of the current density $\mathbf{j}=\rho d\mathbf{x}/dt$). The *equations of the potentials* (cf. Eq. (10.20))

$$\triangle\mathbf{A}-\frac{1}{c^2}\frac{\partial^2\mathbf{A}}{\partial t^2}=-\mu_0\mathbf{j},\quad \triangle\phi-\frac{1}{c^2}\frac{\partial^2\phi}{\partial t^2}=-\frac{1}{\epsilon_0}\rho=-\mu_0c^2\rho, \tag{17.72}$$

together with the *Lorenz gauge* (10.19), i.e.

$$\boldsymbol{\nabla}\cdot\mathbf{A}+\frac{1}{c^2}\frac{\partial\phi}{\partial t}=0, \tag{17.73}$$

and the four-vector

$$(A^\mu)=\left(\frac{1}{c}\phi,\mathbf{A}\right)\equiv(A^0,\mathbf{A}), \tag{17.74}$$

can be written in the form

$$\partial^\rho\partial_\rho A^\mu=\mu_0J^\mu,\quad -\Box\equiv\partial^\mu\partial_\mu=\frac{1}{c^2}\frac{\partial^2}{\partial t^2}-\triangle, \tag{17.75}$$

with the Lorenz condition written

$$\partial_\mu A^\mu=0. \tag{17.76}$$

The *field strengths* $\mathbf{E}$ and $\mathbf{B}$ are given by (cf. (7.33))

$$\begin{aligned}\mathbf{E}&:=(E_1,E_2,E_3)=-\nabla\phi-\frac{\partial\mathbf{A}}{\partial t}=-c\left(\boldsymbol{\nabla}A^0+\frac{\partial\mathbf{A}}{\partial x_0}\right),\\ \mathbf{B}&:=(B_1,B_2,B_3)=\boldsymbol{\nabla}\times\mathbf{A}.\end{aligned} \tag{17.77}$$

Recalling Eqs. (17.63) and (17.74), this means

$$\begin{aligned}
\frac{1}{c}E_1 &= -\frac{\partial A_0}{\partial x^1} - \frac{\partial A^1}{\partial x_0} = -\partial^0 A^1 + \partial^1 A^0, \\
\frac{1}{c}E_2 &= -\frac{\partial A_0}{\partial x^2} - \frac{\partial A^2}{\partial x_0} = -\partial^0 A^2 + \partial^2 A^0, \\
\frac{1}{c}E_3 &= -\frac{\partial A_0}{\partial x^3} - \frac{\partial A^3}{\partial x_0} = -\partial^0 A^3 + \partial^3 A^0, \\
B_1 &= \frac{\partial A^3}{\partial x^2} - \frac{\partial A^2}{\partial x^3} = -(\partial^2 A^3 - \partial^3 A^2), \\
B_2 &= \frac{\partial A^1}{\partial x^3} - \frac{\partial A^3}{\partial x^1} = -(\partial^3 A^1 - \partial^1 A^3), \\
B_3 &= \frac{\partial A^2}{\partial x^1} - \frac{\partial A^1}{\partial x^2} = -(\partial^1 A^2 - \partial^2 A^1).
\end{aligned} \tag{17.78}$$

The similar form of the expressions for $E_i/c$ and $B_j$ suggests the *introduction of the field tensor* $F_{\mu\nu}$, i.e. an antisymmetric tensor of the second rank with 6 independent components, i.e.

$$\begin{aligned}
F^{\mu\nu} &\equiv \frac{\partial A^\nu}{\partial x_\mu} - \frac{\partial A^\mu}{\partial x_\nu} \equiv \partial^\mu A^\nu - \partial^\nu A^\mu \\
&= \begin{pmatrix} 0 & -E_1/c & -E_2/c & -E_3/c \\ E_1/c & 0 & -B_3 & B_2 \\ E_2/c & B_3 & 0 & -B_1 \\ E_3/c & -B_2 & B_1 & 0 \end{pmatrix}.
\end{aligned} \tag{17.79}$$

One should note that here $\mathbf{E}/c$ and $\mathbf{B}$ appear, and not $\mathbf{E}/c$ and $\mathbf{H}$. The corresponding tensor with lower indices is obtained as

$$\begin{aligned}
F_{\mu\nu} &= \eta_{\mu\rho} F^{\rho\gamma} \eta_{\gamma\nu} \\
&= \begin{pmatrix} 0 & E_1/c & E_2/c & E_3/c \\ -E_1/c & 0 & -B_3 & B_2 \\ -E_2/c & B_3 & 0 & -B_1 \\ -E_3/c & -B_2 & B_1 & 0 \end{pmatrix} = -F_{\nu\mu}.
\end{aligned} \tag{17.80}$$

The tensor $F_{\mu\nu}$ is also an antisymmetric tensor. Furthermore we observe: $F_{\mu\nu}$ follows from $F^{\mu\nu}$ by the substitution $\mathbf{E} \to -\mathbf{E}$. Another quantity which is needed is the so-called *dual field tensor* $\mathcal{F}^{\mu\nu}$ defined by the relation[9]

$$\mathcal{F}^{\mu\nu} \equiv {}^*F^{\mu\nu} = \frac{1}{2}\epsilon^{\mu\nu\rho\sigma}F_{\rho\sigma} = \begin{pmatrix} 0 & -B_1 & -B_2 & -B_3 \\ B_1 & 0 & E_3/c & -E_2/c \\ B_2 & -E_3/c & 0 & E_1/c \\ B_3 & E_2/c & -E_1/c & 0 \end{pmatrix}, \quad (17.81)$$

where

$$\begin{aligned} \epsilon^{\mu\nu\rho\sigma} &= +1 \quad \text{for } \mu\nu\rho\sigma \text{ an even permutation of } 0,1,2,3, \\ &= -1 \quad \text{for } \mu\nu\rho\sigma \text{ an odd permutation of } 0,1,2,3, \\ &= 0, \quad \text{if two indices are equal.} \end{aligned} \quad (17.82)$$

(Note: "*Self-duality*" means ${}^*F^{\mu\nu} = F^{\mu\nu}$, *anti-self-duality* means ${}^*F^{\mu\nu} = -F^{\mu\nu}$, neither of which apply here). The dual field tensor $\mathcal{F}^{\mu\nu}$ is obtained from $F^{\mu\nu}$ by the substitution

$$\frac{1}{c}\mathbf{E} \to \mathbf{B}, \quad \mathbf{B} \to -\frac{1}{c}\mathbf{E}. \quad (17.83)$$

We see: In four-form notation the electromagnetic field is no longer described by two separate vectors, but by one antisymmetric tensor of the second rank. In a Lorentz transformation, i.e. transformation from the reference frame of one observer to that of another observer moving relative to the first with uniform velocity, the components $E_i, B_i$ are mixed. Thus e.g. the pure field $\mathbf{E}/c$ of a static charge in one frame becomes a mixture of $\mathbf{E}'/c$ and $\mathbf{B}'$ in the other frame.

The Lorentz transformation of the field tensor $F_{\mu\nu}$ is given by the transformation of a second rank tensor (see also below):

$$F'_{\mu\nu} = l_\mu{}^\sigma l_\nu{}^\rho F_{\sigma\rho} \quad (17.84)$$

with the special Lorentz transformation

$$(l_\mu{}^\nu) \overset{(17.45)}{=} \begin{pmatrix} \gamma & \gamma\beta & 0 & 0 \\ \gamma\beta & \gamma & 0 & 0 \\ 0 & 0 & 1 & 0 \\ 0 & 0 & 0 & 1 \end{pmatrix} \quad (17.85)$$

[9]Note that in three dimensions (with $i,j,k = 1,2,3$, $g = \det(g_{ij}) = 1$) the dual of the 2-form $F_{ij}$ is the 1-form $B^k$, i.e. $B^i = \epsilon^{ijk}F_{jk}/2\sqrt{g}$, and so $B^1 = F_{23}, B^2 = F_{31}$ and $B^3 = F_{12}$.

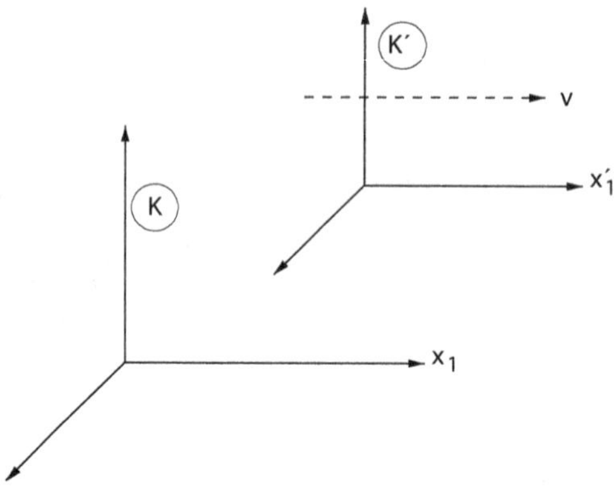


**Fig. 17.3.** Reference frames $K, K'$ in uniform translational motion with respect to each other.

called a "*boost*" along the $x_1$-axis with velocity $c\beta$ from reference frame $K(E, B)$ to reference frame $K'(E', B')$ indicated in Fig. 17.3.

The relation (17.84) defines a tensor of the second rank. That the quantity defined by the expression

$$F_{\mu\nu}(x) := \partial_\mu A_\nu(x) - \partial_\nu A_\mu(x) \tag{17.86}$$

is a covariant tensor of the second rank, can be shown by starting from the relation which defines a covariant vector (i.e. tensor of the first rank), and that means the way this transforms under the appropriate transformations, in the present case of Lorentz transformations. Thus we recall the transformation of a covariant four-vector $A_\mu(x)$:

$$A_\nu(x') = \frac{\partial x^\mu}{\partial x'^\nu} A_\mu(x). \tag{17.87}$$

In order to verify Eq. (17.84), we have to show that with the transformation

$$x'_\mu = l_\mu{}^\nu x_\nu, \tag{17.88}$$

the relation (17.87) implies

$$A_\mu(x') = l_\mu{}^\nu A_\nu(x). \tag{17.89}$$

Since from Eq. (17.42) $l^\kappa{}_\mu l_\kappa{}^\nu = \delta^\nu_\mu$, we deduce from Eq. (17.88):

$$x_\mu = \delta^\nu_\mu x_\nu = l^\kappa{}_\mu l_\kappa{}^\nu x_\nu = l^\kappa{}_\mu x'_\kappa = x'_\kappa l^\kappa{}_\mu, \tag{17.90}$$

or raising the index with the metric tensor:

$$x^\mu = \eta^{\mu\rho} l^\kappa{}_\rho x'_\kappa = \eta^{\mu\rho} l^\kappa{}_\rho \eta_{\kappa\nu} x'^\nu, \tag{17.91}$$

so that

$$\frac{\partial x^\mu}{\partial x'^\nu} = \eta^{\mu\rho} l^\kappa{}_\rho \eta_{\kappa\nu} = l_\nu{}^\mu, \tag{17.92}$$

and hence from Eq. (17.87):

$$\begin{aligned} A_\nu(x') &= \frac{\partial x^\mu}{\partial x'^\nu} A_\mu(x) = \eta^{\mu\rho} l^\kappa{}_\rho \eta_{\kappa\nu} A_\mu(x) = A^\rho(x) l^\kappa{}_\rho \eta_{\kappa\nu} \\ &= \eta_{\nu\kappa} l^\kappa{}_\rho A^\rho(x) = l_{\nu\rho} \eta^{\rho\kappa} A_\kappa(x) = l_\nu{}^\kappa A_\kappa(x), \end{aligned} \tag{17.93}$$

i.e. Eq. (17.87) implies Eq. (17.89) which had to be shown.

We now consider the covariant field strength tensor (where $\partial_{[\mu} A_{\nu]} := \partial_\mu A_\nu - \partial_\nu A_\mu$)

$$\begin{aligned} F_{\mu\nu}(x') &= \partial_{[\mu} A_{\nu]}(x') = \frac{\partial}{\partial x'^\mu} A_\nu(x') - \frac{\partial}{\partial x'^\nu} A_\mu(x') \\ &\overset{(17.87)}{=} \frac{\partial}{\partial x'^\mu}\left[A_\beta(x)\frac{\partial x^\beta}{\partial x'^\nu}\right] - \frac{\partial}{\partial x'^\nu}\left[A_\beta(x)\frac{\partial x^\beta}{\partial x'^\mu}\right] \\ &= \left(\frac{\partial A_\beta(x)}{\partial x^\gamma}\frac{\partial x^\gamma}{\partial x'^\mu}\right)\frac{\partial x^\beta}{\partial x'^\nu} + A_\beta(x)\frac{\partial^2 x^\beta}{\partial x'^\mu \partial x'^\nu} \\ &\quad - \left(\frac{\partial A_\beta(x)}{\partial x^\gamma}\frac{\partial x^\gamma}{\partial x'^\nu}\right)\frac{\partial x^\beta}{\partial x'^\mu} - A_\beta(x)\frac{\partial^2 x^\beta}{\partial x'^\nu \partial x'^\mu} \\ &= \frac{\partial x^\gamma}{\partial x'^\mu}\frac{\partial x^\beta}{\partial x'^\nu}\partial_{[\gamma} A_{\beta]}(x) \\ &\overset{(17.92)}{=} l_\mu{}^\gamma l_\nu{}^\beta \partial_{[\gamma} A_{\beta]}(x) \end{aligned} \tag{17.94}$$

in agreement with Eq. (17.84).

In multiplying matrices we have to keep in mind that Lorentz indices disappear through contraction. Consider first with the help of Eqs. (17.45) and (17.80):

$$\begin{aligned} (L)_{\mu\rho} := (l_\mu{}^\sigma F_{\sigma\rho}) &\overset{(17.45)}{=} \begin{pmatrix} \gamma & \gamma\beta & 0 & 0 \\ \gamma\beta & \gamma & 0 & 0 \\ 0 & 0 & 1 & 0 \\ 0 & 0 & 0 & 1 \end{pmatrix} \begin{pmatrix} 0 & E_1/c & E_2/c & E_3/c \\ -E_1/c & 0 & -B_3 & B_2 \\ -E_2/c & B_3 & 0 & -B_1 \\ -E_3/c & -B_2 & B_1 & 0 \end{pmatrix} \\ &= \begin{pmatrix} -\gamma\beta E_1/c & \gamma E_1/c & \gamma E_2/c - \gamma\beta B_3 & \gamma E_3/c + \gamma\beta B_2 \\ -\gamma E_1/c & \gamma\beta E_1/c & \gamma\beta E_2/c - \gamma B_3 & \gamma\beta E_3/c + \gamma B_2 \\ -E_2/c & B_3 & 0 & -B_1 \\ -E_3/c & -B_2 & B_1 & 0 \end{pmatrix}. \end{aligned} \tag{17.95}$$

The last matrix defines the quantity $L$ which will appear in the following. Next we consider the quantity $L$ multiplied by the matrix (17.85) from the right:

$$(L)_{\mu\rho}(l_\nu{}^\rho) = \begin{pmatrix} -\gamma\beta E_1/c & \gamma E_1/c & \gamma E_2/c - \gamma\beta B_3 & \gamma E_3/c + \gamma\beta B_2 \\ -\gamma E_1/c & \gamma\beta E_1/c & \gamma\beta E_2/c - \gamma B_3 & \gamma\beta E_3/c + \gamma B_2 \\ -E_2/c & B_3 & 0 & -B_1 \\ -E_3/c & -B_2 & B_1 & 0 \end{pmatrix}$$
$$\times \begin{pmatrix} \gamma & \gamma\beta & 0 & 0 \\ \gamma\beta & \gamma & 0 & 0 \\ 0 & 0 & 1 & 0 \\ 0 & 0 & 0 & 1 \end{pmatrix}$$
$$= \begin{pmatrix} 0 & E_1/c & \gamma E_2/c - \gamma\beta B_3 & \gamma E_3/c + \gamma\beta B_2 \\ -E_1/c & 0 & \gamma\beta E_2/c - \gamma B_3 & \gamma\beta E_3/c + \gamma B_2 \\ -E_2/c + \gamma\beta B_3 & -\gamma\beta E_2/c + \gamma B_3 & 0 & -B_1 \\ -\gamma E_3/c - \gamma\beta B_2 & -\gamma\beta E_3/c - \gamma B_2 & B_1 & 0 \end{pmatrix}. \tag{17.96}$$

This expression which is the right hand side of Eq. (17.84), is to be identified with the left hand side of Eq. (17.84), i.e. as in Eq. (17.80) but with primed quantities:

$$F'_{\mu\nu} = \begin{pmatrix} 0 & E'_1/c & E'_2/c & E'_3/c \\ -E'_1/c & 0 & -B'_3 & B'_2 \\ -E'_2/c & B'_3 & 0 & -B'_1 \\ -E'_3/c & -B'_2 & B'_1 & 0 \end{pmatrix}, \tag{17.97}$$

so that by comparison of elements of Eq. (17.96) with corresponding elements of Eq. (17.97) we obtain the following transformation equations:

$$\begin{aligned} E'_1 &= E_1, & cB'_1 &= cB_1, \\ E'_2 &= \gamma(E_2 - c\beta B_3), & cB'_2 &= \gamma(cB_2 + \beta E_3), \\ E'_3 &= \gamma(E_3 + c\beta B_2), & cB'_3 &= \gamma(cB_3 - \beta E_2). \end{aligned} \tag{17.98}$$

It can be shown,[10] that in the general case (i.e. for the constant velocity $\mathbf{v}$ with arbitrary orientation) these equations are:

$$\mathbf{E}' = \gamma(\mathbf{E} + \mathbf{v} \times \mathbf{B}) - \frac{\gamma^2}{c^2(\gamma+1)}\mathbf{v}(\mathbf{v} \cdot \mathbf{E}),$$
$$c\mathbf{B}' = \gamma\left(c\mathbf{B} - \frac{\mathbf{v} \times \mathbf{E}}{c}\right) - \frac{\gamma^2}{c(\gamma+1)}\mathbf{v}(\mathbf{v} \cdot \mathbf{B}). \qquad (17.99)$$

For instance for $E'_1$ with $\mathbf{v}$ parallel to $\mathbf{x}_1$, and

$$\beta = v/c \quad \text{and} \quad \beta^2 = (\gamma^2 - 1)/\gamma^2, \qquad (17.100)$$

one obtains the relations:

$$E'_1 = \gamma(E_1 + 0) - \frac{\gamma^2}{\gamma+1}\beta^2 E_1 = \frac{\gamma^2 + \gamma - \gamma^2\beta^2}{\gamma+1}E_1 = \frac{1+\gamma}{\gamma+1}E_1 = E_1. \qquad (17.101)$$

We observe again: $\mathbf{E}$ and $\mathbf{B}$ do not exist independently of each other. A purely electric or purely magnetic field in the reference frame of one observer appears as a mixture of both types of fields in the reference frame of another observer moving relative to the first with uniform velocity $\mathbf{v}$. Thus the subdivision of the electromagnetic field into electric and magnetic components has no fundamental meaning. Which field components appear depends on the choice of reference frame of the observer. We also see that $\mathbf{E}$ and $\mathbf{B}$ are the fundamental quantities and not $\mathbf{E}$ and $\mathbf{H}$.

## 17.6 Transformation from the Rest Frame to an Inertial Frame

As an example we consider the fields observed by an observer in reference frame $K$, when a point charge $q$ passes him in rectilinear motion with uniform velocity $\mathbf{v}$. This example is always impressive, because one starts from the simple Coulomb potential in the rest frame of the charge, then transforms to the other frame and ends up with the law of Biot–Savart in the nonrelativistic limit!

We let $K'$ be the rest frame of the charge. In this case the transformation equations of the fields are inverse to those given above. We obtain the

[10] See e.g. also Jackson [73].

equations we need by making in the above the substitutions

$$\mathbf{E}', \mathbf{B}' \leftrightarrow \mathbf{E}, \mathbf{B} \quad \text{and} \quad \mathbf{v} \to -\mathbf{v}. \tag{17.102}$$

We obtain (since $K$ moves away from $K'$ with velocity $-\mathbf{v}$)

$$\begin{aligned} E_1 &= E_1', & cB_1 &= cB_1', \\ E_2 &= \gamma(E_2' + c\beta B_3'), & cB_2 &= \gamma(cB_2' - \beta E_3'), \\ E_3 &= \gamma(E_3' - c\beta B_2'), & cB_3 &= \gamma(cB_3' + \beta E_2'). \end{aligned} \tag{17.103}$$

Of course, these equations also follow from Eqs. (17.98) by solving for $E_i$, $B_i$. We let $b$ be the shortest distance of the charge $q$ from the observer at point $P$ in $K$ as indicated in Fig. 17.4. At time $t = t' = 0$ we let the origin $O'$ be at $O$ and charge $q$ at the shortest distance from $P$. The coordinates of $P$ as seen from $K'$ are therefore according to Fig. 17.4:

$$x'_1 = -vt', \quad x'_2 = b, \quad x'_3 = 0, \tag{17.104}$$

where

$$r' = \sqrt{x_1'^2 + x_2'^2 + x_3'^2} = \sqrt{b^2 + (vt')^2}. \tag{17.105}$$

We ask: What is $r'$, containing $t'$, expressed in terms of the coordinates of $K(x_1, x_2, x_3, t)$? Thus we require $t' = t'(x_1, x_2, x_3, t)$. This follows from the Lorentz transformation in the direction of $x_1$. Thus from Eqs. (17.58):

$$t' = \gamma\left(t - \frac{vx_1}{c^2}\right) \quad \text{or} \quad t = \gamma\left(t' + \frac{vx'_1}{c^2}\right). \tag{17.106}$$

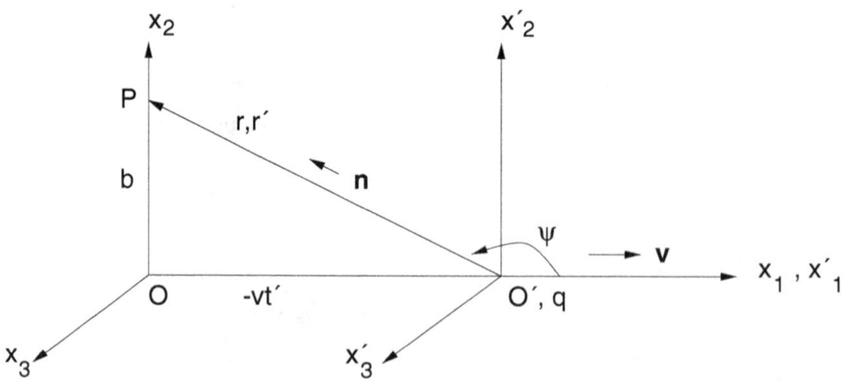


**Fig. 17.4.** Reference frames $K, K'$ with charge $q$ at the origin $O'$ in $K'$.

Since we consider point $P$ fixed in reference frame $K$, we have $x_1 = 0$. This means ($\gamma = 1/\sqrt{1-\beta^2}$)

$$t' = \gamma t \quad (\text{and } x'_1 = -\gamma v t = -v t'). \tag{17.107}$$

In the rest frame $K'$ of the charge $q$ the fields at the point $P$ are:

$$E'_1 = -\frac{q}{4\pi\epsilon_0 r'^2}\left(\frac{vt'}{r'}\right) = -\frac{qvt'}{4\pi\epsilon_0 r'^3} \quad \text{newtons/coulomb}, \tag{17.108}$$

where $E'_1 = \mathbf{E}'_1 \cdot \mathbf{e}'_{x_1}$ and (cf. Fig. 17.4) $-vt'/r' = \cos\psi$. Similarly we have

$$E'_2 = \frac{q}{r'^2}\frac{\cos(\pi/2 - \overline{\pi - \psi})}{4\pi\epsilon_0} = \frac{qb}{r'^3}\frac{1}{4\pi\epsilon_0}, \quad E'_3 = 0 \quad \left(\propto \frac{x'_3}{r'^3}\right), \tag{17.109}$$

the latter would be the tangential component of a radial field. Since the charge is at rest in $K'$, we have no magnetic field there and hence

$$B'_1 = 0, \quad B'_2 = 0, \quad B'_3 = 0. \tag{17.110}$$

We now express these quantities in terms of the coordinates of $K$, i.e. with

$$r' = \sqrt{b^2 + (vt')^2} = \sqrt{b^2 + v^2\gamma^2 t^2}, \quad t' = \gamma t, \tag{17.111}$$

we obtain

$$\begin{aligned} E'_1 &= -\frac{q\gamma v t}{(b^2 + \gamma^2 v^2 t^2)^{3/2}}\frac{1}{4\pi\epsilon_0} \quad \text{newtons/coulomb}, \\ E'_2 &= \frac{qb}{(b^2 + \gamma^2 v^2 t^2)^{3/2}}\frac{1}{4\pi\epsilon_0} \quad \text{newtons/coulomb}, \end{aligned} \tag{17.112}$$

and the others vanish. Substituting these expressions into the transformation Eqs. (17.103), we obtain

$$\left.\begin{aligned} E_1 &= E'_1 = -q\gamma v t/4\pi\epsilon_0(b^2 + \gamma^2 v^2 t^2)^{3/2}, \\ E_2 &= \gamma E'_2 = q\gamma b/4\pi\epsilon_0(b^2 + \gamma^2 v^2 t^2)^{3/2}, \end{aligned}\right\} \frac{E_1}{E_2} = -\frac{vt}{b},$$

$$\begin{aligned} E_3 &= 0, \\ B_1 &= B'_1 = 0, \\ B_2 &= 0, \\ cB_3 &= \gamma\beta E'_2 = \beta E_2. \end{aligned} \tag{17.113}$$

Thus the observer in reference frame $K$ observes a magnetic field. The *relativistic effects* become particularly evident, if $v \to c$. First we see that in the *nonrelativistic limit* $\gamma \to 1$ ($\gamma = 1/\sqrt{1-\beta^2}$):

$$\begin{aligned} cB_3 = \beta E_2 &= \frac{q\gamma\beta b}{(b^2+\gamma^2 v^2 t^2)^{3/2}} \frac{1}{4\pi\epsilon_0}, \\ &\overset{\gamma\to 1}{\simeq} \frac{q\beta b}{(b^2+v^2t^2)^{3/2}} \frac{1}{4\pi\epsilon_0}, \end{aligned} \tag{17.114}$$

which can be rewritten as

$$\begin{aligned} cB_3 &= \frac{q}{c}\frac{vb}{r^3}\frac{1}{4\pi\epsilon_0} = \frac{q}{c}\frac{vr\sin(\pi-\psi)}{4\pi\epsilon_0 r^3} \\ &= \frac{q}{c}\frac{vr\sin\psi}{4\pi\epsilon_0 r^3} = \frac{q}{c}\frac{vr\sin(\mathbf{v},\mathbf{r})}{4\pi\epsilon_0 r^3}, \\ c\mathbf{B} &= q\frac{\mathbf{v}\times\mathbf{r}}{r^3}\frac{c\mu_0}{4\pi} \end{aligned} \tag{17.115}$$

with $c^2 = 1/\epsilon_0\mu_0$. Here, $-\mathbf{r}$ is the vector $\mathbf{PO'}$ described earlier as $\mathbf{r}$. We recognize the result (when divided by $c$) as the *law of Biot–Savart* for the magnetic field of a moving charge, i.e. our earlier Eq. (5.20) but infinitesimally,

$$d\mathbf{B} = \frac{\mu_0 I d\mathbf{s}(\mathbf{r}')\times(\mathbf{r}-\mathbf{r}')}{4\pi|\mathbf{r}-\mathbf{r}'|^3} \quad \text{teslas}, \quad I = \frac{dq}{dt}, \quad \mathbf{v} = \frac{d\mathbf{s}}{dt}. \tag{17.116}$$

Thus the transformation laws of the Special Theory of Relativity supply not only relativistic corrections, but also the Biot–Savart law. However, it is not correct to conclude, it would be possible to derive the magnetic field and magnetic force, as well as the full set of Maxwell equations from the Coulomb law and Special Relativity. We referred to this already in Chapter 1.[11]

Next we have a closer look at *the longitudinal field* $E_1$ of Eq. (17.113). Considered as a function of $vt$ this has the form shown in Fig. 17.5 for $v$ very small ($\beta \simeq 0$) and very large ($\beta \lesssim 1$). The expression is[12]

$$E_1 = -\frac{q\gamma vt}{4\pi\epsilon_0(b^2+\gamma^2v^2t^2)^{3/2}} \quad \text{newtons/coulomb}. \tag{17.117}$$

The maxima and/or minima of $E_1$ are given by

---

[11]See e.g. Jackson [73], Section 12.2, p. 578.

[12]See e.g. Panofsky and Phillips [111], their Eq. (19.30), p. 350, for additional discussion.

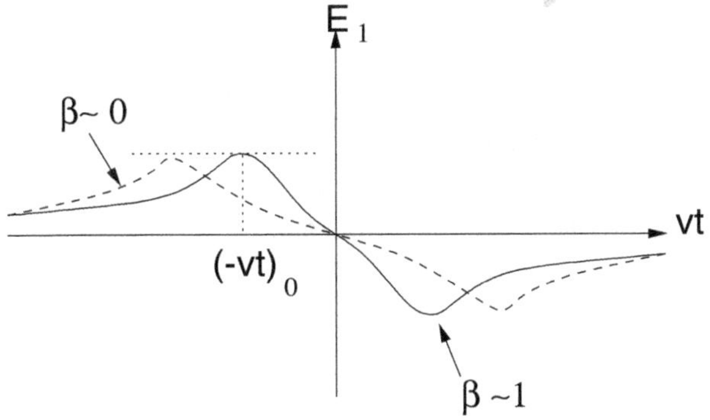


**Fig. 17.5.** $E_1$ for limiting cases of $\beta$.

$$\frac{d}{dvt}E_1 = \frac{d}{dvt}\left[-\frac{q\gamma vt}{4\pi\epsilon_0(b^2+\gamma^2v^2t^2)^{3/2}}\right] = 0, \tag{17.118}$$

i.e.

$$\frac{1}{(b^2+\gamma^2v^2t^2)^{3/2}} - \frac{3}{2}\frac{2\gamma^2v^2t^2}{(b^2+\gamma^2v^2t^2)^{5/2}} = 0,$$

$$b^2+\gamma^2v^2t^2-3\gamma^2v^2t^2 = 0, \quad b^2 = 2\gamma^2v^2t^2, \tag{17.119}$$

or

$$(vt)_0 = \pm\frac{1}{\sqrt{2}}\frac{b}{\gamma}, \quad \gamma^2 = \frac{1}{1-\beta^2}. \tag{17.120}$$

For $\beta \lesssim 1$, we see that $(vt)_0$ is small; for $\beta \approx 0$, we see $(vt)_0 \sim \pm b/\sqrt{2} \sim$ is large, as shown in Fig. 17.5. The value of $E_1$ at $(vt)_0 \sim \pm b/(\sqrt{2}\gamma)$ is

$$(E_1)_0 = \mp\sqrt{\frac{4}{27}\frac{q}{b^2}\frac{1}{4\pi\epsilon_0}} \quad \text{newtons/coulomb.} \tag{17.121}$$

This expression is independent of $v$; hence the maxima have the same height in Fig. 17.5. We see, that $E_1$ of Eq. (17.117) is an odd function of $vt$. In an observation in the time interval $\Delta t = (-t_0, t_0)$ the effects average out and $E_1$ is not observed at $P$. The observer at $P$ therefore observes practically only the transverse field of the components $E_2, cB_3 = \beta E_2$, which for $\beta \approx 1$ are not only mutually transverse but also almost equal in modulus. If we look at the field $E_2$ as a function of $vt$, we see that this has the following form:

$$E_2 = \frac{q\gamma b}{4\pi\epsilon_0(b^2+\gamma^2v^2t^2)^{3/2}} \quad \text{newtons/coulomb.} \tag{17.122}$$

For the limiting cases the behavior is sketched in Fig. 17.6. We see that $E_2$ has the shape of a "*pulse*".[13] Thus the field of a charged particle moving at relativistic speed and observed at the point $P$ in reference frame $K$ is equivalent to that of an electromagnetic pulse. This principle is exploited in attempts to observe heavy magnetic poles. In such experiments an atom is used to feel the effect of a passing magnetic pole (see also Chapter 21).

Finally we consider the spatial distribution of the fields with reference to the instantaneous position of the charge. With $\mathbf{E} = (E_1, E_2)$ we have the ratio

$$\frac{E_1}{E_2} = -\frac{vt}{b}. \tag{17.123}$$

From this and the force diagram we infer that the electric field has the direction of the vector $\mathbf{n}$ along $r$, with $\mathbf{E} \parallel \mathbf{n}$ and (see Fig. 17.7),

$$\cos\psi = \frac{\mathbf{n}\cdot\mathbf{v}}{v} = -\frac{vt}{r}. \tag{17.124}$$

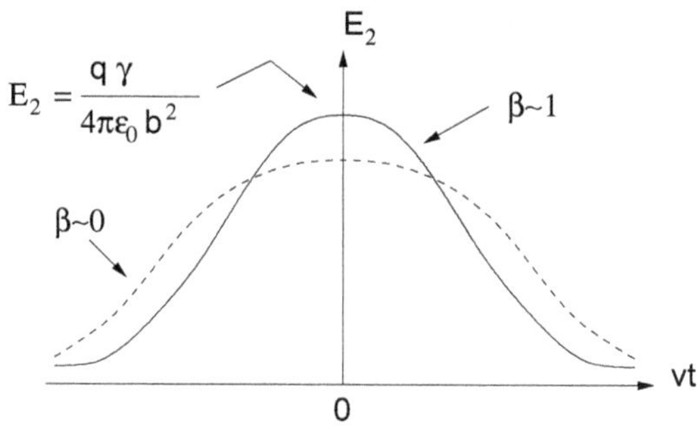


**Fig. 17.6.** $E_2$ in limits of $\beta$ small and large.

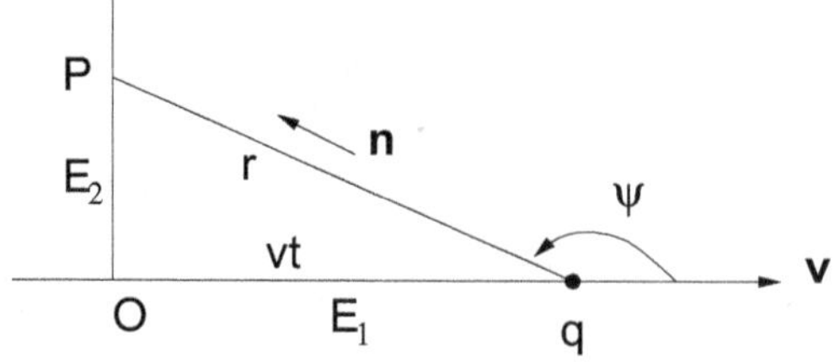


**Fig. 17.7.** The direction of the electric field.

[13] Panofsky and Phillips [111], p. 350.

But now

$$\sin\psi = \frac{b}{r} \tag{17.125}$$

and $(vt)^2 + b^2 = r^2$, so that

$$1 - \beta^2\sin^2\psi = 1 - \beta^2\frac{b^2}{r^2} = \frac{r^2 - \beta^2 b^2}{r^2} = \frac{b^2 + (vt)^2 - \beta^2 b^2}{r^2}, \tag{17.126}$$

and hence

$$\begin{aligned} 1 - \beta^2\sin^2\psi &= \frac{b^2(1-\beta^2) + (vt)^2}{r^2} = \frac{b^2/\gamma^2 + (vt)^2}{r^2} \\ &= \frac{b^2 + \gamma^2(vt)^2}{\gamma^2 r^2}. \end{aligned} \tag{17.127}$$

Thus

$$\begin{aligned} \mathbf{E} = (E_1, E_2) &= \frac{q\gamma(-vt, b)}{4\pi\epsilon_0(b^2 + \gamma^2(vt)^2)^{3/2}} \\ &= \frac{q\gamma\mathbf{r}}{4\pi\epsilon_0(\gamma^2 r^2)^{3/2}(1 - \beta^2\sin^2\psi)^{3/2}} \\ &= \frac{q\mathbf{r}}{4\pi\epsilon_0(\gamma^2 r^3)(1 - \beta^2\sin^2\psi)^{3/2}}. \end{aligned} \tag{17.128}$$

Thus the observed radial field at $P$ has an isotropic distribution only for $\beta \sim 0$, i.e. at a very small velocity of the passing charge as illustrated in Fig. 17.8. At $\psi = 0$ and $\pi$, the field $\mathbf{E}$ is

$$\mathbf{E} = \frac{q\mathbf{r}}{4\pi\epsilon_0\gamma^2 r^3} = \frac{(1-\beta^2)q\mathbf{r}}{4\pi\epsilon_0 r^3} \to 0 \quad \text{for} \quad v \to c. \tag{17.129}$$

At $\psi = \pm\pi/2$, the field $\mathbf{E}$ is

$$\mathbf{E} = \frac{q\gamma^3\mathbf{r}}{4\pi\epsilon_0\gamma^2 r^3} = \frac{1}{4\pi\epsilon_0\sqrt{1-\beta^2}}\frac{q\mathbf{r}}{r^3} \to \infty \quad \text{for} \quad v \to c. \tag{17.130}$$

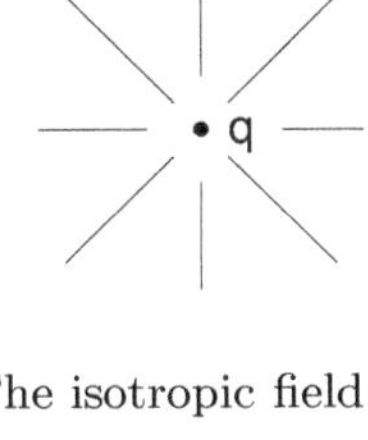


**Fig. 17.8.** The isotropic field for $\beta = v/c \sim 0$.

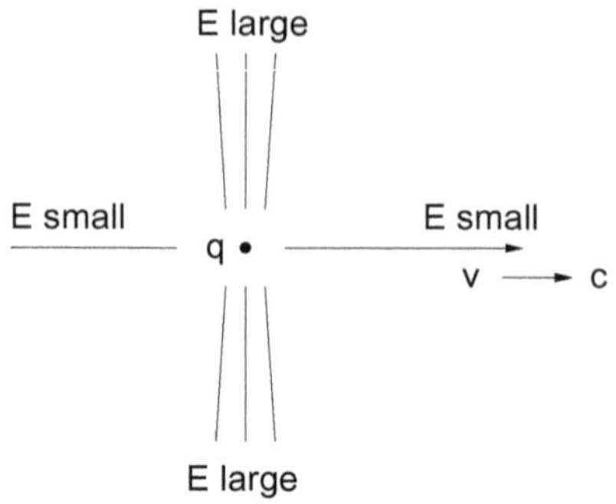


**Fig. 17.9.** The observed electric field at $P$ at high velocity $v \sim c$.

We see therefore: Whereas the electric field is isotropic in the case of the static (or almost static) charge, this is no longer the case for the field of a charge moving with velocity $v$; in the latter case the longitudinal component of the electric field (i.e. along the direction of motion of the particle) is smaller than the transverse component, and this effect becomes more and more pronounced as the velocity approaches that of light. This is illustrated in Fig. 17.9.

## 17.7 Covariant Form of Maxwell's Equations

Our next step is to re-express Maxwell's equations in terms of the field strength tensor $F_{\mu\nu}$. Later we shall derive these equations from an action integral. Maxwell's equations as we obtained them in Chapter 7 are

$$\begin{aligned} \boldsymbol{\nabla} \cdot \mathbf{E} &= \frac{\rho}{\epsilon_0} \text{ (Gauss)}, \\ \boldsymbol{\nabla} \times \mathbf{B} - \frac{1}{c^2}\frac{\partial \mathbf{E}}{\partial t} &= \mu_0 \mathbf{j} \text{ (Ampère–Maxwell)}, \\ \boldsymbol{\nabla} \cdot \mathbf{B} &= 0 \text{ (no magnetic poles)}, \\ \boldsymbol{\nabla} \times \mathbf{E} + \frac{\partial \mathbf{B}}{\partial t} &= 0 \text{ (Faraday)}. \end{aligned} \tag{17.131}$$

We now demonstrate that we obtain these equations by forming $\partial_\alpha F^{\alpha\beta}$ and $\partial_\alpha \mathcal{F}^{\alpha\beta}$, where $\mathcal{F}^{\alpha\beta}$ is the dual tensor defined by Eq. (17.81). In fact, we obtain (to be verified below):

$$\partial_\alpha F^{\alpha\beta} = \mu_0 J^\beta, \quad \partial_\alpha \mathcal{F}^{\alpha\beta} = 0, \tag{17.132}$$

for the four-current $J^\beta$ defined below. We now verify these equations. Considering the left hand side of the first equation, we have

$$\partial_\alpha F^{\alpha\beta} = \left(\frac{\partial}{\partial x^0}, \nabla\right) \begin{pmatrix} 0 & -E_1/c & -E_2/c & -E_3/c \\ E_1/c & 0 & -B_3 & B_2 \\ E_2/c & B_3 & 0 & -B_1 \\ E_3/c & -B_2 & B_1 & 0 \end{pmatrix}, \tag{17.133}$$

and hence

$$\begin{aligned} \partial_\alpha F^{\alpha\beta} &= \left(\frac{1}{c}\boldsymbol{\nabla}\cdot\mathbf{E}, -\frac{1}{c}\frac{\partial E_1}{\partial x^0} + \frac{\partial B_3}{\partial y} - \frac{\partial B_2}{\partial z},\right. \\ &\quad \left. -\frac{1}{c}\frac{E_2}{\partial x^0} - \frac{\partial B_3}{\partial x} + \frac{\partial B_1}{\partial z}, -\frac{1}{c}\frac{\partial E_3}{\partial x^0} + \frac{\partial B_2}{\partial x} - \frac{\partial B_1}{\partial y}\right) \\ &= \left(\frac{1}{c}\nabla\cdot\mathbf{E}, -\frac{1}{c}\frac{\partial \mathbf{E}}{\partial x^0} + \boldsymbol{\nabla}\times\mathbf{B}\right). \end{aligned} \tag{17.134}$$

Setting this equal to

$$\mu_0 J^\beta = \mu_0(c\rho, \mathbf{j}), \tag{17.135}$$

and identifying components, we obtain (with $\mu_0\epsilon_0 c^2 = 1$)

$$\boldsymbol{\nabla}\cdot\mathbf{E} = \frac{\rho}{\epsilon_0}, \quad -\frac{1}{c}\frac{\partial \mathbf{E}}{\partial x^0} + \boldsymbol{\nabla}\times\mathbf{B} = \mu_0\mathbf{j}, \tag{17.136}$$

in agreement with the first two of the above Maxwell equations.

Similarly

$$\partial_\alpha \mathcal{F}^{\alpha\beta} = \left(\frac{\partial}{\partial x^0}, \boldsymbol{\nabla}\right) \begin{pmatrix} 0 & -B_1 & -B_2 & -B_3 \\ B_1 & 0 & E_3/c & -E_2/c \\ B_2 & -E_3/c & 0 & E_1/c \\ B_3 & E_2/c & -E_1/c & 0 \end{pmatrix} \tag{17.137}$$

and hence

$$\begin{aligned} \partial_\alpha \mathcal{F}^{\alpha\beta} &= \left(\boldsymbol{\nabla}\cdot\mathbf{B}, -\frac{\partial B_1}{\partial x^0} - \frac{1}{c}\frac{\partial E_3}{\partial y} + \frac{1}{c}\frac{\partial E_2}{\partial z}, -\frac{\partial B_2}{\partial x^0} + \frac{1}{c}\frac{\partial E_3}{\partial x} - \frac{1}{c}\frac{\partial E_1}{\partial z},\right. \\ &\quad \left. -\frac{\partial B_3}{\partial x^0} - \frac{1}{c}\frac{\partial E_2}{\partial x} + \frac{1}{c}\frac{\partial E_1}{\partial y}\right) \\ &= \left(\boldsymbol{\nabla}\cdot\mathbf{B}, -\frac{\partial \mathbf{B}}{\partial x^0} - \frac{1}{c}\boldsymbol{\nabla}\times\mathbf{E}\right). \end{aligned} \tag{17.138}$$

Equating this to zero, we obtain the other two Maxwell equations. The second of Eqs. (17.132) can also be written:

$$\partial_\alpha \epsilon^{\alpha\beta\rho\sigma} F_{\rho\sigma} = 0. \tag{17.139}$$

Using the total antisymmetry of the Levi-Civita symbol the left side can be written:

$$\begin{aligned}\epsilon^{\alpha\beta\rho\sigma}\partial_\alpha F_{\rho\sigma} &= -\epsilon^{\beta\alpha\rho\sigma}\partial_\alpha F_{\rho\sigma} = -\frac{\epsilon^{\beta\alpha\rho\sigma}}{3}[\partial_\alpha F_{\rho\sigma} + \partial_\rho F_{\sigma\alpha} + \partial_\sigma F_{\alpha\rho}] \\ &= -\frac{\epsilon^{\beta\alpha\rho\sigma}}{3}[\partial_\alpha\partial_\rho A_\sigma - \partial_\alpha\partial_\sigma A_\rho \\ &\quad + \partial_\rho\partial_\sigma A_\alpha - \partial_\rho\partial_\alpha A_\sigma + \partial_\sigma\partial_\alpha A_\rho - \partial_\sigma\partial_\rho A_\alpha] = 0. \end{aligned} \tag{17.140}$$

Solely from the antisymmetry

$$F_{\mu\nu} = -F_{\nu\mu}$$

we obtain

$$\partial_\alpha F_{\beta\gamma} + \partial_\beta F_{\gamma\alpha} + \partial_\gamma F_{\alpha\beta} = 0. \tag{17.141}$$

This equation is automatically valid for the antisymmetric tensor $F_{\mu\nu}$ and represents an identity which is called *Jacobi identity* or *Bianchi identity.* The equations $\partial_\alpha \mathcal{F}^{\alpha\beta} = 0$ follow from this, and are therefore sometimes referred to as examples of equations of motion which are not derived as Euler–Lagrange equations. Equations (17.132) are Maxwell's equations in covariant form, i.e. in form-invariant formulation, which means they transform under Lorentz transformations like Lorentz or Minkowski 4-vectors.

## 17.8 Covariantized Newton Equation of a Charged Particle

Our objective now is to re-express the equation of motion of a Newtonian particle with charge $q$ in the presence of an electromagnetic field (i.e. the Lorentz force) in covariant form, i.e. the equation we encountered several times earlier, cf. e.g.

$$\frac{d\mathbf{p}}{dt} = q(\mathbf{E} + \mathbf{v} \times \mathbf{B}) \quad \text{newtons.} \tag{17.142}$$

One should note that the left side of this equation consists of the vector $\mathbf{p}$ preceded by the 0-component or $d/dt$ of the 4-vector $\partial_\mu$. It is therefore necessary to change something in order to achieve the same transformation property on the left of the equation as on the right. This is done by replacing $t$ by an invariant quantity. In order to generalize the *Newton equation*

$$\frac{d}{dt}\mathbf{p} = \mathbf{F}, \quad \frac{d}{dt}p^i = F^i \tag{17.143}$$

relativistically, one introduces therefore the so-called *eigentime* $\tau$, defined by

$$(ds)^2 = c^2(dt)^2 - (d\mathbf{x})^2 = c^2(dt)^2\left[1 - \frac{1}{c^2}\left(\frac{d\mathbf{x}}{dt}\right)^2\right] = c^2(d\tau)^2. \tag{17.144}$$

Here $\tau$ is the time, measured by a clock fixed in the particle, i.e. in its rest frame. This is the reason that $\tau$ is called the *eigentime of the particle.* With

$$\mathbf{v} = \frac{d\mathbf{x}}{dt} \tag{17.145}$$

(which is constant for inertial frames), this is the important relation which describes the so-called "*clock paradox*", i.e.[14] (with $\beta = v/c$)

$$d\tau = \frac{dt}{\gamma} = \sqrt{1-\beta^2}dt, \quad \gamma^2 = \frac{1}{1-\beta^2}, \quad 1 \leq \gamma \leq \infty. \tag{17.146}$$

This relationship expresses, so to speak, that a stationary clock runs faster than a moving clock (in the sense of observed time intervals), a phenomenon known as "*time dilation*" (cf. the discussion in Jackson [73] with detailed explanation of experimental verifications). Since $(ds)^2$ is Lorentz invariant, then also $(d\tau)^2$ or $\sqrt{(d\tau)^2} = d\tau$. This suggests one to generalize the Newton equation as $(\alpha = 0, 1, 2, 3)$

$$\frac{d}{d\tau}(mu_\alpha) = K_\alpha \equiv \text{ Minkowski force} \tag{17.147}$$

(definition). Here $u_\alpha$ is the 4-*velocity* (also to be understood as tangential vector with respect to the Minkowski manifold)

$$u^\alpha = \frac{dx^\alpha}{d\tau} \tag{17.148}$$

with (cf. Eq. (17.146))

$$u_i := -\frac{dx_i}{\sqrt{1-\beta^2}dt} = -\frac{v_i}{\sqrt{1-\beta^2}} \tag{17.149}$$

[14]For curvilinear coordinate systems the corresponding equation is $d\tau = \sqrt{g^{\rho\kappa}dx_\rho dx_\kappa}/c$.

and

$$u_0 = \frac{dx_0}{d\tau} = \frac{cdt}{\sqrt{1-\beta^2}dt} = \frac{c}{\sqrt{1-\beta^2}}. \tag{17.150}$$

The 4-*form equation* (17.147) can now be written

$$\frac{d(mu_\alpha)}{\sqrt{1-\beta^2}dt} = K_\alpha. \tag{17.151}$$

The spatial part implies

$$-\frac{d}{dt}\left(m\frac{v_i}{\sqrt{1-\beta^2}}\right) = \sqrt{1-\beta^2}K_i. \tag{17.152}$$

Momentum conservation for $K_i = 0$ implies the identification

$$\text{momentum}: \quad p_i = \frac{mv_i}{\sqrt{1-\beta^2}}. \tag{17.153}$$

Hence also for the three-force:

$$F_i := -\sqrt{1-\beta^2}K_i. \tag{17.154}$$

The significance of the time component of the 4-form equation (17.147) can be seen as follows. We multiply the equation by $u^\alpha$:

$$u^\alpha\frac{d}{d\tau}(mu_\alpha) = u^\alpha K_\alpha = \frac{d}{d\tau}\left(\frac{1}{2}mu^\alpha u_\alpha\right). \tag{17.155}$$

However,

$$u^\alpha u_\alpha = (u_0)^2 - (u_i)^2 = \frac{c^2}{1-\beta^2} - \frac{v^2}{1-\beta^2} = \frac{c^2-v^2}{1-\beta^2} = c^2. \tag{17.156}$$

Hence with Eq. (17.155)

$$0 = u^\alpha K_\alpha, \quad \text{i.e.} \quad u_0K_0 = u_iK_i, \tag{17.157}$$

or

$$\frac{c}{\sqrt{1-\beta^2}}K_0 = \frac{v_i}{\sqrt{1-\beta^2}}\frac{F_i}{\sqrt{1-\beta^2}},$$

i.e.

$$K_0 = \frac{1}{c}\frac{\mathbf{v}\cdot\mathbf{F}}{\sqrt{1-\beta^2}}. \tag{17.158}$$

Thus the fourth component of Eq. (17.151) becomes:

$$\frac{d(mc/\sqrt{1-\beta^2})}{\sqrt{1-\beta^2}dt} = \frac{1}{c}\frac{\mathbf{v}\cdot\mathbf{F}}{\sqrt{1-\beta^2}},$$

i.e.

$$\frac{d}{dt}\left(\frac{mc^2}{\sqrt{1-\beta^2}}\right) = \mathbf{v}\cdot\mathbf{F}, \tag{17.159}$$

i.e.

$$\frac{dT}{dt} = \mathbf{v}\cdot\mathbf{F}, \quad T = \frac{mc^2}{\sqrt{1-\beta^2}}. \tag{17.160}$$

Since $\mathbf{v}\cdot\mathbf{F} =$ work per unit time (or power), $T$ is the *total energy*, i.e.

$$T = \frac{mc^2}{\sqrt{1-\beta^2}} = mc^2 + \frac{1}{2}mv^2 + O\left(\frac{1}{c^2}\right). \tag{17.161}$$

Then we complete Eq. (17.153) with

$$p_\alpha = mu_\alpha \tag{17.162}$$

with ($T$ being the energy in the laboratory frame)

$$p_0 = mu_0 = \frac{mc}{\sqrt{1-\beta^2}} = \frac{T}{c}. \tag{17.163}$$

For $v \ll c$, Eq. (17.160) reproduces Newton's equation of motion:

$$\frac{d}{dt}\left(\frac{1}{2}m\mathbf{v}^2\right) = \mathbf{v}\cdot\mathbf{F}, \quad \text{or} \quad \frac{d}{dt}(m\mathbf{v}) = \mathbf{F}. \tag{17.164}$$

From Eqs. (17.162) and (17.156), we obtain in addition

$$p^\alpha p_\alpha = m^2u^\alpha u_\alpha = m^2c^2, \tag{17.165}$$

and so

$$p_0^2 - \mathbf{p}^2 = m^2c^2, \quad \text{or} \quad \left(\frac{\mathrm{T}}{\mathrm{c}}\right)^2 = \mathbf{p}^2 + \mathrm{m}^2\mathrm{c}^2. \tag{17.166}$$

For $\mathbf{p} = 0, T \equiv E$ this entails the well known *Einstein formula* $E = mc^2$.

Finally we consider the Minkowskian generalization of the Lorentz force. We define the four-vector $K^\alpha$ as follows and use Eq. (17.79):

$$\begin{aligned} K^\alpha &\equiv qF^{\alpha\beta}u_\beta \\ &= q\begin{pmatrix} 0 & -E_1/c & -E_2/c & -E_3/c \\ E_1/c & 0 & -B_3 & B_2 \\ E_2/c & B_3 & 0 & -B_1 \\ E_3/c & -B_2 & B_1 & 0 \end{pmatrix}\begin{pmatrix} \frac{c}{\sqrt{1-\beta^2}} \\ \frac{-v_x}{\sqrt{1-\beta^2}} \\ \frac{-v_y}{\sqrt{1-\beta^2}} \\ \frac{-v_z}{\sqrt{1-\beta^2}} \end{pmatrix} \end{aligned} \tag{17.167}$$

and so

$$K^\alpha = \frac{q}{\sqrt{1-\beta^2}}\begin{pmatrix} \mathbf{E}\cdot\mathbf{v}/c \\ E_1 + v_yB_3 - v_zB_2 \\ E_2 + v_zB_1 - v_xB_3 \\ E_3 + v_xB_2 - v_yB_1 \end{pmatrix} = \frac{q}{\sqrt{1-\beta^2}}\begin{pmatrix} \mathbf{E}\cdot\mathbf{v}/c \\ \mathbf{E} + \mathbf{v}\times\mathbf{B} \end{pmatrix}. \tag{17.168}$$

The four-dimensional equation (17.147),

$$\frac{dp^\alpha}{d\tau} = \frac{d(mu^\alpha)}{d\tau} = K^\alpha \tag{17.169}$$

together with the derivatives

$$\frac{dp^\alpha}{d\tau} = \frac{1}{\sqrt{1-\beta^2}}\frac{d}{dt}(p_0, \mathbf{p}) = \frac{1}{\sqrt{1-\beta^2}}\frac{d}{dt}\left(p_0, \frac{m\mathbf{v}}{\sqrt{1-\beta^2}}\right) \tag{17.170}$$

on the left hand side and expression (17.168) on the right, yields the equations

$$\frac{d\mathbf{p}}{dt} \equiv \frac{d}{dt}\left(\frac{m\mathbf{v}}{\sqrt{1-\beta^2}}\right) = q(\mathbf{E} + \mathbf{v}\times\mathbf{B}), \tag{17.171}$$

and

$$\frac{dp_0}{dt} = q\frac{\mathbf{E}\cdot\mathbf{v}}{c}, \quad \text{i.e.} \quad \frac{d}{dt}\left(\frac{T}{c}\right) = q\frac{\mathbf{E}\cdot\mathbf{v}}{c}, \tag{17.172}$$

and so

$$\underbrace{\frac{dT}{dt}}_{\text{power}} = q\underbrace{\frac{\mathbf{E}\cdot d\mathbf{x}}{dt}}_{\text{work/time}}. \tag{17.173}$$

We see that the 4-form

$$K^\alpha = qF^{\alpha\beta}u_\beta \tag{17.174}$$

is the Lorentz force supplemented by the power as the additional fourth component.

In the case of the macroscopic Maxwell equations we have to distinguish between $(\mathbf{E}, \mathbf{B})$ and $(\mathbf{D}, \mathbf{H})$. Evidently we need only to make the following substitutions in the above equations: $\mathbf{E} \to \mathbf{D}, \mathbf{B} \to \mathbf{H}$. The equations then describe *macroscopic averages of the atomic properties.* In this reformulation, the polarization $\mathbf{P}$ and magnetization $-\mathbf{M}$ can be combined like $(\mathbf{E}, \mathbf{B})$ and $(\mathbf{D}, \mathbf{H})$ to form an antisymmetric tensor of the second rank, and acquire their physical significance as macroscopic averages of atomic properties in the rest frame of the medium.[15]

We add a comment. In the force-free case

$$\frac{du^\alpha}{d\tau} = 0, \quad \text{i.e.} \quad \frac{d^2x^\alpha}{d\tau^2} = 0 \tag{17.175}$$

in agreement with $d^2x^\alpha/d\tau^2 = 0$ for constant velocity of a particle in an inertial frame. In curvilinear coordinate systems we have (also for $K = 0$) $d^2u^\alpha/d\tau^2 \neq 0$, and one says, the coefficients of the metric act as potentials of fictitious forces.[16]

## 17.9 Examples

**Example 17.1 (Gauge invariance).** Is the theory defined by the Lagrangian

$$L = \frac{1}{2}mu_\alpha u^\alpha + eu^\alpha A_\alpha \tag{17.176}$$

gauge invariant?

**Solution:** The gauge transformation is

$$A_\alpha \to A'_\alpha = A_\alpha - \partial_\alpha\chi, \quad \chi = \chi(x^\beta) \quad \text{arbitrary.} \tag{17.177}$$

Hence the variation of the Lagrangian is with (see Eq. (17.148)) $u^\alpha = dx^\alpha/d\tau$

$$\delta L = eu^\alpha\delta A_\alpha = -eu^\alpha\partial_\alpha\chi = -e\frac{dx^\alpha}{d\tau}\frac{\partial\chi}{\partial x^\alpha} = -e\frac{\delta\chi}{\delta\tau}. \tag{17.178}$$

[15]See e.g. Panofsky and Phillips [111], pp. 334–337.
[16]See e.g. Müller–Kirsten [100].

The answer is therefore: Yes, because (with $\chi = 0$ at $\tau_1, \tau_2$)

$$\text{action} \quad S = \int_{\tau_1}^{\tau_2} L d\tau \quad \text{and} \quad \delta S = -e \int_{\tau_1}^{\tau_2} d\tau \frac{\delta \chi}{\delta \tau} \to 0. \tag{17.179}$$

**Example 17.2 (Addition of velocities[17]).** Let $K, K', K''$ denote reference frames with parallel $z$-axes. Frame $K'$ moves with velocity $v$ away from $K$, and $K''$ moves parallel to these with velocity $u$ away from $K'$. What is the relativistic law of addition of velocities (i.e. obtain the velocity $w$ of $K''$ with respect to that of $K$)?

**Solution:** For a change we use the Lorentz transformation with Euclidean metric in the form $X' = AX$, where $X = (x, y, z, ict)$ and (cf. Eq. (17.61))

$$A \equiv A_v = \begin{pmatrix} 1 & 0 & 0 & 0 \\ 0 & 1 & 0 & 0 \\ 0 & 0 & \cosh\phi & i\sinh\phi \\ 0 & 0 & -i\sinh\phi & \cosh\phi \end{pmatrix}, \tag{17.180}$$

with (cf. Eqs. (17.55)–(17.57))

$$\cosh\phi = 1/\sqrt{1-\beta^2}, \quad \sinh\phi = \beta/\sqrt{1-\beta^2}, \quad \beta = v/c. \tag{17.181}$$

Performing two consecutive Lorentz transformations along parallel $z$-axes, we obtain

$$A_w = A_u A_v = \begin{pmatrix} 1 & 0 & 0 & 0 \\ 0 & 1 & 0 & 0 \\ 0 & 0 & \cosh(\phi+\phi') & i\sinh(\phi+\phi') \\ 0 & 0 & -i\sinh(\phi+\phi') & \cosh(\phi+\phi') \end{pmatrix}$$

$$\equiv \begin{pmatrix} 1 & 0 & 0 & 0 \\ 0 & 1 & 0 & 0 \\ 0 & 0 & \cosh\overline{\phi} & i\sinh\overline{\phi} \\ 0 & 0 & -i\sinh\overline{\phi} & \cosh\overline{\phi} \end{pmatrix}, \tag{17.182}$$

[17] This problem was already solved by Einstein in his famous paper of 1905, see Einstein [44].

where the parameters $\phi, \phi', \overline{\phi}$ correspond to the velocities $v, u, w$. We obtain therefore $\overline{\phi} = \phi + \phi'$, where

$$\cosh\overline{\phi} = \frac{1}{\sqrt{1 - w^2/c^2}}, \quad \cosh\phi = \frac{1}{\sqrt{1 - v^2/c^2}}, \quad \cosh\phi' = \frac{1}{\sqrt{1 - u^2/c^2}}. \tag{17.183}$$

With the relation

$$\cosh\overline{\phi} \equiv \cosh(\phi + \phi') = \cosh\phi\cosh\phi' + \sinh\phi\sinh\phi' \tag{17.184}$$

and the relations (17.181) it follows that

$$\begin{aligned}\frac{1}{\sqrt{1 - w^2/c^2}} &= \frac{1}{\sqrt{1 - v^2/c^2}}\frac{1}{\sqrt{1 - u^2/c^2}} + \frac{uv/c^2}{\sqrt{1 - v^2/c^2}\sqrt{1 - u^2/c^2}} \\ &= \frac{1 + uv/c^2}{\sqrt{1 - u^2/c^2}\sqrt{1 - v^2/c^2}}.\end{aligned} \tag{17.185}$$

Squaring and taking the reciprocal and later the square root, we obtain

$$\frac{w^2}{c^2} = 1 - \frac{(1 - u^2/c^2)(1 - v^2/c^2)}{(1 + uv/c^2)^2} = \frac{(u/c + v/c)^2}{(1 + uv/c^2)^2}, \tag{17.186}$$

and hence

$$w = \frac{u + v}{1 + uv/c^2} \quad \text{meters/second.} \tag{17.187}$$

We see: For $uv \ll c^2$ the result is simply the sum as in the familiar nonrelativistic considerations.

**Example 17.3 (The twin problem, or clock paradox).** A and B are twin brothers. On their 21st birthday A leaves his twin brother B on earth and flies for 7 years in the direction of $z$ (7 years measured on A's own watch). The velocity $v$ of A relative to that of the earth is $\tilde{\gamma}c$, the fraction $\tilde{\gamma}$ of the velocity of light $c$. After 7 years A reverses his direction and returns to earth with the same speed. How old are A and B then?

**Solution:** A is evidently 21 + 7 + 7 = 35 years old according to his watch (frame $K$). We calculate the age of B (as measured by B on his watch, frame $K'$) as follows. Reversing the Lorentz transformation of Eq. (17.59) but here along the $z$-axis, i.e.

$$x' = x, \quad y' = y, \quad z' = z\cosh\phi - ct\sinh\phi, \quad t' = -\frac{z}{c}\sinh\phi + t\cosh\phi, \tag{17.188}$$

we obtain

$$x = x', \quad y = y', \quad z = z'\cosh\phi + ct'\sinh\phi, \quad t = \frac{z'}{c}\sinh\phi + t'\cosh\phi, \tag{17.189}$$

where with $\beta = v/c$ we have $\cosh\phi = 1/\sqrt{1-\beta^2} = 1/\sqrt{1-\tilde{\gamma}^2} > 1$. We now use the last of Eqs. (17.189): $t = (z'/c)\sinh\phi + t'\cosh\phi$. Since A is at rest in his frame $K'$ with primed coordinates, we have $z' = 0$. Then $t$, the time measured by B on earth in frame $K$, is for $t' = 7$ years: $t = 7\cosh\phi$ years, i.e.

$$t = \frac{7}{\sqrt{1-\tilde{\gamma}^2}} \quad \text{years}. \tag{17.190}$$

The age of B is therefore at the time of return of A

$$= 21 + \frac{2 \times 7}{\sqrt{1-\tilde{\gamma}^2}} \quad \text{years}.$$

If $\tilde{\gamma} = 4/5$, the age of B follows as:

$$21 + \frac{2\times 7}{\sqrt{1-(4/5)^2}} = 21 + \frac{2\times 7\times 5}{3} = 21 + \frac{70}{3} \simeq 44 \text{ years}. \tag{17.191}$$

Thus 35 years and 44 years are the time intervals which A and B measure on their respective watches. Hence in this way, as remarked above and loosely speaking, the stationary clock appears to run faster than a moving clock.

**Example 17.4 (Lifetimes of particles).** Establish the lifetime of a particle in the laboratory frame for the cases:

(a) a particle with constant velocity $v$, and
(b) a particle with arbitrary velocity $v$.
(c) Is the photon a stable particle?

**Solution:**
(a) and (b) The spacetime distance $ds$ between the spacetime point 1 of creation of the particle and the spacetime point 2 of its decay is given by

$$ds^2 = c^2dt^2 - d\mathbf{x}^2 = c^2dt^2\left\{1 - \left(\frac{d\mathbf{x}}{c\,dt}\right)^2\right\}. \tag{17.192}$$

The invariant eigentime or proper time of the particle is the time $\tau$ given by $ds^2 = c^2 d\tau^2$. Thus in case (a) the lifetime of the particle measured in the laboratory frame is $(t_2 - t_1)$ given by

$$\tau_2 - \tau_1 = (t_2 - t_1)\sqrt{1-\beta^2}, \quad v^2 = \beta^2 c^2, \tag{17.193}$$

and in case (b)

$$\tau_2 - \tau_1 = \int_{t_1}^{t_2} dt \sqrt{1-\beta(t)^2}. \tag{17.194}$$

(c) Suppose the photon were unstable. Then the only invariant way to define its lifetime would be as that in its rest frame. However, since the velocity of the photon is constant and unequal to zero, it has no rest frame. This can also be seen as follows. The lifetime in the rest frame is:

$$\tau_2 - \tau_1 = (t_2 - t_1)\sqrt{1-\beta^2}, \tag{17.195}$$

where $t_2 - t_1$ is observed as infinite in the laboratory frame and the square root is zero, and hence the product indeterminate. Thus the question is meaningless. This applies to all particles with mass zero. (See also the remarks at the beginning of this chapter).

**Example 17.5 (The dipole in a magnetic field).** The charge $+q$ of two charges $+q, -q$, both initially at point $\mathbf{a}$, is moved in the $(x, y)$-plane through the magnetic field $F_{i3}$ to the point $\mathbf{b}$.

(a) With the gauge choice $\partial_3(A_0, \mathbf{A}) = 0$ determine the mechanical momentum $p_3$ of the system in the direction 3.
(b) Is this momentum independent of the path from $\mathbf{a}$ to $\mathbf{b}$?
(c) Show that the momentum $p_3$ is cancelled by a corresponding momentum of the electromagnetic field.

**Solution:**
(a) Starting from the Lorentz force $\mathbf{F} = d\mathbf{p}/dt = q d\mathbf{x}/dt \times \mathbf{B}, d\mathbf{p} = q d\mathbf{x} \times \mathbf{B}$, we have

$$p_3 = +q\int_{\mathbf{a}}^{\mathbf{b}} F_{3i} dx^i = -q\int_{\mathbf{a}}^{\mathbf{b}} F_{i3} dx^i. \tag{17.196}$$

Now,

$$\begin{aligned}\int_{\mathbf{a}}^{\mathbf{b}} F_{i3} dx^i &= \int_{\mathbf{a}}^{\mathbf{b}} (F_{13} dx^1 + F_{23} dx^2) \\ &= \int_{\mathbf{a}}^{\mathbf{b}} (\partial_1 A_3 - \partial_3 A_1) dx^1 + \int_{\mathbf{a}}^{\mathbf{b}} (\partial_2 A_3 - \partial_3 A_2) dx^2. \end{aligned} \tag{17.197}$$

With the condition to be assumed, $\partial_3(A_0, \mathbf{A}) = 0$, we obtain from Eq. (17.196):

$$p_3 = -q\int_{\mathbf{a}}^{\mathbf{b}} \boldsymbol{\nabla} A_3 \cdot d\mathbf{x} = -q\int_{\mathbf{a}}^{\mathbf{b}} dA_3 = -q(A_3(\mathbf{b}) - A_3(\mathbf{a})) \tag{17.198}$$

and hence the mechanical momentum is

$$p_3 = q(A_3(\mathbf{a}) - A_3(\mathbf{b})), \tag{17.199}$$

depending only on the endpoints $\mathbf{a}$ and $\mathbf{b}$.
(b) We consider the curl of $d\mathbf{p}$:

$$\boldsymbol{\nabla} \times d\mathbf{p} = \boldsymbol{\nabla} \times \{qd\mathbf{x} \times \mathbf{B}\} = q[\boldsymbol{\nabla} \cdot \mathbf{B}d\mathbf{x} - (\boldsymbol{\nabla} \cdot d\mathbf{x})\mathbf{B}] = 0, \tag{17.200}$$

since $dF = \partial_1 F_{23} + \partial_2 F_{31} + \partial_3 F_{12} = 0$, i.e. here $\boldsymbol{\nabla} \cdot \mathbf{B} = 0$, and $d(\boldsymbol{\nabla} \cdot \mathbf{x}) = 0$. As a result of this condition $p_3$ is independent of the path from $\mathbf{a}$ to $\mathbf{b}$ (compare with the definition of a conservative force in mechanics).
(c) The field momentum of the electromagnetic field has been defined in Section 7.7 as

$$\mathbf{P}^{\text{field}} = \int dV(\mathbf{D} \times \mathbf{B}) \quad \text{newton} \cdot \text{seconds}. \tag{17.201}$$

Thus we have in the present case for this quantity:

$$P_3^{\text{field}} = \epsilon_0 c\int dV F_{0i}F_{3i} = -\epsilon_0 c\int dV F_{0i}\partial_i A_3 = \epsilon_0 c\int dV A_3\partial_i F_{0i}. \tag{17.202}$$

In the last step the integral over a total divergence is zero. In the next step we use Eq. (17.59), i.e. the Maxwell equation $\nabla \cdot \mathbf{D} = \rho$ in the form

$$\partial_i F_{0i} = q\mu_0 c\{\delta(\mathbf{x} - \mathbf{b}) - \delta(\mathbf{x} - \mathbf{a})\}$$

together with $\epsilon_0\mu_0 c^2 = 1$, so that

$$P_3^{\text{field}} = \epsilon_0 c\mu_0 cq\int dV A_3\{\delta(\mathbf{x} - \mathbf{b}) - \delta(\mathbf{x} - \mathbf{a})\} = q(A_3(\mathbf{b}) - A_3(\mathbf{a})), \tag{17.203}$$

which had to be shown. Thus, the mechanical momentum (17.199) determined by the Lorentz force, is cancelled by a corresponding negative momentum (17.202) of the field. This example was motivated by a paper of Taylor [147]. But see also Examples 7.2 and 21.2.

**Example 17.6 (Orthogonality expressed invariantly).** Show that when $\mathbf{E} \cdot \mathbf{B} = 0$, also $F_{\mu\nu}\mathcal{F}^{\mu\nu} = 0$.

**Solution:** Using Eq. (17.80) for $F_{\mu\nu}$ and Eq. (17.81) for $\mathcal{F}^{\mu\nu}$ and multiplying one finds

$$F_{\mu\nu}\mathcal{F}^{\mu\nu} = \frac{1}{c}\mathbf{E}\cdot\mathbf{B}. \tag{17.204}$$

## 17.10 Exercises

Exercises on Special Relativity can be found in many books. The text of The Physics Coaching Class [150], for instance, contains 54 solved problems on Special Relativity.[18]

**Example 17.7 (Addition of velocities).** In Example 17.2 we used a Euclidean metric. Redo the problem using the Minkowski metric (17.19).

**Example 17.8 (Decay of a nucleus).** A nucleus at rest with mass $M$ decays into two parts with rest masses $m_1$ and $m_2$. Calculate the energies $E_1$ and $E_2$ of the decay products with masses $m_1$ and $m_2$ in terms of $M, m_1, m_2$ and $c$.[19]

**Example 17.9 (The relativistic aberration law).** A transverse wave of frequency $\nu = \omega/2\pi$ propagates in a direction of angle $\alpha$ with the $x$-axis of inertial frame $K$ of the source (in applications this source can, for example, be the end of a moving stick from which light is scattered). The source moves in the $x$-direction at speed $v < c$ towards an observer with rest frame $K'$, the $x'$-axis being parallel to the $x$-axis, as indicated in Fig. 17.10. Determine the frequency $\nu'$ observed in $K'$ and its angle of observation.

(Answers: $\nu' = \gamma(1-\beta\cos\alpha)\nu$, $\cos\alpha' = (\cos\alpha-\beta)/(1-\beta\cos\alpha)$, where $\beta = v/c$ and $\gamma = 1/\sqrt{1-\beta^2}$).

**Example 17.10 (The Doppler effect).** Light is emitted from a source in the inertial frame $K'$ and observed in frame $K$, with frame $K'$ moving at speed $v$ away from $K$ ($\beta = v/c$), the axes being parallel. Also $\nu$ is the frequency observed by the observer in frame $K$ and $\nu'$ is the frequency of the light emitted by the source in frame $K'$. Derive the relation between $\nu$ and $\nu'$ when

(a) $K, K'$ approach each other,
(b) $K, K'$ recede from each other,
(c) $K, K'$ pass each other in perpendicular positions, as indicated in Fig. 17.11.

---

[18]See also Müller–Kirsten [101].

[19]For the worked solution see Müller–Kirsten [101], Example 12.4.

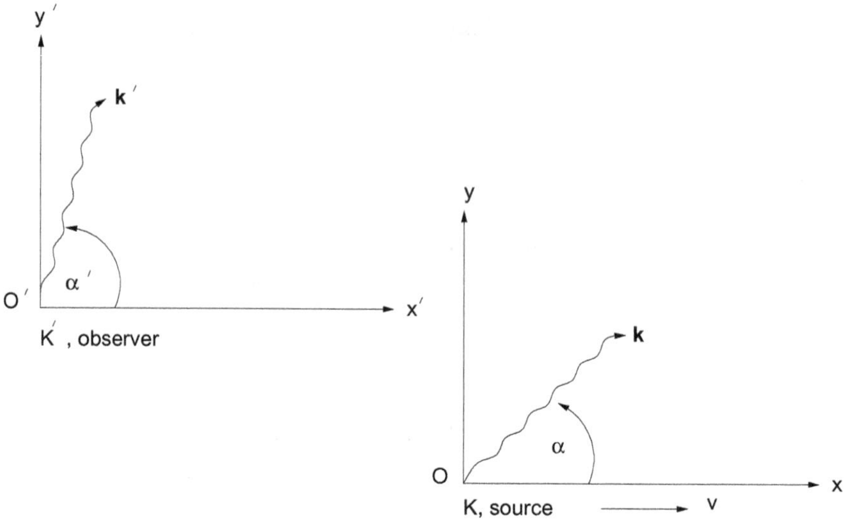


**Fig. 17.10.** Light leaving frame $K$ which moves relative to frame $K'$.

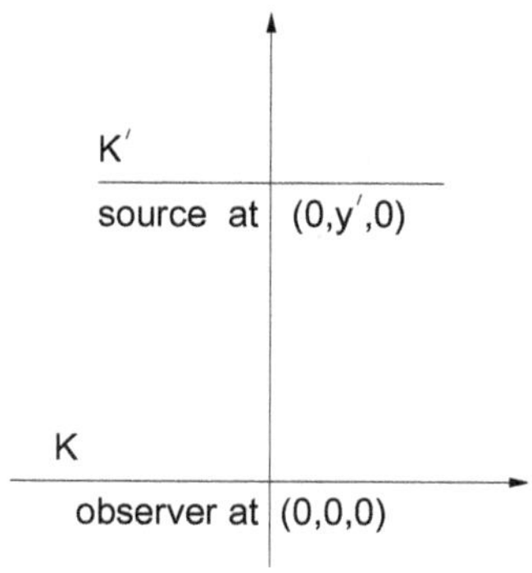


**Fig. 17.11.** Source and observer passing at right angles.

(Answers: (a) $\nu = \nu'\sqrt{(1+\beta')/(1-\beta')}$, (b) $\nu = \nu'\sqrt{(1-\beta')/(1+\beta')}$, (c) $\nu = \nu'/\sqrt{1-\beta^2}$). [20]

**Example 17.11 (Electron–positron annihilation).** A positron $e^+$ is annihilated by colliding with an electron $e^-$ which is almost at rest in the laboratory frame, i.e.

$$e^+ \text{ (fast) } + e^- \text{ (at rest) } \longrightarrow \text{ photons } \gamma. \tag{17.205}$$

[20]For the worked solution see Müller–Kirsten [101], Example 12.3.

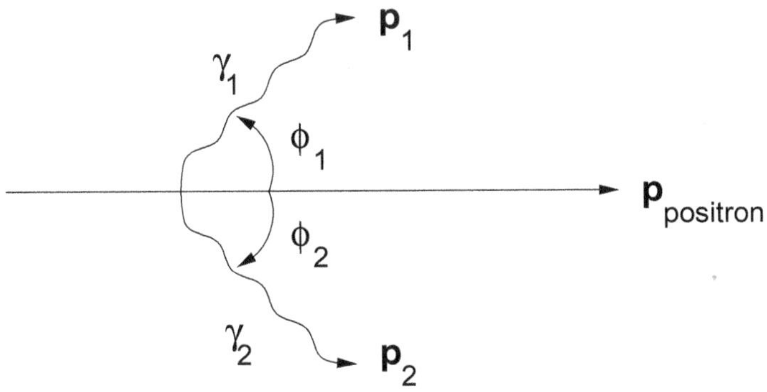


**Fig. 17.12.** Electron–positron annihilation with production of two photons.

(a) Consider the collision in a reference frame in which for the momenta of the initial particles $\mathbf{p}_{e^+} + \mathbf{p}_{e^-} = 0$. What is the smallest number of photons that must be created? (b) Considering as in Fig. 17.12 the case $e^+ + e^- \longrightarrow \gamma_1 + \gamma_2$, derive an expression for the energy of one of the $\gamma$-rays $\gamma_1, \gamma_2$ in the laboratory frame ($\mathbf{p}_{e^-} = 0$); the expression will involve the angle $\phi_1$ between $\mathbf{p}_e$ and $\mathbf{p}_1$, the latter being the momentum of $\gamma_1$. (c) Finally obtain the largest and smallest $\gamma$-ray energies in the laboratory frame.[21]

**Example 17.12 (Effective duration of collision).** In Section 17.6 the passage of a charge $q$ past an observer at closest distance $b$ was considered. Show that the effective duration of their encounter ("collision") is $\sim b/\gamma v$, where $v$ is the velocity of the charge.

**Example 17.13 (Momentum transferred to a target particle).** Suppose the observer of Section 17.6 is replaced by a target particle of mass $M$ and charge $Q$ (like a nucleus). The momentum transfer $\triangle\mathbf{p}$ is given by

$$\triangle\mathbf{p} = \int_{-\infty}^{\infty} Q\mathbf{E}dt \quad \text{newton} \cdot \text{seconds}. \tag{17.206}$$

Show that

$$\triangle\mathbf{p} = \triangle\mathbf{p}_\perp, \quad \triangle p_\perp = \frac{2qQ}{4\pi\epsilon_0 bv}. \tag{17.207}$$

Thus this is independent of masses. What is the energy $\triangle E$ transferred to the target particle (assumed initially at rest)?

(Answer: $2(qQ)^2/Mb^2v^2(4\pi\epsilon_0)^2$ joules).

[21]For the worked solution see Müller–Kirsten [101], Example 12.5.

**Example 17.14 (No momentum conservation in one photon pair creation).** Show that in electron-positron pair creation by a single photon the momentum is not conserved.
[Hint: Consider conservation of energy $h\nu$, and conservation of momentum, $(h\nu)^2 \leq (p_+ + p_-)^2 c^2$].

**Example 17.15 (Two particles combining to one).** Two identical particles of rest mass $m_0$ travel in mutually orthogonal directions with the same speed $c/2$. They collide to form a single particle. What is the rest mass of this particle?

(Answer: $(14/3)^{1/2} m_0$).

**Example 17.16 (Relative velocity of two photons).** Two photons (speed $c$) are sent in opposite directions along a line. What is their relative velocity with respect to some inertial frame?

(Answer: $c$).

**Example 17.17 (Another electron-positron annihilation).** A positron travelling with velocity $3c/5$ is annihilated in a collision with a stationary electron yielding two photons which emerge in opposite directions along the line of the incoming positron. Show that the photons have energies $3mc^2/4$ and $3mc^2/2$ where $m$ is the mass of each of the oppositely charged particles.

## Chapter 18

# The Lagrange Formalism for the Electromagnetic Field

## 18.1 Introductory Remarks

In this chapter we derive Maxwell's equations as the equations of motion of the electromagnetic field in spacetime in the customary way as the Euler–Lagrange equations derived from an action integral. We discuss *gauge transformations, transversality and masslessness of the electromagnetic field* and touch briefly (restricted by our predominantly classical topic here) its *spin*. Finally examples are given to illustrate the diversity of some recent explorations motivated by the Lagrangian of Maxwell electrodynamics.

## 18.2 Euler–Lagrange Equation

We define as *Lagrangian density* the functional

$$\mathcal{L}(A_\mu(x), \partial_\mu A_\nu(x)).$$

We note that $\mathcal{L}$ is *not* $\mathcal{L}(A_\mu(x), \partial_\mu A_\nu(x), t)$; rather, the time dependence is assumed to be contained in the vector potential $A_\mu(x)$. Here $A_\mu(x)$ is the electromagnetic 4-potential describing the local electromagnetic field. We define as *Lagrangian* the volume integral

$$L(t) = \int d^3x \mathcal{L}(A_\mu(x), \partial_\mu A_\nu(x)) \tag{18.1}$$

and as *action* or *action integral*

$$S = \int_{t_1}^{t_2} dt L(t). \tag{18.2}$$

We demand that $S$ or rather the Lagrangian density be a Lorentz invariant (recall we had verified in Eq. (17.68) the invariance of the spacetime (Minkowski space) volume element). The time dependence of $L$ is implicitly contained in the fields. Our aim is to derive the Maxwell equations in analogy to equations of motion in classical mechanics. Hence we construct an action to which we apply a *Hamilton's principle.* This means we demand that in varying the action by varying the fields $A_\mu(x)$ and their derivatives $\partial_\mu A_\nu(x)$ at fixed endpoints, *this action is to remain stationary.* Thus we demand

$$0 = \delta S = \delta \int_{t_1}^{t_2} dt d^3x \mathcal{L}(A_\mu(x), \partial_\mu A_\nu(x)) \tag{18.3}$$

with

$$\delta A_\mu = 0 \quad \text{at} \quad t = t_1, t_2 \quad \text{and} \quad A_\mu(x) \to 0 \quad \text{for} \quad |\mathbf{x}| \to \infty. \tag{18.4}$$

Hence

$$\begin{aligned} 0 &= \int_{t_1}^{t_2} dt d^3x \left[\frac{\partial \mathcal{L}}{\partial A_\mu}\delta A_\mu + \frac{\partial \mathcal{L}}{\partial(\partial_\mu A_\nu)}\delta(\partial_\mu A_\nu)\right] \\ &= \int_{t_1}^{t_2} dt d^3x \left[\frac{\partial \mathcal{L}}{\partial A_\mu}\delta A_\mu + \frac{\partial \mathcal{L}}{\partial(\partial_\mu A_\nu)}\partial_\mu(\delta A_\nu)\right]. \end{aligned} \tag{18.5}$$

As in classical mechanics one varies with respect to a parameter $\alpha$, whose values label the various paths in configuration space, so that

$$\delta(\partial_\mu A_\nu) = \frac{\partial}{\partial\alpha}(\partial_\mu A_\nu(x,\alpha))\delta\alpha = \partial_\mu(\delta A_\nu). \tag{18.6}$$

After a partial integration the variation of $S$ is seen to be

$$0 = \int_{t_1}^{t_2} dt d^3x \left[\frac{\partial \mathcal{L}}{\partial A_\nu} - \partial_\mu\left(\frac{\partial \mathcal{L}}{\partial(\partial_\mu A_\nu)}\right)\right]\delta A_\nu + \int dS_\mu \left[\frac{\partial \mathcal{L}}{\partial(\partial_\mu A_\nu)}\delta A_\nu\right]. \tag{18.7}$$

Here $S_\mu$ is a 3-dimensional pseudo-plane orthogonal to $x_\mu$. If the surface $S_\mu$ is time-like, i.e. an area orthogonal to the time-axis, the second integral vanishes, because $\delta A_\nu$ is zero at the integration limits $t = t_1, t_2$. But also the non-timelike surface-area contributions vanish by extending the volume of integration far enough so that all field components and their derivatives vanish on the boundary. (This is a plausible and in field theory frequently made assumption on the basis of the localizability of the fields, i.e. finiteness of the velocity of light, $c$). Since the components $A_\mu$, or rather their variations $\delta A_\nu$, are independent of one another, we obtain the *Euler–Lagrange equations*

$$\frac{\partial \mathcal{L}}{\partial A_\mu} = \partial_\nu \left( \frac{\partial \mathcal{L}}{\partial(\partial_\nu A_\mu)} \right). \tag{18.8}$$

These equations are *Lorentz covariant*, provided $\mathcal{L}$ is a Lorentz invariant.

## 18.3 Symmetries and Energy-Momentum Tensor

In classical mechanics a very generally valid theorem known as *Noether's theorem* is well known. This theorem says: Every continuous symmetry (group) of $\mathcal{L}$ (i.e. invariance of $\mathcal{L}$ under the transformations of the group) has as its consequence a conservation law. With the help of the Euler–Lagrange equations (18.8) we can immediately derive one conservation law or conserved current. We have

$$\delta\mathcal{L} = \frac{\partial \mathcal{L}}{\partial A_\mu}\delta A_\mu + \frac{\partial \mathcal{L}}{\partial(\partial_\mu A_\nu)}\delta(\partial_\mu A_\nu). \tag{18.9}$$

Using the Euler–Lagrange equations (18.8), we have

$$\delta\mathcal{L} = \partial_\nu \left( \frac{\partial \mathcal{L}}{\partial(\partial_\nu A_\mu)} \right) \delta A_\mu + \frac{\partial \mathcal{L}}{\partial(\partial_\mu A_\nu)}\delta(\partial_\mu A_\nu) = \partial_\mu \left( \frac{\partial \mathcal{L}}{\partial(\partial_\mu A_\nu)}\delta A_\nu \right). \tag{18.10}$$

Consider now a transformation depending on some parameters, with an infinitesimal "*form transformation*" $\delta_P$ of the field, i.e. in the present case of $A_\mu$, defined by the relation

$$\delta_P A_\mu := A'_\mu(x) - A_\mu(x) \tag{18.11}$$

(note the argument $x$ is the same in both quantities on the right). Then, if in the corresponding variation of $\mathcal{L}$ without the use of the equations of motion one has

$$\delta_P \mathcal{L} = \partial_\mu V^\mu, \quad V^\mu \quad \text{some} \quad \text{4-vector}, \tag{18.12}$$

one says the transformation represents a continuous symmetry of the theory. Setting $\delta A_\mu = \delta_P A_\mu$ and $\delta\mathcal{L} = \delta_P \mathcal{L}$, one has

$$\partial_\mu j^\mu = 0 \tag{18.13}$$

with

$$j^\mu = \frac{\partial \mathcal{L}}{\partial(\partial_\mu A_\nu)}\delta_P A_\nu - V^\mu. \tag{18.14}$$

Then $j^\mu$ is called a *conserved current* or *Noether current*.[1] There are cases for which $V^\mu \neq 0$ (e.g. in the case of translations, rotations), and cases for which $V^\mu = 0$ (in the latter case of gauge transformations). A simple example for the latter is $\mathcal{L} = \phi^*\phi$, the star meaning complex conjugation. The infinitesimal transformation $\delta\phi = i\epsilon\phi$ ($\epsilon$ real) implies $\delta_P\mathcal{L} = (-i\epsilon\phi^*)\phi + \phi^*(i\epsilon\phi) = 0$, i.e. according to Eq. (18.12) $\partial_\mu V^\mu = 0$ and so $V^\mu = 0$. One says, the transformation is an "*internal symmetry*". Equation (18.13) expresses the conservation of the current $j^\mu$. In Example 18.1 the derivation of the conservation law associated with translations is demonstrated in the case of the simpler theory of a scalar field $\phi(x)$. Instead of deriving the currents we are interested in with lengthy calculations from the above expression,[2] we simply define certain quantities here and then verify that they are currents. We define as the so-called *energy-momentum tensor* the quantity

$$T^{\mu\nu} := \frac{1}{c\mu_0}\left[-\eta^{\mu\nu}\mathcal{L} + \frac{\partial\mathcal{L}}{\partial(\partial_\mu A_\rho)}\partial^\nu A_\rho\right]. \tag{18.15}$$

The constant in front is introduced to ensure agreement of the field momentum with our earlier definition in Section 7.7.2. It follows that

$$\begin{aligned}
c\mu_0\partial_\mu T^{\mu\nu} &= -\partial^\nu\mathcal{L} + \partial_\mu\left(\frac{\partial\mathcal{L}}{\partial(\partial_\mu A_\rho)}\partial^\nu A_\rho\right)\\
&= -\partial^\nu\mathcal{L} + \partial_\mu\left(\frac{\partial\mathcal{L}}{\partial(\partial_\mu A_\rho)}\right)\partial^\nu A_\rho + \left(\frac{\partial\mathcal{L}}{\partial(\partial_\mu A_\rho)}\right)\partial_\mu\partial^\nu A_\rho\\
&= -\partial^\nu\mathcal{L} + \frac{\partial\mathcal{L}}{\partial A_\rho}\partial^\nu A_\rho + \frac{\partial\mathcal{L}}{\partial(\partial_\mu A_\rho)}\partial^\nu(\partial_\mu A_\rho)\\
&= -\partial^\nu\mathcal{L} + \partial^\nu\mathcal{L} = 0,
\end{aligned} \tag{18.16}$$

where we used in an intermediate step the Euler–Lagrange equations (18.8). The result represents a conservation law or equation of continuity. If we integrate the equation with respect to space coordinates at a fixed time, we obtain

$$\int d^3x\partial_\mu T^{\mu\nu} = 0, \tag{18.17}$$

[1]This procedure, attributed to R. Jackiw, was emphasized in Fleming [56].
[2]See e.g. Schweber [133], pp. 207–211.

and hence

$$\partial_0 \int d^3x T^{0\nu} + \int d^3x \partial_i T^{i\nu} = 0. \tag{18.18}$$

The second contribution is the volume integral of a total divergence, which with Gauss' divergence theorem can be converted into a surface integral, and this vanishes on account of a localization of the fields in a finite part of space (i.e. the fields decrease to zero on the boundary). Thus we obtain

$$0 = \frac{d}{dt} \int d^3x T^{0\nu}, \quad \nu = 0, 1, 2, 3. \tag{18.19}$$

For a further evaluation of this expression in electrodynamics, we need to know the components $T^{0\nu}$, which therefore have to be determined.

We define as *4-momentum density* $p^\mu$ the 4-vector

$$p^\mu = T^{0\mu} = \frac{1}{c\mu_0}\left[-\eta^{0\mu}\mathcal{L} + \frac{\partial \mathcal{L}}{\partial(\partial_0 A_\rho)}\partial^\mu A_\rho\right], \tag{18.20}$$

and as *4-momentum* $P^\mu$ the spatial integral of the density,

$$P^\mu = \int d^3x p^\mu = \frac{1}{c\mu_0}\int d^3x[-\eta^{0\mu}\mathcal{L} + \pi^\rho \partial^\mu A_\rho]. \tag{18.21}$$

In this expression, $\pi^\rho$ is the density of the *conjugate momentum*

$$\pi^\rho = \frac{\partial \mathcal{L}}{\partial(\partial_0 A_\rho)} \tag{18.22}$$

(defined in analogy to the conjugate momentum in classical mechanics i.e. to $p = \partial L(q,p)/\partial \dot{q}$). The time component of $p^\mu$ is

$$p^0 = \frac{1}{c\mu_0}[-\mathcal{L} + \pi^\rho \partial^0 A_\rho]. \tag{18.23}$$

This expression represents a *Legendre transform* and we set

$$\mathcal{H} = \mathcal{H}(A_\mu, \pi_\mu) = cp^0 = \frac{1}{\mu_0}[\pi^\rho \partial^0 A_\rho - \mathcal{L}], \tag{18.24}$$

and we shall see that this is the energy density or *Hamilton density* in agreement with our earlier expression, i.e. the density contained in $W$ of Eq. (7.65). We observe that contrary to $\mathcal{L}$, $\mathcal{H}$ is *not a Lorentz invariant*, but the 0-component of a 4-vector! For the further evaluation of these expressions we require an explicit expression for the *Lagrangian density*. We then expect that we obtain for the momentum density the field density and for the Hamilton density the energy density of the electromagnetic field — both of which we encountered in Section 7.7.

**Example 18.1 (Translations as continuous symmetry of a Lagrangian).** Show that the translations $x'^{\mu} = x^{\mu} + \epsilon^{\mu}$ ($\epsilon^{\mu}$ a constant 4-vector) represent a continuous symmetry of the Lagrangian of a scalar field $\phi(x)$ with density

$$\mathcal{L}(\phi, \partial_\mu \phi) = \frac{1}{2}\partial^\mu \phi \partial_\mu \phi - \frac{1}{2}m^2\phi^2. \tag{18.25}$$

**Solution:** According to the discussion around Eq. (18.12), we have to show that

$$\delta_P \mathcal{L} = \partial_\mu V^\mu, \tag{18.26}$$

and have to determine $V^\mu$. Since $\phi(x)$ is a scalar, we have $\phi'(x') = \phi(x)$ (recall that scalars, vectors, tensors in some space are defined by the way they transform under the transformations of this space). But also for infinitesimal $\epsilon^\mu$ (by Taylor expansion)

$$\phi'(x') = \phi'(x) + (x'^{\mu} - x^\mu)\partial_\mu \phi(x) = \phi'(x) + \epsilon^\mu \partial_\mu \phi(x). \tag{18.27}$$

(It suffices to consider the infinitesimal transformation, since the finite transformation can be generated from a sequence of infinitesimal transformations). Hence (recall Eq. (18.11)) the *form variation* $\delta_P \phi \equiv \phi'(x) - \phi(x)$ is with $\phi'(x') = \phi(x)$ in the case of translations given by

$$\delta_P \phi(x) = \phi'(x) - \phi(x) = -\epsilon^\mu \partial_\mu \phi(x). \tag{18.28}$$

With this we obtain

$$\delta_P \mathcal{L} = (\partial^\mu \phi \partial_\mu - m^2\phi)\delta_P \phi = (\partial^\mu \phi \partial_\mu - m^2 \phi)(-\epsilon^\rho \partial_\rho \phi) \tag{18.29}$$

or

$$\delta_P \mathcal{L} = -\epsilon^\rho \partial_\rho \left[\frac{1}{2}\partial^\mu \phi \partial_\mu \phi - \frac{1}{2}m^2 \phi^2\right] \equiv \partial_\mu V^\mu. \tag{18.30}$$

Hence in the present case of translations

$$V^\mu = -\epsilon^\mu \mathcal{L}. \tag{18.31}$$

This result was obtained without recourse to an equation of motion. Applying Eq. (18.14) in the present case, we have with Eq. (18.28):

$$j^\mu = \frac{\partial \mathcal{L}}{\partial(\partial_\mu \phi)}\delta_P \phi - V^\mu = -\epsilon^\nu \left\{\frac{\partial \mathcal{L}}{\partial(\partial_\mu \phi)}\partial_\nu \phi - \eta^\mu_\nu \mathcal{L}\right\}, \tag{18.32}$$

and so $\partial_\mu j^\mu = 0$ can be written $\partial_\mu T^{\mu\nu} = 0$, where

$$T^{\mu\nu} = \text{const.}\left\{\frac{\partial \mathcal{L}}{\partial(\partial_\mu \phi)}\partial^\nu \phi - \eta^{\mu\nu}\mathcal{L}\right\}, \tag{18.33}$$

which is the corresponding energy-momentum tensor. For the transformations of the full Poincaré group the corresponding considerations are somewhat more involved.

## 18.4 The Lagrangian of the Electromagnetic Field

We have seen previously (cf. Eq. (17.132)) that two of the 4 Maxwell equations can be combined in the following covariant form

$$\partial_\mu F^{\mu\nu} = \mu_0 J^\nu. \tag{18.34}$$

We saw (cf. the discussion after the Jacobi identity (17.141)) that the other two Maxwell equations follow from the Jacobi identity or are trivial consequences of the antisymmetry of the field strength tensor $F^{\mu\nu}$. We now search for an expression of $\mathcal{L}$, from which Eq. (18.34) can be derived as Euler–Lagrange equation. We consider the following expression and we shall see that this yields the desired equations:

$$\mathcal{L}(A_\nu, \partial_\mu A_\nu) = -\frac{1}{4}F^{\mu\nu}F_{\mu\nu} - \mu_0 J^\nu A_\nu. \tag{18.35}$$

We see immediately that the expression is Lorentz invariant. With some thought one will realize that (apart from a possible additional term) there is hardly any possibility to construct some other expression under the conditions it has to satisfy (only first derivatives, Lorentz invariance, gauge invariance). Consider the expression:

$$\begin{aligned}&\frac{\partial}{\partial(\partial_\nu A_\mu)}\left\{\frac{1}{4}(\partial^\rho A^\kappa - \partial^\kappa A^\rho)(\partial_\rho A_\kappa - \partial_\kappa A_\rho)\right\}\\&= \frac{1}{4}F^{\rho\kappa}(\delta^\nu_\rho \delta^\mu_\kappa - \delta^\nu_\kappa \delta^\mu_\rho) + \frac{1}{4}(\delta^{\rho\nu}\delta^{\mu\kappa} - \delta^{\nu\kappa}\delta^{\mu\rho})F_{\rho\kappa}\\&= \frac{1}{4}[F^{\nu\mu} - F^{\mu\nu} + F^{\nu\mu} - F^{\mu\nu}] = -\frac{4F^{\mu\nu}}{4} = +F^{\nu\mu}.\end{aligned} \tag{18.36}$$

Hence the *equation of motion*

$$\frac{\partial \mathcal{L}}{\partial A_\mu} = \partial_\nu\left(\frac{\partial \mathcal{L}}{\partial(\partial_\nu A_\mu)}\right) \tag{18.37}$$

implies the relation

$$-\mu_0 J^\mu = -\partial_\nu (F^{\nu\mu}), \quad i.e. \quad \partial_\nu F^{\nu\mu} = \mu_0 J^\mu \tag{18.38}$$

in agreement with the desired equation (18.34).

Next we calculate the *density of the canonical momentum*, i.e.

$$\pi^\mu \overset{(18.22)}{=} \frac{\partial \mathcal{L}}{\partial(\partial_0 A_\mu)} = -F^{0\mu} = F^{\mu 0}, \quad \pi^0 = 0, \quad \pi^i = F^{i0}. \tag{18.39}$$

According to the expression for $F^{\mu\nu}$ in Eq. (17.79),

$$F^{i0} = \frac{E_i}{c}. \tag{18.40}$$

The product $F^{\mu\nu}F_{\mu\nu}$ in the Lagrangian density can also be re-expressed in terms of the field strengths with the help of the matrices (17.79) and (17.80). Thus ("Tr" meaning "*trace*")

$$\begin{aligned}
-\sum_{\mu,\nu} F^{\mu\nu}F_{\mu\nu} &= \mathrm{Tr}(F^{\mu\nu}F_{\nu\mu}) = \mathrm{Tr}\left[\begin{pmatrix} 0 & -E_1/c & -E_2/c & -E_3/c \\ E_1/c & 0 & -B_3 & B_2 \\ E_2/c & B_3 & 0 & -B_1 \\ E_3/c & -B_2 & B_1 & 0 \end{pmatrix}\right. \\
&\qquad \left. \times \begin{pmatrix} 0 & E_1/c & E_2/c & E_3/c \\ -E_1/c & 0 & -B_3 & B_2 \\ -E_2/c & B_3 & 0 & -B_1 \\ -E_3/c & -B_2 & B_1 & 0 \end{pmatrix}\right] \\
&= \frac{\mathbf{E}^2}{c^2} + \left(\frac{E_1^2}{c^2} - B_3^2 - B_2^2\right) + \left(\frac{E_2^2}{c^2} - B_3^2 - B_1^2\right) \\
&\quad + \left(\frac{E_3^2}{c^2} - B_2^2 - B_1^2\right) \\
&= 2\left(\frac{\mathbf{E}^2}{c^2} - \mathbf{B}^2\right).
\end{aligned} \tag{18.41}$$

Hence the Lagrangian density is also (without the $J^\nu$-term)

$$\mathcal{L} = \mathcal{L}(E_i, B_j) = \frac{1}{2}\left(\frac{\mathbf{E}^2}{c^2} - \mathbf{B}^2\right). \tag{18.42}$$

We note the analogy with the simple Lagrangian in classical mechanics: $L =$ kinetic energy minus potential energy, and compare also with the energy density (see below).[3]

We consider now the case of the *free electromagnetic field*, i.e. $J^\nu = 0$. Then

$$\begin{aligned}\pi^\rho \partial^0 A_\rho &= \pi^0 \partial^0 A_0 + \pi^i \partial^0 A_i = 0 - \pi^i \partial^0 A^i = -F^{i0} \partial^0 A^i \\ &\stackrel{(18.40)}{=} -\frac{1}{c} E_i \partial^0 A^i = \frac{1}{c} E_i \left(-\frac{1}{c}\frac{\partial A^i}{\partial t}\right) = \frac{1}{c^2} E_i (E_i - \nabla^i \phi), \end{aligned} \tag{18.43}$$

where we used Eq. (17.77) in the last step. Thus in vector form

$$\pi^\rho \partial^0 A_\rho = \frac{\mathbf{E}^2}{c^2} - \frac{1}{c^2}\mathbf{E}\cdot\boldsymbol{\nabla}\phi = \frac{\mathbf{E}^2}{c^2} - \frac{1}{c^2}[\boldsymbol{\nabla}\cdot(\mathbf{E}\phi) - \phi\boldsymbol{\nabla}\cdot\mathbf{E}], \tag{18.44}$$

where the last term vanishes on account of $J^\nu = 0$ (free field, no source). We obtain therefore for the energy density $\mathcal{H}$ (cf. Eqs. (18.20) and (18.23))[4]

$$\begin{aligned}\mathcal{H} = cp^0 = cT^{00} &= \frac{1}{\mu_0}[\pi^\rho \partial^0 A_\rho - \mathcal{L}] \\ &= \frac{1}{\mu_0}\left[\frac{\mathbf{E}^2}{c^2} - \frac{1}{c^2}\boldsymbol{\nabla}\cdot(\mathbf{E}\phi) - \frac{1}{2}\left(\frac{\mathbf{E}^2}{c^2} - \mathbf{B}^2\right)\right] \\ &= \frac{1}{\mu_0}\left[\frac{1}{2}\left(\frac{\mathbf{E}^2}{c^2} + \mathbf{B}^2\right) - \boldsymbol{\nabla}\cdot\left(\frac{\mathbf{E}\phi}{c^2}\right)\right] \quad \text{joules/meter}^3. \end{aligned} \tag{18.45}$$

(Compare the first contribution with $\mathcal{L}$ above). For the energy of the electromagnetic field we obtain by integrating over all space

$$\begin{aligned}E_{\text{field}} &= \int d^3x \mathcal{H} = \frac{1}{2\mu_0}\int d^3x \left(\frac{\mathbf{E}^2}{c^2} + \mathbf{B}^2\right) = \frac{1}{2\mu_0}\int d^3x(\epsilon_0\mu_0\mathbf{E}^2 + \mathbf{B}^2) \\ &= \frac{1}{2}\int d^3x(\mathbf{E}\cdot\mathbf{D} + \mathbf{H}\cdot\mathbf{B}) \quad \text{joules}, \end{aligned} \tag{18.46}$$

where with the help of Gauss' divergence theorem the contribution of the divergence again vanishes for localized fields (as here assumed).[5]

---

[3]Observe also that, with $c^2\epsilon_0\mu_0 = 1$, one can define $\mathbf{D}$ as $\mathbf{D} = \partial\mathcal{L}/\partial(\mu_0\mathbf{E})$ and similarly $\mathbf{H} = -\partial\mathcal{L}/\partial(\mu_0\mathbf{B})$.

[4]Observe the factor $1/\mu_0$. This is the price we have to pay for using the practical MKSA units. In the Gaussian system of units $\epsilon_0$ and $\mu_0$ are taken as 1 and dimensionless. See Eqs. (1.21)–(1.23).

[5]We can always add a divergence to the Lagrangian density so that the Lagrangian density is not uniquely defined.

In an analogous way we can calculate the *field momentum* $P^i$, i.e. (cf. Eq. (18.21))

$$\begin{aligned} c\mu_0 P^i &= c\mu_0 \int d^3x p^i = \int d^3x[-\eta^{0i}\mathcal{L} + \pi^\rho \partial^i A_\rho] = \int d^3x \pi^\rho \partial^i A_\rho \\ &\overset{(18.39)}{=} \int d^3x(F^{\rho 0})\partial^i A_\rho = \int d^3x F^{j0}\partial^i A_j \\ &= \int d^3x F^{j0}(\partial^i A_j - \partial_j A^i) + \int d^3x F^{j0}\partial_j A^i, \end{aligned} \tag{18.47}$$

and hence

$$\begin{aligned} c\mu_0 P^i &= -\int d^3x F^{j0}(\partial^i A^j - \partial^j A^i) - \int d^3x F^{j0}\partial^j A^i \\ &\overset{(18.40)}{=} -\int d^3x \frac{E_j}{c}(\partial^i A^j - \partial^j A^i) - \int d^3x \partial^j (F^{j0} A^i), \end{aligned} \tag{18.48}$$

where, as above, we have

$$\partial^j (F^{j0} A^i) = (\partial^j A^i) F^{j0} + A^i \partial^j F^{j0} \tag{18.49}$$

with $\partial^j F^{j0} = \boldsymbol{\nabla}\cdot\mathbf{E}/c = \rho/(\epsilon_0 c) = 0$. But $E_j(\partial^i A^j - \partial^j A^i)$ is element "$i$" of the product (cf. the submatrix in Eq. (17.79))

$$\begin{aligned} &\begin{pmatrix} E_1 & E_2 & E_3 \end{pmatrix} \begin{pmatrix} 0 & -B_3 & B_2 \\ B_3 & 0 & -B_1 \\ -B_2 & B_1 & 0 \end{pmatrix} \\ &\qquad = (E_2B_3 - E_3B_2, E_3B_1 - E_1B_3, E_1B_2 - E_2B_1) \\ &\qquad = \mathbf{E}\times\mathbf{B}. \end{aligned} \tag{18.50}$$

We obtain therefore

$$P^i = -\frac{1}{c^2\mu_0}\int d^3x(\mathbf{E}\times\mathbf{B})^i - \frac{1}{c^2\mu_0}\int d^3x \boldsymbol{\nabla}\cdot(\mathbf{E}A^i), \tag{18.51}$$

or with arguments as before

$$\mathbf{P} \equiv (P_1, P_2, P_3) = \frac{1}{c^2\mu_0}\int d^3x(\mathbf{E}\times\mathbf{B}) = \int d^3x(\mathbf{D}\times\mathbf{B}). \tag{18.52}$$

For $P_0, P_i$ of Eqs. (7.65) and (7.83) we obtain therefore the expected expressions (18.46) and (18.52), although those for the densities differ from the earlier expressions by divergence contributions.

## 18.5 Gauge Invariance and Charge Conservation

We know from Noether's theorem that every continuous symmetry of the Lagrangian implies a conservation law. Thus invariance under spatial translations implies conservation of momentum and so on, as is familiar from classical mechanics. An additional aspect of electrodynamics is its invariance under gauge transformations. The associated conserved quantity is the *electric charge*. This is the aspect we investigate next.

In general one distinguishes between *global gauge invariance* — corresponding to $x$-independent transformations — and *local gauge invariance* — corresponding to $x$-dependent transformations. In the context of our considerations here only the electromagnetic potential $A_\mu$ and the current $J_\mu = (c\rho, -\mathbf{j})$ appear, not however single electrons and positrons other than classical pointlike charges. In order to take electrons and positrons properly into account we would have to construct the complete Lagrangian for charged objects, or rather fields, in interaction with the electromagnetic fields, and that would mean the Lagrangian of *quantum electrodynamics.* The corresponding Lagrangian density has the form

$$\mathcal{L}(A_\mu, \partial_\nu A_\mu, \psi, \partial_\mu \psi) = \overline{\psi}(i\gamma^\mu \partial_\mu - e'\gamma^\mu A_\mu - m)\psi - \frac{1}{4}F^{\mu\nu}F_{\mu\nu}, \qquad (18.53)$$

where $\psi$ is the field of the electron, the so-called *Dirac field*, a 4-component column matrix called *spinor*, and $\gamma^\mu, \mu = 0,1,2,3$, are corresponding $4 \times 4$ matrices called *Dirac matrices*, which obey a Clifford algebra. We do not enter into a discussion of specific properties of the electron field here, as this belongs into the realm of quantum electrodynamics. However, we want to consider the gauge invariance of the Lagrangian. To this end we observe that $\mathcal{L}$ is invariant under the following local gauge transformation

$$\psi \to \psi' = e^{i\theta(x)e'}\psi, \quad \overline{\psi} \to \overline{\psi}' = e^{-i\theta(x)e'}\overline{\psi},$$
$$A_\mu \to A'_\mu = A_\mu - \partial_\mu \theta(x). \qquad (18.54)$$

The invariance depends crucially on the fact, that $\partial_\mu\theta(x)$-contributions of the free Dirac field ($A_\mu = 0$) are cancelled by contributions derived from the interaction with the electromagnetic field (physically this implies that it is not possible to separate the electromagnetic field from the electron). This means

$$\mathcal{L} = -\frac{1}{4}F^{\mu\nu}F_{\mu\nu} - \mu_0 J^\nu A_\nu \qquad (18.55)$$

is gauge invariant only if we add to $\mathcal{L}$ the free Lagrangian of those particles that give rise to the current $J^\nu$ or if $J^\nu$ is a conserved current, $\partial_\nu J^\nu = 0$, and we ignore a divergence term which, integrated over the spatial volume, gives zero: $\partial_\mu(J^\mu\theta) = J^\mu\partial_\mu\theta + (\partial_\mu J^\mu)\theta$. The *invariance of* $F^{\mu\nu}F_{\mu\nu}$ is easily verified:

$$\begin{aligned} F^{\mu\nu} \to F^{\mu\nu\prime} &= \partial^\mu A^{\nu\prime} - \partial^\nu A^{\mu\prime} \\ &= \partial^\mu(A^\nu - \partial^\nu\theta(x)) - \partial^\nu(A^\mu - \partial^\mu\theta(x)) = F^{\mu\nu}. \end{aligned} \tag{18.56}$$

The invariance of the Lorentz gauge condition $\partial_\mu A^\mu(x) = 0$ requires

$$\partial^\mu\partial_\mu\theta(x) = 0. \tag{18.57}$$

We assume therefore that the Lagrangian density $\mathcal{L}$ is gauge invariant, and return to Noether's theorem. The *simplest way* to obtain the corresponding conservation law is by differentiating the equation of motion (here a kind of trick), i.e. differentiating Eq. (18.38),

$$\partial_\nu(\partial_\mu F^{\mu\nu} - \mu_0 J^\nu) = 0. \tag{18.58}$$

Since (with $\partial^\mu\partial_\mu = -\Box$)

$$\partial_\nu\partial_\mu F^{\mu\nu} = \partial_\nu(\partial_\mu\partial^\mu A^\nu - \partial_\mu\partial^\nu A^\mu) = -\Box\partial_\nu A^\nu + \Box\partial_\mu A^\mu = 0, \tag{18.59}$$

it follows that

$$\partial_\nu J^\nu = 0 \quad \text{(current conservation)}. \tag{18.60}$$

The charge $q$ is given by

$$q = \int d^3x\rho = \frac{1}{c}\int d^3xJ^0, \tag{18.61}$$

i.e. (recall Eq. (17.71), where $J^\mu = (c\rho, \mathbf{j})$)

$$\begin{aligned} \partial_0 q &= \frac{1}{c}\int d^3x\partial_0 J^0 = -\frac{1}{c}\int d^3x\partial_j J^j = -\frac{1}{c}\int d^3x\boldsymbol{\nabla}\cdot\mathbf{j} \\ &= -\frac{1}{c}\int d\mathbf{F}\cdot\mathbf{j} = 0, \end{aligned} \tag{18.62}$$

if we assume (as frequently explained above) that the fields are localized (having a finite velocity of propagation). This result, $\partial_0 q = 0$, is described as *charge conservation*, since $q = \text{const}$.

## 18.6 Lorentz Transformations and Associated Conservation Laws

The conservation laws which result from the invariance of the Lagrangian under translations and spatial rotations (i.e. conservation of energy, momentum and angular momentum) are familiar from mechanics (with the difference that here (e.g.) the mechanical momentum has its counterpart in the field momentum, and that one has to distinguish between momentum and momentum density). A natural question is therefore now: What are the conserved quantities resulting from invariance with respect to the special Lorentz transformations? We do not enter into detailed field calculations here, and instead resort to a *plausibility argument.*

The $x$-component of the conserved angular momentum $\mathbf{L}$ is

$$L_x = yp_z - zp_y. \tag{18.63}$$

In the case of a special Lorentz transformation along the $x$-axis the correspondingly conserved quantity is evidently

$$l_x = xp_0 - x_0p_x, \quad x_0 = ct, \quad p_0 = \mathcal{H}/c, \tag{18.64}$$

or more generally

$$\mathbf{l} = \mathbf{r}p_0 - x_0\mathbf{p}. \tag{18.65}$$

Since we have here field densities, the corresponding conservation law is

$$0 = \frac{d}{dx_0}\int d^3x[\mathbf{r}p_0 - x_0\mathbf{p}], \quad \int d^3x\mathbf{p} = \frac{d}{dx_0}\int d^3x\mathbf{r}p_0 \tag{18.66}$$

with e.g. $\mathbf{p}$ the field momentum density. With this we construct

$$\frac{\int \mathbf{p}d^3x}{\int p_0d^3x} = \frac{1}{\int p_0d^3x}\frac{d}{dx_0}\int \mathbf{r}p_0d^3x = \frac{d}{dx_0}\left(\frac{\int \mathbf{r}\mathcal{H}d^3x}{\int \mathcal{H}d^3x}\right) \equiv \frac{d}{dx_0}\mathbf{R}. \tag{18.67}$$

The vector $\mathbf{R}$ is (for obvious reasons) called "*center of mass of the field*". Thus if we visualize the localized field like a cloud travelling in space, this vector would be that of its center of mass; this is usually not of much interest and explains why this is rarely mentioned.

## 18.7 Masslessness of the Electromagnetic Field

Our next aim is to demonstrate that the electromagnetic field (and so the photon) is massless. In principle this is already evident by looking at the Lagrangian and observing that this does not contain a mass term, i.e. $m^2 A_\mu A^\mu$. A term of this type would violate the gauge invariance of the Lagrangian. We can also argue as follows without reference to the Lagrangian. We consider the Maxwell equations rewritten in the form of equations of motion of the free electromagnetic field ($J_\mu = 0$), i.e.

$$\partial^\mu F_{\mu\nu} = 0, \tag{18.68}$$

or (important: we do not yet choose a gauge fixing condition!)

$$(\partial_\mu \partial^\mu) A_\nu - \partial_\nu (\partial^\mu A_\mu) = 0. \tag{18.69}$$

For the solution we make the ansatz

$$A_\mu(x) = \epsilon_\mu e^{ip_\nu x^\nu}. \tag{18.70}$$

The vector $\epsilon_\mu$ is the *polarization vector*. Inserting this into Eq. (18.69), we obtain

$$0 = [-p_\mu p^\mu \epsilon_\nu + p_\nu (p^\mu \epsilon_\mu)] e^{ip_\nu x^\nu}. \tag{18.71}$$

Since the solution $A_\mu$ is not to be zero, we have

$$(p_\mu p^\mu)\epsilon_\nu = (p^\mu \epsilon_\mu) p_\nu. \tag{18.72}$$

Either

$$(a) \quad p_\mu p^\mu \neq 0, \quad \text{or} \quad (b) \quad p_\mu p^\mu = 0. \tag{18.73}$$

*In case* (a) we have

$$\epsilon_\nu = \left(\frac{p^\mu \epsilon_\mu}{p^\mu p_\mu}\right) p_\nu, \quad \text{i.e.} \quad \epsilon_\nu \propto p_\nu. \tag{18.74}$$

Hence the solution is

$$A_\nu = \alpha p_\nu e^{ip_\mu x^\mu}, \quad \alpha = \text{const.} \tag{18.75}$$

Such solutions are considered to be *trivial*: They can be "*gauged away*" (see below) and are therefore not observable.

"*Gauging away*": The gauge transformation is

$$A_\nu \to A'_\nu = A_\nu + \partial_\nu \chi. \tag{18.76}$$

Since $\chi$ is arbitrary, we can set

$$\chi = i\alpha e^{ip_\mu x^\mu}, \qquad \partial_\nu \chi = -\alpha p_\nu e^{ip_\mu x^\mu} \tag{18.77}$$

and so, since from Eq. (18.75) $A_\nu = \alpha p_\nu e^{ip_\mu x^\mu}$,

$$A'_\nu = \alpha p_\nu e^{ip_\mu x^\mu} - \alpha p_\nu e^{ip_\mu x^\mu} = 0. \tag{18.78}$$

Thus, these solutions do not yield anything. *Hence we have case* (b), i.e. $p_\mu p^\mu = 0$, and so from Eq. (18.72),

$$0 = (p^\mu \epsilon_\mu) p_\nu, \quad \text{i.e.} \quad p^\mu \epsilon_\mu = 0, \tag{18.79}$$

since in general $p_\nu \neq 0$.

We conclude therefore: The solutions are

$$A_\nu = \epsilon_\nu e^{ip_\mu x^\mu} \quad \text{with} \quad p_\mu p^\mu = 0 \quad \text{and} \quad p^\mu \epsilon_\mu = 0. \tag{18.80}$$

The condition $p_\mu p^\mu = 0$ says: The mass of the field is zero (compare with Eq. (17.166): $(T/c)^2 = \mathbf{p}^2 + m^2c^2$). The condition is also described as *mass shell condition*, implying that the momentum is physical (in field theory intermediate states may have unphysical momenta).

We recognize in the condition $p^\mu \epsilon_\mu = 0$ our earlier condition (10.19), i.e. $\partial_\mu A^\mu = 0$, which we called *Lorenz gauge* (fixing condition). We shall see below, that with the help of these two conditions we can eliminate the unphysical components of the vector potential.

## 18.8 Transversality of the Electromagnetic Field

We now demonstrate explicitly that two of the four components of the 4-potential $A_\mu$ can be gauged away, i.e. that the free electromagnetic field is transverse (to the direction of propagation). We choose again

$$\chi = i\alpha e^{ip_\mu x^\mu} \tag{18.81}$$

($\alpha$ will be chosen suitably later). Then with Eq. (18.70):

$$\begin{aligned} A'_\nu &= A_\nu + \partial_\nu \chi = \epsilon_\nu e^{ip_\mu x^\mu} - \alpha p_\nu e^{ip_\mu x^\mu} \\ &= (\epsilon_\nu - \alpha p_\nu) e^{ip_\mu x^\mu} \equiv \epsilon'_\nu e^{ip_\mu x^\mu} \end{aligned} \tag{18.82}$$

with

$$\epsilon_\nu \to \epsilon'_\nu = \epsilon_\nu - \alpha p_\nu. \tag{18.83}$$

This means, with the gauge transformation we also change the polarization vector. We select a wave travelling in the direction of $z$:

$$p_\mu = (\omega/c, 0, 0, k), \quad (\mathbf{p} \equiv (p^i),\ \mathbf{x} = (x^i)), \tag{18.84}$$

where the mass shell condition (18.80) implies

$$p_\mu p^\mu = 0 \;\Rightarrow\; \omega^2 - c^2k^2 = 0, \qquad \omega = ck. \tag{18.85}$$

Since $p^\mu \epsilon_\mu = 0$, we have

$$\frac{\omega}{c}\epsilon_0 - k\epsilon_3 = 0, \quad \text{so that} \quad \epsilon_0 = \epsilon_3. \tag{18.86}$$

It follows that

$$A_\nu = \epsilon_\nu e^{ip_\mu x^\mu} = (\epsilon_3, \epsilon_1, \epsilon_2, \epsilon_3)e^{i\omega(ct-z)/c}. \tag{18.87}$$

However, we can still perform a gauge transformation, i.e. we can replace $\epsilon_\nu$ by $\epsilon'_\nu$, $\epsilon'_\nu = \epsilon_\nu - \alpha p_\nu$, i.e.

$$\epsilon'_\nu = \left(\epsilon_3 - \alpha\frac{\omega}{c}, \epsilon_1, \epsilon_2, \epsilon_3 - \alpha k\right). \tag{18.88}$$

*Choosing* now

$$\alpha = \frac{\epsilon_3}{k}, \tag{18.89}$$

we have

$$\epsilon'_\nu = (0, \epsilon_1, \epsilon_2, 0), \tag{18.90}$$

i.e. the components $\epsilon_0, \epsilon_3$ have completely disappeared and have no physical significance. The part $\epsilon_0$ of $A_\mu$ is described as the *scalar part* ("scalar photon"), the other $\epsilon_3$ part as the *longitudinal part.* We see that only the two transverse parts $\propto \epsilon_1, \epsilon_2$ have physical significance. This result is a consequence of the mass shell condition and the gauge fixing condition.

## 18.9 The Spin of the Photon

We now have a closer look at the polarization vector. We put

$$\begin{pmatrix}\epsilon_0\\ \epsilon_1\\ \epsilon_2\\ \epsilon_3\end{pmatrix} = \tilde{\epsilon}_0\begin{pmatrix}1\\0\\0\\0\end{pmatrix} + \tilde{\epsilon}_3\begin{pmatrix}0\\0\\0\\1\end{pmatrix} + \underbrace{\tilde{\epsilon}_+\begin{pmatrix}0\\1\\-i\\0\end{pmatrix}}_{(\alpha)} + \tilde{\epsilon}_-\begin{pmatrix}0\\1\\i\\0\end{pmatrix}, \tag{18.91}$$

where

$$\tilde{\epsilon}_{\pm} = \frac{1}{2}(\tilde{\epsilon}_1 \pm i\tilde{\epsilon}_2). \tag{18.92}$$

(Since $\mathbf{E} = -\partial\mathbf{A}/\partial t - \boldsymbol{\nabla}\phi$, i.e. $\mathbf{E} \parallel -\mathbf{A}$, the waves $\epsilon_{\pm}\exp[i\omega(ct - z)/c]$ are circularly polarized). Actually $\tilde{\epsilon}_{\mu} \equiv \epsilon_{\mu}$, but we want to distinguish between the two more clearly. We want to examine the solution of a wave travelling in the direction of $z$:

$$A_{\mu} = \epsilon_{\mu}e^{ip^{\rho}x_{\rho}} = \epsilon_{\mu}e^{i\omega(ct-z)/c} \quad \text{or} \quad A^{*}_{\mu} = \epsilon^{*}_{\mu}e^{-i\omega(ct-z)/c}. \tag{18.93}$$

Naturally $A_{\mu}(x)$ can be decomposed exactly like $\epsilon_{\mu}$. Consider an ordinary rotation about the $z$-axis, i.e. a special case of a homogeneous Lorentz transformation,

$$x'_{\mu} = l_{\mu}{}^{\nu}x_{\nu} \tag{18.94}$$

with rotation operator $R_{\theta}$ for rotation $\theta$ about the $z$-axis:

$$R_{\theta}: \quad l_{\mu}{}^{\nu} = \begin{pmatrix} 1 & 0 & 0 & 0 \\ 0 & \cos\theta & \sin\theta & 0 \\ 0 & -\sin\theta & \cos\theta & 0 \\ 0 & 0 & 0 & 1 \end{pmatrix}. \tag{18.95}$$

Here $\theta$ is the angle of the rotation in the $(x, y)$-plane. Then

$$A'_{\nu} = l_{\nu}{}^{\mu}A_{\mu} = l^{\mu}_{\nu}\epsilon_{\mu}\exp(ip^{\rho}x_{\rho}) \equiv \epsilon'_{\nu}\exp(ip^{\rho}x_{\rho}), \tag{18.96}$$

where

$$\epsilon'_{\nu} = l^{\mu}_{\nu}\epsilon_{\mu}. \tag{18.97}$$

Thus the rotation affects only the polarization vector (cf. Eq. (18.91)):

$$\tilde{\epsilon}_0 \begin{pmatrix} 1 \\ 0 \\ 0 \\ 0 \end{pmatrix} \to \tilde{\epsilon}'_0 = l^{\mu}_{0}\epsilon_{\mu} = \tilde{\epsilon}_0 \left[ \begin{pmatrix} 1 & 0 & 0 & 0 \\ 0 & \cos\theta & \sin\theta & 0 \\ 0 & -\sin\theta & \cos\theta & 0 \\ 0 & 0 & 0 & 1 \end{pmatrix} \begin{pmatrix} 1 \\ 0 \\ 0 \\ 0 \end{pmatrix} \right]_0 = \tilde{\epsilon}_0 \begin{pmatrix} 1 \\ 0 \\ 0 \\ 0 \end{pmatrix}. \tag{18.98}$$

Similarly

$$\tilde{\epsilon}_3 \begin{pmatrix} 0 \\ 0 \\ 0 \\ 1 \end{pmatrix} \to \tilde{\epsilon}_3 \begin{pmatrix} 0 \\ 0 \\ 0 \\ 1 \end{pmatrix}. \tag{18.99}$$

However,

$$\begin{aligned} \tilde{\epsilon}_\pm \begin{pmatrix} 0 \\ 1 \\ \mp i \\ 0 \end{pmatrix} &\to \tilde{\epsilon}_\pm \begin{pmatrix} 1 & 0 & 0 & 0 \\ 0 & \cos\theta & \sin\theta & 0 \\ 0 & -\sin\theta & \cos\theta & 0 \\ 0 & 0 & 0 & 1 \end{pmatrix} \begin{pmatrix} 0 \\ 1 \\ \mp i \\ 0 \end{pmatrix} \\ &= \tilde{\epsilon}_\pm \begin{pmatrix} 0 \\ \cos\theta \mp i \sin\theta \\ -\sin\theta \mp i\cos\theta \\ 0 \end{pmatrix} = \tilde{\epsilon}_\pm e^{\mp i\theta} \begin{pmatrix} 0 \\ 1 \\ \mp i \\ 0 \end{pmatrix}. \end{aligned} \tag{18.100}$$

If we consider $R_\theta$ as rotation operator, we have the equation called "*eigenvalue equation*":

$$R_\theta A_\mu = \text{eigenvalue} \times A_\mu, \tag{18.101}$$

where the eigenvalue is a number. The rotation operator is given by

$$R_\theta = e^{i\theta I_z} \tag{18.102}$$

where $I_z$ is the $z$-component of the (eigen)-angular momentum operator ("spin operator"). Then in a rotation about the $z$-axis Eqs. (18.98)–(18.100) are:

$$\begin{aligned} R_\theta A_0 &= 1 \text{ times } A_0 \\ &\equiv e^{i\theta I_z} A_0 \quad \text{with eigenvalue of } I_z = 0, \quad \text{(to give 1)} \\ R_\theta A_3 &= 1 \text{ times } A_3 \\ &\equiv e^{i\theta I_z} A_3 \quad \text{with eigenvalue of } I_z = 0, \quad \text{(to give 1)} \\ R_\theta A_\pm &= e^{\mp i\theta} A_\pm \\ &\equiv e^{i\theta I_z} A_\pm \quad \text{with eigenvalue of } I_z = \mp 1 \quad (\text{to give } \mathrm{e}^{\mp i\theta}). \end{aligned} \tag{18.103}$$

It follows therefore that for a scalar or longitudinal field or photon the spin projection on the $z$-axis (i.e. in the direction of propagation) is zero, whereas the field of the transverse photon has the spin projections $S_z = \pm 1\hbar$. Since only the transverse components are physical, i.e. observable, it is evident that the photon has spin $1(\hbar), s = 1\hbar$ ($\mathbf{S}^2$ has eigenvalue $s(s+1)\hbar^2, s = maximal\, s_z, S_z \equiv I_z$). The photon never has spin projection 0 in the direction of the momentum vector, i.e. in the direction of propagation. One says therefore, the photon has *helicity* $\pm 1$.

## 18.10 Examples

The following examples are motivated by the Lagrangian of electrodynamics, but they also lead somewhat beyond the considerations of primary interest here and, maybe, illustrate the fascinating use of electrodynamics for the exploration of a plethora of other issues.

**Example 18.2 (The Chern–Simons term and Rydberg atoms).** Verify the gauge invariance of the following term, called *Chern–Simons term*, which one could add to a Maxwell Lagrangian density:

$$T_{\text{CS}} := \epsilon^{\lambda\mu\rho} A_\lambda \partial_\mu A_\rho \tag{18.104}$$

(up to a divergence). Explore the simple choice $A_i = x_i$ with $i = 1, 2$, and $A_0 = 0$.

**Solution:** With the gauge transformation $A_\mu \to A_\mu + \partial_\mu \chi$, and using the antisymmetry of the Levi-Civita tensor, the term $T_{\text{CS}}$ becomes

$$T'_{\text{CS}} = \epsilon^{\lambda\mu\rho}(A_\lambda + \partial_\lambda \chi)\partial_\mu(A_\rho + \partial_\rho \chi) = T_{\text{CS}} + \triangle, \tag{18.105}$$

where

$$\begin{aligned}
\triangle &= \epsilon^{\lambda\mu\rho} A_\lambda \partial_\mu \partial_\rho \chi + \epsilon^{\lambda\mu\rho}\partial_\lambda \chi \partial_\mu (A_\rho + \partial_\rho \chi) \\
&= 0 + \epsilon^{\lambda\mu\rho}\partial_\lambda \chi \partial_\mu A_\rho + \underbrace{\epsilon^{\lambda\mu\rho}\partial_\lambda \chi \partial_\mu \partial_\rho \chi}_{0} \\
&= \epsilon^{\lambda\mu\rho}(\partial_\lambda(\chi \partial_\mu A_\rho) - \chi \partial_\lambda \partial_\mu A_\rho) \\
&= \epsilon^{\lambda\mu\rho}\partial_\lambda(\chi \partial_\mu A_\rho) = \partial_\lambda(\chi \epsilon^{\lambda\mu\rho}\partial_\mu A_\rho) \\
&= \partial_\lambda(\chi V^\lambda), \\
V^\lambda &= \epsilon^{\lambda\mu\rho}\partial_\mu A_\rho,
\end{aligned} \tag{18.106}$$

which had to be shown (gauge invariance up to a total divergence).

In the special case of $\lambda, \mu, \rho = 0, 1, 2$ with gauge choice $A_0 = 0$ and $i, j = 1, 2$ and $m$ the mass of the atom, we have

$$T_{\rm CS} = -\epsilon_{ij} A_i \partial_0 A_j. \tag{18.107}$$

Selecting the simple case of $A_i = \alpha x_i, \alpha = \text{const.}$, so that $\partial_0 A_j = \alpha \dot{x}_j / c = \alpha p_j / mc$, which relates the electric field to momentum, we have

$$T_{\rm CS} = \frac{\alpha^2}{mc} \epsilon_{ij} p_i x_j. \tag{18.108}$$

This type of term appears in studies of atomic physics where the Hamiltonian of a Rydberg atom of mass $m$ in the presence of electric and magnetic fields (the latter along $x_3$) is written as

$$\begin{aligned} H &= \frac{1}{2m}\left(p_i + \frac{1}{2} g \epsilon_{ij} x_j\right)^2 + \frac{1}{2}\kappa {x_i}^2 \\ &= \frac{1}{2m}{p_i}^2 + \frac{1}{2m} g \epsilon_{ij} p_i x_j + \frac{1}{2}\left(\kappa + \frac{g^2}{4m}\right){x_i}^2, \end{aligned} \tag{18.109}$$

where $g$ and $\kappa$ are constants and $\sum_i \epsilon_{ij}\epsilon_{ik} = \sum_i \epsilon^T_{ji}\epsilon_{ik} = \delta_{jk}$. The central term in the last expression is seen to be of Chern–Simons type. (Observe the difference between the Hamiltonian here and that of Eq. (19.7) to be discussed later. The minimal coupling to $A_i$ discussed there (cf. the end of Section 19.2.1) corresponds here effectively to the coupling to the dual of $A_i$). The so-called *Rydberg atom* is an atom regarded as one with a permanent dipole moment resulting from a single electron in a shell far away from the filled electron shells enclosing the nucleus, so that the structure of the latter can be ignored.

**Example 18.3 (6-dimensional gauge field theory).** In a (1+5)-dimensional gauge field theory the gauge field is the tensor field $A_{\mu\nu} = -A_{\nu\mu}, \mu, \nu = 0, 1, 2, \ldots, 5$, and the field tensor is

$$F_{\mu\nu\lambda} = \partial_\mu A_{\nu\lambda} + \partial_\nu A_{\lambda\mu} + \partial_\lambda A_{\mu\nu}. \tag{18.110}$$

(a) Verify the invariance of the field strength $F_{\mu\nu\lambda}$ with respect to gauge transformations

$$\delta A_{\mu\nu} = A'_{\mu\nu} - A_{\mu\nu} = \partial_\mu \chi_\nu - \partial_\nu \chi_\mu. \tag{18.111}$$

(b) Derive the Maxwell equations from the action integral

$$S = -\frac{1}{3!}\int d^6x F_{\mu\nu\lambda} F^{\mu\nu\lambda}. \tag{18.112}$$

**Solution:**
(a) Trivial.
(b) Set

$$E_{ij} := F_{0ij}, \quad B_{ij} := \frac{1}{3!}\epsilon_{ijklm}F^{klm}. \tag{18.113}$$

The equations of motion are

$$(a)\ \partial^{\mu}F_{\mu\nu\lambda} = 0, \quad (b)\ \epsilon^{\mu\nu\lambda\rho\sigma\delta}\partial_{\lambda}F_{\rho\sigma\delta} = 0. \tag{18.114}$$

Consider (a):

$$\partial^{\mu}F_{\mu i0} = 0, \quad \partial^{0}F_{0i0} + \partial^{j}F_{ji0} = 0, \quad F_{00i} = \partial_0 A_{0i} + \partial_i A_{00} + \partial_0 A_{i0} = 0, \tag{18.115}$$

so that

$$\partial_i E_{ij} = 0 \quad \text{and} \quad \partial^0 F_{0jk} + \partial^i F_{ijk} = 0, \tag{18.116}$$

i.e.

$$\partial_0 E_{jk} = -\partial^i F_{ijk} \overset{\text{see below}}{=} -\frac{1}{2}\partial^i \epsilon_{ijklm}B^{lm}. \tag{18.117}$$

Verification:

$$\partial^i \epsilon_{ijklm}B^{lm} = \partial^i \epsilon_{ijklm}\frac{1}{3!}\epsilon^{lmabc}F_{abc} = \partial^i \frac{1}{3!}\left(\underbrace{\epsilon_{ijklm}\epsilon^{lmabc}}_{2!\mathrm{Sgn}\begin{bmatrix} i & j & k \\ a & b & c\end{bmatrix}}\right)F_{abc}$$

$$= \partial^i \frac{1}{3!}2!3!F_{ijk} = 2!\partial^i F_{ijk}. \tag{18.118}$$

The other equations follow similarly from (b). (For the $\epsilon$ contractions and related topics see e.g. Felsager [51], p. 354).

**Example 18.4 (Born–Infeld theory).** The static, nonlinear BI Lagrangian, here simplified to that involving only the electric field $\mathbf{E} = -\boldsymbol{\nabla}\phi$

in a flat space, and with $c = 1$, hence $\epsilon_0\mu_0 = 1$ (hence we state no units here), is given by

$$L_{\rm BI} = \int d\mathbf{r}\mathcal{L}_{\rm BI}, \quad \mathcal{L}_{\rm BI}(\phi, \boldsymbol{\nabla}\phi) = 1 - \sqrt{1-(\boldsymbol{\nabla}\phi)^2} - 4\pi e\phi\delta(\mathbf{r}). \qquad (18.119)$$

Show that the energy of the charge $e$ at the origin is given by

$$H_{\rm BI} = 4\pi\epsilon_0(3.09112)e^{3/2}, \qquad (18.120)$$

i.e. is finite, and not infinite like the energy of a static point-charge, i.e.

$$\frac{1}{2}\int d\mathbf{r}(\boldsymbol{\nabla}\phi)^2, \qquad (18.121)$$

in Maxwell electrodynamics.

**Solution:** In an attempt to avoid the infinite self-energy of a point charge in Maxwell theory, Born and Infeld[6] arrived at the above Lagrangian which is now named after them and plays an important role in String Theory. In fact, $\mathcal{L}_{\rm BI}|_{e=0} = \mathcal{L}_{b=1}$, where

$$\mathcal{L}_b = b^2\left[1 - \sqrt{1-\frac{2}{b^2}\mathcal{L}_M}\right] \stackrel{b^2\to\infty}{=} \mathcal{L}_M, \qquad (18.122)$$

and $\mathcal{L}_M$ is the Maxwell Lagrangian density, in the present case for $\mathbf{B} = 0$ and $c = 1$ simply $\mathcal{L}_M = \mathbf{E}^2/2$.

The Euler–Lagrange equation (18.8) gives the equation corresponding to the Gauss equation in Maxwell's electrodynamics:

$$\partial_i\left(\frac{\partial_i\phi}{\sqrt{1-(\partial_j\phi)^2}}\right) = -4\pi e\delta(\mathbf{r}). \qquad (18.123)$$

Integrating over a sphere of radius $r$ and using the Gauss divergence theorem, this becomes

$$\int\frac{\boldsymbol{\nabla}\phi}{\sqrt{1-(\phi'(r))^2}}\cdot d\mathbf{F} = -4\pi e, \quad \text{or} \quad \frac{\phi'(r)}{\sqrt{1-(\phi'(r))^2}}4\pi r^2 = -4\pi e, \qquad (18.124)$$

i.e.

$$\frac{\phi'(r)}{\sqrt{1-(\phi'(r))^2}} = -\frac{e}{r^2}. \qquad (18.125)$$

[6] Born and Infeld [19]; see also Jackson [73], p. 10 and Gibbons and Rasheed [59]. The Born–Infeld theory has also been discussed in this context by Anderson [3], p. 99.

Solving this equation for $\phi'(r)$, the solution is then seen to be

$$\phi'(r) = -\frac{1}{\sqrt{1+r^4/e^2}}, \quad \text{and hence} \quad \phi_{\rm BI}(\mathrm{r}) = \int_{\mathrm{r}}^{\infty} \frac{\mathrm{dx}}{\sqrt{1+\mathrm{x}^4/\mathrm{e}^2}}. \tag{18.126}$$

With the help of Tables of Integrals we can evaluate this function at the origin and obtain $\phi_{\rm BI}(0)$. We rewrite this integral as

$$\phi_{\rm BI}(0) = e\int_0^{\infty} \frac{dx}{\sqrt{e^2+x^4}} = \sqrt{e}\int_0^{\infty} \frac{dx}{\sqrt{1+x^4}}. \tag{18.127}$$

The integral on the right can be evaluated as an elliptic integral using formula 3.166(10) of Gradshteyn and Ryzhik [60], p. 262. But it is easier to use formula 3.241(4), p. 292, which (plugging in our values of their parameters) gives immediately:

$$\int_0^{\infty} \frac{dx}{\sqrt{1+x^4}} = \frac{1}{4}\frac{\Gamma(1/4)\Gamma(1/4)}{\Gamma(1/2)} \equiv \frac{1}{4}B[1/4,1/4], \tag{18.128}$$

where $\Gamma(n) = (n-1)!$. Evaluating the gamma functions or factorials of the beta function $B$, one finds that $\phi_{\rm BI}(0) = 1.854074677e^{1/2}$. This was the intention of Born and Infeld, to obtain this result finite in a covariant theory.

Defining (with $\mathbf{E} = -\boldsymbol{\nabla}\phi$)

$$\mathbf{D} = \frac{1}{\mu_0}\frac{\partial \mathcal{L}_{\rm BI}}{\partial \mathbf{E}} = \frac{1}{\mu_0}\frac{\mathbf{E}}{\sqrt{1-\mathbf{E}^2}}, \tag{18.129}$$

we have (cf. Eq. (18.15) or Eq. (18.45) with Eq. (18.43), here with $c=1$)

$$\mu_0 T^{00} = F^{0i}\frac{\partial \mathcal{L}_{\rm BI}}{\partial F^{0i}} - \mathcal{L}_{\rm BI} = \mu_0 \mathbf{E}\cdot\mathbf{D} - \mathcal{L}_{\rm BI} = \frac{1}{\sqrt{1-\mathbf{E}^2}} - 1 + 4\pi e\phi\delta(\mathbf{r}). \tag{18.130}$$

The energy of the charge is then found to be after some nontrivial integration which we give below:

$$H_{\rm BI} = \int d\mathbf{r} T^{00} = 4\pi\epsilon_0(3.09112)e^{3/2}. \tag{18.131}$$

The steps leading to this result (with $c = 1, \mu_0\epsilon_0 = 1$) are the following. We have, with $\phi' = \phi'_{\rm BI}(r) = -1/\sqrt{1+r^4/e^2}$ and $d\mathbf{r} = 4\pi r^2 dr$,

$$\begin{aligned}\mu_0 H_{\rm BI} &= \int d\mathbf{r}\left[\frac{1}{\sqrt{1-(\phi'(r))^2}} - 1 + 4\pi e\phi(r)\delta(\mathbf{r})\right]\\ &= 4\pi e\phi(0) + \int 4\pi r^2 dr\left[\frac{1}{r^2}\sqrt{e^2+r^4} - 1\right]\\ &= 4\pi e\phi(0) + 4\pi\int_0^\infty dr[\sqrt{e^2+r^4} - r^2]. \qquad (18.132)\end{aligned}$$

Setting $r = x\sqrt{e}$, we obtain (with $\phi(0) \equiv \phi_{\rm BI}(0)$ as given above Eq. (18.129)):

$$\begin{aligned}\mu_0 H_{\rm BI} &= 4\pi e\phi(0) + 4\pi e^{3/2}\int_0^\infty dx[\sqrt{1+x^4} - x^2]\\ &= 4\pi e^{3/2}(1.854074677) + 4\pi e^{3/2}(1.2360498) = 4\pi e^{3/2}(3.09112),\end{aligned} \qquad (18.133)$$

where the integral has been evaluated numerically with MAPLE.

**Example 18.5 (Non-gauge invariant solutions of Maxwell's equations[7]).** A type of nonlinear, non-gauge invariant solutions $W_{\mu\nu}$ of Maxwell's equations has been found and investigated by Schiff [130]. These are given by the mixed tensor $W_{\mu\nu}$ defined by

$$W_{\mu\nu} := F_{\mu\nu} + \sqrt{2}g^{-1}A_\mu A_\nu, \quad (g = \text{const.}). \qquad (18.134)$$

Show that $W_{\mu\nu}$ satisfies everywhere the conditions

$$^*W_{\mu\nu}W^{\mu\nu} = 0, \quad W_{\mu\nu}W^{\mu\nu} = 0, \qquad (18.135a)$$

which define it as a "*null field*". Show that static, radially symmetric solutions of the homogeneous part of the non-gauge invariant equation

$$\partial_\mu W^{\mu\nu} = \mu_0 j^\nu \qquad (18.135b)$$

are given by

$$A_r = \frac{k_0}{r}, \quad k_0 = \text{const.}, \qquad (18.136)$$

[7]Schiff [130].

and

$$\phi_1 = \frac{b}{r} \quad \text{and} \quad \phi_2 = \frac{c_2}{r^\gamma}, \quad \text{where} \quad \gamma = \sqrt{2}g^{-1}k_0, \quad b, c_2 = \text{const.} \tag{18.137}$$

What is the total charge $Q = \int d\mathbf{r}\rho$? Finally solve the null condition with $\mathbf{B} = 0$ for the Coulombic solution $\phi_1$ and determine the charges $b$.

**Solution:** The original starting argument of Schiff [130] is simple. Since for a point charge $q$ at rest the Coulomb potential $\phi$ and electric field strength $E$ are given by

$$\phi = k\frac{q}{r}, \quad E = k\frac{q}{r^2}, \tag{18.138}$$

one has $E^2 = \phi^4/k^2q^2$, or (cf. Eqs. (17.80) and (17.74))

$$A_0 = \frac{\phi}{c}, \quad {F_{i0}}^2 = \frac{E^2}{c^2} = \frac{\phi^4}{k^2q^2c^2} = \frac{{A_0}^4}{g^2}, \quad g^2 = \frac{k^2q^2}{c^2}. \tag{18.139}$$

The covariant generalization is seen to be

$$F_{\mu\nu}F^{\mu\nu} = -\frac{2}{g^2}(A^\mu A_\mu)^2, \quad A^\mu A_\mu = \frac{\phi^2}{c^2} - \mathbf{A}^2, \tag{18.140}$$

as one can verify by reduction to the case above. For $\mathbf{B} = 0$ (i.e. $F_{ij} = 0$ in the quantity on the left) this null field condition is

$$\frac{\mathbf{E}^2}{c^2} = \frac{1}{g^2}\left(\mathbf{A}^2 - \frac{\phi^2}{c^2}\right)^2. \tag{18.141}$$

Writing the relation as

$$F_{\mu\nu}F^{\mu\nu} + \frac{2}{g^2}(A^\mu A_\mu)^2 = 0, \tag{18.142}$$

we see that — due to opposite symmetries of $F_{\mu\nu}$ and $A_\mu A_\nu$ (under interchange of $\mu$ and $\nu$) — we can write it as $W_{\mu\nu}W^{\mu\nu} = 0$ with $W_{\mu\nu}$ as defined in Eq. (18.134). Multiplying $W_{\mu\nu}$ by the four-dimensional Levi-Civita tensor, we see that its dual satisfies $^*W_{\mu\nu} = {}^*F_{\mu\nu}$, so that with $^*F_{\mu\nu}F^{\mu\nu} = 0$ ($\mathbf{E} \perp \mathbf{B}$), we obtain $^*W^{\mu\nu}W_{\mu\nu} = 0$. Thus $W_{\mu\nu}$ satisfies the conditions of a null field as defined above.

The static, radially symmetric part of the homogeneous equation $\partial_\mu W^{\mu\nu} = 0$ with $A = A_r(r)$ and so $\mathbf{B} = 0$ is

$$\boldsymbol{\nabla} \cdot (A\mathbf{A}) = 0, \tag{18.143}$$

which is solved by $A = k_0/r$, since with the divergence expressed in spherical polar coordinates, Eq. (2.22a),

$$\nabla \cdot (A\mathbf{A}) = \nabla A \cdot \mathbf{A} + A\nabla \cdot \mathbf{A} = -\frac{k_0 \mathbf{r}}{r^3} \cdot \frac{k_0 \mathbf{r}}{r^2} + \frac{k_0}{r}\frac{1}{r^2}\frac{\partial}{\partial r}\left(r^2 \frac{k_0}{r}\right) = 0. \tag{18.144}$$

The zero component of $\partial_\mu W^{\mu\nu} = \mu_0 j^\nu$ is (with $W_{0i} = F_{0i} - \sqrt{2}(\mathbf{A})_i A_0/g$) the Gauss law

$$\nabla \cdot \left(\mathbf{E} - \frac{\sqrt{2}}{g}\mathbf{A}\phi\right) = \mu_0 c^2 \rho(r), \quad \text{or}$$

$$-\frac{1}{r^2}\frac{\partial}{\partial r}\left(r^2 \frac{\partial \phi}{\partial r}\right) - \frac{\sqrt{2}k_0}{g}\nabla \cdot \left(\frac{\mathbf{r}}{r^2}\phi\right) = \mu_0 c^2 \rho(r), \tag{18.145}$$

which with

$$\begin{aligned}\nabla \cdot \left(\frac{\mathbf{r}}{r^2}\phi\right) &= (\nabla\phi) \cdot \left(\frac{\mathbf{r}}{r^2}\right) + \phi\nabla \cdot \left(\frac{\mathbf{r}}{r^2}\right) \\ &= \frac{1}{r}\frac{\partial \phi}{\partial r} + \frac{\phi}{r^2}\frac{\partial}{\partial r}\left(r^2 \frac{r}{r^2}\right) = \frac{\phi}{r^2} + \frac{1}{r}\frac{\partial \phi}{\partial r}\end{aligned} \tag{18.146}$$

can be written (as shown by Schiff [130])

$$-\frac{1}{r}\frac{d^2(r\phi)}{dr^2} - \gamma \frac{1}{r^2}\frac{d(r\phi)}{dr} = \mu_0 c^2 \rho(r), \tag{18.147}$$

where $\gamma = \sqrt{2}k_0/g$. Irrespective of the value of $\gamma$ the homogeneous part has solutions for $r\phi = \text{const.}$, i.e. $\phi_1 = b/r$, $b$ being a Coulombic charge, but for $\gamma \neq 1$ also solutions $\phi_2 = c_2 r^{-\gamma}$, which are "*confinement potentials*" for $\gamma \leq -1$ (see Example 2.8).

Integrating the Gauss law (18.145), one obtains in the usual way the total charge (with $\epsilon_0 \mu_0 c^2 = 1$)

$$Q = \int d\mathbf{r}\rho(r) = \epsilon_0 \int d\mathbf{F} \cdot \left(\mathbf{E}_1 - \frac{\sqrt{2}}{g}\mathbf{A}\phi_1\right) \tag{18.148}$$

(the index referring to solution $\phi_1$), i.e.

$$Q = \epsilon_0 \int r^2 d\Omega \frac{1}{r^2}\left(b - \frac{\sqrt{2}}{g}\frac{\gamma g}{\sqrt{2}}b\right) = \epsilon_0 4\pi b(1-\gamma). \tag{18.149}$$

Inserting the Coulombic solution $\phi_1$ into the null field condition (18.141), i.e.

$$\frac{E}{c} = \pm\frac{1}{g}\left(A^2 - \frac{\phi^2}{c^2}\right), \tag{18.150}$$

one obtains

$$\frac{1}{r^2}\left(\frac{b^2}{c^2} \pm g\frac{b}{c} - \frac{\gamma^2 g^2}{2}\right) = 0 \tag{18.151}$$

with solutions

$$\frac{b}{c} = \pm\frac{g}{2}[1 \pm \sqrt{1 + 2\gamma^2}]. \tag{18.152}$$

This determines the charge $b$ in the potential $\phi_1$. In the papers of Schiff [130] these conditions are further extended to the confinement potentials $\phi_2$, and it is shown that fractional charges such as those attributed to quarks can be obtained. This is an interesting observation, although, of course, this is not the acceptable overall theory. Further aspects are considered in Example 18.7.

**Example 18.6 (Quantum Hall effect).** Starting from the Lagrangian density $\mathcal{L}$ of the electromagnetic field (cf. Eq. (18.42)) in a dielectric with polarization energy $U(\mathbf{E}, \mathbf{B})$,

$$\mathcal{L}(\mathbf{E}, \mathbf{B}) = \frac{1}{2}\left(\frac{\mathbf{E}^2}{c^2} - \mathbf{B}^2\right) - U(\mathbf{E}, \mathbf{B}), \quad \text{with} \quad U := -g(\mathbf{E}\cdot\mathbf{B}), \tag{18.153}$$

show that if the induced surface charge is $ne$, i.e. an integral multiple $n$ of the elementary charge $e$, and the magnetic flux is $\Phi = m\Phi_0$, i.e. an integral multiple of the magnetic flux quantum $\Phi_0 = h/e$ (in MKSA units), that then the surface *Hall conductivity*,[8] $\sigma_H$, defined by

$$\mathbf{j}_M = \sigma_H \mathbf{E}, \tag{18.154}$$

[8] For a recent discussion with references to earlier literature see Avron, Osadchy and Seiler [6]. The standard definition of the magnetic flux quantum in SI units is $\Phi_0 = h/2e$ (as in Appendix B). In c.g.s. units it is $hc/2e$. This is for superconductors where the charge carriers are Cooper pairs of charge $2e$ (ignoring the sign of the charge). The above article considers single electrons, thus the elementary quantum of magnetic flux there is defined as $\Phi_0 = hc/e$ in c.g.s. units. The author is indebted to Dr. Taylor, National Institute of Standards and Technology, USA, for clarifying correspondence.

is given by

$$\sigma_H = \frac{e}{\Phi_0}\frac{n}{m}(\text{length})^{-1} = \frac{e^2}{hc}\frac{n}{m}(\text{length})^{-1}. \tag{18.155}$$

The Hall conductivity plays an important role in the *quantum Hall effect*[9] (this example does not claim to explain this, instead is only an exercise in considerations around it).

**Solution:** We had the equations (cf. (4.112) and (6.99))

$$\mathbf{D} = \epsilon_0\mathbf{E} + \mathbf{P}, \quad \mathbf{H} = \frac{\mathbf{B}}{\mu_0} - \mathbf{M}, \tag{18.156}$$

and recalling remarks after Eq. (18.42),

$$\therefore \mathbf{D} = \frac{1}{\mu_0}\frac{\partial\mathcal{L}}{\partial\mathbf{E}} \quad \text{implies} \quad \mathbf{P} = -\frac{1}{\mu_0}\frac{\partial U}{\partial\mathbf{E}}, \tag{18.157}$$

and correspondingly

$$\mathbf{H} = -\frac{1}{\mu_0}\frac{\partial\mathcal{L}}{\partial\mathbf{B}} \quad \text{implies} \quad \mathbf{M} = -\frac{1}{\mu_0}\frac{\partial U}{\partial\mathbf{B}}. \tag{18.158}$$

For the given expression of $U$ in Eq. (18.153) these relations imply

$$\mathbf{P} = \frac{g}{\mu_0}\mathbf{B}, \quad \mathbf{M} = \frac{g}{\mu_0}\mathbf{E}. \tag{18.159}$$

Since we had for induced charge and current densities (cf. Eqs. (4.113) and (6.98))

$$\rho_P = -\boldsymbol{\nabla}\cdot\mathbf{P}, \quad \mathbf{j}_M = \boldsymbol{\nabla}\times\mathbf{M}, \tag{18.160}$$

we have $\rho_P = 0$ (since $\boldsymbol{\nabla}\cdot\mathbf{B} = 0$), but (cf. (4.113)) the surface charge density and the magnetic current density resulting from polarization are

$$\sigma_P = \mathbf{P}\cdot\mathbf{n} = \frac{g}{\mu_0}\mathbf{B}\cdot\mathbf{n}, \quad \mathbf{j}_M = \boldsymbol{\nabla}\times\mathbf{M} = \frac{g}{\mu_0}\boldsymbol{\nabla}\times\mathbf{E}, \tag{18.161}$$

where $\mathbf{n}$ is a unit vector pointing out of the surface or interface area $\mathbf{F}$ with, say, $F = L^2$ and circumference $L_0$. Then, using the theorem of Stokes,

$$j_M F = \frac{g}{\mu_0}\int_F \boldsymbol{\nabla}\times\mathbf{E}\cdot d\mathbf{F} = \frac{g}{\mu_0}\int_0^L \mathbf{E}\cdot d\mathbf{l} = \frac{g}{\mu_0}EL_0. \tag{18.162}$$

[9]See von Klitzing, Dorda and Pepper [157].

Hence, cf. Eq. (18.154),

$$j_M = \frac{g}{\mu_0}\frac{EL_0}{F} \equiv \sigma_H E, \quad \therefore \sigma_H = \frac{g}{\mu_0}\frac{L_0}{F}. \tag{18.163}$$

The quantity $\sigma_H$ is the interface *Hall conductivity*. Considering the surface charge density $\sigma_P$, we have

$$\text{surface charge} \;\equiv Q = \int \sigma_P dF = \frac{g}{\mu_0}\int \mathbf{B}\cdot d\mathbf{F} = \frac{g}{\mu_0}\Phi, \tag{18.164}$$

where $\Phi$ is the magnetic flux through the surface. Setting $Q = ne$ and $\Phi = m\Phi_0$, we have

$$g = \frac{\mu_0 e}{\Phi_0}\frac{n}{m} \quad \text{and so} \quad \sigma_H = \frac{e}{\Phi_0}\frac{n}{m}\frac{L_0}{F}. \tag{18.165}$$

In strong magnetic fields $\mathbf{B}$ the Hall conductivity is observed to be quantized.

## 18.11 Exercises

**Example 18.7 (The Lagrangian of Schiff's non-gauge invariant theory).** A Lagrangian containing only the field $A_\mu$ in the theory of Schiff [130] cannot be obtained for the derivation of Eq. (18.135b), i.e. $\partial_\mu W^{\mu\nu} = \mu_0 j^\nu$, in Example 18.5. In order to achieve this one has to introduce an auxiliary vector field $V_\mu$ as Schiff [130] has shown with

$$G_{\mu\nu} := \partial_\mu V_\nu - \partial_\nu V_\mu. \tag{18.166}$$

The Lagrangian density $\mathcal{L}$ is then defined by

$$\begin{aligned}&\mathcal{L}(A_\mu, V_\mu, \partial_\mu A_\nu, \partial_\mu V_\nu)\\ &:= -\frac{1}{2}\left[\frac{1}{2}G_{\mu\nu}F^{\mu\nu} - \frac{\sqrt{2}}{g}V_\nu\partial_\mu(A^\mu A^\nu) + \mu_0 V_\nu j^\nu + \mu_0 A_\nu {}^*j^\nu\right],\end{aligned} \tag{18.167}$$

where $j^\nu$ and ${}^*j^\nu$ are currents. Show that the Euler–Lagrange equations are

$$\partial_\mu\left[F^{\mu\nu} + \frac{\sqrt{2}}{g}A^\mu A^\nu\right] = \mu_0 j^\nu, \tag{18.168a}$$

$$\partial_\mu G^{\mu\nu} - \frac{\sqrt{2}}{g}A^\mu(\partial_\mu V^\nu + \partial^\nu V_\mu) = \mu_0 {}^*j^\nu. \tag{18.168b}$$

For further aspects of the theory we refer to the papers of Schiff [130].

**Example 18.8 (Relativistic equation of motion).** Obtain the equation of motion of a particle of rest mass $m_0$ and charge $e$ in an electromagnetic field with potentials $\phi$ and $\mathbf{A}$ from the Lagrangian

$$L = -m_0c^2\sqrt{1-\mathbf{v}^2/c^2} - e\phi + e\mathbf{v}\cdot\mathbf{A}. \tag{18.169}$$

Compare with Eqs. (17.171) and (7.121).

**Example 18.9 (Lorentz invariants yes or no?).** Is $2(\mathbf{B}^2 - \mathbf{E}^2/c^2)$ invariant with respect to Lorentz transformations? And how about $\mathbf{E}\cdot\mathbf{B}$?

(Answer to both: Yes).

Chapter 19

# The Gauge Covariant Schrödinger Equation and the Aharonov–Bohm Effect

## 19.1 Introductory Remarks

In this chapter we first establish the *Schrödinger equation for a charged particle* moving in an unspecified electromagnetic field. We demonstrate the covariance of this equation under gauge transformations which ensures that the probability density is independent of the choice of a particular vector potential. We then consider the special case of a homogeneous solenoidal magnetic field restricted to a small domain behind a diaphragm with a double slit and with an electron source on the other side. It is then explained why the resulting interference pattern observed on a screen (on the solenoid side of the slits and some distance away from it) is displaced when the current in the solenoid is switched on or off. This *Bohm–Aharonov effect* or *Aharonov–Bohm effect* discovered by Bohm and Aharonov [13] around 1959 is a quantum mechanical effect resulting from the phase of the Schrödinger wave function, and may be looked at as experimental evidence of the vector potential. *The considerations here supplement our considerations of solenoidal fields* in Chapters 5 and 8, where, in particular, we investigated the electromagnetic vector potential **A** in the neighborhood of a long, thin solenoid (in practice provided, for instance, by certain magnetized iron crystals). The Aharonov–Bohm effect is clearly a fundamentally significant phenomenon. The effect has been further explored in numerous gauge field theoretical contexts and has been verified experimentally. The most remarkable aspects of the effect are summarized at the end of this chapter.

## 19.2 Schrödinger Equation of a Charged Particle in an Electromagnetic Field

### 19.2.1 *Hamiltonian of a charge in an electromagnetic field*

We return once again to the nonrelativistic equation of motion of a particle with mass $m$ and charge $e$ in an electromagnetic field — this time, however, with a view to the formulation of the corresponding Schrödinger equation. Previously we wrote the former equation as a Newton equation with the Lorentz force as external force, i.e. Eq. (7.121),

$$\frac{d}{dt}(m\dot{\mathbf{r}}) = e(\mathbf{E} + \dot{\mathbf{r}} \times \mathbf{B}), \quad |\dot{\mathbf{r}}| \equiv |\mathbf{v}| \ll c. \tag{19.1}$$

This equation can also be derived as in ordinary classical mechanics from the following Lagrangian (with $\mathbf{r} \equiv (q_1, q_2, q_3)$):

$$L(q, \dot{q}) = \frac{1}{2}m\dot{q}_i^2 + e(\dot{q}_i A_i(q,t) - \phi(q)), \tag{19.2}$$

as we verify now. With the Lagrangian we can then (via a Legendre transform) construct the Hamiltonian as a prerequisite for the transition to the quantum mechanical Schrödinger equation. From $L$ we obtain

$$\frac{\partial L}{\partial \dot{q}_i} = m\dot{q}_i + eA_i, \quad \frac{\partial L}{\partial q_j} = e\dot{q}_i\frac{\partial A_i}{\partial q_j} - e\frac{\partial \phi}{\partial q_j}, \tag{19.3}$$

and the Euler–Lagrange equation of motion

$$\frac{d}{dt}\left(\frac{\partial L}{\partial \dot{q}_j}\right) - \frac{\partial L}{\partial q_j} = 0 \tag{19.4}$$

yields

$$m\ddot{q}_j + \frac{d}{dt}(eA_j) = e\dot{q}_i\frac{\partial A_i}{\partial q_j} - e\frac{\partial \phi}{\partial q_j}, \tag{19.5}$$

where $d/dt$ is the total time derivative, i.e.

$$\frac{d}{dt} = \frac{\partial}{\partial t} + \dot{q}_i\frac{\partial}{\partial q_i}, \tag{19.6}$$

so that

$$m\ddot{q}_j + e\frac{\partial A_j}{\partial t} + e\frac{\partial A_j}{\partial q_i}\dot{q}_i = e\dot{q}_i\frac{\partial A_i}{\partial q_j} - e\frac{\partial \phi}{\partial q_j}, \tag{19.7}$$

and hence

$$m\ddot{q}_j = e\left(-\frac{\partial A_j}{\partial t} - \frac{\partial \phi}{\partial q_j}\right) + e\left(\frac{\partial A_i}{\partial q_j}\dot{q}_i - \frac{\partial A_j}{\partial q_i}\dot{q}_i\right). \tag{19.8}$$

However, we know that

$$\mathbf{E} = -\frac{\partial \mathbf{A}}{\partial t} - \boldsymbol{\nabla}\phi \quad \text{newtons/coulomb} \quad \text{and} \quad \mathbf{B} = \boldsymbol{\nabla} \times \mathbf{A} \quad \text{teslas,} \tag{19.9}$$

and thus (note that in the present context we are only concerned with 3-dimensional quantities and therefore do not have to distinguish between upper and lower indices)

$$\begin{aligned} \mathbf{v} \times \mathbf{B} = \mathbf{v} \times (\boldsymbol{\nabla} \times \mathbf{A}) &= \boldsymbol{\nabla}(\mathbf{v} \cdot \mathbf{A}) - (\mathbf{v} \cdot \boldsymbol{\nabla})\mathbf{A} \\ &= \frac{\partial}{\partial \mathbf{r}}(\mathbf{v} \cdot \mathbf{A}) - \left(\mathbf{v} \cdot \frac{\partial}{\partial \mathbf{r}}\right)\mathbf{A}, \end{aligned} \tag{19.10}$$

i.e.

$$(\mathbf{v} \times \mathbf{B})_j = \frac{\partial}{\partial x_j}(v_i A_i) - v_i \frac{\partial}{\partial x_i} A_j, \tag{19.11}$$

where

$$\frac{\partial}{\partial x_j}(v_i A_i) = v_i \frac{\partial A_i}{\partial x_j} + A_i \frac{\partial v_i}{\partial x_j} = v_i \frac{\partial A_i}{\partial x_j} \tag{19.12}$$

and hence

$$m\ddot{q}_j = eE_j + e(\mathbf{v} \times \mathbf{B})_j \quad \text{newtons,} \tag{19.13}$$

as had to be shown, i.e. that the Euler–Lagrange equation is identical with Newton's equation.

One should note: The *conjugate momentum* is $p_i = \partial L/\partial \dot{q}_i$; this is also the *canonical momentum* in the Hamilton formalism. We have (cf. Eq. (19.3))

$$\mathbf{p} = m\mathbf{v} + e\mathbf{A}. \tag{19.14}$$

However, the so-called *mechanical momentum* is $m\mathbf{v}$ or

$$m\mathbf{v} = \mathbf{p} - e\mathbf{A} \equiv m\dot{\mathbf{q}}. \tag{19.15}$$

This distinction is very important.

We obtain the *Hamiltonian* $H(q,p)$ as in mechanics with a Legendre transform, i.e. by defining

$$\begin{aligned} H(q,p) &:= p_i \dot{q}_i - L(q,\dot{q}) = p_i \dot{q}_i - \frac{1}{2} m \dot{q}_i^2 - e(\dot{q}_i A_i - \phi) \\ &= \frac{p_i}{m}(p_i - eA_i) - \frac{1}{2}\frac{m}{m^2}(p_i - eA_i)^2 + e\phi - \frac{eA_i}{m}(p_i - eA_i), \end{aligned} \tag{19.16}$$

or

$$\begin{aligned} H(q,p) &= \frac{1}{2m}(p_i - eA_i)^2 + e\phi = \frac{1}{2m}(\mathbf{p} - e\mathbf{A})^2 + e\phi \\ &= \left[\frac{1}{2m}(m\mathbf{v})^2 + e\phi\right]_{m\mathbf{v}\to\mathbf{p}-e\mathbf{A}} . \end{aligned} \tag{19.17}$$

The contribution $-e\mathbf{A}$ is seen to be the effect of the Lorentz force expressed in terms of the vector potential. The substitution

$$\mathbf{p}_{\text{mech}} = m\mathbf{v} \to \mathbf{p} - e\mathbf{A} \tag{19.18}$$

for the ordinary momentum in mechanics is frequently described as *minimal electromagnetic coupling.*

### 19.2.2 *The gauge covariant Schrödinger equation*

We now construct the time-independent Schrödinger equation and explain, what is meant by "*gauge covariance*". We saw above: For a particle with charge $e$ the relation between mechanical momentum and conjugate momentum is changed by the effect of the Lorentz force, i.e. the momentum $\mathbf{p}$ becomes $\mathbf{p} - e\mathbf{A}$. Correspondingly we have in nonrelativistic quantum mechanics the substitution

$$\mathbf{p}_{\text{mech}} \to -i\hbar\boldsymbol{\nabla} - e\mathbf{A} = -i\hbar\left(\boldsymbol{\nabla} - \frac{ie}{\hbar}\mathbf{A}\right) \tag{19.19}$$

and $(\boldsymbol{\nabla} - (ie/\hbar)\mathbf{A})$ is called *gauge covariant derivative.*

The *canonical variables* required for quantization are the spatial Cartesian coordinates $x_i$ and the appropriate *canonical momenta* $p_i$. In the present case *quantization* requires the replacement of $\mathbf{p}$ by the operators $\boldsymbol{\nabla}$, and to consider all quantities as operators in the space of states $\psi$. The corresponding Schrödinger equation in position space and with eigenvalue $E$ is then (cf. Hamiltonian (19.17))

$$-\frac{\hbar^2}{2m}\left(\boldsymbol{\nabla} - \frac{ie}{\hbar}\mathbf{A}\right)^2\psi = (E - e\phi)\psi. \tag{19.20}$$

One expects a sensible formulation of the theory to be such that Schrödinger equations with different potentials $\mathbf{A}$ are "equivalent" in the sense that they yield the same observable results, which depend only on the probability density $|\psi|^2$. To achieve this, the wave function $\psi$ has to be transformed along with the potential in the gauge transformation from one potential $\mathbf{A}$

to a different one. Thus, with the gauge transformation (8.38) of the vector potential,

$$\mathbf{A} \to \mathbf{A}' = \mathbf{A} + \boldsymbol{\nabla}\chi \tag{19.21}$$

(where our time-independent considerations leave $\phi$ unchanged, see the 4-dimensional transformation (10.11)), we also transform the phase of the wave function by writing

$$\psi \to \psi' = \exp\left[\frac{ie\chi}{\hbar}\right]\psi. \tag{19.22}$$

This equivalence (to be verified below) is described as *gauge covariance.* We now verify that these transformations convert a Schrödinger equation in $\psi'$ and $\mathbf{A}'$ into one in $\psi$ and $\mathbf{A}$ or vice versa. We start from the Schrödinger equation in primed quantities:

$$-\frac{\hbar^2}{2m}\left(\boldsymbol{\nabla} - \frac{ie}{\hbar}\mathbf{A}'\right)^2\psi' = (E - e\phi)\psi'. \tag{19.23}$$

Now (observe in the second line on the right the first $\boldsymbol{\nabla}\chi$ originating from the wave function $\psi'$ implying that the gauge transformation also changes the momentum $\mathbf{p}$ to $\mathbf{p}' = \mathbf{p} + e\boldsymbol{\nabla}\chi$),

$$\begin{aligned}\left(\boldsymbol{\nabla} - \frac{ie}{\hbar}\mathbf{A}'\right)\psi' &= \left(\boldsymbol{\nabla} - \frac{ie}{\hbar}(\mathbf{A} + \boldsymbol{\nabla}\chi)\right)e^{ie\chi/\hbar}\psi \\ &= e^{ie\chi/\hbar}\left(\boldsymbol{\nabla} + \frac{ie}{\hbar}\boldsymbol{\nabla}\chi - \frac{ie}{\hbar}(\mathbf{A} + \boldsymbol{\nabla}\chi)\right)\psi \\ &= e^{ie\chi/\hbar}\left(\boldsymbol{\nabla} - \frac{ie}{\hbar}\mathbf{A}\right)\psi,\end{aligned} \tag{19.24}$$

i.e.

$$\left(\boldsymbol{\nabla} - \frac{ie}{\hbar}\mathbf{A}'\right)^2\psi' = e^{ie\chi/\hbar}\left(\boldsymbol{\nabla} - \frac{ie}{\hbar}\mathbf{A}\right)^2\psi. \tag{19.25}$$

This implies the equation in $\mathbf{A}, \psi$:

$$-\frac{\hbar^2}{2m}\left(\boldsymbol{\nabla} - \frac{ie}{\hbar}\mathbf{A}\right)^2\psi = (E - e\phi)\psi. \tag{19.26}$$

Since the only physically observable quantity in $\psi$ is the probability density $|\psi|^2$, and this remains unchanged under the gauge transformation, we see that all gauge potentials, which are related by gauge transformations, describe the same physical state. (A wave function is a probability amplitude; the phases of wave functions have a *relative* significance, not an absolute significance, as quantum mechanical interference experiments show). In quantum mechanics the wave function of a free particle of mass $m$, energy $E = p^2/2m$ and momentum $p$ is given by

$$\psi(x,t) = \exp\left[\frac{i}{\hbar}(px - Et)\right]. \tag{19.27}$$

Recall briefly how quantum mechanics enters here. *Canonical quantization* implies, that the Cartesian quantities $x_i$ and $p_i$ (as operators in the space of states) obey the following commutator algebra:

$$[x_i, x_j] = 0, \quad [p_j, x_k] = -i\hbar\delta_{jk}, \quad [p_j, p_k] = 0. \tag{19.28}$$

These operator relations are satisfied by the position space representation $p_j = -i\hbar\partial/\partial x_j$ of the momentum operator. Now assume $E = p^2/2m + V(x)$. Then, classically in the non-free case (i.e. potential $V(x) \neq 0$) $p^2/2m = E - V(x)$ and quantum mechanically

$$-\frac{\hbar^2\boldsymbol{\nabla}^2}{2m}\psi = (E - V(x))\psi, \tag{19.29}$$

where $\psi(x)$ is a wave function. In general one can separate the angles and one obtains a one-dimensional differential equation analogous to the Schrödinger equation in one dimension (with centrifugal term). An approximation $\psi_0$ (for $V \neq 0$, exact for $V = 0$) of $\psi$ is obtained from the classical $p = \sqrt{2m(E - V(x))}$ as solution of

$$-i\hbar\frac{d}{dx}\psi_0 = \sqrt{2m(E - V(x))}\psi_0, \tag{19.30}$$

i.e.

$$\int\frac{d\psi_0}{\psi_0} = \int\frac{i}{\hbar}\sqrt{2m(E - V(x))}dx \tag{19.31}$$

or

$$\psi_0 = \exp\left[\frac{i}{\hbar}\int\sqrt{2m(E - V(x))}dx\right] = \exp\left[\frac{i}{\hbar}\int p(x)dx\right]. \tag{19.32}$$

Considering now the case of a particle travelling in the Maxwell field $\mathbf{A}$, then according to the above prescription of minimal coupling, we have to replace the free particle momentum $\mathbf{p}$ by $\mathbf{p} - e\mathbf{A}$, so that the wave function has the form

$$\psi = \exp\left[\frac{i}{\hbar}\int \mathbf{p}\cdot d\mathbf{r} - \frac{ie}{\hbar}\int \mathbf{A}\cdot d\mathbf{r}\right]. \tag{19.33}$$

The contribution

$$-\frac{ie}{\hbar}\int \mathbf{A}\cdot d\mathbf{r} \tag{19.34}$$

is the effect of the Lorentz force in the quantum mechanical case. We can consider it as the corresponding quantum mechanical law. It plays an important role in the interpretation of the Aharonov–Bohm effect. One should note that the expression $\oint \mathbf{A}\cdot d\mathbf{r}$ and hence the wave function $\psi$ (for integration around a closed path) is invariant under nonsingular gauge transformations, since with the gauge transformation (19.21):

$$\oint \mathbf{A}'\cdot d\mathbf{r} = \oint \mathbf{A}\cdot d\mathbf{r} + \oint \boldsymbol{\nabla}\chi\cdot d\mathbf{r} = \oint \mathbf{A}\cdot d\mathbf{r}, \tag{19.35}$$

because

$$\oint \boldsymbol{\nabla}\chi\cdot d\mathbf{r} = \oint d\chi = 0. \tag{19.36}$$

## 19.3 The Aharonov–Bohm Effect

We restrict ourselves here largely to a qualitative treatment.[1] Electrons with momentum $p = h/\lambda$ are sent through a diaphragm or partition with two slits as indicated in Fig. 19.1. If the difference $a$ in the lengths of the two possible

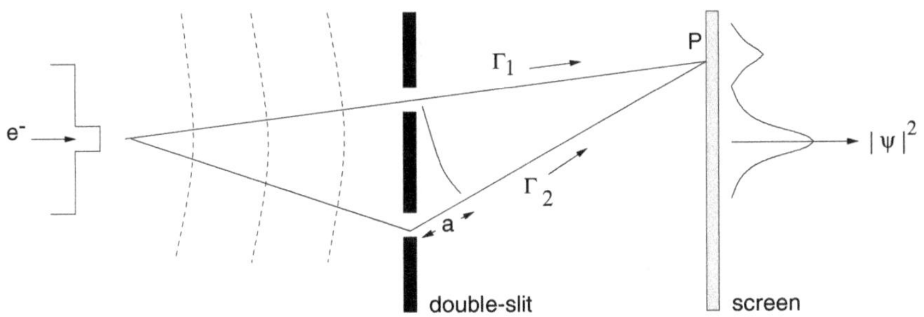


**Fig. 19.1.** The two paths $\Gamma_1, \Gamma_2$ of the electron wave and their interference spectrum.

[1]For detailed literature see e.g. Peshkin and Tonomura [114].

paths $\Gamma_1, \Gamma_2$, is of the order of the wavelength $\lambda$ of the waves one expects and observes an interference pattern on the screen behind the slits (due to constructive or destructive interference). We write the difference between the phases of the wave function of the electrons due to the different paths

$$\triangle \left( \frac{1}{\hbar} \int \mathbf{p} \cdot d\mathbf{r} \right)_{p=\hbar k} = ka \equiv \delta. \tag{19.37}$$

At a maximum $P$ as indicated in Fig. 19.1 we have the effect of the superposition of two waves which have a phase difference $ka$. If we arrange the experiment such that electrons can only pass through one slit, then $\delta = 0$, because $a = 0$ (as well as $\psi = 0$). In this case we observe no interference and the conditions of quantum mechanics are not provided.

The present problem with (as we shall see) a homogeneous magnetic field vertically out of the plane of the diagram of Fig. 19.1 in a small domain behind the slits is seen to have *cylindrical symmetry.* This suggests one to use cylindrical coordinates $r, \theta, z$. In these coordinates,[2] with field components

$$\mathbf{A} = \mathbf{e}_\theta A, \quad \mathbf{B} = \boldsymbol{\nabla} \times \mathbf{A} \parallel \mathbf{e}_z, \tag{19.38}$$

the gradient is given by

$$\boldsymbol{\nabla} \Leftrightarrow \mathbf{e}_r \left\{ \frac{\partial}{\partial r} \right\} + \mathbf{e}_\theta \left\{ \frac{1}{r} \frac{\partial}{\partial \theta} \right\} + \mathbf{e}_z \left\{ \frac{\partial}{\partial z} \right\}, \tag{19.39}$$

and the divergence by

$$\boldsymbol{\nabla} \cdot \mathbf{C} = \frac{1}{r} \frac{\partial}{\partial r}(rC_r) + \frac{1}{r} \frac{\partial C_\theta}{\partial \theta} + \frac{\partial C_z}{\partial z}. \tag{19.40}$$

Hence

$$\boldsymbol{\nabla} \cdot \boldsymbol{\nabla} = \frac{1}{r} \frac{\partial}{\partial r} \left\{ r \frac{\partial}{\partial r} \right\} + \frac{1}{r} \frac{\partial}{\partial \theta} \left\{ \frac{1}{r} \frac{\partial}{\partial \theta} \right\} + \frac{\partial}{\partial z} \left\{ \frac{\partial}{\partial z} \right\}. \tag{19.41}$$

For the chosen vector potential $\mathbf{A}$ in the direction of $\mathbf{e}_\theta$ (in the outside region around the solenoid as we saw in Chapter 8) we have

$$\begin{aligned} \left( \boldsymbol{\nabla} - \frac{ie}{\hbar} \mathbf{A} \right)^2 &= \left( \boldsymbol{\nabla} - \frac{ie}{\hbar} A \mathbf{e}_\theta \right) \cdot \left[ \frac{\partial}{\partial r}, \frac{1}{r} \frac{\partial}{\partial \theta} - \frac{ie}{\hbar} A, \frac{\partial}{\partial z} \right] \\ &= \frac{1}{r} \frac{\partial}{\partial r} r \frac{\partial}{\partial r} + \frac{1}{r} \frac{\partial}{\partial \theta} \left( \frac{1}{r} \frac{\partial}{\partial \theta} - \frac{ie}{\hbar} A \right) \end{aligned}$$

[2]See Eqs. (2.21a) and (2.21b), or e.g. Magnus and Oberhettinger [90], p. 145.

$$
\begin{aligned}
&+\frac{\partial}{\partial z}\left(\frac{\partial}{\partial z}\right)-\frac{ie}{\hbar}A\left(\frac{1}{r}\frac{\partial}{\partial\theta}-\frac{ie}{\hbar}A\right)\\
&=\frac{\partial^2}{\partial r^2}+\frac{1}{r}\frac{\partial}{\partial r}+\left(\frac{1}{r}\frac{\partial}{\partial\theta}-\frac{ie}{\hbar}A\right)^2+\frac{\partial^2}{\partial z^2}\\
&=\frac{\partial^2}{\partial r^2}+\frac{1}{r}\frac{\partial}{\partial r}+\frac{1}{r^2}\left(\frac{\partial}{\partial\theta}-\frac{ie2\pi r}{h}A\right)^2+\frac{\partial^2}{\partial z^2},
\end{aligned}
\tag{19.42}
$$

where we switched from $\hbar$ to $h = 2\pi\hbar$. But now

$$
\oint_{C(F)}\mathbf{A}\cdot d\mathbf{l} = A2\pi r = \int(\boldsymbol{\nabla}\times\mathbf{A})\cdot d\mathbf{F} = \int_{F(C)}\mathbf{B}\cdot d\mathbf{F} \equiv \Phi,
\tag{19.43}
$$

where $\Phi$ is the magnetic flux through the surface $F$. Hence with

$$
\alpha := -\frac{e2\pi rA}{h} = -\frac{e\Phi}{h} = -\frac{e\Phi}{2\pi\hbar}
\tag{19.44}
$$

the expression (19.42) becomes

$$
\left(\boldsymbol{\nabla}-\frac{ie}{\hbar}\mathbf{A}\right)^2 = \frac{\partial^2}{\partial r^2}+\frac{1}{r}\frac{\partial}{\partial r}+\frac{1}{r^2}\left(\frac{\partial}{\partial\theta}+i\alpha\right)^2+\frac{\partial^2}{\partial z^2}.
\tag{19.45}
$$

The corresponding Schrödinger equation for scattering in two dimensions is

$$
\left(\boldsymbol{\nabla}-\frac{ie}{\hbar}\mathbf{A}\right)^2\psi = -\frac{2mE}{\hbar^2}\psi \equiv -k^2\psi,
\tag{19.46}
$$

or, with $\psi(r,\theta,z)\to\psi(r,\theta)$ and $\theta$ as in Fig. 19.2,

$$
\left[\frac{\partial^2}{\partial r^2}+\frac{1}{r}\frac{\partial}{\partial r}+\frac{1}{r^2}\left(\frac{\partial}{\partial\theta}+i\alpha\right)^2+k^2\right]\psi(r,\theta) = 0.
\tag{19.47}
$$

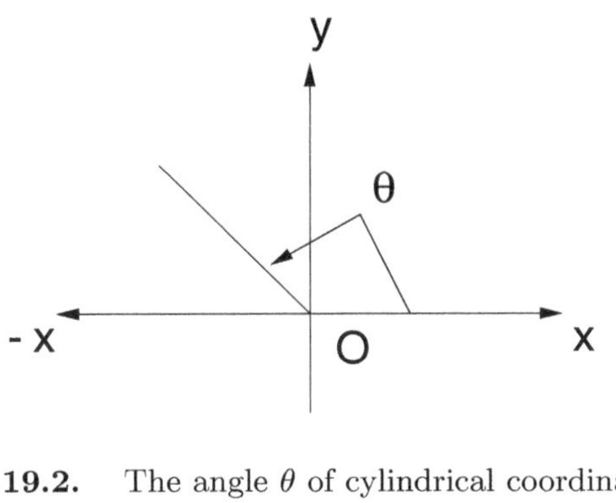


**Fig. 19.2.** The angle $\theta$ of cylindrical coordinates.

Setting $\psi(r,\theta) \propto e^{im\theta} = e^{im(\theta+2\pi)}$, $m = 0, \pm 1, \pm 2, \ldots$, we obtain

$$\left[\frac{\partial^2}{\partial r^2} + \frac{1}{r}\frac{\partial}{\partial r} - \frac{(\alpha+m)^2}{r^2} + k^2\right]\psi(r,\theta) = 0. \tag{19.48}$$

Comparison with the Bessel equation

$$\frac{d^2 Z_\nu}{dz^2} + \frac{1}{z}\frac{dZ_\nu}{dz} + \left(1 - \frac{\nu^2}{z^2}\right) Z_\nu = 0 \tag{19.49}$$

shows that the general solution of the Schrödinger equation has the form

$$\psi = \sum_{m=-\infty}^{\infty} e^{im\theta}\left[a_m J_{|m+\alpha|}(kr) + b_m J_{-|m+\alpha|}(kr)\right], \tag{19.50}$$

where $a_m, b_m$ are arbitrary constants and $J_\nu, J_{-\nu}$ are Bessel functions. Since the wave function has to be regular at $r = 0$, all coefficents $b_m$ must be zero, i.e.

$$\psi(r,\theta) = \sum_{m=-\infty}^{\infty} a_m e^{im\theta} J_{|m+\alpha|}(kr). \tag{19.51}$$

This is the small-$r$ expansion. The Bessel functions can also be expressed as the sum of two Hankel functions valid at large $r$, i.e.

$$J_\nu(z) = \frac{1}{2}(H_\nu^{(1)}(z) + H_\nu^{(2)}(z)), \tag{19.52}$$

where for $|z| \to \infty$ (see books on Special Functions, like Magnus and Oberhettinger [90])

$$H_\nu^{(1)}(z) = \sqrt{\frac{2}{\pi z}} e^{i(z-\nu\pi/2-\pi/4)}(1 + O(z^{-1})), \tag{19.53}$$

$$H_\nu^{(2)}(z) = \sqrt{\frac{2}{\pi z}} e^{-i(z-\nu\pi/2-\pi/4)}(1 + O(z^{-1})). \tag{19.54}$$

Hence for $r \to \infty$ the wave function (19.51) can be written

$$\begin{aligned}\psi(r,\theta) = \frac{1}{2}\sum_{m=-\infty}^{\infty} a_m e^{im\theta}\sqrt{\frac{2}{\pi kr}} \\ \times \left\{e^{i(kr-|m+\alpha|\pi/2-\pi/4)} + e^{-i(kr-|m+\alpha|\pi/2-\pi/4)}\right\}(1 + O(1/r)).\end{aligned} \tag{19.55}$$

The wave function $\psi(r,\theta)$ can now be subdivided for large values of $r$ into an incident or ingoing wave (let's say from the right) and a scattered or outgoing wave whose asymptotic behavior defines the scattering amplitude $f(\theta)$, i.e.

$$\psi(r,\theta) = \psi_{\rm in} + \psi_{\rm out} = \underbrace{\langle e^{-ikr\cos\theta}\rangle}_{e^{-ikx}} + \psi_{\rm out} \tag{19.56}$$

(the braces $\langle\ldots\rangle$ are meant to indicate that the enclosed expression is not the strictly correct free wave, as is explained below) with (for $r \to \infty$ we deduce from the differential equation or the above Hankel functions that $\psi \sim e^{ikr}/\sqrt{r}$)

$$\psi_{\rm out} = \frac{1}{r^{1/2}} e^{ikr} f(\theta)(1 + O(1/r)). \tag{19.57}$$

We expand the ingoing wave $\psi_{\rm in}$ in terms of Bessel functions.

We take this expansion from the literature,[3] i.e.

$$\begin{aligned} e^{-ikr\cos\theta} &= \sum_{n=-\infty}^{\infty} (-i)^n J_n(kr) e^{in\theta} \\ &\simeq \sum_{n=-\infty}^{\infty} (-i)^n e^{in\theta} \left(\frac{2}{\pi kr}\right)^{\frac{1}{2}} \cos\left(kr - n\frac{\pi}{2} - \frac{\pi}{4}\right). \end{aligned} \tag{19.58}$$

We substitute Eqs. (19.58) and (19.57) into Eq. (19.56) and obtain

$$\begin{aligned} \psi(r,\theta) \simeq & \sum_{n=-\infty}^{\infty} \frac{1}{2}(-i)^n e^{in\theta} \left(\frac{2}{\pi kr}\right)^{1/2} \\ & \times \left\{ e^{i(kr - n\pi/2 - \pi/4)} + e^{-i(kr - n\pi/2 - \pi/4)} \right\} + \frac{e^{ikr}}{r^{1/2}} f(\theta). \end{aligned} \tag{19.59}$$

Comparison of the coefficients of the second exponentials with the factor $\exp(-ikr)$ in this expression and those in Eq. (19.55), i.e.

$$\begin{aligned} \psi(r,\theta) \simeq & \sum_{m=-\infty}^{\infty} \frac{1}{2} a_m e^{im\theta} \left(\frac{2}{\pi kr}\right)^{1/2} \\ & \times \left\{ e^{i(kr - |m+\alpha|\pi/2 - \pi/4)} + e^{-i(kr - |m+\alpha|\pi/2 - \pi/4)} \right\}, \end{aligned} \tag{19.60}$$

then yields

$$(-i)^m e^{-i(-m\pi/2)} = a_m e^{i|m+\alpha|\pi/2}, \tag{19.61}$$

---

[3] Gradshteyn and Ryzhik [60], p. 973.

i.e.

$$a_m = e^{-i|m+\alpha|\pi/2}. \tag{19.62}$$

Comparison of the coefficients of $e^{+ikr}$ yields the scattering amplitude $f(\theta)$,

$$\begin{aligned} f(\theta) = &\sum_{m=-\infty}^{\infty} \frac{1}{2} a_m e^{im\theta} \left(\frac{2}{\pi k}\right)^{1/2} e^{-i(|m+\alpha|\pi/2+\pi/4)} \\ &- \sum_{m=-\infty}^{\infty} \frac{1}{2}(-i)^m e^{im\theta} \left(\frac{2}{\pi k}\right)^{1/2} e^{-i(m\pi/2+\pi/4)}, \end{aligned} \tag{19.63}$$

i.e.

$$\begin{aligned} f(\theta) &= \sum_{m=-\infty}^{\infty} \left(\frac{1}{2\pi k}\right)^{1/2} e^{im\theta} \left\{a_m e^{-i(|m+\alpha|\pi/2+\pi/4)} - e^{-i\pi m/2} e^{-i(m\pi/2+\pi/4)}\right\} \\ &= \sum_{m=-\infty}^{\infty} \frac{1}{\sqrt{2\pi i k}} e^{im\theta} e^{-i\pi m} \left\{e^{-i|m+\alpha|\pi+i\pi m} - 1\right\}, \end{aligned} \tag{19.64}$$

and hence

$$f(\theta) = \frac{1}{\sqrt{2\pi i k}} \sum_{m=-\infty}^{\infty} e^{im(\theta-\pi)}(e^{2i\delta_m} - 1) \tag{19.65}$$

with the *scattering phase*

$$\delta_m = -\frac{\pi}{2}|m+\alpha| + \frac{\pi}{2}|m|. \tag{19.66}$$

We see therefore, that for $\alpha \neq 0$ the scattering phase $\delta_m \neq 0$. Actually the ingoing wave in Eq. (19.56) is in the present case wrong. In the usual case of the scattering off a potential like the Coulomb potential (as in the familiar case of 3-dimensional scattering theory), one assumes that the potential vanishes for $r \to \infty$. Here we consider the scattering of electrons (or the electron wave) in the field of an infinitely long solenoid, i.e. off magnetic flux lines which do not return. Aharonov and Bohm showed in their work,[4] that

[4]For the original papers see Bohm and Aharonov [13].

the ingoing wave therefore carries a phase factor, i.e. that Eq. (19.56) must be written

$$\psi(r,\theta) = e^{-i\alpha\theta(x,y)}e^{-ikr\cos\theta} + \psi_{\text{out}}. \tag{19.67}$$

This change is by no means self-evident. Indeed the calculation of the scattering amplitude $f(\theta)$ here depends on the order in which the sum $\sum_m$ and the limit $\lim_{r\to\infty}$ are taken. Aharonov and Bohm used in their complicated calculation the order $\lim_{r\to\infty}\sum_m$, whereas in the above partial wave expansion these steps are taken in the opposite order. It is found that the summed expressions for $f(\theta)$ differ in forward direction by a contribution proportional to $\delta(\theta-\pi)$. We do not enter into a deeper analysis of these mathematical details here.[5]

We now consider the experimental setup shown schematically in Fig. 19.3 and proposed by Aharonov and Bohm with a solenoid (infinitely long, so that the field outside is zero) immediately behind the double slit and perpendicular to the diagram. In practice, in real experiments, the solenoid is replaced by microscopically thin magnetized iron crystals. One can see that with the current in the solenoid switched on, i.e. the magnetic flux, the phase of the electron wave function changes by an amount given by[6]

$$\frac{\pi}{2}\alpha \overset{(19.44)}{=} \frac{\pi}{2}\left(-\frac{e\Phi}{2\pi\hbar}\right) = -\frac{\pi}{2}\frac{e}{2\pi\hbar}\oint \mathbf{A}\cdot d\mathbf{l}, \tag{19.68}$$

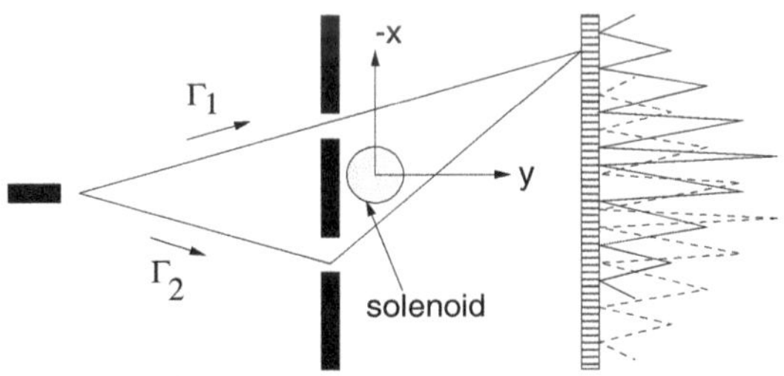


**Fig. 19.3.** Displacement of the interference spectrum when a current flows in the solenoid.

[5]The partial wave treatment can be found in particular in the work of Ruijsenhaars [127]. The discussion on the order of $\sum_m$ and $\lim_{r\to\infty}$ can be found in Hagen [67]. Further discussions are given in the book of Peshkin and Tonomura [114].

[6]In some considerations the wave function with phase factor $\exp\{ie\int_\Gamma \mathbf{A}\cdot d\mathbf{l}/\hbar\}$ is not unique, since several revolutions around the field $\mathbf{B}$ are possible. This problem can be circumvented by summing over arbitrarily many revolutions or by using only uniquely defined wave functions (see e.g. the discussion in the last paragraph of Berry [10]).

which may be rewritten

$$\frac{\pi}{2}\alpha = -\frac{\pi}{2}\frac{e}{2\pi\hbar}\left[\int_{\Gamma_1}\mathbf{A}\cdot d\mathbf{l} - \int_{\Gamma_2}\mathbf{A}\cdot d\mathbf{l}\right]. \tag{19.69}$$

Here $\Gamma_1$ and $\Gamma_2$ are the two paths from the electron source through the two slits to the interference point on the screen. The interference spectrum suffers a corresponding shift as indicated in Fig. 19.3. This effect is known as *Aharonov–Bohm effect.*[7] The effect has been confirmed experimentally[8] and plays an important role in numerous theoretical considerations.

What makes the Aharonov–Bohm effect so remarkable is:

1. that it is a purely quantum mechanical (wave function) effect,
2. that the electrons travel only in such domains in which **E** and **B** are zero, and
3. (as we saw in Chapter 8) that also the vector potential **A** can be almost everywhere zero, i.e. where the singular gauge transformation is defined. However, there is no need to perform a singular gauge transformation. Nonetheless the singularity is crucial for the phenomenon to occur. As we saw in Section 8.3 as a result of the vector potential there is a singularity inside the closed path $\Gamma = \Gamma_1 - \Gamma_2$ and an associated discontinuity in the coordinate space. The latter implies that the space of the path is not simply connected. The phase that the electron wave function receives as a consequence of the nonzero magnetic flux is also an example of the (more general) Berry phase.[9] Thus, in quantum mechanics the vector potential has physical significance and becomes effectively observable in the shift of the interference spectrum, i.e. in an indirect way.

**Example 19.1 (Eigenvalues of momentum).** For the case that $\alpha$ is proportional to the magnetic flux in the direction of $z$ as in the above, and for propagation of electrons in the direction of $x$, so that the wave function $\psi$ of the electrons of mass $m$ is

$$\psi = e^{-i\alpha\theta(x,y)}e^{-ikx}, \quad \cot\theta = \frac{x}{y}, \tag{19.70}$$

[7]The original papers have been cited above. A readable account of the Aharonov–Bohm effect can be found in Felsager [51], pp. 49–55.
[8]An early experimental verification can be found in Möllenstaedt and Bayh [98]. A more recent confirmation has been reported in Tonomura *et al.* [149].
[9]Berry [10].

verify the momentum eigenvalue equations

$$mv_x\psi = -\hbar k\psi, \quad mv_y\psi = mv_z\psi = 0. \tag{19.71}$$

**Solution:** We have the operator relation (19.18),

$$m\mathbf{v} = -i\hbar\boldsymbol{\nabla} \pm e\mathbf{A}, \tag{19.72}$$

and magnetic flux through a circular area of radius $\rho$,

$$\Phi = \int_F \mathbf{B}\cdot d\mathbf{F} = \int_F \boldsymbol{\nabla}\times\mathbf{A}\cdot d\mathbf{F} = \oint_C \mathbf{A}\cdot d\mathbf{l} = A(\rho)2\pi\rho, \tag{19.73}$$

so that $\alpha$ of Eq. (19.44) is

$$\alpha = \pm\frac{e\Phi}{2\pi\hbar} = \pm\frac{eA(\rho)2\pi\rho}{2\pi\hbar} = \pm\frac{eA(\rho)\rho}{\hbar}. \tag{19.74}$$

Hence

$$mv_x\psi = \left(-i\hbar\frac{\partial}{\partial x} \pm eA_x\right)e^{-i\alpha\theta(x,y)}e^{-ikx} \tag{19.75}$$

with, cf. Fig. 19.4,

$$\cot\theta = \frac{x}{y}, \quad \theta = \cot^{-1}\frac{x}{y},$$

$$\frac{d}{dx}\theta = \frac{d}{dx}\cot^{-1}\frac{x}{y} = -\frac{y}{x^2+y^2} = -\frac{y}{\rho^2} = -\frac{\sin\theta}{\rho}. \tag{19.76}$$

From Eq. (19.74) it follows that

$$mv_x\psi = \left(-i\hbar(-i\alpha)\frac{\partial\theta(x,y)}{\partial x} - i\hbar(-ik) \pm eA_x\right)e^{-i\alpha\theta(x,y)}e^{-ikx}. \tag{19.77}$$

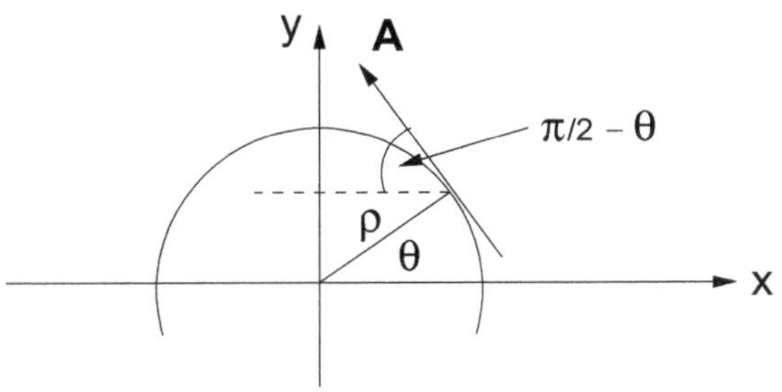


**Fig. 19.4.** The direction of **A** with respect to the angle $\theta$.

This equation, i.e.

$$mv_x\psi = \left(-\alpha\hbar\frac{\partial\theta}{\partial x} - \hbar k \pm eA_x\right) e^{-i\alpha\theta(x,y)}e^{-ikx}, \tag{19.78}$$

may now be rewritten as

$$\begin{aligned} mv_x\psi &= \left(-k\hbar + \alpha\hbar\frac{\sin\theta}{\rho} \pm eA_x\right) e^{-i\alpha\theta}e^{-ikx} \\ &\overset{(19.74)}{=} \left(-k\hbar \pm eA(\rho)\rho\frac{\sin\theta}{\rho} \pm eA_x\right) e^{-i\alpha\theta}e^{-ikx} \\ &= (-k\hbar \pm eA(\rho)\sin\theta \pm eA_x)e^{-i\alpha\theta}e^{-ikx} \\ &= -k\hbar e^{-i\alpha\theta}e^{-ikx} = -k\hbar\psi, \end{aligned} \tag{19.79}$$

where in an intermediate step we replaced $A_x$ by $-A(\rho)\cos(\pi/2-\theta) = -A(\rho)\sin\theta$. In an analogous way one can verify that

$$mv_y = 0, \quad \text{and} \quad mv_z = 0. \tag{19.80}$$

## 19.4 Exercises

**Example 19.2 (Derivation of current from continuity equation).** Use the time-dependent Schrödinger equation (i.e. that with $E$ in Eq. (19.26) replaced by $i\hbar\partial/\partial t$) and the equation of continuity (5.12), i.e.

$$\frac{\partial\rho}{\partial t} + \boldsymbol{\nabla}\cdot\mathbf{j} = 0, \tag{19.81}$$

where now $\rho = \psi^*\psi$, and show that the current density $\mathbf{j}$ is given by

$$\begin{aligned} \mathbf{j} &= -\frac{i\hbar}{2m}\left\{\psi^*(\boldsymbol{\nabla}\psi) - \psi(\boldsymbol{\nabla}\psi^*) - \frac{2ie}{\hbar}\psi^*\psi\mathbf{A}\right\} \\ &= -\frac{i\hbar}{2m}\{\psi^*(\mathbf{D}\psi) - \psi(\mathbf{D}\psi)^*\}, \end{aligned} \tag{19.82}$$

where

$$\mathbf{D} = \boldsymbol{\nabla} - \frac{ie}{\hbar}\mathbf{A}. \tag{19.83}$$

**Example 19.3 (Cooper current).** Show that the current $\mathbf{j}$ of Example 19.2 can be re-expressed in the form quoted in Example 7.13, i.e. as

$$\mathbf{j} = \rho_0[\boldsymbol{\nabla}\varphi(\mathbf{r},t) - e\mathbf{A}/\hbar] \quad \text{amperes/m}^2, \tag{19.84}$$

where $\varphi(\mathbf{r},t)$ is the phase of the wavefunction,

$$\psi(\mathbf{r},t) = \sqrt{|\rho(\mathbf{r},t)|}e^{i\varphi(\mathbf{r},t)}, \tag{19.85}$$

and $\rho_0 = \text{const.}$

**Example 19.4 (String of magnetic flux).** Consider a string of magnetic flux. In going once round the string the phase of a wave function makes an integral (quantized) jump. Does this jump imply that the flux is quantized?

(Answer: No. Otherwise the Aharonov–Bohm effect would disappear).[10]

**Example 19.5 (Expulsion and trapping of magnetic flux).** Assuming constancy of the *generalized angular momentum* $\mathbf{L}$,

$$\mathbf{L} = \mathbf{r} \times (m\mathbf{v} + e\mathbf{A}), \tag{19.86}$$

show that in the case of a simply connected body (e.g. a solid sphere) the magnetic flux is excluded (Meissner effect explanation of F. London, cf. Example 7.13), but in the case of a non-simply connected body (e.g. a solid ring) the magnetic flux is trapped. (This is a more difficult problem. For the solution with ample explanation and illustration see Portis [121], pp. 275–282).

**Example 19.6 (Proportionality between current and A).** The general definition of velocity $\mathbf{v}$ is $\mathbf{v} = \partial T/\partial \mathbf{p}$, where $\mathbf{p}$ is the momentum and $T$ the kinetic energy, here $T = (\mathbf{p} - e\mathbf{A})^2/2m$, of a single electron of mass $m$. Show that in the case of $n$ electrons per unit volume and total momentum $\mathbf{P} = 0$, the current density $\mathbf{j}$ is given by $\mathbf{j} = -(e^2n/m)\mathbf{A}$. How do you explain the fact that the physical quantity on the left is proportional to the non-gauge-invariant quantity $\mathbf{A}$ on the right?
[Hint: Recall that the momentum changes also under a gauge transformation].

[10]For discussion see Felsager [51], p. 502.

Chapter 20

# Quantization of the Electromagnetic Field and the Casimir Effect

## 20.1 Introductory Remarks

The crux of the *Casimir effect*[1] can already be explained with the help of the simple harmonic oscillator. In the case of the one-dimensional oscillator defined on $x \in [-\infty, \infty]$ one obtains the ground state energy $E_0 = \hbar\omega/2$, also called zero point energy, where $\omega$ is the frequency of the oscillator. If the oscillator is restricted to a domain $x \in [-a, a]$, $a$ large but finite, i.e. if the wave function $\psi$ is subjected to the boundary condition $\psi(\pm a) = 0$ (infinitely high potential walls), the eigenvalues naturally change and become something like[2]

$$E_0 = \frac{1}{2}\hbar\omega(1 + O(1/a)), \tag{20.1}$$

so that for $a \to \infty$ the zero point energy $E_0 = \frac{1}{2}\hbar\omega$ is regained. Thus with the boundary condition one obtains a contribution $\propto 1/a$ (or similar) to the energy, and hence a

$$\text{force} = -\frac{\partial}{\partial a}(\text{energy}) \propto \frac{1}{a^2}. \tag{20.2}$$

Thus as a consequence of the boundary conditions, quantum mechanics implies a force proportional to $1/a^2$. The additional force derived from such a boundary-dependence in the case of quantized electrodynamics (in which

[1]The first monograph on this topic is that of Milton [95].

[2]We write simply $O(1/a)$ as an example. In most cases the change is stronger or even exponentially small.

conductors play the role of the walls at $x = \pm a$ in the above) is referred to as *Casimir-effect.*

In the following we introduce first the simple canonical quantization of the $n$-dimensional harmonic oscillator. This method is then used to perform an analogous quantization of the electromagnetic field which naturally implies divergences if the field is visualized as providing harmonic oscillators at every point in space. Thus the calculation requires the definition of an artificial but sensible regularization procedure which leaves a physically sensible result. The method is first applied to the academic example of electrodynamics in one space dimension but with boundary conditions. Thereafter the three dimensional case is treated along similar lines, and it is shown how the Casimir force arises. Our considerations here can be regarded as a natural supplement to those of Chapter 14 on wave guides and resonators.

## 20.2 Quantization of the $n$-Dimensional Harmonic Oscillator

In the canonical quantization of the $n$-dimensional harmonic oscillator one usually proceeds as follows. The Hamiltonian of the problem to be quantized can be taken as (with a suitable choice of coefficients which are irrelevant here, hence we state no units here)

$$H(q,p) = \sum_{i=1}^{n} \frac{1}{2}(p_i^2 + q_i^2) \equiv \sum_{i=1}^{n} \frac{1}{2}(p_i^2 + \omega_i^2 q_i^2)|_{\omega_i=1}. \tag{20.3}$$

Canonical quantization implies, that we associate with every dynamical variable $q_i, p_i$ (Hermitian) operators $\hat{q}_i, \hat{p}_i$, which obey the following algebra:

$$[\hat{q}_i, \hat{q}_j] = 0, \quad [\hat{p}_i, \hat{p}_j] = 0, \quad [\hat{p}_i, \hat{q}_j] = -i\hbar\delta_{ij} \tag{20.4}$$

(in the following we usually set $\hbar = 1$), where the generalized coordinates $q_i$ have to be Cartesian coordinates (i.e. $q_1 = x, q_2 = y, q_3 = z$). The momentum operator $\hat{p}_i$ has in position space the differential operator representation $-i\hbar\partial/\partial q_i$. Next one defines the "*annihilation operators*" $\hat{a}_i$ and "*creation operators*" $\hat{a}_i^\dagger$ by the relations

$$\hat{a}_i := \frac{1}{\sqrt{2}}(\hat{q}_i + i\hat{p}_i), \quad \hat{a}_i^\dagger := \frac{1}{\sqrt{2}}(\hat{q}_i - i\hat{p}_i), \tag{20.5}$$

and it is readily verified that in view of the relations (20.4) these satisfy the following commutator relations

$$[\hat{a}_i, \hat{a}_j] = 0, \quad [\hat{a}_i^\dagger, \hat{a}_j^\dagger] = 0, \quad [\hat{a}_i, \hat{a}_j^\dagger] = \delta_{ij}. \tag{20.6}$$

With a little algebra one can verify that $H(q,p)$ becomes the *Hamilton operator* given by

$$\hat{H}(\hat{q},\hat{p}) = \sum_{i=1}^{n}\left(\hat{a}_i^{\dagger}\hat{a}_i + \frac{1}{2}\right) \equiv \sum_{i=1}^{n}\omega_i\left(N_i + \frac{1}{2}\right)\bigg|_{\omega_i=1}. \tag{20.7}$$

The operator

$$\hat{N}_i = \hat{a}_i^{\dagger}\hat{a}_i \tag{20.8}$$

is called *particle number operator*. In the case of the harmonic oscillator we do not have real particles and therefore one uses instead the terms "*quasi-particles*" and "quasi-particle operators". The "*vacuum state*" or "*ground state*" $|0\rangle$ with wave function (i.e. in $q_i$-representation)

$$\psi_0 \equiv \langle q_i|0\rangle \tag{20.9}$$

is defined by the condition of "all annihilation operators applied to the vacuum give zero", i.e.

$$\hat{a}_i|0\rangle = 0 \quad \text{for all} \quad i = 1,2,\ldots,n. \tag{20.10}$$

In this state the system has an energy given by (with $\hbar = 1$)

$$E_0 = \langle 0|\hat{H}(\hat{q},\hat{p})|0\rangle = \frac{1}{2}\sum_{i=1}^{n}\omega_i, \tag{20.11}$$

called "*zero point energy*". It is an amusing exercise to find a way to Maxwell's equations with the help of the commutator relations (20.4), as in the following example.

**Example 20.1 (Feynman's "proof" of Maxwell's equations).** Assume a particle has mass $m$ and coordinates $x_j (j = 1,2,3)$ and velocity $\dot{x}_j$. Also assume the velocity satisfies Newton's equation $m\ddot{x}_j = F_j(x,\dot{x},t)$, and the coordinates and velocity components satisfy the commutator relations

$$[x_j, x_k] = 0, \quad m[x_j,\dot{x}_k] = i\mu_0\hbar\delta_{jk}. \tag{20.12}$$

Then show that there are fields $\mathbf{E}_i(x,t)$ and $\mathbf{H}_j(x,t)$, which satisfy the Lorentz-force equation $F_j = E_j + \epsilon_{jkl}\dot{x}_k H_l$ and the Maxwell equations $\nabla\cdot\mathbf{H} = 0$, $\mu_0\partial\mathbf{H}/\partial t + \nabla\times\mathbf{E} = 0$. The remaining two Maxwell equations (with $\epsilon_0\mu_0 c^2 = 1$) $\nabla\cdot\mathbf{E} = \rho/\epsilon_0$, $-\epsilon_0\partial\mathbf{E}/\partial t + \nabla\times\mathbf{H} = \mathbf{j}$, then simply define the external charge and current densities $\rho$ and $\mathbf{j}$.

**Solution:** The solution can be found in the paper of Dyson[3] and is therefore not reproduced here. This publication gave rise to further discussion in the literature.[4]

## 20.3 Hamiltonian of the Gauge Field

Our next objective is to derive the Hamiltonian density of the free electromagnetic field in analogy to the Hamiltonian of the $n$-dimensional harmonic oscillator as a necessary step towards its quantization. For reasons of simplicity we use from now on again the same symbols $q, p$ for c-number quantities as well as for operators $\hat{q}, \hat{p}$, since it is clear from the context which case is meant. Indeed, in the case of the free electromagnetic field we can proceed parallel to the case of the harmonic oscillator. The Lagrangian density $\mathcal{L}$ and the Lagrangian $L$ are, as we saw in Chapter 18, given by

$$\mathcal{L} = -\frac{1}{4} F^{\mu\nu} F_{\mu\nu} \quad \text{and} \quad L = \int d^3x \mathcal{L}. \tag{20.13}$$

We have four dynamical (i.e. time-dependent) variables $A_\mu, \mu = 0, 1, 2, 3$. The Euler–Lagrange equations are, as we saw in Chapter 18,

$$\partial_\mu \left( \frac{\partial \mathcal{L}}{\partial(\partial_\mu A_\nu)} \right) - \frac{\partial \mathcal{L}}{\partial A_\nu} = 0, \tag{20.14}$$

i.e.

$$\partial_\mu F^{\mu\nu} = 0 \quad \left( \frac{\partial \mathcal{L}}{\partial A_\nu} = 0 \right). \tag{20.15}$$

These equations separate into (for $\nu = i, 0$):
(1)

$$\partial_0 F^{0i} = -\partial_j F^{ji} \quad (\partial \mathcal{L}/\partial A_i = 0) \tag{20.16}$$

(because of the time derivative this is an *equation of motion*), and
(2)

$$\partial_k F^{k0} = 0 \quad (F^{00} = 0, \quad \partial \mathcal{L}/\partial A_0 = 0) \tag{20.17}$$

(this is a *constraint*).

---

[3]Dyson [43].
[4]See e.g. Farquhar [50] and Vaidya and Farina [153].

We note that (2) results from the variation with respect to $A^0$. If one would impose from the very beginning the gauge fixing condition $A^0 = 0$, then (2), and hence (see below) the Gauss law would be lost!

The momenta canonically conjugate to the components $A_\mu$ are given by (again as we saw earlier, cf. Eq. (18.39))

$$\pi_k := \frac{\partial \mathcal{L}}{\partial(\partial_0 A^k)} = F_{k0} = \partial_k A_0 - \partial_0 A_k. \tag{20.18}$$

The result

$$\pi_0 := F_{00} = 0 \tag{20.19}$$

is an additional constraint. From (2), and the definition of the momentum density components given by Eq. (20.18), we obtain

$$\partial_k \pi_k = 0. \tag{20.20}$$

This is the condition $\nabla \cdot \mathbf{E} = 0$, which is the *Gauss law* (with no source since we consider the "free electromagnetic field"). We observed earlier (see Eq. (18.42)), that the Lagrangian density can be expressed as

$$\mathcal{L} = \frac{1}{2}\left(\frac{\mathbf{E}^2}{c^2} - \mathbf{B}^2\right). \tag{20.21}$$

The quantities entering here do not involve $A_0$. This suggests that we proceed as follows. We add the Gauss law with a Lagrangian multiplier $A^0$ to $\mathcal{L}$, so that (with — as we saw — $\partial\mathcal{L}/\partial A^0 = 0 = \nabla \cdot \mathbf{E}/c$)

$$\mathcal{L} = \frac{1}{2}\left(\frac{\mathbf{E}^2}{c^2} - \mathbf{B}^2\right) + A^0 \nabla \cdot \frac{\mathbf{E}}{c}. \tag{20.22}$$

The canonical Hamiltonian density $\mathcal{H}$ is as usual given by a Legendre transform for which (cf. Eq. (18.45))

$$\begin{aligned}
\mu_0 \mathcal{H} &= \pi^\rho \partial^0 A_\rho - \mathcal{L} = -\pi_k(\partial_0 A_k) - \mathcal{L} \\
&\overset{(20.18)}{=} -\pi_k(-\pi_k + \partial_k A_0) - \frac{1}{2}(\pi_k^2 - B_k^2) - A_0(\partial_k \pi_k) \\
&= \frac{1}{2}(\pi_k^2 + B_k^2) - \pi_k(\partial_k A_0) - A_0(\partial_k \pi_k) \\
&= \frac{1}{2}(\pi_k^2 + B_k^2) - \partial_k(\pi_k A_0).
\end{aligned} \tag{20.23}$$

Here $k = 1, 2, 3$ for the three spatial Cartesian coordinates, summation being understood. Hence if we integrate over all space

$$H = \frac{1}{\mu_0}\int d^3x\mathcal{H} = \frac{1}{\mu_0}\int d^3x\frac{1}{2}(\pi_k^2 + B_k^2). \tag{20.24}$$

We now choose the gauge $A_0 = 0$ and $\nabla \cdot \mathbf{A} = 0$, so that (see Eq. (10.15))

$$\Box\mathbf{A} = 0. \tag{20.25}$$

The condition $A_0 = 0$ suggests itself, since $\pi_0 = 0$; i.e. the 0-degree of freedom is completely eliminated. The other constraint $\nabla \cdot \mathbf{A} = 0$ together with Eq. (20.25) then ensures that only two components of $\mathbf{A}$ are independent.

## 20.4 Quantization of the Electromagnetic Field

Our aim in this section is to obtain quantization conditions of the electromagnetic field in analogy to the quantization conditions (20.6) of the multi-dimensional harmonic oscillator. Our procedure is similar to that of Section 24.4 of Panofsky and Phillips [111]. Like these authors we also have to carry along the constant $\mu_0$ or $\epsilon_0$ (with $\epsilon_0\mu_0 c^2 = 1$), but we use some different definitions. Equation (20.25) allows us to write the solutions

$$\mathbf{A} \propto \boldsymbol{\epsilon}_{k\lambda}e^{i(\mathbf{k}\cdot\mathbf{x}-\omega t)} \quad \text{(or } \sin, \cos). \tag{20.26}$$

For these fields Eq. (20.25) implies the dispersion relation

$$\omega = kc \equiv \omega_k. \tag{20.27}$$

The solutions of $\Box\mathbf{A} = 0$ and $\Delta\mathbf{A} = -\mathbf{k}^2\mathbf{A}$ are therefore periodic and we can write them as Fourier expansions (with $q_{k,\lambda}(t) = q_{k,\lambda}(0)e^{-i\omega t}$)

$$\begin{aligned}\mathbf{A}(\mathbf{x}, t) &= \frac{c}{(2\pi)^{3/2}}\sum_\lambda\int d^3k\boldsymbol{\epsilon}_{k,\lambda}e^{i\mathbf{k}\cdot\mathbf{x}}q_{k,\lambda}(t),\\ \boldsymbol{\pi}(\mathbf{x}, t) &= \frac{1}{(2\pi)^{3/2}}\sum_\lambda\int d^3k\boldsymbol{\epsilon}_{k,\lambda}e^{i\mathbf{k}\cdot\mathbf{x}}p_{k,\lambda}(t),\end{aligned} \tag{20.28}$$

where $\lambda = 1, 2$ label the two possible, mutually orthogonal directions of polarization with

$$\boldsymbol{\epsilon}_{k,\lambda}\cdot\boldsymbol{\epsilon}^*_{k,\lambda'} = \delta_{\lambda,\lambda'}. \tag{20.29}$$

The symbols $\sum\int$ indicate summation over discrete parameters ($\lambda$) and integration over continuous parameters (the components of $\mathbf{k}$). The vector $\mathbf{k}$ here is continuous. We obtain its discretization in a large volume

$V \to \infty$ with the 3-dimensional generalization of the one-dimensional case of Eq. (3.133). Thus $\mathbf{k} = (2\pi/V^{1/3})(l\mathbf{e}_x + m\mathbf{e}_y + n\mathbf{e}_z)$, where $l, m, n$ are positive or negative integers, and $\int d^3k \leftrightarrow \sum_{\mathbf{k}}/V$. The condition $\boldsymbol{\nabla} \cdot \mathbf{A} = 0$ implies

$$\boldsymbol{\epsilon}_{k,\lambda} \cdot \mathbf{k} = 0. \tag{20.30}$$

Then with

$$\frac{1}{(2\pi)^3} \int d^3x e^{i\mathbf{k}\cdot\mathbf{x}} = \delta(\mathbf{k}), \tag{20.31}$$

in the second step (in the following) we have

$$\begin{aligned}
&\int \boldsymbol{\pi}^2 d^3x \\
&= \frac{1}{(2\pi)^3} \int d^3x \sum_\lambda \int d^3k \sum_{\lambda'} \int d^3k' \boldsymbol{\epsilon}_{k,\lambda} \cdot \boldsymbol{\epsilon}_{k',\lambda'} e^{i(\mathbf{k}+\mathbf{k}')\cdot\mathbf{x}} p_{k,\lambda}(t) p_{k',\lambda'}(t) \\
&= \sum_{\lambda,\lambda'} \int d^3k \int d^3k' \boldsymbol{\epsilon}_{k,\lambda} \cdot \boldsymbol{\epsilon}_{k',\lambda'} \delta(\mathbf{k}+\mathbf{k}') p_{k,\lambda}(t) p_{k',\lambda'}(t) \\
&= \sum_{\lambda,\lambda'} \int d^3k \boldsymbol{\epsilon}_{k,\lambda} \cdot \boldsymbol{\epsilon}_{-k,\lambda'} p_{k,\lambda}(t) p_{-k,\lambda'}(t).
\end{aligned} \tag{20.32}$$

We demand that $\mathbf{A}$ and $\boldsymbol{\pi}$ are real and as operators Hermitian (since they represent quantum mechanical observables). It follows that for $\boldsymbol{\pi}(\mathbf{x}, t)$ (observe that $\sum_k$ contains the summation over positive as well as negative components of $\mathbf{k}$, which is used in the last step in the following)

$$\begin{aligned}
\boldsymbol{\pi}(\mathbf{x}, t) &= \frac{1}{(2\pi)^{3/2}} \sum_\lambda \int d^3k e^{i\mathbf{k}\cdot\mathbf{x}} \boldsymbol{\epsilon}_{k,\lambda} p_{k,\lambda}(t) \\
&= \frac{1}{(2\pi)^{3/2}} \sum_\lambda \int d^3k e^{-i\mathbf{k}\cdot\mathbf{x}} \boldsymbol{\epsilon}^*_{k,\lambda} p^*_{k,\lambda}(t) \quad (= \boldsymbol{\pi}^*(\mathbf{x}, t)) \\
&= \frac{1}{(2\pi)^{3/2}} \sum_\lambda \int d^3k e^{+i\mathbf{k}\cdot\mathbf{x}} \boldsymbol{\epsilon}^*_{-k,\lambda} p^*_{-k,\lambda}(t),
\end{aligned} \tag{20.33}$$

i.e. (comparison of the first line with the third)

$$\boldsymbol{\epsilon}_{k,\lambda} p_{k,\lambda}(t) = \boldsymbol{\epsilon}^*_{-k,\lambda} p^*_{-k,\lambda}(t), \tag{20.34}$$

or

$$\boldsymbol{\epsilon}^*_{k,\lambda} p^*_{k,\lambda}(t) = \boldsymbol{\epsilon}_{-k,\lambda} p_{-k,\lambda}(t). \tag{20.35}$$

Returning to Eq. (20.32) we have

$$\int \pi^2 d^3x \overset{(20.32)}{=} \sum_{\lambda,\lambda'} \int d^3k \boldsymbol{\epsilon}_{k,\lambda} \cdot \boldsymbol{\epsilon}_{-k,\lambda'} p_{k,\lambda}(t) p_{-k,\lambda'}(t)$$

$$\overset{(20.34)}{=} \sum_{\lambda,\lambda'} \int d^3k \boldsymbol{\epsilon}_{k,\lambda} \cdot \boldsymbol{\epsilon}^*_{k,\lambda'} p_{k,\lambda}(t) p^*_{k,\lambda'}(t)$$

$$\overset{(20.29)}{=} \sum_{\lambda} \int d^3k p^*_{k,\lambda}(t) p_{k,\lambda}(t). \tag{20.36}$$

In order to be able to handle the other terms in $H$ involving $B_k^2$ in an analogous manner, we first have to rewrite these. It can be shown, as we do below, that

$$\int d^3x (\boldsymbol{\nabla} \times \mathbf{A}) \cdot (\boldsymbol{\nabla} \times \mathbf{A}) = -\int d^3x \mathbf{A} \cdot (\triangle \mathbf{A}). \tag{20.37}$$

We use Gauss' divergence theorem,

$$\int d^3x \boldsymbol{\nabla} \cdot \mathbf{D} = \int \mathbf{D} \cdot d\mathbf{F}. \tag{20.38}$$

Here we set

$$\mathbf{D} := \mathbf{A} \times (\boldsymbol{\nabla} \times \mathbf{A}). \tag{20.39}$$

Then

$$\boldsymbol{\nabla} \cdot \mathbf{D} = \boldsymbol{\nabla} \cdot \{\mathbf{A} \times (\boldsymbol{\nabla} \times \mathbf{A})\}. \tag{20.40}$$

We use (see Appendix C)

$$\boldsymbol{\nabla} \cdot (\mathbf{A} \times \mathbf{B}) = \mathbf{B} \cdot (\boldsymbol{\nabla} \times \mathbf{A}) - \mathbf{A} \cdot (\boldsymbol{\nabla} \times \mathbf{B}), \tag{20.41}$$

so that

$$\boldsymbol{\nabla} \cdot \{\mathbf{A} \times (\boldsymbol{\nabla} \times \mathbf{A})\} = (\boldsymbol{\nabla} \times \mathbf{A}) \cdot (\boldsymbol{\nabla} \times \mathbf{A}) - \mathbf{A} \cdot (\boldsymbol{\nabla} \times \{\boldsymbol{\nabla} \times \mathbf{A}\}). \tag{20.42}$$

Recalling the relation "*curl curl* = *grad div minus div grad*", we have

$$\boldsymbol{\nabla} \times (\boldsymbol{\nabla} \times \mathbf{A}) = \boldsymbol{\nabla}(\boldsymbol{\nabla} \cdot \mathbf{A}) - \triangle \mathbf{A}. \tag{20.43}$$

Hence

$$\boldsymbol{\nabla} \cdot \{\mathbf{A} \times (\boldsymbol{\nabla} \times \mathbf{A})\} = (\boldsymbol{\nabla} \times \mathbf{A}) \cdot (\boldsymbol{\nabla} \times \mathbf{A}) - (\mathbf{A} \cdot \boldsymbol{\nabla})(\boldsymbol{\nabla} \cdot \mathbf{A}) + \mathbf{A} \cdot (\triangle \mathbf{A}). \tag{20.44}$$

We insert this relation in Eq. (20.40) and obtain:

$$\begin{aligned}\int d^3x \boldsymbol{\nabla}\cdot\mathbf{D} &= \int d^3x(\boldsymbol{\nabla}\times\mathbf{A})\cdot(\boldsymbol{\nabla}\times\mathbf{A})\\ &\quad -\int d^3x(\mathbf{A}\cdot\boldsymbol{\nabla})(\boldsymbol{\nabla}\cdot\mathbf{A})+\int d^3x\mathbf{A}\cdot(\triangle\mathbf{A})\\ &\overset{(20.39)}{=}\int \mathbf{A}\times(\boldsymbol{\nabla}\times\mathbf{A})\cdot\mathbf{dF}\\ &= \int d\mathbf{F}\cdot\{A_i\boldsymbol{\nabla}A_i-(\mathbf{A}\cdot\boldsymbol{\nabla})\mathbf{A}\}. \end{aligned} \tag{20.45}$$

The right side is an integral over the surface area of the volume $V$ of integration on which $\mathbf{A}\to 0$. Hence this integral vanishes. With the gauge fixing condition $\boldsymbol{\nabla}\cdot\mathbf{A}=0$ another contribution drops out, and we obtain the result (20.37), i.e.

$$\int d^3x(\boldsymbol{\nabla}\times\mathbf{A})\cdot(\boldsymbol{\nabla}\times\mathbf{A}) = -\int d^3x\mathbf{A}\cdot(\triangle\mathbf{A}). \tag{20.46}$$

Hence (see remarks after Eq. (20.27))

$$\begin{aligned}\int d^3xB_k^2 &= \int d^3x(\boldsymbol{\nabla}\times\mathbf{A})\cdot(\boldsymbol{\nabla}\times\mathbf{A}) = -\int d^3x\mathbf{A}\cdot(\triangle\mathbf{A})\\ &= k^2\int d^3x\mathbf{A}\cdot\mathbf{A}\end{aligned} \tag{20.47}$$

since $\triangle\mathbf{A} = -k^2\mathbf{A}$. Repeating the procedure of above (i.e. now with $\mathbf{A}$ of Eq. (20.28), before we used $\boldsymbol{\pi}$) we obtain evidently with Eq. (20.27), $\omega_k^2 = k^2c^2$,

$$\int d^3xB_k^2 = \sum_\lambda\int d^3k\omega_k^2q_{k,\lambda}^*(t)q_{k,\lambda}(t). \tag{20.48}$$

Thus the expression for the Hamiltonian (20.24) becomes

$$H = \frac{1}{2\mu_0}\sum_\lambda\int d^3k[p_{k,\lambda}^*(t)p_{k,\lambda}(t)+\omega_k^2q_{k,\lambda}^*(t)q_{k,\lambda}(t)]. \tag{20.49}$$

The next step is quantization. We *demand* as quantization conditions at equal times $t$ (with transition from c-number valued $A_i, \pi_j$ to operators, and in analogy with the harmonic oscillator)

$$\begin{aligned}&[A_i(\mathbf{x},t), A_j(\mathbf{x}',t)] = 0, \quad [\pi_i(\mathbf{x},t),\pi_j(\mathbf{x}',t)]=0,\\ &[\pi_j(\mathbf{x},t), A_i(\mathbf{x}',t)] = -\frac{i\mu_0\hbar}{c}\delta_{ij}\delta(\mathbf{x}-\mathbf{x}'),\end{aligned} \tag{20.50}$$

or discretized as here

$$[q_{k,\lambda}(t), q_{k',\lambda'}(t)] = 0, \quad [p_{k,\lambda}(t), p_{k',\lambda'}(t)] = 0,$$
$$[p_{k,\lambda}(t), q_{k',\lambda'}(t)] = -i\mu_0\hbar\delta_{k,k'}\delta_{\lambda,\lambda'}. \tag{20.51}$$

We now define annihilation and creation operators

$$a_{\lambda,k} = \sqrt{\frac{\omega_k}{2}}\left(q_{k,\lambda} + \frac{i}{\omega_k}p_{k,\lambda}\right), \quad a^\dagger_{\lambda,k} = \sqrt{\frac{\omega_k}{2}}\left(q_{k,\lambda} - \frac{i}{\omega_k}p_{k,\lambda}\right) \tag{20.52}$$

with

$$[a_{\lambda,k}, a_{\lambda',k'}] = 0, \quad [a^\dagger_{\lambda,k}, a^\dagger_{\lambda',k'}] = 0, \quad [a_{\lambda,k}, a^\dagger_{\lambda',k'}] = \mu_0\hbar\delta_{\lambda,\lambda'}\delta_{k,k'}, \tag{20.53}$$

so that in analogy with the case of the harmonic oscillator

$$H = \sum\nolimits_\lambda \int d^3k\omega_k\left(\frac{a^\dagger_{k,\lambda}a_{k,\lambda}}{\mu_0} + \frac{\hbar}{2}\right). \tag{20.54}$$

(The factor $\mu_0$ could be avoided by combining $\sqrt{\mu_0}$ and $\sqrt{\epsilon_0}$ suitably with $q_{k,\lambda}$ and $p_{k,\lambda}$).

The *vacuum state* $|0\rangle$ or ground state of the quantized electromagnetic field is defined by

$$a_{\lambda,k}|0\rangle = 0, \quad \text{all} \quad k, \lambda. \tag{20.55}$$

The discretization $\int d^3k \to \sum_{\mathbf{k}} \sim \sum_{l,m,n}$ makes the uncountable number of space points a countable number. We thus have a complete analogy to the case of the multi-dimensional harmonic oscillator. The energy of the system in this state is the zero point energy

$$E_0 = \langle 0|H|0\rangle = \frac{1}{2}\sum_{k,\lambda}\hbar\omega_k. \tag{20.56}$$

Since we have to sum here over a countable number of infinitely many values of $\mathbf{k}$, the sum is (different from the case of the $n$-dimensional harmonic oscillator) infinite. This result seems very natural because in field theory we have effectively at every point in space one or more oscillators, each of which contributes to the zero point energy. If the physical field is to satisfy boundary conditions at boundaries, these change the spectrum, i.e. the values of $\omega_k$ and hence the zero point energy. For this reason it is plausible to define the finite *physical vacuum energy* as a difference in zero point energies, i.e. as

$$E_{\text{vac}} = E_{0,\text{with boundary}} - E_{0,\text{without boundary}}. \tag{20.57}$$

If $E_{\text{vac}} \neq 0$, this effect or the corresponding force has to be attributed to the presence of boundaries. The first investigation of this effect was published by Casimir in 1948.[5] Casimir considered the zero point energy of the electromagnetic field between two infinitely extended parallel perfectly conducting metal plates a distance $d$ apart, and showed that as a result of the quantum nature of the field an attractive force $\propto -1/d^4$ acts between the plates. This can also be viewed as a pressing together of the plates from the field outside.

## 20.5 One-Dimensional Illustration of the Casimir Effect

In his original work (see above) Casimir considered two *uncharged, parallel, perfectly conducting* metal plates, and showed that an attractive force proportional to $-1/d^4$ acts between the plates, where $d$ is the distance between the plates. We now want to consider explicitly the analogous calculation for the academic, though simpler, 1-dimensional case. We shall see that it is not enough to define the vacuum energy as a difference. In order to obtain well defined, i.e. finite expressions, one must also carry out a *regularization* of the expressions. In order to justify this regularization physically, we recall that at high frequencies of the radiation, $\omega > \omega_c$ ($\omega_c$ a critical or so-called cutoff frequency) the metal behaves like a dielectric and hence is practically transparent to the radiation.[6] This means that the eigenfrequencies of very high modes ($\omega_n > \omega_c$) remain unaffected by the boundaries and hence do not contribute to the vacuum energy.

We consider a "one-dimensional box" of length $a$, i.e. a box with cross section practically zero (cf. the rectangular resonator we discuss later), between two conductor planes, and we let the one-dimensional separation of these be between $x = 0$ and $x = a$. Moreover, we imagine a third and parallel conductor plane at a farther distance $L - a$ as indicated in Fig. 20.1. This additional plate serves the purpose of permitting the construction of the difference of energies referred to above. At the metal boundaries the electromagnetic field in the intermediate space must satisfy boundary conditions.

In the present idealized one-dimensional case this means, we have to select conditions that correspond to the boundary conditions we discussed

[5] Casimir [26]. A very readable review paper is that of Plunien, Müller and Greiner [117] which we use in the remaining parts of this chapter.
[6] See the discussion at the end of Section 9.4.

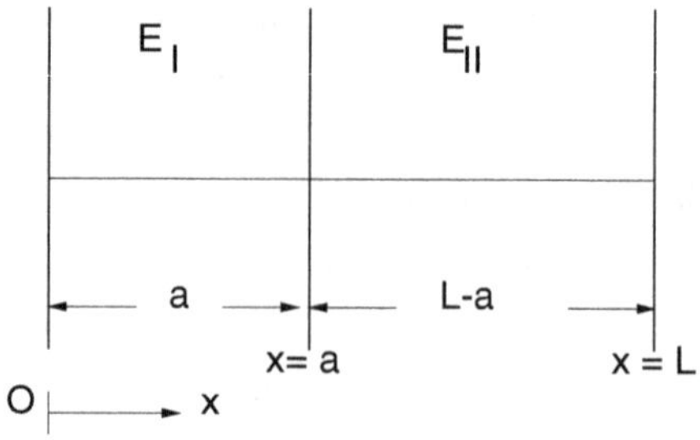


**Fig. 20.1.** The three conducting plates.

at length in Chapter 14, Section 14.3, i.e.

$$\mathbf{n} \cdot \mathbf{B} = 0, \quad \mathbf{n} \times \mathbf{E} = 0. \tag{20.58}$$

Since the radiation is confined to the one-dimensional space between the "walls" (i.e. endpoints), this implies, that we have to demand periodicity, so that we obtain stationary waves.

This means we have instead of the vector potential $\mathbf{A}(\mathbf{x}, t)$ only a scalar field $A(x, t)$ (i.e. we have the case of only $\mathbf{E} = -\boldsymbol{\nabla}\phi, \phi = A$) with $A(x, t) = 0$ at $x = 0$ and $x = a$. Since

$$\Box A(x, t) = 0, \quad A(x, t) \propto e^{i\omega t} \sin kx \tag{20.59}$$

with

$$A(x, t)|_{x=0,a} = 0 \quad \to \quad ka = n\pi, \quad n = 0, \pm 1, \pm 2, \ldots. \tag{20.60}$$

Thus the momentum is discretized. This is why we considered immediately discretized momenta above. It follows that $\omega = kc$ leads to the eigenfrequencies

$$\omega_n = \frac{n\pi}{a} c. \tag{20.61}$$

The vacuum energy in the domain I in Fig. 20.1 is therefore (since the Hamilton operator is positive, semidefinite, only values $n \geq 0$ make sense)

$$E_{\mathrm{I}}(a) = \frac{1}{2} \sum_{k,\lambda} \omega_k = \frac{1}{2} \sum_{n>0} \frac{n\pi}{a} \tag{20.62}$$

with $\hbar = c = 1$. The summation over $\lambda$ yields in general the factor 2, in view of the two possible directions of polarization, here in the one-dimensional case, of course, not. This means we obtain $E_{\mathrm{I}}(a) = \frac{1}{2} \sum_{n>0} n\pi/a$, an absolutely divergent series! We assume, as argued at the beginning of this section, that high eigenfrequencies, i.e. high frequency modes, do not contribute significantly. This implies that the sum does not have to be extended up to $n = \infty$. For this reason we introduce a parameter $\alpha$ as a cutoff factor, and write

$$E_{\rm I}(a) \to E_{\rm I}(a,\alpha) = \frac{1}{2}\sum_{n=1}^{\infty}\left(\frac{n\pi}{a}\right)e^{-\alpha(n\pi/a)} = -\frac{\partial}{\partial\alpha}\frac{1}{2}\sum_{n=1}^{\infty}e^{-\alpha(n\pi/a)}. \quad (20.63)$$

The sum is now a geometric progression which is found to give

$$E_{\rm I}(a,\alpha) = -\frac{\partial}{\partial\alpha}\frac{1}{2}\frac{1}{1-e^{-\alpha(\pi/a)}} = \frac{\pi}{2a}e^{-\alpha\pi/a}[1-e^{-\alpha\pi/a}]^{-2}. \quad (20.64)$$

Expanding the exponentials we have

$$\begin{aligned} E_{\rm I}(a,\alpha) &= \frac{\pi}{a}\left[1-\alpha\left(\frac{\pi}{a}\right)+\frac{1}{2}\alpha^2\left(\frac{\pi}{a}\right)^2-\cdots\right]\frac{1/2}{\alpha^2(\pi/a)^2} \\ &\quad\times\left[1+\alpha\left(\frac{\pi}{a}\right)+\alpha^2\left(-\frac{2}{6}+\frac{3}{2^2}\right)\left(\frac{\pi}{a}\right)^2-\cdots\right], \end{aligned} \quad (20.65)$$

and hence

$$\begin{aligned} E_{\rm I}(a,\alpha) &= \frac{1/2}{\alpha^2(\pi/a)}\left[1+\alpha^2\left(\frac{\pi}{a}\right)^2\left(\frac{1}{2}+\frac{5}{12}-1\right)+\cdots\right] \\ &= \frac{1/2}{\alpha^2(\pi/a)}\left[1-\frac{\alpha^2}{12}\left(\frac{\pi}{a}\right)^2+\cdots\right]. \end{aligned} \quad (20.66)$$

Thus the result becomes

$$E_{\rm I}(a,\alpha) = \frac{1}{2}\left[\left(\frac{a}{\pi}\right)\frac{1}{\alpha^2}-\frac{1}{12}\left(\frac{\pi}{a}\right)+O(\alpha^2)\right]. \quad (20.67)$$

For the domain II in Fig. 20.1, the corresponding expression is (replacing in Eq. (20.67) $a$ by $L-a$)

$$E_{\rm II}(L-a,\alpha) = \frac{1}{2}\left[\left(\frac{L-a}{\pi}\right)\frac{1}{\alpha^2}-\frac{1}{12}\left(\frac{\pi}{L-a}\right)+O(\alpha^2)\right]. \quad (20.68)$$

Hence for $L \gg a$

$$E_{\rm I}(a,\alpha)+E_{\rm II}(L-a,\alpha) = \frac{1}{2}\left[\left(\frac{L}{\pi}\right)\frac{1}{\alpha^2}-\frac{1}{12}\left(\frac{\pi}{a}\right)-\frac{1}{12}\left(\frac{\pi}{L}\right)+O(\alpha^2)\right]. \quad (20.69)$$

We see that this sum removed the contribution $a/2\pi\alpha^2$ in Eq. (20.67), but produced the contribution $L/2\pi\alpha^2$. We now construct the procedure which allows the removal of the wall at $x = L$ to infinity without, however, producing a new contribution $\propto 1/\alpha^2$. This is achieved by another pair

**Fig. 20.2.** The subtraction procedure.

of walls as depicted in Fig. 20.2 with $L$ thereafter taken to infinity. In other words, we single out the finite boundary effect by subtracting the corresponding expression with $a$ replaced by $L/\eta, \eta > 1$ as indicated in Fig. 20.2 (we demand $\eta > 1$, so that $L - L/\eta \neq 0$). The additional "walls" (endpoints) are then removed by taking the limit $L \to \infty$. This means we consider the difference

$$[E_{\rm I}(a,\alpha) + E_{\rm II}(L-a,\alpha)] - [E_{\rm III}(L/\eta,\alpha) + E_{\rm IV}(L-L/\eta,\alpha)]. \qquad (20.70)$$

Here, according to Eq. (20.67),

$$E_{\rm III}(L/\eta,\alpha) = \frac{1}{2}\left[\left(\frac{L/\eta}{\pi}\right)\frac{1}{\alpha^2} - \frac{1}{12}\left(\frac{\pi}{L/\eta}\right) + O(\alpha^2)\right] \qquad (20.71)$$

and

$$E_{\rm IV}(L-L/\eta,\alpha) = \frac{1}{2}\left[\left(\frac{L-L/\eta}{\pi}\right)\frac{1}{\alpha^2} - \frac{1}{12}\left(\frac{\pi}{L-L/\eta}\right) + O(\alpha^2)\right]. \qquad (20.72)$$

Hence for $L \to \infty$

$$E_{\rm III} + E_{\rm IV} = \frac{1}{2}\left(\frac{L}{\pi}\right)\frac{1}{\alpha^2} + O\left(\frac{1}{L}\right) + O(\alpha^2). \qquad (20.73)$$

Thus finally

$$\lim_{\alpha\to 0, L\to\infty}[(E_{\rm I} + E_{\rm II}) - (E_{\rm III} + E_{\rm IV})] = -\frac{\pi}{24a}. \qquad (20.74)$$

Hence there remains a finite energy or potential

$$V(a) = -\frac{\pi}{24a} \quad \text{newton} \cdot \text{meters} \qquad (20.75)$$

with the repulsive force

$$F = -\frac{\partial V}{\partial a} = +\frac{\pi}{24a^2} \quad \text{newtons}. \qquad (20.76)$$

This quantum mechanical effect which yields a force between the walls, is called *Casimir-effect*.

## 20.6 The Three-Dimensional Case

In the original 3-dimensional problem considered by Casimir the geometry is analogous to that of the previous case, but now areas in the $(x, y)$-plane represent metal plates. This geometry reminds us immediately of that of resonators, which we discussed in Chapter 14, where boundary conditions resulted from the plates at both ends of an otherwise open, cylindrical wave guide. From the considerations there we obtain (see e.g. Eq. (14.287)) for the case when the interior is the vacuum, and for $-L/2 \leq x, y \leq L/2$ as in Figs. 20.3 and 20.4 (instead of $0 \leq x, y < L$), so that with $L \to \infty$ these walls can be removed (for details see Plunien *et al.* [117]),

$$\frac{\omega^2}{c^2} = \left(\frac{n\pi}{a}\right)^2 + k_\perp^2, \quad n = 0, \pm 1, \pm 2, \ldots, \tag{20.77}$$

or (see Section 14.10)

$$\frac{\omega^2}{c^2} = \left(\frac{n\pi}{a}\right)^2 + \left(\frac{n_x\pi}{L/2}\right)^2 + \left(\frac{n_y\pi}{L/2}\right)^2, \tag{20.78}$$

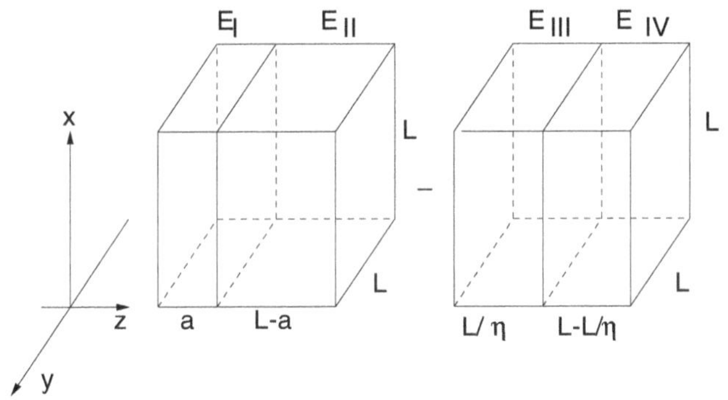


**Fig. 20.3.** The subtraction procedure in the 3-dimensional case.

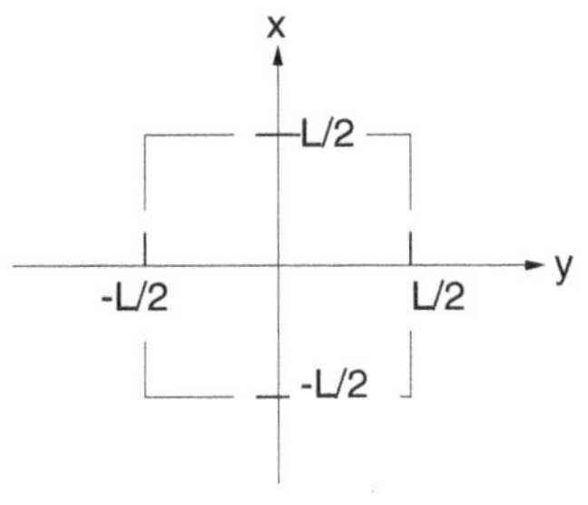


**Fig. 20.4.** The cross section.

considering a resonator of volume $L^2a$. In the present case $n_x, n_y$ are not fixed as integers, since we allow the corresponding areas to extend to infinity (with $L$). The summations over $n_x, n_y$ must therefore be replaced by integrations. Because

$$k_x = \frac{n_x\pi}{L/2}, \quad dn_x = \frac{L}{2\pi}dk_x, \quad dn_x dn_y = \left(\frac{L}{2\pi}\right)^2 dk_\perp^2, \tag{20.79}$$

and the sum

$$E_0 = \frac{1}{2}\sum_{\lambda,k} \hbar\omega_k \tag{20.80}$$

yields

$$E_{\mathrm{I}}(a) = \frac{\hbar}{2}\sum_{\lambda,n} c \int dn_x dn_y \sqrt{\left(\frac{n\pi}{a}\right)^2 + \left(\frac{n_x\pi}{L/2}\right)^2 + \left(\frac{n_y\pi}{L/2}\right)^2}. \tag{20.81}$$

We can write this

$$E_{\mathrm{I}}(a) = \frac{1}{2}\sum_{\lambda,n} \hbar c \left(\frac{L}{2\pi}\right)^2 \int dk_\perp^2 \sqrt{\left(\frac{n\pi}{a}\right)^2 + k_x^2 + k_y^2}. \tag{20.82}$$

Separating the $n = 0$ part and introducing the cutoff exponential, this becomes

$$E_{\mathrm{I}}(a) = \frac{\hbar c}{2}\sum_{\lambda} \left(\frac{L}{2\pi}\right)^2 \int d^2k_\perp$$
$$\times \left[\underbrace{\sqrt{k_x^2 + k_y^2}e^{-\alpha k_\perp^2}}_{n=0 \text{ part}} + 2\sum_{n=1}^{\infty} e^{-\alpha k_\perp^2}\sqrt{\left(\frac{n\pi}{a}\right)^2 + k_x^2 + k_y^2}\right]. \tag{20.83}$$

Here $\sum_\lambda$ is the sum over the two possible directions of polarization. Since the integral is in any case divergent, we have again introduced a frequency cutoff $\alpha$, which will be sent to zero at the end of the calculation. The arguments now run parallel to those of the previous case. Finally, one obtains the energy

$$E(a) = \lim_{L\to\infty}[\{E_{\mathrm{I}}(a) + E_{\mathrm{II}}(L-a)\} - \{E_{\mathrm{III}}(L/\eta) + E_{\mathrm{IV}}(L - L/\eta)\}]$$
$$= -\frac{L^2\pi^2}{720}\frac{\hbar c}{a^3} \text{ joules}, \tag{20.84}$$

with the attractive force per unit area (vacuum pressure) (i.e. divided by the area $L^2$)[7]

$$\begin{aligned} P_C &= -\frac{\partial E(a)}{\partial a} = -\frac{\pi^2}{240}\frac{\hbar c}{a^4_{\text{meter}}} \text{ newtons/meter}^2 \\ &= -\frac{1.30 \times 10^{-26}}{a^4_{\text{cm}}} \text{ dynes/cm}^2. \end{aligned} \tag{20.85}$$

Thus this is a force pressing the perfectly conducting plates together. The earliest experimental investigation of the Casimir effect was carried out by Sparnaay.[8] Sparnaay investigated the effect in the domain

$$0.5\mu \lesssim a \lesssim 2\mu \ \ (1\mu = 10^{-6} \text{ cm}) \tag{20.86}$$

and observed a force, which did not contradict Casimir's original theoretical findings. By now the Casimir effect has been investigated in a large number of field theoretical models and has been verified experimentally. One of the most recent experiments on the verification of the Casimir effect is that of Lamoreux,[9] who achieved agreement with theoretical prediction to an error of at most 5%. In recent years the Casimir effect has become a very active area of highly specialized research and therefore will not be treated in more detail in the present context.[10] The case in which the metal plates of the above example are replaced by concentric spherical shells has also been treated in the literature.[11] We add, that external boundary conditions like those we considered here, play a role analogous to that of external fields, which supply observable effects via quantum fluctuations.

In recent years the Casimir effect has found interest and relevance in numerous directions. For reviews see in particular Bordag [16] and Bordag *et al.* [17], Mostepanenko *et al.* [99], but also Levin and Micha [86].

---

[7]See e.g. Plunien, Müller and Greiner [117] or Fierz [55]. See also Robinson [124], p. 203. A more accurate calculation which leads to a slightly different numerical factor has been given by Fierz [55]. As Robinson remarks, the effect should not be confused with the van der Waals attraction between any two solid objects which obeys a different power law, although the two effects are connected. See comments here after Eq. (3.188).

[8]Sparnaay [142].

[9]Lamoreux [83].

[10]Some references from which also earlier literature can be traced back are Barton and Eberlein [7] and Pendry [112]. These also treat the case with motion, which is the reason that the second of these papers also discusses the concept of "quantum friction". A critical evaluation of various aspects of calculations of the Casimir effect can be found in an article by Hagen [68].

[11]Boyer [21].

A particularly interesting development is the consideration of the reversed (repulsive) Casimir effect by Leonhardt *et al.* [85].

**Example 20.2 (The fine-structure constant).** For a perfectly conducting sphere of radius $a$ the attractive Casimir energy has been calculated[12] to be

$$E_C(a) = -\frac{1}{8\pi a}\hbar c. \tag{20.87}$$

Assuming that the charge of the electron were uniformly distributed over the surface of a sphere of radius $a$, calculate the energy of this electron's repulsive Coulomb force as that of a spherical shell. Assuming that these energies cancel each other, calculate the value of the dimensionless fine-structure constant $\alpha = e^2/4\pi\epsilon_0\hbar c$.

**Solution:** We obtain the Coulomb energy from the formula for the energy of charge distributions, i.e.

$$W = \frac{1}{2}\int d\mathbf{r}\rho(\mathbf{r})\phi(\mathbf{r}), \quad \text{with} \quad \rho(\mathbf{r}) = e\delta(\mathbf{r}-\mathbf{a}), \quad \phi(\mathbf{r}) = \frac{e}{4\pi\epsilon_0 r}, \tag{20.88}$$

so that

$$W = \frac{1}{2}\int d\mathbf{r}e\delta(\mathbf{r}-\mathbf{a})\frac{e}{4\pi\epsilon_0 a}, \quad \text{i.e.} \quad W = \frac{e^2}{8\pi\epsilon_0 a}. \tag{20.89}$$

The corresponding force is

$$-\frac{\partial W}{\partial a} = +\frac{e^2}{8\pi\epsilon_0 a^2}, \quad \text{repulsive.} \tag{20.90}$$

We set

$$E_C(a) + W = 0, \quad \frac{\hbar c}{8\pi a} = \frac{e^2}{8\pi a\epsilon_0}, \quad \text{i.e.} \quad \frac{e^2}{\epsilon_0} = \hbar c. \tag{20.91}$$

Then (with $\hbar = c = 1$)

$$\alpha = \frac{e^2}{4\pi\epsilon_0} = \frac{1}{4\pi} = \frac{1}{12.56}. \tag{20.92}$$

Hence the value obtained here for the fine-structure constant is too large by a factor of 10 (the correct value being $\sim 1/137$).

[12] Milton [94].

## 20.7 Exercises

**Example 20.3 (Casimir pressure).** Calculate the Casimir pressure for the plate separations $d = 190\,\mathrm{nm}$ and $11\,\mathrm{nm}$.

(Answers: $1\,\mathrm{Pa}, 100\,\mathrm{kPa}$).[13]

**Example 20.4 (Finite range (boundary) normalization[14]).** As an illustration of the finite range or boundary normalization referred to in Section 20.1 consider the following two eigenvalue problems:

$$(a) \qquad \left(\frac{d^2}{dz^2} + 6\,\mathrm{sech}^2 z - n^2\right)\psi_0(z) = 0 \tag{20.93}$$

with $n^2 = 1$ and $\psi_0(\pm\infty) = 0$, and

$$(b) \qquad \left(\frac{d^2}{dz^2} + 6\,\mathrm{sech}^2 z - n^2\right)\tilde{\psi}_0(z) = -\tilde{\omega}_0^2\tilde{\psi}_0(z) \tag{20.94}$$

with $n^2 = 1$ and $\tilde{\psi}_0(\pm T) = 0$. Show that for $T$ large

$$\tilde{\omega}_0^2 \simeq 24\,e^{-2T}. \tag{20.95}$$

[Hint: Multiply (a) by $\tilde{\psi}_0(z)$, (b) by $\psi_0(z)$, subtract and integrate].

[13] Note, 1 nanometer being $10^{-9}$ meter, Pa meaning pascal (see Appendix B, Table B.1).

[14] See Müller–Kirsten [103], pp. 638–640.

# Chapter 21

# Duality and Magnetic Monopoles

## 21.1 Symmetrization of the Maxwell Equations

A conspicuous and eye-catching aspect of the four Maxwell equations is a certain symmetry they exhibit between electric and magnetic fields. But it is also obvious that this symmetry is violated if one looks at charge and current densities. This symmetry is described as *duality*. In this chapter we consider attempts to gain some understanding of these observations. Maxwell had banished isolated magnetic charges called poles from his equations since these had never been observed, the existence of two oppositely charged poles in magnets having been established already in the 13th century by the monk Peregrinus who also introduced the description of north and south poles.

In our treatment of the Aharonov–Bohm effect we also encountered singular fields. A deeper investigation into such fields leads to the topic of magnetic monopoles, i.e. to that of single magnetic poles. The full symmetrization of the Maxwell equations also requires the introduction of these monopoles.[1]

We first recapitulate the Maxwell equations with new notation for charge and current densities to which we affix an index "$e$" for "electric":

$$\begin{aligned} \nabla\cdot\mathbf{D} &= \rho_e, \qquad \nabla\cdot\mathbf{B} = 0, \\ \nabla\times\mathbf{H} &= \mathbf{j}_e + \frac{\partial\mathbf{D}}{\partial t}, \qquad \nabla\times\mathbf{E} = -\frac{\partial\mathbf{B}}{\partial t}, \end{aligned} \tag{21.1}$$

and the equation of continuity is

$$\frac{\partial\rho_e}{\partial t} + \nabla\cdot\mathbf{j}_e = 0. \tag{21.2}$$

---

[1] Some of the few introductory texts which also treat magnetic monopoles are Lorrain and Corson [89], see index there, and Konopinski [79], Chapter 6.

Thus, as far as these Maxwell equations are concerned, there are electric charges as sources of the electric field, but no magnetic charges or poles as sources of the magnetic field, which are here provided by electric currents. The equation $\boldsymbol{\nabla}\cdot\mathbf{B} = 0$ has no source term. This asymmetry of the equations is not appealing. It is therefore a natural step to introduce magnetic poles and to completely symmetrize the Maxwell equations and to explore the consequences which result from this. For symmetry reasons it is helpful in this case to use the formulation in terms of the MKSA-units as we do here, and not Gaussian units (but, of course, this is not essential). Thus we symmetrize the equations by adding magnetic charge and current densities with an index "$m$". The equations are then, $\rho_m$ having the dimensions of coulombs/meter$^3$ (C/m$^3$), $j_m$ those of amperes/meter$^2$ (A/m$^2$),

$$\begin{aligned} \boldsymbol{\nabla}\cdot\mathbf{D} &= \rho_e, \qquad \boldsymbol{\nabla}\cdot\mathbf{B} = \rho_m\mu_0 c, \\ \boldsymbol{\nabla}\times\mathbf{H} &= \mathbf{j}_e + \frac{\partial\mathbf{D}}{\partial t}, \qquad -\boldsymbol{\nabla}\times\mathbf{E} = \mathbf{j}_m\mu_0 c + \frac{\partial\mathbf{B}}{\partial t}. \end{aligned} \tag{21.3}$$

Here $\mu_0$ is the magnetic permeability of the vacuum.

Previously we encountered the following quantities which were already symmetric:

(a) Energy density (cf. Eq. (7.65))

$$u = \frac{1}{2}(\mathbf{E}\cdot\mathbf{D} + \mathbf{H}\cdot\mathbf{B}) \quad \text{joules/meter}^3, \tag{21.4}$$

(b) Poynting vector (cf. Eq. (7.64))

$$\mathbf{S} = \mathbf{E}\times\mathbf{H} \quad \text{watts/meter}^2, \tag{21.5}$$

(c) Maxwell stress tensor and field momentum density (cf. Eq. (7.81))

$$\begin{aligned} T_{ij} &= E_i D_j + H_i B_j - \frac{1}{2}u\delta_{ij} \quad \text{joules/meter}^3, \\ \mathbf{p}^{\text{field}} &= \mathbf{D}\times\mathbf{B} \quad \text{newton}\cdot\text{seconds/meter}^3. \end{aligned} \tag{21.6}$$

However, the Lorentz force $\mathbf{F}_e$ must be supplemented by a force $\mathbf{F}_m$ to yield the pair of symmetrized forces.

(d) Symmetrized forces

$$\mathbf{F}_e = e(\mathbf{E} + \mathbf{v}\times\mathbf{B}), \quad \mathbf{F}_m = g\mu_0 c(\mathbf{H} - \mathbf{v}\times\mathbf{D}) \quad \text{newtons} \tag{21.7}$$

with magnetic charge density $\rho_m$ defined by

$$\rho_m = g\delta(\mathbf{r}) \quad \text{coulombs/meter}^3. \tag{21.8}$$

This system of equations is now completely symmetric. The symmetry can be expressed in terms of a transformation called *duality transformation* of the fields and charge or pole densities, i.e. the relations (with $\epsilon_0\mu_0 c^2 = 1$)

$$\begin{aligned}\mathbf{E} \to \mathbf{E}' &= \mathbf{E}\cos\phi + c\mathbf{B}\sin\phi,\\ c\mathbf{B} \to c\mathbf{B}' &= -\mathbf{E}\sin\phi + c\mathbf{B}\cos\phi\end{aligned} \tag{21.9}$$

and

$$\begin{aligned}\rho_e \to \rho_e' &= \rho_e\cos\phi + \rho_m\sin\phi,\\ \rho_m \to \rho_m' &= -\rho_e\sin\phi + \rho_m\cos\phi.\end{aligned} \tag{21.10}$$

Setting $\phi = 0$, one has $\rho_e' = \rho_e$, i.e. charges $q_e' = q_e \equiv e$ and the magnetic charges or pole strengths are correspondingly $q_m' = q_m \equiv 0$, no single pole known. Setting $\phi = \pi/2$, one has $q_e' = q_m \equiv 0$ and $q_m' = -q_e = -e$. The question is: Is a certain angle $\phi$ for all particles a universal entity? Within the limits of present-day experiments $\phi = 0$ is the correct choice and in agreement with experiment. From this point of view the symmetry of classical electrodynamics is no argument for or against the existence of single magnetic poles. The question is not, do all particles in Nature have the same ratio $\rho_m/\rho_e = \tan\phi$ (in this case $\rho_m' = 0$ and we could choose units suitably, so that the usual Maxwell equations remain valid). The decisive question is really: Are there particles with different ratios $q_m/q_e$?

**Example 21.1 (Magnetic pole and electric charge at rest).** Show that a system consisting of a magnetic pole $g$ and a charge $e$, at rest at time $t = 0$, is always at rest.

**Solution:** We consider the expressions (21.7) for the forces acting on the pole and the charge. In $\mathbf{F}_e$ the vectors $\mathbf{E}$ and $\mathbf{B}$ are fields of the monopole, which act on the charge. Since the monopole is at rest the field $\mathbf{E}$, which would be generated by the monopole's motion (cf. Eq. (21.3)), is zero. Also since $\mathbf{v} = 0$, we have $\mathbf{v} \times \mathbf{B} = 0$. Hence $\mathbf{F}_e = 0$. Similarly we have $\mathbf{F}_m = 0$.

**Example 21.2 (Separating the charge-pole pair).** The magnetic pole $g$ of a dyon, i.e. charge-pole pair, with negligible binding energy is moved away from the charge $e$ to a point $\mathbf{x}'$. Show that the mechanical momentum $\mathbf{p}^{\rm mech}$ of the pole $g$ is exactly cancelled by the field momentum. (Compare with Examples 7.2 and 17.5).

**Solution:** The Lorentz force acting on a pole $g$ is given by Eq. (21.7), $\mathbf{F}_m = -g\mu_0 c\mathbf{v} \times \mathbf{D}$. Suppose we give the pole mechanical momentum $\mathbf{p}^{\rm mech}$ by pulling it away from the charge with a force equal and opposite to that of the Lorentz force. Then Newton's equation of motion for the pole is with $\mathbf{v} = d\mathbf{x}/dt$ and $F_{0k} = E_k/c$, where $F_{\mu\nu}$ is the field tensor (17.80),

$$\frac{dp_i^{\text{mech}}}{dt} = g\mu_0\epsilon_0 c^2\, \epsilon_{ijk}\frac{dx_j}{dt}F_{0k} \quad \text{newtons.} \tag{21.11}$$

In moving the pole it generates field momentum which has to be such that the total momentum is conserved, i.e. here equal to that initially, i.e. zero. Thus

$$p_i^{\text{mech}} = g\epsilon_{ijk}\int_0^{x_j'} dx_j\; F_{0k} \quad \text{newton}\cdot\text{seconds.} \tag{21.12}$$

Since $\nabla\cdot\mathbf{B} = \mu_0 cg\delta(\mathbf{r})$, we have $\int dV\nabla\cdot\mathbf{B} = \int dV\mu_0 cg\delta(\mathbf{r}) = \mu_0 cg = \int \mathbf{B}\cdot d\mathbf{a} = \int B_l da_l$, where $da_l$ is an element of area with normal in the direction $l$, and $dx_j da_l = \delta_{jl}dV$. Hence, replacing $g$ in Eq. (21.12) with this relation,

$$\begin{aligned} p_i^{\text{mech}} &= \frac{1}{\mu_0 c}\epsilon_{ijk}\int dx_j\, F_{0k}B_l da_l = \frac{1}{\mu_0 c}\epsilon_{ijk}\int dV F_{0k}B_j \\ &= \epsilon_0 c\epsilon_{ijk}\int dV F_{0k}B_j = -\int dV(\mathbf{D}\times\mathbf{B})_i = -P_i^{\text{field}}. \end{aligned} \tag{21.13}$$

It follows that the total momentum vanishes,

$$p_i^{\text{mech}} + P_i^{\text{field}} = 0, \tag{21.14}$$

as in the original state of the system.

**Example 21.3 (Angular momentum of separated charge-pole pair at rest).** Consider a magnetic pole $g$ at the origin $\mathbf{r} = 0$ and a charge $q$ at $\mathbf{r} = r_0\mathbf{e}_z$ at rest and the distance $r_0$ apart along the $z$-axis. Show that the system has a nonvanishing angular momentum only along the line drawn from the pole to the charge and determine this angular momentum.[2]

**Solution:** The angular momentum $\mathbf{L}$ is given by

$$\mathbf{L} = \int d\mathbf{r}\mathbf{l}, \quad \mathbf{l} = \mathbf{r}\times\mathbf{p}^{\text{field}}, \quad \mathbf{p}^{\text{field}} = \mathbf{D}\times\mathbf{B} \quad \text{newton}\cdot\text{seconds/meter}^3. \tag{21.15}$$

[2]See also Felsager [51], p. 487.

In the present case we have explicitly

$$\mathbf{B} = \frac{\mu_0 c g \mathbf{r}}{4\pi r^3} \quad \text{teslas}, \quad \mathbf{D} = \frac{\mathrm{q}(\mathbf{r} - \mathbf{r}_0)}{4\pi|\mathbf{r} - \mathbf{r}_0|^3} \quad \text{coulombs/meter}^2. \tag{21.16}$$

The vector $\mathbf{D} \times \mathbf{B}$ is perpendicular to the $z$-axis which points from the pole to the charge at $z = r_0$, as shown in Fig. 21.1. At fixed angle $\theta$ between $\mathbf{r}$ and $\mathbf{r}_0 = r_0\mathbf{e}_z$ but varying azimuthal angle $\varphi$ in the $(x, y)$-plane the point $\mathbf{r}$ traces a circle with $\mathbf{D} \times \mathbf{B} \propto (\mathbf{r} - \mathbf{r}_0) \times \mathbf{r}$ along its tangential direction $\mathbf{e}_\varphi$. Thus the geometry is analogous to that of a circular current with the Biot–Savart field to be determined at the origin. The components perpendicular to the $z$-axis cancel each other, so that only a $z$-component remains, in the present case $L_z$. We obtain this as follows.

We have

$$l_z = [\mathbf{r} \times (\mathbf{D} \times \mathbf{B})] \cdot \mathbf{e}_z = (\mathbf{r} \cdot \mathbf{B})D_z - (\mathbf{r} \cdot \mathbf{D})B_z. \tag{21.17}$$

Inserting the explicit expression for $\mathbf{B}$ we obtain

$$L_z = \int d\mathbf{r} l_z = \frac{\mu_0 c g}{4\pi} \int d\mathbf{r} \frac{1}{r^3}[r^2(\mathbf{D} \cdot \mathbf{e}_z) - (\mathbf{r} \cdot \mathbf{D})(\mathbf{r} \cdot \mathbf{e}_z)]. \tag{21.18}$$

With the help of the formula (see Appendix C)

$$\boldsymbol{\nabla}(\mathbf{u} \cdot \mathbf{v}) = (\mathbf{v} \cdot \boldsymbol{\nabla})\mathbf{u} + (\mathbf{u} \cdot \boldsymbol{\nabla})\mathbf{v} + \mathbf{v} \times (\boldsymbol{\nabla} \times \mathbf{u}) + \mathbf{u} \times (\boldsymbol{\nabla} \times \mathbf{v}), \tag{21.19}$$

we have (setting $\mathbf{u} = \mathbf{r}/r$, $\mathbf{v} = \mathbf{e}_z$)

$$\boldsymbol{\nabla}\left(\frac{\mathbf{r} \cdot \mathbf{e}_z}{r}\right) = (\mathbf{e}_z \cdot \boldsymbol{\nabla})\left(\frac{\mathbf{r}}{r}\right) + \mathbf{e}_z \times \left(\boldsymbol{\nabla} \times \frac{\mathbf{r}}{r}\right) \tag{21.20}$$

implying (since the expression in the bracket on the far right vanishes)

$$\boldsymbol{\nabla}\left(\frac{\mathbf{r} \cdot \mathbf{e}_z}{r}\right) = \frac{1}{r^3}[r^2\mathbf{e}_z - \mathbf{r}(\mathbf{r} \cdot \mathbf{e}_z)]. \tag{21.21}$$

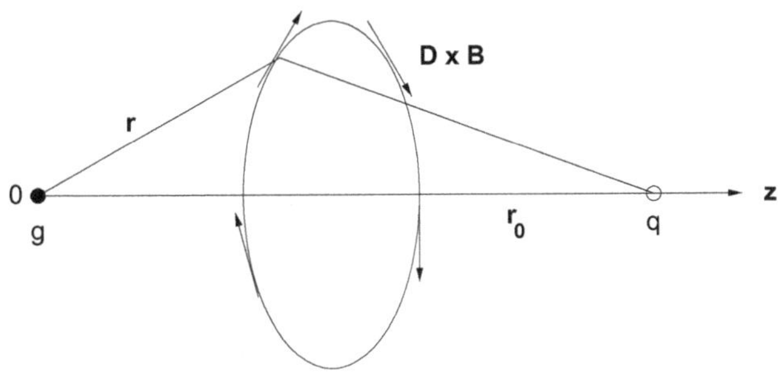


**Fig. 21.1.** Field momentum density $\mathbf{D} \times \mathbf{B}$ in the $(x, y)$-plane.

Hence

$$\mathbf{D}\cdot\boldsymbol{\nabla}\left(\frac{\mathbf{r}\cdot\mathbf{e}_z}{r}\right)=\frac{1}{r^3}[r^2(\mathbf{D}\cdot\mathbf{e}_z)-(\mathbf{D}\cdot\mathbf{r})(\mathbf{r}\cdot\mathbf{e}_z)], \tag{21.22}$$

and so

$$L_z=\frac{\mu_0 cg}{4\pi}\int d\mathbf{r}\mathbf{D}\cdot\boldsymbol{\nabla}\left(\frac{\mathbf{r}\cdot\mathbf{e}_z}{r}\right). \tag{21.23}$$

But (cf. Appendix C, $\boldsymbol{\nabla}(u\mathbf{v})=\boldsymbol{\nabla}u\cdot\mathbf{v}+u\boldsymbol{\nabla}\cdot\mathbf{v}$)

$$\mathbf{D}\cdot\boldsymbol{\nabla}\left(\frac{\mathbf{r}\cdot\mathbf{e}_z}{r}\right)=\boldsymbol{\nabla}\cdot\left(\frac{\mathbf{r}\cdot\mathbf{e}_z}{r}\mathbf{D}\right)-\frac{\mathbf{r}\cdot\mathbf{e}_z}{r}\boldsymbol{\nabla}\cdot\mathbf{D}. \tag{21.24}$$

Ignoring again the total derivative with the usual reasoning we obtain therefore

$$L_z=-\frac{\mu_0 cg}{4\pi}\int d\mathbf{r}\frac{\mathbf{r}\cdot\mathbf{e}_z}{r}\boldsymbol{\nabla}\cdot\mathbf{D}\quad \mathrm{kg}\cdot\mathrm{m}^2/\mathrm{s}. \tag{21.25}$$

Here $\boldsymbol{\nabla}\cdot\mathbf{D}=q\delta(\mathbf{r}-\mathbf{r}_0)$, so that

$$L_z=-\frac{\mu_0 cg}{4\pi}\int d\mathbf{r}\frac{\mathbf{r}\cdot\mathbf{e}_z}{r}q\delta(\mathbf{r}-\mathbf{r}_0)=-\frac{\mu_0 cgq}{4\pi}\quad \mathrm{kg}\cdot\mathrm{m}^2/\mathrm{s}. \tag{21.26}$$

This expression is seen to involve already the product $qg\mu_0/2\pi$, which will appear later in Dirac's charge quantization condition. In fact, the latter follows from the present result with the half-integral quantization condition $|L_z|=n\hbar/2$, $n$ an integer.

## 21.2 Quantization of Electric Charge

In 1931 Dirac considered the question: Is the existence of single magnetic poles, called "monopoles", compatible with the principles of quantum mechanics? His conclusion was: Yes, provided the elementary electric charge $e$ and the magnetic charge $g$ satisfy the condition[3]

$$\frac{eg}{\hbar}=\frac{2\pi n}{\mu_0 c},\quad (n \text{ an integer}). \tag{21.27}$$

The basic idea of this connection can be explained in the following elementary way (easiest possibility to make the quantization condition (21.27) plausible). We let $g$ coulombs be the magnetic charge located at the origin of coordinates

[3]Observe that the quantity $eg$ on the left is given in coulombs$^2$ since on the right $1/\mu_0$ is given in amperes$^2$/newton; cf. Appendix B, Table B.2.

which, similar to the Coulomb potential, generates at a point $\mathbf{r}$ the magnetic flux density of

$$\mathbf{B}(\mathbf{r}) = \frac{g\mu_0 c\mathbf{r}}{4\pi r^3} \quad \text{teslas.} \tag{21.28}$$

We now allow a particle with electric charge $e$ — as illustrated in Fig. 21.2 — to travel along a straight line with shortest distance $b$ from the origin and with velocity $\mathbf{v} = v\mathbf{e}_z$ past the pole charge $g$. On its flight this particle experiences the Lorentz force (pointing out of the paper in Fig. 21.2)

$$\mathbf{F} = F_y\mathbf{e}_y = evB_x\mathbf{e}_y = \frac{e\mu_0 cg}{4\pi}\frac{vb\mathbf{e}_y}{(b^2+v^2t^2)^{3/2}} \quad \text{newtons.} \tag{21.29}$$

Thereby the particle receives an impulse with momentum transfer

$$\underbrace{\triangle P_y = \int_{-\infty}^{\infty} F_y dt}_{dP_y/dt=F_y} = \frac{egvb\mu_0 c}{4\pi}\underbrace{\int_{-\infty}^{\infty}\frac{dt}{(b^2+v^2t^2)^{3/2}}}_{2/vb^2} = \frac{eg\mu_0 c}{2\pi b}. \tag{21.30}$$

This momentum transfer leads to a change in angular momentum $L_z$ along the direction of flight given by

$$\triangle L_z = b\triangle P_y = \frac{eg\mu_0 c}{2\pi} \quad \text{kg m}^2/\text{s}. \tag{21.31}$$

According to quantum mechanics orbital angular momentum is always quantized, so that

$$L_z = n\hbar, \quad n = 0, \pm 1, \pm 2, \pm 3, \ldots. \tag{21.32}$$

It follows that

$$\frac{eg\mu_0 c}{2\pi} = n\hbar, \quad g = n\frac{2\pi\hbar}{e\mu_0 c}, \tag{21.33}$$

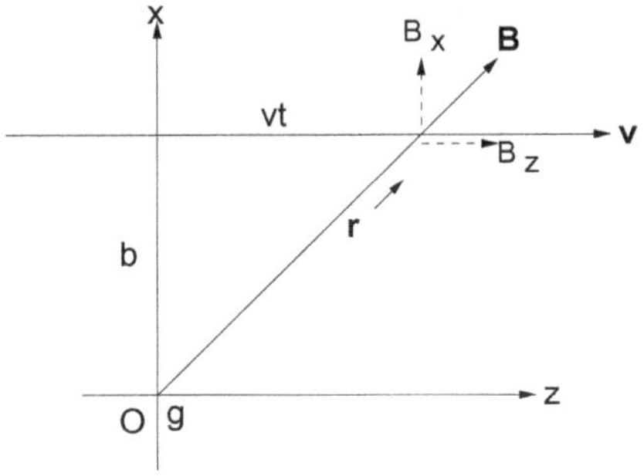


**Fig. 21.2.** Velocity $\mathbf{v}$ of the electric charge in passing the magnetic pole at the origin $O$.

i.e.

$$\frac{eg}{\hbar} = \frac{2\pi n}{\mu_0 c}. \tag{21.34}$$

This implies that

$$\frac{e}{4\pi} = \frac{n}{2}\left(\frac{\hbar}{g\mu_0 c}\right) \quad \text{coulombs} \tag{21.35}$$

must be quantized, i.e. the electric charge can appear only in positive or negative integral multiples of the unit $4\pi\hbar/2g\mu_0 c$. The charge is therefore quantized as a consequence of the quantization of angular momentum.

One should note: For this result to be physical, it suffices that a single magnetic pole exists somewhere in the universe, i.e. one we can use for the above argument.

A presumably more respectable derivation of the result (21.27) is based like Dirac's original derivation on the uniqueness of the Schrödinger wave function $\psi$, which we encountered already in Chapter 19. To be able to proceed to this, however, we first need to take a closer look at the field of the monopole.

The coupling constant of electromagnetic fields is given by the dimensionless so-called *fine-structure constant*

$$\alpha := \frac{e^2}{4\pi\hbar c\epsilon_0} \simeq \frac{1}{137}. \tag{21.36}$$

With the above result the corresponding "*magnetic fine-structure constant*" is the dimensionless quantity (using $\epsilon_0\mu_0 c^2 = 1$)

$$\begin{aligned} \frac{\mu_0 c}{4\pi\hbar} g^2 &= c\frac{\mu_0(\mu_0\epsilon_0 c^2)}{4\pi\hbar} g^2 = c\frac{\mu_0(\mu_0\epsilon_0 c^2)}{4\pi\hbar}\left(\frac{2\pi\hbar n}{e\mu_0 c}\right)^2 \\ &= \frac{4\pi\hbar c\epsilon_0}{e^2}\left(\frac{n^2}{4}\right) \simeq \frac{137 n^2}{4} \overset{n=1}{=} 34.25, \end{aligned} \tag{21.37}$$

for $n = 1$. Thus, instead of Eq. (21.36) one has

$$\beta := c\frac{g^2\mu_0}{4\pi\hbar} \simeq 34.25 \tag{21.38}$$

as dimensionless coupling of the electromagnetic field to the magnetic pole, which is seen to have a much larger value. One can say, a unit magnetic charge exerts a force equivalent to that of an electrically charged nucleus with $Z = 137/4$ protons.

## 21.3 The Field of the Monopole

If we want to take magnetic poles in a Lagrangian formalism into account, we immediately face problems. If we insist, that the field strengths $F_{\mu\nu}$ are defined by gauge potentials, we can do this only at the price of admitting singularities in the gauge potentials, i.e. the "*Dirac strings*". Consider once again the pure monopole field[4] (21.28), i.e.

$$\mathbf{B} = \frac{g\mu_0 c\mathbf{r}}{4\pi r^3}, \quad \mathbf{r} = (x, y, z), \quad r = \sqrt{x^2 + y^2 + z^2}. \tag{21.39}$$

This field is singular at the origin (like the Coulomb field). The field $\mathbf{B} = \boldsymbol{\nabla} \times \mathbf{A}$ can be obtained from the vector potential (Cartesian components written as column elements)

$$\mathbf{A} = \frac{\mu_0 c g}{4\pi} \begin{pmatrix} y/r(r-z) \\ -x/r(r-z) \\ 0 \end{pmatrix} = -\frac{g\mu_0 c}{4\pi}\frac{(1+\cos\theta)}{r\sin\theta}(\sin\varphi, \cos\varphi, 0) \tag{21.40}$$

(the first expression in Cartesian coordinates, the second in spherical polar coordinates). For instance, $B_x = \partial_y A_z - \partial_z A_y$,

$$\begin{aligned} B_x &\propto \frac{\partial}{\partial y}(0) + \frac{\partial}{\partial z}\left(\frac{x}{r(r-z)}\right) = -\frac{x\partial r/\partial z}{r^2(r-z)} - \frac{x(\partial r/\partial z - 1)}{r(r-z)^2} \\ &= -\frac{xz/r}{r^2(r-z)} - \frac{x(z/r-1)}{r(r-z)^2} = -\frac{xz/r - x}{r^2(r-z)} = \frac{x}{r^3}, \end{aligned} \tag{21.41}$$

*q.e.d.* We observe that $\mathbf{A}$ is singular at the origin $r = 0$ (like $\mathbf{B}$), but also along the positive $z$-axis, $z = r \geq 0$ (different from $\mathbf{B}$), or for $\theta = 0$. This second singularity from $r = 0$ along the positive $z$-axis is described as *Dirac string* (this is a singularity and is not to be confused with the strings of String Theory!). One may note: At $\theta = \pi$ we have

$$\begin{aligned} \sin\theta &= \sin((\theta - \pi) + \pi) \simeq -(\theta - \pi), \\ \cos\theta &= \cos((\theta - \pi) + \pi) \simeq -1 + \frac{1}{2}(\theta - \pi)^2, \end{aligned} \tag{21.42}$$

so that at $\theta = \pi$ the factor in Eq. (21.40)

$$\frac{1+\cos\theta}{\sin\theta} \propto (\theta - \pi) = \text{finite}. \tag{21.43}$$

[4]The topic we discuss here has also been treated in Fraas [57], p. 173.

Thus the vector potential is singular at the "*north pole*" where $\theta = 0$, but regular at the "*south pole*" where $\theta = \pi$. With a gauge transformation one can find another and different vector potential which is regular at the north pole, but singular at the south pole; in the overlap region around the equator the potentials are proportional to one another. In this way, i.e. by patching together different vector potentials, as in Section 21.4, one can avoid the Dirac string (singularity). The price one has to pay is this patching together of different potentials.[5]

## 21.4 Uniqueness of the Wave Function

We now use two vector potentials, $\mathbf{A}_1, \mathbf{A}_2$, on a sphere $S^2$ surrounding the magnetic charge $g$ at the center. We choose the vector potentials such that one can be continued smoothly into the other in the equatorial region. We choose the potential $\mathbf{A}_1$ to be regular on the northern hemisphere $S^2_+$ (with Dirac string along the negative $z$-axis through the south pole), and $\mathbf{A}_2$ correspondingly regular on the southern hemisphere $S^2_-$. We thus have for one and the same system two Schrödinger equations with different gauge potentials $\mathbf{A}_1, \mathbf{A}_2$ and $\phi_1, \phi_2$, and so with Hamilton operators

$$\hat{H}_1 = -\frac{\hbar^2}{2m}\left(\boldsymbol{\nabla} - \frac{ie}{\hbar}\mathbf{A}_1\right)^2 + e\phi_1, \quad \hat{H}_2 = -\frac{\hbar^2}{2m}\left(\boldsymbol{\nabla} - \frac{ie}{\hbar}\mathbf{A}_2\right)^2 + e\phi_2. \tag{21.44}$$

The two gauge potentials are related by the gauge transformation

$$\mathbf{A}_1 \to \mathbf{A}_2 = \mathbf{A}_1 + \boldsymbol{\nabla}\chi, \tag{21.45}$$

however, the function $\chi$ is then singular along the $z$-axis (i.e. along the Dirac strings) (otherwise the left side would possess the singularity but not the right side). The necessary associated transformation of the wave function $\psi$ for the gauge covariance of the Schrödinger equation, $\hat{H}\psi = E\psi$, is (as we saw in Chapter 19),

$$\psi \to \psi' = \exp\left(\frac{ie\chi}{\hbar}\right)\psi, \quad \chi = \chi(\mathbf{r}, t), \tag{21.46}$$

i.e. in the present case

$$\psi_1 \to \psi_2 = \exp\left(\frac{ie\chi}{\hbar}\right)\psi_1. \tag{21.47}$$

[5]The well known reference for this is: Wu and Yang [151]. The Dirac string is also discussed in the book of Felsager [51], p. 476. The same book, p. 486, also discusses the angular momentum as a consequence of the monopole field. See also Witten [165], pp. 424–425.

The uniqueness of the wave function in the nonsingular domain requires therefore, that after one closed orbit about the string this wave function has the same value, i.e.

$$\frac{e}{\hbar}\oint_C d\chi = 2\pi n, \quad n = 1, 2, 3, \ldots. \tag{21.48}$$

The left hand side is really the discontinuity across the singularity of the function $\chi$. We compute this as follows. We have

$$\begin{aligned}\frac{2\pi n\hbar}{e} &= \oint_C d\chi = \oint_C \boldsymbol{\nabla}\chi \cdot d\mathbf{l} = \oint_C (\mathbf{A}_2 - \mathbf{A}_1)\cdot d\mathbf{l} \\ &= \int_{S^2_+} (\boldsymbol{\nabla}\times\mathbf{A}_2)\cdot d\mathbf{F} - \int_{S^2_-} (\boldsymbol{\nabla}\times\mathbf{A}_1)\cdot d\mathbf{F},\end{aligned} \tag{21.49}$$

and therefore

$$\begin{aligned}\frac{2\pi n\hbar}{e} &= \int_{S^2_+} \mathbf{B}\cdot d\mathbf{F} - \int_{S^2_-} \mathbf{B}\cdot d\mathbf{F} = \int_{S^2} \mathbf{B}\cdot d\mathbf{F} \\ &= \int_{S^2} \frac{g\mu_0 c\, dF}{4\pi r^2} = g\mu_0 c.\end{aligned} \tag{21.50}$$

Hence

$$\frac{2\pi\hbar n}{e} = g\mu_0 c, \quad \frac{eg}{\hbar} = \frac{2\pi n}{\mu_0 c}. \tag{21.51}$$

We see here that the magnetic charge provides the discontinuity across the Dirac string singularity. The Dirac string itself, however, has only a virtual, not a real significance.

There is another, more complicated argument, which leads to the same result.[6] This argument is closely related to the arguments involved in the explanation of the Aharonov–Bohm effect. There we saw that when the current in the solenoid is switched on, the interference pattern observed on the screen undergoes a displacement depending on the magnetic flux through the solenoid. The interference pattern results from interference of electron waves along the paths $\Gamma_1, \Gamma_2$ passing through the two slits, i.e. from the phase factors

$$\exp\left(\frac{ie}{\hbar}\int_{\Gamma_2}\mathbf{A}\cdot d\mathbf{l}\right)\exp\left(-\frac{ie}{\hbar}\int_{\Gamma_1}\mathbf{A}\cdot d\mathbf{l}\right) = \exp\left(\frac{ie}{\hbar}\oint_{\Gamma}\mathbf{A}\cdot d\mathbf{l}\right). \tag{21.52}$$

[6] A detailed treatment can be found in the book of Felsager [51], p. 501.

Here

$$\oint_\Gamma \mathbf{A}\cdot d\mathbf{l} = \int \mathbf{B}\cdot d\mathbf{F} \tag{21.53}$$

is the magnetic flux through the solenoid. Thus the entire system consists of the charged particle (electron) and the magnetic field. Consider again the case of an electron and a magnetic pole. These generate the electromagnetic field $F_{\mu\nu}$, which, as we saw, has to be singular and cannot be derived from a single global vector potential. However, one can imagine an additional tensor field $S_{\mu\nu}$ and can construct it such that the sum of both, $F+S$ with $\int_{\mathbf{S}^2} F = g\mu_0 c$, $\int_{\mathbf{S}^2} S = -g\mu_0 c$, is generated by a still singular $\mathbf{A}$. We take $\int_{\mathbf{S}^2} S$ to be the magnetic flux through $\mathbf{S}^2$ due to the Dirac string. Then we have

$$\oint_\Gamma \mathbf{A}\cdot d\mathbf{l} = \int_{\mathbf{S}^2} F + \int_{\mathbf{S}^2} S. \tag{21.54}$$

This separation is now inserted into the phase factor (21.52). However, the string-part $S$ may not contribute to this, since otherwise the interference would depend on its position in space. It follows therefore that we must have

$$\exp\left(\frac{ie}{\hbar}\int_{\mathbf{S}^2} S\right) = 1, \tag{21.55}$$

and this implies

$$\frac{eg}{\hbar} = \frac{2\pi n}{\mu_0 c}. \tag{21.56}$$

Finally we add yet another derivation of the Dirac charge quantization condition which, in view of its brevity or even the associated geometry of Fig. 21.3, may be described as a "*nutshell proof*".[7] We saw in Chapter 19 that for a particle with charge $e$ moving in a magnetic field (here that of a pole $g$), the *Lorentz force* appears in the quantum mechanical wave function as the phase factor

$$\exp\left(\frac{ie}{\hbar}\oint_\Gamma \mathbf{A}\cdot d\mathbf{l}\right). \tag{21.57}$$

We also saw that the vector potential cannot be gauged away in view of its Dirac string singularity extending, as shown in Fig. 21.3, from the pole through the north pole to infinity. However, we can proceed as follows.

[7]Polchinski [119], p. 148.

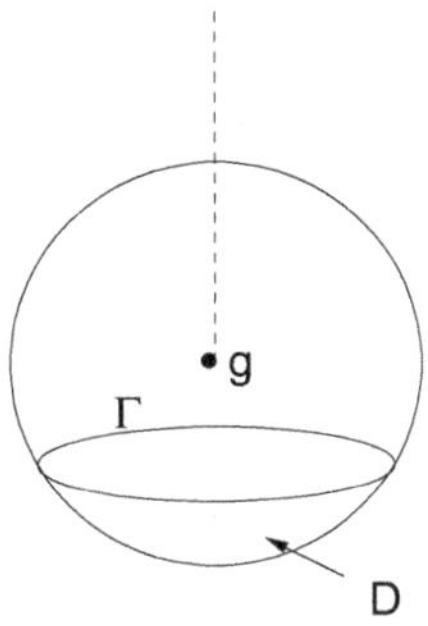


**Fig. 21.3.** The Dirac string emanating from the magnetic pole $g$, and the orbit $\Gamma$ as boundary of area $D$.

The closed path $\Gamma$ is the boundary of the part $D$ of the surface of a sphere around the south pole. Pulling the path $\Gamma$ across the equator and contracting it to a small circle around the Dirac string through the north pole, the area of $D$ becomes that of the sphere and we can write the phase factor

$$\exp\left(\frac{ie}{\hbar}\oint_{\mathbf{S}^2}\mathbf{B}\cdot d\mathbf{F}\right) = \exp\left(\frac{ie}{\hbar}g\mu_0 c\right). \tag{21.58}$$

Since there may not be any trace of the Dirac string, this factor must be 1 giving again

$$\frac{e}{\hbar}g\mu_0 c = 2\pi n, \quad n \text{ integral.} \tag{21.59}$$

## 21.5 Regularization of the Monopole Field

Finally we have a closer look at the vector potential (21.40) in order to gain some physical understanding of the Dirac string, i.e. what we would have to do in order to remove its singular behavior. We shall see that a thin solenoid along the position of the string will achieve this. To this end we regularize the vector potential (21.40), by replacing in it $r$ by $\sqrt{r^2+\epsilon^2}$, where $\epsilon$ is small. We rename the correspondingly modified fields $\mathbf{A}, \mathbf{B}, \mathbf{E}$ respectively $\mathbf{A}_\epsilon, \mathbf{B}_\epsilon, \mathbf{E}_\epsilon$. With this modification the field $\mathbf{A}_\epsilon$ is well defined in the whole of space. Since this regularization does not affect the time dependence, we have $\mathbf{E}_\epsilon = 0$. We maintain, of course, $\mathbf{B}_\epsilon = \nabla\times\mathbf{A}_\epsilon$ and see, that changes ensue. First we replace in $\mathbf{A}$ of Eq. (21.40) $r$ by $\sqrt{r^2+\epsilon^2}$ and expand the expressions up to the leading contributions in powers of $\epsilon^2$. With this we obtain

$$\mathbf{A}_\epsilon = \frac{\mu_0 cg}{4\pi}\begin{pmatrix} (y/r(r-z))\{1-\epsilon^2(2r-z)/2r^2(r-z)\} \\ -(x/r(r-z))\{1-\epsilon^2(2r-z)/2r^2(r-z)\} \\ 0 \end{pmatrix}. \tag{21.60}$$

These expressions *seem* to be singular at $r = z$. But consider once again the starting expressions, by taking for instance, the $x$-component of $\mathbf{A}_\epsilon$:

$$A_{\epsilon x} \propto \frac{y}{\sqrt{r^2+\epsilon^2}(\sqrt{r^2+\epsilon^2}-z)}. \tag{21.61}$$

This expression is regular at $r = z$. We avoid these problems in the following by replacing the factor $(r-z)$ where it seems to be problematic, by $(\sqrt{r^2+\epsilon^2}-z)$. With this we calculate the components of $\mathbf{B}_\epsilon = \boldsymbol{\nabla}\times\mathbf{A}_\epsilon$ and obtain

$$B_{\epsilon x} = \frac{\mu_0 cg}{4\pi}\frac{x}{r^3}\left(1-\frac{3\epsilon^2}{2r^2}\right), \quad B_{\epsilon y} = \frac{\mu_0 cg}{4\pi}\frac{y}{r^3}\left(1-\frac{3\epsilon^2}{2r^2}\right),$$
$$B_{\epsilon z} = \frac{\mu_0 cg}{4\pi}\frac{z}{r^3}\left(1-\frac{3\epsilon^2}{2r^2}\right) - \frac{\mu_0 gc}{4\pi}\frac{\epsilon^2(2r-z)}{r^3(r-z)^2}. \tag{21.62}$$

We now see that the $z$-component displays an additional contribution of the order of $\epsilon^2$, i.e. a $\mathbf{B}$-field along the string. We have a closer look at this expression, i.e. at

$$w_\epsilon := \frac{\epsilon^2(2r-z)}{r^3(\sqrt{r^2+\epsilon^2}-z)^2}, \tag{21.63}$$

in which we replaced $(r-z)$ as explained above. We choose a point $z_0 > 0$ along the $z$-axis and explore the behavior of $w_\epsilon$ in its neighborhood, i.e. around $r \sim z_0$. First we have with $\rho^2 = x^2+y^2$

$$\begin{aligned} r - z_0 \to \sqrt{r^2+\epsilon^2} - z_0 &= \sqrt{\rho^2+z_0^2+\epsilon^2} - z_0 \\ &\simeq z_0\left(1+\frac{\rho^2+\epsilon^2}{2z_0^2}\right) - z_0 = \frac{\rho^2+\epsilon^2}{2z_0}. \end{aligned} \tag{21.64}$$

Also

$$(2r - z_0) = (r - z_0) + r \to \frac{\rho^2+\epsilon^2}{2z_0} + r. \tag{21.65}$$

With these expressions we obtain

$$w_\epsilon \to \frac{\epsilon^2\{(\rho^2+\epsilon^2)/2z_0 + r\}(2z_0)^2}{r^3(\rho^2+\epsilon^2)^2} \overset{r\sim z_0}{=} \frac{4\epsilon^2\{1+(\rho^2+\epsilon^2)/2z_0^2\}}{(\rho^2+\epsilon^2)^2}$$
$$= \frac{4\epsilon^2(2z_0^2+\rho^2)}{2z_0^2(\rho^2+\epsilon^2)^2} \simeq \frac{4\epsilon^2}{(\rho^2+\epsilon^2)^2}, \tag{21.66}$$

and hence with Eq. (21.62)

$$B_{\epsilon z} = \frac{\mu_0 cg}{4\pi}\frac{z}{r^3} - \mu_0 cg\frac{\epsilon^2}{\pi(\rho^2+\epsilon^2)^2}. \tag{21.67}$$

We now use the 2-dimensional representation (2.71) of the delta function which we found in Example 2.6, i.e.

$$\lim_{\epsilon\to 0}\frac{\epsilon^2}{\pi(\epsilon^2+\rho^2)^2} = \delta(x)\delta(y), \tag{21.68}$$

and obtain

$$\lim_{\epsilon\to 0} B_{\epsilon z} = \frac{\mu_0 cg}{4\pi}\frac{z}{r^3} - \mu_0 cg\delta(x)\delta(y)\theta(z). \tag{21.69}$$

Here we have multiplied the second contribution by the step function since the string runs along the positive $z$-axis. The magnetic field concentrated in the Dirac string can therefore be written (there is only a $z$-component)

$$\mathbf{B}_{\text{string}} = -\mu_0 cg\delta(x)\delta(y)\theta(z)\mathbf{e}_z \quad \text{teslas}. \tag{21.70}$$

The divergence of this expression is

$$\boldsymbol{\nabla}\cdot\mathbf{B}_{\text{string}} = -\mu_0 cg\delta(x)\delta(y)\delta(z). \tag{21.71}$$

On the other hand the divergence of the monopole part of the field $\mathbf{B}$, i.e. the first part on the right of (21.69) and the corresponding components contained in Eq. (21.62), is

$$\boldsymbol{\nabla}\cdot\mathbf{B}_{\text{monopole}} = \mu_0 cg\delta(x)\delta(y)\delta(z). \tag{21.72}$$

We see that the sum of the two sources cancels exactly as a consequence of our regularization of the originally singular vector potential. Thus $\boldsymbol{\nabla}\cdot\mathbf{B}_\epsilon = 0$.

Finally we calculate the curl of the string field. We obtain evidently

$$\boldsymbol{\nabla}\times\mathbf{B}_{\text{string}} = -\mu_0 cg\begin{pmatrix}\delta(x)\delta'(y)\\ -\delta'(x)\delta(y)\\ 0\end{pmatrix}\theta(z) \equiv \mu_0\mathbf{j}_{\text{string}} \tag{21.73}$$

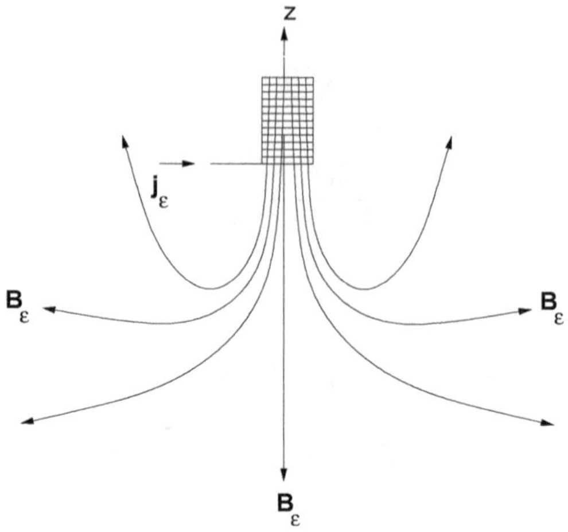


**Fig. 21.4.** The regularized field **B** of a monopole.

with a singular current density. We can visualize the current associated with this current density in the $(x, y)$-plane as a current in a very narrow, semi-infinite solenoid which supplies the magnetic field concentrated along the $z$-axis. This situation together with the appropriate field **B** of Eq. (21.62) are illustrated in Fig. 21.4. The regularization we explored above is therefore equivalent to introducing this solenoidal flux along the $z$-axis, so that the effect of the original singularity is cancelled.

## 21.6 Concluding Remarks

In the above we investigated a few important elementary aspects related to the introduction of classical magnetic monopoles, and other consequences of the symmetrization of Maxwell's equations. The experimental search for evidence of magnetic monopoles received considerable impetus in 1982 with the possibility that an appropriate event had been observed, at least it was not easy at the time to propose some other explanation for the observations of Cabrera.[8] A review of the experimental situation thereafter, which has hardly changed since then, and which also recapitulates the historical development and cites extensive literature can be found in the paper of Fryberger.[9] The principle which is used in experimental attempts to observe a monopole, is similar to that discussed in Section 17.6, i.e. that a monopole passing an atom

[8]Cabrera [24].
[9]Fryberger [58].

at some distance generates a pulse of magnetic flux which can be observed. However, due to conservation of magnetic charge, magnetic monopoles must be created in pairs — as e.g. in the so-called Drell–Yan pair production process [40]. A complicated MACRO-detector search for magnetic monopoles was carried out throughout the 1990s. It found none but set stringent limits. The final results have recently been published.[10] Monopoles which play an important role in field theories, including grand unified theories and string theory, as topologically nontrivial field configurations are of a completely different nature. However, seen from far away, they assume the structure of the Dirac monopole discussed above.[11] The question of whether magnetic monopoles really exist is therefore the remaining big puzzle of electrodynamics. Lately, in spite of the earlier discouraging attempts, a new initiative has been undertaken to search for the existence of monopoles. This initiative resulted in the CERN Research Board approval in December 2009 of the MoEDAL (Monopole and Exotics Detector) experiment whose prime motivation is to search for the direct production of the magnetic monopole at the Large Hadron Collider (LHC).[12] This experiment is still in progress. Thus in view of pressure particularly from theoretical considerations this search has not been given up. In fact, reversing Dirac's argument and saying that quantized electric charge requires the existence of magnetic monopoles, Polchinski [119] went still further and said "the existence of magnetic monopoles seems one of the safest bets that one can make about physics not yet seen".

## 21.7 Exercises

**Example 21.4 (Do magnetic forces perform work?).** A charge $q$ moves a distance $d\mathbf{l} = \mathbf{v}dt$ in the presence of a magnetic field $\mathbf{B}$. Show that $dW_{\text{magnetic}} = 0$, i.e. magnetic forces do no work. (This important observation is given particular emphasis in the book of Griffiths [63], p. 207).

**Example 21.5 (Force of a monopole on a charge e).** Assume a particle exists with a magnetic monopole strength $g$, and assume a point charge $e$ moves with velocity $\mathbf{v}$ relative to the monopole. Show that if the origin of coordinates is on the monopole, the charge detects its presence

[10] Ambrosio *et al.* [2].
[11] See, for instance, Goddard and Olive [62].
[12] Pinfold [116].

through a force

$$\mathbf{F} = eg\frac{\mathbf{v} \times \mathbf{r}}{r^3}\frac{\mu_0 c}{4\pi} \quad \text{newtons.} \tag{21.74}$$

Does the total field angular momentum $\mathbf{L}$ in the static situation depend on $\mathbf{r}$?

**Example 21.6 (Energy acquired by a magnetic monopole).** A magnetic charge $q_M$ situated in a magnetic field $B$ is subjected to the force $q_M B/\mu_0$ newtons. What is the energy acquired by a magnetic monopole in a field of 10 teslas over a distance of 0.16 meters?

(Answer: 66 GeV, 1 GeV$=10^9$ eV).[13]

**Example 21.7 (Dimensional verification of Dirac condition).** Verify that the Dirac quantization condition (21.27) is dimensionally correct.

**Example 21.8 (The monopole inducing a current).** Show — using the symmetrized Maxwell equations — that a magnetic monopole passing through a loop of wire induces in the latter a net current.
[Note: This is effectively the reverse of Example 5.7. Such a test would provide unambiguous evidence of the monopole but in practice one would use a superconducting loop to obtain a sustainable current].

[13]Lorrain and Corson [89], p. 214.

## Appendix A

# The Delta Distribution

The *delta distribution*[1] or delta function, as it was originally called, is a singular function beyond the realm of classical analysis. As we observed in the text, in physical contexts the delta distribution arises in the consideration of point charges, and similarly in that of mass points. The density of a unit charge or unit mass at (say) $x = 0$ which is written $\delta(x)$ is everywhere zero except at $x = 0$, where it is so large that the total charge, i.e. its integral over all space, becomes 1, i.e.

$$\delta(x) = 0 \quad \text{for} \quad x \neq 0, \qquad \int_{-\infty}^{\infty} \delta(x)dx = 1. \tag{A.1}$$

No function of classical analysis has such properties since for any function which is everywhere zero except at one point, the integral must vanish (irrespective of the concept of the integral). As a further example, which leads to a singular function, we consider the case of two charges of opposite signs with intensities $\pm 1/\epsilon$ and located at the points $x = 0$, $x = \epsilon$. The density distributions of the charges are $\delta(x)/\epsilon, -\delta(x - \epsilon)/\epsilon$. In the limit $\epsilon \to 0$ the charges approach each other, with the product of intensity $1/\epsilon$ and mutual separation $\epsilon$, i.e. the dipole moment, remaining constant. One thus obtains a dipole with density

$$\delta'(x) = \lim_{\epsilon \to 0} \left( \frac{\delta(x) - \delta(x - \epsilon)}{\epsilon} \right). \tag{A.2}$$

This limiting value is undefined and does not exist in the context of classical analysis. However, in the theory of distributions developed by L. Schwartz, the delta function, its derivative $\delta'(x)$ and similar quantities, find exact

[1]We follow here Güttinger [65].

definitions which, moreover, permit application of customary operations of classical analysis. Physical considerations point the way to how best to proceed in developing such a theory. In order to localize an object with a certain density distribution, one observes its reaction to some action of a test object. If there is no reaction, one concludes that the object was not present in the tested domain. We can characterize the density distribution of the object by $f(x)$, and that of the test object by $\varphi(x)$. Then the product $f(x)\varphi(x)$ represents the result of the test for the spatial point $x$, because $f(x)\varphi(x) = 0$ if not simultaneously $f(x) \neq 0$, $\varphi \neq 0$, i.e. if there is no encounter of the object with the test object. The integral $\int f(x)\varphi(x)dx$ then describes the result of the testing or scanning procedure over the entire space. If we perform this procedure with different test objects, i.e. with different test functions $\varphi_i(x), i = 1, 2, \ldots$, we obtain as the result over the whole of space different numbers corresponding to the individual $\varphi_i$. These $\varphi$-dependent numbers are written

$$f\langle\varphi\rangle = \int f(x)\varphi(x)dx. \tag{A.3}$$

If $f\langle\varphi\rangle = 0$ for all continuous differentiable functions $\varphi(x)$, then $f(x) = 0$. In general one expects that a knowledge of the numbers $f\langle\varphi\rangle$ and the test functions $\varphi(x)$ characterizes the function $f(x)$ itself provided the set of test functions is complete. In this way one arrives at a new concept of functions: Instead by its values $y = f(x)$, the function $f$ is now determined by its action on all *test functions* $\varphi(x)$. One calls $f$ a *functional*: The functional $f$ assigns every test function $\varphi$ a number $f\langle\varphi\rangle$. The functional is therefore the mapping of a space of functions onto a space of numbers. $f\langle\varphi\rangle$ is the value of the functional at the "point" $\varphi$. The concept of a functional permits us to define objects which are not functions in the sense of classical analysis. Consider e.g. the delta function. This is now defined as the functional $\delta\langle\varphi\rangle$ which assigns every test function $\varphi(x)$ a number, in this case the value of the test function at $x = 0$, i.e.

$$\delta\langle\varphi\rangle = \varphi(0), \tag{A.4}$$

where with Eq. (A.3)

$$\delta\langle\varphi\rangle = \int \delta(x)\varphi(x)dx. \tag{A.5}$$

The result of the action of the delta function on the test function $\varphi(x)$ is the number $\varphi(0)$. Writing this $\int \delta(x)\varphi(x)dx$ is to be understood only symbolically. The example of the delta function demonstrates that a function does not have to be given in order to be able to define a functional.

To ensure that in the transition from a function $f(x)$ defined in the classical sense to its corresponding functional $f\langle\varphi\rangle$ no information about $f$ is lost, i.e. so that $f\langle\varphi\rangle$ is equivalent to $f(x)$, the class of test functions has to be sufficiently large. This means: If the integral $\int f(x)\varphi(x)dx$ is to exist even for a function $f(x)$, which increases with $x$ beyond all bounds, then the test functions must decrease to zero sufficiently rapidly for large values of $x$ (how, depends of course on the given physical situation). However, it is clear that the space of test functions must contain those functions $\varphi$, which vanish beyond a bounded and closed domain, because these correspond to the possibility, to measure mass distributions that are restricted to a finite domain. The functions $\varphi$ must also have a sufficiently regular behavior so that the integral of Eq. (A.3) exists. One demands continuous differentiability of any high order. Thus we define $D$ as the space of all infinitely often differentiable functions $\varphi(x)$, which like all their derivatives $\varphi^{(n)}(x), n = 0, 1, 2, \ldots$, vanish at $x = \pm\infty$. The functionals $f\langle\varphi\rangle$ defined on $D$ are called *distributions*.

We expect also that certain continuity properties of the function $f(x)$ are reflected in the associated functional. The reaction of a mass distribution $f(x)$ to a test object $\varphi(x)$ is the weaker, the weaker $\varphi(x)$ is. It makes sense, therefore, that if a sequence of test functions $\varphi_\nu(x)$ as well as the sequences of their derivatives of arbitrary order $\varphi_\nu^{(n)}(x)$ converge uniformly towards zero, then also the sequence of numbers $f\langle\varphi_\nu\rangle$ converges towards zero.

The linearity

$$f\langle\varphi_1 + \varphi_2\rangle = f\langle\varphi_1\rangle + f\langle\varphi_2\rangle \tag{A.6}$$

follows from the superposition principle. The derivative of a distribution $f\langle\varphi\rangle$ is defined by the following equation:

$$f'\langle\varphi\rangle = -f\langle\varphi'\rangle. \tag{A.7}$$

This definition is natural. Because if we associate with the function $f(x)$ the distribution

$$f\langle\varphi\rangle = \int_{-\infty}^{\infty} dx f(x)\varphi(x), \tag{A.8}$$

then the derivative $f'(x)$ is assigned the functional

$$f'\langle\varphi\rangle = \int_{-\infty}^{\infty} dx f'(x)\varphi(x). \tag{A.9}$$

Partial integration of this integral yields, since $\varphi(\pm\infty) = 0$,

$$f'\langle\varphi\rangle = [f(x)\varphi(x)]_{-\infty}^{\infty} - \int_{-\infty}^{\infty} dx f(x)\varphi'(x) = -f\langle\varphi'\rangle \tag{A.10}$$

as in Eq. (A.7). Equation (A.7) defines the derivative of the functional $f\langle\varphi\rangle$, even if no function $f(x)$ exists, which defines the functional. For instance, we have in the case of the delta function with Eqs. (A.7) and (A.4):

$$\delta'\langle\varphi\rangle = -\delta\langle\varphi'\rangle = -\varphi'(0). \tag{A.11}$$

Formally one writes

$$\begin{aligned}\int_{-\infty}^{\infty} dx\delta'(x)\varphi(x) &= [\delta(x)\varphi(x)]_{-\infty}^{\infty} - \int_{-\infty}^{\infty} dx\delta(x)\varphi'(x)\\ &= -\delta\langle\varphi'\rangle = -\varphi'(0).\end{aligned} \tag{A.12}$$

For an infinitely often differentiable function $g(x)$ we have evidently

$$(g \cdot f)\langle\varphi\rangle = f\langle g\varphi\rangle, \tag{A.13}$$

so that

$$(x\delta)\langle\varphi\rangle = \delta\langle x\varphi\rangle = [x\varphi]_{x=0} = 0 \cdot \varphi(0) = 0, \tag{A.14}$$

or

$$\int x\delta(x)\varphi(x)dx = 0, \quad x\delta(x) = 0. \tag{A.15}$$

Thus formally one can operate with the delta function like with a function of classical analysis. As another example we consider the relation

$$f(x)\delta(x) = f(0)\delta(x). \tag{A.16}$$

According to Eq. (A.13) we have:

$$\begin{aligned}\int f(x)\delta(x)\varphi(x)dx &= \delta\langle f\varphi\rangle = [f(x)\varphi(x)]_{x=0}\\ &= f(0)\varphi(0) = f(0)\delta\langle\varphi\rangle\\ &= \int f(0)\delta(x)\varphi(x)dx,\end{aligned} \tag{A.17}$$

as claimed with Eq. (A.16). Formal differentiation of the relation (A.16) gives

$$f(x)\delta'(x) = f(0)\delta'(x) - f'(x)\delta(x). \tag{A.18}$$

One can convince oneself that this formal relation also follows from the defining equation (A.3). In the special case $f(x) = x$ we obtain the useful result

$$x\delta'(x) = -\delta(x). \tag{A.19}$$

A function which is important in applications is the *Heaviside* or *step function*, which is defined as follows:

$$\theta(x) = \begin{cases} 1 & \text{for} \quad x > 0, \\ 0 & \text{for} \quad x < 0. \end{cases} \tag{A.20}$$

From Eq. (A.20) we derive the important relation

$$\theta'(x) = \delta(x). \tag{A.21}$$

In order to prove this relation, we associate with the step function the functional

$$\theta(x)\langle\varphi\rangle = \int_{-\infty}^{\infty} \theta(x)\varphi(x)dx = \int_0^{\infty} \varphi(x)dx. \tag{A.22}$$

For the derivative we have according to Eq. (A.7):

$$\begin{aligned} \theta'(x)\langle\varphi\rangle &= -\theta(x)\langle\varphi'\rangle = -\int_{-\infty}^{\infty} dx\theta(x)\varphi'(x) \\ &= -\int_0^{\infty} dx\varphi'(x) = \varphi(0) - \varphi(\infty) \\ &= \varphi(0) = \delta(x)\langle\varphi\rangle, \end{aligned} \tag{A.23}$$

or symbolically

$$\theta'(x) = \delta(x), \tag{A.24}$$

which is Eq. (A.21).

Fourier integrals were formally introduced in Chapter 3. We abstain here from delving into rigorous proofs. We recall that the function represented by the Fourier series is piece-wise smooth. The Fourier integrals of Chapter 3 therefore supply us an integral representation of such a piece-wise smooth function. The Fourier theorem says: If $f(x)$ is a piece-wise continuous function and if the integral $\int_{-\infty}^{\infty} |f(x)|dx$ exists, and if $f(x)$ is defined at a discontinuity $x_0$ by the relation

$$f(x_0) = \frac{1}{2}[f(x_0 - 0) + f(x_0 + 0)], \tag{A.25}$$

then the following representation of the function $f(x)$ exists:

$$f(x) = \frac{1}{2\pi}\int_{-\infty}^{\infty} dk \int dy f(y) e^{ik(y-x)}. \tag{A.26}$$

In order to verify (A.26), we consider first the function

$$\delta_\kappa(x) = \frac{1}{2\pi}\int_{-\kappa}^{\kappa} dk e^{ikx} = \frac{1}{\pi}\frac{\sin \kappa x}{x}. \tag{A.27}$$

The behavior of this function for $\kappa > 0$ but not too large is illustrated in Fig. A.1. We observe that as the central maximum $\kappa/\pi$ grows with $\kappa \to \infty$, the zeros at $n\pi/\kappa$ converge toward the origin and we have

$$\delta(x) = \lim_{\kappa\to\infty} \delta_\kappa(x), \tag{A.28}$$

because $\delta(x)$ is defined by the properties $\delta(x) = 0$ for $x \neq 0$, $\delta(x) = \infty$ for $x = 0$. From Eqs. (A.27) and (A.28) we obtain the important integral representation of the delta function

$$\delta(x) = \frac{1}{2\pi}\int_{-\infty}^{\infty} dk e^{ikx}. \tag{A.29}$$

From Eqs. (A.27) and (A.28) we also obtain

$$\delta(x) = \delta(-x). \tag{A.30}$$

The representation (A.29) also proves the relation (A.26) claimed by Fourier's theorem. Because if we perform the integration with respect to

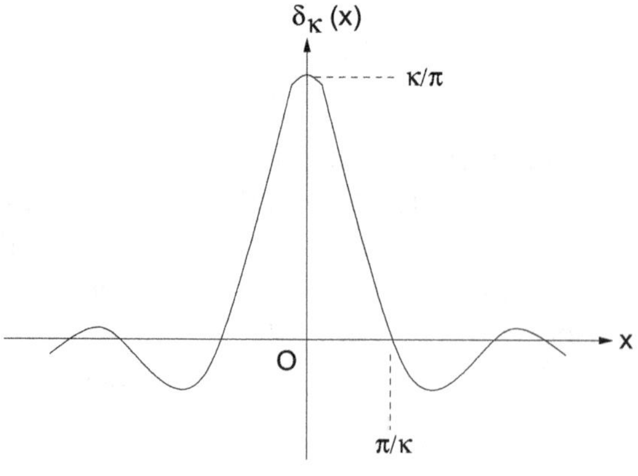


**Fig. A.1.** The function $\delta_\kappa(x)$.

$k$ on the right hand side, we obtain

$$\int dy f(y)\delta(y-x) = f(x), \tag{A.31}$$

as claimed.

In the following example we use yet another representation of the delta function which is very similar to the representation (A.28) (effectively the square of the latter). However, in this example our interest is focussed on the explicit use of a test function and the consequent necessity to evaluate the integral defining the distribution.

**Example A.1 (Explicit application of test functions).** The delta function can be represented by the limit

$$\delta(x) = \lim_{t\to\infty} \delta_t(x), \tag{A.32a}$$

where

$$\delta_t(x) := \frac{1}{\pi}\frac{\sin^2 xt}{x^2 t} = \begin{cases} t/\pi & \text{for} \quad x \to 0, \\ \leq 1/\pi x^2 t & \text{for} \quad x \neq 0. \end{cases} \tag{A.32b}$$

Use the method of test functions to verify the above result in the sense of distribution theory.

**Solution:** For test functions $\phi(x)$ the delta distribution is defined by the functional

$$\int \delta(x)\phi(x)dx = \phi(0). \tag{A.33}$$

In the case of the representation given by Eqs. (A.32a) and (A.32b), this is the relation

$$\lim_{t\to\infty} \frac{1}{\pi}\int_{-\infty}^{\infty} dx \frac{\sin^2 xt}{x^2 t}\phi(x) = \phi(0). \tag{A.34}$$

Thus we have to verify that the right hand side of this equation is obtained with a suitable choice of test functions $\phi(x)$. We choose the test function

$$\phi(x) = e^{-|x|}, \quad \phi(0) = 1. \tag{A.35}$$

This is a test function $\phi(x) \in S(\mathbb{R})$, which is the set of those test functions which together with all of their derivatives fall off at infinity faster than any inverse power of $|x|$ (the symbol $S$ being derived from the name of Schwartz referred to above). We therefore have to evaluate the integral

$$I = \int_0^\infty dx \frac{\sin^2 xt}{x^2} e^{-x}. \tag{A.36}$$

We do this with the help of Tables of Integrals.[2] We have

$$\begin{aligned}\int_0^\infty e^{-|p|x} \sin ax \sin bx \frac{dx}{x^2} &= \frac{p}{4} \ln \frac{p^2+(a-b)^2}{p^2+(a+b)^2} + \frac{a}{2} \tan^{-1} \frac{2pb}{p^2+a^2-b^2} \\ &\quad + \frac{b}{2} \tan^{-1} \frac{2pa}{p^2+b^2-a^2},\end{aligned} \tag{A.37}$$

so that for $a = b = t, p = 1$

$$I = -\frac{1}{4} \ln(1+4t^2) + t \tan^{-1}(2t). \tag{A.38}$$

Then

$$\begin{aligned}\lim_{t\to\infty} \frac{1}{\pi} \int_{-\infty}^\infty dx \frac{\sin^2 xt}{x^2 t} e^{-|x|} &= 2 \lim_{t\to\infty} \left[ -\frac{1}{4t\pi} \ln(1+4t^2) + \frac{t}{t\pi} \tan^{-1}(2t) \right] \\ &= \frac{2}{\pi} \tan^{-1}(\infty) = \frac{2}{\pi}\frac{\pi}{2} = 1.\end{aligned} \tag{A.39}$$

Hence the result is $\phi(0) = 1$ as expected.

Finally we consider the following two examples. The first of these deals with the Fourier transform, but contains in addition a further useful representation of the delta function. In the second example we consider the first of the relations (2.44), which remains to be verified.

**Example A.2 (Determine the Fourier transform of $e^{-ax^2}$ for $a > 0$).** The function required is

$$g(k) = \int_{-\infty}^\infty dx e^{-ax^2} e^{ikx}. \tag{A.40}$$

**Solution:** In general one uses for the evaluation of such an integral the method of contour integration. In the present case, however, a simpler

[2] See Gradshteyn and Ryzhik [60], formula 3.947, p. 491.

method suffices. Differentiation of Eq. (A.40) with respect to $k$ yields the same as

$$g'(k) = -\frac{i}{2a}\int_{-\infty}^{\infty} dx \frac{d}{dx}(e^{-ax^2})e^{ikx}. \tag{A.41}$$

Partial integration of the right hand side implies

$$g'(k) = \frac{i}{2a}\int_{-\infty}^{\infty} ike^{ikx}e^{-ax^2}dx = -\frac{k}{2a}g(k). \tag{A.42}$$

Thus $g(k)$ satisfies the differential equation

$$g'(k) + \frac{k}{2a}g(k) = 0. \tag{A.43}$$

Simple integration yields $g(k) = Ce^{-k^2/4a}$ with the constant $C = g(0)$. Since

$$g(0) = \int_{-\infty}^{\infty} dxe^{-ax^2} = \sqrt{\frac{\pi}{a}}, \tag{A.44}$$

we obtain

$$g(k) = \sqrt{\frac{\pi}{a}}e^{-k^2/4a}. \tag{A.45}$$

From Eqs. (A.40) and (A.45) we obtain a representation of the delta function:

$$\delta(k) = \lim_{a\to 0}\frac{1}{2\pi}\int_{-\infty}^{\infty} dxe^{-ax^2}e^{ikx} = \lim_{a\to 0}\frac{1}{2\sqrt{\pi a}}e^{-k^2/4a} = \lim_{\epsilon\to 0}\frac{e^{-k^2/\epsilon^2}}{\epsilon\sqrt{\pi}}. \tag{A.46}$$

**Example A.3 (Another important property of the delta function).** Verify the relation

$$\delta[(x-a)(x-b)] = \frac{1}{|a-b|}[\delta(x-a) + \delta(x-b)] \tag{A.47}$$

for $a \neq b$. The formula can also be written

$$\delta[g(x)] = \sum_n \frac{1}{|g'(x_n)|}\delta(x - x_n), \tag{A.48}$$

where $g(x_n) = 0, g'(x_n) \neq 0$.

**Solution:** We start with the integral (for $\epsilon > 0$)

$$\frac{1}{2\pi}\int_{-\infty}^{\infty} dke^{-\epsilon|k|}e^{ikx} = \frac{1}{\pi}\int_0^{\infty} dke^{-\epsilon k}\cos kx = \frac{1}{\pi}\frac{\epsilon}{\epsilon^2+x^2}. \tag{A.49}$$

From this we obtain as representation of the delta function:

$$\delta(x) = \lim_{\epsilon\to 0}\frac{1}{2\pi}\int_{-\infty}^{\infty} dke^{-\epsilon|k|}e^{ikx} = \lim_{\epsilon\to 0}\frac{1}{\pi}\frac{\epsilon}{\epsilon^2+x^2}. \tag{A.50}$$

With the help of this representation of the delta function one can show by partial fraction decomposition that the above relation (A.47) holds for $a \neq b$. Thus consider the sum

$$\begin{aligned}
&\frac{\epsilon}{(x-a)^2+\epsilon^2} + \frac{\epsilon}{(x-b)^2+\epsilon^2} \\
&= \frac{1}{2i}\left[\frac{1}{x-a-i\epsilon} - \frac{1}{x-a+i\epsilon} + \frac{1}{x-b-i\epsilon} - \frac{1}{x-b+i\epsilon}\right] \\
&= \frac{1}{2i}\left[\frac{(a-b+2i\epsilon)}{[(x-a)-i\epsilon][(x-b)+i\epsilon]} - \frac{(a-b-2i\epsilon)}{[(x-a)+i\epsilon][(x-b)-i\epsilon]}\right].
\end{aligned} \tag{A.51}$$

For $\epsilon$ small and $a \neq b$ this is

$$\begin{aligned}
&\frac{(a-b)}{2i}\left[\frac{1}{(x-a)(x-b)+i\epsilon(b-a)} - \frac{1}{(x-a)(x-b)-i\epsilon(b-a)}\right] \\
&= \frac{\epsilon(a-b)^2}{(x-a)^2(x-b)^2+\epsilon^2(a-b)^2} = |a-b|\left[\frac{\epsilon'}{(x-a)^2(x-b)^2+\epsilon'^2}\right],
\end{aligned} \tag{A.52}$$

where we set $\epsilon' = |a-b|\epsilon$. Hence

$$\frac{1}{\pi}\left[\frac{\epsilon}{(x-a)^2+\epsilon^2} + \frac{\epsilon}{(x-b)^2+\epsilon^2}\right] = \frac{1}{\pi}|a-b|\left[\frac{\epsilon'}{(x-a)^2(x-b)^2+\epsilon'^2}\right]. \tag{A.53}$$

Again using Eq. (A.50) and taking the limit $\epsilon \to 0$ on both sides, we obtain the result Eq. (A.47).

## Appendix B

# Units and Physical Constants

The units used in the text are those internationally agreed on and called SI units or SIU, *Systeme International d'Unité*, or often as earlier also the MKSA system of units, depending on meter, kilogramme, second and ampere. The international abbreviation is SI (cf. references below). One distinguishes between two types of units, those of the first type being the *fundamental units*, of which the relevant ones for our purposes here are meter (m), kilogramme (kg), second (s), ampere (A) and kelvin (K), the last for thermodynamical temperature. All other units, like newton, hertz, etc. are *derived SI units*. The most important derived units for our purposes here, and possibly related uses, are summarized in Table B.1, especially as a reference of internationally agreed symbols for the various units.[1] According to international agreement, symbols for units which are proper names, as in Table B.1, are written with a first capital letter, their special names are written with small letters (e.g. hertz); plurals of unit names are formed according to the usual rules of grammar with irregular exceptions (hertz, siemens). Symbols for physical quantities are set in italic (sloping) type, while symbols for units are set in roman (upright) type.[2] Multiplication of units is indicated by a raised dot (Table B.1) or a space between the units (cf. again Nelson [105]). Note that in the past $\Omega^{-1}$ was occasionally described as mho with the MKSA unit of conductivity 1 mho/m. Also note that capacitance is American usage for capacity.

[1]Extracted (with permission gratefully granted by the author and with notification to the Rights Office, AIP) from Nelson [105].
[2]For all of these conventions see in particular Nelson [105].

**Table B.1.** Some derived SI units and their symbols.

| Quantity | Special Name | Unit Symbol | Equivalent |
|---|---|---|---|
| plane angle | radian | rad | $m/m = 1$ |
| solid angle | steradian | sr | $m^2/m^2 = 1$ |
| speed, velocity | | | $m/s$ |
| acceleration | | | $m/s^2$ |
| angular velocity | | | $rad/s$ |
| angular acceleration | | | $rad/s^2$ |
| frequency $\nu = \omega/2\pi$ | hertz | Hz | $s^{-1}$ |
| force | newton | N | $kg \cdot m/s^2$ |
| pressure, stress | pascal | Pa | $N/m^2$ |
| work, energy, heat | joule | J | $N \cdot m$, $kg \cdot m^2/s^2$ |
| impulse, momentum | | | $N \cdot s$, $kg \cdot m/s$ |
| power | watt | W | $J/s$ |
| electric charge | coulomb | C | $A \cdot s$ |
| electric potential, emf | volt | V | $J/C, W/A$ |
| resistance | ohm | $\Omega$ | $V/A$ |
| conductance | siemens | S | $A/V$, $\Omega^{-1}$ |
| magnetic flux | weber | Wb | $V{\cdot}s$ |
| inductance | henry | H | $Wb/A$ |
| capacitance | farad | F | $C/V$ |
| electric field strength | | | $V/m$, $N/C$ |
| magnetic flux density | tesla | T | $Wb/m^2$, $N/(A \cdot m)$ |
| electric displacement | | | $C/m^2$ |
| magnetic field strength | | | $A/m$ |
| radioactivity | becquerel | Bq | $s^{-1}$ |

Table B.2 summarizes important fundamental constants [3] (for frequent usefulness with inclusion of the standard value of the gravitational acceleration $g_n$).

In many cases the Gauss units mentioned in Chapter 1 are still in use. In Table B.3 the factors for the appropriate conversions to Gaussian units of equations given in the text in MKSA units are summarized (Jackson [73] uses Gaussian units and therefore gives a table in the reverse direction). Units for force and other non-electromagnetic quantities remain unaffected.

[3]Extracted (with permission gratefully granted by the authors and with notification to the Rights Office, AIP) from Mohr and Taylor [97]. Note that in our brief extract some names of quantities have been shortened and values have been reduced to fewer decimal places.

**Table B.2.** Fundamental physical constants (relevant for electrodynamics and related areas).

| Quantity | Symbol | Value | Unit |
|---|---|---|---|
| speed of light in vacuum | $c$ | 299 792 458 | $\mathrm{m\,s^{-1}}$ |
| magnetic constant | $\mu_0$ | $4\pi \times 10^{-7}$ | $\mathrm{N\,A^{-2}}$ |
| | | $= 12.566370614\ldots \times 10^{-7}$ | $\mathrm{N\,A^{-2}}$ |
| electric constant $1/\mu_0 c^2$ | $\epsilon_0$ | $8.854187817\ldots \times 10^{-12}$ | $\mathrm{F\,m^{-1}}$ |
| vacuum impedance $\sqrt{\mu_0/\epsilon_0}$ | $Z_0$ | $376.730\ldots$ | $\Omega$ |
| Newton grav. constant | $G$ | $6.673\ldots \times 10^{-11}$ | $\mathrm{m^3\,kg^{-1}\,s^{-2}}$ |
| Planck constant | $h$ | $6.626068\ldots \times 10^{-34}$ | $\mathrm{J\,s}$ |
| Planck constant | $h$ | $4.135667\ldots \times 10^{-15}$ | $\mathrm{eV\,s}$ |
| $h/2\pi$ | $\hbar$ | $1.054571\ldots \times 10^{-34}$ | $\mathrm{J\,s}$ |
| Planck mass $\sqrt{\hbar c/G}$ | $m_P$ | $2.176\ 7 \times 10^{-8}$ | kg |
| Planck length $\hbar/m_P c$ | $l_P$ | $1.616\ 0 \times 10^{-35}$ | m |
| Planck time $l_P/c$ | $t_P$ | $5.390\ 6 \times 10^{-44}$ | s |
| elementary charge | $e$ | $1.602\ 176 \times 10^{-19}$ | C |
| magnetic flux quantum $h/2e$ | $\Phi_0$ | $2.067\ 833 \times 10^{-15}$ | Wb |
| Bohr magneton $e\hbar/2m_e$ | $\mu_B$ | $927.400\ 899 \times 10^{-26}$ | $\mathrm{J\,T^{-1}}$ |
| Bohr magneton $e\hbar/2m_e$ | $\mu_B$ | $5.788\ 381\ 7 \times 10^{-5}$ | $\mathrm{eV\,T^{-1}}$ |
| fine-structure constant $e^2/4\pi\epsilon_0\hbar c$ | $\alpha$ | $7.297\ 352 \times 10^{-3}$ | |
| $1/\alpha$ | $\alpha^{-1}$ | $137.035\ 999\ 76\ldots$ | |
| electron mass | $m_e$ | $9.109\ 381\ 88 \times 10^{-31}$ | kg |
| $m_e$ equivalent energy | $m_e c^2$ | $8.187\ 104 \times 10^{-14}$ | J |
| $m_e$ equivalent energy | $m_e c^2$ | $0.510\ 998\ 9$ | MeV |
| electron charge/mass | $-e/m_e$ | $-1.758820 \times 10^{11}$ | $\mathrm{C\,kg^{-1}}$ |
| grav. acceleration | $g_n$ | $9.806\ 65$ | $\mathrm{m\,s^{-2}}$ |

## Examples

1. MKSA: $\nabla \cdot \mathbf{D} = \rho$ ,

   $\therefore$ Gauss: $\nabla \cdot \sqrt{\frac{\epsilon_0}{4\pi}}\mathbf{D} = \sqrt{4\pi\epsilon_0}\rho$,

   i.e.

$$\nabla \cdot \mathbf{D} = 4\pi\rho.$$

2. MKSA: $\nabla \times \mathbf{E} = -\frac{\partial \mathbf{B}}{\partial t}$,

   $\therefore$ Gauss: $\nabla \times \frac{\mathbf{E}}{\sqrt{4\pi\epsilon_0}} = -\frac{\partial}{\partial t}\sqrt{\frac{\mu_0}{4\pi}}\mathbf{B}$,

   i.e.

$$\nabla \times \mathbf{E} = -\sqrt{\mu_0\epsilon_0}\,\mathbf{B} = -\frac{1}{c}\frac{\partial \mathbf{B}}{\partial t}.$$

**Table B.3.** Conversion from MKSA to Gaussian units.

| Quantity in MKSA Units | $\to$ | Quantity in Gaussian Units |
|---|---|---|
| $1/\sqrt{\mu_0\epsilon_0}$ | $\to$ | $c$ |
| $(\mathbf{E},\phi,V)$ | $\to$ | $\frac{1}{\sqrt{4\pi\epsilon_0}}(\mathbf{E},\phi,V)$ |
| $\mathbf{D}$ | $\to$ | $\sqrt{\frac{\epsilon_0}{4\pi}}\mathbf{D}$ |
| $\rho,q,\mathbf{j},I,\mathbf{P}$ | $\to$ | $\sqrt{4\pi\epsilon_0}(\rho,q,\mathbf{j},I,\mathbf{P})$ |
| $\mathbf{B}$ | $\to$ | $\sqrt{\frac{\mu_0}{4\pi}}\mathbf{B}$ |
| $\mathbf{H}$ | $\to$ | $\frac{1}{\sqrt{4\pi\mu_0}}\mathbf{H}$ |
| $\mathbf{M}$ | $\to$ | $\sqrt{\frac{4\pi}{\mu_0}}\mathbf{M}$ |
| $\sigma$ | $\to$ | $4\pi\epsilon_0\sigma$ |
| $\epsilon$ | $\to$ | $\epsilon_0\epsilon$ |
| $\mu$ | $\to$ | $\mu_0\mu$ |
| $(R,Z)$ | $\to$ | $\frac{1}{4\pi\epsilon_0}(R,Z)$ |
| $L$ | $\to$ | $\frac{1}{4\pi\epsilon_0}L$ |
| $C$ | $\to$ | $4\pi\epsilon_0 C$ |

3. MKSA: $\boldsymbol{\nabla}\times\mathbf{H}=\mathbf{j}+\frac{\partial}{\partial t}\mathbf{D}$,

$\therefore$ Gauss: $\boldsymbol{\nabla}\times\frac{\mathbf{H}}{\sqrt{4\pi\mu_0}}=\sqrt{4\pi\epsilon_0}\,\mathbf{j}+\frac{\partial}{\partial t}\sqrt{\frac{\epsilon_0}{4\pi}}\mathbf{D}$,

i.e.

$$\boldsymbol{\nabla}\times\mathbf{H}=4\pi\sqrt{\mu_0\epsilon_0}\,\mathbf{j}+\frac{\partial}{\partial t}\sqrt{\epsilon_0\mu_0}\,\mathbf{D}=\frac{4\pi}{c}\mathbf{j}+\frac{1}{c}\frac{\partial\mathbf{D}}{\partial t}.$$

4. MKSA: $\mathbf{F}=\rho\mathbf{E}+\mathbf{j}\times\mathbf{B}$,

$\therefore$ Gauss: $\mathbf{F}=\sqrt{4\pi\epsilon_0}\rho\frac{\mathbf{E}}{\sqrt{4\pi\epsilon_0}}+\sqrt{4\pi\epsilon_0}\,\mathbf{j}\times\sqrt{\frac{\mu_0}{4\pi}}\mathbf{B}=\rho\mathbf{E}+\frac{1}{c}\mathbf{j}\times\mathbf{B}$,

i.e.

$$\mathbf{F}=\rho\left(\mathbf{E}+\frac{\mathbf{v}\times\mathbf{B}}{c}\right).$$

These equations show that the factor $c$ always appears whenever a magnetic (or electric) quantity is related to an electric (magnetic) quantity respectively, as we observe this also in the field tensor $F_{\mu\nu}$ (see e.g. Eq. (17.80)).

Thus the factor $c = 1/\sqrt{\epsilon_0\mu_0}$ appears in the transition from purely electric units (long ago called e.s.u.), to purely magnetic units (long ago called e.m.u.). In the Gaussian (c.g.s. = cm gm sec) system of units (see Appendix in Jackson [73]), as in the original e.s.u. ('s' for 'static') the quantities charge, current, potential (voltage) and capacity are given in *statcoulomb, statampere, statvolt* and *statfarad*. With $c = 3 \times 10^{10}$ cm/sec ('3' as good approximation for 2.997 924 58...) one defines

$$1\ \mathrm{C} = 3 \times 10^9\ \text{e.s.u.} \equiv 3 \times 10^9\ \text{statcoulombs}, \quad 1\ \mathrm{A} = 3 \times 10^9\ \text{statamperes}.$$

Then, using the equivalents of Table B.1,

$$\begin{aligned} 1\ \mathrm{V(volt)} &= \frac{1\ \mathrm{J(joule)}}{1\ \mathrm{C(coulomb)}} = \frac{10^7\ \mathrm{erg}}{3 \times 10^9\ \text{e.s.u.}} \\ &= \frac{1}{300}\frac{\mathrm{erg}}{\text{e.s.u.}} = \frac{1}{300}\ \text{statvolt}, \\ 1\ \mathrm{F(farad)} &= \frac{1\ \mathrm{C(coulomb)}}{1\ \mathrm{V(volt)}} = \frac{3 \times 10^9\ \text{statcoulombs}}{1\mathrm{V}} \\ &= \frac{3 \times 10^9\ \text{statcoulombs}}{(1/300)\ \text{statvolt}} = 9 \times 10^{11}\ \text{statfarads} \end{aligned}$$

with

$$1\ \text{statfarad} = 1\ \text{statcoulomb}/1\ \text{statvolt}.$$

For the electric field strength $E$ one has therefore

$$E: \qquad 1\ \mathrm{V \cdot m^{-1}} = \frac{1}{3} \times 10^{-4}\ \text{statvolt} \cdot \mathrm{cm}^{-1}.$$

For the electric displacement **D** one obtains:

$$\begin{aligned} D: \qquad 1\ \mathrm{C \cdot m^{-2}} &= \frac{3 \times 10^9\ \text{statcoulombs}}{10^4\ \mathrm{cm}^2} \\ &= 3 \times 10^5\ \text{statcoulombs} \cdot \mathrm{cm}^{-2}. \end{aligned}$$

For resistance $R$:

$$\begin{aligned} R: \qquad 1\Omega = 1\ \mathrm{V}/1\ \mathrm{A} &= \frac{\text{statvolt}}{300 \times 3 \times 10^9\ \text{statamperes}} \\ &= \frac{1}{9} \times 10^{-11}\ \text{statvolt sec/statcoulomb} \\ &\quad (\text{or sec/statfarad or sec} \cdot \mathrm{cm}^{-1}), \end{aligned}$$

and in the case of the conductivity $\sigma$:

$$\sigma: \qquad 1\,\Omega^{-1}\,\mathrm{m}^{-1} = 9\times 10^9 \text{ statamperes cm}^{-1}/\text{statvolt(or sec}^{-1}).$$

The old e.m.u. (electromagnetic units) are still occasionally mentioned, e.g. also in Jackson [73], and one finds these in older texts. We therefore mention these briefly, since some aspects of their definition are not without interest, even today. These units were introduced via the formula for the field strength $H$ at the center of a single circular conductor with radius $r$ and current $I$, i.e. the expression $H \propto 2\pi I/r$, in which the proportionality constant is chosen as 1, i.e. with

$$H = \frac{2\pi I}{r}.$$

For $r = 1\,\mathrm{cm}, I =$ unit of current in e.m.u. (in old literature also called "abampere") one has $H = 2\pi$ oersted, with the unit oersted given in dyne per magnetic pole. (Thus in the case of the e.m.u. the unit of current is defined first, whereas in the case of the e.s.u. the unit of charge is defined first). Using the expressions $I = dq/dt, dq = Idt$, current $I$ is given in abampere and $dt$ in seconds, and charge $dq$ in abcoulomb, where 1 abcoulomb is defined as that amount of charge which is transported in one second by a current of one abampere.

The relation between e.m.u. and e.s.u. can be determined experimentally. The easiest way is with the help of a parallel plate condenser of known mass. Let $q_s$ be the charge in e.s.u. and $q_m$ its value in e.m.u. What one is looking for is the constant $c$ in the relation $q_m = cq_s$. The capacity of the condenser in vacuum is (let us say)

$$C_s = \frac{q_s}{V_s} \text{ e.s.u.} \quad \text{or} \quad C_m = \frac{q_m}{V_m} \text{ e.m.u.}$$

Since in the case of both kinds of units the energy is measured in ergs, one can equate them, i.e.

$$\frac{1}{2}q_sV_s = \frac{1}{2}q_mV_m \quad \text{or} \quad \frac{q_s}{q_m} = \frac{V_m}{V_s} = \frac{q_m}{q_s}\frac{C_s}{C_m}.$$

If we replace here the ratio of the potentials by that of the capacities from the preceding relation, we obtain

$$\frac{q_s}{q_m} = \sqrt{\frac{C_s}{C_m}}$$

and hence

$$c = \sqrt{\frac{C_m}{C_s}}.$$

The value of $C_s$ can now be given directly in centimeters for the parallel plate condenser (cf. Example 2.9), and $C_m$ can be determined experimentally, e.g. with the help of a ballistic galvanometer. This interesting experiment was performed in 1907 by Rosa and Dorsey[4] and yielded to a high degree of precision for the constant $c$ the numerical value of the velocity of light. Hence

$$q \text{ e.m.u.} = 3 \times 10^{10}\, q \text{ e.s.u.},$$

or

$$1 \text{ C} = 3 \times 10^{9} \text{ e.s.u.} = 10^{-1} \text{ e.m.u.}$$

In this way we obtain the relationships between the systems of units given in Table B.4 (here only for a few typical examples). The derivation of such

**Table B.4.** Relations between e.s.u. and e.m.u.

| Quantity | e.s.u. | | e.m.u. | MKSA |
|---|---|---|---|---|
| charge | $3 \times 10^9$ | = | $10^{-1}$ | C |
| current | $3 \times 10^9$ | = | $10^{-1}$ | A |
| potential | 1/300 | = | $10^8$ | V |
| electric field strength | 1/300 | = | $10^6$ | $\text{V}\cdot\text{m}^{-1}$ |
| resistance | $1/9 \times 10^{-11}$ | = | $10^{-9}$ | Ω |
| capacity | $9 \times 10^{11}$ | = | $10^{-9}$ | F |
| inductance | $1/9 \times 10^{-11}$ | = | $10^9$ | H |

[4]See e.g. Lemon and Ference [84], p. 385, and Bleaney and Bleaney [12], pp. 159–160, p. 228.

**Table B.5.** Energy equivalences.

| | Relevant Unit | | | |
|---|---|---|---|---|
| | J | kg | $m^{-1}$ | Hz |
| 1 J | $(1\ J) = 1$ | $(1\ J)/c^2 = 1.112\ 650 \times 10^{-17}$ | $(1\ J)/hc = 5.034\ 117\ 6 \times 10^{24}$ | $(1\ J)/h = 1.509\ 190\ 5 \times 10^{33}$ |
| 1 kg | $(1\ kg)c^2 = 8.987\ 551\ 787 \times 10^{16}$ | $(1\ kg) = 1$ | $(1\ kg)c/h = 4.524\ 439 \times 10^{41}$ | $(1\ kg)c^2/h = 1.356\ 392\ 77 \times 10^{50}$ |
| 1 $m^{-1}$ | $(1\ m^{-1})hc = 1.986\ 445 \times 10^{-25}$ | $(1\ m^{-1})h/c = 2.210\ 218\ 6 \times 10^{-42}$ | $(1\ m^{-1}) = 1\ m^{-1}$ | $(1\ m^{-1})c = 299\ 792\ 458$ |
| 1 Hz | $(1\ Hz)h = 6.626\ 068\ 76 \times 10^{-34}$ | $(1\ Hz)h/c^2 = 7.372\ 495\ 78 \times 10^{-51}$ | $(1\ Hz)/c = 3.335\ 640\ 95 \times 10^{-9}$ | $(1\ Hz) = 1$ |
| 1 K | $(1\ K)k = 1.380\ 650\ 3 \times 10^{-23}$ | $(1\ K)k/c^2 = 1.536\ 180\ 7 \times 10^{-40}$ | $(1\ K)k/hc = 69.503\ 56$ | $(1\ K)k/h = 2.083\ 664\ 4 \times 10^{10}$ |
| 1 eV | $(1\ eV) = 1.602\ 176 \times 10^{-19}$ | $(1\ eV)/c^2 = 1.782\ 661\ 7 \times 10^{-36}$ | $(1\ eV)/hc = 8.065\ 544\ 77 \times 10^{5}$ | $(1\ eV)/h = 2.417\ 989\ 49 \times 10^{14}$ |

relations can, for instance, also be found in electrodynamics exercises in the literature.[5]

Finally we collect in Table B.5 some useful values of energy-equivalences for conversion from one unit to another.[6]

[5]See e.g. Booth and Nicol [15], pp. 506–507.

[6]Extracted (with permission gratefully granted by the authors and with notification to the Rights Office, AIP) from Mohr and Taylor [97]. Note that in our brief extract some precise values have been reduced to fewer decimal places.

Appendix C

# Formulas

## C.1 Vector Products

The following relations are frequently used in the text and are therefore summarized here[1]:

$$\mathbf{A}\cdot(\mathbf{B}\times\mathbf{C}) = (\mathbf{A}\times\mathbf{B})\cdot\mathbf{C},$$
$$(\mathbf{P}\times\mathbf{Q})\times\mathbf{R} = (\mathbf{R}\cdot\mathbf{P})\mathbf{Q} - (\mathbf{R}\cdot\mathbf{Q})\mathbf{P},$$
$$\mathbf{P}\times(\mathbf{Q}\times\mathbf{R}) = (\mathbf{P}\cdot\mathbf{R})\mathbf{Q} - (\mathbf{P}\cdot\mathbf{Q})\mathbf{R}.$$

For vectorial derivatives one has in particular: *curl curl* = *grad div* − *div grad*, i.e.

$$\boldsymbol{\nabla}\times(\boldsymbol{\nabla}\times\mathbf{A}) = \boldsymbol{\nabla}(\boldsymbol{\nabla}\cdot\mathbf{A}) - (\boldsymbol{\nabla}\cdot\boldsymbol{\nabla})\mathbf{A}.$$

In addition:

$$\boldsymbol{\nabla}\times(\mathbf{u}\times\mathbf{v}) = \mathbf{u}(\boldsymbol{\nabla}\cdot\mathbf{v}) - \mathbf{v}(\boldsymbol{\nabla}\cdot\mathbf{u}) + (\mathbf{v}\cdot\boldsymbol{\nabla})\mathbf{u} - (\mathbf{u}\cdot\boldsymbol{\nabla})\mathbf{v},$$
$$\boldsymbol{\nabla}(\mathbf{u}\cdot\mathbf{v}) = (\mathbf{v}\cdot\boldsymbol{\nabla})\mathbf{u} + (\mathbf{u}\cdot\boldsymbol{\nabla})\mathbf{v} + \mathbf{v}\times(\boldsymbol{\nabla}\times\mathbf{u}) + \mathbf{u}\times(\boldsymbol{\nabla}\times\mathbf{v}),$$
$$\boldsymbol{\nabla}\cdot(u\mathbf{v}) = \boldsymbol{\nabla}u\cdot\mathbf{v} + u\boldsymbol{\nabla}\cdot\mathbf{v},$$
$$\boldsymbol{\nabla}\times(u\mathbf{v}) = \boldsymbol{\nabla}u\times\mathbf{v} + u\boldsymbol{\nabla}\times\mathbf{v},$$
$$\boldsymbol{\nabla}\cdot(\mathbf{u}\times\mathbf{v}) = \mathbf{v}\cdot(\boldsymbol{\nabla}\times\mathbf{u}) - \mathbf{u}\cdot(\boldsymbol{\nabla}\times\mathbf{v}).$$

The following relations hold

$$\text{curl grad}\, V = 0 \quad \text{and} \quad \text{div curl}\,\mathbf{F} = 0,$$

[1] A suitable reference for these formulas and more is Weatherburn [159].

i.e.

$$\nabla \times \nabla V = 0 \quad \text{and} \quad \nabla \cdot (\nabla \times \mathbf{F}) = 0.$$

## C.2 Integral Theorems

Gauss' divergence theorem:

$$\int_V \nabla \cdot \mathbf{E}\, dV = \int_{F(V)} \mathbf{E} \cdot d\mathbf{F},$$

and Stokes' circulation theorem:

$$\int_F \nabla \times \mathbf{E} \cdot d\mathbf{F} = \oint_{C(F)} \mathbf{E} \cdot d\mathbf{s}.$$

# References


[1] A.C. Aitken, *Determinants and Matrices* (Oliver and Boyd, 1954).

[2] M. Ambrosio *et al.*, *Eur. Phys. J.* **C25** (2002) 511 and **C26** (2002) 163; see also *CERN Courier* **43**(4) (2003) 23.

[3] N. Anderson, *The Electromagnetic Field* (Plenum Press, 1968).

[4] N. Ashcroft and N.D. Mermin, *Solid State Physics* (Holt, Rinehart and Winston, 1976).

[5] S.S. Attwood, *Electric and Magnetic Fields* (Dover Publications, 1967).

[6] J.E. Avron, D. Osadchy and R. Seiler, A topological look at the quantum Hall effect, *Physics Today* (August 2003) 38.

[7] G. Barton and C. Eberlein, *Ann. Phys.* (N.Y.) **227** (1993) 222.

[8] W.E. Baylis, *Electrodynamics, A Modern Geometric Approach* (Birkhäuser, 1999). This text adopts instead of the traditional covariant approach the vector approach which uses Clifford's geometric algebras. As the author says (p. xiii): "The cost has been a layer of abstraction from familiar physical quantities and their interrelationships that has impeded adoption by the broader physics community".

[9] G. Bekefi and A.H. Barrett, *Electromagnetic Vibrations* (PIT Press, Cambridge, Mass., 1977).

[10] M.V. Berry, *Proc. R. Soc. Lond.* **A392** (1984) 45.

[11] S.K. Blau, E.I. Guendelman and A.H. Guth, *Phys. Rev.* **D 35** (1987) 1747.

[12] B.I. Bleaney and B. Bleaney, *Electricity and Magnetism*, 3rd ed. (Oxford University Press, 1976).

[13] D. Bohm and Y. Aharonov, Significance of electromagnetic potentials in the quantum theory, *Phys. Rev.* **115** (1959) 485; Further considerations on electromagnetic potentials in the quantum theory, *Phys. Rev.* **123** (1961) 1511.

[14] B. Bolton, *Electromagnetism and Its Applications* (van Nostrand Reinhold, 1980).

[15] E.H. Booth and P.M. Nicol, *Physics, Fundamental Laws and Principles with Problems and Worked Solutions*, 12th ed. (Australasian Medical Publishing Co., 1952).

[16] M. Bordag, The Casimir effect 50 years later, in *Proceedings of 4th Workshop on Quantum Field Theory* (World Scientific, 1999).
[17] M. Bordag *et al.*, *Advances in the Casimir Effect* (Oxford University Press, 2009).
[18] M. Born and E. Wolf, *Principles of Optics*, 4th ed. (Pergamon, 1970).
[19] M. Born and L. Infeld, *Proc. R. Soc.* **A144** (1934) 425.
[20] F. Bopp, *Z. Phys.* **169** (1962) 45.
[21] T.H. Boyer, *Phys. Rev.* **174** (1968) 1764.
[22] C.J. Bradley, *Electrostatics and Current Flow* (Mathematical Institute, Oxford, 1974).
[23] K.E. Bullen, *Introduction to the Theory of Mechanics* (Science Press, Sydney, 1951).
[24] B. Cabrera, *Phys. Rev. Lett.* **48** (1982) 1378.
[25] H. Carprasse, *An Elementary Method to Evaluate Certain Principal Value Integrals* (Universite de Liege, 1973), Unpublished.
[26] H.B.G. Casimir, *Proc. Kon. Ned. Akad. Wet.* **51** (1948) 793.
[27] L.G. Chambers, *An Introduction to the Mathematics of Electricity and Magnetism* (Chapman and Hall, 1973).
[28] R.G. Chambers, *Proc. R. Soc.* **A65** (1952) 458.
[29] D.K. Cheng, *Field and Wave Electromagnetics* (Addison–Wesley, Reading, 1983).
[30] W.B. Cheston, *Elementary Theory of Electric and Magnetic Fields* (J. Wiley and Sons, 1964).
[31] B.H. Chirgwin, C. Plumpton and C.M. Kilmister, *Elementary Electromagnetic Theory*, Vol. I (Pergamon Press, 1971).
[32] P.C. Clemmow, *An Introduction to Electromagnetic Theory* (Cambridge University Press, 1973).
[33] D.M. Cook, *The Theory of the Electromagnetic Field* (Prentice-Hall, 1975).
[34] W.N. Cottingham and D.A. Greenwood, *Electricity and Magnetism* (Cambridge University Press, 1991).
[35] C.A. Coulson and T.J.M. Boyd, *Electricity*, 2nd ed. (Longman, 1979).
[36] B.G. Dick, *Am. J. Phys.* **41** (1973) 1289.
[37] R.B. Dingle, *Electricity and Magnetism*, Mimeographed Lecture Notes (University of Western Australia, 1956).
[38] R.B. Dingle, *Physica* **19** (1953) 311.
[39] R.B. Dingle, *Lectures on Contour Integration, Fourier, Laplace and Mellin Transforms and their Applications*, Notes Prepared by E.J. Moore and Doreen Arndt (University of Western Australia, 1957).
[40] S.D. Drell and T.M. Yan, *Phys. Rev. Lett.* **25** (1970) 316, 902.
[41] C.V. Durell and A. Robson, *Advanced Algebra*, Vol. II (G. Bell and Sons, 1950), Chapter XI: Difference Equations, pp. 226–252.
[42] H.B. Dwight, *Tables of Integrals and Other Mathematical Data*, 3rd ed. (Macmillan, 1957).
[43] F.J. Dyson, *Am. J. Phys.* **58** (1990) 209.
[44] A. Einstein, *Ann. Phys.* **17** (1905) 891.

[45] A. Einstein, *The Meaning of Relativity* (Chapman and Hall, 1967).
[46] G. Epprecht, H. Carnal, E. Schanda, H. Severin, H. Bremmer and D.J.R. Stock, *Theorie der Elektromagnetischen Wellen* (Birkhäuser, 1969).
[47] Staff of Research and Education Association, *The Electromagnetic Problem Solver* (SREA, Revised 1988).
[48] A. Erdélyi, W. Magnus, F. Oberhettinger and F.G. Tricomi, *Higher Transcendental Functions*, Vols. I, II, III (McGraw–Hill, 1955).
[49] L. Eyges, *The Classical Electromagnetic Field* (Addison–Wesley, 1972).
[50] I.E. Farquhar, *Phys. Lett.* **A151** (1990) 203.
[51] B. Felsager, *Geometry, Particles and Fields* (Odense University Press, 1981).
[52] L.B. Felsen and N. Marcuvitz, *Radiation and Scattering of Waves* (Prentice-Hall, 1973).
[53] R.L. Ferrari, *An Introduction to Electromagnetic Fields* (Van Nostrand Reinhold Co., 1975).
[54] V.C.A. Ferraro, *Electromagnetic Theory* (The Athlone Press, University of London, 1954).
[55] M. Fierz, *Helv. Phys. Acta* **33** (1960) 855.
[56] H. Fleming, *Noether Theorems*, University of Sao Paulo Report IFUSP/P-517 (1985), Unpublished.
[57] H. Fraas, Magnetische Monopole, *Phys. unserer Zeit* **15** (1984) 173.
[58] D. Fryberger, *Magnetic Monopoles*, Stanford Linear Accelerator Report, SLAC-PUB-3535 (1984), Invited Paper at *1984 Applied Superconductivity Conference*, San Diego, California, unpublished.
[59] W. Gibbons and D.A. Rasheed, *Nucl. Phys.* **B454** (1995) 185.
[60] I.S. Gradshteyn and I.M. Ryzhik, *Table of Integrals, Series and Products* (Academic Press, 1965).
[61] W. Greiner, *Theoretische Physik, Vol. 3: Klassische Elektrodynamik*, 3. Auflage (H. Deutsch, 1982).
[62] P. Goddard and D. Olive, *Rep. Prog. Phys.* **41** (1978) 1357.
[63] D.J. Griffiths, *Introduction to Electrodynamics,* 3rd ed. (Prentice-Hall, 1999).
[64] R. Guenther, *Modern Optics* (Wiley, 1990).
[65] W. Güttinger, *Fortschr. Phys.* **14** (1966) 483.
[66] E. Hagen and H. Rubens, *Ann. Phys.* **11** (1903) 873.
[67] C.R. Hagen, *The Aharonov–Bohm Scattering Amplitude*, University of Rochester Report UR-1103 (1989).
[68] C.R. Hagen, Casimir energy for spherical boundaries, hep-th/9902057.
[69] G.P. Harnwell, *Principles of Electricity and Magnetism* (McGraw–Hill, 1949).
[70] A.C. Hewson, *An Introduction to the Theory of Electromagnetic Waves* (Longman, 1970).
[71] J. Hilgevoord, *Dispersion Relations and Causal Description* (North–Holland Publishing Co., 1962).
[72] T. Holstein, *Phys. Rev.* **82** (1951) 1427.
[73] J.D. Jackson, *Classical Electrodynamics*, 2nd ed. (Wiley, 1975).
[74] E. Jahnke and F. Emde, *Tables of Functions*, 4th ed. (Dover Publications, 1945).

[75] R. Jost, *Elektrodynamik* (Verlag der Fachvereine Zürich, 1975).
[76] T.J. Killian, *Nucl. Instrum. Methods* **176** (1980) 355.
[77] C. Kittel, *Introduction to Solid State Physics* (Wiley and Sons, 1972).
[78] D. Kleppner, *Physics Today* (October 1990) 9.
[79] E.J. Konopinski, *Electromagnetic Fields and Relativistic Particles* (McGraw–Hill, 1981).
[80] H.A. Kramers, *Collected Scientific Papers* (North–Holland Publishing Co., 1956).
[81] R. Kronig, *J. Opt. Soc. Am.* **12** (1926) 547.
[82] D.F. Lawden, *Electromagnetism* (George Allen and Unwin, 1973).
[83] S.K. Lamoreux, *Phys. Rev. Lett.* **78** (1997) 5.
[84] H.B. Lemon and M. Ference, *Analytical Experimental Physics* (University of Chicago Press, 1943).
[85] U. Leonhardt *et al.*, Quantum levitation by left-handed metamaterials *New J. Phys.* **9** (2007) 254.
[86] F.S. Levin and D.A. Micha, *Long-Range Casimir Forces* (Plenum Press, 1993).
[87] L.R. Lieber, *The Einstein Theory of Relativity* (Dobson, 1949).
[88] Y.K. Lim, *Introduction to Classical Electrodynamics* (World Scientific, 1986).
[89] P. Lorrain and D.R. Corson, *Electromagnetism: Principles and Applications*, 2nd ed. (W.H. Freeman, 1990).
[90] W. Magnus and F. Oberhettinger, *Formulas and Theorems for the Functions of Mathematical Physics* (Chelsea Publ. Co., 1954).
[91] L.M. Magid, *Electromagnetic Fields, Energy and Waves* (John Wiley, 1972).
[92] L.C. Maier and J.C. Slater, *Field Strength Measurements in Resonance Cavities*, *J. Appl. Phys.* **23** (1952) 68.
[93] A.A. Michelson and E.W. Morley, *Am. J. Sci.* **34** (1887) 333; reprinted in L.P. Williams (ed.), *Relativity Theory: Its Origin and Impact on Modern Thought* (Wiley, 1968). See also their paper in *Phil. Mag.* **24** (1887). A detailed discussion of the experiment can be found in Schwartz [132], pp. 105–110.
[94] K.A. Milton, *Ann. Phys.* (N.Y.) **127** (1980) 49.
[95] K.A. Milton, *The Casimir Effect: Physical Manifestation of Zero-Point Energy* (World Scientific, 2002).
[96] H.R. Mimno, *Rev. Mod. Phys.* **9** (1937) 1–43, and *Ergebnisse der exakten Wissenschaften* **17** (Springer, 1938).
[97] P.J. Mohr and B.N. Taylor, *The Fundamental Physical Constants*, *Physics Today* **56** (August 2003) BG6.
[98] G. Möllenstaedt and W. Bayh, The continuous variation of the phase of electron waves in field free space by means of the magnetic vector potential of a solenoid, *Phys. Blätter* **18** (1962) 299.
[99] V.M. Mostepanenko *et al.*, *The Casimir Effect and Its Applications* (Clarendon Press, 1997).
[100] H.J.W. Müller–Kirsten, *Classical Mechanics and Relativity* (World Scientific, 2008).
[101] H.J.W. Müller–Kirsten, *Classical Mechanics and Relativity* 2nd ed. (World Scientific, 2024).

[102] H.J.W. Müller–Kirsten, *Basics of Statistical Physics* 3rd ed. (World Scientific, 2022).
[103] H.J.W. Müller–Kirsten, *Introduction to Quantum Mechanics: Schrödinger Equation and Path Integral* (World Scientific, 2006).
[104] M. Nelkon, *Electricity, An SI Advanced Level Course* (Edward Arnold, 1970).
[105] R.A. Nelson, Guide for metric practice, *Physics Today* **56** (August 2003) BG 15.
[106] H.C. Ohanian, *Classical Electrodynamics* (Allyn and Bacon, 1988).
[107] L. Onsager, *J. Am. Chem. Soc.* **58** (1936) 1486.
[108] J.R. Oppenheimer, *Lectures on Electrodynamics* (Gordon and Breach, 1970). This book considers also basic quantum effects.
[109] L. Page and N.I. Adams, *Electrodynamics* (Dover Publications, 1965), pp. 154–155.
[110] L. Palumbo and V.G. Vaccaro, *Wake Field Measurements*, Frascati-Report LNF-89/035(P) (1989).
[111] W.K.H. Panofsky and M. Phillips, *Classical Electricity and Magnetism*, 2nd ed. (Addison-Wesley, 1962).
[112] J.B. Pendry, *J. Phys. Condens. Matter* **9** (1997) 10301.
[113] R. Penrose, *The Road to Reality, A Complete Guide to the Laws of the Universe* (Vintage Books, 2004).
[114] M. Peshkin and A. Tonomura, *The Aharonov–Bohm Effect*, Lecture Notes in Physics, Vol. 340 (Springer, 1989).
[115] E.G. Phillips, *Functions of a Complex Variable* (Oliver and Boyd, 1954).
[116] J. Pinfold, MoEDAL becomes the LHC's magnificent seventh, *CERN Courier* **50** (4) (May 2010) 19.
[117] G. Plunien, B. Müller and W. Greiner, *Physics Reports* **134** (1986) 87.
[118] B. Podolsky and K.S. Kunz, *Fundamentals of Electrodynamics* (M. Dekker, 1969).
[119] J. Polchinski, *String Theory*, Vol. II (Cambridge University Press, 1998).
[120] J. Polchinski, *Int. J. Mod. Phys.* **A19** (2004) (51) 145.
[121] A.M. Portis, *Electromagnetic Fields* (John Wiley, 1978).
[122] A. O'Rahilly, *Electromagnetic Theory*, Vols. I, II (Dover Publications, 1965).
[123] G.E.H. Reuter and E.H. Sondheimer, *Proc. R. Soc.* **A195** (1948) 336.
[124] F.N.H. Robinson, *Macroscopic Electromagnetism* (Pergamon, 1973).
[125] F.N.H. Robinson, *Electromagnetism* (Clarendon Press, 1973).
[126] V. Rossiter, *Electromagnetism* (Heyden, 1979).
[127] S.N.M. Ruijsenhaars, *Ann. Phys.* (N. Y.) **146** (1983) 1.
[128] J.J.G. Scanio, *Am. J. Phys.* **41** (1973) 415.
[129] S.A. Schelkunoff, *Electromagnetic Waves* (D. van Nostrand, 1960).
[130] H. Schiff, *Can. J. Phys.* **47** (1969) 2387; Quark-like potentials in an extended Maxwell theory, hep-th/0308091; *Classical Quarks in Dual Electromagnetic Fields*, University of Alberta Report (2010), to be published in *Int. J. Phys.* (2011).
[131] F. Schwabl, *Quantenmechanik* (Springer, 1998).

[132] M. Schwartz, *Principles of Electrodynamics* (McGraw–Hill, 1972).
[133] S. Schweber, *An Introduction to Relativistic Quantum Field Theory* (Harper and Row, 1961).
[134] T. Shintake, Transient wave analysis program using wave equation of vector potential, Paper Contributed to *1984 Linear Accelerator Conference*, May 1984, Darmstadt, Germany, and KEK Report 84–8.
[135] H. Sipila, V. Vanha-Honko and J. Bergqvist, *Nucl. Instrum. Methods* **176** (1980) 381.
[136] J.C. Slater and N.H. Frank, *Electromagnetism* (Dover Publications, 1969).
[137] J.C. Slater, *Microwave Transmission* (Dover Publications, 1959).
[138] J.C. Slater, *Microwave Electronics* (D. Van Nostrand, 1950).
[139] A. Sommerfeld, *Elektrodynamik*, 3rd ed., revised and supplemented by F. Bopp and J. Meixner (Akademische Verlagsgesellschaft, 1961).
[140] A. Sommerfeld, *Z. Phys.* **47** (1928) 1–32.
[141] A. Sommerfeld and H.A. Bethe, *Elektronentheorie der Metalle* (Springer, 1933).
[142] M.J. Sparnaay, *Physica* **XXIV** (1958) 751.
[143] L. Spitzer, *Physics of Fully Ionized Gases* (Interscience Publishers, 1956).
[144] S.G. Starling and A.J. Woodall, *Physics* (Longmans, Green and Co., 1950).
[145] E. Stiefel, *Methoden der Mathematischen Physik II* (Verlag der Fachvereine, Zürich, 1980).
[146] J.A. Stratton, *Electromagnetic Theory* (McGraw-Hill, 1941).
[147] W. Taylor, JHEP **0007** (2000) 039.
[148] J.J. Thomson, *Recent Researches in Electricity and Magnetism* (Clarendon Press, 1893).
[149] A. Tonomura, O. Noboyuki, T. Matsuda, T. Kawasaki, J. Endo, S. Yano and H. Yamada, Evidence for Aharonov–Bohm effect with magnetic field completely shielded from electron wave, *Phys. Rev. Lett.* **56** (1986) 729.
[150] The Physics Coaching Class, University of Science and Technology of China, *Problems and Solutions on Mechanics* (World Scientific, 1994).
[151] T.T. Wu and C.N. Yang, *Phys. Rev.* **D12** (1975) 3845.
[152] G. Tyras, *Radiation and Propagation of Electromagnetic Waves* (Academic Press, 1969).
[153] A. Vaidya and C. Farina, *Phys. Lett.* **A153** (1991) 265.
[154] I. Veit, *Technische Akustik*, 6th ed. (Vogel Buchverlag, 2005).
[155] M. Visser, *Phys. Lett.* **A 139** (1989) 99.
[156] W. Voigt, *Gött. Nachr.* (1887) 45.
[157] K. von Klitzing, G. Dorda and M. Pepper, *Phys. Rev. Lett.* **45** (1980) 494.
[158] G.N. Watson, *Theory of Bessel Functions*, 2nd ed. (Cambridge University Press, 1952).
[159] C.E. Weatherburn, *Advanced Vector Analysis* (C. Bell and Sons, 1947).
[160] W. Weihs and G. Zech, *Numerical Computation of Electrostatic Fields in Multiwire Chambers*, Report University of Siegen (May 1989).
[161] V.F. Weisskopf, *The Formation of Cooper Pairs and the Nature of Superconducting Currents*, CERN-Report 79–12.

[162] E.T. Whittaker and G.N. Watson, *A Course of Modern Analysis*, 4th ed. (Cambridge, 1958).
[163] A.H. Wilson, *The Theory of Metals*, 2nd ed. (Cambridge University Press, 1953).
[164] E. Witten, *Adv. Theor. Math. Phys.* **2** (1998) 253; hep-th/9802150.
[165] E. Witten, *Nucl. Phys.* **B223** (1983) 422.
[166] E. Witten, *Nucl. Phys.* **B249** (1985) 557.
[167] M. Zahn, *Electromagnetic Field Theory* (Wiley, 1979).
[168] J.M. Ziman, *Principles of the Theory of Solids* 2nd ed. (Cambridge University Press, 1972).

# Index

**K**



**L**



**M**

**Y**



**Z**

www.ingramcontent.com/pod-product-compliance
Lightning Source LLC
Chambersburg PA
CBHW081547260326
41744CB00006B/265